SCIENCE ON A MISSION

Science on a Mission

HOW MILITARY FUNDING
SHAPED WHAT WE DO AND DON'T
KNOW ABOUT THE OCEAN

+ + + + + + + + + + + + + + + +

NAOMI ORESKES

THE UNIVERSITY OF CHICAGO PRESS
CHICAGO AND LONDON

The University of Chicago Press, Chicago 60637
The University of Chicago Press, Ltd., London
© 2021 by The University of Chicago
Published 2021
Paperback edition 2022
Printed in the United States of America

31 30 29 28 27 26 25 24 23 22   1 2 3 4 5

ISBN-13: 978-0-226-73238-1 (cloth)
ISBN-13: 978-0-226-82400-0 (paper)
ISBN-13: 978-0-226-73241-1 (e-book)
DOI: https://doi.org/10.7208/chicago/9780226732411.001.0001

Library of Congress Cataloging-in-Publication Data

Names: Oreskes, Naomi, author.
Title: Science on a mission : how military funding shaped what
    we do and don't know about the ocean / Naomi Oreskes.
Description: Chicago : University of Chicago Press, 2021. |
    Includes bibliographical references and index.
Identifiers: LCCN 2020020806 | ISBN 9780226732381 (cloth) |
    ISBN 9780226732411 (ebook)
Subjects: LCSH: Oceanography—Research—Finance. |
    Oceanography—Research—United States. | Military
    research—United States.
Classification: LCC GC58 .O744 2020 | DDC 333.91/640973—
    dc23
LC record available at https://lccn.loc.gov/2020020806

♾ This paper meets the requirements of ANSI/NISO
Z39.48-1992 (Permanence of Paper).

*To all the oceanographers, geophysicists and marine geologists—
past and present—who have worked to help us understand the sea
around us, and to the archivists who have made it possible to tell
their stories.*

*To Ronald Rainger and David van Keuren, who helped to pioneer the
field of history of oceanography but found themselves prematurely
sailing the seas of eternity.*

And to Ken, who has surfed all the waves I have generated.

The growth of the profession of science to its present dimensions is not a sign of a spontaneous increase in the number of individuals gifted with natural curiosity, but of the realization of the value that science can bring to those who finance it.... Science uses curiosity, it needs curiosity, but curiosity did not make science.

J. D. BERNAL, *The Social Function of Science* (1939)

If oratorios could kill, the Pentagon would long ago have supported musical research.

ERWIN CHARGAFF, *Heraclitean Fire: Sketches of a Life before Nature* (1978)

It's amazing to me that the public is viewed as being so stupid that we would believe that the Navy is suddenly concerned about global warming.

MICHELLE WATERS, public comment C-36 on Marine Mammal Research Program Draft Environmental Impact Statement (1995)

Contents

Introduction

"There's no such thing as tainted money, except 't ain't enough"—so a senior colleague told me many years ago, when I was first raising money for my own scientific research.[1] Like many jokes, this one contains a germ of truth and masks an anxiety: some money *is* tainted. And it suggests a serious question: what difference does it make who pays for science? Many scientists would say none at all. If scientists seek to discover fundamental truths about the world, and if they do so in an objective manner using well-established methods, then how could it matter who foots the bill? History, alas, suggests that it does matter. Few patrons have ever supported science for the love of knowledge alone; most have had orthogonal (or at least oblique) motivations, be they prestige, power, or the solution of practical problems, and the available evidence suggests that those motivations make a difference.[2] On the positive side, patrons can encourage scientists to attend to neglected questions, consider matters from new angles and perspectives, or try a new approach. In medical research, we have seen how patients have positively influenced researchers who previously neglected important questions.[3] Historians of technology have shown how the demands of industry and commerce can stimulate scientific innovation.[4]

On the negative side, however, the interests of patrons may cause scientists to focus on immediate answers to pressing problems at the expense of fundamental understanding (which, as we shall see in this book, many Cold War oceanographers feared would happen to their field). The pressure of external deadlines can cause scientists to take shortcuts, make mistakes, or miss important elements of a problem. The needs of funders may also introduce bias into the design of scientific studies, as when scientists funded by the chemical or plastics industries choose test animals known to be insensitive to the potential effects of concern, or into the data interpretation, as when scientists funded by the tobacco industry fail to find the adverse effects of smoking that their independent colleagues find.[5] Most worrisome, the demands of patrons may grossly distort science, as when, under the pressure of the Soviet government in the 1930s, Trofim Lysenko rejected advances in modern genetics in favor of an empirically inadequate theory of environmentally dominated inheritance, development, and growth, and used his political power to discredit colleagues who disagreed with him.[6]

Most scientists would like to think that these sorts of problems and pitfalls are rare and that scientists—both individually and collectively—are sufficiently smart and self-aware to recognize and avoid them. We like to think of Lysenko as the exception that proves the rule—a grotesque intellectual expression of the broader horrors of Stalinism.[7] Indeed, it is easy to compartmentalize problematic cases as anomalous, to assume that they are exceptions, or to think that they apply only in narrow domains where the interests of sponsors are overt or extreme. Sadly, these assumptions have been shown not to hold. There is empirical evidence to demonstrate that scientists have been overly optimistic about their ability to maintain their intellectual integrity, particularly in cases where the desiderata of their funders are obvious, as with tobacco or pharmaceutical research.

This is not to suggest that the interests of funders are necessarily at odds with those of the funded. In many cases scientists and their patrons have a shared interest in gaining knowledge, which enables them to work productively together. Under such circumstances, the positive impact of funding is obvious—scientists get to do work they want to do. Any negative impact, however, is subtle and harder to discern. Oceanography during the Cold War is a case in point.

The Transformation of American Ocean Science

Before World War II, no American earned a living as an oceanographer investigating the oceans beyond coastal waters. Two institutions encompassed American oceanography—the Woods Hole Oceanographic Institution in Massachusetts and the Scripps Institution of Oceanography in La Jolla, California—and both were small, young, and poorly funded.[8] In fact, neither was really "oceanographic," because most of their scientists did their work at the seashore or in small boats that plied coastal waters. Woods Hole had no year-round staff and no regular external funding.[9] Scripps had a scientific staff of eleven, kept barely afloat by the institution's modest endowment income and funds from the University of California that covered salaries and operating expenses but left little for research.[10] In 1940, Scripps launched its first expedition, to the Gulf of California, with a hard-won grant of $10,000 from the Geological Society of America. Money was so tight that the institution's director implored his scientists not to stay away even a day longer than planned, for "we shall find ourselves in a deep pit when you return.... Funds are pitifully low."[11]

The same was true for marine geology and geophysics. The structure and composition of the ocean basins, pertinent to global tectonics, was of enormous interest to geologists, but a lack of access to the deep sea meant the topic was more speculation than investigation.[12] In 1935, Lehigh University geophysicist Maurice Ewing helped to invent the field of marine geophysics by applying seismic techniques developed for shallow oil exploration to study the Earth's deep crust. Ewing completed the first comprehensive geophysical study of the structure of a continental margin, but he hadn't a clue what to make of his results.[13] Conventional wisdom posited a sunken Paleozoic continent off the east coast of North America to account for thick Paleozoic sequences in the Appalachians, but the data revealed not a trace of it.[14] Ewing turned to his Lehigh colleague, geology professor Benjamin L. Miller, but he couldn't make sense of the results, either.[15] Miller supposed, tentatively (and in hindsight wrongly), that "somewhere in the Atlantic Ocean, there [must be] extensive Paleozoic strata."[16]

Within just a few years, matters would change dramatically. As war spread in Europe and a US entry appeared likely, American military planners recognized that this second world war would not be like the first. Throughout human history, warfare had taken place on two-dimensional battlefields—the surface of the land or the sea (or, in the case of U-boats, just barely beneath that surface). The impending war would be fought not only on those battlefields but also in the air above and the sea below them. After World War II, the newly formed US Air Force would look to the skies, the upper atmosphere, and even outer space as theaters of warfare; the Navy would look to the deep sea. The earth sciences—particularly physical oceanography and marine geophysics—would become crucial for antisubmarine warfare, weather and surf forecasting, undersea communications, navigation, air-sea rescue, vessel design and testing, submarine-based ballistic missile launching, the tracking of atomic bomb fallout, and a number of other operational ambitions and concerns. As Roger Revelle put it in 1947 in a report to Navy officials, during the war "a knowledge of oceanography was proved essential," and in the future an "increased emphasis on subsurface warfare in which a thorough knowledge of the medium is of prime importance" would lead to an even greater "requirement for oceanographic information." And there was scarcely an aspect of the ocean that was not operationally relevant: "The Navy, which operates on, under, and over the sea will be improved in effectiveness and striking power by precise knowledge concerning every aspect of the oceans."[17] Revelle was a scientist trying to make the case for his science, but relevant Navy officials and political leaders apparently agreed.

It is well known that the war economy provided unprecedented levels of research funding for American physics, particularly for the development of weapons systems. What is less known is how much was invested in understanding the environments in which those weapons systems would operate.[18] It is not that the US military suddenly discovered the value of earth science: meteorology, economic geology, geodesy, and cartography had long associations with terrestrial military campaigns, and navies had long recognized the operational value of various forms of oceanographic data.[19] Rather, it is that warfare itself was changing in a manner that required new kinds of scientific information, some of which could be obtained only with the help of innovative scientific research and nearly all of which increasingly seemed imperative.[20] The result, as historian Jacob Hamblin has put it, was that over the next half century, oceanographic science became "unsurpassed in its interconnections with the American military-industrial complex."[21]

With the expansion of submarine and antisubmarine warfare, questions of the internal configuration and conditions of the deep ocean would no longer be the domains of science fiction writers and their imaginations or scientists and their speculations; they would be domains of knowledge essential to military operations. During the Cold War interest in the deep sea would intensify as the submarine-launched ballistic missile became an arm of the nuclear triad, and as the US Navy built a global listening system to detect Soviet submarines carrying ballistic missiles of their own. It would not suffice simply to put sophisticated weapons on submarines. It would also be necessary to understand the environments through which those submarines would have to travel and from which those weapons would be launched.[22]

The result was an influx of money and logistical support that transformed American oceanography and marine geophysics and led to remarkable growth of scientific knowledge about the oceans, the seafloor, and the life associated with those domains. Scientists answered questions in which they had long been interested and also discovered some entirely new things. The Cold War was a not just a period of unprecedented growth in American oceanographic science; it was also a period of unprecedented growth in oceanographic knowledge.

Not surprisingly, when oceanographers look back on the Cold War, they tend to see it as a golden age, a time when they had both funding and freedom and used them to great effect. In the spring of 2000, the US Office of Naval Research (ONR) sponsored a series of colloquia across the United States to celebrate that history. As part of the effort, the ONR commissioned oral history interviews with oceanographers and geophysicists whose research it had supported. Most of them sang the ONR's praises, largely because of its support for basic research. Douglas Inman, a senior researcher at Scripps, put it this way: "They *were* the basic science supporters in this country. If you look at who supported basic science ... after World War II, it was ONR."[23]

The view that the US Navy, and particularly the ONR, freely supported scientific investigations without regard to military utility has long been widely held. In 1948, for example, in the aftermath of political attacks on physicist Edward Condon, a writer in *Fortune* alleged a crisis in American science caused by increasing restrictions on scientists' activities, decreasing intellectual freedom, and a lack of moral and financial support for basic research. The bright spot in this otherwise bleak landscape, the writer claimed, was the ONR, which "stepped into the breach created by the delay in establishing a civilian National Science Foundation, and was generously supporting pure research, with no strings attached, and a maximum of freedom for working scientists."[24] This early representation became the standard view, which over time developed into a prevailing narrative that Navy funding did not affect the science, except to make it possible. But was it true?

As is typically the case, historical attention suggests a more complex situation.[25] In his 1990 history, political scientist Harvey Sapolsky suggested that broad-ranging support of diverse research was characteristic only of the ONR's first few years, before the creation of the US National Science Foundation and at a time when few officials in the Navy were paying much attention to what the ONR program directors were doing.[26] Moreover, there is something peculiar about the claim that the ONR supported basic research without regard to salience—if this were true, it would stand at odds with its legal mandate to support research on behalf of the Navy mission. For ONR officials to have supported "pure research, with no strings attached," would have meant that they were not actually doing their job; they might even have been guilty of misappropriating federal funds. Wouldn't it make more sense to assume that they funded research that matched their goals, or that Navy funding involved various constraints, some innocuous but others perhaps not? After all, as biochemist Erwin Chargaff quipped in the 1970s, "If oratorios could kill, the Pentagon would long ago have supported musical research."[27]

Moreover, the ONR was one of several Navy bureaus that supported oceanographic and marine geophysical research during the Cold War. The other offices included the Bureau of Ships, the Bureau of Naval Weapons, the Hydrographic Office, and the Chief of Naval Operations.[28] It would have been inconsistent with the goals—indeed, the legal obligations—of these bureaus to spend resources on activities irrelevant to their mission. When we broaden our compass to examine Navy support for research in general, we find a far more complex and thought-provoking story than the one that has been told to date.

Why Oceanography?

Historians of physics have long been interested in the impact of military funding on their science in the Cold War, and their work provides us with some orientation and guidance. In the mid-1980s, the impact of military patronage was addressed head-on by two of the most influential historians of modern physics: Daniel J. Kevles and Paul Forman. Forman suggested that military funding had dramatically altered the character of physics, causing its practitioners to drift from an earlier goal of fundamental understanding of the laws of nature toward a science of "gadgeteering" that was preoccupied with technical prowess.[29] Kevles disagreed. He acknowledged the reality and significance of the military's pervasive patronage but insisted that American physicists had "retained control of their intellectual agenda."[30] He also insisted that there is little sense in arguing about what scientists might have done in a different world: History is the story of what has happened, not what might have happened. In any case there is no essential definition of what constitutes physics—physics is what physicists do. So Forman's claim that physics was "distorted" by Cold War concerns is not—and could not be—a historical claim.[31]

Kevles's argument had intuitive appeal to many historians of science, who reject essentialist notions of science and believe their job to be describing the world as it is, not as it might, should, or could have been.[32] "Science" as a category is both flexible and the subject of ongoing diminution and augmentation, so the observation that science changed during the Cold War—or during any particular historical period we might examine—is not by itself profound. Moreover, thanks to the work of Kevles and Forman, as well as that of many other historians since the 1980s, the idea that American science changed dramatically in the second half of the twentieth century is no longer novel, either.[33] What is of interest—and still not entirely resolved—is how it changed, why it changed in those particular ways, and how those changes were productive of our current states of knowledge and of ignorance. These are the questions I take up in this book.

Since the debates of the 1980s, historians of science have greatly broadened their outlook; there is now a robust literature on the history of the diverse sciences during the Cold War, both in the United States and elsewhere.[34]

Nearly all historians agree that American science took a dramatic turn during and after World War II, a turn that in myriad ways changed the priorities and perspectives of scientific communities. Toward the end of World War II and throughout the Cold War, the US government poured unprecedented amounts of money and levels of logistical support into American science. Scientists often focus their attention on the National Science Foundation and National Institutes of Health as their most important patrons, but during the Cold War in many domains the lion's share of the support came through the armed services. Much of this was to support areas of science, such as oceanography, that had been poorly funded or even had scarcely existed before. So this influx of funding mattered profoundly. One important way in which it mattered involves the question of the direction of science and who or what determines that direction.

Most oceanographers welcomed the wartime infusion of military funding, correctly anticipating the opportunities that it would create, the work it would enable. But, as the stories told in these pages will show, paths of inquiry were also shifted—if not entirely altered—and sometimes blocked. Moreover, many scientists were concerned that an expanded relationship with the US Navy would cause them to lose control of their science—the very thing that Forman argues happened in physics.[35] As World War II gave way to the Cold War and the US Navy became the principal patron of American oceanography, these concerns did not go away. Quite the contrary. They continued to express themselves in terms of anxieties—often privately felt, sometimes publicly expressed—and occasional open conflict over secrecy, the right to publish, and, above all, the question of who was setting the research agenda. The Cold War was a golden age of science in terms of the abundance of work that was made possible by generous financial and logistical support, but it was also a period of deep anxiety and profound conflict over the purpose and character of American science. These anxieties and conflicts—both personal and professional—animate the stories told here.

Throughout the Cold War and even after it was over, the oceanographers whose stories are told in this book insisted that they were doing "basic research," and it is true that for the most part they were not trying to solve specific operational problems. However, we will see in these pages that the Navy supported oceanographic work not *qua* basic research but because it was salient to specific problems the Navy was trying to solve. Even before the Cold War ended, scientists found that when those problems were solved—or resolved by other means—Navy support weakened and sometimes ended entirely. Research that was not salient was not funded, and therefore not done, unless other sources of support could be found. When the Cold War did end—and with it the geopolitical context that had justified the support they had been receiving and motivated much of the work they were doing—oceanographers found themselves scrambling to resituate themselves and their science, a project that did not always go well.

Forman and Kevles were right in what they affirmed but wrong in what they denied. Oceanographers, like physicists, were by and large grateful to have the abundant funding that made it possible to answer important and long-standing questions in earth science. In this sense, the Cold War was indeed a golden age for American oceanography. But many oceanographers were deeply concerned about matters of intellectual control and worried that they might lose—or were already losing—control of their science. Oceanographers worked to take advantage of the opportunities represented by military funding to improve their understanding of the natural world, but at the same time they struggled to preserve the degree of autonomy that most of them believed was essential to the pursuit of knowledge.[36] How did their goals intersect and interact with the aspirations of their funders, and how did they negotiate or adjust those goals when they needed to? The history presented here challenges the conventional dichotomies of autonomy and capture, intellectual integrity and corruption, pure and tainted. We will see that military support for oceanography and marine geophysics was both enabling and constricting. It resulted in the creation of important domains of knowledge, but it also created significant, lasting, and consequential domains of ignorance.

Scope of This Book

The primary focus of this book is on three institutions—the Scripps Institution of Oceanography, the Woods Hole Oceanographic Institution, and the Lamont Geological Observatory—that were the leading centers for oceanography and marine geophysics in the United States at the time (and remain among the most important institutions of their kind in the world). Roger Revelle would overstate the case when he claimed after World War II that Woods Hole and Scripps were the "only two institutions that have contributed in any way to military oceanography," but they (and Lamont) did receive the greatest abundance of military funding, and a rich archival reservoir has made it possible to reconstruct what they were doing in (what I hope is) convincing detail.[37] My concern is not with Scripps, Woods Hole, or Lamont as institutions per se, but as sites where both seminal scientific breakthroughs and painful contestations occurred. I therefore also include a discussion of the work at Princeton University of Harry Hess, who made key contributions to the theory of plate tectonics but also played a major role in challenging Navy data classification.

My focus on these particular institutions should not be read as disparaging work done elsewhere. I have not, for example, looked at oceanographic research within the Navy, such as at the Hydrographic Office—although the Naval Research Laboratory makes several appearances. I discuss only in passing other academic institutions where oceanographic work was already being done in the 1930s, such as the University of Washington, or that became important centers for oceanography in the 1960s, such as Oregon State University and the University of Rhode Island. Centers of oceanographic work outside the United States play a role in the story only insofar as they highlight important points about American oceanography by comparison. In this sense, this book is not a history of oceanography; rather, it uses oceanography to address the question of the impact of funding on the subject, scope, and tenor of scientific work.

I have not attempted to explore the Navy "side" to my story in an equivalent manner as the scientific "side." Some readers may feel that I have given short shrift to the military perspective on the collaborations that drive this story, but that analysis has already been undertaken by Navy historian Gary Weir.[38] That said, I have endeavored not to make the Navy monolithic, stressing throughout that the military side of this story is not only—and perhaps even not even primarily—represented by the Office of Naval Research.[39] However, what mattered most for scientists was not what their military patrons really needed—if such a thing could be ascertained—but what scientists thought they needed and, more important, what scientists could persuade them to fund.[40]

Finally, I have not attempted to compare the impact of Navy funding with other patrons of earth science, such as industry. If I had, this book would have threatened (even more than it has) to spiral out of control. In any case, it would have been a different book. In the conclusion, I do offer some general conclusions about what made the Navy a good patron from the perspective of many earth scientists, and even, to a certain extent, from my own perspective.

Throughout this narrative, I have tried to understand how scientists took positive advantage of the opportunities presented to them in the Cold War and how they navigated the twin challenges of military expectations and military secrecy. Above all, I have tried to determine whether Navy patronage affected the content of the scientific work that was done and, if so, how. I want to show what scientists learned—and did not learn—on the Navy's dime, and what difference it made that it was the Navy, and not some other patron, who paid for this work. To the extent that I am concerned with balance, it is that I am equally interested in the production of knowledge and of ignorance.

The book also seeks to explain why, throughout this time period, oceanographers downplayed the impact of Navy funding. For, despite saying generally good things about the Navy as a patron, oceanographers have tended to de-emphasize the Navy role, as if it could just as well have been the Forest Service or the Post Office that funded them. During the ONR's anniversary, scientists rarely identified anything particular about the Navy that they considered salient—other than the claim that the Navy funded basic research.

The stories told here challenge that framework. I argue that the Navy role was highly salient. Most obviously, Navy priorities largely set the research agenda. The chapters that follow provide concrete examples of projects that were rejected because they did not fit "the mission profile," including some that in hindsight are of obvious societal importance and intellectual significance. We will see how Navy priorities focused attention on particular natural phenomena that in some cases inspired productive lines of thinking and investigation, but in other cases thwarted them. We will also see how Navy control of information created large domains of classified knowledge not available to scientists who did not have a "need to know." And we will see how fifty years of Navy sponsorship had cultural consequences that affected what scientists could do when the Cold War ended.

Because some scientists might misread my intent, let me stress this: I am not saying that the oceanographers and geophysicists in this story were bad human beings or were necessarily ill-motivated. This book is not an exposé. Readers may conclude from chapter 8 that Charles Hollister broke the law when he used government funds to lobby for deep-sea disposal of nuclear waste, or from chapter 9 that scientists at Scripps made serious errors of judgment when they discounted both public opinion and the views of other scientific experts to push aggressively for a project that they wanted but that others found problematic. For the most part, however, the scientists in this history did not do anything wrong: no clinical subjects were exploited, no higher animals were sacrificed.[41] To the extent that the American people knew about this work at the time, they had little complaint with it. On the contrary, during the Cold War, the need for an expansive, sophisticated military presence in the global oceans was broadly accepted, and some scientists were proud to acknowledge the link between their scientific work and the geopolitical exigencies of the Cold War.[42] But many scientists downplayed— and some even lied about—the military linkages, insisting that the Navy was supporting them to do basic research even when that was manifestly not the case.

Why did these scientists feel the need to insist that they were doing basic science, to downplay the interconnections between their research and military matters, and, above all, to insist that they had not lost control of their intellectual agenda? I suggest that they were, in fact, quite worried that if they had not yet lost control of their science, they well might.

Chapter 1 begins by investigating a disturbing incident in the 1930s. Most people today are likely to assume that if anyone were to object to military funding of scientific research, it would be political liberals; certainly that was the case during the Vietnam War years. But in the late 1930s, a group of conservative faculty at Scripps objected to Navy funding on the grounds that it would threaten the autonomy of science. These men were part of the so-called freedom-in-science movement, which opposed government funding or direction of science as socialistic, a threat to scientists' intellectual autonomy, and a threatening impediment to scientific progress. The conservatives lost this debate and were proved wrong: government funding flowed, and it did not lead to the socialization of science. But the conflict did create a lasting schism between the conservative faculty and the institution's Norwegian director, Harald Sverdrup, who a few years later was accused by the same faculty of being a Nazi. This led to Sverdrup's being denied security clearance and prevented from working on wartime classified military projects, undermining his leadership position at the institution. It also raised troubling questions as to whether he would be able to direct its programs after the war, since many of those programs would be at least partly classified or would rely on classified data. As a result, Sverdrup left both the United States and the field of oceanography. He had welcomed military funding as good for his science, but it proved bad for him personally.

Chapter 2 dives into the question of whether scientific patronage can affect not just the questions asked but also the answers obtained, tracking the development of one of the most important theoretical advances of twentieth-century oceanography: the Stommel-Arons model of deep-ocean circulation. Before midcentury, it was a matter of considerable debate as to whether there were deep-ocean currents and, if so, what forces could drive them. The question was answered affirmatively by Henry Stommel's pathbreaking work with physicist Arnold Arons on deep-ocean circulation driven by density gradients and the rotation of the Earth. Many senior scientists point to this work as exemplary of basic scientific research supported by the US Navy. However, Stommel's work was not independent of military concerns. Not only was it linked to problems in sonar transmission, but the central insight on which it was based—the existence of the thermocline—was a direct outcome of operational work. If there is a relation to be discerned here between basic and applied science, it is the inverse of what is often asserted: basic science did not lead to application in this case; rather, an operational problem led to a fundamental scientific insight. Had Stommel not been paying particular attention to the thermocline—had he not been troubled by its existence in a way that few, if any, oceanographers before him had been—he would not have framed the problem in the way that he did. A specific operational problem led him to attend to something that others working on deep circulation had ignored, and this led to the insight that became the basis of his theoretical breakthrough.

Chapter 3 continues investigation of the issue of scientific autonomy through the story of the "Palace Revolt," a faculty mutiny at Woods Hole in the early 1960s. The revolt was triggered by the very developments that the apprehensive scientists at Scripps had feared in the 1930s: the role of the Navy in setting the research agenda and the prioritization of Navy needs over the interests of fundamental research. Feeling that their director was too responsive to Navy needs and too little invested in basic science, a group of Woods Hole faculty demanded that the trustees ask for his resignation. Their demand was rebuffed, the mutinous faculty driven out, and research at Woods Hole continued to be heavily directed toward science that, in the director's words, "fit the mission profile."

Chapters 4 and 5 address the question of secrecy. It has long been an article of faith among many scientists that research must be free and open to operate or, at least, to operate optimally. Yet historically, a great deal of science has been done in secret. This was particularly true in the Cold War. Did it matter? In chapter 4, we follow Harry Hess, the Princeton professor credited with developing the concept of seafloor spreading, the crucial idea that laid the foundations for the theory of plate tectonics. Hess felt that secrecy was stymying his science, and he worked his Navy contacts to try to get key data declassified, but without success. In chapter 5, I follow marine geologists Bruce Heezen and Bill Menard, who were both thwarted by security restrictions in trying to interpret key data from the seafloor.

Hess feared that military secrecy would be bad both for military officers who would not know what useful information existed and for scientists who did not have access to information that could advance their field. But he went further than arguing that secrecy was problematic in principle: he claimed that science was being impeded in practice, that advances were not being made because of military-imposed secrecy. Of course, it is impossible to prove that an advance was not made that under other circumstances could or would have been made, but in these two chapters, I argue that Hess was right: Navy secrecy stood in the way of the emergence of modern global tectonic theory.

The impact of secrecy also helps to explain comments that appear otherwise astonishing. In 1966, on the eve of the plate tectonics revolution, George Woolard, president of the American Geophysical Union who was also working closely with the US Air Force on gravity measurements related to missile guidance, complained that earth science was "in a bad way." Writing to National Aeronautics and Space Administration (NASA) director James Webb, Woolard griped that scientists needed an "earth program" comparable to the space program. He bemoaned that geoscientists had a bad history of "mixing fact with fiction in studying the earth, [and placing] too much significance ... on limited data."[43] Woolard lodged this complaint at the very moment that data collection in earth science had reached a historical zenith. There was no dearth of data, but, for the reasons that Hess bemoaned, there was a startling dearth of knowledge about and access to that data, even among leaders of geophysical science.[44] That dearth, I argue, had both social and epistemic consequences.

Chapter 6 and 7 continue the argument that Navy patronage impeded scientific investigations that were not seen as pertinent to Navy needs. In chapter 6, I explore the history of the deep submersible research vessel *Alvin*, which has long been touted for its role in "basic science," particularly the discovery of deep-sea hydrothermal vents and the remarkable biotic communities they sustain. But *Alvin* was not developed as a research vessel. It was developed to satisfy the demand for deep-submergence capacity to assist salvage operations and to develop a long-range active listening system to detect Soviet submarines. Although it may be hard to believe in hindsight, during the planning stages few Woods Hole scientists could imagine much scientific use for a deep-submergence vessel. But after *Alvin* became a research vessel, its early history and role in classified projects were whitewashed. Thus, I argue, contra earlier accounts, that scientists did not "paint their projects blue," taking basic science and pretending it had military relevance. Rather, they "painted their projects white," cloaking military projects under the cover of basic research. Chapter 7 explores how, even once *Alvin* became a research vessel, its agenda was still largely set by military demands, and projects that did not fit the mission profile were rejected even when they were of profound scientific interest.

Chapters 8 and 9 consider what happened to oceanographers as military funding began to wane and they needed to find new patrons and a new context of motivation for their work. Chapter 8 follows the work of Charles

Hollister, a marine geologist and dean of graduate studies at Woods Hole. When funding dried up for his research on deep-sea sedimentology, Hollister turned to the problem of radioactive waste disposal. For some years the US Department of Energy funded him to study the deep sea as a potential nuclear waste repository, but when the US government decided to focus instead on land-based disposal, he refused to accept that decision. Hollister became an outspoken political advocate for deep-sea disposal and an opponent of Yucca Mountain (the land-based alternative designated by the US Congress), writing opinion pieces, lobbying members of Congress, and appearing on television with the conservative pundit William F. Buckley to make his case.

The final empirical chapter, chapter 9, considers the problem of the legacies of Cold War funding through the lens of a sad saga that took place at the end of the Cold War. In the early 1990s, a group of oceanographers proposed a clever but controversial project: Acoustic tomography of ocean climate. It was clever because the scientists realized that military underwater listening systems developed in the Cold War could be used to determine whether the world ocean—and by implication, the entire globe—was warming and thereby prove the reality of climate change. It was controversial because they failed to attend to the fact that the sound transmissions they proposed using had the potential to disrupt a number of forms of marine life, including several species of endangered whales. When cetacean biologists and whale aficionados raised this concern, the oceanographers responded in an arrogant and dismissive manner. The result was a painful, prolonged, and expensive conflict in which the oceanographers—trying to shift their attention from military to civilian projects—found themselves distrusted by civilians who questioned their motivations and doubted that the military tigers had changed their stripes. I argue here that motivations matter in framing not only what scientists decide to do but also in how they are viewed by others, and therefore whether they are seen as trustworthy.

In the conclusion, I return to the motivating question of this book: What difference does it make who pays for science? The short answer is: a lot.

The Production of Knowledge and Ignorance

In my prior work on the history of debates over continental drift, plate tectonics, and anthropogenic climate change, I have been primarily interested in the production of scientific knowledge. I have queried how scientists decide when they have enough evidence of sufficient quality to say that a scientific question has been answered, as well as how they judge what constitutes "evidence" and "quality."[45] Here, I am interested in how military funding affected which questions scientists believed needed answering in the first place and how they went about answering them. I argue that, while Navy funding produced a great deal of scientific knowledge, it was also productive of considerable ignorance, not only by bringing some questions to the fore and pushing others aside, but also by structuring how scientists thought about the ocean and what they even thought the ocean was. The military context of motivation led oceanographers to view the ocean primarily as a medium through which sound was transmitted and men and machines would travel, and not as an abode of life. This, I argue, had significant, lasting consequences. Thus, I offer this work as a contribution both to the history of science—the study of the production of knowledge—and to agnotology, the study of the production of ignorance.

To write this book has required delving into the history of these scientists' work in considerable detail. As my longtime colleague Steven Shapin has put it (in a different context and with no pun intended), "The very possibility that history can contribute to the general understanding of human behavior arises from the depth of detail it can dredge up from the past."[46] This book has involved much dredging. Each chapter offers a detailed account of one particular discovery, research program, conflict, or failure in order to understand why the work was undertaken or the conflict erupted, and how and why it succeeded or failed.

I have attempted to tell these scientists' stories in their rich complexity in order to recover the texture of their lives *qua* scientists trying to understand complex phenomena that are vexingly difficult to access without expensive equipment and instrumentation, which in turn makes their work impossible without deep-pocketed patrons. This book is not one story, but a set of intercalated stories about a group of men and women (though mostly men) who worked to satisfy these patrons and sustain support for their research while maintaining their focus on the natural world and their own vision of what it meant to be a scientist. That challenge was not a trivial one: sometimes they succeeded, sometimes they did not. Sometimes they were happy; sometimes they suffered.

Every history is a history of choices made and not made, paths taken and forgone. Every history of science is a history both of knowledge produced and of ignorance sustained. The world as we know it is one of many possible worlds; it is the task of the historian to understand how and why it came to be this particular world.[47] Although the chapters in this book may be read individually, they are intended to be read collectively as narrative history, with each chapter illuminating one or more aspects of Cold War oceanography. Some chapters focus on a particular individual, and others on a particular development, discovery, or problem. Collectively, they attempt to paint a landscape of scientific work in a particular time and place. But while the time is the Cold War and the place is America, it is, more broadly, illustrative of the landscapes in which scientists strive to make durable discoveries about the planet we live on in a social and political world that is at least as complex and difficult to understand as the natural one.[48]

1 *The Personal, the Political, and the Scientific*

For American oceanographers, the Cold War arguably began during World War II, when federal funds began to flow freely into academic science for the first time in US history.[1] In time, the Navy would come to be a trusted and valued partner—praised by oceanographers for supporting their work generously and with relatively few constraints—and the Cold War would be viewed as a golden age in oceanography because of the enabling effects of Navy support.[2] But the beginnings of Navy-academic relations were rough: leading scientists had doubts about the wisdom of hitching their intellectual horse to the armed services wagon. Chief among the issues raised by the wartime funding for ocean studies was the foundational question of the purpose of scientific research. Who would set the institution's research priorities? Was science to be driven by the interests and curiosity of scientists or by the expectations of their patrons?

This was not a new question—scientists had pondered it before—but it gained a new urgency as Navy funds began to compete with and then eclipse other sources. If scientists accepted Navy funding, to what degree would the Navy set the research agenda? Would the Navy direct their work?[3] At the Scripps Institution of Oceanography in La Jolla, California, the issue came to the fore in the late 1930s. When the Navy approached Scripps to do work on its behalf, the institution's brilliant and ambitious director, Harald Sverdrup, welcomed the prospect of a new and generous source of badly needed funds. But some of his staff feared that Navy patronage would alter the purpose of their research and cede control to outside forces and extrascientific considerations.

This led to a major dispute, first about the role of Navy funding specifically, but then about larger questions of loyalty, intentions, and purpose. A debate that began about the politics of science ultimately became deeply and damagingly personal.

Figure 1.1 Harald Ulrik Sverdrup, 1938.
Image by Eugene Cecil LaFond, from the Eugene Cecil LaFond Papers, Special Collections and Archives, UC San Diego. Reprinted with permission of UCSD.

Tensions over Money, Priorities, and Privileges

When Harald Sverdrup (1888–1957) accepted the Scripps directorship in 1936, it was widely considered a coup for the institution. His name had come up repeatedly during the search process, but few believed that Sverdrup would be willing to leave his native Norway.[4] After all, Sverdrup was much more famous than Scripps was (fig. 1.1).

Sverdrup was one of the leaders of the new field of dynamic oceanography, in which the principles of dynamic meteorology, which had revolutionized that field, were being applied to oceanography.[5] Sverdrup had studied in Leipzig, Germany, with the founder of dynamic meteorology—Vilhelm Bjerknes—and then traveled with the great polar explorer Roald Amundsen on the Norwegian North Polar Expedition (1918–1925), better known as the *Maud* expedition, after the name of the ship.[6] Using the principles of dynamic oceanography, Sverdrup had used the *Maud* data to analyze ocean-atmosphere interactions.[7] Although his contributions rested firmly on the theoretical training he had obtained from Bjerknes, Sverdrup valued above all the "extensive and varied field practice [obtained] during my many years in the arctic ... which brought me into the closest possible contact with nature."[8] In the Arctic, Sverdrup had come to believe in the necessity of collecting one's own data, without which there was no way to understand the "accuracy of various methods and the ... errors which would have to be considered" in interpreting them.[9]

But the opportunities at Scripps to collect one's own data at sea were scant (and to do so in the Arctic, nonexistent). Indeed, the scientific opportunities at Scripps were scant, tout court. Founded in 1903 as the laboratory of the Marine Biological Association of San Diego, Scripps Institution of Oceanography had twice changed its name: first in 1912, when it became part of the University of California, to the Scripps Institution for Biological Research, and then in 1925 to the name that would stick. But the latter change was mostly aspirational: no one in La Jolla was doing serious work at sea, and no one was teaching dynamic oceanography.[10] Sverdrup was hired to change that.

As director, Sverdrup found himself spending most of his time and energy worrying about money. In this regard, affairs at Scripps had changed little in the previous twenty years. During the early 1920s, director William Ritter (1856–1944) had relied heavily on the contributions of philanthropist Ellen Browning Scripps (1836–1932) and her younger brother E. W. (1854–1926). From 1903 to 1912, the Scripps siblings had provided nearly all operating funds for the biological station; after it became part of the University of California, they continued to match the state's support. But although Ellen had decided as early as 1906 to create an endowment (in the long run becoming the institution's largest donor), in the early years it was E.W. who gave the most money and had the most influence over the institution. He never created an income-producing endowment, though, preferring to support requests on a case-by-case basis.[11]

Things had changed somewhat by the mid-1930s: Sverdrup could count on a regular budget for salary and operational expenses from the university. But despite a bit of help from the Rockefeller Foundation, funds for research were extremely limited.[12] For specific needs, Sverdrup had to appeal to University of California president Robert Sproul (1891–1975), or to the latest Scripps scion, at that time Robert "Bob" Paine Scripps. And the institution was growing faster than its budget was. Money was a not just a source of worry but also a source of conflict, linked as it was to the problem of priorities.

Sverdrup had been hired with a mandate to take Scripps to sea and to coordinate and balance the overall research effort. Since the institution already had a strong presence in biology, this meant strengthening the physical sciences.[13] Sverdrup referred to this vision as "balanced" oceanography, but the biologists saw it differently; they perceived the increased emphasis on physical oceanography as a shift away from the institution's earlier focus on marine life. In one of his first staff meetings in September 1936, Sverdrup tried to reassure his colleagues, suggesting that he "would have preferred that our name had been the Scripps Institution of Marine Sciences."[14] But this was slightly disingenuous. On more than one occasion Sverdrup made clear that—like many scientists trained in physics, in this case, the physics of the atmosphere—he considered the physical sciences to be more fundamental than the biological ones. Also like many physical scientists, he knew scant biology. In a lecture to the California Alumni Association in April 1938, he explained that although both physics and biology were part of oceanography, physical oceanography was the foundation of the discipline, because physical conditions ultimately controlled life. One could not understand plankton, for example, without understanding the circulation that controlled the nutrients on which they depended. As for his own knowledge of biology, when he came to a slide of a zooplankton, he bantered, "Don't ask me what the creatures are called!"[15]

Two Scripps biologists were particularly apprehensive about what Sverdrup's leadership would mean: Denis L. Fox (1901–1983) and Claude E. ZoBell (1904–1989). Fox's field was the biochemistry of marine organisms, and his focus the comparative biochemistry of carotenoids—the yellow, orange, and red pigments found in plants and animals. His degree in chemistry was from UC Berkeley; his PhD in biochemistry from Stanford. He had come to Scripps in 1931 as instructor of physiology of marine organisms (fig. 1.2); ZoBell was hired a month later and soon became Fox's closest colleague.[16]

Figure 1.2 Denis L. Fox, 1935.
Image by Eugene Cecil LaFond, from the Eugene Cecil LaFond Papers, Special Collections and Archives, UC San Diego. Reprinted with permission of UCSD.

Figure 1.3 Claude E. ZoBell, 1952.

Image by the Scripps Photographic Library, from the Scripps Institution of Oceanography Photographs, Special Collections and Archives, UC San Diego. Reprinted with permission of UCSD.

Born in Utah and raised in Idaho in modest circumstances, ZoBell received his bachelor's and master's degrees in bacteriology from Utah State University before completing a PhD at Berkeley.[17]

Fox and ZoBell were both laboratory scientists with a biochemical focus; neither was an oceanographer. Their materials came from the sea, but they did not go to sea. In the tradition of Scripps in its early years, and of many marine biological stations, they collected samples on the pier or the beach or from colleagues who went to sea.[18] This wasn't just a matter of activities but also of identity: photographs from the 1930s almost invariably show ZoBell in a white lab coat, usually near a microscope (fig. 1.3), and he viewed his role at Scripps as building the experimental program.[19] His activity report for the year ending March 1936 stressed his efforts to build a "broad and firm foundation for future experimental work."[20]

Fox and ZoBell were also men of firm views. Fox's concerns revolved around the institution and he sought on many occasions to ensure it ran with decorum.[21] He wrote numerous memoranda: about tatty sign plates, proper conventions for numbering courses, and the need to enforce rules against sample collecting on the beach. He once complained to Sverdrup about a lack of proper stationery.[22] He was especially fussy about academic titles and on more than one occasion asked to change his.[23]

ZoBell's concerns were broader: he was dedicated to the ideal of science as a public good.[24] When his work in bacteriology of sediments led to a method to improve secondary recovery of petroleum, he took out a patent not to protect his own financial interest but to ensure that the benefits would be "dedicated to the public on a royalty-free basis."[25] But ZoBell was concerned about the institution, too. In his files he kept a reprint of a 1937 article by Scripps's first physical oceanographer, George McEwen (1882–1972), which told the story of how Scripps had grown from its roots as a small biological research station to encompass the whole of oceanography.[26] McEwen explained that at first the institution focused explicitly on biological matters, and physical conditions were studied secondarily, in relation to the life forms that they affected. But that soon led the institution into physical oceanography and meteorology. From there it was a short step to "recognition of the essential unity, physical, chemical, and biological, in the field of oceanography [that] has served to unite the interests of all."[27] This was the optimist's view; the pessimist's was that biological matters were being downgraded, even marginalized.

Sverdrup was not uninterested in biology. His germinal work *The Oceans* (coauthored with chemical oceanographer Richard Fleming [fig. 1.4] and with Scripps's first biological oceanographer, Martin Johnson) was subtitled "Their Physics, Chemistry, and General Biology," and seven of its twenty chapters were dedicated to the latter.[28] But it was clear from the book's contents—as well as from Sverdrup's comments—that the order of the sciences in the title reflected a view of their importance. Moreover, the book presented some areas of biology as part of oceanography while downplaying or entirely dismissing others.[29] Plankton garnered considerable attention because, as the authors explained, one can investigate the dynamics of water masses from the distribution of plankton, and conversely, one needs to understand those dynamics in order to understand plankton. This made plankton an oceanographic topic. Marine microbiology and biochemistry were another matter: they could be studied more or less independently of the oceans. Fish were omitted entirely, presumably for the same reason. Oceanographers would later quip that meteorologists don't study birds, so why should oceanographers study fish?[30]

Figure 1.4 Richard Fleming, 1935, coauthor with Harald Sverdrup and Martin Johnson of *The Oceans.*
Image by Eugene Cecil LaFond, from the Eugene Cecil LaFond Papers, Special Collections and Archives, UC San Diego. Reprinted with permission of UCSD.

Sverdrup confronted this tension directly in his plan for use of the new oceanographic vessel, the *E. W. Scripps*.[31] He thought that the ship's main efforts should be on the systematic study of a small geographic region in which all the represented branches of the institution would cooperate. This would include "large-scale ecological studies," as well as bottom topography, sedimentation, and the study of the water masses and their movements, to engage "nearly all the members of the present staff."[32] Laboratory work would be subsumed under this larger oceanographic program.[33] Sverdrup wanted biological experiments to be motivated by and to serve problems encountered at sea, but not to have a life of their own. He explained: "Laboratory experiments ... will more and more be designed directly towards explaining features encountered in the ocean.... At an oceanographic institution the work at sea and the information obtained from the sea has to form the basis for all research."[34]

Fox and ZoBell were unhappy at being told that their research programs had to change, not because they were wrong or bad scientifically but because they did not fit in. And as time went on, the pressure to fit in increased. In his annual report of 1940, Sverdrup reiterated his expectation that Fox's and ZoBell's research could be better integrated into the overall program of the institution. "Several staff members have continued projects which have been in progress over a period of years, some of which are independent in their nature," Sverdrup wrote, "others of which deal with questions which sooner or later will find their places in the common program. Of the latter I wish particularly to mention Mr. ZoBell's studies in bacteriology ... and Mr. Fox's studies in the biochemistry of marine organisms."[35]

There is no reason to think that Sverdrup did not respect the quality of the work that Fox and ZoBell were doing, but there is reason to think that he viewed it as peripheral.[36] This led to difficulties: awkward incidents and disagreements plagued Sverdrup's relations with them and contributed to an overall atmosphere of distrust. When World War II broke out, Navy projects at Scripps became more important, and the FBI began security investigations of the scientists who would do the Navy work, this atmosphere of distrust would create serious problems. The years 1936 to 1941 were ones of normal academic disagreements; 1941 to 1943 saw the distrust generated by those disagreements transmogrified into a matter of national security. The personal, the political, and the scientific would become entangled.

Trouble at Scripps

In one of Sverdrup's first meetings with his new staff, ZoBell complained about inequity between the physical and the biological research programs. If the biologists needed equipment, such as nets, the costs came out of their divisional funds, but equipment for physical sampling, such as Nansen bottles or thermometers, were purchased with funds allotted for boat work. Sverdrup told ZoBell he would try to ensure the biologists had the equipment they needed, but privately he was irritated, writing in a memo to the file, "I did not consider this a matter of such importance that it would be necessary to discuss it at a staff meeting."[37]

ZoBell was also troubled by his title, assistant professor of marine microbiology in charge of the Biological Program, the latter part to acknowledge the administrative work he handled. ZoBell found the title awkward and asked if it might be changed to chairman of the Biological Group. Sverdrup agreed and forwarded the request to University of California president Robert Sproul, who declined the request on the grounds that only departments had chairs. ZoBell received a perfunctory letter informing him that his title had been changed to assistant professor of marine microbiology.[38] ZoBell was livid because the change made it appear that he had been demoted; he wrote back immediately to insist it not be implemented in the forthcoming university bulletin. However, when the bulletin next came out, it was just as he had anticipated.[39] Sverdrup finally found a title with which ZoBell was content— assistant to the director—but it took more than a year to resolve, and considerable ink and effort were spent reassuring ZoBell that there had been no ill intent.

Research support, titles—the obvious remaining issues for any academic were salary and promotion. ZoBell had been hired as an instructor in January 1932, but because of the financial strictures caused by the Great Depression, in 1933 the university denied him a promotion to assistant professor. In 1934, both Fox and ZoBell were recommended for promotion, but finances were again cited to deny them.[40] In December 1938, Sverdrup recommended them for promotion to associate, but the recommendation was rejected on the grounds that the men had scarcely been assistant professors! Sverdrup came to the men's defense, insisting, quite fairly, that they not be penalized because their earlier promotions had been delayed for fiscal reasons.[41] But the request was turned down again.

Fox and ZoBell did not know the efforts Sverdrup had made on their behalf—Sverdrup was a man to keep his own counsel—but they correctly perceived that they were not his favorites. That distinction went to his two young instructors in oceanography, Richard Fleming (1909–1989) and Roger Revelle (1909–1991). Although Sverdrup never wrote anything bad about the biologists, he often praised them faintly on the grounds that he did not know their fields. Of ZoBell, he wrote: "He is, I understand, highly thought of as a teacher.... He works very methodically and is able to present his research briefly and clearly." Of Fox, he is "a good lecturer and I believe a good teacher.... [He] is highly rated in his special field." In contrast, Sverdrup wrote of Revelle's "unusually clear mind ... and great ability as an independent worker." Fleming belonged to "the group of young men at the Institution, who during the last years have brought new lines of thought and new ideas and whose services in the future can be expected to be of increasing value."[42] And it was these men—particularly Roger Revelle—who would help Sverdrup build the institution, with the help of the US Navy.

Government Support versus Freedom of Science

Like all seafaring organizations, the US Navy had to deal with "fouling"— the problem of ships' hulls being damaged by marine organisms. Since 1932, ZoBell and his Scripps colleague W. E. Allen had been in contact with the Navy Bureau of Construction and Repair, which wanted to engage Scripps scientists in antifouling studies.[43] In 1935, Navy representatives invited ZoBell and Allen to submit a research proposal on the "mechanics of the action of toxics in preventing attachments" of barnacles, mussels, and other marine organisms; in the spring of 1937, Scripps received its first grant, of $450.[44] ZoBell and Allen would work with a representative of the Navy's Bureau of Ships named W. Forest Whedon.[45]

By December 1939, the project was thriving.[46] In June 1940, the budget was dramatically increased: to $2,270, with additional monthly billing for secretary services up to $1,200.[47] Compared to the institutional budget of $98,500, this was modest but measurable.[48] Within a few months, the budget mushroomed to $12,000 per year.[49] The expanded antifouling project was a watershed for Scripps in two respects: It was larger than any previous outside grant or contract, and it exceeded the yearly grant from the Ellen Browning Scripps Foundation, which at that time was about $10,000 per year.[50]

The Navy had become the leading external patron of oceanographic research at Scripps, and its money had come easily, with the exchange of a just a handful of letters. Sverdrup was relieved to have such significant funding flowing in, helping to cover staff salaries and badly needed research assistantships for graduate students. In exchange, Scripps had to provide only space, expertise, and a promise that "in the case of the [Navy] research assistants, the regular staff members will have daily contact with their work."[51] But trouble was brewing.

As the project expanded, Fox and ZoBell began to express reservations, particularly about who was actually in charge. Fox was troubled by Whedon's role; he felt strongly that a faculty member should supervise the project. Sverdrup had described the contract as supporting the salaries of staff who were assisting Whedon, who had also been given an academic title. This implied (to Fox, at least) that Whedon was in charge.[52]

In a staff meeting in February 1942, Sverdrup insisted that the project was a university project because all the participants were employed by the university, albeit "on special funds." But this ignored Whedon, who employed by the Navy, not the university, and Fox and ZoBell were not assuaged.[53] Sverdrup assured them that the students involved would report to their academic supervisor, who in turn "takes over responsibility of tendering report to Bureau of Ships via W[hedon]." Fox and ZoBell were not convinced.[54] In their view, graduate students were working for the Bureau of Ships.

Another issue was secrecy and publication. Would the Navy allow their results to be published? What would happen to students working on the project if they were not able to publish their findings? In his notes on the staff meeting, Fox recorded: "Publishability of research? [Sverdrup] thinks so with scientific results, for scientific world, with permissions of Navy. Can apply only to what student is doing for Bureau of Ships. If student thesis applies directly to the fouling problem, permission to publish is [definitely] necessary."[55] The scientists would have to find a way to segregate the "scientific" work from the "Navy" work—to create a kind of intellectual apartheid. How would they do that? The distinction proposed was whether the results applied directly or only indirectly to the fouling problem; the latter could be freely published. Either way, the Navy's permission would be needed.

These concerns were formalized in a memo signed by Fox, ZoBell, and the three other professors, who insisted that the head of any student's research committee always be "an academic staff member."[56] ZoBell followed up the next day with a more pointed memorandum, underscoring the urgency for academic research to remain under academic direction: "[The project] places the institution in an embarrassing position to expect its staff members to serve under the direction of one who is neither an academic staff member nor of recognized scientific standing. By the same token, we would object to working under the direction (either implicit or implied) by the hireling of any other governmental bureau or agency."[57] ZoBell and Fox already saw their autonomy shrinking as Sverdrup built on his vision of an integrated oceanographic program; now it would shrink further. Moreover, ZoBell argued that the separation of research into "academic" and "Navy" domains was hardly unproblematic. There was no sharp line to be drawn between them, and the threat of loss of control, even over the academic portions, was real: "The work of several staff members is already concerned with sedentary organisms. If we accept ... student assistants on any of these projects, the results can be construed as belonging to the Navy, and publication may require the formal approval of the Navy."[58]

In retrospect, ZoBell's insight sounds like a harbinger of complaints lodged in the 1960s against military funding of university research from the left wing of the American political spectrum or in the 2000s against corporate funding.[59] Like later opponents of military funding of academic research, ZoBell's opposition was based on a commitment to academic freedom and concerns over the twin threats of secrecy and state control. However, unlike that of later critics, Fox and ZoBell's stance emerged not from the left of the political spectrum but from the right.

Both Fox and ZoBell were members of the Society for Freedom in Science (SFS), founded in 1939 in opposition to the left-wing British Association of Scientific Workers. The latter advocated state direction of science for the public good, whereas the former was dedicated to ensuring liberty in scientific research, by which they meant the freedom of scientific workers to choose their own research topics and "to work separately or in collaboration as they may prefer."[60] In a later letter to Fox, SFS secretary John R. Baker lamented the growing influence of Marxism among European scientists. "The movement against freedom is now very powerful in Britain," he wrote in 1944, "being repeatedly and loudly voiced by first-rank scientists (e.g. Bernal, Haldane, Hogben, Watson Watt)."[61]

Explicitly anti-Marxist, the SFS took its intellectual lead from British philosopher and professor of physical chemistry at Manchester University Michael Polanyi, whose reprints were distributed to members. Polanyi critiqued state direction of science for the common good as meretricious: its intent, to serve society, was good, but its effect, to damage science, would be bad. Moreover, it was premised on a logical error: eliding the distinction between pure and applied science. For Polanyi, these were different enterprises, requiring different intellectual stances. Applied science advanced by finding new ways to attack old problems. It required a willingness to try approaches lacking obvious relation to what had gone before, and it flourished when interacting with ideas generated in the outside world. Pure science, in contrast, advanced by increments, each step building on the one before, which required isolation and detachment from the world at large. It had to be allowed to follow its own internal dictates, which accounted for the academic tradition of withdrawal from the "real" world: "Scientific thought can only flourish in a scientific atmosphere," Polanyi wrote. "Academic seclusion represents, therefore, an indispensable framework for a single-minded application to systematic science."[62] Applied science could be directed, but pure science needed to be left alone.[63]

Fox had taken Polanyi's concerns to heart. Along with the Polanyi reprints, he kept in his files a clipping from the *San Diego Union Tribune* that quoted Polanyi stating that, in the context of Marxism, war, and emigration, Britain and the United States were the "keepers of science." It was up to British and American scientists to carry the torch of intellectual freedom.[64]

Consistent with these principles, ZoBell withdrew from the antifouling project and, during the war, worked on a project funded by the American Petroleum Institute on the role of bacteria in petroleum generation.[65] In contrast, war changed Fox's mind. In July 1943, he declared himself "ready and willing to contribute to the Navy's project ... any services and consultation which our training and experience might afford of value."[66] Following Sverdrup's suggestion, the antifouling project was divided into two parts; paint (Navy) and organisms (academic), with Navy funds paying for the former and university funds for the latter.[67] In practice, this became a formality: Fox was soon testing the antifouling efficacy of mixtures of hydrophobic materials and detergents applied to planks of wood hung off the Scripps pier; his reports and letters reveal no indication as to which parts of the project belonged to academia and which to the Navy.[68] But while he may have come to terms with the Navy, he had not come to terms with Sverdrup.

World War II Comes to Scripps

By 1940, the antifouling studies were only a small part of the war work that oceanographers were undertaking on behalf of the US Navy.[69] A new civilian scientific agency, the National Defense Research Committee (NDRC), was interested in the role that oceanographers, physicists, and geophysicists could play in subsurface warfare. By the summer of 1941, the NDRC had established two new laboratories. One of these, to be run by the University of California, would be located at Point Loma, fifteen miles south of Scripps. Its focus would be the application of underwater acoustics to antisubmarine warfare. John T. Tate (1889–1950), vice chairman of NDRC Division C, wanted Sverdrup to head the program.[70]

Sverdrup began to work two or three days a week at the Point Loma laboratory—by then renamed the UC Division of War Research—supervising a small oceanographic division. He and Richard Fleming conducted a variety of studies tracking currents and water masses off the California coast, developing instruments for measuring temperatures at sea, and analyzing records from the bathythermograph (BT), a new instrument for measuring temperature as a function of depth in order to enhance sonar capability and improve ocean acoustic surveillance of enemy submarines.[71]

Meanwhile, Sverdrup had also been asked to develop a program to train Army Air Force (AAF) meteorologists for war duty. Joseph Kaplan (1902–1991), chairman of the Physics Department at the University of California at

Los Angeles (UCLA), had long argued that the AAF needed an officer corps trained in weather forecasting. The Nazis had sent Luftwaffe personnel to Norway for meteorological training; two of the scientists who had helped to educate the Germans were Vilhelm Bjerknes's son Jacob (1897–1975) and Jürgen Holmboe (1902–1979), both of whom had fled Norway after the occupation and landed at UCLA. By June 1941 Kaplan had them teaching for the Allied cause. By summertime, he had convinced Sverdrup to join them.[72]

Sverdrup had agreed to train forecasters because he had been excluded from more sensitive (and more scientifically rewarding) work on underwater sound. In fact, when the Point Loma facility was launched in July, a Military Intelligence Division (MID) investigation had also been launched. By September the FBI had concluded that Sverdrup should not be trusted with state secrets. A confidential memo from the War Department to Lieutenant Colonel J. Edgar Hoover (1895–1972) of the FBI summarized: "Mr. Sverdrup has a fine scientific record ... but, in view of conflicting reports as to his loyalty, there is considerable doubt as to the advisability of his employment on the National Defense Program."[73] J. Edgar Hoover concurred.

When interviewed by the FBI, the vast majority of Sverdrup's associates had firmly defended his integrity and loyalty. At the California Institute of Technology, a prominent physicist pointed out that Sverdrup's visibility in the international scientific community made it impossible for any serious character flaw to have remained undisclosed.[74] In La Jolla, the manager at the local branch of Bank of America was certain that Sverdrup could be "relied on as to discretion, loyalty, and integrity."[75] The lawyer for the Scripps' estate noted that at the time of the German invasion of Norway, Sverdrup was "grief-stricken over the plight of his country and country-men."[76] Lieutenant Colonel Kenneth C. Masteller of Camp Callan, California—a temporary Army base established in the hills above Scripps that would later become the campus of UC San Diego—had arranged for Sverdrup to speak of his Arctic voyages at the Camp Callan Service Club, and found him cooperative, loyal to America, and harboring "great resentment toward Germany for that country's invasion of Norway."[77]

Why, then, would anyone doubt Sverdrup's integrity or question his loyalty to the Allied cause? The answer came in an anonymous letter postmarked in New York City, reporting several persons as "suspected Nazi sympathizers"— among them Harald Sverdrup. The letter claimed that Sverdrup was a friend of Sven Hedin, the Swedish geographer, cartographer, and travel writer who mapped Marco Polo's Great Silk Road for the Chinese government and actively supported the Nazis. Given the intimate size of Scandinavian intellectual circles in the 1930s—particularly Scandinavian explorers—it is likely that Sverdrup did know Hedin. But this was not the only charge.

Based on information gleaned from people in La Jolla, George L. Shea, a commander in the Civilian Conservation Corps in Sacramento, California, accused Sverdrup of harboring communist sympathies. Among other things, he alleged that Sverdrup had "surrounded himself with ... personnel, who, in their actions and statements, have indicated that they had advanced communistic attitudes." Shea was particularly troubled by Lodewyk Lek (1905–1968), a Dutch-Jewish oceanographer who had come to the United States as a refugee from fascist Europe.[78]

Born in the Netherlands to a wealthy diamond merchant, Lek had participated in the Dutch *Snellius* expedition (1929–1930) and then studied in Berlin under the great Alfred Defant (1884–1974), Germany's leading expert in ocean-atmosphere dynamics.[79] Like so many European Jews, Lek saw his plans derailed by political events, and he fled for the United States in 1936, applying for citizenship that year. In March 1939, he asked Sverdrup if he might work at Scripps (he also had a brother in La Jolla). Consistent with a policy of encouraging scientific visitors, Sverdrup said yes.[80]

Lek's academic credentials and oceanographic experience made it easy for Sverdrup to welcome him. His independent wealth no doubt helped as well—Lek had not asked for a salary—but for Shea that was grounds for suspicion. So were Lek's personal habits. Besides being "provided with an exceptionally large amount of money," Shea noted, Lek "has had at his residence ... cars containing strange people from nearly half the states in the union coming to his house in the early evening, and remaining until late at night."[81] According to one neighbor, "various activities at the house, and the secretive manner of conducting these visits, have indicated that Dr. Lek was covering up activities he did not wish known." In a previous residence, Lek had (allegedly) boarded up the lower portion of his house, and "late at night there were people in that lower section of the house, and apparently some type of activity going on."[82] Besides late-night visits from strange men, Lek and his brother were also accused of taking pictures with "high powered cameras of the type used by our military intelligence" and of sending out "a volume of correspondence."[83]

Lek and his French wife were said to be unsympathetic to the Allied cause. When the French-born wife of a retired US Army colonel called at the Lek's house and mentioned the "calamity that had befallen both Holland and France," Mrs. Lek (allegedly) replied that the French had got what they deserved. Shea concluded, "With both Sverdrup and Lek, there always seems to be extreme glee at the Axis successes."[84] Shea admitted that Sverdrup had never actually said anything disloyal, but there was an explanation for that: "He is too shrewd to say much of anything to anyone except those people that he definitely knows are dissatisfied with our American ideas."[85]

Shea's case was one of guilt by association, leavened with nativism: Lek was foreign and suspicious, Sverdrup was foreign and associated with Lek, ergo Sverdrup was suspicious.[86] Evidently oblivious of the internationalism of 1930s science—and ignoring the fact that people have long been attracted to the beauty of a place whose name means "the jewel"—Shea noted gravely that La Jolla, "for some strange reason, has had a large number of persons of foreign extraction."[87]

These accusations derived from hearsay, but they were seconded by a man who was in a position to know: a Scripps staff assistant named Stanley Chambers (1890–1967). Born in Nova Scotia, Chambers served in the Canadian Army in World War I; in later life he liked to be called "Captain," after his military title. After immigrating to the United States in 1922, he settled in Coronado, California, and in 1924 was hired by George McEwen to take temperatures and current measurements off the Scripps pier.

Shea had accused Sverdrup of communist sympathies, but Chambers was convinced that Sverdrup was a fascist. In a Military Intelligence Division interview, he insisted that Sverdrup was "pro-Nazi" but "too intelligent to make any public or semi-public avowal of his beliefs."[88] It was Sverdrup's wife, Gudrun, who had let slip, making "utterances in both his presence and that of Mrs. Chambers of decided sympathy for the Nazi movement." Chambers was unequivocal that Sverdrup "should not be trusted with information vital to this country's defense because of his decided sympathies with Germany and the Nazi movement."[89] He found it significant that Sverdrup had spent World War I in Leipzig and claimed that on the night the Germans had begun their invasion of Norway, Sverdrup stated "that Hitler would show the world something."[90] Moreover, one incident that occurred at Scripps before the war had suggested to at least some people at Scripps that Sverdrup harbored Nazi sympathies.

When Fox did not receive his promotion to associate professor in 1938, he did get approval for a much-desired sabbatical, which he arranged to take in Cambridge, England.[91] Shortly before he and his family were to leave, there was a going-away party at which Sverdrup staged a little joke, presenting Fox with a "letter" dated Berlin, July 15, 1938. It expressed regret that the eminent "cognoscente of organic pigments" had decided to take his sabbatical in England, and implored Fox to come to Germany instead (fig. 1.5): "Your beautiful investigations of the pigments of organisms interest us particularly because we see therein a new method with the help of which the race

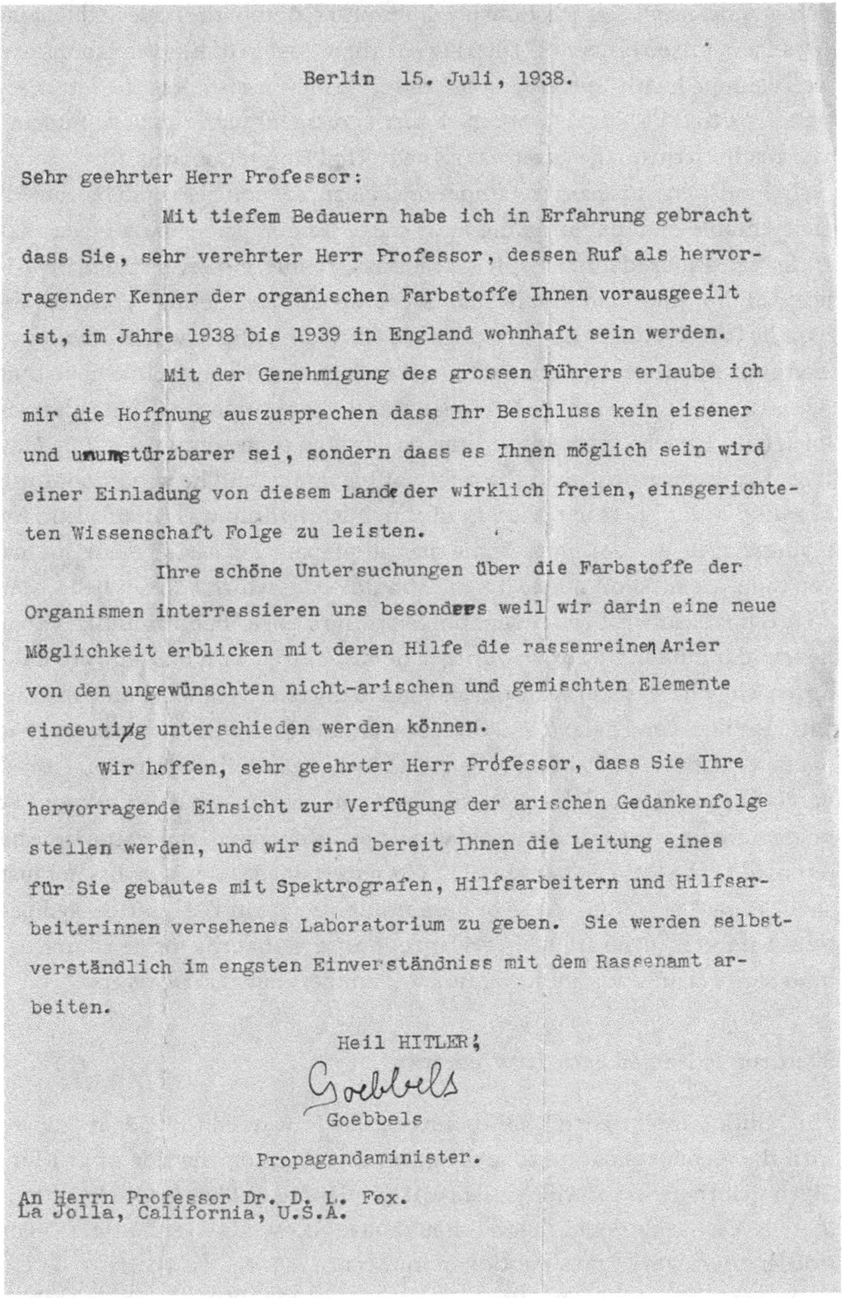

Berlin 15. Juli, 1938.

Sehr geehrter Herr Professor:

        Mit tiefem Bedauern habe ich in Erfahrung gebracht dass Sie, sehr verehrter Herr Professor, dessen Ruf als hervorragender Kenner der organischen Farbstoffe Ihnen vorausgeeilt ist, im Jahre 1938 bis 1939 in England wohnhaft sein werden.

        Mit der Genehmigung des grossen Führers erlaube ich mir die Hoffnung auszusprechen dass Ihr Beschluss kein eiserner und unumstürzbarer sei, sondern dass es Ihnen möglich sein wird einer Einladung von diesem Lande der wirklich freien, einsgerichteten Wissenschaft Folge zu leisten.

        Ihre schöne Untersuchungen über die Farbstoffe der Organismen interessieren uns besonders weil wir darin eine neue Möglichkeit erblicken mit deren Hilfe die rassenreinen Arier von den ungewünschten nicht-arischen und gemischten Elemente eindeutig unterschieden werden können.

        Wir hoffen, sehr geehrter Herr Professor, dass Sie Ihre hervorragende Einsicht zur Verfügung der arischen Gedankenfolge stellen werden, und wir sind bereit Ihnen die Leitung eines für Sie gebautes mit Spektrografen, Hilfsarbeitern und Hilfsarbeiterinnen versehenes Laboratorium zu geben. Sie werden selbstverständlich im engsten Einverständniss mit dem Rassenamt arbeiten.

                Heil HITLER!

                Goebbels

                Propagandaminister.

An Herrn Professor Dr. D. L. Fox.
La Jolla, California, U.S.A.

Figure 1.5 The "Goebbels" letter.

Reprinted with permission of UCSD.

pure Aryans may be unquestionably differentiated from the undesirable non-Aryan and mixed elements." Offering Fox directorship of his own laboratory, well equipped with "spectrographs, and ... female assistants," the letter assured Fox that he would "work in closest agreement with the Race-Bureau." Written in German, the letter was signed: "Heil Hitler! Goebbels."[92]

In hindsight, it is hard to fathom how Nazi race policies could be considered suitable material for going-away party humor, particularly given that Hitler had annexed Austria only four months before. However, like many educated Europeans (and Americans), Sverdrup admired German culture; perhaps he found it difficult to imagine that Nazi race policies would amount to anything. In a letter to Fox later that year, Sverdrup expressed his hope that "the sky will clear and that a European war will not come at all. I am an optimist, but I have had some difficulties in clinging to my optimism."[93] By May 1939, Fox was worried about the safety of his family, and he wrote to inform Sverdrup of his plans in the event of war, but Sverdrup was complacent, replying that he was "happy to know that all preparation can be made so that you can leave on short notice if a war should come, which I do not believe."[94]

It is not clear what Fox thought of Sverdrup's "joke," but it is unlikely that he found it humorous. Fox was British: he had moved with his family to California when he was four years old and had many relatives in England. In 1939, with war between England and Germany looming (as well as the threat that his hard-won sabbatical might be cut short), it is unlikely that Fox looked upon the Nazi threat with equanimity. As time went on, whatever humor Fox might have seen in the letter would likely have evaporated, and by the time he returned to the United States in 1939, Fox entertained serious doubts not just about Harald Sverdrup's sense of humor but also about the political sympathies behind it.[95] The available evidence indicates that the inside source behind Shea's claims was not just Captain Chambers but also Denis Fox.

Sverdrup Is Denied Security Clearance

The Military Intelligence Division report was forwarded to J. Edgar Hoover with the recommendation that the government err on the side of caution. Given that there was a "wide variance of opinions as to [Sverdrup's] loyalty ... there is reasonable doubt that he should have access to secret matters."[96] For Hoover, there was no presumption of innocence.

As word spread that Sverdrup's clearance had been denied, leading scientists tried to get the decision reversed. John Tate wrote to A. H. van Keuren (1899–1962), the assistant chief of the Bureau of Ships, insisting that Sverdrup was "'practically indispensable," with "better qualifications for assisting us in oceanographic problems than any person living today."[97] The director of the Carnegie Institution of Washington pointed out that Sverdrup would not have been offered the position of Scripps director had many others not held him in the same high regard. Robert Sproul insisted that Sverdrup was "not only pro-America but anti-Nazi," while the university's great E. O. Lawrence insisted that Sverdrup held an "intense hatred of totalitarian governments." Moreover, America needed him: "His knowledge and experience are vital to the submarine detection program. It would be no less than a calamity to lose his services at this time."[98] At Point Loma, Captain Wilbur J. Ruble, director of the Navy Radio and Sound Laboratory, described Sverdrup not only as the most valuable scientist at Point Loma but also "the only indispensable one."[99]

Ruble proposed a compromise: have Sverdrup supervise the work but have the data analysis done by "a loyal American citizen, Lieutenant Roger Revelle." Revelle—who would go on to become one of the most important geoscientists of his generation—was already working at the Radio and Sound Lab and had been noticed by both scientific and military leaders.[100] Sverdrup could have a desk in the same room as Revelle, and at the end of the day Revelle could put the classified data in Navy safes.[101]

Admiral Van Keuren forwarded these materials to the Secretary of the Navy, requesting "reconsideration of the clearance of Dr. Sverdrup in order that he may be employed in connection with secret and confidential matters." Van Keuren agreed that Sverdrup was "a distinguished scientist" and that there appeared to be "no question as to his loyalty to this country."[102] Vern Knudsen (1893–1974), the UCLA physicist who was supervising the scientific work at Point Loma, also wrote to the Navy Secretary, insisting that Sverdrup's services were of "vital importance ... in fact, irreplaceable."[103] Knudsen meanwhile was lining up the heavy artillery: Vannevar Bush (1890–1974), head of the US Office of Scientific Research and Development, direct adviser to President Franklin Roosevelt, and chief architect of the US military-scientific mobilization. On October 25, Bush wrote to Dr. Jerome Hunsaker (1886–1984), the first Navy coordinator for research and development, supporting Sverdrup's presence at Point Loma.[104]

Sverdrup was meanwhile taking steps of his own to remedy the situation. Although he had originally promised to stay at Scripps only three years, when war broke out, Sverdrup concluded that his future was in the United States, and in June 1940 he applied for citizenship.[105] Thinking this might help his case, he cabled Tate to tell him that he had filed his "declarations of intention ... with the Superior Court of San Diego on June 11 1940," and that naturalization could "take place after June 11 1942."[106]

But the MID and the FBI had already drawn their conclusions; the Army now ratified them. A memo written in September summarized their view in a single sentence: "After a careful review of all reports and records available to this office, there appears to be evidence, as shown in the attached report, which indicates reason to question the loyalty of the person named below: Harald Ulrik Sverdrup."[107]

None of the scientists at the NDRC knew the reasons behind Sverdrup's difficulties. Indeed, their proposed solutions reveal how little they knew. Thinking that the troubles stemmed from Sverdrup having relatives in occupied Norway, Knudsen proposed to Tate that Sverdrup sign a "statement or oath to the effect that whatever pressure may be exerted upon his relatives in Norway, he will not violate any confidences which would be injurious to the welfare of the United States."[108] Or perhaps a solution was to be found in the fact that Sverdrup had a half brother, Leif, who had come to the United States in 1914, fought in World War I, and was "now a colonel in the United States Engineer Corps … serving somewhere in the Pacific."[109] (Leif Sverdrup [1898–1976] was in fact one of General Douglas MacArthur's chief engineers.) If there were concerns with relatives in occupied Norway, Knudsen wrote, "it would seem appropriate to ask that Sverdrup's case be considered in the light of action which has already been taken in behalf of his brother Leif."[110] Knudsen also took the case to the US National Academy of Sciences, where President Frank Jewett assured him that Sverdrup would soon be cleared.[111]

An Expanding Inquiry

The files are silent as to the weight they gave to the various testimonies; one might suppose that the FBI would have weighed the testimony of distinguished scientists who knew Sverdrup well more heavily than that of minor figures who scarcely knew him at all. But there was a source of trouble close to home. While nationally prominent scientists were testifying on Sverdrup's behalf, the FBI had gone to Scripps and found the two people who knew him on a day-to-day basis and were willing to speak against him: Denis Fox and Claude ZoBell.

In 1941–1942, the FBI undertook an expanded investigation, which produced reports from four anonymous informants in La Jolla: W, X, Y, and Z. From the particulars it is clear that W and X were Shea and Chambers, respectively, whose earlier comments were reiterated in greater detail.[112] New in this report were informants Y and Z, both identified as Scripps faculty. Informant Y was particularly sensitive to Sverdrup's lack of sympathy for the English; informant Z had known Sverdrup for five or six years. Later documents identify them by name: Y was Denis Fox, and Z was Claude ZoBell.

To Fox it was clear that Sverdrup and his wife were Nazi sympathizers, "to say the least," but now that war was on, they were "cagey." However, "every once in a while these feelings got the best of them, resulting in these outbreaks."[113] One outbreak occurred at the time of the invasion of Norway, when Sverdrup "turned white and stated, 'G— D—it. I hate the English and hope Germany wins the war.'" Sverdrup, Fox claimed, "always took the attitude that the British were trying to spread the war to Norway, and has often indicated his dislike for the English." He "expressed no sympathy for the Poles, and asserted that it is no one's business what Germany does in Europe, and furthermore most intelligent people in the world realize that Germany runs Europe, anyway."[114]

Another outburst occurred when the son of a staff member asked Sverdrup how Norwegian science would be affected.[115] "Lots of Norwegian scientists have no objection to German scientists in Norway," Sverdrup allegedly replied. Gudrun volunteered that "Hitler would take all of Europe and the United States, and we would have the good old German government here." Regarding Japan, "Imagine a little country like Japan giving this country a beating," he (ostensibly) declared.[116]

ZoBell had little concrete to offer—indeed, one of his complaints was that he saw Sverdrup only infrequently—but he took as suspicious the fact that Sverdrup had scarcely mentioned the war since its outbreak, even though he had had many opportunities to do so. ZoBell found this "most unusual, particularly in view of the fact that [his] wife ... was reportedly very pro-Nazi in her conversations with her acquaintances."[117]

FBI agents also interviewed one Scripps faculty member who defended Sverdrup in great detail. Providing more than fifteen single-spaced pages of testimony, this informant knew Sverdrup intimately and made a point of explaining how Sverdrup's personal and scientific philosophy might be misconstrued if taken out of context. Sverdrup, he explained, was a man who believed that all men needed "something besides themselves to work for," to have a purpose larger than themselves, to be part of a larger whole. For Sverdrup that larger whole was science—particularly international science. Although he believed in democracy, he was not entirely uncritical of the American version of it because he found it overly individualistic. Although it was true that Sverdrup had spent four years in Germany, he found most German scientists "pompous." His loyalty was not to Germany as a nation or culture but to the man he had gone there to work with: Vilhelm Bjerknes, a man of absolute integrity, who was "reported to be the center of anti-German intrigue in Norway at the present time despite the fact that he [was] ... past 75 years of age." Above all, Sverdrup was a man whose "life was his reputation as a scientist, and [he] would not do anything injurious to his reputation." He did not care about money and had no incentive to be an agent of a foreign power unless he believed in the principles of that power.[118]

Only one man in America knew Sverdrup that well, and probably only one man at Scripps at the time had that level of political and philosophical perspicacity: Roger Revelle. Revelle was well connected in La Jolla, so it fits that this report yielded a fresh set of leads there, all of whom seconded his favorable views. One informant suggested that people were misinterpreting Sverdrup's homesickness, pointing out that when he first came to the United States "it had been very difficult for [Sverdrup] to adjust himself to the [contrasting] nature of the Norwegian and American people." A second suggested that, like many Europeans and more than a few Germans, Sverdrup had at first been sympathetic to some aspects of the Nazi program but had soon changed his mind: "When Dr SVERDRUP first came here he ... thought that HITLER had done some good for the German people since [he] realized the chaos in which the German people were left after the last World War and also considering the organic character of the German people." But this changed after the invasion of Norway. Both Sverdrup and his wife, as well as their relatives in Norway, were now all opposed to Hitler and considered the collaborationist president of Nazi-occupied Norway, Vidkun Quisling (1887–1945), "a very low type of person." A third allowed that they were irritated by the Sverdrups, explaining that everyone in La Jolla had become "oversaturated" with the way they "talked concerning the greatness of Norway," but that did not mean they were Nazis.[119]

FBI officials were delving deeply into Sverdrup's views and eliciting more reflective comments than before, but the net result was no more consistent. Some said Sverdrup was definitely loyal, others that he was definitely suspicious. Much of the information collected was hearsay; the sum was highly contradictory.[120] However, a new charge was added, a deeply serious one that spoke directly to the question of whether Sverdrup could be trusted with military secrets.[121] It was about Sverdrup's role in the antifouling studies. Noting that the work at Scripps had begun before the war, Chambers claimed that Sverdrup had "ordered research along this line to be discontinued and that the members of the Institute should turn their attention to experiments with the California sardine." If these studies had resulted in the development of any preventative, that would have been "of the greatest military value."[122] (This, of course, was untrue: the complaint against Sverdrup was not that he had blocked the studies but that he had *pushed* them over faculty objections.)

Finally, the FBI questioned Sverdrup himself, asking about his family background and education, his time in Germany, his beliefs, his relationship with

Quisling (they had both attended the Norwegian War Academy), and his half brother Leif. Sverdrup had apparently gleaned—perhaps from Revelle—that his comments about the German invasion of Norway were an issue, because the two men told an identical story. On hearing that Quisling was to head the occupational government, Sverdrup had indeed said, "Thank the Lord," not because he approved of Quisling, but because he knew that Quisling had no following.[123] Sverdrup acknowledged that at first it was difficult for him and Gudrun to adjust to their new home, but they had come to like America and planned to stay. Sverdrup closed the interview by volunteering that "no amount of pressure on his relatives in Norway would in any way affect any confidential matter or information which he may have in his possession."[124]

The FBI was nothing if not thorough: agents were also working the case in Chicago, Boston, Los Angeles, San Francisco, and Seattle. Most of these leads turned up little new information; the exception came from the Boston field office.[125] Agents there reported on interviews with a biologist and former Woods Hole director in Cambridge, Massachusetts, as well as with the institution's current director. The former could only have been Harvard professor Henry Bigelow (1879–1967); the latter was Columbus O'Donnell Iselin (1904–1971).[126]

Bigelow recalled that about a year and a half before, when military work was getting under way, "certain individuals ... questioned the advisability of placing Sverdrup in the position of authority and confidence in these matters which involved secret scientific methods and formulas which in reality are the property of the Navy department." Being himself highly patriotic, Bigelow insisted that he would have recalled had Sverdrup ever said anything even faintly disloyal—and he recalled nothing. He did recollect some trouble at Scripps "regarding a certain type of work sponsored by the Navy which involved employees who were said to have had Communistic sympathies," but Bigelow was sure that Sverdrup "had nothing whatsoever to do with this particular group who were charged with those attitudes of a 'pink' nature." Indeed, it was his impression that Sverdrup had been instrumental in cleaning it up.[127] Bigelow concluded by noting that Sverdrup had told him that relatives had been imprisoned in Norway: in the past this would have been a source of shame, but "today it is an honor to be in prison in Norway."[128]

Like Bigelow, Iselin was mostly but not entirely reassuring. He agreed that Sverdrup's respect for Germany did not extend to political developments there, but he did wish that Sverdrup were a US citizen. Iselin (wrongly) thought that Sverdrup had not applied for citizenship and attributed this to "Sverdrup's long-standing intent to return one day to Norway, and become director of the Oceanographic Institute at Bergen." Iselin may also have been the source of Bigelow's comments about communists and pacifists, for he recounted that some students coming to Woods Hole from Scripps had expressed the feeling that if Sverdrup "were a good American citizen, he would have made some active effort to banish Communistic or Pacifist sympathy prevalent among the employees of the Institute." This was, he alleged, why the Navy department had chosen to conduct its oceanographic work at Point Loma rather than Scripps. Iselin suggested that it was possible that Sverdrup, being essentially apolitical and focused on science, was unaware of the situation, but he would feel better if Sverdrup "were to make application for American citizenship."[129]

By February, the Navy had reached its decision, based on the FBI and MID reports. Navy summaries acknowledged that the "motivations and reliability" of the informants were not known, that much of what was being reported was rumor rather than evidence, and that the diverse accusations were contradictory. Nevertheless, the negatives were weighed more heavily than the positives. On February 11, 1942, the Navy recommended that Sverdrup be denied employment on classified matters. On February 23, the chief of naval operations wrote to the Navy Secretary that he was "unable to recommend" that Sverdrup be cleared. Three days after that, the Navy Secretary formally refused consent for Sverdrup's employment.[130]

The report concluded: "Reliable information reports that subject has made remarks which indicate sympathy with the Axis cause and that he is a close friend of a prominent Swedish Nazi [geographer Sven Hedin]. There is considerable doubt as to the desirability of disclosing to him vital or secret information relative to the National Defense Program."[131] On March 1, 1942, Sverdrup was informed that he would be denied further access to Point Loma.

Weather Yes, Acoustics No

Over the following few months, Sverdrup worked with Joseph Kaplan on the weather prediction program—training officers to make forecasts—but this was teaching, not research, and Sverdrup continued to look for ways to contribute his research expertise. In early May, he wrote to Vilhjalmur Stefansson (1879–1962), the polar explorer by then living in New York City, suggesting that they might offer their special knowledge of the polar seas to military leaders. Stefansson empathized, writing that it was "a crying shame that men like you ... are not being employed to the full in ... the war effort, especially as you are ... eager to contribute. I myself am being employed, in the sense that I send one memorandum after another to Washington."[132] Stefansson subsequently arranged a meeting on Arctic problems with members of the

US military's Office of the Coordinator of Information, which Sverdrup attended, but nothing came of it.[133]

Like Stefansson, Sverdrup kept himself employed, and the next day submitted a detailed, two-page letter to the Bureau of Aeronautics proposing a research program considering whether a man lost at sea could survive on plankton. Can plankton "be used by man for nourishment?" he wondered. "Would the salt adhering to it be harmful? Would the plankton themselves be poisonous? And what kind of net would be needed to catch them?" Sverdrup hypothesized that even if plankton were not particularly nutritious, or if not enough could be caught "to sustain a man for any length of time, it is believed that the small amounts caught will be important and that in all cases the psychological effect will be beneficial." The bureau declined to pursue the idea.[134]

Throughout this time, Sverdrup continued to believe that the obstacle to his obtaining clearance was his family in Norway. Writing to Sproul in June 1942, he noted: "You may have learned that … I had to leave the project at Point Loma, because Navy regulations do not permit a person with relatives in an occupied country to work with classified material. My case is not an isolated one, but I much regret the enforcement of the rule."[135] Sverdrup still did not know the true nature of the charges against him.

The Navy, of course, was not the only armed service, and in 1942 the Weather Directorate of the Army Air Force—the sponsors of the officer training program—invited Sverdrup to direct a meteorological study for military purposes. It would have four parts:

1. Preparation of isothermal charts of the world's oceans
2. Preparation of monthly average isothermal charts for specific oceanic regions, particular coastal regions, or "any part of the world as required to meet military needs and priorities"
3. Special studies of combined meteorological and oceanographical phenomena
4. Specific forecasts of combined meteorological and oceanographical conditions for specific military operations[136]

The special studies and forecasts became the most important part of the project: using meteorological data to predict surf conditions on the beaches at sites of proposed amphibious assaults (including, ultimately, the beaches of northern France on D-Day). Sverdrup would run the project and the Army would provide funds for up to fifteen (human) computers and clerks, a contingency fund of $5,000, and money for Sverdrup to go to Washington four times in the fiscal year 1942–1943. Sverdrup would be assisted by Lek, who, though an alien, could be hired "on a temporary basis," and by his newest protégé, Austrian-born Walter Munk (1917–2019), a brilliant graduate student who would become one of the leading physical oceanographers of his generation. Ten days later, a document was issued granting Sverdrup "restricted consent on oceanographic studies only."[137] The US military had found a way to employ Harald Sverdrup—as well as two other talented foreign-born scientists—by focusing on work that required less access to highly classified information than submarine acoustics did. But the story does not end there.

A Third Round of Investigations

In May, Robert Sproul received a letter from George Turner, a staff member at UCLA, transmitting information from informants wishing "to be of direct service to the nation." The three-page letter outlined thirteen points of concern about Sverdrup and Scripps, information said to come from "about 5 on the faculty" of whom "two are well-known, [and] have talked with the government agents from time to time."[138]

Turner's letter was a mix of old and new accusations, some consistent with the facts, some patently wrong. The letter repeated old charges about the comings and goings of Lek (who was accused of doing research on morphine!), about Sverdrup's Germanophilia and Anglophobia, and about Gudrun Sverdrup's "open admiration for Hitler." In exemplary circular reasoning, the visits from the FBI were cited as evidence that something at Scripps was amiss.[139] Most pointedly, the letter expanded on the earlier accusation that Sverdrup had impeded war work at Scripps: "At least two important undertakings, apparently very useful to the federal government, have been slowed down or side-tracked: the study of fouling organisms and sea-weed [for agar]. Progress in these investigations has been brought to a stand-still.... [An] American Inst[itute] of Petroleum's offer of $36,000 for the investigation of pigmented marine muds [sic] seems also to have been involved in a policy of delay. In the matter of fouling organisms, the Bureau of Aeronautics has been interested."[140]

This was totally garbled. The petroleum project had nothing to do with "pigmented muds"—it was a study of the role of bacteria in petroleum formation and recovery, the project ZoBell moved onto after he quit the antifouling work (and which would later make his name).[141] It was Fox and ZoBell who expressed reservations about the antifouling project and Sverdrup who had promoted it. As for agar, that was a project begun in earnest only in 1942–1943, and Sverdrup had endorsed it, too.[142] And the antifouling studies had nothing to do with the Bureau of Aeronautics. Perhaps Fox and ZoBell were defensive about their earlier position and sought to turn the tables on Sverdrup, or perhaps Turner was a moron. But what did sound like a plausible complaint from Fox and ZoBell was the additional assertion that Sverdrup's leadership was autocratic.

Turner recounted the incident in which Fox and ZoBell had opposed Sverdrup on the organization of the antifouling study. Noting (correctly) that several faculty members had signed a protest over the authority given to Whedon, the complaint asserted that Sverdrup had ridden roughshod over the faculty. "The evening of Halloween, in the presence of nine men, including Navy officers, the director ... declared Whedon to be the unanimous choice of the faculty." Why was this relevant? Because, he alleged, Sverdrup's autocratic style could "facilitate subversion."[143]

Sverdrup had been working intensively with Richard Fleming and Martin Johnson (1893–1984) on what would become their acclaimed textbook, *The Oceans*, and the faculty complained that "it has been increasingly difficult to have stenographic aid, as it is almost exclusively used to prepare the volumes being written or collaborated on by the director. As many as 4 stenographers and typists are solely occupied with this task over long periods."[144] It was unclear why this was relevant—other than perhaps to prove that Sverdrup ran roughshod over his faculty—but Turner's offered two further pieces of circumstantial evidence that Sverdrup was engaged in questionable activities. One was an incident in which Sverdrup discovered a janitor in his office, examining the contents of the wastepaper basket. When Sverdrup demanded to know what the man was doing, he confessed that the FBI had instructed him to do this. Sverdrup "quickly dumped the papers into the fireplace where they burned.... The attendant resigned some weeks later." The second incident involved telephones. Turner asserted that there had been "considerable listening by the telephone, made possible by the connections. It is said to come directly from the office of the director."[145]

Finally, there was one new charge: Prentice Hall, the publisher of Sverdrup's other forthcoming book, *Oceanography for Meteorologists* (an outgrowth of the work with Kaplan), had contacted Turner with a request for an aerial photograph of the UCLA campus.[146] Turner had "some recollection of this name in connection with un-American propaganda" and contacted Army intelligence. He was instructed not to send the photographs and subsequently found out that Prentice Hall was the subject of an FBI investigation. "It may or may not be coincidence," he noted ominously, "that an important work linking oceanography with meteorology should be courted by this publisher. Galley proofs now at the institution might be of service to the enemy."[147] Turner had little or no direct knowledge of events at Scripps (or Prentice Hall), but equally clearly someone had been talking. Only five people had been involved in the dispute over Whedon, and only one—Denis Fox—was the sort to worry about stenography; it seems likely that Fox was deliberately spreading complaints and casting suspicion on Sverdrup.[148] Two further events confirm that the source of Turner's information was Fox and ZoBell.

In June, ZoBell wrote directly to Sproul, questioning Lek's employment at Scripps and stating that he was an "unsavory character," an alien "not in sympathy with our defense program."[149] Three weeks later, Fox wrote to the FBI. The letter, dated July 9, 1942, was forwarded to the Navy from the FBI with "the latest pertinent information" about Sverdrup; it involved aerial photographs. The FBI summarized: "Dr. D. L. Fox, of the Scripps Institution of Oceanography ... advised that the subject had received a communication from the Norwegian embassy in Washington, D.C., and that he and his wife left Washington on 6/19/42. SVERDRUP allegedly indicated his intention of making only a short trip, but Dr. Fox advised that both the subject and his wife had recently procured passport pictures. It is also reported from another source that the subject is in possession of aerial photographs of the San Diego coast."[150] Earth scientists routinely obtain and use aerial photographs, but Sverdrup, possessing photos of the Southern California coast, was meeting with representatives of German-occupied Norway. Perhaps he was planning to leave the country? He had gone at the request of the Norwegian embassy, but "the purpose of the trip ... was unknown [and] a mystery surrounded the whole situation." Fox claimed that Sverdrup had made arrangements for a "whole year in case of his absence [and it] is not known when or if the subject will ever return to the Institute."[151] A related memo, written by the Department of Justice to J. Edgar Hoover on August 15, confirmed that Sverdrup had the aerial photographs of San Diego in his possession when he went to Washington.[152]

And there was a new charge, this one to do with Walter Munk. Sverdrup had invited Munk to work on the AAF meteorology project, but some felt that Munk was "not particularly professionally qualified for the task required." The same objection was raised against Lek: "The proposed employment of these two men (Munk and Lek) was to be under a subterfuge so as they would be paid from the funds of the Institute rather than the War Department funds so as not to bring them under the jurisdiction of the War Department."[153] Who were the staff who deemed Lek and Munk unqualified to work on the AAF project? Stanley Chambers, Denis Fox, and Claude ZoBell.

On November 15, 1942, Sverdrup's partial clearance was revoked and his employment with the AAF terminated. A confidential memorandum on that same date states, "No further investigation is contemplated by this Division."[154] But in January 1943, investigations would begin yet again, although this time the results would be more favorable to Sverdrup.

Roger Revelle to the Rescue

Throughout the summer and early autumn of 1942, Sverdrup had worked on the AAF project producing isotherm charts of the world's oceans and initiating research on combined meteorological and oceanographic phenomena—work that was important, but hardly commensurate with his scientific standing. He also taught aerology courses for AAF officers and contributed to Kaplan's training program at UCLA.[155] With his termination from the weather research, however, he was left working only in a teaching and training capacity.

Meanwhile, Revelle was working inside the Navy, where he was gaining influence and developing connections. Throughout 1941–1942, Revelle had been at work on submarine acoustics at the Navy Radio and Sound Laboratory, where there had been various difficulties between Navy personnel and civilian scientists. In response, the Navy coordinator for research and development, Admiral Julius A. Furer, decided to create a new billet in the Hydrographic Office for an officer to coordinate civilian and military oceanographic research. In late 1942, Revelle was appointed to that position (fig. 1.6).

Meanwhile, as the United States took the offensive in the war, the Joint Chiefs of Staff Committee on Meteorology began to worry about wave and surf forecasting for beach landings. Responsibility for developing forecasting methods lay with the AAF; in early 1943 the AAF's results were presented to a new oceanography subcommittee of the Committee on Meteorology, with Revelle in attendance.[156]

Revelle emphasized that the Navy as much as the Army had an interest in wave and surf forecasting. Indeed, he and many others considered it odd that the Army had taken the lead on the issue, anticipated troop beach landings notwithstanding. Revelle took advantage of discrepancies between predictions developed by the AAF and those of British Admiralty to suggest that Sverdrup—with his deep theoretical knowledge of air-sea interaction—could develop a scientifically grounded forecasting system that would yield better results than the empirical methods on which the AAF was relying. Revelle provided the committee with a copy of *The Oceans*, noting its status as the only complete textbook on the subject in English. In other words, the mil-

Figure 1.6 Roger Revelle.
Image by Eugene Cecil LaFond, from the Eugene Cecil LaFond Papers, Special Collections and Archives, UC San Diego. Reprinted with permission of UCSD.

itary had ignored the expert in its midst who could solve one of their most pressing problems.[157]

Revelle won the day: in May 1943 Sverdrup was appointed chief civilian scientist on the wave-forecasting project.[158] It was not just that Revelle had convinced the Army not to ignore the talent they had before them; Revelle had also finally convinced the Navy that the complaints against Sverdrup were spurious. Naval Intelligence issued a new report, one that for the first time contained positive testimonials from within the Navy.

Revelle had been waging his campaign to clear Sverdrup's name since the previous year. In March 1942, a remarkable new FBI report had been submitted, describing Sverdrup as "an outstanding advocate of the Governmental form of this country and to be absolutely reliable, honest, and trustworthy." The report went on: "Unusually discreet. Wife and daughter are reported to be great 'fans' for this country and Government. Informants advise no grounds for suspicion of un-American or subversive activities."[159] The long report, which explained the importance of Sverdrup's scientific work and extolled his character and personality, could have been compiled only by someone who was close to Sverdrup but also understood the US military. Just one person fit that bill: Roger Revelle.

Sverdrup was now described as a man who had early on experienced a "complete conversion ... to the American life and ways of living" and who "after his stay here for one year in 1924 or 1925 ... built himself an American house in Norway, completely American in all details." He was "of unimpeachable character ... very much Americanized and devoted to the principles of this country."[160] The negative rumors about Sverdrup were dismissed as the groundless gossip of idle La Jolla housewives. Ellen Revelle, the unimpeachable hostess and scion of the original Scripps family, had invited the FBI agents into her home and insisted that she and Roger "knew of absolutely not one statement or action" by the Sverdrups "which could be conceivably interpreted as being un-American or indicating sympathies with any foreign power ... Dr. Sverdrup and Mrs. Sverdrup were *not* pro-Nazi."[161]

Evidently, the Revelles had realized that their earlier tactic—of trying to account for statements that might seem on the face of it suspicious—was inadequate, and so they shifted their strategy to defend Sverdrup unequivocally and in every respect. No explanations, no apologies, just unconditional support, down to the character of the Sverdrups' house. The report also included a new set of interviews, begun in January 1943 and continued through the early spring. Sverdrup was anti-Nazi and always had been. Chambers was jealous of Lek because Lek was rich. Sverdrup did know Quisling—they had been classmates at the War Academy in Norway—but it was well known that "Sverdrup viewed Quisling with disgust." Regarding England and Germany, Sverdrup had resigned himself to the occupation of his home country and accepted that one had to consider the presence of both British and German forces as matters of "military necessity." This could be misinterpreted as suggesting Nazi sympathies, but that would be entirely wrong: Sverdrup's "hatred for the Nazis had been intensified since the death of his brother who was serving in the British army."[162] This was a reference to Sverdrup's younger brother Einar, director of the Norwegian Spitsbergen Coal Company, which the Norwegians had evacuated and blown up in 1941. In 1942, Einar was asked by the British to lead a party of free Norwegian forces to reopen and defend the mines. Einar agreed, but when his ship entered the harbor, the Germans bombed and sank it. Sverdrup family members later recalled that Leif was angry that the British had not provided better air cover for the assault.[163]

Revelle had come to understand that military officials were not particularly concerned with Sverdrup's friends and family in Norway; rather, it was his associations with foreign scientists in the United States, particularly those with German or Japanese connections, that troubled them. This included Lek (who was Dutch-Jewish), but Naval intelligence had come to believe there was reason to doubt the "motives or reliability" of the individuals who had made the charges against Lek. A more serious concern was Sverdrup's connection with the Japanese oceanographer, Koji Hadaka, of the Imperial Marine Observatory: Hadaka had come to Berkeley for a conference in the summer of 1939 and visited Scripps after the meeting. No doubt thanks to Revelle's counsel, Sverdrup understood that this was a concern, and when interviewed again, he responded directly to the questions about his associations with foreign scientists. He claimed he had known no Japanese scientists prior to the Berkeley meeting but had exchanged reprints with them precisely because their work was of military importance to the United States.[164]

Revelle's access to information through his position at the Hydrographic Office was critical to Sverdrup's ability to respond to charges that had never been presented. But perhaps most important was the effect of Revelle's efforts within the military establishment, for officials in the Bureau of Ships now came to Sverdrup's defense. An undated letter to the chief of naval operations from the Bureau of Ships acknowledged that the "existence of reports adverse to the loyalty and integrity of Dr. Harald U. Sverdrup has again been brought to the attention of the Bureau of Ships," but the military dismissed these reports as unfounded. The bureau had known Sverdrup since 1940 and had already indicated its confidence in him by employing him on several projects. The letter also called attention to the "numerous signed statements from persons of unquestioned integrity which are favorable to Dr. Sverdrup's character" and insisted that the derogatory reports against him were "without foundation." The letter concluded: "The Bureau of Ships hereby reiterates its statement that there is no reason to question Dr. Sverdrup's loyalty or integrity."[165]

On August 29, 1943, a letter to J. Edgar Hoover from the Office of the Chief of Naval Operations summarized the Office of Naval Intelligence files, concluding that there was no reason to exclude Sverdrup from the forecasting effort.[166] After twenty-five months of investigations, the matter was finally settled. The Navy had confronted a wide range of conflicting allegations and concluded they were unfounded. From that point until the end of the war, both Sverdrup and Munk would work on meteorological projects for the Army and the Navy without further ado. By September, they had developed a new theoretical model for forecasting sea and surf conditions, one that Navy officers were satisfied was far superior to previous techniques.[167] The value of the method in launching massive beach attacks was demonstrated at D-Day and in several other amphibious operations.[168]

Was Sverdrup Disloyal—And Does It Matter?

Harald Sverdrup was a man who kept his own counsel and left no diary of his Scripps years, so we will never know exactly what he felt and thought during this time. We do know that what he did not say was held against him. We also know that there were plenty of other scientists to work on acoustics. Does it matter that he was denied security clearance for two years, or that ultimately he worked on forecasting rather than acoustics?

Certainly, it mattered to him. Sverdrup wanted to contribute his expertise to the war effort, and it frustrated him that he was closed out of the most important efforts. His letters to Stefansson, his ideas about using plankton to save downed airmen, and, above all, his decision to become a US citizen speak to his desire to aid in the war effort and contribute to the liberation of his homeland. In later years, a nephew recalled that Sverdrup was "very upset" that his capacities had not been used to their full extent and felt that "he had much to contribute." George Sverdrup explained: "The fact that Norway had been invaded and his family together with the others there were having such a difficult time added to the distress [and it] didn't help to see Leif having such a prominent part in the war effort. Knowing Harald, I'm sure there was not so much envy as frustration."[169]

Snippets from Sverdrup's personal letters betray this frustration. In a letter to Revelle in July 1943, when the sea surf and swell project was just getting under way, Sverdrup noted the difficulties he was still having: "I still have to write you personal letters because I do not know what official relations I have with the Navy."[170] When Sverdrup was elected to the US National Academy of Sciences in 1945 and Columbus Iselin wrote to congratulate him, Sverdrup replied that he hoped "as a member of the Academy I may be more useful in the post-war period."[171] Sverdrup was useful during World War II, but much of what he was asked to do was less than the maximum use of his talents. In a letter to meteorologist Carl-Gustaf Rossby, Sverdrup allowed that training weathermen was all right as an "emergency measure," but he was certainly not enamored of it.[172] Some of the work Sverdrup was asked to do—for example, compiling ocean isotherms—could have been done by an undergraduate.

Perhaps more important, the clearance problem undermined Sverdrup's own sense of leadership. In June 1942, he was trying to decide whether to stay in La Jolla to work on the AAF project or to go to Washington to work on behalf of the Norwegian embassy. This, in fact, was the reason for his trip to the embassy that had generated so much suspicion. In a letter to Sproul, he explained that he thought he could go; because everyone at Scripps was engaged in war work, his own presence was "not essential."[173] It is hard to imagine that Sverdrup was untroubled by being inessential.

Sverdrup's troubles clearly mattered to the institution he served and the science he helped to build, because soon after the war ended—and despite having finally obtained US citizenship—he left them both. In 1948, Sverdrup returned to Norway to head the Norsk Polarinstitutt. He continued to be highly active in administrative, educational, and humanitarian affairs: he led a Norwegian-Swedish-British expedition to Antarctica in 1951, organized the Norwegian famine aid to India in 1952, and, as dean of the natural sciences at Oslo University, spearheaded a major educational reform to reorganize Norwegian higher education on the American (rather than German!) model. As an educator, he became an advocate for the importance of the humanities in tackling problems which science had "no claim to answer."[174] But as William Nierenberg (1919–2000), who became Scripps's seventh director in 1965, later put it, his scientific work was "not of the intensity or depth of what had occupied him at Scripps."[175] Once back in Norway, Sverdrup made few further scientific contributions to oceanography.

Nierenberg attributed Sverdrup's departure to his desire to have a greater influence on international affairs, his low academic salary in the United States, his homesickness, and his wife's concern that their daughter not become "Americanized."[176] Historian Robert Marc Friedman, the biographer of Vilhelm Bjerknes (1862–1951), affirms that Sverdrup was deeply homesick throughout his stay in the United States and eager to return to Norway, particularly if he could help the country rebuild after the devastating Nazi occupation. Moreover, Friedman notes, Sverdrup had a "thick skin and developed very early an ability to absorb major disappointments and worse.... He experienced an even more bitter disappointment with the Norwegian government not fulfilling the promises for a research-based polar institute, but instead of complaining and accusing, he simply and quietly accepted other tasks that he felt were worthwhile."[177] If he felt he could continue to be productive in the United States, it seems unlikely that homesickness and patriotism alone would have caused Sverdrup to give up his hard-won US citizenship and reverse his studied decision to stay. And his salary in Norway would not have been higher than it was in California.

It seems more likely that Sverdrup used homesickness as an excuse, rather than confront the reality that his leadership had been undermined (fig. 1.7). After all, the principal argument for remaining in the United States would have been the far greater opportunities to continue to pursue oceanography—and to build Scripps as the great scientific institution that it

Figure 1.7 Harald Sverdrup, portrait by Paul Williams, 1946, around the time of his decision to leave the United States and return to Norway.

From the Scripps Photographic Library Special Collections and Archives, UC San Diego, reprinted with permission of UCSD.

indeed later became. But those opportunities, Sverdrup knew, would for him be constrained, because the end of the war did not mean the end of security concerns. On the contrary, it was clear that those concerns would loom even larger and likely prove even more demeaning.

Even before the war was over, scientists and Navy officers were taking steps to ensure continued Navy support into peacetime. Sverdrup, Iselin, Revelle, Fleming, and other leaders of American oceanography had no desire to return to impecunious prewar conditions; they wanted the relationship not merely to be maintained but also to grow and prosper.[178] Following Vannevar Bush's September 1944 announcement of the termination of the Office of Scientific Research and Development, the Navy organized several conferences to discuss the completion and disposition of NDRC subsurface warfare projects.[179]

At one meeting in La Jolla in January 1945, Sverdrup, Fleming, Iselin, and physicist Lyman Spitzer (1914–1997), who had worked during the war on sonar (and would later champion the Hubble Space Telescope) discussed the important role that oceanography could play in surface and subsurface warfare. Among other things, they recommended that several UC Division of War Research oceanography projects be transferred to civilian laboratories.[180] In at least two reports proposing postwar oceanographic work that could be done at Scripps, Sverdrup constructed tables specifying the Scripps research programs in one column and their military value in another to underscore the close linkage between oceanographic science and military operations (fig. 1.7).[181] In various ways, civilian scientists made clear they wanted to maintain the continuity of scientific research and "insure continuous flow of information to the Navy."[182]

Moreover, oceanographers could not "go back to their ships," as physicists who had worked on wartime projects could go back to their laboratories, because before the war they scarcely had any ships. Without Navy support, there would be little to go back to. Oceanographers wanted Navy support to continue, and the Navy had made it clear that it wanted to strengthen and extend the navy-oceanography link. Much of this work would be classified, so it would have been essential for Sverdrup to know whether the problems of the war years were fully resolved. Given that he had spent 1943–1945 on military work, it might have seemed that they were. The reality was otherwise—and Sverdrup knew it.

First, there was the question of leadership. Sverdrup's forecasting method had helped to enable beach landings under favorable conditions—a central military need based on oceanography and meteorology—but it nevertheless placed him on the periphery of military oceanographic research. Most of his colleagues and students had spent the war years on acoustics, and after the war, acoustic research would be of prime importance. Would Sverdrup be allowed to participate? Would he be a full partner? Could he be a leader? Even if the answers to the first two questions were yes, the answer to the third was probably no; he was behind in his knowledge and understanding. No doubt he could have caught up, but having spent the war on the sidelines, it was unlikely he would take a leadership role. If the answers to all three questions were no, then this meant a thoroughgoing marginalization.

Second, clearance was not a yes-no proposition (any more than it is today), and there were persistent boundaries to Sverdrup's participation in military-scientific projects. For example, in 1943, when Sverdrup's Navy clearance was finally being approved, Revelle recommended the development of new manuals to aid submariners. Traditionally, the Hydrographic Office had produced "Sailing Directions"—pamphlets to provide mariners with information on surface waters. In light of the increasing importance of submarine warfare in the Pacific, Revelle called for the production of "Submarine Supplements to the Sailing Directions" with detailed information on subsurface conditions. As the Pacific War intensified, the production of submarine supplements became a top priority for the Hydrographic Office. Sverdrup played a leading role in compiling the supplements, but this work underscored the limitations to his security clearance, because he was not involved in the data collection— only its compilation.[183] He received the temperature data from Richard Fleming, who was collecting the data aboard ships and submarines—and training Navy personnel to do so as well—and then passing on the results.[184]

Sonar information also came to Sverdrup indirectly. At Woods Hole, Columbus Iselin and geophysicist Maurice Ewing (1906–1974) had compiled prediction manuals describing the bearing of ocean conditions on evasive tactics, including the greatest distance at which submarines could be evaded and detected (see chapter 2).[185] But Sverdrup could work neither directly with sonar nor on Navy vessels that carried sonar equipment, so he had to rely on others to compile and present him with the necessary data.[186] This might seem a small thing—nowadays few senior scientists collect their own data in the field—and some scientists might have been pleased to be supplied with large amounts of data that they did not have to take the time to collect. But Sverdrup was not such a scientist; he held to the principle of collecting his own data in order to fully understand it.

By late 1944, agreements had been reached whereby much of the data on temperature profiles and bottom sediments would be transferred to Scripps and compiled by a team (of mostly women) there. Existing contracts with the Hydrographic Office and the Bureau of Ships would also be extended, but provisions still had to be made for work in the subsurface domain. The Navy, the NDRC, and UC agreed that the UC Division of War Research would be re-configured as a new facility, the Marine Physical Laboratory (MPL), to be run jointly by the university and the Bureau of Ships.[187] But Sverdrup's suggestion that the new laboratory be made part of a new university-wide Institute of Geophysics was rejected; MPL was established instead at Point Loma on the grounds of the defunct Navy Radio and Sound Laboratory.[188] Scripps could contribute to MPL activities through instrument development and related oceanographic work, but for the most part there were few direct ties between Scripps and MPL during Sverdrup's years. It was only in 1948, when Revelle returned from the Navy and MPL director Carl Eckart (1902–1973) became the new Scripps director, that a much closer relationship began to develop.[189]

Sverdrup and Scripps were also hampered by security restrictions in the area of deep-sea exploration and research in radioactivity and oceanography. Deep-sea dredging and sonar tests had demonstrated the existence of guyots—flat-topped seamounts—leading to renewed interest in the structure of the ocean floor.[190] Meanwhile, wartime development of seismic refraction techniques and bathymetry provided new means to study its structure (see chapters 4 and 5).[191] Although the scientific promise of these techniques was evident, they required use of classified instrumentation and access to ships, submarines, and strategic locations.[192] The same was true of work that was expanding as a result of the atomic bomb.

The use of atomic bombs in Japan in 1945 had created immediate consternation within the Navy, whose leaders scrambled to demonstrate the Navy's continued relevance in the atomic age. In the autumn of 1945, the US government approved a joint Army-Navy project, Joint Task Force 1—better known as Operation Crossroads—to assess the effects of Pacific atomic bomb tests.[193] Revelle was put in charge of the oceanography section, asked to evaluate the waves generated by the blast and their effect on the organisms and geology of Bikini Atoll. But Sverdrup was not invited to participate in Operation Crossroads, or in it its follow-up, the Bikini Scientific Resurvey of 1947.[194]

Beyond Sverdrup's limited participation in these major postwar initiatives, evidence suggests that the political economy of postwar oceanography weighed on his mind. He expressed concerns over secrecy, publication, and the costs of military support—indeed, some of the very concerns that Fox and ZoBell had expressed at a time that likely seemed a lifetime away.[195]

In the fall of 1945, when the Navy Office of Research and Inventions—soon to become the Office of Naval Research (ONR), much praised for the support it rendered—began sending officers to discuss possible research contracts, Sverdrup's feelings were mixed. In November, he told a Berkeley colleague categorically that Scripps would not accept any contract that restricted freedom in any phase of its work. In contrast, writing to UC vice president C. A. Dykstra just a month later, he maintained that "key men in any research group need full cognizance of possible military applications of their work. That will stimulate their research and draw attention to other new opportunities."[196] Sverdrup was clearly conflicted, having experienced firsthand the constraints and—in his personal case—indignities of military work, but equally witnessing the military and scientific significance of the work done. Military work had created major new opportunities for oceanographers, and the Navy had become the source of all the most significant of those opportunities.[197]

The Navy's limitations on Sverdrup's access to subsurface work may have had implications for the new Institute of Geophysics being developed at UCLA and later expanded to the La Jolla campus. The institute brought together the scientists who had worked on the officer training program—Jacob Bjerknes, Jürgen Holmboe, and Joseph Kaplan—to offer new opportunities for research and education in meteorology and ocean-atmosphere interactions, but they could offer little in the way of marine geophysics, in part because of security issues.[198]

One letter proves that Sverdrup's problems had not gone away and that he knew it.[199] In 1946, Richard Fleming accepted a position with the Hydrographic Office as chief of the Division of Oceanography, where he would have a substantial influence on Navy support for oceanographic research. In August 1947, he wrote to Sverdrup to ask whether Scripps could prepare reports on bottom sediments for the Navy. The work could be done, he suggested, under an existing Hydrographic Office contract on submarine geology.[200] At the bottom of the letter, however, Fleming penciled in a handwritten note: "This work would probably have to be classified, as the areas and the publications are CONFIDENTIAL. This may cause you some difficulties. —Richard."[201]

Sverdrup must have been uncertain how to reply, because he delayed uncharacteristically in responding. Two months later, Fleming reminded him that they had "never decided" on the submarine geology program; he offered to visit in November to discuss it. If the visit occurred, there is no record of it. Perhaps that is because, just around that time, Sverdrup made the decision to return to Norway.[202] Did Sverdrup leave because of the security issue? We may never know for sure, but consider this: How could Sverdrup continue to lead Scripps if he were unable to participate in its major research initiatives? How could he judge his faculty's work—their prospects for tenure and promotion—if he were unable to read all their publications?

Sverdrup's decision to leave threw Scripps into disarray. Simmering tensions broke out into open dispute over who would be the next director (a dispute in which Fox and ZoBell would again play a major role). After a brief interregnum under physicist Carl Eckart, Revelle became director. His close links with the Navy appear to have been part of his appeal to both university administrators and Navy patrons; over the following decade Revelle would move the institution even farther in the direction of Navy sponsorship than Sverdrup had ever envisaged.[203]

Conclusion

American science in the nineteenth and early twentieth century was not independent of the federal government, and many leaders of American science had links to the armed services of one sort or another.[204] But World War II represented a new level of involvement and unprecedented funding by the federal government, particularly by its military and security agencies.[205] This had significant ramifications for the direction of, and at times even control over, scientific work.

Henry Bigelow perhaps went too far when he argued that scientific projects funded by the Navy are "in reality the property of the Navy department." Then again, perhaps not. When oceanographers joined military projects during the war and looked to the Navy to be their patron after the war, they were implicitly agreeing to a significant degree of Navy control over projects, publications, and personnel. The disputes at Scripps over the antifouling studies and the exclusions of Harald Sverdrup and Walter Munk from particular projects were troubling to these men and to their colleagues. But the larger historical point is not about the impacts of security restrictions on one man or several: it is about the impact of Navy patronage on their scientific discipline and the direction in which it became oriented.[206]

In the decades following the war, American oceanographers would strive to put oceanography on a firm financial footing without sacrificing their intellectual integrity and autonomy. In 1948, it was not clear that they could or would succeed. One thing, however, *was* clear. American oceanography in the future would be very different from what it had been in the past, and the US Navy would be central to that difference.

When Sverdrup left the United States, Scripps in particular and American oceanography in general lost an exceptional leader: a man of profound integrity and broad scientific vision. They lost a leader committed to fieldwork and the experience of nature in its rawest forms. And they lost a committed internationalist who knew and worked with scientists from around the globe, for whom science was always a collective and never just a personal project.

When Harald Sverdrup went back to Norway, American science was the poorer for it.

The lasting irony of this story is that the two men who did the most to damage Harald Sverdrup's reputation—Denis Fox and Claude ZoBell—opposed him on the issue of military sponsorship of science. Sverdrup wanted military contracts for Scripps; Fox and ZoBell believed it threatened the freedom and autonomy of scientific research. Who was right? How much control did the US Navy exert on American oceanography in the years to come? What difference *does* it make who funds science? These questions guide the remainder of this book.

2 *Seeing the Ocean through Operational Eyes: The Stommel-Arons Model of Abyssal Circulation*

Some scientists at Scripps in the 1930s were worried that Navy concerns might overwhelm their research agenda, but others welcomed financial support that they viewed as enabling them to do "their" science. What neither group seems to have anticipated was that focusing attention on operational problems might lead them, for better or worse, to view the ocean in a different way. And they certainly did not anticipate that looking at the ocean in a different way, because of operational concerns, would lead to a major theoretical breakthrough. But at Woods Hole in the 1950s, this is what happened. The operational problem was the effect of the thermocline on sonar; the breakthrough was the understanding of deep-ocean circulation.

Throughout history, naval warfare had mostly been a matter of ships at sea—which is to say ships *on* the sea—and this had made the matters of winds, waves, tides, and currents of great salience.[1] But these phenomena had been interpreted as essentially surficial; from a naval standpoint the deep sea was largely irrelevant. Scientists, in contrast, had long been motivated to understand the deep sea but for the most part lacked the means to do so. This did not prevent them from trying; nor did it inhibit their theorizing. In the nineteenth century there had been a great debate about the origins of currents: whether they were a surface phenomenon related to wind, or a deeper-seated phenomenon perhaps related to variations in temperature and salinity. Some military hydrographers, notably the nineteenth-century hydrographer Matthew Fontaine Maury, had engaged in these debates, but most Navy officers felt they had more pressing things to worry about.

In the mid-twentieth century, this changed. The impact in World War II of subsurface warfare—and the emerging notion after the war that in the not-too-distant future nuclear-powered submarines would carry nuclear-powered weapons—made the conditions of the deep sea suddenly very pertinent. Indeed, in a future nuclear Navy, they might become decisive. For the first time

in history, scientists and military officers found themselves with a shared interest in the deep sea, particularly the interrelated questions of the origin of
(surface) currents and whether there might be deep-ocean circulation.

From the military standpoint, an element of particular concern was sound
and its potential role in subsurface warfare.[2] Radio waves do not propagate
through water, so for a fully submerged vessel sound was the only potential
means of communication. It was also a potential means to detect the enemy
or be detected by him. As early as World War I, scientists in Germany, Russia, and the United States had studied underwater sound transmission for
its pertinence to submarine warfare, but no effective means of underwater
communication or detection had been developed.[3] In any event, early U-boats
could not stay submerged for very long. Things changed with the invention
of the snorkel and overall improvements in submarine design, which permitted submarines to go deeper and stay submerged longer. The topic of subsurface communication and detection became a major focus of the US National
Defense Research Committee during World War II, inspired by the increased
efficacy and exigencies of submarine warfare.[4]

One key scientific advance was the establishment that the speed of sound
in water depends on the water's density, and that sound waves, like light
waves, are refracted as they move through zones of different density. This
results in the production of acoustic "shadow zones"—regions where sound
waves do not penetrate—thus possibly allowing a submarine to avoid detection. And the density of water is in part a function of temperature and salinity. Access to accurate data on temperature and salinity changes with depth
could therefore make the difference between hiding successfully from an enemy and being torpedoed.

Sound refraction also creates acoustic "channels" in which sound waves
travel efficiently—enabling effective long-distance communication. Because
sound channels are caused by refraction, understanding them, like the shadow
zones, also required knowing the spatial distribution of temperature and salinity. And so, starting during World War II and continuing into the Cold War,
the Navy invested considerable resources into understanding temperature
and salinity variation with depth. Key to this was the bathythermograph (BT),
an innovative instrument that recorded temperature variation with depth.[5]
The BT was used during World War II to make hundreds of thousands of high-
accuracy temperature-depth profiles—typically down to about six hundred to
nine hundred feet—referred to as bathythermograms or BT profiles.[6]

Before the development of the BT there were scant data on subsurface conditions, and of very variable quality. Now, for the first time, oceanographers
had reliable, geographically extensive data on the conditions of the ocean at

greater depth.[7] This meant they could begin to revisit the long-standing questions of the driving force of ocean circulation and whether the great ocean currents were only a surface phenomenon or extended to depth. Nine hundred feet was not nearly the entire ocean depth, of course, but by the 1950s, means to explore the deepest portions of the sea were beginning to emerge. In 1955, Naval Electronics Laboratory oceanographer Robert Dietz (1914–1995)—later known for his contributions to plate tectonics (chapter 5)—was particularly excited about the bathyscaphe *Trieste*, a deep-sea submersible vessel invented in the late 1940s by explorer Auguste Piccard (and later made famous by Jacques Cousteau). Writing to Scripps colleague Henry W. "Bill" Menard (1920–1986), he suggested they might use the bathyscaphe to measure deep currents, "one of the most critical unknowns of physical oceanography."[8]

Deep-ocean interest was reinforced in the 1950s by the emergence of nuclear waste as a political and scientific problem.[9] To many people, the ocean seemed a good place to dump high-level radioactive waste, but this rested on the assumption that there were no deep currents, or only very sluggish ones, so that radioactive waste put at the bottom of the sea would remain there.[10] If strong currents swept the seafloor, then the wisdom of disposing of dangerous wastes in the deep sea was doubtful at best (chapter 8).[11]

Thus in the 1950s, both the US Navy and the US Atomic Energy Commission invested substantial resources in the study of the deep sea. The result of this confluence of scientific, military, and political interest was a comprehensive research program at four academic institutions—the Lamont Geological Observatory (later Lamont-Doherty, today the Earth Institute) at Columbia University, the Scripps Institution of Oceanography, the Woods Hole Oceanographic Institution, and the University of Washington—which revolutionized the understanding of deep oceanographic processes. At Lamont, oceanographer Bruce Heezen (1924–1977) and his graduate student Charles Hollister (1936–1999) demonstrated the existence of "contourites"—sedimentary deposits formed by currents running along the topographic contours of the continental shelf. Hollister later demonstrated the existence of benthic storms capable of scouring the ocean beds and redepositing large amounts of sediment in very short periods of time.[12] At Scripps, Walter Munk analyzed friction and wind data to lay the basis for the contemporary understanding of coupled ocean-atmosphere circulation.[13] And at Woods Hole, Henry Melson Stommel (1920–1992) and Arnold Arons (1916–2001) constructed the first generally accepted mathematical model of abyssal circulation, which demonstrated that there is deep-sea circulation and that density differences are central to it.[14]

The Stommel-Arons model of abyssal circulation still provides key elements of our basic theoretical framework of ocean circulation: density-driven circulation combined with the Earth's rotation. Cold-water sinks in relatively limited areas, but there is an upward flow of water throughout the abyssal ocean. Density-driven circulation combined with the Earth's rotation also explains the existence of western boundary currents and the phenomenon known as "westward intensification"—the strengthening of the western arms of ocean currents. The Stommel-Arons model predicted that these currents—already recognized at the time in some of the world's oceans—had to be a general phenomenon, an insight often cited by oceanographers as a rare example of a successful prediction in their field.[15]

But this profound scientific breakthrough did not come simply from thinking harder or more clearly. Nor did it come from collecting more or better data. It came from Henry Stommel's paying attention to a natural phenomenon—the thermocline—that other oceanographers had mostly ignored but whose understanding had proved critical to successful sonar operation. To appreciate his breakthrough, we need to step back into the long history of debate about ocean circulation.

The Problem of Ocean Circulation

For centuries, scientists and mariners sought to understand ocean circulation, particularly the great surface currents that were crucial to sailing ships.[16] Historically, there were two major schools of thought: one, that winds were the driving forces of ocean currents, and the other, that density differences were. The equatorial current—widely recognized in the sixteenth century as traveling from east to west—was obviously consistent with prevailing easterly winds, which could be explained by the rotation of the Earth.

Less easy to decipher was the Gulf Stream. In the eighteenth century, it was known that ships could make the trip across the Atlantic fastest if they went out along low latitudes and returned at higher ones; some interpreted this as a northward deflection of the equatorial current when it collided with the Americas on the western side of the Atlantic. However, Benjamin Franklin had shown that the surface waters in the Gulf Stream were warmer than in the adjacent waters, suggesting that temperature variations (and the density differences they produced) might be causal. It also meant that temperature measurements could be used to reveal currents, providing a substantial incentive for mariners to make such measurements.[17] But they were vexing to do: thermometers generally broke when submerged to any extent or overestimated temperature because of pressure on the bulb, and a thermometer lowered to great depth would in any case re-equilibrate as it was retrieved.

Various devices were invented and approaches tried—such as wrapping thermometers in lamb's wool—but most were more ingenious than successful.[18]

In the late eighteenth century it was discovered that Antarctic waters were colder at the surface than at depth—the reverse of what had been seen everywhere else. In 1787, the Irish chemist Richard Kirwan (1733–1812) proposed that sinking of these cold (and therefore dense) surface waters could drive deep circulation. Benjamin Thompson, Count Rumford (1753–1814), took up the idea, and in the work that is generally credited as the first developed concept of density-driven circulation, proposed that cold air blowing over the polar regions cooled the sea's surface layer; this cold, dense water would sink, displacing warmer water at depth and forcing it toward the equator. Equatorial waters would then flow along the surface to replace the sunken polar waters, creating a giant convective system. Rumford's ideas were embraced by the great Alexander von Humboldt (1769–1859) and by Emil von Lenz (1804–1865).[19] In 1845 Lenz proposed a simple and influential model in which cold deep water derived from the polar regions merged in two major zones of vertical upwelling on either side of the equator (fig. 2.1).[20]

In North America, the concept of density-driven circulation was championed by Matthew Fontaine Maury (1806–1873), the controversial superintendent of the US Naval Observatory who would resign his service in 1861 to join the Confederacy (President Andrew Johnson pardoned him in 1868). Maury's treason has made him a contested figure in history, but in his day he was arguably the world's leading hydrographer. His 1855 *Physical Geography of the*

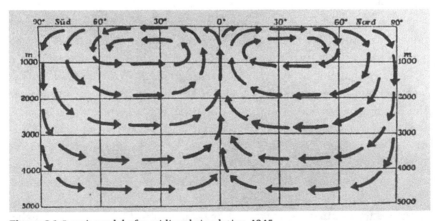

Figure 2.1 Lenz's model of meridional circulation, 1845.
From E. Lenz, "Bermerkungen über die Temperatu des Weltmeeres in verschiedenen Tiefen," *Bulletin de la Classe physico-mathématique de l'Académie impérial des Sciences de Saint-Pétersbourg* 5 (1845): 67–74.

Sea provided the first comprehensive scientific discussion of the deep sea, as well as the first proposal for density-driven circulation to be widely noticed in the United States.[21]

Maury focused attention on a key difference between the ocean and the atmosphere. In the atmosphere warm air rises and cold air sinks, driving atmospheric circulation. However, in the ocean, ceteris paribus, the warmest waters are at the surface, a thermodynamically stable configuration. Ergo, no circulation, but for one thing: salinity. Therefore, the salt content of the ocean—known to vary substantially in some surface waters—must be a factor in ocean circulation. In hindsight, Maury's important contribution was to call attention to the haline component of what we now call the thermohaline circulation, but it was not viewed that way at the time. In the United States, his work was discredited by his treason; in Britain it ran up against James Rennell (1742–1830) and the 4° theory and then subsumed in a bitter scientific feud known as the Carpenter-Croll debate.[22]

James Rennell, the 4° Theory, and the Carpenter-Croll Debate

James Rennell was the leading British geographer of his day; his 1832 *Investigation of the Currents of the Atlantic Ocean* provided the first detailed and systematic treatment of Atlantic surface currents.[23] His work persuaded most of his readers that the major currents, like the Agulhas and Gulf Stream, were driven by winds modified by interaction with land masses with density differences playing little if any role.[24]

Why did Rennell discount density effects? At the time, most people assumed that seawater would behave like fresh water, experiencing a density maximum near 4°C (39°F)—and most deepwater measurements came out around that value—so it began to be assumed that all deep water was 4°C; historian Margaret Deacon has labeled this the "4° theory." If the 4° theory were correct—and deep waters were more or less uniformly at their density maximum—then there would be no significant differences to drive circulation.[25]

Meanwhile, scientists were also debating whether there was—or even could be—life in the deep sea. Most natural historians tended to assume that without light and nutrients there couldn't be life at depth; this was broadly linked to a static picture of the deep sea.[26] Matters changed in the late 1850s when a broken cable retrieved from 1,200 fathoms during the laying of the transatlantic telegraph line was found to have marine animals clinging to it.[27] In August 1868, British naturalist C. Wyville Thomson (1830–1882), professor of natural history at Queens' College, Belfast, and William B. Carpenter

(1813–1885), vice president of the Royal Society (which supported the venture), sailed on HMS *Lightning* to look for life in the deep sea.

Thomson and Carpenter found that the distribution of temperature with depth was not as expected. Not only did temperature frequently fall below 4°; it also varied dramatically, sometimes over short distances. Deducing that earlier workers must have failed to account for pressure effects on their measurements, they contacted the firm that manufactured the Admiralty's meteorological instruments. The result was the Miller-Casella protected thermometer, in which an outer bulb filled with alcohol protected an inner mercury-filled measuring device.[28] With the new thermometer in hand, Thomson and Carpenter went back to the North Sea on HMS *Porcupine*.[29] The 1869–1870 voyage disproved the 4° theory, showed that there was life in the deep sea, and convinced Carpenter that there was density-driven circulation.

In the *Lightning* observations, Thomson and Carpenter had found warm water—in the upper 40s Fahrenheit—at depths as great as five hundred fathoms (approximately three thousand feet). Yet only a few miles away, temperatures fell to 32°F. Associated with these different thermal regimes were different sets of fauna; this indicated that the temperature contrasts were not instrumental artifacts. On *Porcupine*, they replicated this result. In each region, depth had little effect on temperature—the cold regions were cold throughout, and the warm regions warm throughout—and each thermal region had a distinctive faunal community. This suggested that the measurements were revealing different water masses, at different temperatures, with different nutrients.[30] Following Rumford, Humboldt, Lenz, and Maury before him, Carpenter became an advocate of density-driven deep circulation. Still, most British scientists remained loyal to Rennell, whose earlier arguments were being defended by geologist James Croll (1821–1890).[31]

Croll is a hero to many climate scientists for his early advocacy of the link between orbital variations and Earth climate cycles—an idea later developed by Milutin Milankovitch (1879–1958), whose name it now carries—but Croll rejected the density-driven circulation that is now thought to play a significant role in climate.[32] Croll had shown that the Ice Age was not a single episode of glaciation but a series of glacial and interglacial periods; he proposed that they were caused by variations in the eccentricity of Earth's orbit, which altered the climate by altering the pattern of wind-driven circulation. To give up the theory of wind-driven circulation was thus to threaten his account of climate change. Croll let things get personal, viewing Carpenter as an enemy whom he attacked in a series of acrimonious and dismissive articles in *Nature* and elsewhere, published in the early to mid-1870s.[33] Scientific opinion remained focused primarily on wind-driven circulation.

The Scandinavian School

In the late nineteenth century, things began to shift as physical oceanography received new impetus from Arctic explorers, particularly the Scandinavians Fridtjof Nansen and Roald Amundsen. In the *Fram* expedition (1893–1896), Nansen completed the first pack-ice drift across the Arctic and discovered that the direction of the ice drift diverged from prevailing wind directions by 20° to 40°. Suspecting the Coriolis force—articulated by French natural philosopher Gaspard Coriolis (1792–1843) but as yet unapplied to ocean dynamics—Nansen placed the problem before a brilliant Swedish oceanography student, Vagn Walfrid Ekman (1874–1954). Ekman demonstrated mathematically that the drift of the *Fram* was consistent with wind-driven currents modified by the Coriolis effect, an effect that now bears his name: the Ekman spiral.[34] This pathbreaking work—with its attention to the geostrophic effects (i.e., the interaction between pressure gradients and the effects of Earth's rotation)—became a cornerstone of the new dynamic oceanography, further developed by Bjørn Helland-Hansen (1877–1957), Johan Sandström (1874–1947), and Harald Sverdrup.

The foundation of dynamic oceanography was an analogy between the atmosphere and the ocean. Sverdrup's mentor Vilhelm Bjerknes had demonstrated that atmospheric air masses moved under the effects of baroclinicity: differences in pressure controlled primarily by temperature and modified by the effect of Earth's rotation. By analogy, the ocean could be understood as water masses moving under the influence of pressure differences that were caused by temperature, salinity, or both, and modified by the Earth's rotation. This defined a research program, and the Bergen school—as it came to be known—developed the equations needed to calculate ocean dynamics.[35] However, researchers lacked the data needed to solve them. Advances would require improved instrumentation to obtain better data. This came (in large part) in the form of the Nansen bottle.[36]

In response to the long-standing difficulty of obtaining accurate deepwater temperature measurements, Nansen had invented a self-sealing insulated bottle to collect deepwater samples and measure their temperatures in situ with a pressure-protected thermometer. Early results of dynamic calculations, based on measurements obtained with Nansen bottles, suggested that density-driven circulation might not be inconsequential, particularly in the middle layers. Meanwhile, the question of deep circulation had been taken up by German-speaking scientists in a series of late nineteenth- and early twentieth-century expeditions: *Gazelle* (1874–1876), *Valdivia* (1898–1899), *Gauss* (1906–1907), *Deutschland* (1911–1912) and, above all, the German Atlantic Expedition of 1925–1927, better known by the name of the ship *Meteor*.[37]

The *Meteor* expedition was the brainchild of Austrian oceanographer Alfred Merz (1880–1925), the third director of the Institut für Meereskunde in Berlin, founded in 1900 (and closed in 1946, having been destroyed during World War II). Merz had worked with the Imperial Navy during World War I preparing tidal charts to aid in submarine warfare. Merz was a passionate advocate of systematic data collection as essential to advancing theoretical understanding; most of his early work consisted of detailed time-series measurements on lakes, which he considered a model for open-ocean processes. He also closely followed the Bergen school.[38] Merz's dream was an extended German expedition to the Pacific and Atlantic Oceans—along the lines of the HMS *Challenger* expedition—but the fiscal realities of the 1920s limited his ambitions. He achieved instead a focused study on the South Atlantic, emphasizing bathymetric, temperature, and salinity measurements. Sadly, Merz contracted pneumonia on the first leg of the expedition and died in Buenos Aires at the age of forty-five; the work was continued by his student George Wüst (1890–1977).[39] In later years, Wüst summarized the intellectual state of affairs at the time: there was empirical evidence of deep circulation, but theoretical understanding "remained very schematic."[40]

Educated in Berlin in the early 1910s, Wüst had traveled to Bergen 1913 to study with Helland-Hansen, then returned to Berlin to be Merz's assistant. In preparation for the *Meteor* work, Wüst and Merz compiled and analyzed data from prior expeditions, particularly the *Deutschland*. In 1921, Wilhelm Brennecke had published the results of that expedition, which showed that the deep Atlantic was filled with cold, dense waters.[41] He suggested that these derived from two discrete polar regions: the Weddell Sea (adjacent to the West Antarctic Ice Sheet) and the North Atlantic above 60° latitude, with the former being colder and denser (fig. 2.2).

Brennecke's model was similar to Lenz's, but it had greater subtlety and detail. The Antarctic bottom waters, for example, did not rise back to the surface at the equator, as Lenz had supposed, but appeared to penetrate north as far as 10° to 20°, slipping below the North Atlantic deep layer. Similarly, the cold waters from the North Atlantic penetrated south past the equator at depths of 1,000–3,000 meters.[42] Brennecke established the principal patterns of meridional circulation in the Atlantic and identified what we now call Antarctic Bottom Water and North Atlantic Deep Water, but, as historian Eric Mills has noted, rather than see themselves as building on what Brennecke had done, Merz and Wüst saw themselves as refuting it.[43]

Part of the problem was jealousy: Brennecke published his results in

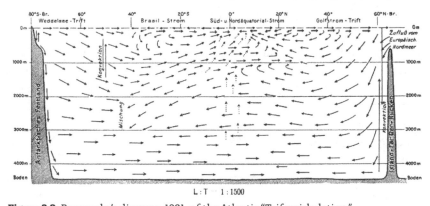

Figure 2.2 Brennecke's diagram, 1921, of the Atlantic "Teifenzirkulation."

From Wilhelm Brennecke, "Die ozeanographischen Arbeiten der Deutschen Antarktischen Expedition 1911–1912," *Aus dem Archiv der Deutschen Seewarte* 39, no. 1 (1921): 216.

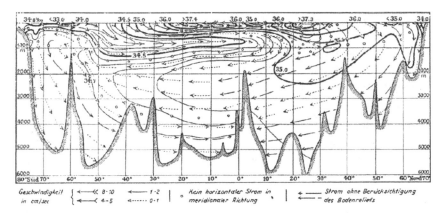

Figure 2.3 Merz and Wüst's model of ocean circulation.

From Alfred Merz and Georg Wüst, "Die atlantische vertikalzirkulation," 1922, from *Die Atlantische Vertikalzirkulation. Zeitschrift der Gesellschaft für Erdkunde zu Berlin, Jahrgang 1922*, 1–35.

1921—a decade after the completion of the *Deutschland*—just as Merz and Wüst were completing their magnum opus, *Der atlantische Vertikalzirkulation*, published in 1922.[44] Not prepared to let Brennecke steal their thunder, Merz and Wüst argued that most of the results emphasized by Brennecke could be extracted from the data of the *Challenger* expedition. Moreover, they challenged Brennecke's picture of ocean circulation by calculating the expected pressure field from the distribution of temperature and salinity (but without geostrophic corrections). Their conclusion? A general northward circulation above 1,000 meters and a southward one below about 1,400 meters (fig. 2.3). But that was all. The pressure gradients did not appear to be sufficient to drive large-scale circulation.[45]

As with Carpenter and Croll, the debate between Wüst and Brennecke became bitter. In 1922 the editor of the *Zeitschrift der Gesellschaft für Erdkunde* was so fed up that he declared the controversy closed; he would publish nothing further on it.[46] But there was a deeper problem, which Eric Mills has noted: Merz's and Wüst's analysis was born obsolete. They had calculated a static pressure distribution, ignoring the geostrophic effects that Ekman and others had demonstrated.[47] Moreover, their work was superseded by British oceanographer George Deacon (1906–1984), who in 1937 published *Hydrology of the Southern Ocean*, based on both the German work and his own data collected on the British *Discovery* expeditions, particularly *Discovery II*'s 1932–1933 circumnavigation of Antarctica.[48] Deacon explicitly proposed Antarctic Bottom Water originating in the Weddell Sea as a major component of a global circulation. Then World War II put the whole matter on hold.

Ocean Circulation after World War II

When European oceanographers regrouped after World War II, the question of ocean circulation was high on their agenda, particularly for Alfred Defant (1884–1974), who had succeeded Merz as director of the Berlin Institute. An Austrian trained in geophysics and meteorology, Defant became the leading German-speaking oceanographer of his generation and a staunch advocate of the dynamic method.[49]

In his 1945 text *Physical Oceanography*, Defant reviewed the previous hundred years of thinking on density-driven circulation and explained why, despite some evidence, arguments for density gradients as a large-scale driving force remained unconvincing. Germans, he argued, were still thinking primarily in terms of the Lenz model, which had assumed "an upward directed component near the equator" to explain the presence of cold near-surface waters there.[50] This was too simplistic. Defant underscored that it was a mistake to press the analogy between ocean and atmosphere too strongly (for the reason Maury had earlier stressed): the atmosphere was warmed from *below*, a thermodynamically unstable situation that caused warm air masses to rise and drive convective overturn, but the ocean was warmed from *above*, a thermodynamically stable configuration. As Defant put it, "Conditions are not favorable for the development of powerful [density-driven ocean] circulation systems. In any case they can be only of small vertical extent and they will be entirely incapable of filling the whole of the oceanic space from the poles to the equator."[51]

Although his argument began along the same lines as Maury's, it diverged on the matter of salinity effects, again for thermodynamic reasons. Increase in salinity acts in concert with temperature decrease: circulation is favored when cold waters are salty and warm waters are fresh. But on Earth, evaporation produces warm, salty waters, so the thermal and haline effects act in opposing ways.[52] Defant did not ignore the possibility that in polar regions, waters could be so cold as to be dense enough to sink, with or without increases in salinity, but like most others before him, he thought that the effects would be minor and local. Drawing on the meteorological tradition that had inspired dynamic oceanography—and picturing the ocean as a mirror image of the atmosphere—he distinguished between the oceanic "troposphere," or the warm surface layer strongly influenced by the atmosphere, and the oceanic "stratosphere," composed of nearly uniform masses of cold, deep water below. He allowed that there must be some sinking of very cold polar waters "due to their low temperature and in spite of their low salinity," and that sinking probably "reaches down to great depths."[53] But how widespread was this effect? And how great were the depths to which the polar waters would sink? The data were inadequate to answer these questions, and so, in the absence of data to demonstrate an effect, Defant assumed there was none.[54]

Defant's views were shared by Harald Sverdrup, who by 1942 was working with Defant's former student, Lodewyk Lek (chapter 1).[55] In *The Oceans*, Sverdrup characterized the state of knowledge of deep circulation as one of ignorance. There were some data to indicate density differences at depth and some to indicate currents but not enough to pin down the relationship, if any, between them. "It is impossible to tell," he concluded, "whether the distribution of density causes the currents or the currents cause the distribution of density.[56] Like Defant (and countless other scientists before and after him), Sverdrup took the absence of evidence as evidence of absence and concluded that any deep currents were likely to be "so weak that they are negligible."[57]

Sverdrup calculated the degree of accuracy that would be needed to determine if density differences were sufficient to drive deepwater flow; his answer was temperature measurements accurate to 0.02° C and salinity measurements accurate to 0.02 percent. For near-surface samples, this had been achieved in the 1930s with reversing thermometers, and in the 1940s with the BT, but accurate deepwater temperature measurements and high-precision salinity remained elusive. Moreover, full understanding would require extensive measurements in time and space to allow for the full range of regional and seasonal variability. Sverdrup concluded that resolution would await much more extensive and more accurate data collection. He was wrong. After nearly two centuries of measuring, theorizing, and, in some cases, bitter argumentation, what made the difference in understanding ocean circulation was not more data but a different way of thinking about the problem.

And that different thinking was inspired by a specific World War II operational problem: the effect on sonar of the ocean thermocline.

Henry Stommel and the Problem of Abyssal Circulation

Brilliant, creative, funny, original, and unpretentious, Henry "Hank" Melson Stommel was a man who teemed with ideas and possessed a unique ability to get to the physical essence of a problem. An unscientific poll of oceanographers in 1979 rated him the most famous oceanographer in the world, primarily for his work on abyssal circulation.[58]

In a letter nominating Stommel for an honorary degree in 1959, one colleague noted that he had the virtually singular distinction of making a theoretical prediction in oceanography confirmed by observation: the prediction of a deep boundary current on the western side of the Atlantic moving toward the equator—essentially a countercurrent to the Gulf Stream—verified in 1957.[59] As Stommel's collaborator Arnold Arons later noted, it was common for oceanographers to devise a posteriori explanations for observed phenomena, but almost unknown for them to predict an unobserved one.[60] Among some oceanographers today, the lore is that Stommel is the *only* oceanographer to do this.[61]

Like many smart people, Stommel was independent minded, strong willed, and more than occasionally impatient—a price he felt "one pays for intelligence and imagination."[62] He never hesitated to refuse service on a committee he considered pointless, or to express his distaste for actions he considered foolish or misguided. He frequently turned down administrative appointments and felt that scientists made such poor administrators that Woods Hole should establish a training course for careers in geophysics administration—akin to hospital administration—so the work could be done by competent professionals rather than incompetent scientists.[63] In his files, Stommel kept several copies of a painting of Michael Faraday declining the presidency of the Royal Society.

Stommel also lived with myriad tensions and contradictions. He expressed misgivings over the application of science to warfare, yet his fame was based on work funded almost entirely by the US Navy. Philosophically, he was committed to personal autonomy and viewed scientific research as an expression of individual curiosity and creativity, yet he became a driving force behind several highly political, big-science projects of the 1970s.[64] He accepted that large-scale data-gathering enterprises were necessary—indeed, he believed that all previous oceanographic advances had been rooted in new and better data—yet he worried that large enterprises were risky to individual autonomy and creativity. His scientific hero (besides Faraday) was John Swallow

(1923–1994), the gentle and modest inventor of the neutral-buoyancy float, a "full sized observer in the old-fashioned one-man way."[65] Yet Stommel's most important contribution would be based not on observation, but on insight.

As strong and perhaps contradictory as his views were, they rarely generated ill will, perhaps because of his wry sense of humor. In the 1960s, when a bitter dispute broke out between faculty and Woods Hole director Paul Fye (1913–1988), Stommel tried to make the faculty case with satire (chapter 3). In 1990, when he thought Woods Hole was tipping too far in the direction of computation at the expense of fieldwork, he proposed a new design for the institution's letterhead: replacing the graceful research vessel *Atlantis* with an ungainly computer monitor (fig. 2.4).[66]

Woods Hole Oceanographic Institution
Woods Hole, MA 02543
Phone: (508) 548-1400
Telex: 951679

June 14, 1990

Drs George Grice, Judith McDowell, Werner Deuser, Joseph Pedlosky:

Thank you for sending the packet including the memo to Senior Scientists of February 14th and the letter from Lane, Frenchman and Associates of May 8.

My first surprise is the low position of ships on the totem pole. Taking the two documents together, I deduce that ships are now regarded to be of lower priority than facilities including a softball field, basketball court, tennis court, volleyball sandpit, paddle tennis court and jogging paths.

My second surprise was the suggestion that the Institution needs to "create a new image". I myself rather like the old Atlantis as a symbol. If something more in accord with our new sense of priorities is desired, perhaps something like the above sample letterhead would do.   .

Yours sincerely,

Henry Stommel

Henry Stommel

Figure 2.4 Stommel's proposed new letterhead for the Woods Hole Oceanographic Institution.

From Henry Melson Stommel Papers, 1946–1996, MC-06, "Correspondence, 1990," Data Library and Archives, Woods Hole Oceanographic Institution. Reprinted with permission from Woods Hole Oceanographic Institution.

Stommel built his exceptional career on no formal education in oceanography. Growing up in Brooklyn, New York, where his mother settled after divorcing his father, he attended a public school populated largely by the children of Jewish and Scandinavian immigrants, where he saw that "application to school work was our only escape from uncertain employment."[67] He spent a year at the prestigious Townsend Harris High School, a competitive public high school in Queens, before the family moved to Long Island. He earned a scholarship to Yale, receiving his BS in astronomy in 1942. For the next two years, he concurrently worked at the Yale Observatory and took courses at the Divinity School, deciding between a career dedicated literally to the heavens or only metaphorically so.

In 1944, he was offered a position at Woods Hole, where money was pouring in for military oceanographic work and qualified scientists were in short supply. He joined as a research associate working with geophysicist Maurice Ewing (1906–1974), analyzing bathythermograms.[68] Ewing was proud of the concrete value of this work for wartime military operations; later, as director of the Lamont Geological Observatory, he would foster extensive ties with the Navy (chapter 5).[69] But Stommel had misgivings, recalling later that he "was troubled by the morality of killing in war." He rationalized his involvement in antisubmarine warfare as "the least immoral of any military application of science that I could do," insofar as it "did not involve civilian targets."[70]

After the war, Stommel would make his name with an analysis of western boundary currents. His 1948 paper "On the Westward Intensification of Wind Driven Currents" demonstrated that westward intensification was a direct consequence of the variation of the Coriolis effect with latitude. This insight explained the strength and location of the Gulf Stream and predicted a deep countercurrent traveling toward the equator under the Gulf Stream. This prediction was confirmed in 1957 by Stommel's hero, John Swallow, along with Valentine Worthington (1920–1995), using Swallow's neutrally buoyant floats to make the needed measurements at the appropriate depth.[71]

In Arons's view, the paper "launched a new episode in physical oceanographic work" that emphasized the connection of the world's oceans at depth and the fact that the atmosphere and ocean form a coupled resonant system.[72] It certainly launched Stommel's career. "Though it all seems too obvious and compelling in retrospect it is well to note that Bjerknes, Ekman, Defant, Sverdrup and [Carl-Gustaf] Rossby had not perceived the connection with westward intensification in a bounded basin. It was the youthful Henry Stommel who did," Arons recalled.[73] Ekman had incorporated the Coriolis effect into his analysis of the relation between winds and currents, but no one had incorporated its variation with latitude, even though the latter was well known. Moreover, this would be just the first step in a comprehensive, path-breaking reconsideration of the problem of ocean circulation. To understand how the youthful Henry Stommel—a onetime aspiring astronomer without a PhD—was able to solve a problem that had eluded the oceanographic greats, we need to look more closely at Woods Hole in the years just after World War II.

Postwar Woods Hole and Its "Basic Task"

The immediate postwar years were an anxious time for all American scientists as they wondered whether wartime military largesse would continue (and some hoped it would not), but the anxiety was particularly acute for oceanographers. Few Americans before the war had ever made professional careers in the field. The issues that plagued Sverdrup at Scripps—whether a continued wartime largesse would help or hinder him personally—also affected Stommel. Lacking a PhD, he particularly wondered whether he would be able to pursue a career in an expanded, more professionalized environment. In 1950, he asked Columbus Iselin for advice: should he go to Scripps for an advanced degree? Iselin replied: "The essence of your problem is whether or not a married man has any business fooling around in an obscure field like oceanography. Whoever heard of a professor of physical oceanography, except in Germany? Things may be different in the future, but past performance would indicate that your children are bound to starve."[74]

Iselin thought the future would be different, but he also thought that Stommel did not need a PhD to be part of it because he had already learned "on the job." Moreover, at Scripps he would always be "one step behind Walter Munk," who was working on a parallel track but with superior mathematical skills.[75] It would be better to stay at Woods Hole, Iselin advised, and wait for a university position to open: "Have a talk with ... anybody else whose opinion you value, but don't sign yourself up as a student at Scripps. What can they teach you?"[76]

Evidently, Stommel believed the folks at Scripps could teach him something, because he wrote to Sverdrup asking to be taken on as a PhD student. Sverdrup rejected him. Stommel thought Sverdrup disliked the fact that he had written a popular book (*Science of the Seven Seas*); more likely Sverdrup rejected him because he (Sverdrup) had already made the decision to return to Norway (chapter 1). In any case, Stommel did not gather a favorable impression when he visited. He greatly respected Munk, but that was about it: "All the other people here who are actually looking at and thinking about ocean circulation ... are junior assistants of the Tuna Commission or something," he concluded dismissively.[77]

Stommel stayed at Woods Hole, and his children did not starve. On the contrary, he became one of the principal beneficiaries of the "basic task contract," which in 1950 had already been supporting his research for four years. Indeed, less than a year after the end of World War II, Woods Hole signed a contract with the Office of Naval Research to continue investigations begun during the war on oceanographic questions relevant to Navy concerns.[78] Officially the title was "military defense oceanography"; Woods Hole staff called it the "basic task." Its stated goal was to increase "our knowledge of oceanographic phenomena." In practice the contract formalized the relationship forged between Woods Hole and the Navy during the war: one of mutual interest in shared concerns but with a primary focus on aspects of physical oceanography relevant to military operations.[79] For the following two decades, military defense oceanography would be the basic task of Woods Hole.

If oceanographers saw the US Navy as a good patron—and despite the troubles at Scripps during the war, nearly all of them did—the basic task contract illustrates why. The Navy wanted to increase oceanographic knowledge and so did oceanographers. It seemed natural, and even obvious, that the relationship forged in war should be continued in peace to mutual benefit. If Woods Hole would change as a result, those changes would be for the better, as the institution would be able to pursue oceanographic investigations with a vigor and consistency that was previously impossible.

As historians Susan Schlee and Gary Weir have documented, Navy funds transformed Woods Hole just as they transformed Scripps. Both institutions had been established by private philanthropy, and at both government funds had come to eclipse endowment resources. At Woods Hole in 1951, for example, institutional research funds amounted to $140,000; government grants and contracts (almost entirely from the armed services) amounted to $1.5 million. Like Scripps, Woods Hole was now a different place. A previously unimaginable level of funding would make it possible to do science that had long been imaginable but not achievable.[80]

The Navy was inevitably more interested in some aspects of oceanography than others; not all oceanography could be characterized as "military defense oceanography."[81] Three areas received particular attention: dynamic oceanography, surface and internal waves, and abyssal circulation. By 1957, Stommel was in charge of the Internal Waves and Abyssal Circulation Group. He summarized his progress in a report to the ONR: "Considerable progress has been made toward developing a theory of the abyssal circulation.... A general scheme showing the distribution of deep water currents on a world-wide basis has been developed theoretically."[82] This must rank as one of the great understatements in the history of science, for 1957 was the year in which Stommel began to solve a problem that had vexed scientists for centuries.

The Stommel-Arons Model of Abyssal Circulation

In October 1957, Henry Stommel reviewed the state of knowledge of abyssal circulation in the pages of *Nature*. Stommel attributed the impasse on the issue of deep-ocean circulation to insufficient empirical evidence: "So very little positive knowledge exists about the nature of the deep circulation that it is possible to maintain entirely different points of view."[83] Russian scientists, he noted, believed "that the abyssal circulation is essentially produced by the wind ... and that the deep flows are almost always in the same direction as surface currents."[84] It was, of course, not only Russians who held those views, but Stommel was perhaps trying to soften the blow to his Anglophone colleagues of what he was about to say: it was no longer reasonable to hold such views. Referring to himself in the third person, Stommel explained why, providing in a single paragraph the theoretical idea that would alter scientific thinking:

> These views are quite the opposite of those expressed by Stommel, who regards the deep circulation as an essentially heat-driven and salt-driven engine, and [that] in general ... the surface and deep flows are in opposite directions.... This theory predicts that in the deep water the mean meridional components of velocity are everywhere directed poleward except in the western boundary streams, which connect the slow central flow to regions of formation of deep water.[85] It is thus possible to predict the deep counter-currents under the Gulf Stream. Similarly, we can expect to find a very narrow intense northward current between 3,000 and 5,000 m. over the Tonga-Kermadec Trench, which feeds the entire Pacific with deep water. One hopes that the adherents of these two rival theories of abyssal circulation will be able to develop them quantitatively and in sufficient detail so that future direct measurements ... will effect a clear-cut decision in favor of one or the other. It is in this way that science progresses.[86]

Everyone accepted that there was wind-driven horizontal displacement of water masses, tied to Earth's rotation. But was there also a significant vertical component? That was unresolved. Rather than perpetuate the impasse, Stommel took a novel approach. He said, in effect, let us assume that there is. We can to do this by conceptualizing the ocean as composed of two layers, across which there is vertical transport. If we do, we find that the rest of the ocean circulation falls out surprisingly simply. In a long paper published the same year, "Survey of Ocean Current Theory," he explained. The key conceptual innovation was "dividing the ocean into two layers by a *level surface* at mid-depth, say 1500 or 2000m and specifying the vertical mass transport across this level surface as a function of geographical position.... A pattern of geostrophic flow can be constructed in each layer to absorb (or give up) the water required for the vertical transport across the level surface, and a western boundary current fitted where needed."[87] In other words, forget for a moment about the winds. Forget about the downward motion of cold polar waters in the North Atlantic and the Weddell Sea. Imagine instead a general upward vertical diffusion throughout the abyssal ocean: if one presumes that there is such vertical transport, then one can calculate the currents required to accommodate it.

Stommel was flipping conventional wisdom on its head. When most oceanographers thought about vertical transport, they imagined it downward—the sinking of dense water in a few unusually cold areas. Stommel was turning this literally upside down and imagining a general upward diffusion throughout the abyss. He illustrated this in a series of sketches showing flow controlled only by vertical transport, flow controlled only by horizontal transport, and flow controlled by a combination of both. The latter, he argued, was closest to the empirical evidence. This suggested that it was unhelpful to think of the ocean as having two *different* forms of circulation—wind driven and density driven. It was more useful simply to envisage circulation, in which the horizontal effects were in part dependent on the vertical transport, and vice versa. Crucially, if one specified the horizontal transport, the rest of the picture would be "determined."[88] Put another way, "arbitrary assumptions about the vertical velocity structure in the ocean are attended by the most far-reaching implications concerning the form of the horizontal current pattern" (figs. 2.5 and 2.6).[89]

Stommel disavowed any aspiration to have his conceptual model interpreted as a theory: "These considerations, and the pictures derivable from them, are not *theories*," he insisted. "They merely provide an interpretive tool, and indicate what flow patterns must be associated with various distributions of vertical transport across level surfaces."[90] Stommel no doubt wished to deflect potential criticism that his model was mathematically underdeveloped, and over the following ten years he worked with colleagues to dignify his model sufficiently to call it a theory. More than that, he worked to show that it was *true*.

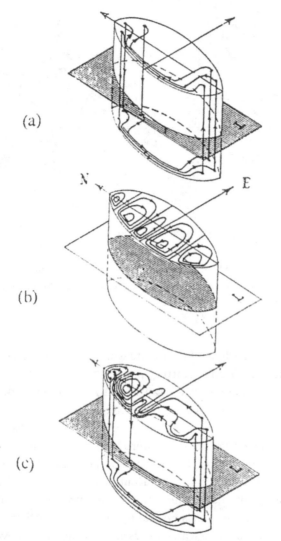

Figure 2.5 Stommel's figure 9, from his "A Survey of Ocean Current Theory."
From H. A. Stommel, "A Survey of Ocean Current Theory," *Deep Sea Research* 4, no. 3 (1957): 160.
Reprinted with permission from Elsevier.

From an Interpretive Tool to a Theory of the Ocean

How did Stommel transform his speculative concept into a persuasive and accepted scientific theory? His first step was to test it experimentally. In an article published in *Tellus* in 1958 and coauthored by Arons and Woods Hole colleague Alan Faller (b. 1929), Stommel pursued his concept with a simple

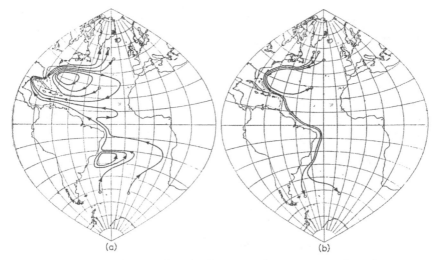

(a)   (b)

Figure 2.6 Stommel's figure 10, from his "A Survey of Ocean Current Theory."
From H. A. Stommel, "A Survey of Ocean Current Theory," *Deep Sea Research* 4, no. 3 (1957): 160.
Reprinted with permission from Elsevier.

experimental setup, designed to simulate what he saw as the key element of ocean circulation: fluids on a rotating sphere.

The relevant parameters were established by his earlier work: that ocean circulation could be described as a simplified regime in which "(a) the flow in the whole layer is steady and geostrophic, except and only (b) at the western boundary where a narrow, intense western boundary current is permitted to depart markedly from geostrophy; and moreover (c) this system, which would otherwise be at rest, is driven by a distribution of sources and sinks of fluids (various driving agents such as the wind can be expressed in terms of source and sink distributions, so this is no real restriction)."[91] In other words, if the ocean is a dynamic system in which fluids are moving, those movements can be analyzed in terms of sources of water (where water flows into a region) and sinks (where water flows away).

The use of the term *sink* in this context can be confusing; in everyday life, a sink is something in which water drains down. For readers familiar with contemporary concepts of thermohaline circulation (in which cold waters sink in polar regions), the term reinforces the commonsense impression that in a sink waters move downward. Stommel is not using the term in that way. Rather, in this discussion a source is anything that brings fluid into any area under consideration, and a sink is anything that removes it. (This is comparable to contemporary discussions of sources and sinks of carbon dioxide in the atmosphere.) Therefore, westward-flowing currents are a source of ocean water into the western sides of basins and eastward-flowing currents are a sink. The same reasoning applies in the vertical dimension. In the Weddell Sea, cold polar waters move down into the abyssal ocean: this is a source of cold, deep water. Upward vertical diffusion is a source of cold water into the upper reaches of the seas. Thus, wind is either a source or sink, depending on which way the wind drives fluids, and the same is true for density differences. Moreover, these driving agents may operate in the vertical or the horizontal dimension.

Why was it important to build a physical model, particularly given the obvious complaint that a bench-scale model could hardly reproduce the complexities of the world ocean? Stommel explained that some of the implications were so "contrary to intuition that it would be reassuring if they could be partially tested by model experiments."[92] The models were not intended to replicate the real ocean, nor could they prove that the ocean behaved in a particular way. But they could increase the plausibility of the proposed conceptual scheme. Implicit was that they might be more convincing to field-based oceanographers than mathematical formalisms would be.[93]

Before constructing the actual experiment, or even doing the math, Stommel knew what he expected to find: "Before we had carried out any mathematical analysis or experimental work, we could see that our regime led qualitatively to some rather remarkable deductions."[94] It was as if he had already done the experiments in his head and found that "the only permissible circulation pattern ... is one in which zonal geostrophic flow extends from the source to the western boundary, thence down the western boundary in a narrow, intense boundary current to the radius of the sink, and thence, in another zonal geostrophic flow to the sink."[95] Moreover, the idea that the correct solution was the simplest one was not, in this case, correct. If the Earth did not rotate, water would simply flow directly from sources to sinks, wherever they were. But the Earth does rotate, and the flow path that results is "very circuitous."[96] In fact, it is about as circuitous as possible, tracking three sides of a closed form (fig. 2.7).

One could also imagine the source at a particular locale, such as the apex of a radial sector (e.g., the pole). A western boundary current is again created, only in this case water flows back toward the source throughout the basin and the return flow is distributed. Finally, imagine a source at the western edge of the rim. In this case, the interior flow is again throughout the basin and in toward the center, as one might expect, but the return flow is again along the western boundary. In other words, the strongest current is toward the source—not what one might intuitively suspect. Crucially Earth's

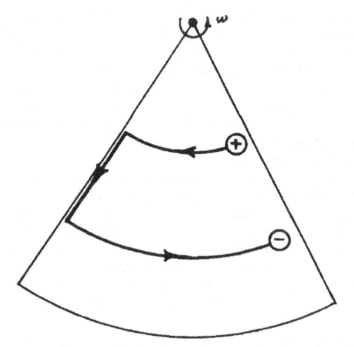

Figure 2.7 Stommel, Arons, and Faller's figure 2, illustrating how, due to the Earth's rotation, the flow path from a source to a sink in a bounded basin is exceedingly indirect. If Earth did not rotate, fluids would flow directly from source to sink. But Earth does rotate, and so the fluids follow a "circuitous" path, tracking three sides of the closed form and yielding a western boundary current.

From Henry Stommel, A. B. Arons, and A. J. Faller, "Some Examples of Stationary Planetary Flow Patterns in Bounded Basins," *Tellus* 10, no. 2 (1958): 181 (fig. 2). Reprinted in accordance with Taylor and Francis open-access policies.

rotation leads to a western boundary current *irrespective of the location of sources and sinks* (figs. 2.8 and 2.9).

To test their thought experiment, Stommel, Arons, and Faller set up the physical experiment, using a design proposed by their Woods Hole colleague, William von Arx (1917–1999) (chapter 3). They began with a seven-foot-diameter rotating tank (fig. 2.10). To simulate the effect of a radially shaped ocean basin, they used wooden partitions to isolate a pie-shaped sector, with a width of 60°. The tank contained a set of quarter-inch slots at various positions in the walls—so that mass could be introduced or removed—and covered so as to prevent wind stress on the surface and simulate only the effects of sources and sinks, which is to say, only the effects of density-driven circulation. Mass could be added from a large can of dyed water connected to the main tank by a tube; a small can of additional water was connected to the large can to maintain the hydraulic head.

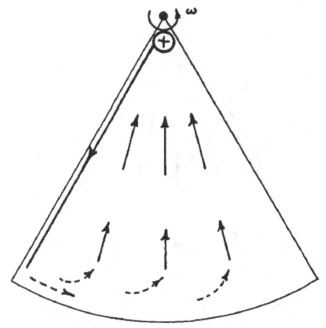

Figure 2.8 Stommel, Arons, and Faller's figure 3, illustrating the flow pattern in a bounded basin, assuming a source near the pole. Note that again the flow path is circuitous, and again a western boundary current is created. In this case the return flow is distributed throughout the basin. Note the possibly counterintuitive result that the return flow is toward the source.

From Henry Stommel, A. B. Arons, and A. J. Faller, "Some Examples of Stationary Planetary Flow Patterns in Bounded Basins," *Tellus* 10, no. 2 (1958): 181 (fig. 3). Reprinted in accordance with Taylor and Francis open-access policies.

With this setup, they mimicked the conditions in each of three idealized cases. First, they created a source and sink at an arbitrary point at the eastern side of the basin. Then they created a source at the pole with a sink in the basin center; then a source at the southwest corner with a sink at the pole; and then a source at the southwest corner with a sink in the basin center (fig. 2.11). The results were as predicted: in every case, the basin exhibited a western boundary current with return flow, and in the latter two cases toward the source. Conceptualization, mathematical analysis, and physical experiments aligned.

How did these experiments relate to the whole ocean? In some sense they merely confirmed what was already accepted about surface circulation, but the territory Stommel had staked was not the surface but the abyss. He therefore had to return to the question of vertical transport, and for this he returned to the first topic in oceanography he had confronted when he joined

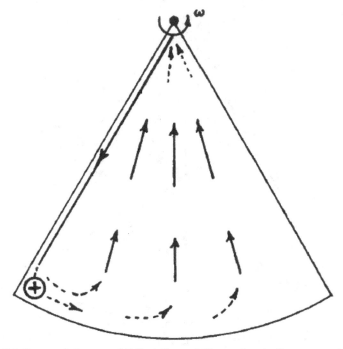

Figure 2.9 Stommel, Arons, and Faller's figure 4, illustrating the flow pattern in a bounded basin assuming a source at the western edge of the basin, close to the equator. Note that the flow pattern is exactly the same as in figure 2.8. Thus, contrary to what one might expect, the flow pattern is independent of the location of the source. Crucially, in all cases Earth's rotation yields a western boundary current irrespective of the location of the source and sink.

From Henry Stommel, A. B. Arons, and A. J. Faller, "Some Examples of Stationary Planetary Flow Patterns in Bounded Basins," *Tellus* 10, no. 2 (1958): 181 (fig. 4). Reprinted in accordance with Taylor and Francis open-access policies.

Maurice Ewing at Woods Hole in 1944: the thermocline. Its existence was a perplexing aspect of ocean dynamics and critical, as well, to the operation of submarines.

The Ocean Thermocline and Vertical Diffusion

Dynamic oceanographers had analyzed the ocean by drawing heavily on analogies to the atmosphere, but (as Maury and Defant had stressed) there was one crucial difference: the atmosphere is warmed from below, which fosters convection, but the ocean is warmed from above, which hinders it. Convection can therefore drive circulation in the atmosphere and foster mixing in a manner that in the ocean it cannot. But there is a significant complication. If

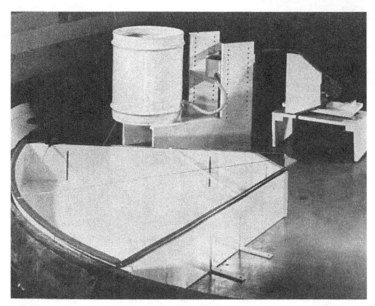

Figure 2.10 Stommel, Arons, and Faller's figure 6, showing their experimental setup. Note the fluid "source" in the upper portion of the photograph.

From Henry Stommel, A. B. Arons, and A. J. Faller, "Some Examples of Stationary Planetary Flow Patterns in Bounded Basins," *Tellus* 10, no. 2 (1958): 184 (fig. 6). Reprinted in accordance with Taylor and Francis open-access policies.

the ocean were entirely thermodynamically stable, and therefore entirely un-mixed, the temperature would drop gradually with depth. But that is not the case. In the ocean, the upper 50 to 200 meters are generally warm (typically 5°C–30°C), and the deep ocean is close to freezing (the source of the confusion behind the 4° theory). Between them is a zone of rapid temperature drop known as the thermocline.

The thermocline had been sedulously studied during World War II for its importance to antisubmarine warfare, which lay in the crucial fact that the velocity of sound is density dependent. As temperature drops rapidly through the thermocline, the density of water increases and sound waves are refracted, just as light is refracted as it travels through media of different density. From a military perspective, this was a fact of surpassing significance, because refraction of sound waves through the thermocline produces acoustic shadow zones where submarines may rest undetected. Moreover, while temperature falls with depth, salinity generally increases (because surface layers are diluted by rainfall); these countervailing effects produce a zone of minimum velocity—the "sound channel"—in which sound waves can be very effectively transmitted (fig. 2.12).[97]

Figure 2.11 Stommel, Arons, and Faller's figures 7–10, photographs of the flow patterns in the physical experiments.

From Henry Stommel, A. B. Arons, and A. J. Faller, "Some Examples of Stationary Planetary Flow Patterns in Bounded Basins," *Tellus* 10, no. 2 (1958): 185–86 (figs. 7–10). Reprinted in accordance with Taylor and Francis open-access policies.

The sound channel proved effective in determining the position of a submerged submarine or an airman downed at sea (the airman could release a small explosive and the sound waves detected, even at great distance) and became the basis of a new military navigational system—Sound Fixing and Ranging, or SOFAR.[98] After the war, this became the basis of the US Sound Surveillance System, or SOSUS, the system of underwater cables and hydrophones used to detect and track Soviet submarines throughout the

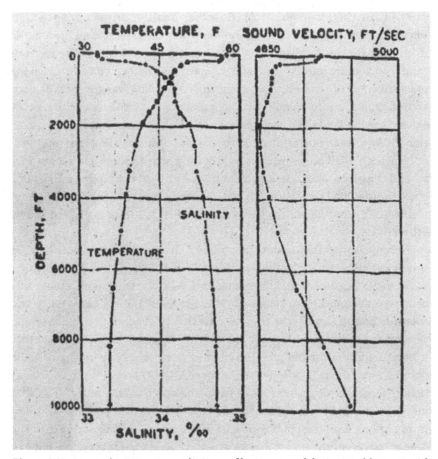

Figure 2.12 A typical temperature-salinity profile, as prepared during World War II, with computed sound velocities showing a steep temperature drop just below the surface layer and a corresponding drop in velocity. Note the velocity minimum at 2,000 feet, created by the countervailing effects of decreasing temperature and increasing salinity.

From Carl Eckart, 1946 (reprinted 1968), "The Refraction of Sound," in *Principles and Applications of Underwater Sound*, NDRC Division 6, 11 (fig. 1).

Cold War (chapter 9).[99] It would not be unreasonable to say that, from an operational standpoint, the analysis of the thermocline and its effects on sound transmission was the most important wartime development in oceanographic science.

Woods Hole had been a center for temperature and salinity studies on behalf of antisubmarine warfare; the thermocline was the focus of Maurice Ewing's work when Stommel joined him in 1944. The bathythermograph, the instrument used during the war to create temperature-depth profiles, was developed at Woods Hole, one of two US centers for the collection and analysis of bathythermographic data (the other was the University of California Division of War Research, where Sverdrup's troubles arose, as recounted in chapter 1). The detailed study of the thermocline was a major focus of military-scientific interest in oceanography and an early domain in which the practical value of theoretical scientific understanding was cogently demonstrated. (It also became a domain of women's work, as male adventurers considered the landlocked task of data compilation too dull to merit their attention.)[100] By the end of World War II, the thermocline was one of the most well documented of all ocean features, as BTs were sent to ships and submarines around the globe. Many thousands of BT records were compiled into charts, atlases, and sailing directions.[101] It is perhaps not surprising, then, that Stommel focused attention on the thermocline. What is surprising is that so few others did.

Stommel called the thermocline "one of the mysteries of the sea," for it was not at all obvious why there should be one.[102] Why wasn't the transition between surface and deeper layers gradual? Why was the water immediately below the thermocline so cold, even in the tropics? Defant, Lenz, Wüst, and Sverdrup had all argued that any adequate theory of ocean circulation had to account for the presence of cold water at depth, but Stommel insisted that it had to account for the presence of cold water nearly everywhere. As he pointed out, "the really deep water, between 4,000 feet and the bottom at 18,000 feet, is close to freezing, even in the tropics. This must mean that most of the water in the oceans flows [toward the middle and low latitudes] from regions near the poles. However, there are no strong currents in the deeper regions of the ocean. The great wind-driven currents are confined to the upper quarter of its depth."[103] The presence of cold water throughout the ocean meant that water must be being moved from the poles to the rest of the abyssal ocean, but how? Clearly something was missing from the prevailing thinking.

Here, Stommel's independence served him well. Most colleagues had only praise for Sverdrup's magnum opus, *The Oceans*, but Stommel felt that it was entirely too definitive. He took particular exception to its preface, in which Sverdrup and his coauthors argued that it was better to present a clear interpretation, even if wrong, then a muddle of conflicting speculations. Stommel strongly disagreed. He felt that this approach discouraged the reader from thinking independently and offered young scientists "few hints that there was much left to do."[104] How would a student know which of these clear statements were *actually* definite?[105] Stommel particularly objected to this line: "At the risk of premature generalization, we have preferred definite statements ... to conflicting interpretations, believing that the treatment selected would be more stimulating." Stommel considered that "deadly."[106] Oceanographers often referred to *The Oceans* as "the Bible," and Stommel found it just about as dull. And to his thinking, no scientific work should be viewed as sacred.

Like Sverdrup a decade before, Stommel recognized that the deep currents were too weak to measure by current meter, and in any case, there was no way to get instruments to the required depths. The answer would have to be theorized. In a characteristic pattern, Stommel at first outlined a qualitative solution and then developed it quantitatively. In his 1957 report challenging Russian thinking, Stommel had already answered the question. The key assumption—laid out in this early sketch—was vertical transport. But it was not just any form of vertical transport; it was upward vertical transport throughout the ocean. Stommel's analysis began by inviting the reader to presume upward vertical flow throughout the abyss.

What was the justification for this radical presumption? And why did Stommel make this presumption where no one else had? The answer can be found buried in the middle of a paragraph in the middle of that original 1957 comment:

> The general picture emerging from ... various descriptive studies is that the general flow of water below 2,000-m depth in the ocean is directed from quite limited areas in the northern North Atlantic and from the shallow shelf areas surrounding Antarctica into the Indian and Pacific Oceans. The regions where water is supplied by sinking to the level beneath 2,000 m. are doubtless quite small, and probably intermittent. On the other hand, the upward flow of water out of the deep ocean evidently occurs in a very widespread manner all over the ocean, and *it is presumably this upward flow which prevents the hot water at the surface from gradually working down into the deep water* (even at the tropics the warm surface layers are relatively shallow and the temperature of the bottom waters is less than 3°C).[107]

The key insight here—made seemingly in passing—is the requirement of upward flow throughout the ocean to explain the thermocline. Without upward flow, there would be no thermocline. The existence of the thermocline is the key fact underpinning the entire model. Stommel had referred to his presumption of upward vertical flow as arbitrary—"arbitrary assumptions about the vertical velocity structure in the ocean"—but it was not arbitrary at all. It was based on demanding: Why is there a thermocline? And how it is sustained?

Instead of asking, as others had done, why deep water is so cold, Stommel asked how the warm surface layer stays warm. His answer was "*upward*

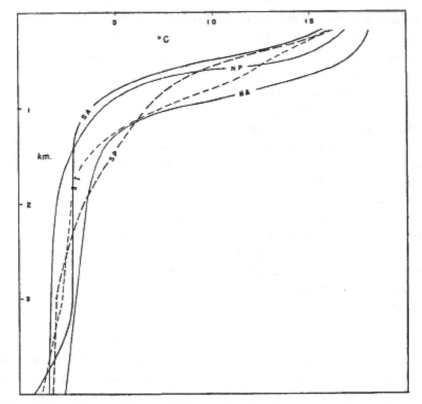

Figure 2.13 Robinson and Stommel's figure 3, from their "The Oceanic Thermocline and the Associated Thermohaline Circulation."

From Allan Robinson and Henry Stommel, "The Oceanic Thermocline and the Associated Thermo-haline Circulation," *Tellus* 11, no. 3 (1959): 305 (fig. 3). Reprinted in accordance with Taylor and Francis open-access policies.

flow which prevents the hot water at the surface from gradually working down into the deep water."[108] In short, the fact of the thermocline meant that there had to be upward flow to maintain it. Here was the explanation simultaneously for the characteristics of both the deep ocean and the thermocline: the upwelling of water throughout the abyssal ocean. Upward diffusion both mixed the abyssal ocean and prevented surface waters from mixing; this created and sustained the observed thermal structure. The Stommel-Arons model hung on the existence of the thermocline. Put another way, it hung on *attending* to the thermocline where others had not (fig. 2.13).

Stommel developed these ideas mathematically in an article coauthored with Woods Hole colleague Allan Robinson (1933–2006), published in *Tellus* in 1959. The authors began by noting that, since the acrimonious and unproductive Carpenter-Croll debate, many authors had approached the topic of ocean circulation with trepidation; many textbooks dealt with the subject only indirectly. And because it was easier to calculate the (large) wind stress than the (small) density effects, researchers had tended to minimize the significance of the latter. It was time to change that.[109] Rather than assume that density differences were insufficient to have an effect, they would assume there was an effect and calculate what it had to be.

The parameters of their mathematical model followed their physical one: a bounded basin driven by a meridional temperature gradient at the surface on a rotating sphere, represented mathematically by a Coriolis parameter varying with latitude. Solving the relevant equations, Stommel and Robinson were able to show that a deep circulation could be generated by the observed meridional differences in surface temperatures, from equator to poles, and that there must be an upward vertical component of flow to maintain the thermocline. Moreover, one could calculate its value: 3.1×10^{-5} cm/sec. Not a large number, to be sure, but in terms of ocean dynamics and integrated over the entire ocean, a significant one. In a separate letter in *Deep Sea Research* that same year, Stommel calculated that integrated value: 90×10^6 m³/sec. *That* was a large number.[110] The upward flow was the sink that permitted the downward advection of dense waters elsewhere. The "sink" consisted of water moving upward, which permitted the polar regions to be a "source" of cold water to the abyssal sea.

Stommel accepted the empirical evidence collected by Wüst, Deacon, and others that deep waters were created only in two areas: the North Atlantic and the Weddell Sea.[111] This suggested a picture of abyssal circulation dramatically different from the earlier models of Humboldt, Lenz, and Brennecke: it was not a conveyor belt, on which coherent water parcels sank at the poles, maintained their identity, and rose at the equator. Rather, parcels lost their identity—like the dye of his experiments—as the water within them diffused throughout the entire abyssal ocean (fig. 2.14). That there were only two small sources of cold water and the entire abyssal ocean was the sink might have seemed a bold—even crazy—claim. But the thermocline was the proof, because without the abyssal sink there would be no thermocline.

Stommel and Robinson's claim that the ocean was mixed was not just an assumption—it was based on temperature and salinity data—but at the same time, they had no way to measure the mixing of the ocean. In their model they included a mixing parameter, admittedly not based on any measured property. As Stommel explained, the model contains "a parametric treatment of the mixing processes embodied in an "eddy thermometric conductivity parameter χ. This parameter is assumed to be constant over the

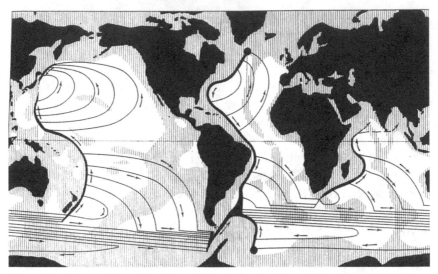

Figure 2.14 Stommel's model of abyssal circulation based on the dynamics of a fluid on a rotating sphere. Two point sources—the North Atlantic and the Weddell Sea—and a distributed "sink" (the entire ocean)—are connected by western boundary currents. The distributed sink is required to maintain the thermocline; the western boundary currents are required to connect the focused sources with the distributed sinks. The picture Stommel drew in a horizontal plane based on the tank experiments—of a concentrated western boundary current and a distributed return flow—could be applied to the vertical plane as well.

From H. A. Stommel, "The Abyssal Circulation," *Deep-Sea Research* 5 (1959): 82 (fig. 2). Reprinted with permission from Elsevier.

entire ocean basin.... Thus we envisage the ocean as being slowly and evenly stirred by some physical process which we cannot specify."[112] Stommel did not attempt to obscure the fact that his thermal circulation worked only if one presumed the ocean were mixed. This was rather different from what most others had assumed—(and indeed, different from misunderstandings of thermohaline circulation prevalent today). In Stommel's calculations, it is not the density differences that mix the ocean, but the mixing of the ocean that permits small density differences to have an effect.

The illustration in figure 2.14, published by Stommel in 1959 (in yet another paper with the title "The Abyssal Circulation"), contains all the important elements of what would come to be known as the Stommel-Arons model, but Stommel was dissatisfied by its essentially qualitative nature. "The above presentation is in the nature of a tour-de-force," he wrote, suggesting that it was more a feat of ingenuity than of labor: "One cannot pretend that it describes the abyssal circulation accurately in detail."[113] For that he turned to Arnold Arons.

Quantifying the Stommel-Arons Model

Stommel was sensitive about the gaps in his formal training and addressed them by finding collaborators with complementary skills. Allan Robinson was one; Arnold Arons was another. A physical chemist by training, Arons had come to Woods Hole during the war when his PhD adviser, Harvard chemist E. Bright Wilson (1908–1992), suggested he join the Underwater Explosive Research Laboratory at Woods Hole (for more on Wilson, see chapter 3). Arons worked on a variety of problems related to the propagation of shock waves through water, culminating in the analysis of wave propagation at the atomic bomb blasts at Bikini Atoll. In 1946, he became a professor of physics, first at the Stevens Institute of Technology and then at Amherst College, where he was renowned as a teacher who insisted on connecting mathematical formulas to the ideas they represented. Throughout his career, he maintained close relations with Woods Hole, spending summers there throughout the 1950s and gradually shifting his interests away from shock waves in the oceans to the oceans themselves. In 2000, Frank Press (1924–2020), president of the National Academy of Sciences and science adviser to President Jimmy Carter, recalled that Arons was an inspiration to him of "how to be a good scientist."[114]

Arons had come to Woods Hole—and to know Henry Stommel—primarily because of the atomic bomb. In 1950, he had been invited to participate in studies sponsored by the Naval Ordnance Laboratory on the dispersal of "falling particles in a liquid"—nuclear fallout—and the question of whether dense particles (from an underwater bomb blast) could trigger the movement of density currents.[115] This project led to discussions with Stommel and William von Arx about the work of Dutch geologist Philip Kuenen (1902–1976), whose recent experiments had demonstrated the role of density currents in creating submarine canyons (chapter 4).[116] Arons had also contributed to a highly classified project investigating the base surge produced by "a large explosion in shallow water harbors." In a letter to Woods Hole director Edward "Iceberg" Smith (1889–1961), Arons suggested that Woods Hole take up scale-model experimental studies of the sort developed by Kuenen, which would contribute to understanding both the blast effects in water and the potential cratering effect of "a 20 kiloton charge detonated in about 40 ft of water" (the first bomb dropped on Hiroshima was an approximately fifteen-kiloton-equivalent charge).[117] Around this time, Stommel and Arons began to work together.

The Stommel-Arons model was laid out in a five-part paper published over the course of twelve years. The most critical were the first two parts,

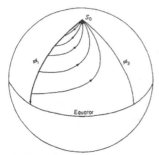

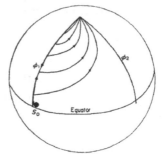

Fig. 6. Circulation pattern in meridionally bounded ocean with concentrated source S_0 at North Pole and a uniformly distributed sink Q_0, such that $S_0 = Q_0 a^2 (\phi_2 - \phi_1)$.

Fig. 8. Circulation pattern in meridionally bounded ocean with concentrated source S_0 (fed by western boundary current from below the equator) and a uniformly distributed sink Q_0 such that $S_0 = Q_0 a^2 (\phi_2 - \phi_1)$.

Figure 2.15 Stommel and Arons's figs. 6 and 8, circulation patterns. Note the match to the experimental results illustrated in this volume's figures 2.8 and 2.9.

From H. A. Stommel and A. Arons, "On the Abyssal Circulation of the World Ocean—I. Stationary Planetary Flow Patterns on a Sphere," *Deep Sea Research* 6 (1960): 146 (fig. 6), 148 (fig. 8). Reprinted with permission from Elsevier.

published in *Deep Sea Research* in 1960.[118] Stommel and Arons provided the mathematical treatment for the concept developed in the rotating tank experiments: a system driven by sources and sinks, in which flow is geostrophic, except at the western boundary, where an intense current develops. The source could be any driving force—including displacement of water masses by wind—but their focus was the "sink," defined as the upward transfer of fluids throughout the abyssal ocean (fig. 2.15).[119]

The model hung on two key points. One, as already noted, was the existence of the thermocline, which justified the assumption of a distributed sink across the whole ocean. The other was the evidence provided by Wüst and others for two sources of cold water: the North Atlantic and the Weddell Sea. Connecting the sources and sinks provided the conceptual picture that they developed mathematically: a treatment that confirmed the requirement for intense western boundary currents and predicted that these currents would develop no matter where the source of the water was. Indeed, one of the most important results of the model is that it does not much matter where the source lies. Whether the source is at the equator or the pole, a western boundary current is required to satisfy the requirements of geostrophy and conservation of mass.

In the second paper, they pursued the question of the residence time of deep waters, a question that greatly concerned the Atomic Energy Commission (see chapter 7).[120] They showed that the residence time of deep waters was likely to range between 200 and 1,800 years—consistent with the diverse conclusions of their colleagues, but not very reassuring if one were hoping to dump radioactive waste in the deep sea.[121] They also addressed the question of the sensitivity of the thermohaline circulation to changes in either the location or the density of the sources of deep water. The answer was not particularly sensitive, because the driving force is not the density differences per se, but the upward flux that maintains the thermocline. The "ultimate cause of the abyssal circulation" is the "thermocline which ... as a result of surface heating and applied wind stress, *demands a certain upward flux at mid-depth*."[122] They concluded: "This way of looking at the abyssal circulation is quite different from the traditional view which implies that the magnitude of the abyssal circulation is determined essentially by the amount of winter cooling at the polar regions and hence that a warming of polar regions of only a few degrees would largely stop the abyssal flow. From our point of view a warming of polar regions of one or two degrees would not affect deep transports except possibly to shift the location of the sources, and to reshuffle the western boundary currents."[123]

With these papers, the tour-de-force had been transformed into a quantitative scientific theory. Attending to the thermocline had led Stommel and Arons to a new and convincing theory of the ocean and, with it, to the resolution of a question that had vexed scientists for centuries. But there is at least one more important contribution that should rightly be viewed as part of this story—not least because of its importance to Stommel's lasting influence in oceanography today: his analysis of whether ocean circulation could display different stable regimes and his conclusion that it could.

In a paper in *Tellus* in 1961—today with more than one thousand citations— Stommel argued that a thermohaline convective system could exhibit two different stable regimes: one in which temperature differences dominated and one in which salinity differences dominated. As Sandström and Defant had stressed, in the ocean temperature and salinity differences work against each other—because warm waters tend to be saltier—and warm surface water overlying cold deep water presents a stable situation that can be altered only if the surface waters become very cold or very salty. For this reason, they (and others) had been skeptical about thermohaline effects. Because heat transfer tends to be faster and more efficient than the processes that change salinity—mainly evaporation, but also mixing, diffusion, and the creation of sea ice—in the natural world, as currently constituted, temperature effects dominate. Thus, it had been reasonably assumed that warm waters would tend to stay near the surface and cold waters to remain at depth. Ergo, no thermohaline circulation.

But one could imagine a different situation in which salinity differences predominated. Moreover, it was a relatively simple matter to calculate the relative effects of temperature and salinity on density to determine how likely such a situation might be. In a series of thought experiments, Stommel imagined adjacent basins with different temperature and salinity conditions and solved the equations representing heat and salinity transfer between them. In the case where two connected vessels were connected by a capillary tube at the bottom (through which deep, dense water could move) and an overflow connector at the top (through which shallow, less dense water could move), there were two stable solutions to the equations: one in which temperature effects dominated and one in which salinity effects dominated. (Philosophers would say the problem was underdetermined, mathematicians that the solutions to the equations were nonunique). This suggested that two different stable regimes could exist in nature: "The fact that even in a very simple convective system ... two distinct stable regimes can occur ... suggests that a similar situation may exist somewhere in nature. One wonders whether other quite different states of flow are permissible in the ocean or some estuaries and if such a system might jump into one of these with a sufficient perturbation. If so, the system is inherently fraught with possibilities for speculation about climatic change."[124]

These possibilities—fraught indeed—are of heightened concern today, as we move into a climatic regime that has not existed on Earth for at least one hundred thousand years.[125] Their recognition is part of the lasting legacy of Henry Stommel (fig. 2.16). Nearly all oceanographers know that. But what most scientists don't know is that this is also part of the lasting legacy of Cold War military defense oceanography. Our modern understanding of deep-ocean circulation was made possible not just by military funding but also by the ways in which operational concerns forced attention to the thermocline. The military didn't simply foot the bill for Henry Stommel's work; military concerns forced scientific attention. Military defense oceanography created a context that led Stommel to pay attention to an aspect of the ocean that just about everyone else had ignored.

Conclusion

Scientists frequently overstate the novelty and significance of their findings—the better to secure tenure and promotion, if not a place in history—but Stommel's and Arons's description of their view as "different from the traditional" was a major understatement. Theirs was one of the major advances of twentieth-century earth science. Their model was radically different from

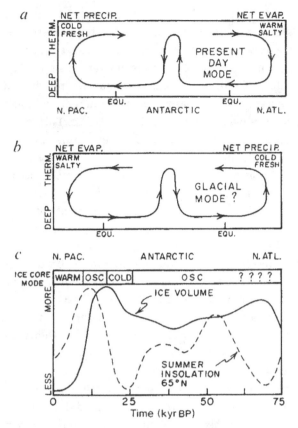

Figure 2.16 Illustration by Wallace Broecker, Dorothy Peteet, and David Rind of the two stable modes of ocean circulation and their possible relation to global climate change, based on the insight of Henry Stommel.

Broecker et al., "Does the Ocean-Atmosphere System Have More Than One Stable Mode of Operation?" *Nature*, May 2, 1985, based on Stommel (1961). Reprinted with permission from Springer Nature.

what had prevailed over the better part of the previous century, and it answered a question that had been debated for longer than that.

The Stommel-Arons collaboration had the effect, as Stommel put it, of turning a concept into a theory, because it put the model on a physical basis—the goal of all dynamic oceanography. It helped to specify in detail predictions that laboratory experiments suggested and to confirm them on physical grounds: poleward flow of interior waters, intense currents along both the western and the southern and northern boundaries, deep currents flowing toward (rather than away from) sources, and western boundary currents not necessarily connected to sinking currents.[126] The flow of water toward sources was counterintuitive, but Stommel and Arons demonstrated that the effects of Earth's rotation required it. Perhaps most important, the predictions were testable, and in the years and decades that followed they were tested and found to be correct. As oceanographer Bruce Warren wrote in his obituary of Arnold Arons in 2001: "It has taken more than thirty years to check the many major ocean basins for the rudimentary Stommel-Arons predictions, and they have always been vindicated.... They constitute the most observationally useful theorizing that anyone has ever done about the large-scale ocean circulation."[127]

The Stommel-Arons model proved powerful not only because it answered a long-standing question in oceanography but also because it made specific, testable predictions about the natural world that turned out to be true. Today, scientists continue to debate the details of the thermohaline circulation, particularly its effect on climate.[128] In discussions of the potential impact of global warming, the effect of thermohaline circulation on the strength and location of the Gulf Stream has come to be a major point of concern, even fear. It has inspired dystopic anxieties and a bad Hollywood film.[129] But while scientists continue to argue about the details, no one disputes that there *is* thermohaline circulation, and for that we largely have Henry Stommel to credit.

The Navy did not force Henry Stommel to think about the thermocline. It did not order or instruct him. No particular pressure was put on him to do the work that he did, much less in the way that he did it. We cannot know whether, in a different context, he might have developed the same theory, albeit perhaps not in the same way, for it is impossible for any historian to explain precisely how and why someone comes up with a new idea or creative solution to a long-standing problem. What we do know is that Navy finance made it possible for Stommel to work on the question of deep-ocean circulation in a sustained and focused manner, creating the conditions that made his work possible, but the influence was not just financial. It was also conceptual. Other oceanographers before him—with access to the same data and better formal training—failed to do what Henry Stommel did. He did it by paying attention to something that other people had ignored. That something was the thermocline, and it came into focus when and where it did because of military defense oceanography.

3 *Whose Science Is It Anyway?*
The Woods Hole Palace Revolt

The years Henry Stommel spent at the Woods Hole Oceanographic Institution in the late 1950s and early 1960s should have been the happiest of his professional life, and in many ways they were; he loved his work and had the resources to do it. But there was a problem, and it was the same one that had troubled scientists at Scripps in the late 1930s and early 1940s: the question of the autonomy of scientific research. The issue expressed itself though money: where it came from, where it should come from, and how it affected the work done at the institution. It erupted at Woods Hole in an episode in the early 1960s that participants in hindsight called the "Palace Revolt."[1]

Throughout the 1950s, the US Office of Naval Research generously supported physical oceanography at Woods Hole and elsewhere, and scientists almost uniformly recall the period as a golden age. Looking back, oceanographers are generous in their praise of the ONR for both its fiscal largesse and its genuine interest in scientific matters.[2] Most ONR program directors were trained as scientists; they understood what their academic colleagues were trying to achieve, and the projects they funded were typically loosely managed. Working with ONR support, Woods Hole scientists by and large had a sense of autonomy. They pursued the questions that interested them, without having to spend undue time or energy writing grants, worrying about funding, or complaining that they did not have enough of it. No wonder they were happy.

The task of raising money at Woods Hole fell largely to the institution's director. Just as Harald Sverdrup had spent time in the late 1930s and early 1940s convincing Navy patrons of the value of oceanographic work at Scripps, director Columbus Iselin did so at Woods Hole. As historian Gary Weir has shown, Iselin—a patrician's son who grew up sailing in New England—had cultivated close ties with Navy officials.[3] Maintaining these ties during and after the war enabled him to expand Woods Hole dramatically and permitted his scientists to do their work with little distraction. Oceanographer George Veronis (1926–2019) recalled, "During my entire stay at Woods Hole we all worked under a large grant that Columbus Iselin had with the ONR, so I never paid much attention to funding."[4] Arnold Arons agreed. "Columbus had ONR in his pocket," he laughed, and if you had an idea, you just "went to Columbus." He would tell you to go ahead, and the money would follow.[5]

Historical evidence suggest that this was not mere lore—Woods Hole scientists in the 1950s rarely had to write a grant proposal. The basic task contract in military defense oceanography was but one example of the large block grants that Iselin negotiated with the ONR, grants that permitted scientists to do their research without distraction. The grants were defined broadly, and because so many aspects of oceanography had potential military pertinence, it was easy for most Woods Hole scientists to fit what they were doing under that rubric. This was particularly so for those working in physical oceanography, meteorology, and air-sea interactions. When Valentine Worthington—a senior scientist in physical oceanography—retired in 1981, he wrote a final report expressing his appreciation for "more than thirty years of almost uninterrupted support."[6]

Stommel also benefited from many years of almost uninterrupted support. We saw in chapter 2 how his attention to the thermocline—a focus of military attention—stimulated a key idea, central to his groundbreaking work on abyssal circulation, so we might presume that his relationship to military defense oceanography was unproblematic. He had a research topic of significance that challenged him intellectually and fit well into an institution trying simultaneously to advance understanding of the oceans and satisfy Navy patrons. The same was true for Veronis, and for William von Arx, the oceanographer who designed the rotating tank that Stommel and colleagues used for their circulation experiments. But these men were not, in fact, professionally content. All three—as well as several of their colleagues—were troubled by what they saw as excessive influence of the Navy on institutional priorities and a drift away from the basic science commitment that had inspired Woods Hole's creation. All three men would play a major role in the Palace Revolt.

The Origins of Woods Hole

In the early 1920s, US scientists from academia and government wanted to expand the institutional basis for oceanographic research in the United States. A series of conversations culminated in 1924 with the Interagency Conference on Oceanography, which brought together representatives of all relevant the major US government agencies and many academics. The following year, Frank Lillie (1870–1947), the biologist who directed the Marine Biological Laboratory at Woods Hole, Massachusetts, began conversations with the Rockefeller Foundation about funding a new institution for oceanography to be built alongside his biological lab.[7]

In 1927, the US National Academy of Sciences established a committee on oceanography to "consider the share of the United States of America in a worldwide program of oceanographic research."[8] Oceanography had been dominated by the British, Germans, and Scandinavians, and it was represented in America only weakly and primarily on the West Coast: at Scripps, which (as we saw in chapter 1) had only just changed its name to indicate its oceanographic aspirations, and at the University of Washington, which would soon rename its marine station at Friday Harbor to the Oceanographic Laboratories at Friday Harbor.

Among other things, the academy's committee recommended the creation of a permanent independent research laboratory on the East Coast to "prosecute oceanography in all its branches"—biological, physical, chemical, and geological. At Scripps this would be called "balanced oceanography"; at Woods Hole it would be called "integrated oceanography." Whatever the name, the intent was clear: to create a prominent academic institution on the East Coast to help build the American scientific presence in oceanography. The scientists' vision was realized in the founding in 1930 of the Woods Hole Oceanographic Institution, next door to the Marine Biological Laboratory.

These scientists wanted their new oceanographic institution to be dedicated to "fundamental" or "basic" science, by which they meant research intended to advance understanding of the ocean without regard to specific utility.[9] Applied oceanographic research, particularly in relation to navigation, already had an American home in government agencies such as the Navy Hydrographic Office and the US Coast and Geodetic Survey, and a European home at several universities and institutions doing work related to fisheries and weather forecasting.[10] Moreover, the Rockefeller Foundation was willing to support basic research—indeed, it was actively doing so across America—so it seemed both desirable and plausible that the new institution focus on basic science.[11]

A $3 million grant from the Rockefeller Foundation, with which the new institution would be launched, was a vigorous endorsement of the scientists' vision. Still, 1930 was a tough year to begin a major initiative in a science as expensive as oceanography, as the United States and the world slid into economic depression. Most of the initial grant went to construction. With only a meager budget for research activities, no year-round staff, and no regular external funding, Woods Hole remained small and, frankly, insignificant throughout the 1930s.[12]

World War II changed the fortunes of Woods Hole as it did for Scripps, as Iselin assured his Navy contacts that his scientists had something to offer them. The institution embarked on diverse projects in underwater sound transmission, photography, and the propagation of explosive charges. As historian Susan Schlee has put it, "A lack of funds had traditionally been the limiting factor for both public and private marine research.... Then, suddenly, that problem was gone."[13]

Moreover, as Schlee has also pointed out, oceanography emerged from the war not merely "a bigger, richer field" but also "one of changed proportions, aims and allegiances."[14] Before the war, the principal applications of the marine sciences were navigation, fisheries, and weather forecasting; during the war it had become clear that far greater resources could be found in support of naval warfare. When the war ended, Woods Hole scientists, like those at Scripps, sought to continue research under Navy auspices. The establishment of the ONR in 1946 facilitated and formalized that goal, and throughout the 1950s Stommel, Veronis, von Arx, and many others at Woods Hole were funded almost entirely by the ONR.

Most Woods Hole scientists were pleased to have the financial and logistical support with which to build their careers as oceanographers. As Iselin had noted, before the war, only in Germany (or perhaps Scandinavia) could an oceanographer have reasonably hoped to feed a family. Things were different now.

The simple fact that they would be able to have careers as oceanographers was obviously a fine thing for the men who had worked on Navy oceanographic projects during the war. But did any of them have misgivings about helping to build or deliver weaponry, or perpetuating the military-scientific alliance? Many physicists had spoken publicly—and in some cases startlingly openly—about their feelings of moral responsibility and even guilt for their role in the destruction wrought at Hiroshima and Nagasaki. Science, some thought, was not meant to be applied to such ends, and they argued for a reconfiguration of postwar science.[15] Others defended what they had done when facing the exigencies of World War II but questioned whether the large military-scientific complex under construction in the 1950s was compatible with scientific ideals. Cornell physicist Philip Morrison, for example, argued, "We cannot tie science to the military and hope to see it used for peace, no matter how ingeniously we write the contracts."[16]

Oceanography was far less publicly visible than nuclear physics, and there was no comparable public conversation about oceanographers' role in the war. Perhaps this is because few people outside of the field knew much about oceanographers' wartime work, so oceanographers were not forced to confront its moral ramifications in the way that physicists were. However, in private, some of them expressed concerns. Henry Stommel had his doubts about applying scientific research to killing; William von Arx had misgivings about the match (or mismatch) between military and scientific leadership. Military leadership was a poor model for science, von Arx suggested, because military leadership was top down and science needed to be bottom up. Science should not proceed "by fiat" but should be driven "by individual enthusiasms among the people who are actually doing the work."[17]

Stommel and von Arx appear to have been in a minority, at least to the extent that they were willing to articulate their unease. Throughout the late 1940s and early 1950s, any real or perceived conflict between military patronage and scientific ideals at Woods Hole remained mostly unexpressed. This began to change, however, in the mid-1950s as the institution searched for a new director and scientists wondered what new policies or philosophies might follow. Whatever freedom and flexibility they had enjoyed in the past, they anticipated changes ahead.

The "Ghost" of Military Justification

The issue of the "freedom of science" was expressed primarily in individualistic terms; for Stommel and his colleagues, *freedom* meant their ability as individuals to determine their own research agendas and have that research unfold according to an interplay of their own interests and the "internal" logic of science.[18] In the mid-1950s, the issue of individual autonomy came to the fore, when Woods Hole meteorologist Joanne Malkus (1923–2010) attempted to recruit a young oceanographer, Kirk Bryan (b. 1929), to join the meteorology group.[19] Bryan would later become internationally known for helping to develop the first general circulation model of the atmosphere; indeed, some consider him the founder of numerical ocean modeling. But Bryan's work had already gained the attention of the Woods Hole group; Malkus wrote to Arnold Arons to enlist his support to recruit him. She stressed that Woods Hole already had an outstanding group in geophysical fluid dynamics—including Stommel, Veronis, and her husband, Willem Malkus (1923–2016)—which Bryan could join and strengthen. In making the case, she was quick to stress that joining this group did not mean that his research priorities would be dictated: "Naturally I am not implying that Dr. Bryan's endeavors would be externally prescribed if he came here—this is antithetical to Woods

Figure 3.1 Portrait of Joanne Malkus, 1960, by Jan Hahn.
© Woods Hole Oceanographic Institution; reprinted with permission.

Hole philosophy (which I sincerely hope will be perpetuated with the new Director)—he would be quite free to permit his interests to evolve in any topic whatsoever."[20]

Malkus was one of only three women scientists at Woods Hole in the 1950s and one of a handful of professional female geophysicists anywhere (fig. 3.1).[21] As an undergraduate, she had entered the University of Chicago in 1940 and joined a meteorological training program established as part of the war effort. After nine months of intensive training, she became an instructor in military weather forecasting, returning to complete her bachelor's degree in 1943 and master's in 1945. While completing her PhD, she taught physics and meteorology at the Illinois Institute of Technology; earned a Guggenheim Fellowship to study cloud physics at Imperial College, London; and joined the scientific staff at Woods Hole in 1951.[22]

Like most women scientists of her generation, Malkus endured discrimination, hostility, and humiliations large and small. Her own thesis adviser—the great meteorologist Carl-Gustaf Rossby—was frankly dubious about a woman geophysicist and told her that she would "look both ridiculous and pathetic if [she] didn't really 'make it big' after creating such a ... spectacle ... and exacting such an unfair sacrifice from [her] husband and child."[23] (Years later, when she had made it "big," Rossby acknowledged that the sacrifices were worth it, but she—thrice divorced—was no longer sure.)[24] Nevertheless, she forged a place for herself in meteorology, helping to explain Atlantic cyclogenesis. Later, as Joanne Simpson, she became the first woman president of the American Meteorological Society. On the occasion of her death, Greg Holland of the National Center for Atmospheric Research offered this assessment: "There is zero doubt that there has never been a more capable woman in meteorology, and she would also be in the top five of all meteorologists in history, no matter the gender."[25]

Malkus's research on cloud formation and cyclogenesis was financed by the armed services for its relevance to military operations and weather forecasting, so she was perhaps sensitive to the perception that her work might be directed by others.[26] If so, it was not because of her gender or because of Rossby's conspicuously qualified support. Alan Faller also raised the question of research autonomy in a conversation with William von Arx in 1957.[27] Faller had just been appointed in the meteorology department at Chicago and sought advice on developing his research program. He had been encouraged by a program officer to apply for an Air Force research contract, but he worried what strings might be attached. "Presumably these funds would be on a very broad basis (not restricted to immediate application) and would therefore be desirable," he wrote in a letter to von Arx.[28] The latter agreed that autonomy was indeed the salient question. However, von Arx turned Faller's concern around, arguing that external support would increase his stature and therefore his autonomy within his department: "You'll have autonomous control of your work and the credit of contributing to the financial well-being of the laboratory."[29] But Faller was not convinced; a few months later he was still pondering the matter: "I am inclined to feel that an Air Force contract would be desirable if it fell into the undefined category of pure research and [if it] could be relied upon for continued support."[30]

Von Arx agreed that these were the crucial ifs. He suggested that perhaps it would be better to get money from the National Science Foundation, which was just starting to expand into oceanography: "If NSF gets money enough for continuing support of research we will be free of the necessity for 'justifying' research in terms of military need. While this has never been troublesome in ONR contracts, it's always there in a ghostly sort of way—and who likes ghosts?"[31]

This was an apt characterization. ONR funding was generous, particularly in physical oceanography, but everyone knew that the Navy expected results. Even if in practice those results were off in the future, at least in principle they had to exist. But naval expectations were not just about long-term promise; the Navy also expected to be able to call on scientists to help with tasks as they arose. In the archives, one finds frequent inquiries to ONR-funded oceanographers from officers in other naval bureaus and bases—such as the Bureau of Ships—soliciting assistance or advice or requesting participation on a committee or advisory panel. One was not merely being paid to do the science of one's choice; one was being paid to be a scientist who was, at least from time to time, on call. In this respect, oceanographers did serve, as sociologist Chandra Mukerji has emphasized, as a reserve labor force.[32] Did this constitute direction? Von Arx felt that it did. Drawing on the then-fashionable work of sociologist David Riesman, he concluded that scientists had become "other-directed."[33] (Stommel would later describe the relationship more ominously as "Kafka-like": someone seemed to be in control, but it was not clear who that someone was nor from whence their power emanated.)[34]

Von Arx may have been unusual; most oceanographers do not seem to have objected to being at least in part other-directed, and many were happy to be of service—recall Denis Fox in 1943 declaring himself "ready and willing." Of course, there are always gaps between what people say and what they do, as well as between what they say in public and feel or express in private. But so far as the evidence permits us to judge, most oceanographers do not seem to have thought that their Navy connections significantly affected the direction of their scientific work. To the contrary, Woods Hole scientists in the 1950s largely felt quite free to follow their interests. In the early 1960s, however, it appeared that might be about to change, and several of them expressed the view that there was, indeed, a problem.

"Need to Know" Science

In 1958, chemist Paul Fye took over as Woods Hole's fifth director. He would serve an unprecedented nineteen years, shepherding an enormous expansion and professionalization of the institution, enabled by Navy support. During the war, Fye had worked in naval ordnance, and in 1942 was recruited to the Underwater Explosives Research Laboratory housed at Woods Hole. His focus was on improving underwater ignition and increasing the explosive power of depth charges, depth bombs, and torpedo warheads. According to Woods

Hole history, the laboratory became an "institution within the Institution," and many scientists who worked there during the war maintained Woods Hole associations afterward.[35] In 1948, Fye joined the US Naval Ordnance Laboratory, where he served as chief of explosives research (1948–1956) and then associate director for research (1956–1958).

With his education as a chemist and experience in ordnance, Fye was perhaps an unlikely candidate to lead an oceanographic institution. With his extensive administrative experience and close links to sources of Navy funding, though, he might be viewed as just the man to develop the institution at a time when funding was becoming both more abundant and more complex and accountability coming more to the fore. As Woods Hole trustee and former president of Bowdoin College James S. Coles (who had also worked during the war at the Underwater Explosive Laboratory) recounted in 1977: "In 1940, the scientific staff [at Woods Hole] had numbered 18 (of whom 15 were part-time). The annual budget was $104,400. The plant consisted of the Bigelow Building, and one small dock for the 142-foot ketch *Atlantis* and the 40-foot workboat *Asterias*. By 1957, there were 65 full-time scientists and technicians, plus 106 support personnel, 8 buildings, 2 research vessels, and 2 docks. Operating costs in 1957 were $2.5 million, and $2.4 million of this came from the Federal Government with the attendant complications of red tape, contracts, and auditors."[36] Given this huge increase in size and budget, as well as the shift from private to public funding, the trustees believed that the institution needed to be put on a more businesslike footing. Fye was hired not for his scientific credentials but for his administrative talent, energy, and decisiveness.

Woods Hole scientists were initially receptive. Some were concerned that he was not an oceanographer nor even an earth scientist, but overall most concurred with von Arx's assessment: "He's a good guy, who knows his own mind—no nonsense—but is fair in that he talks things over with others before coming to any firm conclusions."[37] Joanne Malkus was also receptive at first, telling Arnold Arons that morale at the institution had been lifted by "the choice of the new director."[38] Over the following two years, however, staff opinion changed. Some would come to see him as autocratic and unfair. Von Arx came to view him as a bureaucrat ready to sell the soul of the institution to the US Navy. Malkus, Veronis, and several others would leave Woods Hole in frustration and discontent.[39]

The trustees had appointed Fye in part in the hope that he would increase private philanthropic support, a goal the staff shared (and which, in the long run, he accomplished), but his immediate vision emphasized prompt responsiveness to Navy needs. He articulated this vision under the rubric of what he called "need-to-know" science. This was a peculiar turn of phrase, for in military circles "need to know" generally referred to security: people obtained clearances to access classified information only if they had a demonstrated need to know that information. For scientists, both the term and the attitudes it signified were unpleasantly associated with Los Alamos and its infamous policy of compartmentalization. As the phrase's originator General Leslie Groves explained in his postwar memoir, *Now It Can Be Told: The Story of the Manhattan Project*, "My rule was simple and not capable of misinterpretation—each man should know everything he needed to know to do his job and nothing else."[40] For scientists who believed that creativity necessitated letting one's intellect range widely, the idea was unsettling even in wartime. In peacetime—and in an institution founded to pursue basic research—it was offensive.

Fye's use of the term was also ill chosen, because what he had in mind was different from its Manhattan Project connotations. He was not advocating compartmentalization, but rather that the institution focus its work on topics that that addressed societal needs, concerns, and problems, in this case, the part of society represented by the armed services. Woods Hole should prioritize its research questions according to society's need to know the answers to those questions. In oceanography, need-to-know science was what the Navy needed to know.

Fye articulated his vision in a memorandum addressed to the Woods Hole Executive Committee on January 20, 1961, and later shared with the entire staff and the trustees. It carried the seemingly innocuous title "Policies for the Woods Hole Oceanographic Institution in the 60s." But its length—eight pages, single spaced—indicated that its intent was no mere statement of operating procedures; it was a mission statement for the institution. Its opening made clear that a rift had already developed: "It has become clear within the past year that there are important differences of view within the Institution about [its] future development.... It is therefore necessary at this time to clarify certain basic policies."[41]

Fye placed developments in oceanography in the context of changes in American science, implying that the changes he was proposing were not idiosyncratic—not a matter of preference, personal or institutional—but the direction that history was taking American science. The argument generated so much fury that it is worth quoting at length:

> During the past twenty years we have seen one pastoral province of science after another rudely invaded by an urgent "need-to-know." These pastoral provinces had been fields of inquiry and investigation having little

direct relevance to immediate needs of the world. For example, during the 30s the field of geophysics involving ionospheric measurements was a tranquil area in which investigators conducted their studies in an unhurried fashion with little concern for applications. It had many features of the ideal for academic research—no pressure, no applications visible.

Suddenly, an application of great importance emerged: radar. Quickly the field was invaded by scientists and engineers unknown to the established investigators. Not only were applications quickly made, but understanding in the whole field progressed much more rapidly than before. Thus was created an urgent requirement for knowledge completely foreign to the workers in this field—a "need-to-know" all about the fundamentals of electromagnetic propagation which was real and demanding to a degree far greater than had heretofore been experienced in this field.[42]

Fye's characterization of the history of ionospheric physics was preposterous—incredible to anyone familiar with the history of radio—but while his history was muddled, his point was not. The reference to the "pastoral" provinces of science was an allusion to Sinclair Lewis's best-selling novel of the 1920s, *Arrowsmith*, in which the hero, Martin Arrowsmith, refuses to compromise the intellectual integrity of his research in the interest of more rapid application. Forgoing the worldly rewards that would accompany compromise (as well as the company of women and associated rewards), he retreats with a single male colleague to the pastoral provinces of Vermont, where he will be able to work in unsullied intellectual isolation.[43] Fye's message was unmistakable: pastoral science was a romantic, literary, and, above all, outdated concept.

Fye listed the provinces of science that had been transformed in the twentieth century by the need to know: nuclear physics, astrophysics, and, now, oceanography. He argued that the time lapse between basic discovery and application was becoming shorter, comparing the "52 years [that] elapsed from the discovery in 1830 of the fundamental law of electrical induction by Michael Faraday until the first electric motor was built by Westinghouse" to the "5 years [sic] from fission to A-bomb."[44] With slightly opaque logic, he argued that oceanography had reached the point at which application was coming first: "Today the requirements of the Polaris submarine crew go beyond present-day oceanography and application has run ahead of discovery."[45]

The Polaris program—with its submarine-launched ballistic missiles—was poised to become the third leg in the US nuclear triad. It was also high in the consciousness of naval officers and certain oceanographers for the demands it would place on oceanographic knowledge of the deep sea (chapter 5). Presumably, this was what Fye meant when he argued that application had

run ahead of discovery: Polaris was going ahead despite incomplete knowledge of the ocean floor, needed to site missile launchings. Oceanographers
would have to catch up, give the Navy the information it needed, and do so
quickly.

Many people—then and now, and of diverse political leanings—might
agree with Fye that scientific research should be responsive to the needs of
the society that supports it. Indeed, a distinctive feature of this story is that
World War II saw a reversal of political affiliations on this point: in the 1930s
it was typically leftists, like socialist geneticist J. D. Bernal, who argued that
science should be guided by the needs of society, and rightists, like neoliberal
chemist Michael Polanyi, who argued that science must be left to its guide itself (chapter 1).[46] These debates also invited the question of who was to decide
what society most needed to know. (This was one reason neoliberals wanted
to leave it to "science," lest governments answer the question in ways they did
not like.) Oceanographic knowledge was potentially relevant to many societal
needs, including mineral resource exploitation, food supply, hazardous waste
disposal (and not just nuclear waste), and even recreation, but Fye's sights
were set on a particular category of necessary knowledge: knowledge needed
by the nuclear Navy.

Much of this was related to the emergence of the deep ocean as a theater
of warfare. As Fye reminded his staff, historically "the Navy's primary mission had been considered to be on the surface and [in] the air above it. The
third dimension below the surface was of decidedly secondary importance.
Submarines were primarily commerce raiders and not necessarily better than
other means. Prenuclear submarines were not true submersibles.... Knowledge of conditions in the oceans was [therefore] chiefly useful for sonar
purposes—useful but not vital."[47] Matters had changed dramatically because
of the expectation that nuclear weapons would soon be carried on nuclear-
propelled submarines:

> [It is] affirmed by daily pronouncements by high Navy officials [that] the
> nuclear-propelled undersea craft fills a vital need in our major deterrent
> capability and will become the principal naval vessel of the future. Sud-
> denly there is recognition of a need for great increase in our knowledge
> of the oceans—including everything about them. For the undersea is now
> the new "atmosphere" in which the Navy will be conducting its chief op-
> erations on an increasing scale, in which it is already traveling through
> hitherto inaccessible seas. On a growing scale, the military missions of the
> Navy will be undersea missions because the advantages of concealment
> and surprise [and] of navigation and of smooth operation are overwhelm-

ing in an age of high vulnerability of surface targets to detection and destruction.[48]

That argument alone would have been enough to disturb some Woods Hole
scientists, far as it was from Henry Bigelow's vision of an oceanographic institution dedicated to basic research. But Fye was concerned not only (or even
primarily) with the scientific needs of the Navy but also with its political and
instrumental need to find a mission in the nuclear age:

> The Navy has been searching for a real naval mission ever since the war.
> Bikini spelled the end of an era.... The Navy needed a new medium and it
> has found it within the sea. It has strong Congressional support to develop
> its new role aggressively. Already the undersea ships and armaments have
> outpaced our understanding of the medium in which they function....
> The "need to know' about the oceans has jumped an order of magnitude in
> the past year or two, and *the need will be satisfied in one way or another just as
> it has been in other fields.*[49]

If anyone harbored the idea that this could be viewed as a both-and rather
than an either-or proposition, Fye disabused them, concluding unequivocally
that "the pastoral tranquility of oceanography is a thing of the past."[50]

Fye's history was in several respects questionable: the Navy had decided
years earlier that the deep ocean was its frontier, and Lamont director Maurice Ewing would surely have objected to the characterization of his sonar
work as "useful but not vital." But Fye's larger point was both clear and correct: the Navy was not supporting science because of a generalized conviction of its overall value, the belief that basic research might inadvertently
yield something useful, or because it wished to have scientists available as a
reserve labor force—although these considerations no doubt contributed to
making the case. There was something far more specific, concrete, and immediate at stake: the need for oceanographic information essential to the
Polaris program.

Fye had been active throughout 1960 and early 1961 mobilizing oceanographers to work on Polaris. In a letter to Lamont's assistant director, he emphasized the urgency of the program, which would "accelerate greatly the kind
of research that is urgently needed in support of the fleet ballistic missile
systems."[51] Some scientists and scientific institutions would naturally resist
the changes taking place, but Fye suggested that in doing so they would be
threatening their own futures: "It may be found that the existing institutions
cannot adjust their philosophy or methods to cope with the new dimensions

and pace of this field. To some, the attractions of the pastoral life are irresistible. It is often more difficult to change the philosophy and methods of existing institutions than to build new ones."[52]

Fye was not asking his scientists to change; he was telling them to, and the remainder of his memo was therefore in the form of a directive. Repeatedly using imperative syntax, he spoke of the "important contribution to national and world security that oceanography must make," the pressures that "must and will be satisfied," the new institutions that "must be created," and the "decision of basic philosophy and policy which must now be made." It was not a choice but an instruction, a presentation of the reality to which Woods Hole scientists had to adapt. In principle, Fye concluded, they could try to continue the "traditional methods of individual inquiry and operations," but in practice that "alternative is not really available to us unless we purposefully plan the decline of the Institution we love." He concluded in no uncertain terms: "Not interpreting my mandate to be one of liquidating WHOI, I have no alternative as a conscientious director but to recommend a planned but rapid expansion that will permit the Institution to fulfill its destiny."[53]

Fye acknowledged that greater planning and stricter organization could undermine creativity, but, tellingly, he discussed the issue not in terms of the advance of scientific knowledge, but in terms of weapon development: "Creativity is essentially a personal and private process. The innovative research worker is keenly sensitive to his environment. Even though he has a firm conviction that research will pay big dividends, he can quickly be diverted to areas of low risk and small thinking if management shows more interest in the small gains of soon-to-be obsolete weapon improvement than in free-wheeling endeavors that may produce wholly new and unique weapons systems."[54] These lines were excerpted from an earlier speech that had included a strong defense of creativity and stressed the need for researchers to control their own tools. But in that speech, Fye had been speaking in the context of oceanographic institutions managing their own ships (rather than the Navy managing them), not in the context of individuals choosing and designing their research programs. Freedom in science played no role in Fye's Weltanschauung, and so he concluded by asking his staff for "a broad policy and purpose which rests on a general consensus and which is understood by all."[55] The "consensus" was in fact a directive.

The Scientists' Response

No one contested whether Fye's characterization of Navy needs was correct; the issue was the policy and purposes of Woods Hole. With his discussion of the elements of "concealment and surprise," the "vulnerability of surface targets to detection and destruction," and the imperative of "wholly new and unique weapons systems," Fye sounded more like a Navy officer than a scientist. Where was the discussion of basic science, of the role of the investigator, of what it meant to do scientific research? Where were the discussions of the differences between civilian and military research, basic and applied science, science and engineering? These categories were deeply salient for most Woods Hole scientists, yet in Fye's framework they disappeared. Fye claimed that he cared about basic science, but his message suggested he was indifferent to its pursuit.

No one denied that world had changed; the question was how to respond to that change. Many technical issues related to the Polaris program were already being addressed or could be addressed in military facilities, such as the Naval Research Laboratory and the Navy Electronics Laboratory. Surely Woods Hole's role ought to be different? Fye's directive made it unclear how, though. What did it mean to be an independent scientific research institution if the agenda was set by the needs of the armed services? Just a few years before, in 1955, the Woods Hole Scientific Advisory Committee had warned against the risk of becoming too much the servant of the Navy: "The Institution must not become a satellite Navy Laboratory, which might result in it being put in the position of being controlled by Navy demands for hardware rather than by Navy support for basic research in oceanography."[56] It seemed to many faculty—including Stommel, von Arx, Malkus, and Veronis—that Fye was asking them to do the very thing the committee had warned against.

Columbus Iselin had also been concerned about the direction Woods Hole was taking. After more than a decade of vigorous efforts to bring Navy patronage to Woods Hole, he had come to believe that the Navy orientation was invariably short term and would not serve the interests of oceanographic science in the long term. In a lecture written in July 1958, he argued that the Navy was inevitably more interested in development than research, in getting specific systems built and tasks accomplished rather than learning more about the oceans. This was to be expected—the Navy had a job to do, and that was fine. But oceanographers had a job to do, too, which if not confined to basic research certainly included it. Dependence on Navy funding would inevitably shift the balance in oceanography away from basic and toward applied research. Iselin thought that to some degree it already had, in that "many of the desks in the Navy department that are supporting oceanography are in effect asking for assistance in development programs." When a scientist helped the Navy, he typically did learn "something new in an area of his science," but he seldom had "the opportunity to look at the more subtle and little understood phenomena." The Navy was "pouring in more and more money," but overall basic research was on the decline.[57]

Iselin accepted that the Navy would and should be oceanography's dominant patron for some time to come, because "almost any kind of knowledge about the sea may in the long run turn out to have applications to the Navy's mission."[58] But the research effort needed to be "protected from oceanographic engineering, which is what the Navy mainly wants without understanding the difference."[59] Solving operational problems was simply not science.

Many Woods Hole scientists loved Iselin in part because of their feeling that he *had* protected them against the threat of ocean engineering—the threat that Navy needs would take over their institution. Now Fye was asking them to depend more rather than less on the organization that Iselin had concluded could not be expected to support basic research.[60] For scientists used to doing their research without being bothered by financial and political matters—and who had been taught to believe in the intrinsic value of basic research without having to defend or justify it—this was shocking. Some felt that Fye was asking them to betray their core values and commitments; others, that it was a betrayal of the institutional principles on which Woods Hole had been established. Either way, it was bad.[61]

The scientists discussed at length what to do. They met with the director and with one another, wrote letters and memoranda, held evening meetings, and lobbied the trustees for support. They spoke of anxiety and apprehension, of deep concern and declining morale, of frustrated ambitions, and even of thwarted dreams. According to Stommel's wife, Elizabeth, he could not sleep at night during this period and would come home from work "ranting and raving" about what was happening.[62] Finally, in November, things came to a head, as the scientific staff of Woods Hole formally rebelled. It was the start of the Palace Revolt.[63]

The Palace Revolt

Even before the circulation of the "need-to-know memo," Stommel had been growing uncomfortable with Fye, whom he had initially considered affable, if uninspiring. Now he viewed him as a threat to the institution. In the summer of 1960, Stommel composed a five-page letter of complaint to the three members of the Board of Trustees whom he most trusted: Iselin, geophysicist Athelstan Spilhaus (1911–1998; inventor of the bathythermograph), and Harvard chemist E. Bright Wilson. Stommel requested that Fye be relieved of his authority, suggesting that if the present situation were "allowed to persist indefinitely [it] would have grave consequences." With an emphasis on the past tense to underscore how things had changed, Stommel continued: "My first fifteen years at the Oceanographic Institution were very happy—it was a delightful place to work—and I have a deep respect and affection for my colleagues. Nothing less could have impelled me to compose this missive.[64]

Stommel suggested that Fye was attempting to run Woods Hole as a military outfit rather than a research institution. Among other things, he had appointed as second in command a man whom the staff thought of as "somewhere between a 'cold fish' and a psychopathic personality.'" He had established the Scientific Policy Committee but then proceeded to ignore it and then dissolve it. His behavior and decision-making process had changed "from candid to secret." Fye claimed that the reason for discounting and then disbanding the committee was that he was disappointed by a "lack of 'positive' response; Stommel saw him as rejecting any views that diverged from his own and accused Fye of disbanding the committee "because he fears it threatens his authority."[65]

Fye had implemented a procedure to review scientific progress that had offended just about everyone. His use of phrases like "clean house," "prune," and "remove dead wood" had "resulted in a certain atmosphere of uneasiness," a feeling confirmed by the "inquisitorial" tone of the review process. Woods Hole scientists increasingly had "the feeling that they are working merely in someone else's laboratory."[66] This must have been particularly difficult for Stommel to stomach, as early in his career he had chosen not to work for Maurice Ewing for just that reason: "Doc" Ewing told you what to do, and you were expected to do it.[67]

Many people clashed with Maurice Ewing (chapter 5), but no one ever doubted that he was a brilliant scientist who he had earned the right to direct a research institution. The same could not be said for Fye. Referring to Fye as "A" and the Scientific Policy Committee as "C," Stommel asked for "A" to be replaced, on the grounds that he did not understand the science that he was supposed to be directing and therefore had "no basis for valid value judgments on the type of research being carried on at the institution."[68] He was like "the amateur mountaineer who has dismissed his guides because he does not trust them or because they irritate him." But it was not the guides who were at fault, and thus, "one concludes that A should be replaced."[69] Some staff scientists had already decided to accept other jobs and Stommel predicted "more will go."[70]

Navy support had increased the flow of funds into the institution, but what was the point of money if not used to advance science? Fye claimed to want to put the institution on a more orderly basis and sound financial footings, but Stommel didn't see it that way: "There is a feeling of uncontrolled and unplanned growth: a mere scrambling in Washington to take any and all money from any source, even if it constitutes a prostitution of professional aims and standards."[71] Stommel had said it: Fye was prostituting their science, turning his beloved institution into a whore.

It is not clear whether the trustees responded, but matters definitely worsened after Fye issued his "need-to-know memo." Before the memo, Stommel had addressed his concerns privately to the trustees; now he confronted the director directly. His emotions were running high, but Stommel tried to maintain a dignified tone. He began by suggesting that there were two issues at stake: autonomy and purpose. Some scientists like to work alone and others in teams, Stommel argued, and both should be not just permitted but also encouraged. Similarly, some science is "directly related to social needs [and] some science is 'pure' and pursued for its own sake."[72] Again, both were legitimate. Fye, however, with his pejorative discussion of "pastoralism," seemed to reject both pure science and solitary scientists. Stommel defended both. For Stommel, the pursuit of knowledge was a form of individualism—an expression of curiosity, creativity, and personality. Denigrating one denigrated the other. Not all pure science was done by solitary individuals, Stommel argued, but most of it was, and this should remain the central focus of the institution: pure science, done primarily by individuals (or small groups) who decided for themselves what to do and how to do it. No matter how urgent the needs of the Navy, Stommel insisted, they "do not over-ride the need for general development of the science.... Young scientists are very much aware of this, and tend to shy away from purely applied naval sciences. My own students are very suspicious about the source of money for work at Woods Hole."[73]

Student views are not much preserved in archives (except insofar as those students later become famous teachers, researchers, or administrators), but von Arx's discussions with Faller certainly reveal a concern by one young scientist over how funding (both having and lacking it) might affect his career. Likewise, Malkus's reassurances about autonomy suggest that she believed that if Woods Hole came to be viewed as a place where scientists lacked autonomy—forced to work only on applied projects in large teams—it would negatively affect recruitment.

Stommel noted that two of Scripps's best geophysicists—Walter Munk and Gordon MacDonald (later an adviser to President Richard Nixon)—had made their reputations not on the applied work they had done, however useful that work might have been, but on their contributions to basic science, particularly their seminal textbook, *The Rotation of the Earth*, which "is unconcerned with national defense needs—and … establishes their reputation as none of their applied work can."[74] He continued: "Young men are anxious to establish an individual personally and internationally recognized reputation. There[fore] you also need to reserve a place in your institution for a few outstanding individuals, who are completely on their own, and respected, and some of whom are avowedly 'pastoral.' We are not all hurrying together on a great well-regulated four-lane highway into a glorious future. I think there is still jungle and undergrowth and unexplored territory where we need individual pathfinders on foot."[75] Stommel offered a solution. Divide Woods Hole into two parts: the Division of Defense Studies and Applied Oceanography and the Division of Oceanographic Sciences. The former could be run by "the best engineer you can find … the whole being efficiently organized," but the latter, in separate physical quarters, would be just the opposite: "loosely organized in an academic fashion, so as to accommodate the solitary investigator and also the man who wants a team in pure science."[76] Stommel had no quarrel with engineering—he was happy to let Fye have it—so long as the engineers stayed out of his way and their work was not mistaken for or conflated with science. Science, Stommel insisted, was constituted by individual investigators pursuing research "subject to no pressures or supervision—in the full meaning and spirit of academic life."

In later years, Stommel would spent time in a conventional academic department and develop a much more jaundiced view of the spirit of academic life, but his point did not depend on the actual conditions of scientific life elsewhere. His point was simply that basic science should be pursued at Woods Hole in an academic fashion, and the best way to do that was to support the most creative and productive individuals. "That is what [Roger] Revelle did for [Walter] Munk when he set up the Institute of Geophysics at La Jolla. Support … is what makes scientists productive."[77] Woods Hole had lost Ewing and Rossby, Stommel suggested, because they were not offered sufficient autonomy and support.[78] He continued: "In 1940, W.H.O.I. was too small a place to contain the aspirations and ambitions of Rossby [and] in 1950 W.H.O.I. was too small to hold Maurice Ewing…. Above all, these men wanted personal autonomy."[79]

Fortunately, it was not too late to build an institution that could sustain scientists of their stature, and so Stommel asked Fye to consider not what scientists could do for Woods Hole but what Woods Hole could do for scientists. That, he insisted, was "to offer an outstanding individual complete support, all the machinery, an assured minimum budget, money with no strings, a high degree of autonomy to do his own hiring, and set up his own sub-kingdom, and then not to interfere in any way at all…. Strong individuals …

demand their freedom as a birthright—and all you have to do to keep them is to give it to them."[80] Stommel's arguments would likely have seemed to Fye more utopian than realistic, but the comments about Munk were largely true: Munk had been given an institute, he was given largely free rein, and his international reputation was based far more on his theoretical work than on the applied problems to which he had attended.[81]

For Stommel—as for the biologists who had resisted Sverdrup's vision at Scripps—the crux of the matter was autonomy. This was linked to the motivation for doing scientific research, which was not national defense. (Indeed, Stommel pointed out that many of these men—Rossby, Sverdrup, Munk—were not US citizens!) The motivation for these men was "to build a part of science all their own, in their own image. It is not so much fame, nor power, nor duty, not civic-mindedness, not disinterested curiosity (all of which do play a role) that motivates their work. What makes them go is a deep-seated drive to wrestle personally with the unknown and to create understanding in their own terms where before there were not even questions."[82]

No doubt Stommel was describing himself, but many of his colleagues would have largely agreed. At least they would have hoped that the quest for understanding was the driving force in their scientific life. Thus, Stommel concluded, have two divisions, "separate[d] physically, morally, [and] financially," and allow the scientific division to be filled with "solitary pastoral scientists."[83]

Of Shepherds and Sheep

In his letter to the trustees, Stommel had referred to scientists who had already left, including Willem and Joanne Malkus, who had gone to UCLA for a year and were deciding whether to stay in Los Angeles.[84] Joanne Malkus joined Stommel in his complaint. Like Stommel, she wrote both to Fye and to the trustees. To Fye, Malkus argued that what a particular science needed depended on the nature of that science and its state of development. Given the undeveloped state of oceanographic and atmospheric science, it required an open-ended and relatively unstructured approach to research. Referring to a monograph she had recently completed—and that was soon to appear in the first volume of *The Sea*, a great compendium of oceanographic and geophysical advances edited by Cambridge geophysicist Maurice Hill (1919–1966)—she explained that it was written "with the decision-making problems of responsible persons like yourself clearly in mind."[85]

Her 116-page chapter, titled "Large Scale Interactions," summarized the state of research in air-sea interaction with an emphasis on theoretical advances made during and since World War II. These advances spoke to the issues that concerned Fye: how to solve problems whose answers society needed to know. In meteorology, that need was weather forecasting. The dilemma, however, was tension between the short and long terms. Improvements in empirical weather forecasting might yield short-term benefits but would do little to improve the field in the long term. That would require a more theoretical approach, to strengthen the foundations of meteorology as a basic science.

Theoretical advance required rigor, but empirical work required comprehensiveness, and there was generally a trade-off between the two. Unlike some theoreticians, Malkus was not dismissive of descriptive work, particularly if intended to solve immediate problems, such as weather forecasting. Indeed, she argued that the practical demands of weather forecasting had provided an important stimulus to the science: "The practical importance of the weather and the sea in man's daily life have ... provided motivation and means of attacking problems which pure curiosity-driven science might have avoided as too difficult or too costly." Nevertheless, she believed that the most important advances of her generation had come, and would continue to come, through rigorous, quantitative, basic research focused on fundamentals. This required tempered expectations of rapid practical benefits: "In general ... the more formal and rigorous the approach, the less useful are the results in the practical sense of forecasting or controlling a planetary phenomenon; at present, even the convective motions in a coffee-cup-sized laboratory cell are only on the threshold of tractability. The challenge is to isolate simplified, prototype problems of this sort, whose rigorous treatment provides key insights into the complexity of natural geophysical systems."[86] Malkus acknowledged what all empiricists know—that fundamental principles don't necessarily help you decide if it is going to rain tomorrow, find an ore deposit, or predict an earthquake. But, she argued, at Woods Hole the goal should be to advance science in the long term, and therefore society would be "best served by the deepest emphasis being placed on basic research."[87] And basic science required the right environment, which was "hardly that of engineering firms, the brave new gadget technologies, nor even of a government laboratory, despite all the virtues we know these to possess." Perhaps scientists did not need their environment to be "pastoral," but they certainly needed it to be "scholarly." She concluded: "Of those who best served the 'need to know' in nuclear development, nearly all the outstanding names are those of scholars, of men who rarely left the academic world and who produced their great efforts in a scholarly environment."[88]

Like Fye, Malkus was a better scientist than historian—neither the atomic nor the hydrogen bomb was built in an academic laboratory—but she was an insightful analyst of the trade-off between reductionism and holism, and between theoretical and practical goals in studying complex natural systems. Like Stommel, she envisaged fundamental science as the necessary foundation if science was, ultimately, to serve practical goals.[89] Indeed, this was the standard rationale for federal support of basic science in the period after World War II, the rationale that had underpinned Vannevar Bush's vision for American science and justified the creation of the National Science Foundation: that undirected scientific investigation would advance the practical interests and material welfare of American society. Often referred to in hindsight as "the linear model," it insisted that basic science necessarily preceded "application," because "basic science" was the thing that was being "applied."

Fye's argument, of course, was that an operational program—like Polaris— might identify areas in which the necessary science was lacking. However, if one accepted the linear model, then one could defend basic research simply by arguing that one's science was young and therefore not yet ready for rapid development. Thus Malkus invoked the notion of research and development to suggest that what Fye wanted was development, but the sciences at Woods Hole were not ready for that: "Only the follow-up or development has generally been done by accelerated programs at large laboratories. Oceanography is not ready for the follow-up stage because the fundamental ideas upon which to base such a follow-up have not evolved. It's really that simple."[90]

Malkus's position was neither original nor surprising; it was precisely the sort of claim one could have heard any American scientist making in defense of basic research at this time, and Fye would have heard such arguments many times. What was surprising was the vehemence with which Malkus made her case to the Woods Hole Board of Trustees.

Like Stommel, Malkus had addressed Fye politely and respectfully, but in truth she was livid. She explained why in a letter to E. Bright Wilson, the Harvard chemistry professor and Woods Hole trustee to whom Stommel had also looked for support.[91] Wilson seemed a likely champion, as he was a distinguished chemist who had coauthored a landmark textbook with Linus Pauling on the application of quantum mechanics to chemistry. More than that, he had written a book, *An Introduction to Scientific Research*, that addressed the very issue at stake: the driving force of scientific progress. Wilson placed that drive in the initiative of the individual investigator.

Having worked during the war on microwave spectroscopy and explosives, Wilson was not dismissive or disrespectful of applied work, but like Malkus and Stommel and many others, he considered it a different kind of work. When it came to what he called "fundamental investigation," he was unequivocal in rejecting control, management, and planning. Paralleling the anticommunist arguments of Michael Polanyi and his followers in the British Society for Freedom in Science (chapter 1), Wilson argued that basic science could not be planned. Science was a process of discovering new things: how could one plan to discover what one did not yet know existed? He wrote: "From time to time the proposal is put forward that pure science should be planned 'by some master board of strategists' which would direct workers to those fields where gaps were thought to exist. The utter folly of this idea is apparent to anyone with the slightest knowledge of the history of science. How could any board have directed anyone to discover radio, or X rays, or penicillin at the time no one even suspected that these things existed?"[92] Fye's "need-to-know memo" was certainly suggestive of a master strategy, if not a board of strategists, and the Woods Hole scientists who had read Wilson's book not unreasonably concluded that he shared their view. Moreover, Wilson had described the necessary conditions for successful applied research to thrive and had specifically warned against pitfalls into which Fye seemed to be falling: "It is very easy to develop the habit of making decisions about programs and handing these to subordinates as dictates, without passing on the information upon which the decisions were made. This is not merely bad for morale; it very frequently leads to foolish and useless undertakings which a closer meeting of the minds would have avoided.... The necessity is very great for strong measures to prevent a chasm from growing between the director and the staff."[93] This was precisely the pattern of behavior of which they were accusing Fye.

Assuming, therefore, that Wilson would come to their aid, Malkus wrote openly of her low opinion of Fye. She wrote that she had tried in her letter to him "to emphasize only constructive aspects" rather than argue the details of the science at stake, because he was "not well enough versed in earth sciences to comprehend how far off the mark he is, and [because] his reaction to criticism, actual or imagined, is usually most unhappy and is more likely to give rise to a struggle to win the point or issue involved rather than to think again about it."[94] She considered his administration "an unqualified disaster," and she let loose a barrage of criticism:

[Fye has shown] little interest in learning the types of problems that the marine scientists face or in comprehending the type of personality or mind that is attracted or necessary to face these problems. He is not excited by the oceans nor by oceanographers' excitement with them, but is more impressed by recognition, status, publicity, and the external drapery, not just in addition to, but at the costs of getting on with the work

in oceanography. He has therefore dealt with the staff with misappre-
hension and mistrust, . . . with manipulation insufficiently hidden when
communication failed (as it usually did), with a show of force in the face
of criticism, with stubborn attention to irrelevant details where either a
broad view or "hands off' was in order, and with apathy in important matters requiring directorial action because he failed to recognize them as
important.[95]

Fye was driving scientists away: "The best and most creative staff [have been]
harassed the most, several to the point of no return. The majority of those
who gave W.H.O.I. its quality and reputation of leadership in marine sciences
have already left or are now seriously considering leaving." Reiterating Stommel's point that a scientific institution "is known and evaluated in terms of
its best men and not by the average," Malkus insisted that Fye was driving out
the best, who were more "aware and more disturbed" by what was going on
than their more "pedestrian" colleagues. In any case, the best people "have
the good offers and opportunities to move elsewhere, which the hacks will
not be able to do."[96]

Malkus wrote to Stommel a few days later, optimistic that Wilson would
help, in part because he was "the smartest and most influential" of the trustees and in part because she believed—based on his book—that he shared
their philosophy of science. She wanted Stommel to know what she had said
in case Wilson followed up with him:

I am sure [Wilson] will now try to check up and find out via other people
whether I am just a hysterical female exaggerating the situation or
whether there is a general opinion to this effect and whether there is any
grounds for such an opinion. I thought best that I should be the one to
stick my neck out and make the first move by putting things in writing because I am not, like you, an oceanographer in any way bound to WHOI and
I really shan't be too harmed either personally or professionally if I do not
go back there—although at the moment I surely hope to.[97]

In a memo to colleague Melvin Stern (1929–2010), Stommel summarized the
situation with his characteristic wit. If Fye rejected pastoralism, what did he
want instead? On the bottom of a copy of the need-to-know memo, Stommel
scrawled: "The dictionary does not define pastoral research, but if one may
paraphrase its definition of pastoral drama: A poem, drama, etc. (science?) of
rustic simplicity in which the speakers (scientists?) assume the characteristics of shepherds. Hence, its opposite, to which we are exhorted to aspire is:

'A science ... in which the scientists assume the characteristics of sheep.'"[98]
Neither Stommel nor Malkus were prepared to be sheep.

The Revolt Fails

Even before he had heard from the scientists, Wilson had written to Fye inquiring what the need-to-know memo was supposed to accomplish and correctly anticipating that it would "stir things up a bit and not necessarily with a happy result."[99] Wilson agreed that Woods Hole should continue to expand, but he also thought things were already going fast in that direction and saw little need to press the point. Moreover, he unequivocally defended the idea that individual investigators drove major advances in science:

> I must confess that the flavor of your [memo] rubs me a little bit the wrong way. Perhaps this is because I am an old-fashioned scientific individualist. I still think that this is the way science advances for the most part. Of course, in some fields, such as parts of physics, enormously expensive machines are considered necessary and these can only be operated by teams. But even here it is the lone investigator who makes the really important advances.... The development of radar for wartime purposes was of course a team enterprise, but it was not the engineers and technical people invading a cozy scientific area in a sleeping state, but rather the other way around. It was hordes of pure scientists who were rudely extracted from their pleasant occupation and put to practical work which had been neglected and botched by the practical people.[100]

Wilson acknowledged that oceanography needed "great team efforts with much money and many ships and many routine workers collecting temperature and other measurements."[101] But that wasn't the best role for Woods Hole: "I am very happy to let other institutions expand as much as they want and do as much practical work as they can. My ambitions for the institution are that it will be the place where the basic understanding of the ocean and its problems is achieved, and I am quite positive that this will come from one man's brain and not from sixteen somehow linked together by millions of dollars of government money."[102]

"I don't like the heavy emphasis on application and military application in particular," he continued. "We earn our bread and butter in this way and have to keep doing it both for money and patriotic reasons, but I would hope we do not lose sight of the fact that we are fundamentally a scientific and not an engineering laboratory."[103] Finally, he concluded presciently, "You have been provocative and ... people are going to object to the tone ... these remarks are likely to make people mad." But despite expressing views quite parallel to those of Stommel and Malkus, Wilson took no action on their behalf.

Over the next several months, additional faculty wrote to the trustees to complain about Fye's leadership, organization and vision. By late fall, memos were flying and passions were running high. In November, the senior scientific staff requested a meeting with the Board of Trustees to discuss "the ineffectiveness of the administration and direction of the institution," about which they were "seriously disturbed." The memo was signed by twenty scientists, including several who had strong Navy ties: Richard Backus, Alan Faller, Brackett Hersey, Bostwick Ketchum, William Schevill, George Veronis, Allyn Vine, William von Arx, Vaughan Bowen, and—at the top of the list—Henry Stommel.[104]

The memo was a single sentence long, asking for an audience with the trustees. Prior to the meeting von Arx had prepared a much longer letter detailing the faculty's concerns, but the group had deemed it best to wait and discuss details in person, so von Arx sent the letter to Wilson as an expression of his personal views.

The six-pager outlined four major points of concern. First, the administration seemed detached from the faculty, "totally unaware of their scientific dreams and ambitions." Second, the staff felt that their work was neither understood nor appreciated. Third, the staff felt excluded from institutional decision-making. Here, von Arx quoted directly from a passage in Wilson's book in which he warned of the habit of directors of making decisions without sharing the information on which they based them. Fye had done precisely that, and the result was precisely what Wilson predicted: bad morale. Fourth and finally, Fye was trying to expand the institution in pursuit of "economic bounties" at the expense of the basic science mission. To remain intellectually nimble, von Arx argued, the institution should remain small rather than "allow[ing] ourselves to become overblown with people and to amass equipment that inevitably will steer the course of our efforts towards large-scale undertaking of a survey or developmental nature."[105] He concluded with an explicit rejection of the need-to-know concept and an affirmation of the primacy of basic science: "We think the Institution should be careful to avoid any policy that could tend to divert its primary concern from the accumulation and dissemination of pure knowledge."[106]

Staff discontent was congealing around two issues: management and vision. George Veronis wrote a letter to the chairman of the Board of Trustees, Noel McLean (1907–1984), focusing attention on the first.[107] Veronis repeated the charge that Fye had failed to communicate with the scientific staff, causing apprehension and demoralization. Veronis wrote: "The director ... does not have the competence to run an organization as large as WHOI. The supporting evidence is composed of myriads of small incidents many of which are not serious taken by themselves but the sum of which presents a devastating argument."[108]

Biologist Bostwick Ketchum (1912–1982) wrote to trustee James S. Coles (1913–1996), a physical chemist who had worked with Fye during the war and at the time was president of Bowdoin College. Ketchum viewed himself as an ally of Fye—describing himself as more "moderate" in his views than some—but he, too, had signed the "no-confidence" memo because he believed that the problems had become "more that Paul can cope with alone." Fye had done an excellent job handling the "relationships between the Laboratory and the outside world," Ketchum felt, but matters within the institution had been neglected: "Paul has been so engrossed with external problems that he has lost the intimate contact with the scientists which is necessary in a director of an institution.... [He] must be relieved of some of his responsibilities."[109]

Ketchum agreed with Stommel that there were two different types of work going on at the institution, guided by two different philosophies, which he characterized in terms of free versus directed research.[110] Contrasting his biology group with that of geophysicist Brackett Hersey's in underwater sound, he wrote:

> My group is run as nearly as possible on the academic tradition, with each staff member having complete freedom to develop his own research as he sees fit.... The contrasted philosophy is more or less one of programmatic research, though when this word is used those whom I consider to be in this category violently object. It is clear, however, that Hersey's group, which is the largest in the Institution, is run quite differently from mine. I would certainly object to imposing these ideas on the Institution as a whole, and perhaps he would object to [me] imposing [mine]. This, however, is the fundamental philosophical problem which must be solved.[111]

Von Arx also wrote to McLean, in a four-page, single-spaced letter returning to what he considered the "fundamental philosophical" problem: the freedom of scientists to set their own research agenda. Bluntly, passionately, and at length, von Arx defended a vision of unfettered research conducted by a modest number of investigators and resisting "political, military and industrial coercion."[112]

The institution could of course "expand freely and thus become a large, highly organized research mill with departments and sub-departments very like any government laboratory," but given how many other laboratories were already doing that, the result would almost certainly be "mediocrity." It would also "invite increasing preoccupation with applied research, which is already too much in evidence."[113] The alternative, which von Arx defended, was to "redirect ourselves on a course consistent with our heritage." He wrote:

> We have grown to be respected throughout the world for our leadership in new fields, far beyond the military "need to know," or the clamorings of industry. Since there are already many laboratories being formed or expanded to satisfy the military and industrial needs of our country I think it behooves W.H.O.I. to stand somewhat apart, to re-dedicate itself to academic pursuits. The pursuits are for the greater part the product of individual minds motivated by curiosity and bear no specific relation to other concerns than the general advance of scientific knowledge. Scientific knowledge of a truly basic character is hard to come by.... We should not surrender our position of leadership in its pursuit.[114]

Paul Fye Responds

On December 12, Fye called a meeting to discuss the events of the previous months. He began by acknowledging the discontent, although he claimed that the memo to the trustees had come as a complete surprise, thinking that his efforts to address faculty concerns had been effective. It was time to address them once and for all.

Fye's comments to staff were long and detailed—the transcript ran to thirteen single-spaced pages—but the overall message was simple: he was doing the job that the trustees had hired him to do. If the faculty thought otherwise, they did not understand the realities of the institution. He had been hired by the trustees "with a directive ... to make certain changes," which he described as follows: "to build the necessary organization, to tend toward active building of the professional staff from adequately trained scientists rather than on an apprentice system, to develop proper titles and salary scale for the research staff, to establish reasonable scientific review of our work and our publications and to evolve a far-sighted long-range plan for the development and improvement of the Institution"—in other words, to professionalize the place.[115]

The faculty seemed not to appreciate the multifaceted commitments and responsibilities that had kept him busy during the previous two years, including congressional hearings, serving on government committees, seeking private financing, dealing with customs issues, and more. Staff criticism of the amount of time he spent away from Woods Hole, he insisted, revealed their ignorance of what his job entailed.

To address the need for closer supervision of internal matters, Fye had asked for and been given authorization to create a new position of associate director to help with internal and day-to-day management while he attended to external and long-range issues. In addition, he proposed an internal reorganization, establishing a set of committees to address contested issues: Organization, Professional Status, and Computers. Finally, a staff council would participate in technical decisions and review the work of all other committees. These changes would assist the scientific staff to do their work while he did his.

Fye concluded by reminding the faculty that they shared responsibility for the troubles at Woods Hole — for complaining behind his back at evening meetings, for making contradictory and irreconcilable demands on the director, and for acting intemperately. He asked for self-restraint in the interests of the institution, as well as in the interests of being able to hire the kind of person needed in the new position of associate director: "I ask for restraint not ... for selfish reasons [but] for your own self-preservation. If the turmoil of the past month continues it can lead to a large measure of self-destruction and can do a great deal of harm to the institution. I just suggest we stop rocking the boat until we are in quieter waters."[116] At the end of what must have been a very long speech, he finally addressed what Stommel, Malkus, von Arx, Veronis, and others felt to be the essence of their complaint: the institution's intellectual direction. Fye had focused his comments on organization and communication; he closed with an attempted reassurance that he was not in disagreement over the importance of basic science. On the contrary, the funds he and others had raised were making basic science possible on an unprecedented scale, and the political situation enabled them to do both basic and applied science so long as they "kept their heads" and did not get overly emotional about it:[117]

The kind of science we are all interested in is assured of support for years to come. We can write our own ticket in a manner and with a freedom never before possible. The monies and facilities now available need not, if we keep our heads, pressure us into a position of prostitution of pure science or of our true missions. Over a year ago I wrote as part of a statement for our Development program: "The Woods Hole Oceanographic Institution is, first of all, a laboratory devoted to basic research. This means a concentrated devoted effort on the part of dedicated individuals in the search for new knowledge. Factual knowledge alone is not sufficient — but rather by basic research we mean the pursuit of the kind of new knowledge that enhances our understanding of nature. Pursuit of this kind of understanding is a creative act of the highest type and it is most important the complete freedom and sympathetic support be given those gifted individuals who are capable of carrying on this type of research. We firmly believe that the challenge of the future in oceanography can be met only by research that truly asks the why and how of nature, simply for the purpose of understanding."[118]

Fye had a point—there *was* more money for basic oceanographic research than there had ever been before—but whether his scientists were being given the "complete freedom and sympathetic support" they wanted was precisely the question at hand. That, and whether it was possible not to be forced "into a position of prostitution of pure science or of our true missions" when an interested party was footing the bill.

Prostitution or Pragmatism?

Fye did not say that he made these recommendations for reorganization with the approval of the trustees, but the evidence suggests that he had. Already in the summer, the trustees had discussed the need for an associate director to handle more of the day-to-day operations of the institution. More important, in communications among themselves they made clear that they considered the faculty complaints naïve and overblown.

James Coles, the Bowdoin president, argued to Wilson that Fye had done an excellent job in presenting the institution to the outside world: "This essential work—interpreting science to the lay public and the business man, making future plans and finding funds for support of scientific research—is something which I believe Paul does very well." If there were other areas in which Fye was less able, then the trustees should find some way to "compensate," but they should not overreact. As for the "need-to know memo," Coles addressed it in a postscript, brushing it aside as simply a poor choice of words: "Please overlook the "Need-to-know." This might better be called 'Desire-to-know.' Careful reading is necessary to recognize the basic emphasis of fundamental science."[119]

Coles dismissed the scientists' concerns as sour grapes, focusing particularly on one man, William Schevill (1906–1988), a cetacean biologist who had done extensive work for and with the Navy on mammalian sound: "Just between us girls," he began, presumably referring to the gossipy character of what he was about to say, "my own impression of Schevill has never been high, and I wonder if he might not be one of those men who, frustrated with his own work or career, finds compensation in criticism of the institution."[120]

Trustee Harvey Brooks (1915–2004) was similarly critical of the scientists, whom he considered naïve, if not irresponsible. Writing to McLean in December to discuss Fye's speech to the faculty, Brooks concluded: "It is a pretty impressive document, neither too aggressive nor too apologetic. I am very hopeful that it will have the effect of calling the more sober heads to a real sense of responsibility."[121] And despite what he had written in his book, Wilson was not, in fact, prepared to take the scientists' side. Writing to Ketchum about von Arx's memo, he wrote simply, "I must admit to being a little astounded."[122] The scientists, he felt, had grossly overreacted.

On December 19, the faculty finally had the opportunity to meet with the trustees, but the decision had been all but made. Prior to the meeting, the trustees had told the faculty that their complaints were too various and in some cases too contradictory to be credible; they suggested it would be best if only a handful of faculty attended, with one appointed speaker.[123] The faculty rejected the suggestion—they were not prepared to have only one person represent their complex concerns—but they did prepare a text to be read at the meeting by physical oceanographer Bill Richardson, with shorter, supplemental comments by others.[124]

Perhaps because Fye had focused his comments on communication and organization—or perhaps because the deeper issues seemed too difficult to broach—the prepared text focused on communication: between staff and director and between staff and trustees. In the past the staff had not felt the need to address the trustees directly because the director was adequate to the task, but this was no longer the case. Fye was not only unable to speak on their behalf; he was incompetent to act as their director.[125] Acknowledging that they could not be objective about themselves—and no doubt realizing that it was impolitic to question the trustees' oversight—the faculty moved directly to the question of the director's competence. As oceanographers often did when seeking vision and wisdom, they turned to the words of Roger Revelle.

In 1955 Revelle had been asked for advice about the choice of a new director for Woods Hole. In his long and thoughtful reply, Revelle placed scientific stature above all other considerations: "The number one requirement for your new director is that he should be a first-rate scientist. He should be a welcome participant in scientific discussions anywhere in Europe and America and should be able to furnish an inspiring example of scientific insight to his colleagues at Woods Hole."[126]

The faculty accepted that Fye was not that kind of a leader, but they did expect him to "provide the environment into which scientific leadership could be introduced or from which it would evolve. He had a good deal of personal charm and considerable experience in administrative duties. He liked Woods Hole and seemed generally eager and sincere in his desire to improve the stature of the institution."[127] But he was "miscast" as director and "not big enough for his job":[128]

> He has administered the institution with such a degree of uncertainty and with so little confidence in himself and in members of his staff that the net result has been a general frustration and lowering of morale of everyone. Most of our criticisms of Dr. Fye can be traced to these characteristics of uncertainty and lack of confidence.... Under the best of conditions, when all available information is in, he is indecisive, often taking months to announce a decision which was apparent from the start. When issues are less clear, the uncertainty shows itself in behavior which is characterized by irresponsibility, vacillation, and even duplicity.[129]

This was a strange complaint. Elsewhere, scientists had called Fye arrogant and autocratic; now they were saying he was vacillating and uncertain. Perhaps they were chastened by his speech and accepted that the man was trying to serve their interests. Or perhaps it was a tactical decision to focus on Fye's competence rather than on their real concern—his vision—for conspicuously missing from the prepared group presentation was any discussion of the event that had triggered the rebellion: the need-to-know memo and the perspectives it revealed. Perhaps the faculty had concluded that the question of vision—the raison d'être of an institution such as Woods Hole— was simply too difficult, too contentious, or too political to address. Perhaps competence seemed a more objective ground on which to build a complaint. Whatever the reason, it was a mistake, because the resulting impression was that the two faculty who addressed institutional purpose were offering a minority view.

The first was von Arx, who asked simply for intellectualism to come first: "The world is changing around us, but the nature of scholarship and research is altered only in its outward appearances: in larger, more complex equipment, for example. Unless these tools and the administrative machinery needed to manage and finance them are firmly regarded as accessories to research, they will become our masters."[130] Von Arx reiterated this theme in private the following day to George Veronis: "I do feel that the major issue is a misunderstanding of the laboratory's mission by the trustees, and that this misunderstanding gave us ... Paul Fye."[131]

The other scientist who spoke at the meeting concurred with von Arx, emphasizing that the personal issues about Fye concerned him less than "the issue of over-riding importance," the future direction of the institution. Notes from the meeting do not reveal this speaker's identity, but it was someone who knew Rossby and was determined to protect scientific pastoralism. It is unlikely that this was anyone other than Henry Stommel: "It was Dr. Rossby who described Woods Hole as a 'scientific hotel,' where any capable scientist with an inborn drive to try his hand at solving the ocean's problems is welcome to come and work. Its charm, uniqueness and great scientific productivity lies in the personal independence afforded to its staff. Self-generated efforts flourish in an atmosphere that is somewhat idyllic, remote from external political and military pressures and internal administrative concerns. Such a haven of intellectualism is rare indeed today, even in the universities. I feel that we must struggle to preserve this atmosphere at all costs."[132]

Overall, however, the impression was given that the central complaint was Fye's style, not his goals. Moreover, suggesting that Fye's incompetence arose from his lack of confidence was counterproductive, because the trustees' response was a series of actions designed to bolster his authority. Not only did they decline to dismiss him; they appointed him president of the Woods Hole Corporation. They also made it clear that anyone who was not prepared to accept Fye's leadership should look for work elsewhere:

The Executive Committee [of the trustees] has carefully considered and weighted all that has been said ... and all that has come to its attention. It has done likewise for the program outlined by the Director ... This program and its effective implementation by the Director have the unqualified support and confidence of the Executive Committee. *This is a policy decision....* The Committee expects the full and complete cooperation of the staff.... The Executive Committee is prepared to accept resignations without prejudice of those who feel it will be impossible to cooperate in the future fulfillment of this policy.... Those who cannot in good conscience support the Institution should go elsewhere."[133]

The Palace Revolt had failed.

The End of the Palace Revolt

The Woods Hole trustees had hired Fye knowing full well that he had made his name in explosives, not oceanography, and they had hired him with a mandate to move Woods Hole forward into what they saw as a more modern, more professional form of management. Perhaps, like the leaders of the University of California, they also supported the political goal of a science allied to national-security concerns, particularly as the Cold War became increasingly frigid. In this respect, the scientists had been naïve: the trustees did not care whether Fye was excited by the oceans, they cared whether he could run the institution in a pragmatic manner in line with the demands of the times. Indeed, in notes from the trustees' meeting on the question of who might plausibly serve as the associate director that Paul Fye had asked for, Wilson wrote, "Not von Arx. Idealistic."[134]

In his 1955 letter, Revelle had argued that a good leader must be "a good gambler ... willing to take many chances and have the will to make his important bets come off. Good gambling in this sense is perhaps another term for good judgment. The real test of a politician, a bureaucrat, or a director is the ability to forecast the future, which often means that he works hard enough to make the future come out the way he predicted."[135] Fye was not a great scientist, but he was an effective gambler. He had taken a chance—holding firm in his views in the face of faculty opposition—and he had won.

As for the staff, their losses were clear and devastating. Many were shocked by the speed and severity of the trustees' response. Veronis blamed himself for failing to articulate his concerns clearly: "On that fateful day [December 19] the atmosphere of terrible embarrassment and nervous anxiety paralyzed my own mental processes and generally made the meeting ineffective and valueless. I am sure that no one was functioning normally at that time.... The point I wish to make is this: Because of the emotionally ridden atmosphere we never did communicate with the Committee."[136]

Records reveal only one trustee who shared the scientists' dismay: Alfred C. Redfield (1890–1983). A distinguished biologist, Harvard professor, member of the National Academy of Sciences, and one of the original staff recruited by Henry Bigelow in the 1930s, Redfield was known for discovering that the ratio of major nutrients in plankton is the same as in the oceans in which they live and could therefore be used as proxy for ocean nutrient content. Redfield had missed the December meeting and wrote to McLean that he was "shocked and disillusioned," by both the content and the tone of the letter they had sent. The scientists who had come to speak to the trustees "included some of the most competent, level-headed, and devoted members of the staff," yet their concerns had been dismissed, and they were "presented a take-it-or-leave-it program without any request for comment [and] are magnanimously given the privilege of resigning."[137]

Some of the institution's best scientists had come to the trustees in good faith but were "adroitly sidetracked by an administration more interested in defending its power, past actions, and policies than in benefiting by the experience of its staff." The predictable outcome was "frustration and humiliation," which did not bode well for the trustees' stated goal of permitting "the staff to carry on scientific work under the best possible conditions."[138] Quite the contrary. Some would likely leave. Redfield himself was considering resigning from his trustee position: "I cannot, in good conscience, support an institution which is managed in the spirit of this manifesto, either by defending its policies or encouraging any scientist to become its employee. I would be relieved if you can show me in what manner I misjudge the situation."[139]

In his comments to staff, Fye accepted that he bore responsibility for the state of the institution, but he also insisted that the staff shared the blame: "Those of you who have refused to accept responsibilities when asked, who have not carried through jobs which needed doing, either at all or only with repeated prodding, and those of you who have stirred up conflicts rather than searching for compromises must surely, in a small way, share this failure with me."[140] In his copy of Fye's remarks, von Arx wrote in the margins: "refused only w/ clear good reason, viz Joanne."[141]

"Joanne" was, of course, Joanne Malkus. By the time her promised manuscript appeared in *The Sea* the following year, she had added a dedication that was a scarcely veiled attack on Fye. By implicit comparison to Columbus Iselin, Malkus made clear her opinion that Fye was no friend of science: "This effort is dedicated to Columbus O'D. Iselin, to whom the writer owes her interest in the area and her opportunity to pursue it without restriction. His efforts have provided a stimulating atmosphere of broad inquiry into the earth sciences.... His glorious confidence in us all has given each of us the necessary self-confidence to tackle difficult problems, while at the same time his own humility in the face of the sea's complexity has imbued us with the necessity of continuous self-criticism in our attempts to understand its behavior."[142] The trustees had stated that anyone who was not prepared to accept Fye's leadership should tender his or her resignation, and Joanne and her husband, Willem, did; they made their move to UCLA permanent and did not spend summers at Woods Hole.[143] And they were not the only ones. Melvin Stern—the recipient of Stommel's shepherds and sheep memo—went to the University of Rhode Island. George Veronis went to Yale. William von Arx became a professor at the Massachusetts Institute of Technology.

In 1965, von Arx concluded sadly that the problems at Woods Hole had come to characterize oceanography as a whole. In an essay entitled "A Science in Bondage," he wrote: "The Navy's need-to-know about the ocean environment has prompted the establishment of several government laboratories wholly dedicated to the problems of undersea warfare and many of the larger contracts for research in private laboratories and universities for environmental observations and prediction are funded, and directed, through the military establishment. Some very good science and good scientists are involved in these efforts but the question of truly intellectual desire-to-know research is largely set aside."[144]

As for Henry Stommel, even before the trustees' decision, he had decided to accept an offer from Harvard, although not without deep misgivings. A man who loved Woods Hole—both the institution and the place—it was extremely difficult for him to imagine living and working anywhere else. He accepted the professorship, but Harvard made him miserable. Working in the Division of Engineering and Applied Physics, Stommel found few congenial colleagues. Two years later, he lamented to Allan Robinson: "There is much more to oceanography than mathematics, but I just cannot seem to find anyway to foster interest [at Harvard] in these other areas.... It has been very lonely for me at the Division. I detest living in suburbs. Cambridge is very dismal. Dean Ford is a prick. Did you ever notice how ludicrous it is to walk up and down the halls and see all the professors sitting in their offices with open doors, like expensive whores, or stall-merchants in an Eastern bazaar? And so on and so on, but there it is."[145] After three years, Stommel moved to MIT, but he never adjusted to Cambridge (during his tenure at MIT he lived in Falmouth and commuted). Nonetheless, so long as Fye was at Woods Hole, he would not consider returning. In 1978, Fye finally retired, and Stommel returned to Woods Hole, where he spent the remainder of his career.

In a letter to Wilson in early 1962, Veronis described the preceding two years as "a depressing period for all of us here," especially because "Willem [Malkus] and Hank were the two really shining lights at Woods Hole.[146] In the decades that followed—after Veronis had also left—ocean-atmosphere interactions would be recognized as a crucial domain of scientific investigation, critical both to applied and basic science, but little of this work would be done at Woods Hole. The irony of this was not lost on von Arx, who recalled in 1991 that "meteorology once flourished at WHOI, [with] Veronis, . . . Malkus, et al.... [but that program] is now gone."[147]

At Woods Hole, the question of curiosity versus mission-curiosity science would be settled in favor of the latter, but only over considerable faculty objection, which helps to explain why faculty continued to insist that they were doing curiosity-driven science. Had they not so insisted, they would have had to acknowledge something that they found troubling: that the reality of their science was something different from not only from what they ideally wanted it to be but also from what they thought it needed to be to flourish.

Was Oceanography Distorted?

There has been considerable debate among historians of science over how to think about the role of the US military in the Cold War period. Most of this debate has taken place with reference to physics, where historian Paul Forman has argued that military values and priorities changed the field, pushing it off the path it was on before World War II and promoting a science of gadgetry and technical virtuosity. In response, Dan Kevles argued that "physics is what physicists do—or have done," and that changes in American physics after World War II signified "not the seduction of American physics from some true path but its increased integration as both a research and advisory enterprise into the national-security system."[148]

Kevles's description of physics as increasingly integrated into a national-security system applies a fortiori for physical oceanography, but this does not negate Forman's question: how was physics changed by this new situation? Science is always changing—what we identify in hindsight (and to some extent anachronistically) as "science" in the seventeenth, eighteenth, and even nineteenth centuries was very different from what we came to see as science in the twentieth, and science is evolving still in the twenty-first. It is not essentialist to attempt to identify and evaluate these changes. As historians, we should be interested in how science has changed and why.

In historical assessments of twentieth-century American physics, the complex question of how and why physics changed has sometimes been flattened into the word *distortion*, to the question of whether physics was distorted by its assimilation into the national-security enterprise. In response, some historians might argue that the word *distortion* is inappropriate, even irrelevant. The shape of a living brachiopod may be distorted as its fossilized version is turned into rock, the sound of a human voice may be distorted in a telephone call, and the filters of time distort our memories of events. But the events of history are what they are. They are not distorted versions of some other set of (counterfactual) events. In this sense, Kevles is right that physics is what physicists have done, just as history is whatever has happened.

But there is another way to consider the matter: that the word distortion does make sense in the Cold War historical context, because it was a word—and a concept—that scientists themselves invoked. It is, as sociologists would say, an actor's category, a term that had meaning to the people who used it. American scientists had values, aspirations, and ideals that were challenged and in some cases undermined by the integration of their sciences into the Cold War leviathan. We can talk meaningfully talk about distortion in this sense, because it is in this sense that scientists talked about and *experienced* it.

Oceanography in the mid-twentieth century was changing in ways that some welcomed and others resisted, and, crucially for this argument, that some of those who resisted viewed as distorting. If Henry Stommel was particularly aggrieved by the events of 1961, it was because they were part of a larger pattern that disturbed him. As a member of the Scientific Policy Committee at Woods Hole in 1958—the committee that Fye later disbanded— Stommel had put his name to a memorandum (signed also by Columbus Iselin, Bostwick Ketchum, Brackett Hersey, Joanne Malkus, William von Arx, and, ironically, Fye himself), which explicitly asked whether military concerns were dominating the activities of the institution to an untoward degree. The question under consideration was the expansion of Woods Hole. The committee favored expansion, but only at a pace that did not "exceed [that by] which the excellence of the scientific program can be maintained."[149] Above all, they warned, expansion should not deflect the institution from its basic scientific mission, a deflection that they already to some degree detected: "BE IT RESOLVED that ... *this expansion not be permitted to distort further the emphasis on direct military and development aspects compared with the basic science* and to this end the Trustees shall endeavor to obtain additional financial support for independent institutional work in oceanography."[150]

"To distort further"—those are their words. This perceived distortion was the outcome of an unequal funding environment, with funds flowing freely from the armed services to support work of value to them, but little available for "independent" work.

Historians may argue that "basic" or "pure" science—unaffected by social concerns—has never actually existed, but we cannot deny that the category of basic science had salience for the men and women who invoked it. We may similarly argue that the notion of distortion implies a Platonic ideal that could never exist in practice, but the evidence nonetheless shows that at least some oceanographers *experienced* military patronage as distorting.

Most of the Woods Hole Scientific Policy Committee felt that it was fine to serve the nation's national-security needs so long as those did not crowd out basic science. However, some staff were prepared to go further, arguing that the institution should not pursue military projects at all. In a memo outlining topics for discussion with the new director, William von Arx's number-one item under "policy questions" was "Should totally classified research and development be a permanent part of the institution's scientific program?"[151] Von Arx's use of the word *permanent* acknowledged the extensive military work that had been pursued since World War II and the continued role of

Navy funding, but it implied discontent at the fact that what had begun as a response to national emergency had started to look like a permanent state of affairs.

To these men (and one woman), oceanography was not whatever oceanographers happened—or were pressured—to do; it was a science dedicated to understanding the oceans in their full physical, biological, and chemical complexity. It was the mission that Henry Bigelow laid out in the 1929 report of the National Academy of Science Committee that led to Woods Hole's creation. It was a science that had already thrived in Europe for some time and that they saw themselves as building in the United States. What Paul Fye and his supporters among the trustees wanted was something different. To the scientists of the Palace Revolt, what Fye and the trustees wanted *was* a distortion.

Trustee James Coles had tried to dismiss William Schevill's complaints as sour grapes, but no one could say that of Henry Stommel. Moreover, his work fit Fye's vision of need-to-know science, as did that of von Arx, Malkus, Veronis, and most of the others involved in the Palace Revolt. Nearly all received Navy funding, and nearly all did work relevant to military operations and aspirations. Nor is there any evidence that they were forced to take on tasks against their will. But this raises an important question: if they received Navy support to do work they wanted to do, what exactly was the problem?

One way to answer this question is by comparison with Lamont. No one went to Lamont without knowing, and at least to some degree accepting, the autocratic leadership of Maurice Ewing and his commitment to Navy-scientific collaboration. Lamont was not changed by Navy funding; it was created by it. But Woods Hole, like Scripps, was changed. Many scientists experienced those changes as mostly for the better, but even those who did expressed concerns for the future: that military funding not be permitted to undermine scientific autonomy, a problem that had not yet manifested but that might still. It was about a tipping point that had not yet been reached but they feared soon would be. Their anxiety was anticipatory, and, as von Arx had put it, about issues that were lurking in a "ghostly" way.

When Henry Stommel looked back on his career thirty years later, he concluded that things had been better in the past. Much of the joy and spontaneity was gone, he felt, and scientists were much more directed by external programs than by their own ideas. They had indeed become other-directed, with more money but less freedom. Stommel concluded wistfully: "They left us alone as long as we were poor."[152]

Autonomy was a theme to which Stommel consistently returned. On receiving the William Bowie Medal from the American Geophysical Union in

1982, he emphasized the importance to him of being able to do "unfettered research." In his notes for his acceptance speech, he wrote:

> I must thank my lucky stars for being born in the century and this country where the ~~Federal Government~~ public has taken such a benevolent interest in supporting scientific research. It was only by the support of agencies such as the Office of Naval Research and the National Science Foundation that it was possible for me to spend a lifetime exploring the ocean, on ships, in the laboratory, and in my mind.... Let me affirm that the freedom to work in science on one's own, with congenial colleagues, unfettered by supervision, with a scientific problem in one's mind when he goes to bed and when he awakes next morning, to be able to give undivided attention to unraveling some puzzle of nature, is a privilege beyond compare. I have only one regret: I was born at the wrong time to witness a transit of Venus.[153]

Stommel had sounded a similar note in 1977 when he received the Ewing Medal from the American Geophysical Union. In preparing his acceptance remarks, he struggled with the right way of expressing gratitude for a prize named in honor of a man whose influence he had endeavored to escape. It was not that Stommel disliked Ewing as a man, but he strenuously objected to his approach to science. After several attempts, Stommel found the right words: "'Doc' Ewing ... really was a boss.... Working under such strong leadership is not my style, [so] I found something else to do at Woods Hole as soon as I could."[154] He concluded: "Most of human history has not afforded men much chance to pursue their curiosity—except as a hobby of the rich or within the refuge of the monastery. We can count ourselves fortunate indeed to live in a society and a time where we are actually paid to explore the universe."[155]

Henry Stommel was indeed fortunate, but in part because the bit of the universe that interested him also interested the US Navy. This was not merely luck—after all, he had studied astronomy and divinity and had been drawn into oceanography as part of the war effort. Stommel found important questions to be answered in the science of oceanography, but the fact that he became an oceanographer was a result of the historical circumstances of his life. Moreover, his "luck" was not without its price, and that price, as he noted more than once, was some measure of personal autonomy.

In his later years, Stommel became an active promoter of several large-scale collaborations in oceanography, which he considered necessary to advance the science, but he retained misgivings about the consequent loss of autonomy. In a conversation with MIT professor Carl Wunsch just two years before his death, Stommel asked, "How can scientists find joy in their work if their tasks are set by others? And if they don't find joy, will they put their hearts into their work?" If joy were not sufficient justification for free inquiry, there was also the question of productivity: "History has shown that you don't necessarily increase the food supply by collectivizing the farmers."[156] Three decades after the Palace Revolt, Henry Stommel still believed in pastoral science.

4 *Stymied by Secrecy: Harry Hess and Seafloor Spreading*

Harry Hess was one of the most important geologists of his generation. His concept of seafloor spreading paved the way for the theory of plate tectonics, which in the late 1960s became the first generally accepted global tectonic theory in the history of science.[1] Hess's ideas were inspired by new information about the shape and structure of the seafloor, particularly in the Atlantic, acquired by scientists working under Navy sponsorship. From an empirical perspective, these data were a new element in the debate over crustal mobility, something that had been lacking in earlier discussions of the theory of continental drift.[2] But the Navy did not collect bathymetric data for their relevance to tectonic theories; it collected them for their relevance to navigation, to the locating of hydrophones used to detect Soviet submarines, and to the submarine-launched ballistic missile (SLBM) program—and so the data were highly classified. This troubled Hess, who thought that data classification was stymying his science.

Bathymetry had always been important to mariners, but in the Cold War it took on a new, urgent significance in relation to the nuclear triad. By the late 1950s, engineers had achieved a high degree of accuracy in targeting and guiding intercontinental ballistic missiles, but it was no use knowing the precise location of Moscow—and being able to launch a projectile on a high-precision trajectory to it—if you didn't know where you were when you fired it.[3] If missiles were to be launched from submarines, those boats would have to know just exactly where they were. Missiles might also one day be launched from seafloor installations, which would require detailed data on bottom topography and conditions. It would also require the Navy to occupy these sites; data collection could help the United States to establish "rights" to those locations. A 1963 report on bathymetric investigations in the Indian Ocean—"a potential launch site for a Polaris A3 missile" as well as for bottom-mounted launchers—put it this way:

The acoustical properties and the bathymetry of the ocean floor are dominant factors influencing all undersea warfare operations and plans.... They determine where the deep submersible can hide and where bottom-mounted sea-based deterrent systems can be located.... Possible sea-based deterrent systems located on the ocean floor and the Navy's potential requirement to establish proprietary rights to the ocean floor by occupying it with manned stations require the type of data provided by this program. Questions such as: where can good foundation sites be found, what are the ocean floor's dynamic sedimentary processes, and where can a system on the floor hide, are being answered.[4]

If these data were crucial to military operations, they were also crucial to understanding the structure of the ocean floor and the processes by which it formed, so Harry Hess tried to convince the Navy to declassify them. Given the haphazard nature of scientific discovery, he argued, one needed the input of as many creative minds as possible. By limiting the number of scientists with access to information, classification decreased the chances of a major discovery. In short: secrecy was a prescription for impeding progress. Insofar as science was crucial to the national defense, impeding science impeded defense, too.

Hess was in line with scientists and analysts of science, past and present, who have held the free exchange of information and ideas to be both a requirement and a defining characteristic of modern science. Robert Merton articulated this view in his classic essay "The Normative Structure of Science," in which he argued that both the raw materials and the results of scientific inquiry are communal property, and the handmaiden of community is communication: "The institutional conception of science as part of the public domain is linked with the imperative for communication of findings. Secrecy is the antithesis of this norm; full and open communication its enactment."[5] Merton linked open communication to democracy, which he held to be uniquely conducive to a flourishing spirit of scientific inquiry.[6]

Merton's essay became famous in part because it articulated what everyone believed but rarely bothered to state; the requirement of open communication had been generally taken by scientists to be self-evident. It was understood as the ideal that undergirded scientific societies, with their annual meetings and published proceedings and the peer-review process. It was the feature invoked to distinguish early modern science from hermetic and alchemical traditions, shrouded as they were in secrecy and mystification. It was also often said to distinguish academic from industrial research, cloaked as the latter often was in claims of propriety and trade secrets.[7] Little wonder that as American science in the 1950s was pursued under conditions of secrecy and data confinement, scientists would worry. Hess and others were striving to advance science under conditions they believed were inimical to it. Were they right?

On the face of it, it seems not, for despite his complaints, Hess succeeded in laying the foundations of modern geophysical theory. American geoscientists in the 1950s routinely lamented the security protocols imposed on them in the Cold War even while they accepted military contracts and worked under conditions in which scientific information was subject to governmental review and its dissemination restricted, so perhaps their complaints were pro forma.[8] Perhaps the anticipated difficulties never developed.[9]

Or perhaps Hess and Merton were wrong about the conditions required for science to flourish. After all, despite scientists' widely held conviction that "science flourishes best in conditions of the open and public exchange of ideas, methods, findings and interpretations," as epidemiologist David Michaels has recently put it, historians have found science flourishing under a variety of restricted conditions.[10] Charles Gillispie has documented secret science projects related to munitions and ordnance in revolutionary France (although it appears the projects were more of a secret than the science that drove them).[11] Loren Graham, Sonja Schmid, Sigrid Schmalzer, and others have analyzed the various branches of science that flourished in conditions of constrained communication in Soviet Russia and Maoist China.[12] Peter Westwick and Peter Galison both concluded that American scientists in the Cold War found ways to operate successfully within the constraints of secrecy by holding classified conferences and creating classified peer-reviewed journals.[13] More recently, scholars have noted that many areas of biomedicine and biotechnology operate today under conditions of at least partial trade secrecy.[14] Given this, Michael Aaron Dennis has gone so far as to say that the conventional equation of science with openness—the "unrestricted exchange of information and knowledge without regard for the race, creed, sex, or national origins of those involved in the exchange"—is "patently false."[15] And Sheila Jasanoff has argued that the assumption that secrecy is necessarily bad should itself be interrogated, that openness and transparency in science cannot be treated as absolute goods or assumed to be natural. Like "most good things," she holds, "even scientific openness has to be purposefully cultivated and judiciously deployed in order to serve its intended functions well."[16]

Clearly, there are situations—both military and civilian—in which secrecy is justified. The problem is the difficulty of evaluating just what the impacts of secrecy—or openness—are in any given case. During the Cold War, the University of Chicago sociologist Edward Shils (1910–1995) argued that the effects of military secrecy would take years to be uncovered, if they could be discerned at all: "Only after many years, if ever," he wrote in 1956, "will it be known which crucial bits of knowledge lay hidden by classification and had to wait for a long time to be rediscovered by someone else who then made use of them for some important discovery."[17]

Cold War physicists often claimed they could separate abstract questions about the fundamental constitution of matter and the universe from concrete problems of weapons design and construction, carving out protected epistemic territory around the former.[18] Whether or not the physicists' view can be defended, in oceanography no such distinction can be sustained. As historian Jacob Hamblin has concluded, in oceanography there was little, if any, meaningful separation between military secrets and scientific information.[19] The information about the nature of the seafloor that bore on theoretical questions was the same information that served the Navy's operational interests, and it came in the same form. Given this, it seems implausible that secrecy had no effect.

We have already seen how security considerations affected Harold Sverdrup during World War II (chapter 1), a story that complements the well-known destructive effects of Cold War anxieties on the careers of physicists Robert Oppenheimer and Edward Condon.[20] We have also seen how it created tensions at Scripps, Woods Hole, and Lamont over questions of scientific meaning, purpose, direction, and autonomy (chapters 1 and 3). Clearly secrecy affected scientists, but did it affect science? Did Cold War secrecy alter the development of our knowledge about Earth? To prove that something was not learned is a near-impossible historical task, and to try to explain why something was not discovered risks charges of presentism and whiggishness.[21] Yet it is possible to show that scientists at the time believed that something was being thwarted. It is also possible to show that data were left unexamined and lines of inquiry abandoned because of the prevailing research conditions. As Robert Proctor and Londa Schiebinger have argued, it is not just knowledge that is socially constructed but also ignorance.[22]

To study the production of ignorance—or the obstruction of knowledge, as Hess would likely have preferred—requires both positive and negative considerations. Which ideas were pursued and which ones abandoned? What was funded or not funded? What did not happen that people at the time thought could or should happen?[23] Hess believed that work was being stymied—that advances were not being made—because of military-imposed secrecy. Is it possible that he was right?

Conventional wisdom among earth scientists is that the prodigious amount of data on the shape, structure, and characteristics of the seafloor, collected under the auspices of the military, laid the foundations for plate tectonics.[24] On this view, Hess was wrong: military secrecy either had no adverse effect, or whatever obstacles it created were small in comparison to the overwhelming positive effect of abundant funding and logistical support. In this chapter, I argue that conventional wisdom is wrong and Hess was right: that Navy secrecy stood in the way of the emergence of modern global tectonic theory. To evaluate this possibility, we need to understand the work that Hess and his colleagues were doing on global tectonic questions before conditions of secrecy were imposed.

Seafloor Studies before World War II

Described as a man who thought about geology from the time he woke up in the morning until the time he went to bed at night, Harry Hammond Hess (1906–1969) was one of the most accomplished geologists of the twentieth century. Known today primarily for the theory of seafloor spreading—first presented in his 1962 paper "History of Ocean Basins"—in his day Hess was known as a versatile adept whose career encompassed geology, geophysics, university administration, naval reserves, and government advisory committees; one colleague said that he lived "five lives simultaneously."[25] Besides generating a prodigious output of scientific work in mineralogy, petrology, geophysics, and tectonics, Hess served as chair of the first National Academy of Sciences' Committee on Radioactive Waste Disposal, head of the Earth Sciences Division of the National Research Council, chairman of the Space Science Board, and president of the Geological Society of America. As a Navy reservist he rose to the rank of rear admiral. A brilliant synthesizer, he was credited by colleagues with great intuition and a prodigious capacity for retaining detail whose relevance he would later distill. He was also a skillful politician who knew how to avoid trouble: as chairman of the Princeton geology department from 1950 to 1966, he eschewed arriving before 10:30 a.m., knowing that by late morning most people would have sorted out their petty problems without him.[26]

Hess began his career as a mineralogist and petrologist focused on ultramafics: rare terrestrial rocks that are unusually rich in iron, magnesium, and calcium. Between 1933 and 1942, he published numerous articles on the mineralogy, petrology, and structure of layered mafic intrusions.[27] These igneous intrusions have remarkably distinct and persistent compositional and mineralogical banding, including peridotite layers—iron- and magnesium-rich rocks composed almost entirely of olivine and pyroxene. They also contain layered structures and cross-bedding reminiscent of sedimentary rocks—stimulating questions about how minerals crystallize within magma chambers—as well as economically valuable concentrations of copper, nickel, chromium, and platinum group elements. The Bushveld Igneous Complex in South Africa, for example, holds the world's largest known concentrations of platinum. Layered mafic intrusions could thus be viewed as an ideal object of earth-scientific interest, seamlessly blending intellectual and economic promise.

Many geologists at the time believed that layered intrusions were a microcosm for Earth as a whole. Their upper layers were compositionally similar to common crustal rocks, so they thought perhaps the lower (peridotite) layers would be equivalent to deep crustal or even mantle rocks. Moreover, there was something peculiar about the peridotites in the lower layers. Apart from their occurrence in layered intrusions, peridotites were known only from island arcs and orogenic belts, where they occurred in highly deformed "ophiolite complexes"—lithological suites composed of ultramafics, basalt, and marine sediments—interpreted by some to be remnant pieces of the seafloor. The peridotites in ophiolite complexes were always altered to hydrous mafic minerals, including serpentinite asbestos—hence they were referred to as *serpentinized*—as were most of the peridotites in the layered intrusions. This suggested, as Hess put it in the circumspect manner characteristic of American geologists of his era, "a number of definite relationships between ultramafic intrusions and tectonics."[28]

The link between petrology and tectonics was a second focus of Hess's early career and the area in which he achieved lasting fame: the structure and tectonics of ocean basins. In the 1930s, Hess had been invited to participate in gravity measurements in the Caribbean under the mentorship of the Dutch geodesist Felix Vening Meinesz (1887–1966). Vening Meinesz had achieved world renown with the invention of a novel pendulum gravimeter that could make accurate measurements under unstable conditions, including at sea, and he had demonstrated its efficacy in a series of landmark submarine expeditions to the East Indies in the 1920s.[29] In the 1930s, Hess collaborated with Vening Meinesz to extend this work to the West Indies and use it to think about global tectonics processes.

Terrestrial gravity measurements had revealed that the Earth's continental crust rested in hydrostatic equilibrium on a denser substrate—a condition known as *isostatic equilibrium* or *isostasy*. (This was reflected in the fact that terrestrial gravity is in most places more or less the same, despite significant differences in elevation and the density of local rocks.) However, Vening

Meinesz had discovered that most of the oceanic crust was out of equilibrium. In particular, there was a large, conspicuous belt of negative gravity anomalies—that is to say, a zone in which gravity was lower than normal (−100 milligals to −200 milligals)—parallel and adjacent to the Java Trench on the southern side of the Indonesian Archipelago and curving around the eastern side of Malaysia and the Philippines.[30] It was nearly five thousand miles long but rarely more than sixty miles wide, bordered on either side by smaller positive anomalies.[31] Earthquake epicenters were concentrated along its axis, which lay about one hundred miles inland from a parallel belt of active volcanism; the negative anomaly belt always fell on the seaward side of the region's active volcanoes (fig. 4.1). Clearly, there was a connection between ocean trenches, volcanism, and disturbances of the gravity field, but what, exactly, was that connection?

Supporters of Alfred Wegener's theory of continental drift had an answer: two pieces of crust were converging, with one sliding beneath the other. Vening Meinesz's measurements were consistent with the idea: the negative anomalies could be the result of low-density crustal material being pushed down and displacing the denser substrate below. If true, it would mean a profound rethinking of prevailing geological and geophysical theory. One way to test the idea would be to see if similar relations held in other places.

In 1928, Vening Meinesz was invited under the auspices of the Carnegie Institution of Washington and the US Coast and Geodetic Survey to measure gravity in the Caribbean aboard the USS *S-21* submarine. The expedition was publicized in the *New York Times* and *Scientific American* and widely discussed. The scientists had found a remarkable parallel between East and West Indies: in both regions a belt of negative anomalies in an area of unusually deep water was associated with earthquakes and active volcanism.[32]

The idea for Vening Meinesz's American visit had come from William Bowie (1872–1940), chief of the geodesy division of the US Coast and Geodetic Survey and a leader in the fledgling American Geophysical Union (AGU). At the Survey, Bowie had been involved in terrestrial gravity measurements in aid of geodetic control, showing that gravity measurements could be used to correct for effects that distorted geodetic measurements. But in this era, before offshore oil exploration, there was as yet no recognized practical use for gravity measurements at sea. If marine gravity investigations were to continue, they needed to be in academic hands. Bowie proposed that the work be continued by the Carnegie Institution, using the ship *Carnegie*, but in 1929 the ship burned in a tragic fire. Bowie turned to Princeton professor Richard Field (1885–1961).

Field was chairman of the AGU's Committee on the Geophysical and Geo-

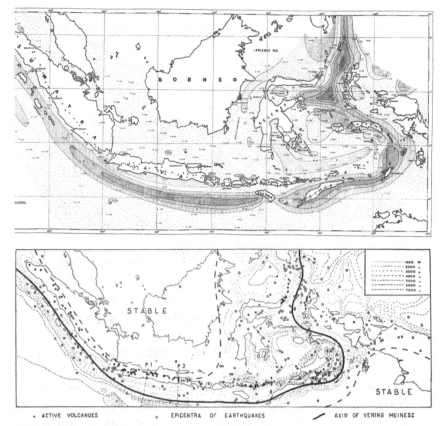

. ACTIVE VOLCANOES   ○ EPICENTRA OF EARTHQUAKES   ⟋ AXIS OF VENING MEINESZ

Figure 4.1 Map from Vening-Meinesz's East Indian work, showing gravity anomalies, trench, earthquake epicenters, and volcanism. (a) Volcanism and negative gravity anomalies; (b) earthquake epicenters.

From F. A. Vening-Meinesz, *Gravity Expeditions at Sea 1923–1930*, vol. 1 (Delft, Netherlands: Geodetic Commission), 1932.

logical Study of the Ocean Basins, and in 1938 he became the AGU's first president. (Previously AGU was a committee of the National Research Council, so its leaders were chairmen, not presidents.) He was also a sedimentary petrologist who since 1927 had been studying reef formation in the Bahamas, where he had used gravity measurements to help interpret subsurface reef structure. Since the early 1930s, he had been active on behalf of the AGU in fostering cooperation between earth scientists and the US Navy Hydrographic Office. With moral support from Bowie and logistical support from the Hydrographic Office, Field invited Vening Meinesz to return to the United States for further gravity work in 1932. The *S-21* expedition had been intended simply to demonstrate the feasibility of gravity measurement at sea; the second expedition would be more scientifically ambitious. Field offered the assistance of a bright young man who had just finished his PhD at Princeton working on ultramafic rocks: Harry Hess.[33]

The Navy-Princeton Gravity Expedition

On the *S-21* expedition, Vening Meinesz had found a strip of negative anomalies north of Puerto Rico and Hispaniola; Vening Meinesz and Hess found that it did not continue westward but stopped abruptly near the eastern edge of Cuba. Over the Bartlett Trough, a region of deep water running west-southwest from the eastern end of Cuba toward the Yucatán, they found a complex pattern of positive and negative anomalies. Over the Bahamas, they found generally negative anomalies. Hess was eager to give these data geological interpretations, but the most important scientific result was not so much the additional data, but the ideas that emerged from the interaction between these two men. Vening Meinesz was struck by the extreme parallel between the East and West Indies results; Hess was struck by the extreme intelligence of Vening Meinesz. Together, they began to develop and promote a geological interpretation of the geophysical data.

Negative gravity anomalies meant concentrations of low-density materials; this could be achieved by a thickening of the low-density crust. Vening Meinesz and Hess proposed a mechanism: a large downfold, generated by surface compression, that displaced the surrounding high-density substrate. Such a downfold would produce a belt of regionally persistent negative anomalies—it would also mean great tectonic forces at play: "A great downward protuberance is developing, which causes a great mass-defect corresponding to the strong negative anomalies." Vening Meinesz wrote: "Knowing that the root must have a cross-section of more than a thousand square kilometers [to explain the measured anomalies], we may realize the magnitude of the phenomenon."[34]

Vening Meinesz initially called this the buckling hypothesis but soon adopted a term suggested by his Dutch colleague, Philip Kuenen (who reportedly got it from the German geologist Eric Haarmann) and called the feature a *tectogene*.[35] Hess embraced the tectogene concept. He read the gravity data as clear evidence of the relation of gravity anomalies to tectonic activity, arguing that the anomalies were caused by the tectogene, which in turn was caused by crustal compression, but what was causing the compression? An obvious answer was continental drift.[36] But American geologists had for the most part rejected that theory, so Hess proceeded cautiously, speaking at this point only of indications, implications, and intimations: "The ... negative strip has important geological implications. It supplies definite indications of a major orogenic belt; and intimates the way in which the earth's crust yields to orogenic forces."[37] To that point, geologists had tried to understand mountains by mapping their visible structures, after they had formed, but geophysical data now offered a means to think about what was happening at depth *while* the mountains might be actually forming.

Vening Meinesz and Hess had found virtually identical patterns on two sides of the globe, and this—along with the scale of the Indonesian anomalies—led Hess to conjecture that they had detected a global process: "This strip was continuous for the length surveyed: 5,000 miles. *It implies that deformation of the earth's crust is taking place in a single and continuous narrow belt....* The terminations of the negative strip were not reached in the region surveyed. Do these strips gradually die out laterally? Do they terminate against faults or shear zones? Or are they perhaps continuous around the world?"[38] In the 1960s, this insight—that tectonism is localized in narrow but continuous zones—would become a linchpin of plate tectonics; these zones came to be understood as plate boundaries. But in the 1930s it was tricky for the young Hess to promote this radical idea. Vening Meinesz was more professionally established, but he was temperamentally more cautious than his younger colleague. Still, he agreed that the observed patterns reflected the local effect of regional stress and that processes involved the movement of large crustal blocks, perhaps even entire continents. He wrote to Hess in March 1933: "The gravity profile ... west of Sumatra [may be explained] in the fact that the direction of stress, exerted by Asia ... is here nearly parallel to the [gravity] strip, [which] causes Sumatra to override the ocean floor a bit, pressing it down."[39]

The scientists had also observed systematic positive gravity anomalies over the Gulf of Mexico, which Vening Meinesz suggested were also related to regional crustal compression—in this case, an increase in density: "I don't see another explanation for them [except] a change of state because of the great lateral compression of the crustal material. This is of course very hypothetic[al]."[40] What did he mean by "great lateral compression of crustal material"? Did he mean continental drift? In published work, Vening Meinesz rhetorically declined to "go into the difficult problem of the cause of these great tectonic phenomena," but then he proceeded to do just that.[41] He was "tempted to look for the cause, or at least one of the causes, of the relative movements in convection currents in the substratum." This was the idea that the substrate between the crust was fluid: not necessarily liquid, but perhaps partially molten or at least sufficiently hot and soft so as to behave in a fluid manner and thus be susceptible to convective heat transfer—the mechanism promoted by British geologist Arthur Holmes (1890–1965) to drive continental drift. But Vening Meinesz pulled back from that temptation, concluding that the "whole question is yet too speculative to allow any definite statement."[42]

In the 1960s, earth scientists would accept mantle convection as the cause of crustal movement, but Vening Meinesz was thinking that tectogenes caused convection rather than the other way around. He assumed that the tectogene brought radioactive crustal material into the less radioactive substrate, which would heat it, triggering convection.[43] However, by 1935, his thinking had changed. In 1934–1935, he completed a monumental expedition supported by the Royal Netherlands Navy—from Holland to Java via Cape Town and West Australia—establishing 226 gravity stations at sea and 20 more on land.[44] He now saw that the positive anomalies measured over the Gulf of Mexico were a general feature of the open oceans; most of the seafloor was characterized by weak positive anomalies.

The quantity and quality of these data seem to have changed his views on the cause-effect relations and strengthened his convictions overall. In a letter to Hess, Vening Meinesz suggested that convection was not just possible but required and might be the cause rather than the effect of the tectogene. Explicitly invoking Arthur Holmes, he implicitly invoked continental drift: "I cannot see another reasonable explanation than by admitting convection-currents in the substratum below the crust, rising below the continents and sinking below the oceans (they may perhaps be caused by temperature differences because of differences of radio-activity, as e.g. Arthur Holmes assumes)."[45]

His reasoning was simple. The positive anomalies indicated denser materials at depth. The simplest explanation for a density contrast was thermal: the denser materials were colder. If they were, then they could be the sinking limbs of a convection current. Indeed, if convection occurred, they would *have* to be the sinking limbs. Because Holmes had invoked convection currents as the mechanism of continental drift, the implication was clear: gravity anomalies reflect tectonic instabilities driven by convection currents, which could drive large-scale crustal motion. In his widely read 1929 paper "Radioactivity and Earth Movements," Holmes had cited the Banda Sea as an area where "subcontinental currents approach and throw the [oceanic] crust into compression."[46] Vening Meinesz now had data consistent with that suggestion.

Tectogenes, Convection Currents, and Continental Drift

In 1934, Vening Meinesz published his magnum opus on the Dutch gravity work, coauthored by J. H. F. Umbgrove (1899–1954), professor of Geology at Delft, and Philip H. Kuenen (1902–1976), geologist on the Dutch Snellius expedition of 1929–1930 and later famous for his work on submarine currents (chapter 5).[47] The Dutch colonial presence in Indonesia had led to considerable detailed geological work on land, and the fifteen-month Snellius expedition had yielded the most detailed regional bathymetric survey completed to date: thirty thousand echo soundings and three thousand wire sounds, accurate to fifty meters, covering an area of sixty thousand square kilometers of the Indonesian Archipelago.[48]

Umbgrove explicitly linked the Indonesian data to continental drift. He summarized the work of B. G. Escher (1885–1967), who had applied Holmes's concept to the Indonesian province using Vening Meinesz's gravity data as a guide. The picture he developed showed large convective cells sinking under the Java trench and generating volcanism in the island arc behind (fig. 4.2).[49] A related perspective was offered by geologist G. A. F. Molengraaff—a well-known proponent of continental drift—who argued on geological evidence that the deep basins and island arcs had formed together as a consequence of regional crustal movement. The surface features—uplifted blocks and down-faulted basins—were the expression of ductile deformation at depth.[50] Umbgrove noted that large-scale underthrusts might be caused by "a North-Western movement of the Australian continent as was supposed by Wegener," stressing that his theory had "been acclaimed by numerous explorers of the East Indian archipelago."[51]

Kuenen, however, was skeptical of Wegener's theory. In Wegener's original version of continental drift, the oceanic crust yielded in the face of drifting continents, implying that it was plastic. In a separate chapter on the relationship between bathymetry and tectonics, Kuenen argued that the bathymetry showed that the oceanic crust had substantial topography, so it must be rigid.[52] Gravity anomalies and earthquake epicenters within the oceanic crust also "point[ed] to a [rigid] crust which is capable to transmit stress [*sic*]."[53] Yet at the same time, Kuenen embraced the idea of convection currents. In hindsight this is tricky to understand: in the 1970s, convection currents would be interpreted as the cause of crustal motions—and leading geologists had already made that argument in the 1920s.[54] But in the 1930s, many scientists viewed convection currents and crustal motion as separate problems; Kuenen was one of them. He believed on geological grounds that the deep troughs

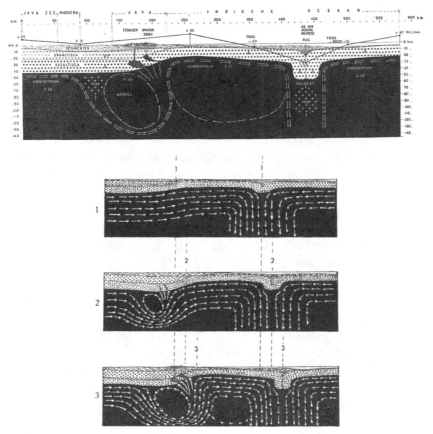

Figure 4.2 Escher's model of convection currents in the Indonesian archipelago.
(a) Overall model of convection related to vulcanism and ocean trenches in Java; (b) detail: "three stages in the evolution of the section through Java and the belt of negative anomalies of gravity."

From B. G. Escher, "On the Relation between the Volcanic Activity in the Netherlands East Indies and the Belt of Negative Gravity Anomalies Discovered by Vening Meinesz," *Proceedings of the Koninklijke Nederlandse akademie van wetenschappen* 36 (1933): 677–85. Reprinted with permission from the Royal Netherlands Academy of Arts and Sciences.

and negative anomalies were associated with compressive folding, and he accepted Vening Meinesz's arguments that convection currents in the substratum could explain them. But this did not, in his view, necessitate continental drift.

In Kuenen's view, two facts of gravity surveys needed explanation: the narrow belts of strongly negative anomalies and the weaker but regionally widespread fields of positive anomalies over the open ocean. He believed the former could be accommodated by the tectogene hypothesis and the latter by a descending convection current.[55] Because the oceanic crust is less radioactive than the continental crust, the substratum below the ocean must also be colder, he argued, on the basis of simple heat conduction. The difference in thermal regimes would set up convective heat transfer from the hot subcontinental regions toward the colder oceanic regions: "If the constitution of the substratum is homogeneous, this temperature difference must cause rising currents below the continents and descending currents below the oceans and so this distribution is in good harmony with the occurrence of fields of positive anomalies in the oceans. This corroboration is a notable support of the convection hypothesis."[56] In short, there would be global-scale convection beneath the crust, driven by the contrast in radiogenic heat between oceans and continents—along the lines that Arthur Holmes had advocated—and this would exert a drag on the crust and produce crustal compression. But while Holmes had linked convection to continental drift, Kuenen did not: "The convection-current ... must exert a viscous drag on the crust and so we have to expect lateral compression in the crust." But he did not assume that this would cause continental-scale crustal displacement.[57] There were now three options on the table: continental drift without convection, continental drift with convection, and convection without continental drift.

The *Barracuda* Expedition, Maurice Ewing, and the Atlantic Ocean

The year 1936 saw the era's third and last Navy-sponsored gravity expedition to the Caribbean region: the Navy–American Geophysical Union Expedition, from November 1936 to January 1937, on the USS *Barracuda*. Once again, the Navy provided the boat, logistical support, and radio time signals for accurate locating of the geophysical stations. The Hydrographic Office handled the organization, the American Philosophical Society provided the funds, and female computers at the Coast and Geodetic Survey computed the isostatic reductions of the gravity data.[58] The geographic focus would be the Lesser Antilles, following a clockwise circuit encompassing the eastern Caribbean southward to Trinidad and Tobago and then west toward Curaçao along the coast of Venezuela.

Hess would serve as lead scientist and supervise the soundings, joined by Maurice Ewing, who would take the gravity measurements. Ewing at the time was assistant professor at Lehigh University, just beginning what would become his pioneering work using seismic techniques developed in the Texas oil fields to study the ocean floor. The expedition thus brought together two of the brightest and most ambitious earth scientists of their generation, but their backgrounds and approaches could scarcely have been more different. Hess was a geologist, schooled in the traditional subjects of mineralogy, petrology, and regional geology but broadening his compass to include marine geophysics and pressing hard on the theoretical implications of his empirical observations. Ewing was a physicist, applying instrumental techniques developed by physicists and engineers to the domain of earth science and pushing the limits of data acquisition, but not particularly knowledgeable about the theoretical arguments taking place in geology at that time.

In 1935, together with polar geophysicist A. P. Crary (1911–1987)—the first man to step on both north and south poles—and geologist H. M. "Mac" Rutherford, Ewing had completed the first comprehensive geophysical study of the structure of a continental margin.[59] With funding from the Geological Society of America, moral support from the American Geophysical Union, and logistical help from the US Coast and Geodetic Survey, they had demonstrated that seismic techniques developed for shallow oil exploration could be used to study the crust in waters up to one hundred fathoms deep. The seismic data had proved capable of distinguishing formations and identifying specific horizons within the sedimentary sequence, and it provided the first quantitative measurement of depth to crystalline basement along the US continental margin—approximately twelve thousand feet at the edge of the continental shelf. It also showed that the depth to basement increased gradually, rather than abruptly or stepwise, suggesting warping and subsidence rather than faulting.[60] Perhaps most important from a tectonic perspective, there was no sign of a sunken Paleozoic continent, which had been invoked to account for thick Paleozoic sequences in the Appalachians that displayed paleocurrent indicators from the east.[61]

Negative results are always ambiguous: either the object or effect does not exist or it does exist but has not yet been found. Ewing and his geological colleagues took the latter view, supposing that somewhere in the Atlantic Ocean there were extensive but as yet unidentified Paleozoic strata.[62] But they could just have legitimately read the data as refuting the theory of sunken continents, because the predicted sediments did not appear to be there.[63] Given that continental drift had been proposed as an alternative to sunken continents, they could also have read the data as confirming continental drift. But in science, social and epistemic convention is to continue to believe what you have always believed until you are compelled to adjust—and Ewing and his colleagues did not feel so compelled, assuming, instead, that the sunken continent was out there somewhere.[64]

In later years, Cambridge marine geophysicist Maurice Hill (1919–1966), son of Nobel laureate A. V. Hill, would write that two outstanding events before World War II formed the basis for the entire postwar development of geophysical exploration of the seafloor: "The development by Professor Vening Meinesz of the submarine pendulum apparatus and the application by Professor Maurice Ewing of the seismic technique to explore marine problems."[65] In Hill's retrospective, the two developments had the common virtue of demonstrating the value and feasibility of geophysical work at sea. But at the time, they were viewed as pointing in divergent theoretical directions: Vening Meinesz's work toward large-scale crustal mobility, Ewing's against it. Hill's Cambridge colleague Sir Edward "Teddy" Bullard (1907–1980) later downplayed the theoretical aspect (some might say theoretical failure) of Ewing's early work, saying that his great achievement was to "show that you could do things at sea that no one had done before."[66]

In the 1960s, Bullard would become an advocate of plate tectonics, but in the 1930s he accepted the interpretation of Ewing's results as unfavorable to any theory of large-scale crustal mobility. In any case, more work was needed. In 1937, Bullard applied to the Royal Society for funds to travel to America to learn Ewing's techniques and apply them on the eastern side of the Atlantic as Ewing had done on the western. Bullard explicitly linked the seismic data to continental drift, but in a negative way. Ewing's work, he suggested, "indicates that the edge of the continental shelf is not a great fault separating the continents from the ocean but is merely the edge of a rubbish dump of unconsolidated sediments. If this is so no extreme horizontal motions of the continents can have occurred."[67]

Not everyone agreed with Bullard's interpretation of the basement data, however, or that unconsolidated sediments were rubbish. Reviewing the proposal, Carnegie Institution of Washington geologist W. B. Wright (1876–1939), one of Wegener's most respected defenders in the United States, challenged the contention that Ewing's data disproved Wegener's hypothesis:

> I think the conclusions to be drawn … are over-stated. Such results would not disprove the horizontal motions postulated by Wegener, but merely show that the rupture, if any, took place along a geosyncline, presumably parallel to the Appalachians, which is where one would expect it to lie. The measurements will not decide whether the oceanic slope beyond the 100 F[atho]m line is of tectonic origin or merely the edge of sedimentation.… I foresee that a claim may some day be made that the Wegener hypothesis has been put to the test of experiment and disproved and wish to state now that I cannot see that the results will be any criterion for or against.[68]

Nevertheless, Wright supported the proposal on empiricist grounds: "They will be none the less interesting.... The measurements are well worth making and will be extraordinarily useful to geologists."[69]

Later in his career, Bullard would be a strong advocate of sticking to one's theoretical guns, but at this early stage in his scientific life—he was thirty—and in a scientific community where empiricist tempers reigned, he deemed it politic to retreat.[70] He responded by disavowing theoretical ambition: "We are not seeking to prove or disprove Wegener, and whether we do so this year or not is a matter of no great importance, but if everything goes well we shall get some facts about the nature of the shelf which I think should be of interest."[71]

The revised proposal was approved and Bullard traveled to the United States in the summer of 1937, where he met both Ewing and Hess. He certainly would have discussed the *Barracuda* results—which Hess and Ewing presented that year at the annual meeting of the AGU and again at a symposium at the American Philosophical Society in Philadelphia in 1938. But he probably would not have garnered a coherent picture, for Hess and Ewing took very different stances toward their data.[72]

Ewing presented the results of fifty-one sea stations and nine harbor stations in a map illustrating the distribution of gravity values. The results confirmed and extended Vening Meinesz's and Hess's earlier work: the narrow belt of intense negative anomalies continued eastward along the north side of Puerto Rico and then curved in a great arc down through Trinidad and Tobago and along the north coast of Venezuela. The anomalies always occurred above or immediately adjacent to the regions of greatest depth: 3,000–4,000 fathoms. As in Indonesia, the Caribbean belt was typically less than 100 miles wide but extended over a great length: in this case more than 1,200 miles. In Ewing's words, "This negative strip is comparable with that found by the epoch-making survey of Vening Meinesz in the East Indies." But he declined to press an interpretation.[73] He expanded on the history of gravity measurement, the instruments and techniques, and the problems of measurement and data reduction, but he offered little in the way of interpretation, concluding weakly: "The large deviations from isostatic equilibrium which exist in the Meinesz belts may be considered as very exceptional features to be explained by stresses in the crust."[74]

Hess's approach was different. He concentrated on linking the new geophysical data to the known geology of the Caribbean and looking for its tectonic significance. In his 1937 AGU presentation, he emphasized the coincidence of maximum negative anomalies with maximum ocean depths and their link to volcanism. The islands interior to the negative belt—Trinidad,

Tobago, Barbados, Martinique, Antigua, St. Kitts, and St. Thomas—formed a zone of active volcanism and recent tectonism: they all contained either volcanic rocks or evidence of deformation of recent sedimentary rocks. Several also contained altered peridotites. Hess linked this to the tectogene hypothesis, suggesting that the peridotites were intrusions that had penetrated the upper crust along fractures generated by the crustal downbuckle: "Serpentinized peridotite intrusions occur all along the great negative strip of the East Indies, and similarly are present in the West Indies.... Where the crust is buckled down into the peridotite substratum, the conditions for the formation of this magma and its migration to the surface through the vertical limbs of the down-buckle are fulfilled."[75] Hess had a theory: in areas of gravity anomalies the crust is compressed, a downbuckle forms, and peridotites are generated and intruded along its limbs.

Hess elaborated these ideas in the paper he presented in Philadelphia in November.[76] His involvement in two major gravity expeditions, his interactions and correspondence with Vening Meinesz, and his personality all seem to have given him the confidence to interpret his findings boldly. He began the Philadelphia paper declaring: "Meinesz's discovery of huge negative anomalies in the vicinity of island arcs is probably the most important contribution to knowledge of the nature of mountain building made in this century.... When the significance of the huge anomalies and the structures they indicate become appreciated, a radical change in some of our ideas on mechanics and processes of mountain building becomes necessary."[77]

Hess stressed that the consistency of the gravity results precluded their arising from instrumental error, observational inaccuracies, or incorrect assumptions about isostatic compensation. "There can be no doubt that the major portion of the anomalies present in the negative strip is due to an abnormal mass distribution in the Earth's crust below the strip ... Meinesz's explanation ... is that the Earth's crust buckles downward in a huge vertical isoclinal fold, and thus the light material of the upper crust might extend downward to a depth of 40 to 60 km."[78] Hess also argued that his interpretation was both unifying and predictive. The tectogene idea resulted in "the coordination of many geologic facts previously merely a collection of observations with no apparent relationship to one another." It "join[ed] together in a single structural entity observations which formerly fit into no definite pattern," and it predicted "relationships that are to be expected (and have since in a number of cases been found)." For all these reasons, Hess was "completely convinced of the soundness of the theory."[79]

But was a giant downbuckle mechanically possible? To answer this, Hess looked to Kuenen, who had agreed that the "combination of the coincidence of both types of profiles and of the geological and seismological evidence leave no doubt about the identity of the phenomenon in the East Indies and the West Indies. So north of Porto Rico [*sic*] we are also led to suppose a downward buckling of the Earth's crust."[80] But was it mechanically plausible that the crust would sustain an enormous downfold, penetrating tens of kilometers into the substrate rather than fracturing in large blocks or crumpling into a sequence of small, superficial folds? Kuenen hoped to answer this question by means of scale models.

Kuenen's Scale Models of the Crust

In a 1936 paper titled "The Negative Isostatic Anomalies in the East Indies (with Experiments)," Kuenen presented the results of an experiment in which he simulated the effects of regional compression to see if they could generate a tectogene.[81] Various workers had built scale models in attempts to re-create fold structures observed at the surface in mountain belts, but Kuenen's work was different: he attempted to re-create a hypothetical feature at depth that could account for measured effects at the surface.[82] He also attempted to scale his model realistically in terms of its aspect ratios and the strength of its component materials. As Hess described it: "The experiments performed by Kuenen are of particular validity because two factors have been taken into consideration which are commonly neglected in experimentation with geologic models. The first is that the strength of materials used has been chosen so as to be of the right order of magnitude for the scale of the model; and second, the crust of the model has been floated on liquid paraffin to supply the zone of no strength beneath the crust."[83]

Kuenen began with a glass aquarium and filled it with various combinations of layers of paraffin, petroleum jelly, mineral oil, and warm water, subject to horizontal compression by means of a plunger at one end. In a series of experiments, which he documented by photographs, he showed that one could generate both underthrusts similar to those observed in mountain belts and downbuckles of the sort postulated by Vening Meinesz. Indeed, his downbuckles were strikingly like the sketches of tectogenes that Hess had drawn before the experiments were performed. Moreover, by placing a weak layer on top, Kuenen could generate small folds on top of a much larger downbuckle, just as one saw small-scale folds on the Earth's surface in mountain belts (fig. 4.3).

Kuenen never claimed that his experiments proved the existence of tectogenes. Rather, he viewed the experiments in heuristic terms, arguing that they were of greatest value "for checking ... theoretical deductions and guid-

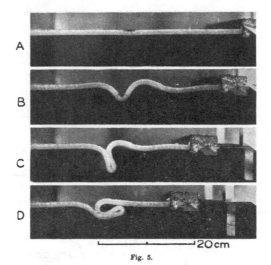

Fig. 5.

Experiment Ic. The model crust in stage A is weighted with a strip of dark paraffin. A geosyncline forms in stage B. In stage C it is closed. In stage D it has grown and floated up, proving the formation of a negative anomaly by the compressive stress. Note the depth of the topographic depression, that is several times the thickness of the crust.

Figure 4.3 Kuenen's experimental model of a how a crustal downbuckle could produce a negative gravity anomaly.

From Ph. H. Kuenen, "The Negative Isostatic Anomalies in the East Indies (with Experiments)," *Leidse geologische mededelingen* 8 (1936): 169–214. Reprinted in accordance with Naturalis Biodiversity Center open-access policies.

ing further research."[84] Hess took a more assertive view. He saw the experiments as a "test of Meinesz's theory" to which they offered "strong support."[85] They could also be used to make sense of prevailing geosyncline theory: the geosynclines observed in nature were the surficial expression of subsurface tectogenes; the geosynclines were sedimentary basins formed at the hinge of the tectogene. As the tectogene buckled downward, the surface was also pulled downward, and so sediments accumulated in what geologists recognized as geosynclines.

In 1938, Hess synthesized this work in a widely cited paper whose title reflected his intellectual starting point but scarcely indicated its scope: "A Primary Peridotite Magma." Ultramafic rocks were generated, he argued, by partial melting of mantle peridotite in zones of crustal buckling, as revealed by gravity anomalies. Crustal compression caused the downbuckle, generating heat and setting up stress fields that resulted in fractures along which the produced melts could migrate (fig. 4.4). Given the striking parallel between East and West Indies—the two places where this problem had been studied in detail—Hess saw no problem in generalizing to the rest of the world:

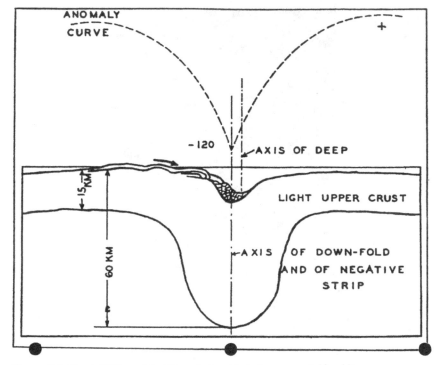

Figure 4.4 Hess's 1938 model of the relationship between crustal buckling, ocean trenches, and negative gravity anomalies.

From H. Hess, "A Primary Peridotite Magma," *American Journal of Science* 35 (1938): 322–44 (fig. 1).

Image © *American Journal of Science*. Reprinted in accordance with stated fair-use policies, available at http://www.ajsonline.org/site/misc/terms.xhtml.

"Apparently all mountain systems have had a tectonic history similar structurally to the present island arcs."[86]

But how were the compressive forces that created the tectogenes themselves created and sustained? Most people, including Kuenen, assumed that the crust was too weak to sustain compressive stresses over geological time. A tectogene, if it could form, would be rapidly heated, melted, and dissipated. Hess needed a driving force that could maintain and sustain regional compression over geological periods. Kuenen's models suggested that tectogenes could form, but they did not demonstrate how they did form. A tectogene was a mechanically possible response to regional compression, but what was the source of compression? Like earlier experimenters with scale models, Kuenen simulated horizontal compression with a plunger at the side of his apparatus without identifying the plunger's real-life referent.[87] Hess believed the solution could been found in the work of David Griggs.

Griggs's Convection Current Model of Continental Drift

In 1938, David Griggs (1911–1974) was a newly appointed professor of geophysics at UCLA. His work focused on the behavior of rocks under the conditions of high temperature and pressure presumed to prevail in the Earth's mantle. As a member of the Society of Fellows at Harvard, he had captured his colleagues' attention by combining mathematical analysis, high-pressure experiments, and bench-scale modeling.

Perhaps influenced by Harvard professor Reginald Daly (1871–1957)—one of North America's few outspoken proponents of continental drift—Griggs took continental drift seriously. Like other American geophysicists, he accepted Harold Jeffreys's criticism that tidal, Coriolis, and *Polflucht* forces were inadequate to drive continental drift. But unlike Jeffreys, Griggs took seriously the quantitative analyses of subcrustal convection provided by Vening Meinesz, as well as by European geophysicists A. L. Hales (1911–2006) and Chaim Pekeris (1908–1993). Griggs believed, as they did, that mathematical analysis could determine whether mantle convection was possible and could be a driving force of crustal motion.

Vening Meinesz and Pekeris had calculated that convection in the Earth's mantle could generate compressive forces of 3,000–5,000 kg/cm^2—more than enough to exceed the laboratory-measured strength of granitic rocks of 2,000–3,000 kg/cm^2—and therefore to deform crustal rocks. It had been shown experimentally that rock strength would increase with increasing confining pressure, but it was also assumed that it would decrease with increasing temperature, so these values were not implausible for deforming near-surface crustal rocks. Hales had calculated that a temperature gradient of only 0.1°/km would be sufficient to set up convection. This could be achieved either by a radioactive heat source at depth or simply if the core were more conductive that the mantle or crust. Assuming convection down to the core boundary at 2,900 km and movement by viscous flow, Pekeris calculated a rate of convective overturn of 5 cm/yr. Assuming convection only down to 1,200 km, Vening Meinesz calculated a rate of 1 cm/yr. Based on his own experimental work demonstrating pseudoviscous flow in limestone—presumably effected by recrystallization, or dissolution and precipitation reactions—Griggs thought the rates of motion would be higher, perhaps 10 cm/yr. Together these studies defined a range: 1–10 cm/yr.[88]

Griggs's analysis was focused on these quantitative constraints and the possibility of episodic convection. What caught Hess's attention was a minor component of Griggs's argument: a tabletop experiment, barely a few inches long. Griggs had read Kuenen and been impressed by his attempt to achieve "dimensional correctness," especially as compared with earlier workers whose choices of materials were at best arbitrary and at worst misleading.[89] Kuenen's model relied on compression from the side; Griggs wanted to test the idea of crustal compression generated by viscous drag from below. Nor did Kuenen's model take into account the time dimension; it was not dynamically similar. Griggs wanted to simulate the effect of convection currents in the mantle on the crust above in a manner that would account for the time-dependent nature of viscous behavior.

Griggs took his lead from the great geologist M. K. Hubbert (1903–1989), who in 1937 had laid out a mathematical approach to model scaling, and from Harvard physicist Percy Bridgman (1882–1961), who had developed the concept of dimensionless products. Defining the model ratios as the ratio of the magnitude of the property in the model to the magnitude of the property in the Earth, Griggs realized that one could create a dimensionless product to determine the scaling factor for time—which in geology was the vexing factor.

Griggs calculated that the time factor in Kuenen's experiments was seventeen orders of magnitude less than in the Earth. This meant that a process requiring ten million years in the Earth would have to occur in the model in one-three-hundredth of a second. A process taking one million years would have to occur in one-three-thousandth of a second. No one could make a model operate that quickly, so the alternative was to scale a different parameter. Length and density could be adjusted only to a limited degree—certainly not by seventeen orders of magnitude. This left only one option: viscosity, which could be varied as required. Griggs explained, "This analysis shows us that by correctly choosing the value of the viscosity we may make a dynamically similar model to operate at any speed we desire, limited only by the viscosity of the materials available."[90] If, for example, one minute in the model were to represent one million years on Earth ($\tau = 1.9 \times 10^{-12}$), and δ is the density ratio between model and Earth, λ the length radio, and τ the time ratio, solving the following equation gives the required viscosity ratio, ψ:

$$\psi = \delta\lambda\tau = 0.35 \times 2 \times 10^{-7} \times 1.9 \times 10^{-12} = 1.3 \times 10^{-19}.$$

In other words: the materials in the model substrate had to have a viscosity nineteen orders of magnitude less than the expected viscosity of the mantle! This result determined Griggs's choice: glycerin. Previous researchers had used sand, mud, plaster, wax, and various other solids or semisolids; Griggs now realized that a realistic mantle model would be liquid, even if the real mantle were not.[91] To build a realistic model, one had to use what seemed to be unrealistic materials.

With this insight, Griggs performed two experiments. In the first, he simulated compression from the side, akin to Kuenen's work and most other earlier experiments, but he got a crucially different result, which was no tectogene, just thickening of the crust immediately in front of the advancing plunger: "The viscous drag of the substratum seems sufficient to prevent the transmission of compressive stresses for long distances through the overriding crust, and causes local thickening of the crust instead."[92] In the second experiment, he simulated the effect of convection currents below. However, it was impractical to achieve a temperature difference comparable to the expected difference between the core and the crust of the Earth—thousands of degrees—so he simulated the effect of convection currents using rotating drums (fig. 4.5).[93] In one simulation, the drums were placed in a small glass boxed filled with glycerin; the crust was simulated by a layer of cylinder oil mixed with fine sawdust. In a another simulation, Griggs used viscous water glass (an aqueous solution of sodium or potassium silicate) and oil mixed with sand, adjusting the proportions to achieve the viscosity required for the size of the model. The drums were rotated by small handheld pulleys.[94]

Griggs's convection model yielded a very different result from previous experiments: underthrusts were developed in which the upper portions of the "crust" passively slid over the lower portions, which thickened and deformed. Griggs correlated these thrusts with structures seen in mountains, such as Alpine nappes. Moreover, if one drum was kept stationary and the other made to rotate fast enough, the oil layer above the rotating drum began to thin and then finally broke, and was "transported into the thickened crust [above the stationary drum]. Finally, the current sweeps all the crustal cover off and piles it up in a peripheral downfold."[95] This, Griggs concluded, was the most important result, for it "opens the attractive possibility of a convection cell covering the whole Pacific basin, comprising sinking peripheral currents localizing the circum-Pacific mountains and rising currents in the center."[96] This was hard to explain in text, so the professor from Los Angeles did something clever: he made a movie, which he presented at the 1938 annual meeting of the Geological Society of America.[97] Harry Hess chaired the session.

In the written version of the paper, published in the *American Journal of Science* the following year, Griggs acknowledged his debt to Hess and Holmes, and especially to Vening Meinesz's "discovery of the great bands of gravity deficiency in the East and West Indies."[98] He saw his work as unifying gravity anomalies, crustal deformation, convection currents, and continental drift.[99] So did Hess. Soon after the meeting, Hess wrote:

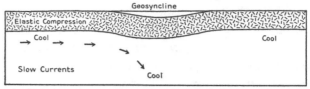

1. First stage in convection cycle — Period of slowly accelerating currents.

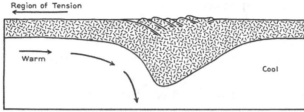

2. Period of fastest currents — Folding of geosynclinal region and formation of the mountain root.

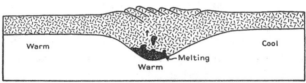

3. End of convection current cycle — Period of emergence. Buoyant rise of thickened crust aided by melting of mountain root.

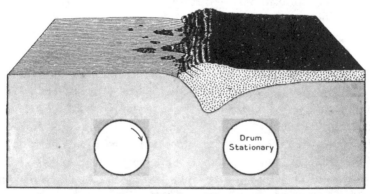

Figure 4.5 David Griggs's 1939 model of mantle convection currents driving crustal motion, illustrating the "hypothetical correlation between phases of the convection-current cycle and phases of the mountain-building cycle." (a) His theoretical picture; (b) stereogram of his dynamically similar physical model, with only one drum rotating, showing development of peripheral tectogene.

From David Griggs, "A Theory of Mountain Building," *American Journal of Science* 237, no. 9 (1939): 611–50 (figs. 15 and 16). Images © *American Journal of Science*. Reprinted in accordance with stated fair-use policies, available at http://www.ajsonline.org/site/misc/terms.xhtml.

Recently an important new concept concerning the origin of the negative [gravity] strip in island arcs has been reported by David Griggs. It is based on model-experiments.... By means of horizontal rotating cylinders, convection currents were set-up in a fluid layer beneath the "crust," and a convection cell was formed. A down-buckle in the crust, similar to that produced in Kuenen's experiments, was developed where two opposing currents meet and plunge downward.... If the buckle were formed entirely by tangential compression in the upper crust, it would be so weak and unstable that it could last for only a very short period of time geologically. [But] by means of convection currents the difficulty of maintaining the [downbuckle] can be removed.[100]

The Revolution That Wasn't

It was the autumn of 1938 and Hess and Griggs had put together the pieces of a global tectonic theory that bore strong resemblance to the arguments made by Alfred Wegener in the 1910s and Arthur Holmes in the 1920s. Their theory, though, had greater emphasis on geophysical evidence of deep crustal structure and offered a quantitatively defensible model for convection currents as the driving force. Its key elements were a layered earth structure with a rigid crust, and convection in the underlying plastic or fluid mantle. The theory accounted for crustal thickening, regional gravity anomalies, deformation and volcanism in island arcs, and—apparently—large-scale crustal motions, which is to say, just about everything that a global tectonic model was expected to account for at the time. It was supported by data on a range of scales, from global geophysics to bench-scale physical modeling; it was consistent with the empirical evidence cited in the 1920s for continental drift; and it was consistent with arguments that had been "in the air" (and the literature) for two decades. Moreover, the values they achieved for the rate of crustal motion, one to ten centimeters per year, were consistent with what would later be demonstrated by other means.

Admittedly, Griggs had not resolved all the prevailing differences of opinion. His model required convection currents to sink into the mantle below the zone of crustal thickening—that is, in the regions of negative anomalies where Hess had put his tectogenes. This was consistent with Hess's requirement for compression and melting in these zones and with the picture drawn by Holmes, but it contradicted Vening Meinesz's and Kuenen's view that convection currents sank beneath the *center* of the oceans. Griggs had begged the question by focusing on the role of downward currents in generating trenches, mountain belts, and a downward pull on the crust, leaving the site of the ascending currents out of his picture. But he agreed with Hess and Holmes that convection currents were most likely rising in the oceans and sinking in regions close to the continental margins, finding the evidence for this in earthquake seismology. Seismologists including Beno Gutenberg and Charles Richter, he noted, "all agree that the foci of deep earthquakes in the circum-Pacific region seem to lie on planes inclined about 45° toward the continents. It might be possible that these quakes were caused by slipping along the convection current surface."[101]

The recognition of dipping zones of deep-focus earthquakes is generally attributed to K. Wadati and Hugo Benioff, who gained attention in Europe and North America after World War II. In the late 1960s, "Benioff zones" came to be understood as the sites of descending crustal plates and descending convection currents—the final, critical piece in the plate tectonic story.[102] But Griggs's comment shows that as early as 1938 geophysicists had recognized these zones, which some of them had interpreted as sites of crustal motion linked to descending convection currents.

The most plausible way to resolve the remaining uncertainties was with more information from the sea. The dearth of data from the marine geophysical environment was why Vening Meinesz, Kuenen, Bullard, Hess, and Ewing had turned to their respective navies for logistical support. Navy support had enabled them to collect data that they would have otherwise been unable to collect, and the navies had imposed no restrictions on where it could be published or with whom it could be discussed. In the 1930s, there had been no reason to suppose that working with navies would be anything other than advantageous. But then the world went to war, and the conditions of scientific work changed dramatically. Secrecy became the order of the day; ideas and data that had been widely discussed among diverse colleagues were then discussed only narrowly, and in the United States, this meant a handful of American geoscientists with security clearances. Meanwhile, a new generation of scientists—mostly trained as physicists—came to work on the seafloor. Did this influx of men and means produce plate tectonic theory, or did it stand in the way of bringing to fruition the work of Vening Meinesz, Hess, and Griggs?

World War II and Pacific Bathymetry

In his work on the *S-21* and the *Barracuda*, Hess had complained that the military chain of command made it difficult for a scientist on board to get things done, so the Navy made Hess a lieutenant in the Naval Reserve. Perhaps this was one of history's small contingencies with large effects, or perhaps Hess would have joined the Navy anyway when hostilities broke out. Either way, on December 8, 1941, he took a morning train to report for active duty.[103] He would serve for the duration of the war.

Hess's first assignment was working on submarine detection in the North Atlantic. Few details of this work are available, but Hess would no doubt have deepened his understanding and appreciation of military exigencies, including the rapidly expanding role of scientific information in submarine warfare. As we have seen in previous chapters, many civilian geophysicists and oceanographers would spend the war working on military projects—at Woods Hole, at Scripps, and at several NDRC facilities dedicated to subsurface warfare—and many came to appreciate Navy operational concerns. But as an active-duty officer, Hess had an experience that was direct and immediate. He was transferred to the Pacific and assigned to the transport ship USS *Cape Johnson*, rose to rank of commander, and participated in several Pacific landings, including Iwo Jima. One of Hess's tasks had scientific meaning as well: echo sounding of the ocean floor.

There was, of course, a long tradition in the Navy of being alert to scientific matters that could aid navigation—tides, currents, depth sounding— but this concern had focused on shallow waters and surface effects. The US Navy had supported gravity work in the 1920s and 1930s because scientists had asked, but there is no evidence that anyone in the service thought gravity data would prove operationally valuable.[104] Echo sounding was different. Naval navigators had always worried about soundings in shallow areas, and increasingly some officers recognized that bathymetry could be used for navigation in the open ocean as well.[105] Echo sounding arguably represented the first form of geophysical data collection about the deep sea that was of direct and substantive interest to both scientists and Navy officers.[106]

Hess had long had an interest in the topography of the seafloor, in part because it was an obvious geological question as to how that topography had formed and in part because he had learned during the *Barracuda* work that the gravity data could not be interpreted without accompanying topographic information.[107] Gravity depends on mass, which in turn depends on topography and rock density. (Rock density could be determined through sampling, and sampling revealed the ocean floor to be almost entirely basalt, so there wasn't much rock density variation, but it was becoming clear that there was a lot of topographic variation.) So after completing the *Barracuda* measurements, Hess and Field asked Hydrographic Office officials to initiate a detailed sounding program in the area. Because bathymetric data were viewed as having a military function, scientific and operational motivations aligned, and the Navy agreed.

In 1940, Hess published a new bathymetric chart of the Caribbean region, based on seventy thousand soundings collected on ninety Navy cruises (plus a much smaller number of additional soundings taken by the Woods Hole research vessel *Atlantis*, HMS *Challenger*, and scattered commercial vessels). This was twice the number of soundings collected on the Snellius expedition. A second paper, "The Floor of the North Pacific Ocean," was published in 1942, based on work presented at the American Geophysical Union in 1940 before Hess had gone to war.

Hess highlighted three major features. First, the North Pacific was surrounded on its northern and western sides by a ring of island arcs and trenches. The trenches were zones of maximum ocean depths—greater than four thousand fathoms—the only place in the Pacific where such great depths were found. Second, the arcs were areas of active earthquakes, volcanism, and serpentinized peridotites, comparable to what he and Vening Meinesz had seen in the East and West Indies. Third was the pattern of structural trends defined by Pacific islands and shallow plateaus. These features were abundant throughout the mid-Pacific, trending east-west across the ocean floor. Some of them had been previously mapped, but the degree of detail made it possible to see the consistent structural trend running across the open ocean (fig. 4.6). It was also possible to see that the Hawaiian Islands were an exception to the rule. Hess could now pursue the question of the character of the Pacific Ocean floor. He appears to have been profoundly happy both to have the opportunity to do science while serving his country and to do military service that served science. Indeed, this was science that would have been nearly impossible to imagine doing before the war—

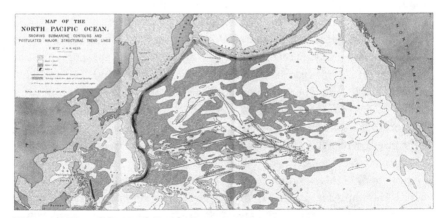

Figure 4.6 Map of the North Pacific Ocean.
From F. J. Betz and H. H. Hess, "The Floor of the North Pacific Ocean," *Geographical Review* 32, no. 1 (1942): 99–116. Reprinted with permission of John Wiley & Sons, Inc.

what historian Graham Burnett has called "royal science"—work whose scope could be achieved only with the support of states or monarchs.[108]

On October 24, 1944, Hess wrote a long letter to his colleagues back at Princeton from the *Cape Johnson*, describing his oddly happy situation:

> We have been taking continuous soundings since we've been out and all sorts of things have come to light. The Captain has become a great sounding enthusiast. I think he is almost convinced we could navigate around the Pacific on sound alone—though I wouldn't go quite as far as that. This is probably the only ship in the US Navy where one can find such phrases as the following in the Captain's night order book: "Steer course 273° true until you hit the 2000 fathom curve then change to 298° and call me when we cross the 500 fathom curve.[109]

Hess also reported the discovery of "an erosion surface (?) submerged about 700 fathoms extending over several million square miles in the west Central Pacific—there is a beautiful accordance of summit levels and many quite extensive flat topped areas at this level."[110] This was apparently Hess's first scientific communication of a discovery for which he would later be celebrated: guyots—flat-topped submerged mountains—named in honor of Princeton's first geology professor, Arnold Guyot. In "Drowned Ancient Islands of the Pacific Basin," published in the *American Journal of Science* in 1946, he documented the existence of more than 160 guyots in the western Pacific between Hawaii and the Marianas. The depth to their tops varied but typically occurred near 750 fathoms below sea level, and they all had a characteristic wide, flat top and gently sloping sides (fig. 4.7). This profile suggested old, eroded volcanoes.

Were guyots drowned volcanic islands? Were they important scientifically? The answer to the latter question was clearly yes, because they spoke to tectonic activity in the oceans. (In later years they would become important militarily as well, when the Navy placed hydrophones on them; see chapters 5 and 6.) The nineteenth-century concept of ocean basins as sunken continents had been largely rejected in both Europe and North America, but it was still unclear how the oceans had formed or whether they were tectonically active. If guyots were drowned volcanoes, this proved that the ocean basins had been active at some point in their history, and their eroded flat tops strongly suggested that they had sunk over the course of geological time—consistent with Darwin's well-known argument that atolls formed as fringing reefs around sinking volcanic islands. Perhaps most important, guyots were a *fact* in a domain that had long been dominated by speculation; Hess believed

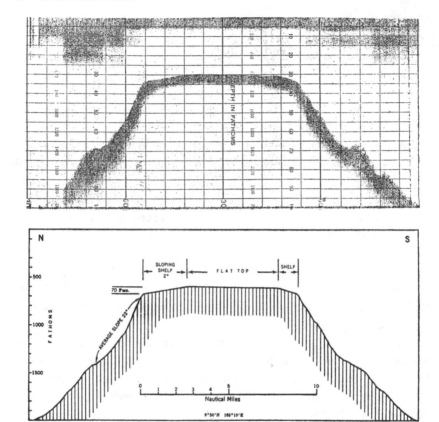

Figure 4.7 (a) Fathometer recorder trace of a typical guyot, 1946; (b) generalized geological sketch, 1946.

From H. H. Hess, "Drowned Ancient Islands of the Pacific Basin," *American Journal of Science* 244 (1946): 772–91. Image © *American Journal of Science*. Reprinted in accordance with stated fair-use policies, available at http://www.ajsonline.org/site/misc/terms.xhtml.

they offered an opportunity to think anew about the ocean basins: "Since it is difficult to discuss any theory of origin of guyots against the background of misconception and ill-founded theories which at present confound geologic literature on ocean basins and the Pacific basin in particular, the writer proposes to wipe the slate clean and start on a new basis."[111]

Before the war, Hess's discussion of gravity anomalies and tectogenes, and of Kuenen's and Griggs's experiments, strongly suggested that he was at least open to the possibility of large-scale crustal movements, perhaps even continental drift. Although Hess rarely referred to the latter explicitly—no doubt because of the hostile reception the idea had received in the United States—and never openly advocated it (a point he emphasized later, to establish his bona fides in taking on the issue in the 1960s), his detailed discussions and emphasis on Griggs's work—even to the point of citing the expected rates of convective motion in the subcrust—at minimum implied its possibility. Moreover, because convection currents were the mechanism invoked by Arthur Holmes to account for drift, any advocacy of convection currents brought Holmes's work to mind, as well as that of Molengraaff and Umbgrove. Hess had spent the prewar years in close communication with Europeans who took seriously the possibility of continental drift—or some theory along those lines—and his writings from that time suggest that he did, too.

By 1946, things had changed: Hess's guyot discussion was framed entirely in the context of a stable ocean. Moreover, the paper strikes a very different tone from his nearly breathless announcement eight years earlier of the need for "a radical change in some of our ideas on mechanics and processes of mountain-building." Hess explicitly denied the possibility that the ocean floor was exposed mantle material, thereby denying the possibility of a weak ocean floor as would be required in Wegener's formulation of drift (though not in Arthur Holmes's). He concluded that the ocean floor was too strong to permit drift—the position advocated by most American geophysicists—and returned to the argument of Vening Meinesz and Kuenen that convection currents descended somewhere in the middle of the ocean basins. To this he added the guyots. If these were former islands, one would expect them to develop fringing reefs, but there were none. Hess therefore concluded that they must have formed before the evolution of reef-building corals—that is, in the Precambrian or early Paleozoic.[112] That meant that the Pacific Ocean basin was very old and that current deformation was restricted to ocean margins. There was downbuckling and deformation along the edges, but the basin itself was static.

These views were aligned with prevailing North American thinking. A static ocean basin with compression along the margins was the essence of geosyncline theory—the dominant American view of continental margins since the mid-nineteenth century. Hess was adding details—literally, attending to the margins—and speculating about mantle processes. But it was hardly the radical change in ideas that before the war he had not only anticipated but announced. The basic framework was entirely compatible with what most geologists already thought. In the absence of the influence of Vening Meinesz and his Dutch colleagues, Hess's thinking had drifted back toward the North American mainstream.

Meanwhile Hess was anticipating the war's end. In 1944 he had written to colleagues: "There are a great many more such interesting facts turning up. I don't know their meaning. I've hardly had time to stop and think about it yet. It will have to be just another one of the many things I am piling up to do 'after the war.'"[113] Hess did indeed pile up a great deal of data, much of which he was able to share with colleagues, and some of which he was able to publish.[114] Yet beneath the surface of his upbeat message was an underlying anxiety: most of the data collected during the war were secret and could not be published.

Hess was looking at data from many sources, but American bathymetry formed the core: he estimated that 95 percent of the data he was using had been obtained by US Navy vessels. It was not merely that the war had provided reason to collect these data but also that it had sent ships to areas far from normal shipping routes.[115] Some of these traversed areas that turned out to be scientifically significant. Navy involvement had made it possible to do things that simply would not otherwise have been done. But what would happen after the war? Could the data be used for scientific purposes? Or openly discussed?

While at war, Hess had written the *Report of the Committee for Geological and Geophysical Study of the Ocean Basins* for the American Geophysical Union. Most of the committee members were either fighting or involved in war work, and Hess noted that "much of this work is secret or confidential and cannot be elaborated on during the war."[116] Ideally, it would be made public once the war ended. But he foresaw two looming problems. The first was declassification: considerable work would be required to declassify data and make it available. The second was access: even if data were not classified, they might simply disappear into the bowels of military bureaucracy: "A great amount of information has been gathered by the Navy which will be of value in the research fields of interest of this committee after the war.... Some concern is felt that information will be lost or mislaid in the course of being assigned to vast government archives at the end of the war."[117]

Postwar Secrecy: The Classification of "First-Order Factual Information"

When Hess's life story was told for the National Academy of Sciences' Biographical Memoir Series, his biographer skipped directly from the discovery of guyots to his later work on Project Mohole (a failed attempt to drill a hole to the Mohorovičič discontinuity, the seismic boundary between the Earth's crust and mantle), on the unit cell dimensions of minerals, on layered mafic intrusions, and, in the 1960s, on the analysis of pyroxenes from lunar samples. This presentation might lead a reader to believe that after the war, Hess returned entirely to civilian life, resuming his geological investigations of minerals, igneous intrusions, and the interior structure of the Earth.[118]

This was untrue, but it fit the standard story, often told about physicists who worked on the Manhattan Project, that when the war ended scientists went "home" to their civilian laboratories and universities to a life of autonomous scientific research. If the military played a role after the war, it was simply as a newly generous source of funding for basic scientific research. But Manhattan Project scientists did not return to the lives they had led before World War II and neither did oceanographers. Hess resumed his research on minerals and layered intrusions, but he also maintained close contacts with the US Navy, particularly the Hydrographic Office, and he dedicated considerable time and energy to a problem of increasing concern: Navy classification of oceanographic data.

Like many American scientists who participated in war work, Hess served in the postwar period on various panels and committees addressing the continuing value of scientific research for military preparedness and operations. One of these panels was the Navy working group on underwater navigation. Following directly from wartime experience, the group was asked to consider all possible scientific techniques that might aid submarine navigation.[119] Navies had long been concerned with improved methods of navigation, but the advent of subsurface warfare had added a new dimension: a submerged submarine can neither see the sky nor receive an electromagnetic transmission. Future submarines would stay submerged longer and their needs for accurate navigational techniques would become more pressing. As the US Navy developed a nuclear fleet, in which submarines might stay submerged indefinitely, the need would become acute.

The committee considered fifteen options, including echo sounding of ocean-floor topography, underwater beacons, free gyroscopes, triangulation from explosive sources at known locations, navigation from variations in Earth's gravitational and magnetic fields, and the latitude effect on cosmic rays. The most likely to be useful in the widest variety of circumstances, they concluded, was topographic navigation. Engineered systems, such as underwater beacons, would require the submarine to be near an installation, but no matter where the submarine was on Earth, the ocean had topography. Finding one's place relative to underwater topographic landmarks was, at least in principle, no different from what men had done for millenia on land and required no novel technology. With the discovery that the ocean had a vast, rich, and complex topography, topographic navigation could be rapidly realized. The Navy agreed and expanded work undertaken during the war. After centuries of looking up to the skies above them for navigation, mariners would now look down to the seafloor below.

The result was a flood of what Hess referred to as "first-order factual in-

formation." Between 1951 and 1963, the Navy accumulated extensive and detailed information on the topography of seafloor, including the mid-ocean
ridges and trenches. Much of this was collected directly by Navy personnel
on Navy ships, but it was also collected by academic scientists working with
Navy support. As scientists at Woods Hole, Scripps, and Lamont launched
major expeditions into the world's oceans, echo sounding became routine.
But there was a major problem: nearly all the data were classified.[120]

Hess argued for the release of these data to the scientific community. His
argument was simple: science helped the US military—something the military implicitly acknowledged through its financial support of science and its
requests to scientists to serve on various relevant panels and committees—
and so the military should help science. "Helping science" required not just
money but also broad dissemination of information. In a long memo titled
"Classification of Factual Material of the Oceans," Hess defined "first-order information" as basic factual data, such as "water depth, temperature gradient,
magnetic field, gravity data, etc., each coupled with a geographic position.
These are the small building blocks on which scientific research depends and
which ultimately may result in important theoretical conclusions."[121] Classification of such basic data presented "some very difficult and not fully recognized problems."[122]

Physicists often argued that they could demarcate basic science from its
military applications; they could, after all, distinguish between the laws of
physics they knew before they arrived in Los Alamos (and which could not be
kept secret) and the bombs they subsequently designed there (whose designs,
at least in principle, could be kept secret). Moreover, no country could go
from the laws of nature to a functional atomic bomb without spending billions on engineering and infrastructure. But oceanographers faced a trickier
situation.[123] They were collecting basic facts about the Earth, such as depth
of the seafloor, distribution of temperature and salinity in ocean waters, and
intensity of the gravity and the magnetic field. These data were not readily
accessible to scientists around the globe without a huge investment on the
part of their sponsors, and, unlike the laws of nature, they could be kept secret until such time as other countries would make such investments. They
were also the building blocks of geophysical theory. The information that scientists needed was the same as what the US Navy needed—and in the same
form. It was what the Soviets needed, too. The data were hard to get, but once
obtained, they could be used immediately, without modification or having to
spend billions on engineering.[124]

Hess did not deny this problem but argued that the benefits to science
were worth the risk to security. He acknowledged that "declassification of in-

formation on soundings would be helpful to a potential enemy of the United States," but the policy of blanket classification was "nevertheless ... highly disadvantageous to our overall national interest. If such a policy were continued for a number of years ... it could well be quite disastrous by greatly retarding the development of several sciences and by limiting to a very few people the scientific know-how in dealing with a number of problems."[125]

This was a familiar argument: that science requires the free flow of information and that society benefits from science. It was a difficult argument to evaluate, based as it was on a predicted gain in scientific knowledge weighed against the loss of information to an enemy. Moreover, one could argue that the relevant American experts had access to the data they needed, as nearly all leading geophysicists and oceanographers had security clearances. Hess anticipated this argument and found a response to it in the military concepts of compartmentalization and the need to know, which might be fine for the military but not for science:

> The argument has been used that all of the information is in fact available to all scientists who "need to know." True, it is available, but for the most part they won't know it exists or what exactly exists. Classified information does not get properly indexed.... But this is not the main difficulty. Those who "need to know" are an infinitely small number of our scientific population. The advances which could be made if the thousands of scientists, who could not prove the need to know, freely had this information are all squelched.... Inasmuch as great discoveries are very rare, we are in effect making almost absolutely sure that they won't be made by the simple procedure of limiting factual information to a very small number of men. Statistically we have a foolproof means of impeding scientific discovery.[126]

This was the crux of the argument. Even if data were in principle available, the labyrinth of Navy bureaus, desks, and departments kept a great deal of information out of sight and therefore out of mind. Moreover, discovery was a fragile and unpredictable matter, so how could one predict in advance which scientists might make good use of the data? The more people with access to data, the greater the odds that someone would use it to good effect.

Hess's assertion that there were thousands of scientists who might use the data might seem like academic hyperbole, but it wasn't. While there were probably only a few hundred American oceanographers and marine geophysicists at the time, there *were* thousands of scientists who could make use of oceanographic data, and they fell into two identifiable categories. One category was geophysicists outside the United States, particularly the Europeans who had worked on marine geophysics before the war. Hess had thrived in his prewar interactions with Vening Meinesz, Kuenen, Umbgrove, and others; his correspondence with Vening Meinesz was the most scientifically detailed and sophisticated of any he ever had. After the war, Hess and Vening Meinesz exchanged the occasional letter, but they lacked the detail and scientific content of their earlier ones. There is no way to prove that secrecy caused the degradation of the quality of their correspondence, but, given Hess's complaints, it seems likely that it played a part. In principle, they could have resumed the conversation, but Hess was no longer free to discuss the data informing his ideas. For a scientist it is difficult to explain what you believe without explaining why you believe it.

A second category was American geologists. While many geophysicists and oceanographers worked closely with the US military in the 1950s and had security clearances—often at fairly high levels—few geologists did.[127] Most land-based geologists working in traditional geological fields—mineralogy and crystallography, igneous and metamorphic petrology, ore deposits, sedimentology, paleontology, structural geology—would not have been thought by the Navy to have a "need to know" about the ocean. Yet from a scientific perspective, they needed to know quite badly: nearly everyone agreed that lack of knowledge about the ocean floor was the principal impediment to the development of global tectonic theory. (This had been widely cited as the major stumbling block in earlier debates over continental drift.) Geologists had played an active role in the creation of Woods Hole and in the transformation of Scripps precisely because they appreciated the importance of oceanography for geology.[128]

Geologists wanted to know whether the ocean basins were different from the land and from each other. How did island arc-trench complexes correlate with deformation in the adjacent continents? How did the terrestrial evidence of orogenic processes and land-sea connections mesh—or conflict—with the evidence obtained at sea? These were critical questions, yet their full consideration was made impossible by the sequestration of marine data.[129] Most of the world's leading experts on tectonic theory did not have access to the ocean data; some did not know they existed. This would have included Hess's own colleagues at Princeton, such as his friend and mentor A. F. Buddington, an expert on continental deep structure and petrology. What might Hess have liked to discuss with Buddington but couldn't?

Hess concluded his long memorandum by acknowledging the legitimacy of some classification but warning of military overreach. He allowed that not all soundings should be declassified, particularly if they would reveal the locations or purposes of military installations, but most of the ocean did not contain such facilities and one could not plausibly claim "that the whole North Atlantic is related to military or naval installations."[130] But the Navy did, in fact, claim that, and nearly all bathymetric data remained classified.

In early 1952, the issue was raised at a meeting of the panel on oceanography of the Department of Defense Joint Chiefs of Staff Research and Development Board, chaired by Roger Revelle. The chief of naval operations had affirmed that all soundings north of 20°S would be formally restricted; Revelle's panel protested. The decision had "resulted in an undue restriction on research activity in the field of basic oceanography by shutting out from such research those scientists and students who would normally engage in research in this field and is detrimental to the effectiveness of the Panel's services." They asked that "the blanket classification be rescinded and replaced by an order classifying only those areas of special strategic significance."[131] The vote was unanimous, albeit with the abstentions of committee members representing the Navy and the Coast Guard.[132]

Revelle summarized the panel's views to Earl Droessler, executive director of the Committee on Geophysics and Geography of the Research and Development Board of the Joint Chiefs of Staff—Revelle's counterpart on the geophysics side—whom he enlisted to present a united front. Revelle noted that he and his colleagues understood, of course, that bathymetric information could be used by America's enemies, but believed that the advantages—both to science and to national defense—of declassification "would, except in special cases, outweigh the advantages to a potential enemy.[133]

Revelle—who had an undergraduate degree in geology—stressed the losses accrued from being intellectually isolated from colleagues in geology. A sounding was just a number at a point in space; scientific judgment was required to interpret it, as well as to extrapolate in areas where available data were thin. For this, oceanographers needed geologists' help: "It is necessary to collate and extrapolate incomplete sounding data using geologic knowledge of the principles underlying the contouring of deep-sea topography. These principles are very inadequately known at the present time, and will best be developed if a wide-spread interest in the topography of the seafloor can be stimulated among competent geologists and geophysicists." This required two things. One, geologists needed to see the data, and two, the scientists doing the work needed to be able to publish their results. Secrecy prevented both.

Restricting data access to just a handful of people also stood in the way of the healthy competition that drove scientific advance, the competitive drive that arose when different scientists were working on the same problem: "Unless there can be competition among scientists to solve the difficult problems involved, it may take twenty years to do what otherwise might be done in five."[134] In physics or chemistry, Revelle suggested, it might be possible to gather the best minds in a few labs, but Earth was a vast and complex domain; the more people who could get at the data, the more likely they could make sense of them. He concluded, "New soundings should be classified only in limited areas (a) which are of great potential importance for Naval Operations and (b) where new knowledge of the seafloor which might be of evident and immediate importance to a potential enemy. Such areas should be carefully delineated after consultation between military personnel and civilian scientists who are aware of the Navy's problems. Perhaps a board of review might be appointed to advise the hydrographer on these matters."[135]

Droessler forwarded Revelle's memo to the chief of naval operations, assuring Revelle that the chief was receptive to his concerns. However, the chief wanted more than abstract arguments; he wanted concrete examples of cases where science had been impeded by classification. Droessler explained that officials at the Hydrographic Office were willing "to discuss de-classification of hydrographic data," but they wanted to be "prepared beforehand by having definite examples of the restriction in research activities and retardation of the advancement of science due to present classification."[136] The Navy wanted evidence.

Lamont was the academic institution with the largest research program collecting and using bathymetric data, so Revelle wrote to Ewing and asked him to state his views "about the classification of deep sea soundings and, if possible, to give me concrete examples of ways in which the Navy's present classification policy has impeded research or teaching."[137] Ewing was at sea and not expected back until mid-May, so the letter was answered by his close colleague (and later associate director of Lamont), J. Lamar (Joe) Worzel (1919–2008).

Of all the Lamont staff, none besides Ewing himself was more closely allied with the Navy, more solicitous of their concerns, or more dependent on them for support than Joe Worzel. (In later years, he would receive a distinguished service citation from Navy Secretary Paul Nitze for his work in the recovery of the lost USS *Thresher* nuclear submarine; chapter 6). Nevertheless, when Revelle asked for examples of classification impeding research, Worzel poured out his frustration over four and a half single-spaced pages.

Worzel documented various specific examples of impacts on individuals, lines of investigation, and the development of new techniques. He discussed Lamont's inability to recruit scientists into areas where publication was difficult, the inconsistent basis on which decisions about publication were made, the logistical bother of handling classified information with its negative impact on research and teaching, and the costs to the Navy itself when data were not shared among its own bureaus. One might imagine that a competitive scientist like Worzel would be pleased to have a data monopoly, but he in fact affirmed Revelle's concern over the inability to discuss information with geologist colleagues who might help him make sense of it.

Worzel began by discussing a specific individual whose work had been blocked by classification: Ivan Tolstoy, who had spent six months preparing a paper based on soundings in the North Atlantic only to be denied permission to publish. The data had been collected on civilian ships and covered no known strategic areas, so the scientists assumed they would get approval to publish. They were mistaken: the Navy classified the paper and refused to approve publication. Tolstoy tried to make the paper acceptable by decreasing the specificity of the data but to no avail: "Several modifications of the paper were attempted generalizing some of the specific information on which we thought the classification was based, but all of these were rejected, and the paper is now in a file sitting idle." This was obviously bad for Tolstoy's career, a point not lost on other scientists, "and we now find it virtually impossible to get any of them to work on topographic problems."[138]

Tolstoy was a rising young star at Lamont, having worked with Ewing in the late 1940s on the mapping of the Mid-Atlantic Ridge. In 1949, Ewing had submitted a request to the Hydrographic Office for charts to support this work, a request that was handled by Richard Fleming (coauthor with Harald Sverdrup of *The Oceans*), who was serving as chief oceanographer of the Hydrographic Office (chapter 1). As a fellow scientist, Fleming was sympathetic; he began his reply by praising Tolstoy for the respect he had garnered at Hydro: "Your Mister Tolstoy has made a very good impression in the Office and … his presence here and his work has, I think, considerably strengthened your case. Having someone working on the data and showing how the soundings can be interpreted and not just plotted on a piece of paper has aroused the interest of the people concerned with the oceanic soundings."[139] But Fleming was unable to help; the request was denied.[140]

Worzel argued that classification didn't just demoralize individuals like Tolstoy, it discouraged entire lines of investigation. For example, the Lamont group had recently developed a new technique of anchoring a buoy in deep water using a radar target, in support of accurate positioning for deep water soundings. Permission was denied to publish on either the data or the technique, so the group "abandoned further developments along these lines, and detailed soundings of significant areas in the deep sea have since been neglected by our group. Not only has this resulted in our neglecting an important phase of the sounding program, but others have not learned of this useful technique."[141]

Investigations of seamounts had also been abandoned. Ever since Hess's guyot discovery, there was great interest to see whether they were unique to the Pacific or a general feature, but secrecy was standing in the way of study. Worzel explained:

> Several of our people have prepared details of a number of seamounts in the Atlantic for publication, only to abandon further projects on similar subjects when the classification of some of these other things has definitely indicated that they would not be cleared. This is plainly interfering with the comparison of seamounts in the various places in the Atlantic and also with those in the Pacific [and] with the development of submarine geology and oceanography, in that the study of the meaning of the shapes of these things which would be made in the process of publication of the work is not being made.[142]

Worzel's own work had been caught in the classification net, too. The Toro Seamount, 350 miles northeast of Bermuda, had been predicted on the basis of topographic echoes and subsequently found where predicted. With preliminary permission from the ONR, Worzel had begun work on an article, but he stopped when Lamont "received a letter from the Hydrographic Office ordering that all charts and papers about it be restricted because of its importance to subsurface navigation. We have recently written a request that this be declassified, as its importance to subsurface navigation is much less than its importance to submarine geology."[143] But the Navy did not weigh the benefit to science against the cost to the Navy; any loss to the Navy, actual or anticipated, was sufficient to justify classification. Moreover, the Navy seemed to apply its standards incoherently: "We cannot believe that this [seamount] *is* important to subsurface navigation, as there are two other seamounts within about 30 miles of this one which are published in public documents and even on Hydrographic Office charts which would serve the purpose equally well, so that we cannot see how one more makes any difference."[144]

The Navy's inconsistent standard was of a piece with a lack of communication among its own departments. The Bureau of Ships had helped to fund the Lamont bathymetric work, so individuals there had written to the Hydrographic Office "to obtain the principles on which sounding matters were classified. They have been unable to get an answer."[145] Worzel did not expect the elimination of classification, but he wanted a consistent standard and to know what it was so he could plan accordingly.

The Toro Seamount was a specific case, but classification also served as a deterrent to investigation of topographic features in general, squashing the spirit of curiosity and the chances for serendipitous discovery. Because it was so hard to publish, "possible submarine features hinted at by various other pieces of data are not investigated by us unless so obviously close to our course that it would be criminal negligence not to investigate them." If a feature was off course, they would simply ignore it, given the low likelihood of being able to write about it, even if it turned out to be significant: "We would much rather utilize the ships' time for other problems on which we can publish." And because classified maps and charts required special handling, they were often deliberately left ashore, so that even when they were "in the vicinity of features requiring further investigation and could have done this without serious interference with our plans, we have neglected to do so." On the whole, Worzel concluded, "our entire program of topographic investigations has been seriously hindered, and most students who could have been interested in this work have by-passed it for other things."[146]

Teaching was also affected, because it was difficult not to talk about what one knew, and trying to do so ended up being both misleading and vexing: "We cannot consider the Atlantic Ocean west of Longitude 37° as very strategic, nevertheless because these are restricted we cannot show them to our classes for discussion, and are forced to show charts which do not include many of the features which we know to exist, and obviously our discussions of the matter are not very intelligible.... This has made it impractical to discuss soundings of ocean depths with large bodies and geologists and geophysicists who are being trained at Columbia."[147]

Most of the Lamont staff at the time were trained in physics or engineering; in later years geophysicist Walter Pitman would stress that when he first went to sea he knew "absolutely nothing about marine science."[148] It is to Worzel's credit that he acknowledged what he didn't know about geology and welcomed help. But classification stood in the way of his getting that help: "We are unable to obtain advice on the meanings of the shapes even from some of our own colleagues in Columbia who have not been cleared by the navy. Their advice could be obtained free and would constitute a great body of authority which we are forced to neglect, as it seems impractical to request that all members of the Geology Department of Columbia University be cleared by the Navy."[149] He concluded: "Those of our people who have done studies on contouring of submarine features say that in general the contour-

ing of features in the Atlantic by untrained geologists seriously violates geologic principles. This is caused by people contouring such features who have
inadequate or no geologic training.... If this is true, obviously this science is
being held back."[150]

The Navy Remains Unmoved

The Hydrographic Office had promised Revelle a review of Navy policy, but
if such a review took place, its only outcome was a clarification of policy, not
a change in it. In a presentation to the Research and Development Board
Panel in February 1953, representatives of the Hydrographic Office enumer-
ated twelve criteria on which soundings might be classified. These included
proximity to war channels and probable war shipping lanes, sonar detection
devices and nonsubmarine sonar contact features and wrecks, features that
would be of value in mine warfare and submarine evasive tactics, features
that would permit enemy evasive tactics, and anything within five hundred
miles of strategic land areas.

The list also included the maddeningly vague items, "configuration of the
bottom" and "bathymetric detail of certain strategic general ocean areas of
use as a system of navigation of a submerged submarine." Officers noted that
matters could be worse: the last item had led only to a restricted classification, "even though operating forces of the Navy have requested higher classification."[151] But there was something far worse in this clarification: although
there were many criteria that justified classification, there was only one official criterion for declassification, and it had nothing to do with science. It
was the release of information on shallow features that could pose a threat
to surface navigation.

Revelle's panel was not happy, and they pressed the point again, noting
that a restricted classification was sufficient to block publication: "Through
a RESTRICTED classification of basic oceanographic data, including bathymetry, many scientific papers were denied publication and distribution; as a
result many critical comments and ensuing evaluations would not be forth-
coming, effectively reducing over-all US oceanographic proficiency. Thus
much of the incentive for research is removed and the attraction of additional scientists to this already underdeveloped field is greatly diminished....
Only by full, unrestricted, and casual scientific research do the really great advances in science occur."[152] The scientists' objections were not based so much
on the formal policies—sweeping though they may have been—but on the
"implementation and interpretation of these policies by the Hydrographic
Office."

The scientists complained that "an interpretation by the office, representing only the military point of view, [did not] give full consideration to scientific benefits." Actually, it gave no consideration to scientific benefits, and so the scientists reiterated an earlier proposal that a joint advisory board "reflecting the scientific point of view" be appointed to "aid the Navy in carrying out a policy of maximum documentary security commensurate with a maximum of scientific, and resulting military, proficiency." The scientific panel concluded by arguing that what was good for American science was good for the American military and therefore good for America: "The opinion of the Panel is that present Navy security policies with regard to classification of all basic oceanographic data ... is unduly restrictive to the advancement of science by over-emphasis of security classification, and therefore detrimental to the military security of the United States."[153]

Revelle's panel appointed a subcommittee to pursue the matter further: its chair was Harry Hess. A month later, Hess wrote his own confidential information memorandum entitled "Classification of Deep Sea Soundings." He stressed that he fully understood the importance of these data to the Navy and was not unmindful of operational concerns. But once again, he tried to insist that there was a trade-off to be made, one in which the benefits of scientific advance should carry some weight: "All new information of oceanic depths undeniably has value to any potential enemy of the United States capable of naval operations. This is a valid reason for some level of classification on all such data. On the other hand, classification of any sort places a serious barrier in the way of scientific research on oceanographic problems and will tend to limit drastically, if widely applied, the total number of scientists actively engaged in such investigations." Part of the problem, he suggested, was that the potential value to the enemy was "self-evident," whereas the potential value to science was harder to articulate and so, without even realizing it, we "may ultimately lose more than we gain.... This does not mean that all soundings should necessarily be non-classified, but rather that sufficiently large areas of the ocean be available to all scientists in this country to permit unhampered basic investigations and free interchange of ideas."[154]

Hess drew on the classic argument from serendipity—that no one could predict where new discoveries would emerge—and added a twist: no one could predict *who* would make those discoveries. It was a utilitarian argument for making data as widely available as possible to ensure the greatest access by the greatest number of individuals: "The way to promote discovery is to have hundreds of scientists working in the field. A few will by chance turn up something quite unexpected.... It is these haphazard discoveries which pay large dividends in the long run. It is these which the Navy must in self-interest facilitate. It is much more important for us to maintain a technical lead in know-how than to hoard basic data."[155]

This unflattering comparison to hoarders might not have endeared Hess to his military counterparts. Nor were Navy officers likely to be impressed by the suggestion that any scientist was in the position to judge what the Navy must do in its own self-interest. But Hess certainly made one inarguable point: more data was being collected than used. Oceanographers working with or in the Navy frequently acknowledged that many research programs were organized more around the accumulation of data than around its use.

In 1951, for example, in a cover letter to a budget request for renewal of the basic task contract at Woods Hole (chapter 2), Director Admiral Edward "Iceberg" Smith summarized the project as "the conduct of basic oceanographic research on a long-term basis." The core activity, however, was not interpretation or explanation of oceanographic data, but its accumulation: "As in the past, the objectives of this work will be to stock-pile basic oceanographic phenomena."[156] What was the point of stockpiling information if you couldn't use it when a relevant question arose?

Even for scientists and officers with clearance, there was the additional problem of knowing what was available and then getting hold of it. People who might make good use of data often did not know it existed: "The Navy today has the majority of the oceanographic scientists in the US working for it through ONR, BuShips, BuOrd, etc., and [scientific] information is available to these men even if it is classified. [But] even among this group classification seriously hampers free interchange of information between widely separated institutions. Not that it cannot be exchanged, but largely because one is not aware that the other has it."[157]

A different man might have been grateful that he had access to a wealth of good data and left it at that. But Hess was courageous as well as generous, and he knew that his credibility with the Navy gave him an opportunity to raise issues that others could not. So for a third time in two years, at the end of 1953, he tried again, this time emphasizing the direct benefits to be had within the Navy from a freer flow of information.

Writing to the Hydrographic Office—the office that had blocked the release of Worzel's data—Hess recalled that in the early years, the office was doubtful that bathymetric charts had any practical use; now the office thought they were so important that no one could have them! But the Hydrographic Office's secrecy was hindering surface vessels' routine use of the charts:

I fought a long battle from about 1932 onwards to convince hydrographers that [bathymetric charts] would be immensely useful to the Navy. When

the hydrographic office started to produce them I felt justifiable pride....
[But] if some way cannot be found to make them readily available to the
fleet, their value is largely lost. It takes years of practice before the bridge
personnel on the ships will use the charts effectively.... This is a plea to
maintain one form of operational efficiency and training at sea against the
sincere viewpoint of men ... to whom security for the information bulks
larger than the operational problems.[158]

Hess also argued not merely for the promise of unexpected discoveries in the
future but also for the specific importance of bathymetric data for tectonic
theory right then. Just as the men at Hydrographic Office desks might not
understand the day-to-day value of bathymetric charts for sailors at sea, so
men at sea did not understand the theoretical value of bathymetric data for
earth scientists. In both cases, the risks of declassification were more easily
perceived than the risks of overclassification. The net result was excess control of information that someone should be using:

Officers with much experience on the operational side may not understand the adverse effects on tectonical developments caused by classification. It is very easy to see that certain types of information would be
useful to an enemy of the United States. The tendency, therefore, is to
classify such information. It is very difficult to see the hazards of with-
holding the same information from one's own people. Only a few experts
familiar with the information, how it can be used, and what essential progress will be impeded by classification can evaluate the case against classification.... The invariable result is that classification of information ... is
the rule.... There must be some way out of this dilemma.[159]

Hess revisited Revelle's suggestion that the Navy create a technical advisory
board to "review and report on cases where differences of opinion on classification exist." But if such a board was ever created, it has left no paper trail.[160]
Nor is there evidence to support Droessler's claim that Navy policies were
flexible and responsive. All the available evidence indicates the opposite: scientists were repeatedly stymied in their efforts to make the deep-sea data
broadly accessible and convinced that this had adverse consequences for
their science.

Throughout these discussions, Maurice Ewing remained largely silent.
When pressed for his views, he declined to criticize his patrons, saying that
"all in all, a very good job is done of handling the difficult problem of classification, and in many cases I do not think that I could do a better job myself."

He claimed even to have been surprised at times by Navy leniency, writing in the mid-1950s: "Through my whole career of working in underwater sound and related problems, I believe that in the cases where the Navy and I have disagreed on questions of classification, I have been more conservative than they—that is, my recommendation was for higher classification than was actually used."[161] But in 1954, even he was driven to complain over the issue that Worzel had raised two years before: the discovery of the Toro Seamount.

Lamont scientists had predicted the existence of the seamount from acoustic evidence, and in March 1952 they found it. But they had "abandoned hope for publication" when the Hydrographic Office immediately classified the soundings and the chart containing them and the inconsistency of military policies reared its head in a particularly galling way. Writing to ONR's chief scientist Emanuel Piore, Ewing explained: "[Worzel] did not make an official request for clearance until a series of papers by J. Northrop of the Hudson Laboratories on similar topographic subjects began to appear. He interpreted the appearance of these articles to mean that policy for clearance of papers of this type had been changed, so he submitted the report … on 19 May 1954 with a request for clearance to publish it.… An article appeared in August 1954 by Northrop describing the same seamount, and shortly thereafter we received word that clearance on Worzel's paper was denied."[162] It was painful to have a major discovery suppressed for security reasons but even worse to see it published by someone else. Ewing continued in uncharacteristically direct language: "This is a very serious situation, for it boils down to the fact that the opportunity to publish is largely governed by luck or by skill in negotiating for clearance." The Hudson Laboratories had been created by Columbia to house certain highly classified projects, including several related to subsurface warfare (see chapter 6), and people who went there understood that the lion's share of their work would be secret. But despite its Navy ties, Lamont was an academic institute inside a university, with graduate students who had to publish if they were to have careers. Ewing had no problem with short-term delays in publication, or even rare cases in which permission was denied entirely in "the interest of national security, but the life blood of our laboratory is the right to publish, and it is my obligation as director to see that a suitable proportion of our work is on subjects where publication is possible."[163]

Piore wrote back defending ONR policy, which, he insisted, was "that publication of research results is vital," and "under normal circumstances the only contractual requirements is that we be sent a manuscript at the same time the manuscript is sent to the publisher." Still, he acknowledged that "normal circumstances" did not always apply, that sometimes information was withheld, and that oceanography presented a particularly thorny situation:

> In some areas of some fields of research it is occasionally necessary to request that a paper be withheld from publication because of the military interest which may be revealed. We have faced this problem of classification of research results in many different fields and usually have been successful in finding a means to protect the scientific priority of the investigator in publication of his results while at the same time protecting the interests of national security. Often this has been a very difficult task, [and we] have had a particularly difficult time in the matter of a classification guide in the field of oceanography and of geophysical investigations of the ocean and ocean bottom which are so closely allied to many problems of great importance in Navy operations and strategy.[164]

This was the essence of the problem: the same confluence of interests that made scientists happy to work with the Navy—that made them feel that the Navy was supporting them to do what they wanted to do—led to conflicts over classification. Piore referred back to Revelle's efforts, noting that the ONR had tried to reach agreement on classification of ocean information with the various Navy agencies, and with the Research and Development Board, but that "no agreement satisfactory to the majority of the scientists ... could be reached." He advised Ewing to continue to work with his office on a case-by-case basis. Where classification was unavoidable, the ONR would try to suggest "a means for downgrading the paper to unclassified by the omission of such portions of it as may reveal military application," and he promised that in the future no paper would be returned as classified without "a specific and detailed statement as to the reasons therefor[e]."[165] For Piore, the solution to classification was either omission or explanation.

Ewing was not mollified. No paper was ever submitted from Lamont, he insisted, unless they had "already debated very seriously whether or not it should be classified in the interests of the country, and have decided that it should not." Given this, it was "extremely disheartening to have [a paper] unfairly classified and denied publication or have it put in cold storage for an indefinite number of months." Ewing suggested that, rather than submit a complete report only to have it classified, scientists should be permitted to submit a "rather full outline, for clearance in principle and for comments and suggestions about things to be avoided or put in," to make the final version publishable.[166] In essence, Ewing was asking the ONR to trust experienced scientists to make the right calls. Indeed, he suggested that, because of his long experience, he was more qualified to pass judgment on matters of national security than the Navy officials. He reminded Piore that "Navy personnel who are dealing with these problems turn over a lot faster than we do. I have been in this business twenty years and most people at those desks in the Navy have not been there three years." He concluded archly, "I think it might be well to tell those people to learn who they are dealing with and deal accordingly."[167]

In 1956 the Eisenhower administration launched a broad review of military classification policies to examine both unauthorized disclosures of information and overclassification. The Committee on Classified Information established by the Department of Defense concluded that military officials tended to overclassify, insofar as there was no penalty for that but potentially great penalties for the reverse. The committee called for a "determined attack" on overclassification "from the Secretary of Defense down." But as former US senator Daniel Patrick Moynihan later noted in his historical account of secrecy in the US government, the defense secretary established no procedures for declassification, no mechanisms for avoiding overclassification, and no penalties for failing to do so.[168] Consistent with this general pattern, at the Hydrographic Office matters remained unchanged.

The importance of bathymetry to navigation was reiterated the following year when a group of Navy officials visited Lamont to discuss the usefulness of various geophysical techniques to aid in "precision navigation techniques." The visitors listened to arguments about gravity and magnetics, but it was precision depth recording on which they were "sold." Ewing wrote in a letter summarizing the visit: "They seem to be completely sold on the usefulness of precision depth sounding.... Their main purpose is to find some geophysical navigation techniques to supplement the dead reckoning (inertial navigation) systems now being developed. It seems that the DR systems, when developed, will only be capable of reliable operation for up to about 36 hours. What they seek, therefore, is to divide up the oceanic operating areas in such a manner that geophysical measurement check points are available every 200–300 miles for recalibration of the DR systems."[169] Bathymetry did this best. That was good for the Navy, but it was bad for science. So once again, the data remained classified.

Two years later, Lamont marine geologist Bruce Heezen wrote to his colleague sedimentologist Kenneth Emery that he, too, had abandoned hope of publishing certain key information because of Navy restrictions. Emery had asked if he had contour charts of the abyssal ocean, to which Heezen replied that he did but "did not dare to submit them" for publication.[170] Two years after that, a colleague at the Naval Electronics Laboratory made a similar inquiry. Heezen's replied flatly: "I have never prepared a bathymetric chart of the entire Atlantic Ocean because of the classification problem."[171]

Hess Returns to Tectonics

Why was Hess so bothered by secrecy when he had access to all the data he wanted? After all, some scientists might relish the competitive advantage of having nearly sole access to reams of propriety data. Hess was a generous man who cared about science as an enterprise as much as he cared about his own career; his efforts were on behalf of earth science as a whole, but the evidence suggests that he also had a personal stake, that when he argued the need for scientific minds to work together, he was arguing, at least in part, for himself. For if we return to the question of Hess's work on tectonic theory, it becomes evident that, from 1939 to 1959, Harry Hess was stuck.

Consider again his 1938 comment: "When the significance of the huge anomalies and the structures they indicate become appreciated, a radical change in some of our ideas on mechanics and processes of mountain-building becomes necessary."[172] In 1938, Hess believed that he and his colleagues were on the verge of a breakthrough, a new paradigm. But in the 1940s and 1950s, he made no progress. In fact, as we have seen, he retreated from his strong advocacy of convection currents to a conventional belief in a stable ocean basin. His thinking had gone from maverick to mainstream.

Hess was well aware of the skepticism and even hostility that had greeted Alfred Wegener's work, and he had perhaps taken the lesson that it was unproductive to push radical ideas in isolation. One needed support, yet support was precisely what Hess lacked after 1939. In the 1920s, continental drift had three active and respected defenders in the United States: Harvard's Reginald Daly, Yale's Chester Longwell, and the Carnegie Institution of Washington's F. E. Wright (the reviewer of Teddy Bullard's grant proposal). By the 1940s, however, Daly was aged and Wright and Longwell had softened their support in the face of collegial opposition. David Griggs had emerged as an important voice in the late 1930s, but after the war he left geophysics to found the RAND Corporation. He maintained close postwar connections with the Department of Defense, including serving in 1951–1952 as chief scientist of the US Air Force, but he no longer worked on tectonics.

In his 1948 paper on the structural features of the western North Pacific, Hess had pointedly declined to discuss "highly speculative relationships such as those bearing on convection currents in the Earth and their possible connection with the formation of the arcs." His reason? "Since the speculative material is bound to be controversial, it is hoped that separate presentation will limit controversy to the speculation without impairing the descriptive portion."[173] Hess's attempt to demarcate description from speculation was understandable but fallible: he presented tectogenes as facts, whereas in hindsight they appear to be speculative (and erroneous). Nevertheless, his comments reveal how careful he felt he needed to be and how great he considered the likelihood that he (or anyone) would be attacked for challenging mainstream thinking. If wider release of relevant data might have helped to garner collegial support, Hess surely would have wanted it. Thus, it seems fair to conclude that, alongside a genuine concern for the progress of his science, Hess's persistent efforts to get oceanographic data declassified were motivated in part by a desire to discuss the data with colleagues and enlist them as allies at least in that discussion, if not in a specific theoretical account.

In 1960–1961, Hess worked with Scripps marine geologist Robert L. Fisher on a chapter on ocean trenches for Maurice Hill's grand compendium, *The Sea* (published in 1963). Mounting seismic evidence pointed to a zone of crustal instability beneath the trenches; Hess now argued that the feature there was not a crustal downbuckle but the crust and upper mantle "moving [downward] together as a unit."[174] This was a critical innovation in his thinking: he had arrived at the concept that would soon be known as subduction. Yet he illustrated the descending crustal slab as vertical—despite the fact that the seismic data revealed a zone dipping at a relatively low angle toward the continent (fig. 4.8). Why did Hess insist that the crust and mantle were moving vertically when the seismic data showed otherwise?

As far back as 1938, Hess had recognized that the pattern of deep-focus earthquakes beneath arc-trench complexes indicated a rigid zone, capable of sustaining brittle deformation, dipping inland from the arc at approximately 45° (rather than distributed vertically below the arc or the trench). In their 1963 article, Hess and Fisher also noted that Hugo Benioff, first in 1949 and then more clearly in 1955, had explicitly "postulated an overthrust mechanism and showed that the deep earthquake foci lie along a plane dipping under Chile at approximately 40°."[175] Why didn't Hess make the connection that Benioff's dipping thrust plane *was* the down-going crustal slab?

The final figure of the article offers the answer: Hess's picture of the down-going slab is as close to his old tectogene as was possible without being the same (fig. 4.9). It is far more detailed—as Hess had by then thought through in much greater detail the rock types involved, the depth at which they would begin to melt, and the generation of his cherished peridotite magmas—but structurally, the picture is nearly the same. The only substantive difference is that the downbuckled slab is permitted to descend indefinitely until it is

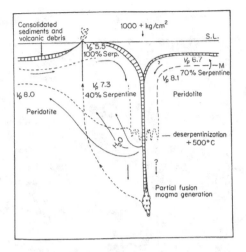

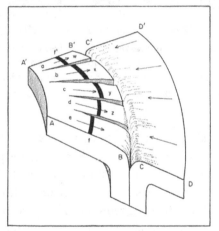

Figure 4.8 (a) Supposed structure for a hypothetical trench-island arc, 1963. Note the inclusion of rock types and seismic velocities. (b)Fisher and Hess model of crustal and mantle buckling to create deep-ocean trenches and generate peridotite magma. Note that the downbuckle is still vertical, as it had been in the 1930s.

From R. L. Fisher and H. H. Hess,"Trenches," in *The Sea: Ideas and Observations on Progress in the Study of the Seas*, ed. M. N. Hill, 411–36 (New York: John Wiley & Sons, 1963), figs. 9 and 10. Reprinted with permission from Elsevier.

rendered unrecognizable by melting or mineralogical change.[176] Moreover, by maintaining a vertical orientation for the compressed crust, the model predicts the occurrence of vertical faults, preserving the conduit for the peridotite intrusions that Hess first proposed in the 1930s. Despite Hess's enormous creativity and courage in promoting seafloor spreading, the major features of his thinking were unchanged since the 1930s.

Figure 4.9 A widely circulated photograph of Harry Hess with his tectonic model, taken in 1961 at Princeton by Fritz Goro. Note how in the schematic sketch to which Hess points, the downgoing crust is vertical, as it was in his models from the 1930s. Reprinted with permission of the LIFE Picture Collection via Getty Images.

In 1960, Hess finally broke through, when he wrote the first draft of his now-classic paper "History of the Ocean Basins" (later published in 1962), proposing the theory of seafloor spreading to account for the structure of the ocean basins and provide the driving force of continental drift.[177] "Mid-ocean ridges ... are interpreted as representing the rising limbs of mantle-convection cells.... The mid-Atlantic Ridge is truly median because each side of the convecting cell is moving away from the crest.... Continents ride passively on convecting mantle."[178] The trenches and island arcs that Hess and others had described from the East and West Indies were the regions where oceanic crust was pressed up against other crustal slabs; the peridotites were slices of ocean floor caught up in the process. The views he had long ago adopted could be put forward again with slight modification. Although Hess did not cite Wegener or Holmes, he did cite Griggs, Kuenen, Umbgrove, and Vening Meinesz. The foundations and inspirations from the 1930s were there for anyone to see.

How did Hess come unstuck? By his own account, two lines of evidence were key. The first was the paleomagnetic work by British geophysicists P. M. S. Blackett, Keith Runcorn, Ted Irving, and others, demonstrating that the continents had moved relative to the magnetic poles and relative to each other. In Hess's words, "One may quibble over the details, but the general picture on paleomagnetism is sufficiently compelling that it is much more reasonable to accept it than to disregard it."[179] This was a major new line of evidence bearing on the question of drift that had not been available in the 1930s. The second was Bruce Heezen and Marie Tharp's demonstration— again in Hess's own words—that "a median graben exists along the crests of the Atlantic, Arctic, and Indian Ocean ridges and that shallow-depth earthquake foci are concentrated under the graben," which had led Heezen and Tharp "to postulate extension of the crust at right angles to the trend of the ridges."[180]

In chapter 5 we examine Heezen and Tharp's work, along with the comparable work done in the Pacific by Bill Menard. This work was based heavily on classified data. For now, two points suffice. First, the paleomagnetic data came from outside the United States and outside domains of classified research. Second, while Heezen and Tharp's data had been available to Hess for some time, Hess's breakthrough came only after the work was published— that is to say, after it was possible to discuss the data and after colleagues were in fact discussing them (including famously, Robert Dietz, who published the seafloor-spreading idea around the same time and very nearly stole Hess's thunder). The evidence does not allow us to say precisely how this mattered to Hess, but evidently it did. Having the data to himself had not sufficed.

This is not really surprising. There is a reason that scientists spend so much time in conferences and workshops and travel the globe to present their work: these are the places where scientific results are mooted. It is not enough for scientists to observe a scientific fact or complete a laboratory experiment. They need to discuss their experiments and observations in order to determine and establish their meaning. As Bruno Latour and others have stressed, in the laboratory, data are just inscriptions; an earthquake is a line of ink on a piece of paper reeling off a seismograph. It is through social processes that the inscriptions acquire meaning.[181] It is through discussions among scientists that a set of seismic data, recording the effects of earthquakes deep in the Earth along a plane dipping toward a continental margin, becomes a Benioff zone, and it is through yet more discussion that it becomes a subduction zone. An instrumental observation—indeed, any observation—is not a fact until a community of experts accepts it as such. Because classification rendered it impossible to discuss so much, many things that might have become facts did not.

The person whose analysis of the adverse impact of classification was probably most accurate was Roger Revelle. In 1952, he argued that classification might cause it to "take twenty years to do what otherwise might be done in five."[182] The story told here suggests that he was too optimistic (Revelle generally was): the resulting work took thirty. Viewed through the long lens of history, thirty years is not a long time, but it certainly is from the perspective of one man's life, or the life of a scientific community. Does this matter? It certainly did to Hess. He believed that work was being stymied by secrecy. The simplest explanation for this belief is that he felt that his own work was stymied by his inability to discuss it freely with colleagues outside the narrow circles of Navy sponsorship and security clearances. Hess thought it was true because it was true *for him*.

Moreover, the reopening of debate over continental drift—and the expansion and transformation of the idea into the theory of plate tectonics—was not triggered by work done under the auspices of the US Navy. The paleomagnetic data to which Hess referred—which demonstrated that the continents were moving relative to the poles—was done in two laboratories in England. One of them was led by the physicist, communist, and Nobel laureate P. M. S. Blackett, who turned to paleomagnetism after World War II because he no longer wanted to work on weaponry.[183] The bathymetric data from the mid-ocean ridges that suggested they were rifts was collected under Navy auspices, but neither of the two American scientists most responsible for the data collection and analysis—Bruce Heezen and Bill Menard—and therefore best placed to use it to answer the global tectonic question, made major contributions to the theory of seafloor spreading (chapter 5). Instead, Frederick Vine and Drummond Matthews in England, working with data independently gathered, confirmed Hess's hypothesis and paved the way for plate tectonics.

Given the thirty years that passed between 1938 and 1968, that the picture of crustal mobility given by plate tectonics in the 1968 was essentially the same as the one Hess and Griggs had drawn in 1938, and that the impetus for reexamining the question of continental drift came largely from outside the United States, the historical evidence supports the conclusion that secrecy did impede scientific work. For the better part of thirty years, the conditions of secrecy imposed on US science during World War II and maintained in the Cold War brought American theoretical thinking about the seafloor to a standstill.

Coda

Harry Hess had warned that it was not just science that would suffer from secrecy; the military would suffer as well. Navy officers were probably doubtful, but history suggests that he was right about this, too. In January 2005, the *New York Times* reported on its front page that a US nuclear submarine moving at top speed had crashed "head-on into an undersea mountain that was not on the charts."[184] One man was killed, sixty were injured (twenty-three seriously), and the commander—who was subsequently relieved of his command—was quoted as saying, "I thought I was going to die." The interior of the boat "looked like a slaughterhouse" as blood splattered on the floor and instrument panels. "Some of the younger sailors said they had not realized how close they had come to dying until they saw the *San Francisco*'s mutilated bow at the dry dock here. 'Your jaw just kind of dropped open, and you wondered why you were still alive.' ... As many as 10 sailors have asked not to return to submarine duty."[185]

The crash was in the western Pacific, southeast of Guam, the very region Hess had patrolled and mapped as a young lieutenant. According to the *Times*, Defense Department officials later reported that a 1999 satellite image revealed the mountain rising to within a hundred feet of the sea's surface. Scripps professor David Sandwell noted that satellite radar data from the 1980s had also indicated a possible undersea mountain at the crash site. But the charts had never been updated.

The initial *Times* report focused on a lack of resources with which to update charts made in the presatellite era, but a later analysis suggested that the situation was even more tragic. The hazard was noted on most charts of the area—but not on the one the captain had in his possession. The information that the captain needed was available, but he had not known to ask for it. It was just as Hess had warned. It is not enough for information to exist. People who need it need to know that it exists.

5 *The Iron Curtain of Classification: What Difference Did It Make?*

Twenty-six years after Harry Hess suggested that tectonic disturbances were focused in narrow belts that might be continuous around the world, this insight would become a cornerstone of plate tectonic theory. Hess had worked on the regions we now understand as convergent plate boundaries. But that was only part of the story. Another part was the regions we now understand as divergent boundaries, and a key player in that story was Lamont marine geologist Bruce Heezen. Working with Lamont director Maurice Ewing and geologist Marie Tharp, Heezen produced the world's first comprehensive map of the Atlantic seafloor, laying the groundwork to confirm the theory of seafloor spreading.[1] The final part of the story involved the transcurrent faults now understood as the places where plates slide past each other, and a key player in that story was Scripps marine geologist William Henry Menard.

Heezen and, to a lesser extent, Tharp were made famous by the reproduction of their map in *National Geographic*.[2] They also became wealthy, collaborating with Menard—who was doing work comparable to theirs in the Pacific—to convert the maps into a children's globe manufactured by the educational toy maker Creative Playthings.[3] In later years, the maps would adorn the walls of geology classrooms and the hallways of earth science departments.

The visual and conceptual centerpiece of the Atlantic physiographic map was the mid-ocean ridge and the dramatic way it was offset by a dense series of east-west bearing faults. The ridge had been recognized since the nineteenth century, but the Atlantic map demonstrated its scale and complexity, and the global map showed its continuity. The North Atlantic ridge was continuous with the South Atlantic ridge, which could be traced into the Antarctic and Indian Oceans and around the Southern Hemisphere into the South Pacific. One Columbia colleague described the map as "unique, and may[be] revolutionary. For the first time, on a single map, one man [*sic*] has attempted to present the fundamental physiographic provinces" of the ocean floor.[4]

To make their maps—and the children's globe—Heezen, Tharp, and Menard had to find a way to deal with the issue of data classification that had vexed Harry Hess. Shortly after the Atlantic map was published, Henry Stommel wrote to congratulate Heezen for running the gauntlet of Hydrographic Office secrecy: "It certainly is nice to see some studies on bathymetry coming out despite the H.O.'s Iron Curtain."[5] How did they manage the challenges of military secrecy? Or did they?

A Scientific Career amid Personal Troubles

Bruce Heezen (1924–1977) received his BA in geology from Iowa State University in 1948 and moved to New York to pursue graduate studies with Maurice Ewing. When Columbia University established the Lamont Geological Observatory in 1949 under Ewing's direction, the timing was perfect for Heezen. He would spend his entire career there, studying seafloor structure and deep-sea sedimentation with abundant support from the Office of Naval Research, the Navy Bureau of Ships, and the American Telephone and Telegraph Company (AT&T).[6]

Heezen was not obviously destined for success. A fraternity boy who enjoyed alcohol more than academics, Heezen had been studying geology and paleontology, anticipating an oil industry career. But his plans changed when he heard Ewing lecture on the geology of the seafloor. He approached Ewing, who invited him to join an upcoming expedition to the Mid-Atlantic Ridge on the research vessel *Atlantis*. This would be a historic cruise: the first major geological expedition to map the ridge in detail, what one geologist would later describe as the beginning of the "road to glory for Doc Ewing."[7]

The *Atlantis* work was a six-week expedition including Ewing, Ivan Tolstoy, geophysicist Gordon Hamilton, and future presidential science adviser Frank Press. *Atlantis* was a Woods Hole ship, and Columbus Iselin had instructed the captain to take particular care in his preparations, as "the scientific importance of the program is greater than any other which you have yet undertaken."[8] At an estimated cost of $26,000, it was also one of the most expensive. But it was money well spent, for the scientists made several major discoveries, including demonstrating the continuity of the ridge throughout the North Atlantic. They also found that the ridge did not follow a smooth curve, snaking down the center of the ocean as drawn in earlier maps, but was composed of distinct linear segments running nearly north-south, with east-west offsets (fig. 5.1).[9] Contrary to plans, however, Heezen did not go. At the eleventh hour, Ewing sent him instead to lead a small cruise surveying the Hudson Canyon off the coast of New York.[10]

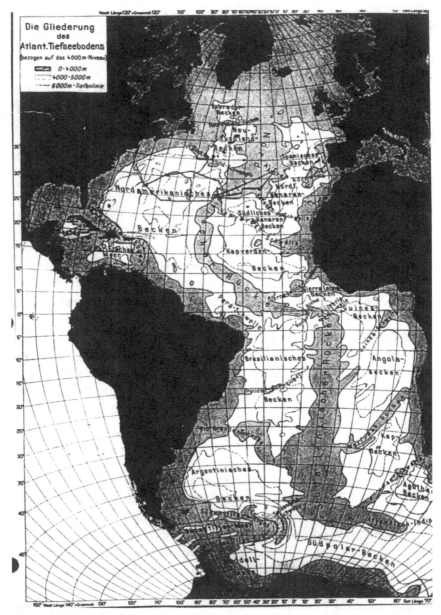

Figure 5.1 The Mid-Atlantic Ridge as mapped before World War II, based primarily on British and German work. This map was found in the logbook of *Atlantis* cruise 150 (1947) and is used as the frontispiece to Ewing and Tolstoy's 1949 work.

From "North Atlantic hydrography and the Mid-Atlantic Ridge," *Geological Society Bulletin* 60, no. 10 (1949): 1527–40. Reprinted with permission of the Geological Society of America, permission conveyed through Copyright Clearance Center, Inc.

The change of plans may have been the result of larger troubles. Heezen enjoyed an active social life, frequenting the theater, concerts, and operas, but he struggled with his coursework, earning mostly Bs. On at least one occasion he received an official university reprimand for unsatisfactory performance.[11] His mediocre classroom work accompanied personal difficulties with love, money, and alcohol. Heezen saved numerous letters from unhappy women he had dropped unceremoniously, including one he owed money.[12] He also saved newspaper clippings on medical advances in treating alcoholism and an uplifting poem on temperance, exhorting its reader to "drink not to elevation."[13] Alcohol may have played a role in an auto accident in December 1951; Heezen was sued for damages resulting from his "negligent operation and control" of his vehicle. The plaintiff's lawyer offered to settle the matter amicably; six months later, Heezen's lawyer was beseeching him to complete and return the relevant papers.[14]

Heezen was sloppy about many things. Joe Worzel chastised him for repeatedly leaving the bathroom at Lamont in disarray and acting as if the cleaning woman were his "personal maid." He reminded Heezen that he was "not in kindergarten anymore" and that "any well brought up person" would not behave this way. Worzel instructed him to modify his behavior "so that more drastic action does not have to be taken."[15]

Heezen may have been depressed. On one occasion he wrote to his parents—in dreadful and erratic handwriting—saying he had dropped plans to go to Bermuda over the holidays, which he described as a "device I had cooked up to engender some enthusiasm in myself. It didn't work." He needed to pass French and German and worried, "Unless I get fired up some more I will never even try." His parents had suggested he take a break, and Heezen reluctantly agreed: "I suppose the suggestion that I come home for a while is good enough.... Hell, I just don't know what to do."[16]

The bright spot in this gloom was going to sea, which Heezen described as his "one way to mental health. Active work on the ship keeps me happy."[17] He soon became known as someone who spent nearly all his time at sea—by his own reckoning as much as nine months out of every year—and the years after World War II were ones in which it was possible for an American oceanographer to do that.[18] Squeezing coursework in between cruises, Heezen completed his PhD in 1952. Just a year later, he earned a listing in *American Men of Science* for demonstrating the existence of turbidity currents, underwater mudflows responsible for the rapid transport of large quantities of coarsegrained land-derived sediments into deep-ocean environments. He had not mapped the Mid-Atlantic Ridge, but he done something perhaps more important: he had prompted reconsideration of the entire question of what occurred in the deep ocean.

The Demonstration of Turbidity Currents

Submarine canyons—deep valleys incised into the continental shelves at their seaward edges—were one of those phenomena, like the ice ages and the rebound of the Fennoscandian Shield, whose existence geologists accepted as facts but whose origins were endlessly argued. When Columbia professor Douglas Johnson wrote a book on the subject in 1939, he outlined no fewer than eleven different hypotheses, explaining, "Few problems in the field of geomorphology have proven so baffling as that of finding a satisfactory explanation for the deeply submerged canyons indenting the seaward margins of many continental shelves. For over half a century the problem has been debated.... Yet there is today no consensus of opinion as to how these remarkable features were produced." The search for a satisfactory explanation had proved "futile."[19]

Johnson was being a bit dramatic. In fact, there were two main theories on how the canyons formed. One was that they were drowned river valleys, carved out during ice-age low-sea-level stands. This had been proposed by Scripps marine geologist Francis Shepard (1897–1985) and was consistent with worldwide evidence of geomorphological effects produced by rising and falling sea level during glacial periods. The other, proposed by Harvard geologist Reginald Daly (1871–1957), America's most prominent defender of continental drift, was that the canyons were scoured by underwater mudflows running down the continental slopes, so-called turbidity currents.[20]

Neither man had direct evidence, but Daly had a strong indirect argument in a type of sedimentary deposit known as greywackes. These were thick sequences of coarse-grained, well-graded but generally poorly rounded sedimentary strata—the sort of clastic deposits one might expect to find in shallow-water environments—interbedded with much thinner layers indicating conditions of very deep water.[21] Greywackes were well described in the stratigraphic record; Daly argued that they were deposited by turbidity currents. Large rivers could bring clastic deposits—mud, sand, and silt—to the edge of the continental shelf, where they could become overpressured, triggering underwater mudslides—turbid mixtures of seawater and sediments that would run down the continental slope. They would be laid down when these mud-choked currents met the abyssal plains. The deposits would be poorly sorted because they came from mudflows, but well graded because they would settle out slowly in the deepwater environment.

This was a clever idea—today it is viewed as right—but most geologists preferred Shepard's hypothesis because they believed that deep sedimentation was dominated by the steady rain of organic debris and very fine-grained clays derived from wind-borne dust. A few samples dredged from oceanographic expeditions in the late nineteenth and twentieth centuries had revealed clastic sediments at abyssal depths, but no one paid much attention to these anomalies. However, in the 1930s, Philip Kuenen (chapter 4) took up Daly's idea, showing experimentally that it was possible for underwater turbid currents to develop and that they would deposit graded beds just like greywackes (fig. 5.2). These experiments were a proof of concept: they showed that turbidity currents could deposit coarse sediments into deepwater environments, but they did not demonstrate that they *did*. However, within a few years, the demand for a mechanism to explain deepwater clastic sediments would grow.

It was well known that underwater telegraph cables in the deep ocean periodically broke, and that sediments dredged by ships attempting cable repairs contained coarse sand and small pebbles. This suggested that some kind of submarine current had transported the materials into the deep ocean, perhaps breaking the cables at the same time. For this reason, AT&T sponsored deep-sea studies at Lamont. As the scientists sampled and photographed the abyssal plains, they found that coarse clastic sediments were common in the abyssal plains adjacent to the continental slopes.[22] Daly's hypothesis was a possible explanation for both the cable breaks and the deep clastic sediments associated with them.

In 1950, Heezen and colleagues published their analysis of deep-sea clastic sediments in the submarine Hudson Canyon. They had completed a topographic and sedimentary survey of the seaward extension of the canyon, which extended for 280 kilometers past the edge of the continental shelf along the southeastward extension of the Hudson River (fig. 5.3). Beyond the detectable edge of the canyon, the topography consisted of a broad plain, gently sloping toward the southeast. Most of the sediments found there were foraminiferal oozes—the remains of marine plankton—but scientists also recovered unconsolidated silts overlying the basement rocks. Photographs revealed an eroded edge, suggesting that erosion was still occurring, even on the continental slope where currents had been assumed to be absent.[23] This result was corroborated by sediment cores brought up by the *Atlantis* that same summer: one sample revealed well-sorted sands, interbedded with clay and globigerina (a type of foraminifera) ooze at a depth of 4,755 meters. This was far below the depths at which most geologists believed that clastic sedimentation occurred; it showed that terrestrial sediments were being transported into abyssal depths as predicted by the turbidity-current hypothesis.

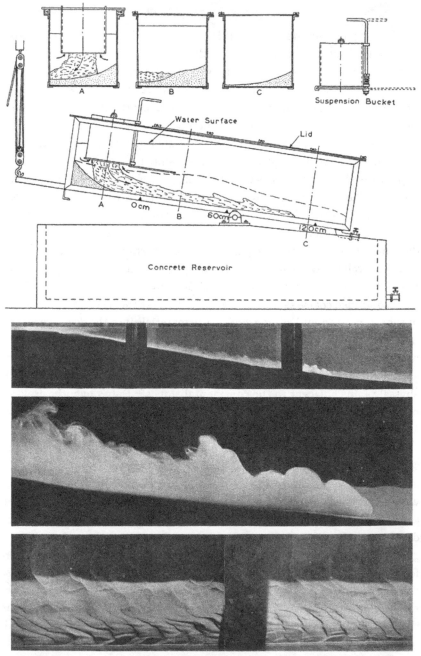

Figure 5.2 Kuenen's experimental model of turbidite formation by underwater mud-flows: (a) illustration of the experimental setup; (b) photographs of the experiment, simulating a submarine turbidity current.

(a) From P. H. Kuenen and C. I. Migliorini, "Turbidity Currents as a Cause of Graded Bedding," *Journal of Geology* 58, no. 2 (1950): 96. Reprinted with permission of University of Chicago Press; permission conveyed through Copyright Clearance Center, Inc. (b) From P. H. Kuenen, "Experiments in Connection with Daly's Hypothesis on the Formation of Submarine Canyons," *Leidse geologische mededelingen* 8 (1937): 327–52. Reprinted in accordance with Naturalis Biodiversity Center open-access policies.

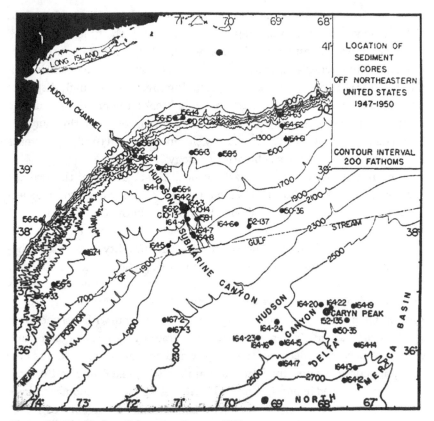

Figure 5.3 The Hudson Submarine Canyon, 1951.
From D. B. Ericson, Maurice Ewing, and Bruce Heezen, "Deep-Sea Sands and Submarine Canyons,"
Geological Society Bulletin 62, no. 8 (1951): 961–66. Reprinted with permission of the Geological Society
of America; permission conveyed through Copyright Clearance Center, Inc.

In a pair of papers published in 1951 and 1952, Heezen and his coauthors
suggested that the clastic sediments at the base of Hudson Canyon were deposited by turbidity currents.[24] Wind could not transport coarse-grained sediments and ice rafting would not explain why the sediments were graded.
Erosion during a prior low-sea-level stand would not explain why the sands
were interbedded with deep-sea clays and oozes. Nor would it explain why
faunal assemblages of the interbedded deposits were contemporaneous with
equivalent deposits on the continental shelf. These observations "demanded
a submarine origin for the canyons."[25]

In a brilliant piece of scientific detective work, Heezen and Ewing showed
that a series of sequential cable breaks along the continental slope following the 1929 Grand Banks earthquake were consistent with an underwater

mudflow traveling rapidly on the upper portion of the continental rise and slowing down where the topography flattened out at the base of the rise.[26] The bedding in the resultant sediments was graded, just as in natural grey-wackes and in Kuenen's experiments, and just as it was at the base of Hudson Canyon. It was a unifying explanation, bringing together seemingly disparate phenomena: an earthquake, a series of cable breaks, the presence of clastic sediments in the abyssal environment, and the origins of submarine canyons. They concluded that "large scale work by slump-generated turbidity currents is a fundamental process in submarine geology."[27]

Heezen had solved a major puzzle of both the structure of the seafloor and the geological record. "It was a happy day for me when I got your letter telling the wonderful news about your findings in the Grand Banks area," Kuenen wrote to him in December 1953. "I had heard so much, not only of doubt, but even cocksure ridicule, concerning the turbidity current hypothesis ... that I was beginning to think that it must be pigheadedness on my part [that I held to it]." One very famous oceanographer had claimed that it was impossible, and "we only have to find the mistake in [our] reasoning. But in spite of much verbiage he could not find it," Kuenen wrote with satisfaction.[28]

The work earned Heezen a position at Lamont as a research associate, where he focused on deep-sea topography and sedimentation. He was better equipped for the task than most at Lamont, who, like Ewing and Worzel, tended to be trained in physics and had little feeling for what submarine geological features might signify (chapter 4), and he found a congenial coworker in geologist Marie Tharp.

Marie Tharp and the Mid-Atlantic Ridge

Marie Tharp (1920–2006) was hired by Ewing as a general research assistant in 1948, the year that Lamont was established and that Heezen began his PhD studies. She had earned a master's degree in geology at the University of Michigan and then worked for Stanolind Oil Company in Tulsa, Oklahoma.[29] Her job at Lamont focused on assisting the (male) graduate students, but within a short time she was working primarily, and then exclusively, for Bruce Heezen.

In 1953, Heezen and Tharp began to analyze Atlantic echo-sounding profiles. Following the then-common pattern, the woman did the detailed labor: the sedulous task of plotting ocean profiles from raw sounding numbers and compiling them into the physiographic map of the seafloor.[30] Tharp also took on Heezen's administrative and secretarial duties, managing his finances, correspondence, and even his health. When he developed a heart murmur in 1960—Heezen would die of a heart attack at age fifty-three—Tharp wrote to his physician to find out what restrictions he should observe. In her letter, Tharp described Heezen as her "'mean ole boss,' best friend and constant companion."[31] In later years, they would live and work together in a house in Nyack, New York, prompting speculation that they were more than just friends.

Continuous sounding had been available since the 1930s, but the data were not highly accurate and were restricted to depths of fewer than 2,000 fathoms (12,000 feet). During the war, both depth and accuracy were greatly improved, and by the end of the war the full range of ocean depths could be explored. Lamont scientists had access to over 150,000 miles of continuous echo-sounding records; Tharp appears to have looked at them all.[32] This work was supported by the ONR and the Bureau of Ships; Heezen and Tharp also had profiles sent to them from Woods Hole, the Hydrographic Office, Bell Laboratories (the research arm of AT&T), and the British Admiralty. A far more detailed picture of the ocean floor began to emerge as Tharp compiled the profiles across the Mid-Atlantic Ridge (fig. 5.4).

The ridge had been known since the *Challenger* expedition (1872–1876), when differences in salinity and temperature across it suggested that it served as a barrier to circulation. Its existence took on heightened significance when geologists began to debate Alfred Wegener's theory of continen-

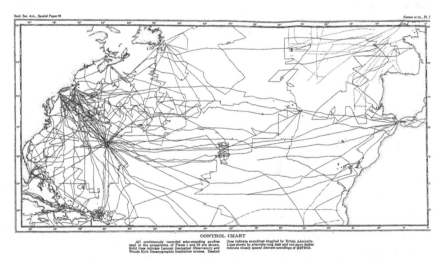

Figure 5.4 Control chart, showing coverage of precision depth recorder profiles, 1959. From B. Heezen, M. Tharp, and M. Ewing, "Special Paper #65: The Floors of the Ocean. I. The North Atlantic," Lamont Geological Observatory, Columbia University, Geological Society of America, Palisades, NY, 1959 (plate 21). Reprinted with permission of the Geological Society of America; permission conveyed through Copyright Clearance Center, Inc.

tal drift. Wegener's supporters interpreted the ridge as the site where Europe and North America had broken apart; British geologist Arthur Holmes suggested that the ridge was a fragment left behind. But in the 1920s, the available picture of the ridge was primitive (see fig. 5.1). It was sharpened in 1938 when scientists on the *Meteor* returned to the site and produced a cross-section of the ridge that revealed "striking depressions" at its crest, a result that was published the following year.[33] This might have been taken as evidence of a rift, but 1939 was not a good year for American scientists to engage with their German colleagues, and little discussion ensued.

After the war, the Mid-Atlantic Ridge was high on Ewing's scientific agenda. In the proposal for the 1947 *Atlantis* work, he noted the relevance of the ridge to the theory of continental drift (albeit confusing Wegener with Wagner): "The geological structure of this ridge is not at all well known, although most of the rocks forming the islands [where the ridge is emergent] are oceanic rather than continental in nature. Proponents of the Wagner [sic] hypothesis of continental drift consider the ridge as a fragment left joined as the continents of the Western hemisphere separated from Europe and Africa."[34] Ewing's primary motivation for mapping the ridge, however, was not to test Wegener's theory; it was to explore for submarine canyons, which, if Daly's hypothesis were correct, implied strong currents that might disrupt cables or other military installations. It was also linked to a major military development: sound fixing and ranging (SOFAR), a technique developed during the war to use sound to determine the position of a submarine or a downed airman at sea (chapter 2). Ewing wrote: "[In] connection with … the SOFAR project in 1945, the USS Muir made a continuous recording of the soundings along the western flank of the ridge for several hundred miles. Extremely rugged topography was revealed, which has been interpreted as indicating the strong probability that submarine canyons similar to those found off the Atlantic coast of the United States have been carved into the flank of the Middle Atlantic Ridge."[35]

Historian Albert Theberge has stressed that the central depression on the Mid-Atlantic Ridge had already been noted in the 1930s by the *Meteor* scientists (and again in 1953 by Maurice Hill), and suggests that Heezen and Tharp took more credit for the discovery than warranted.[36] The issue of credit is further complicated by gender. In later years Heezen would credit Tharp with the recognition that the cleft in the center of the ridge showed that it was a rift, but at the time her contribution was not acknowledged; the first published papers to focus on the cleft and its significance were authored only by Heezen and Ewing (fig. 5.5).[37]

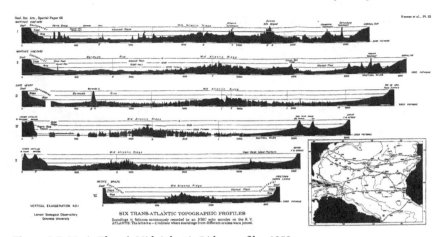

Figure 5.5 Marie Tharp's Mid-Atlantic Ridge profiles, 1959.

From B. Heezen, M. Tharp, and M. Ewing, "Special Paper #65: The Floors of the Ocean. I. The North Atlantic," Lamont Geological Observatory, Columbia University, Geological Society of America, Palisades, NY, 1959 (plate 22). Reprinted with permission of the Geological Society of America; permission conveyed through Copyright Clearance Center, Inc.

Whatever the appropriate apportioning of credit, it is undisputed that Tharp noticed the similarity of the Mid-Atlantic Ridge with profiles across the East African Rift and highlighted the fact that earthquake epicenters on the ridge were clustered in the central 100–120 miles of a complex system that was otherwise 1,000–1,500 miles across (fig. 5.6). The concentration of earthquakes supported the rift interpretation, and if the ridge were a rift, then it might be the location at which two segments of Earth's crust were being pulled apart.

These results were summarized in 1958 in the first release of the physiographic map of the seafloor and then in a 1959 special paper of the Geological Society of America. (Tharp was second author, Ewing third.) The map and paper were widely cited and reproduced and led Heezen and Tharp to produce similar maps for other oceans. Heezen was offered a permanent academic appointment at Columbia as an assistant professor of geology. (He became an associate professor in 1964; Tharp was offered nothing.) Colleagues from around the world wrote to congratulate Heezen (and to a lesser extent Tharp). Many specifically noted the accomplishment of building the unique map based heavily on classified data. As Stommel said, they had negotiated a bureaucracy worthy of its Eastern Bloc counterparts, overcoming the obstacles that had vexed Roger Revelle and Harry Hess. How had they done it?

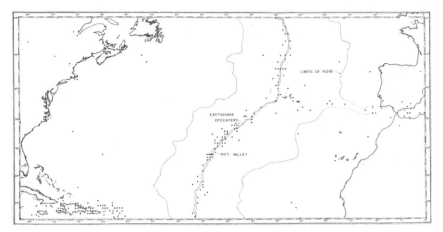

Figure 5.6 Distribution of earthquake epicenters, 1959.

From B. Heezen, M. Tharp, and M. Ewing, "Special Paper #65: The Floors of the Ocean. I. The North Atlantic," Lamont Geological Observatory, Columbia University, Geological Society of America, Palisades, NY, 1959 (plate 29). Reprinted with permission of the Geological Society of America; permission conveyed through Copyright Clearance Center, Inc.

The Solution to Classification: Data Degradation

In February 1954, Harry Hess was revising his lecture notes on submarine geology and wrote to Heezen requesting a figure "showing at least roughly the outlines of the North East Atlantic Plain."[38] Heezen replied that there was none, for there had been no Lamont work on the northeastern Atlantic plain since 1947, and he did not know what data the Hydrographic Office might have. "Since hydro started this dam[n] classification policy," he wrote, "I have not been back to hydro so I don't know what data they might have on it."[39] Actually, Hess clarified, he had meant to ask for a picture of the entire North Atlantic basin. "This would be a big help if you could send me a picture."[40] This idea—to make a picture of the whole North Atlantic—would provide part of the solution to the problem of the "damn" classification. The other part would be what Emanuel Piore suggested to Maurice Ewing: data omission and degradation (chapter 4).

The physiographic maps were based on echograms—continuous linear records made with the sonic depth finder—collected on numerous cruises during and since World War II. The results had been plotted as two-dimensional profiles—cross sections from east to west across the ocean—and then generalized by Tharp into a three-dimensional picture. But the maps were in an important way very different from comparable maps on land: "The shape of all land features is a matter of recorded fact," Tharp explained, but "the preparation of a marine physiographic diagram requires the author to postulate the patterns and trends of the relief on the basis of cross sections and then to portray this interpretation in the diagram."[41]

This was the truth but not the whole truth: the physiographic maps were also a device to evade security restrictions. In fact, they were the solution to the classification problem: a means to present data qualitatively without revealing the location of any particular datum or the quantitative details of any of them. Heezen and Tharp solved the problem of dealing with classified data by downgrading it, so that an overall picture emerged but the details on which it was based did not.

Tharp presented it as a win-win situation: they solved the classification problem and presented their data in a "natural" way. "At that time," she later recounted, "detailed contour maps of the ocean floor were classified by the US Navy, so we had to find another way of presenting our data. Packaging our information in the form of physiographic diagrams not only solved our publication problems but also provided a much more natural picture of the seafloor. The diagrams allowed us, for example, to capture the seafloor's many textured variations, contrasting the smoothness of the abyssal plains with the ruggedness of the mountains along the ridges."[42] They also protected classified information by giving features names other than the ones by which were known in the Navy and, in some cases, by slightly shifting their positions.[43]

This story has been told repeatedly as an example of how earth scientists found ingenious ways to evade security restrictions. One might wonder whether it is a post hoc rational reconstruction, but contemporary accounts support it. In 1960, Heezen wrote to a colleague at the Navy Electronics Laboratory: "Due to the Navy's classification of contour charts, we ... shifted emphasis to physiographic diagrams, which could show more information on texture without incurring the wrath of the classifiers, and for ten years now, this has been the basis of our studies in morphology."[44] Heezen and Tharp did not make physiographic maps because they wanted to; they made them because they had to, or at least because they couldn't think of any other way to convey the information while evading the "iron curtain" of classification.

Invention, however, does emerge from necessity, and the maps were very inventive, striking for their evocative detail and dramatic bird's-eye view of what the oceans would look like were they drained of water. They are much more arresting than a conventional contour map would have been and much easier for a nonexpert to grasp. In later years, when Frank Press had his portrait painted on the occasion of becoming president of the US National Academy of Sciences—the first earth scientist to earn that role—he posed in front

Figure 5.7 Frank Press in official NAS portrait, 1995.
Image © Jon R. Friedman, reprinted with permission.

of the physiographic map of the Atlantic seafloor (fig 5.7). Facing the challenge of security restrictions, Heezen and Tharp had produced something perhaps even more impactful than might otherwise have been the case.

From a scientific perspective, however, there were problems. One problem was that quantitative data been rendered qualitative. By his own admission, Heezen would have preferred to make contour charts, and it is easy to see why: they would have preserved the quantitative character of the data. Field scientists often defend the epistemic value of qualitative information, but no one would have defended the deliberate degradation of data unless there were no alternative.[45] A second problem was that data degradation meant that there was no way for colleagues to know which interpretations were based on rich data, which were interpolated between sparse data points, and which were pure speculation. The maps gave the impression that the entire ocean floor had been mapped, but this was far from the case.[46] Tharp joked about adding mermaids to the map, a reminder that the maps were not merely inventive but in some cases *invented*. Some portions were nearly made up, and colleagues without clearances had no way to determine which those were, or to judge the legitimacy of the interpretations.

Heezen acknowledged this problem, albeit indirectly. In June 1959, two months after *The Floor of the Oceans* was published, Columbus Iselin again raised the issue of the impact on science of data classification. Like Revelle six years before (chapter 4), Iselin wrote to colleagues asking them to "jot down some thoughts on the problem of the classification of deep-sea soundings." Heezen made clear that, despite having found a laudable work-around, he considered the Navy restrictions unreasonable. The topographic data were "vital" to basic research, but blanket classification of marine data left "less than 30% of the earth" available to scientists to investigate freely. Like Revelle, Hess, Iselin, and Ewing, Heezen accepted that the Navy "might have some legitimate reason to classify certain areas"; the question was how much data might be released without compromising US military advantage.

His answer was quite a lot, if one recognized the different scales needed for military operation and scientific investigation. What scientists needed above all was to elucidate the big picture, such "questions of geology as: the origins of the continents and the oceans; continental drift; continental growth; convection currents; global shrinkage; global expansion; the origins of mountain ranges from geosynclines; the origins of petroleum deposits; future mining of the deep sea." A solution, he argued, was to release soundings on a coarse scale—adequate for scientific work but too coarse to give away much to an enemy: "The release of world-wide grids of precision soundings on a 20–40 mile spacing could in no way endanger the military use of topographic detail in navigation and warfare, but would serve the ... demands ... for basic geological investigations."[47]

Heezen also noted that current practice lumped together a variety of different types of data, some of which could be released for scientific use without compromising Navy needs. For example, soundings made before the invention of the precision depth recorder were often of low quality and eventually would likely be discarded, so why not simply declassify them? In fact, it would be better for the Navy if it did, insofar as "the numerous errors and inconsistencies would not be complimentary to the Navy's reputation if workers inexperienced in the use of such unsatisfactory data attempted to draw charts on the basis of this information." Scientists should be given the chance to use the lower-quality data before they were discarded: "It would be a shame if all the compilations were assigned to the furnace still classified, for many of the gross features of topography could be worked out from them even though they are useless for the classified application."[48]

Heezen's point about data quality was not only valid but also personal. All data are not created equal, and scientific interpretation requires assessment of data quality. Scrutiny, evaluation, and testing are the sine qua non of scientific research, but when data are classified, or when data obtained by various means are lumped together, assessment becomes impossible. This was the central dilemma of his physiographic map: few people could say how good the interpretations were, because few had access to the data on which they were based. Most of Heezen's and Tharp's colleagues trusted them, but it clearly bothered Heezen that his own map lumped together the good-quality data with the bad.

On one level, it is not surprising that Heezen's and Tharp's colleagues accepted their maps: only a handful of scientists ever have the time, motivation, and expertise to check their colleagues' work. Unless something very significant is at stake, the institutional rewards for doing so are scant. Moreover, as sociologist Harry Collins has stressed, no replication is ever exact, and Steven Shapin has stressed that all scientific knowledge contains an ineluctable element of trust.[49] But it is one thing to say that scientific processes of replication and peer review are imperfect—that data are not always evaluated as crisply as we might like—and another thing to be in a situation in which data cannot be evaluated. No one claimed that what Heezen and Tharp had done was wrong; the question was which parts of the map were likely to be reliable and which parts less so, and then whether the resulting picture was accurate.

Furthermore, geological science had been moving rapidly and assertively in the direction of greater quantitative specificity, but Heezen and Tharp moved in the opposite direction. The military context of seafloor topographic studies led them to produce something that was less rigorous than it would otherwise have been and that was in some sense old-fashioned. Perhaps more important for the historical argument, it was different from what Heezen himself would have done had he been given a choice or seen an alternative. It was not what Heezen *wanted* to do. It was what, under the circumstances, he had to do.

We know this because earlier in his career Heezen had criticized a colleague for precisely the same maneuver. In 1953, he had written to Robert Dill, an oceanographer at the US Navy Electronics Lab, about a paper Dill had just published in the *Bulletin of the Geological Society of America* on the submarine Monterey Canyon. Heezen queried the contrast between the small number of sounding lines reported in the paper and the degree of detail offered in the contouring of the canyon. "It is obvious," he wrote to Dill, "that you have much more information than you were allowed to show. For my own edification I would like to know how much more control you did have. The trouble with this classification problem is that the foreign reader looking at your map will wonder why you drew it the way you did not knowing how much more information you have but can't tell about because of the naval security problem."[50]

The question of control is a basic scientific one, and from it follow several additional questions: What is this conclusion based on? How much evidence is there? How good is the evidence? How strongly does it support the interpretation? Heezen framed the problem in terms of foreign readers, but the same would have been true of any geologist without clearance, which meant most readers of the *Bulletin*.

Epistemic and Social Difficulties

While Heezen was becoming internationally famous for his Mid-Atlantic Ridge work, in Lamont circles he was becoming infamous for his difficulties dealing with Maurice Ewing. Colleagues often glossed the difficulties as personal—a clash between two strong personalities—but historical evidence suggests that their relations were strained at least in part because of conflicts over military matters. The troubles culminated in the late 1960s, when Ewing attempted to have Heezen dismissed from the university. The documentary record of the conflicts between Heezen and Ewing is so extensive, convoluted, and bitter as to make it difficult to determine what was really at stake. Heezen and Ewing argued about scientific priority, about diction and handwriting, and about Marie Tharp. Some of these difficulties may have been personal, but others were clearly professional. One matter in particular stands out: the handling of classified materials.

Heezen had always been challenged to stay on top of details. Tharp's assistance relieved him of nearly all administrative and secretarial work, but still he struggled. Perhaps he was dyslexic—in one memo he addressed Harry Hess, whom he knew well, as "Harvey"—but he was sloppy, particularly with respect to intellectual boundaries, including the sensitive matter of acknowledging and crediting his colleagues' work.[51] In 1957 he apologized profusely to Philip Kuenen—the man whose experimental work Heezen had spectacularly confirmed in Hudson Canyon—for suggesting that Kuenen had not suspected the "vast significance" of turbidity currents in deep-sea sedimentation, a characterization that was both wrong and insulting.[52] In 1958, his signature work, "The Floor of the Oceans," was nearly rejected by the Geological Society of America because of the disastrous state of the manuscript, which was full of slang, poor-quality illustrations, inconsistent captions and section headings, and numerous "careless mistakes and type errors."[53]

Heezen's carelessness extended as well to university matters. He became persona non grata at the Columbia Geology Department library when at one point he had forty-one items that were years overdue. When Heezen's secretary replied to the library that he would not be able to return the books for another three weeks because he was "extremely busy," the librarian noted dryly "so are we all."[54] On another occasion, Heezen was reprimanded for charging expenses to accounts that he knew had no funds left in them.[55]

Unpaid library fines are not a federal offense, but mishandling classified materials can be. When Ewing discovered that Heezen's sloppiness extended to the handling of classified charts, he wrote a terse letter to "protest the arrangement made about the chart files in the submarine topography room" and the risk it posed:[56] "Some of the files were not locked, and an arrangement was left so that a person without the proper clearance could get access to the keys. This is very serious.... Your general attitude toward the seriousness of this problem is not a credit to you, and could bring serious consequences to the Observatory and to me personally, as well as, of course, to you. This is not a matter to be treated casually.... I realize that it is difficult and vexing and all that, but still it is not to be trifled with."[57] Ewing asked Heezen to acknowledge the letter and to reply "making any comment you care to." If he did, there is no record of it.[58]

Few scientists were as conscientious as Ewing about maintaining security protocols. Roger Revelle, for example, left classified documents under his bed in La Jolla, and Harry Hess jumbled classified reports with reprints, class notes, and ordinary correspondence in his Princeton files.[59] Indeed, on several occasions Ewing boasted to military officials about his institution's aggressive policy on document destruction, which makes it difficult to assess how often arguments erupted on the topic. Still, Heezen seems to have been exceptionally cavalier, and evidence of continued tension can be found, particularly in an incident from 1964.

Heezen had again left classified materials in unsecured conditions and received a terse note of complaint about it from the Lamont security officer. This time Heezen did reply, defending the procedures in his lab against the appearances:

1. The procedure for locking classified charts in map cases has *not* been abandoned.
2. Classified charts are not stored in open areas and are not accessible to anyone that comes into our department.
3. There are over 100 downgraded formerly classified charts that have not been stamped this situation will be remedied within the next few days.[60]

Heezen considered the matter to be merely bureaucratic—that certain downgraded materials had not yet been stamped accordingly—and he asked his secretary to order a rubber stamp reading, "Declassified as per OPNAV INSTRUCTION 3160.6A 20 Jan 1960."[61] But this was in 1964: the instruction to downgrade had been issued four years before.

Ewing was also vexed over the perception that Heezen exaggerated his contributions to work that belonged jointly to various Lamont staff, including Ewing.[62] In 1963, for example, Heezen initiated a plan to collaborate with a marine biologist on an analysis of biological phenomena in underwater photos taken from a Lamont ship. Ewing was furious, believing that Heezen was trying to scoop him.[63] In 1966, Heezen presented at a Moscow conference a paper on the relation between magnetic reversals and the evolution and extinction of life.[64] Paleomagnetism was a major research area at Lamont, but Heezen played no role in it, so why present at an international meeting? While Heezen made promiscuous use of data that staff considered "corporate property," he and Tharp were notoriously reluctant to share their own.[65]

In November 1967, Heezen was suspended from his employment at Lamont; Ewing instructed him and Tharp to vacate their offices, allowing Heezen to retain only minimal space for meeting with his graduate students. (Soon he would ban Heezen from the observatory entirely.) Ewing terminated Tharp's contract, requesting the university to initiate formal disciplinary action against Heezen with the goal of his dismissal. The central issue was Heezen's inappropriate use of Lamont data.

Heezen was required to submit all his manuscripts to Ewing for approval to ensure that he was not inappropriately submitting papers for publication based on other people's work.[66] The senior faculty agreed with Ewing's complaint; the Executive Committee of the Senior Staff summarized their views: "There is no doubt that Professor Heezen has gravely offended numbers of his colleagues by the manner of his use of ideas and data that are wholly or in part the intellectual property of others or of the laboratory as a whole."[67] Heezen had failed to act in accord with "mutual respect and fair play. Greed and unbridled ambition cannot be tolerated."[68]

Heezen considered these actions harassment and censorship, and he penned a bitter, five-page grievance to Columbia's president Grayson Kirk, focusing his vitriol on Ewing. He accused Ewing of attempting to usurp academic tenure, of blackmailing students, and of imposing military-style hierarchy and efficiency at the expense of academic freedom and imagination.

Heezen began with Ewing's treatment of students, alleging that Ewing's militaristic approach had led to the exclusion of students from research programs in favor of "a corps of efficient but often unimaginative technicians who could more easily be controlled than the more gifted but often rebellious graduate students." He claimed that, when he and Ewing had quarreled, Ewing had retaliated by threatening his students' financial support. But the crux of the complaint was that Ewing was a martinet, running Lamont not like the academic institution it was supposed to be but like a branch of the armed services.[69] In the years he had been at Lamont, Heezen claimed, only two staff meetings had ever been held: "I am no student of government but my modest observations indicate that a dictatorial government devoted to the principle in which the means justified the ends is a terrifying opponent to a democratic system.... The Administrative system of Lamont must be seriously reformed in order to insure the academic freedom of the Lamont facility."[70]

Heezen insisted that he respected what Ewing had accomplished—particularly in comparison with Woods Hole and Scripps—and that he did not wish to see Lamont degraded by an excess of committees. But he did wish to see Ewing reined in or driven out: "One principal oceanographic organization in the US suffers from so much democracy that it actually has anarchy, another suffers from fair but dull leadership.... It is time that the University reaffirm its faith in academic freedom of the members of our 'community of scholars' that the temptation to misuse administrative prerogatives in order to gain personal advantage be condemned anew. A firm declaration may suffice for the moment but over the long run a solution must be sought for the underlying administrative disease."[71]

No one would dispute that Ewing ran Lamont with a firm hand. As we saw in chapter 2, when Henry Stommel won AGU's Ewing Medal he expressed discomfort about a medal named for a man under whom he had decided he could not work. Stommel had never worked with anyone who was "so clearly a 'boss' as Doc," the "oceanographic equivalent of General Patton."[72] Elizabeth Stommel put it less charitably: "Ewing ... had an empire, and he had any number of people who were ... willing to sacrifice their own autonomy to do whatever it was that he required." Heezen was not the only one to use the word *dictator* in reference to Ewing; on several occasions Mrs. Stommel did, too.[73] Henry Stommel was fortunate that there were other things to do at Woods Hole, but there was no way to work at Lamont and not be under Ewing's direction and authority.

Academic freedom had also come up at Scripps in the 1930s and at Woods Hole in the 1960s, and at both institutions the central issue was the freedom of scientists to pursue curiosity-driven research. At Lamont, however, the central issue was the management of classified research. Ewing's most urgent grievance against Heezen was his repeated "infractions of established rules." Elsewhere on the Columbia campus, however, students , staff, and other faculty were questioning why the staff at Lamont were doing classified research at all.[74]

Classified Research and the Politics of Neutral Knowledge

Of the three leading US oceanographic institutions—Woods Hole, Scripps and Lamont—the latter was the one in which military patronage was most fully accepted by its scientists and where the fewest severe conflicts and complaints about it arose. Lamont was a bit different from Woods Hole and Scripps: the latter two had been founded in the early twentieth century on private philanthropic support with no military involvement and had diverse emphases under their early directors. Lamont, in contrast, was founded in 1949 and from its inception was strongly linked to Navy projects and funding.[75]

This ready acceptance of Navy support was challenged in 1968 when Columbia pondered whether to eliminate classified research on campus. Because of military secrecy, it was hard to know just how much Navy funding flowed into Lamont, but by some reckonings it was the largest site of classified research on campus, and Lamont's military links became an issue in the student unrest that rocked the university that year. Ewing was deeply disturbed by the events—which recapitulated on a grand scale the dispute he was having with Heezen—and he sought advice from the chair of the geology department, oceanographer and marine geologist Charles L. "Chuck" Drake.[76]

Ewing was not contemplating giving up classified research; he was fathoming how best to defend it against the challenges that were arising on the main campus.[77] Drake suggested that Ewing stress that the choice facing them was not Navy versus no Navy, but rather Navy versus some other external patron. If Navy funding were rebuffed, what would replace it, and would it be better or worse?[78] Drake advised Ewing that he should insist that the alternative would be worse.

Before World War II, Drake suggested, the only significant external patron of the earth sciences was private industry—mostly oil, gas, and mineral companies or individuals who had made fortunes in those industries. This was not actually true—the US Geological Survey and state surveys were also important patrons—but what was true was that after the war, the armed services and the National Science Foundation (NSF) had become the principal funders.[79] Drake believed that the military was just as "enlightened" in its support of basic research as the NSF and had done far more for the earth sciences than private industry; Columbia was hypocritical not to acknowledge this.[80] Without money, there would be no science, and money had to come from somewhere, since universities were "a sink not a source for money": "Any outside source of major funding has the potential for influencing the university or the investigator if he allows it. If the investigator disapproves of the source, he is free to terminate his association. If the university feels that it is being unduly influenced by the outside source, it can refuse to accept the funds, and, if the investigator takes issue with this, he can investigate somewhere else."[81] The key dilemma of directed research, Drake argued, was whether the outside funding source unduly influenced the investigations.

Like Vannevar Bush before him—and likely following Bush's argument—Drake was insistent that science was a free market in which investigators and universities could accept or reject funds as they saw fit. But this was not a realistic appraisal—indeed, it was disingenuous—because in oceanography the Navy was pretty much the only game in town.

In the late 1990s, retired Harvard sedimentologist Raymond Siever (1923–2004) made this point in a volume edited by Noam Chomsky on the Cold War university.[82] Woods Hole (in his opinion) was effectively a "government laboratory that was privately run," and when he arrived in the 1950s, he was told that he should be cleared "as a matter of course."[83] He explained that he did not need security clearance because he was neither doing nor intending to do classified work. He was told that, irrespective of what he was working on, without clearance he would be unable to enter certain buildings and would be entirely "barred from the ships." The centrality of Navy funding would have made anyone who declined to participate an outcast, and the prevalence of projects requiring security clearance effectively made it impossible to operate without it:

A few years later, during a research conference, I was sitting at the bar with a navy program director, and after too many drinks we got into a discussion of the possibility that the Cold War would turn hot. He said that in that event we would need all our oceanographers for war work and when we did, we would know exactly where to go to find them, using a list of all the oceanographers supported by grants. I demurred, saying that I had never been cleared and so would not be eligible for war work—aside from the absurdity of discussing a post-nuclear world in rational terms. He laughed and informed me that whether I knew it or not, I had been investigated and that was already taken care of. So much for trying to stay out of the system.[84]

As for spurning further military funding, many junior scientists were only vaguely aware of where the funding was coming from, anyway. Just as the senior scientists and institution director took care of the clearances, they also took care of the money.[85]

At Lamont and Scripps it was no different. By the late 1950s, the entire world of oceanography, marine geology, and marine geophysics was intercalated with the US Navy: the institutions, the personnel, the money, the instruments, and, above all, the ships. Academic scientists served as ONR program directors, military officials led major scientific institutions, and, as Siever learned to his chagrin, background checks were performed on everyone. Even if a researcher's grants came from the NSF, the ship he traveled on had likely been donated or loaned by the Navy, its operating expenses funded by the ONR, and the instruments developed on Navy grants. At Lamont in 1968, the two research ships, *Vema* and *Robert D. Conrad*, were funded half by the National Science Foundation and half by the Navy.[86] As Siever learned, a scientist without clearance would be unable to board those ships, irrespective of the source of research funds.

In principle, one might decline to be affiliated with military projects; in practice, Siever tried and found it impossible. He concluded, "If I had insisted on having absolutely nothing to do with the [Department of Defense] in any way, however indirect, I would have had to leave oceanography." Drake's claim that scientists were free to decline military funding reminds one of Anatole France's comment that the law in its majestic equality forbids rich and poor alike from sleeping under bridges.

In 1968, there were nine classified projects at Lamont, with a total annual budget of just under $4 million dollars, a bit less than half of Lamont's total scientific research budget at the time.[87] Six were subject only to minor restrictions: three ONR grants for ship support, long-range research in geophysics, and underwater sound in the Arctic; one Air Force grant for seismology, related to nuclear-test-ban verification; and two National Aeronautics and Space Administration (NASA) grants on lunar seismology and lunar heat flow, both of which contained secret information on rocket booster characteristics. Three others involved high-level security: one grant from the Naval Ships Systems Command for underwater sound measurements and acoustic properties of the deep ocean, and two ONR grants, both based at the Columbia University Bermuda Geophysical Station, on underwater sound transmission and explosion effects.

This did not include, however, the Geophysical Field Station itself, run by Gordon R. Hamilton. With an annual budget of $900,000, the Bermuda station was the second-largest project at Lamont, exceeded only by the

ONR grant for general ship support and geophysical research.[88] The station was dedicated to the application of underwater acoustics to antisubmarine warfare, a direct continuation of Ewing's World War II work on underwater sound (chapter 2). The early work there had involved a good deal of basic science, but soon that science was well worked out and the station became an integral part of the US military network for ocean acoustic surveillance. As early as 1951, Hamilton acknowledged this. Writing to administrators at Columbia, he noted that it was an open secret that the station did not, in fact, do basic research. This did not bother Hamilton, but it was adversely affecting recruitment. He was therefore looking for ways to sweeten the pot and asked Columbia to pay moving expenses for scientists recruited to work there: "I stress this because I feel that I must be able to get good men down here. This station is long past the stage where it is doing research in underwater sound transmission for the advancement of science. The objective is detection of Russian submarines at long ranges."[89]

In the 1950s, Columbia had amply supported Hamilton's work, and for more than a decade he worked happily. He was well funded and had Ewing's direct moral, intellectual, and logistical support. He wasn't an outlier, either: his was but one of several laboratories and many projects working directly on issues of military necessity.[90] By 1968, however, the scientific and political landscape had changed. Columbia officials informed Hamilton that his was "the only large organization remaining in the University that engaged primarily in classified research."[91] Hamilton thought he could solve the problem rhetorically, telling Ewing: "For the moment I am shifting the paper work emphasis here from classified to unclassified—basically talking only about the research but omitting the purposes for the research."[92]

Hamilton's strategy was common among those whose work involved classified goals: explain what you are doing but not why. For Hamilton, whether to continue classified research was not a question: he had spent his entire scientific career on military applications of underwater acoustics and had neither the desire nor the ability to change. "My personal intention is to remain in this type of work—my competence is here, I enjoy the work, and the technical competition for funding is less," he wrote.[93] But both Hamilton and Ewing had underestimated the forces gathering at Columbia. Facing mounting pressure, Ewing turned to another of his colleagues, seismologist and Lamont assistant director James Dorman.

Dorman took a hard line, denying that the term *classified research* was even appropriate to describe Lamont activities, because the work was not done in secret and permission to publish was typically granted "without formal application."[94] This was a tortured argument, relying as it did on a very stringent notion of secrecy. It was also false, because (as we have seen) permission to publish was not routine and sometimes not granted at all. But the crux of Dorman's argument was that Lamont's work was needed not just by Lamont but also by the military, the nation, and the world. Scientists engaged with military officials had a salutary influence on those officials, Dorman argued, and both scientists and military officers were doing work that the nation— indeed, the world—needed. If unruly undergraduates or rebellious professors didn't like it, it was because they didn't understand it. Everyone who was "personally familiar with the actual operation of 'classified research' in a university believe[s] it is wise, beneficial, and in the interest of the individual, the University, the scientific community, the nation and in the interest of world peace to participate on their own terms in certain 'classified research.'"[95]

There were "pitfalls" in classified research, to be sure, but scientists were right to "wish to retain ... their enlightened and unselfish influence on military decision-making.... In this situation, the university scientist engaged in 'classified research' reasonably finds that his obligation to his students and his overall job is more important than his commitment to university unity."[96] Like the traveler who thinks that everyone else has an accent, Dorman insisted that his was the apolitical, value-neutral position: "If the University were to announce an official distaste for 'classified research,' this would be a dangerous precedent of taking a political position.... It would be better for universities to maintain the traditional role, which I think is officially nonpartisan but dedicated to furnishing facts and conclusions without comment on their political implications."[97]

Dorman's position was both asymmetrical and incoherent—defending his position as apolitical and at the same time righteous because it served the (political) interests of the nation—but it was not unique. Ewing made a similar argument when he defended his program to Columbia's acting president Andrew Cordier: "The Department of Defense should have the right of our counsel on those problems for which our studies have fitted us to help them, and it is our desire to provide such counsel.... We as individuals, and the University as an institution, have the obligation to bring such matters to the attention of the government no matter whose money or efforts have led to their discovery. In any case, the majority of the senior staff feels that their obligations as US citizens supersede their obligation as Columbia University Staff Members in such matters."[98]

Ewing's representation of the views of his staff was fair. Scientists at Lamont generally considered that being engaged in classified research was more than legitimate; it was beneficent. Had they not felt that way they could not have sustained working there, because whether the politics of classified research were right or wrong, whether restrictions on the work were large or small, and whether "classification" meant what those outside classified circles thought it meant, the reality was that formally classified projects represented the lion's share of the scientific effort at Lamont. As at Woods Hole, anyone who refused to participate would have been isolated from colleagues and probably unable to do his job. But Lamont had an additional component not present at Woods Hole: the need to support students. Most of the financial support for graduate students came through the ONR and other Navy grants.[99] A professor of oceanography who rejected military support would have had a very difficult time funding his graduate students. As far as the historical evidence goes, there never was any such professor.[100]

The exclusion of anyone who had a different view helped to create a mutually reinforcing community of like-minded men who had difficulty comprehending why their colleagues—much less the outside world—had problems with classified research. While the university around them engaged in a deep and wide-ranging discussion, not only over the ethics of classified research but also over the purposes of the university, they engaged only in a narrow defense of their own status quo ante, a defense that was hypocritical and also at times bordered on the absurd. Lamont scientists defended classified research on the grounds of the good values that it served, but when others attempted to invoke an alternative set of values that they likewise believed were good, this was rejected as a "dangerous precedent."[101] They characterized rejection of military funding and classified research as political (and therefore inappropriate) while insisting that maintaining the status quo was entirely apolitical. Left unaddressed was the fact that Cold War military oceanography was implicated in the largest political question of all: which nation and what type of economic and political system should dominate the world?

In hundreds of pages of archival material from this period, it is hard to find evidence that Ewing ever took seriously the arguments of his Columbia colleagues and administrators that there was perhaps some contradiction in pursuing secret science in order to defend the free world, much less in a university dedicated to the free and open exchange of ideas. Ewing appears to have been entirely disdainful of such views.[102] But his disdain did not win the day. Ultimately, the Columbia Faculty Senate would conclude that classified research was not consonant with the university mission and should be removed from campus.[103]

While this larger debate was unfolding, the dean of faculty advised Ewing not to make his stand on Heezen's handling of classified materials, which, while "deplorable," was not going to give Ewing "much mileage" and might backfire: "If the consequence of occasional inept management were dismissal, a lot of us, including the incumbent Vice President, would not be here. While I admire and respect a civilized urbanity of conduct, insolence to the Director is not, I think, actionable.... For all of Heezen's difficulties, it has not been proposed that he is other than a good scientist, heavily immersed in science, with an above par record of supervising the research of PhD candidates, an important part of his role as a teacher."[104] Ewing was gravely disappointed by the university's response, in terms of both the specific case and the larger issues at stake. When the senate recommended the termination of classified research on campus, Ewing made the decision to leave Columbia and return to his native Texas.[105] He died two years later of a massive cerebral hemorrhage.

It would be implausible to blame Heezen's troubles entirely on Navy sponsorship of research. Heezen was a troubled and troubling person in many ways. But the military context of scientific work at Lamont created special social and epistemic difficulties, not just for Heezen but for many scientists. We have seen how both Harald Sverdrup at Scripps and Paul Fye at Woods Hole expected to run their institutions with a high degree of centralized authority—Ewing was not exceptional in this respect. Admittedly, Sverdrup had this expectation before Navy funding became significant at Scripps, and strong leadership was probably necessary in oceanographic institutions where expeditions required tremendous organization and mistakes could cost both money and lives. But it would be implausible to ignore the military dimension, not least because having to maintain military standards of secrecy and security amplified the tendency toward and justification of strong centralized leadership. And this created conflict with scientists who expected a high degree of intellectual freedom, felt they had a right to such freedom, and in many cases believed that freedom was necessary for the advance of science.

Sverdrup had trained at the Norwegian War Academy; Fye had been a military officer in World War II; Ewing had done his first oceanographic work at Woods Hole for the Navy and continued to work closely with Navy patrons throughout his professional life. These men's expectations of how to run an academic institution were conditioned by their military training and experiences, reinforced by the abundant Navy funding with which they built their institutions. In attempting to enforce a certain way of doing things, they ran afoul of their colleagues' academic expectations. Perhaps it is not coincidental that Sverdrup was betrayed by his own faculty, that Fye suffered a mutiny in his ranks, and that Ewing spent years in a futile attempt to control Heezen.[106] Military and academic standards were simply not the same. Moreover, the staff complaints about Heezen's use of data that were considered "corporate property" raised a fundamental question: to whom, exactly, did scientific information collected on the Navy's dime belong? This issue had arisen in the 1930s at Scripps, and it arose in Heezen's relations with colleagues.

British colleagues frequently shared information with American colleagues; Heezen's work on the North Atlantic relied in part on Admiralty information. But on several occasions Heezen declined requests to reciprocate. In 1955, British oceanographer Tony Laughton of the National Institute of Oceanography in Southampton suggested that data be traded between the British and US Hydrographic Offices.[107] Laughton took the liberty of sending Heezen some materials that Maurice Hill had shared with him of a survey of a portion of the Mid-Atlantic Ridge. These were a combination of data from HMS *Challenger*, collected in 1953, and from RRS *Discovery II* in 1954. These materials had been sent with the permission of the British Hydrographic Office on the understanding that any materials sent from Lamont would in turn be shared with them; Heezen wrote back claiming that US officials refused to accede to that plan.[108]

When a colleague from the Universidad Nacional Autónoma de México requested soundings and geophysical data from the continental margins of Mexico, Heezen again refused, answering regretfully six months later, "I am very sorry to inform you that due to conditions beyond our control we will be unable to send you the bathymetric chart of the Pacific Ocean adjacent to Mexico."[109] Heezen similarly turned down requests from scientists at Bell Labs and from other American and foreign colleagues.[110] Whether any of these colleagues might have suggested alternative interpretations of the data can never be answered; none had the opportunity. Nor can we say for sure whether Navy restrictions prevented Heezen from sharing data. It may be that he used them as an excuse, because Heezen was also difficult with Bill Menard, his colleague and counterpart at Scripps who had the same access to classified data as he did.

In the early 1960s, Heezen and Menard had cooperated on the Creative Playthings physiographic globe of the world and corresponded about the interpretation of the data that the globe portrayed. In doing this work, some data were shared; in 1961 Heezen suggested they forge an agreement for further collaborative work. Menard begged off additional work on the grounds of being too busy, but he was willing to share data. "Perhaps I should bow out of the contractual arrangements and merely offer you my data for old times' sake," Menard wrote, suggesting they simply agree to share whatever relevant data they had, since it would benefit them both.[111] He continued: "Now that we are getting into each other's ocean so to speak, I would suggest a general working arrangement to the effect that you should feel free to draw on our soundings [and vice versa]. My experience with your lines off New Zealand indicates that in that region you had little to lose by releasing soundings, and I had a lot to gain."[112] It was the same argument Hess and Revelle had made to the Navy: there was more to gain from data sharing than to lose. But Heezen did not accept Menard's proposal, writing back to thank him for what he had sent but ignoring the data-sharing suggestion.

Heezen seems to have had a generally defensive attitude, as if others would try to exploit sharing to his disadvantage. In 1968, when Menard was elected to the National Academy of Sciences, Heezen wrote to congratulate him but inserted a sour note, complaining that Menard had not kept him fully informed of his work: "In looking through my reprint file I find I have not received reprints from you since the late 50s. Inasmuch as our work overlaps to such an extent I would think it desirable for us to exchange reprints. I enclose a list of my publications."[113] In fact, Menard had sent reprints, scrawling in the margin of this letter: "14 different reprints were sent 7/7/67." He replied to Heezen a week later: "With regard to your reprint file, I am a little nonplussed. According to my index, you have received a copy of every paper I have written including those under joint authorship.... The last batch of papers, numbering 12, were sent to you on 7 July 1967.... Anyway under separate cover I am sending you copies of everything still available. I certainly cannot think of a better place for them to go."[114] Heezen knew that secrecy could work both for and against him: there had been several occasions when he was unable to get data he wanted. In April 1957, a Woods Hole colleague penned an urgent memo to Heezen asking him immediately to return a rough copy of a chart he had lent. The colleague had sent it without checking its classification status and then discovered he should not have shared it; he asked Heezen to return it immediately by registered mail "so as to get me off the hook."[115]

In May 1956, Heezen had complained to Robert Dietz, working at the ONR in London, that a request for telegraph charts (presumably of the North Atlantic) had been denied on the grounds that they were "not available to 'the general public.'"[116] Heezen did not consider himself to be part of the general public and asked Dietz to intercede. That same month, Heezen wrote to Warren Wooster at Scripps asking for his compilation of deepwater soundings, complaining that it was difficult for "an outsider to get a clear picture of what data there is at hand."[117] A clear picture well might have made a difference, because, just at that time, Heezen's view of the tectonic processes controlling the ocean floor were moving in a direction that few colleagues would follow: he was about to adopt the hypothesis of an expanding Earth.[118]

Conceptually, the most important aspect of Heezen's work was the demonstration that the mid-ocean ridges form a continuous belt around the globe. This was the idea Hess had postulated back in 1933: "Deformation of the earth's crust is taking place in a single and continuous narrow belt." After more than twenty years, Hess was returning to this idea and developing the theory for which he would earn lasting fame—that the seafloor splits apart at the mid-ocean ridges, driving the crustal plates apart. Hess's readiness to revisit the arguments of the 1930s rested heavily on new paleomagnetic data coming out of England, which showed that the continents had moved along distinct, independent "apparent polar wandering curves." But his thinking also rested heavily on the conclusion by Heezen and Tharp that the Mid-Atlantic Ridge was a rift. If paleomagnetism proved the continents were moving, then the rift indicated how: by the splitting apart of the seafloor at the mid-ocean ridges. Given the crucial role of Heezen's work in articulating the idea of seafloor spreading, one might have thought that Heezen would have endorsed Hess's views and become one of the revolutionaries of plate tectonics.

That claim has been made. In her retrospective accounts, Marie Tharp has described Heezen as an early advocate of drift: once he recognized that the ridge was a rift, it was clear that the oceans were splitting apart in the middle and continental drift was the inevitable conclusion. In an article co-authored with philosopher Henry Frankel, she claimed that in 1952, when she first showed Heezen the way the rift valley lined up on the Atlantic profiles, he "groaned and said, 'it cannot be. It looks too much like continental drift.'"[119] Upon publication of their physiographic map of the South Atlantic in 1961, she also claimed: "One of our Lamont buddies came up to us and said, 'Now you can hardly avoid the conclusion that you guys believe in continental drift.' Of course he was right."[120]

Sadly, this is a fiction. Heezen *did* avoid the conclusion of continental drift. Throughout the period of the plate tectonics revolution, Heezen was a vocal and consistent opponent of the new paradigm, supporting instead the theory of Earth expansion. Along with several other colleagues at Lamont, he held these views not only in defiance of most informed scientific opinion but also in defiance of a great deal of new evidence. This was possible in part, I suggest, because of a shocking ignorance of what others, outside the world of Navy-funded science, had done and a shocking ignorance of the way the new data resolved the problems of the old debates.

The Closed World of Lamont Science

By the mid-1950s, Heezen and colleagues had traced the seismically active portion of the North Atlantic through the South Atlantic and Indian Ocean, through the Gulf of Aden, and into the East African Rift. Going north, the Mid-Atlantic Ridge could be traced through Iceland into the Arctic, and the ridges of the eastern Indian Ocean could be traced continuously into the South Pacific, the East Pacific Rise, and the Gulf of California and the San Andreas Fault. Hess had been shown to be right: the tectonically active areas of the world ocean formed a continuous belt of deformation around the globe. This raised the obvious question of whether Arthur Holmes had been right, too: the oceans were splitting apart along their medial ridges, which in turn drove the motion of the continents. Tharp had claimed that the conclusion of continental drift could hardly be avoided, and many geologists would go on to link the worldwide pattern of mid-ocean ridges to seafloor spreading and continental drift.[121] But not Heezen.

Heezen knew little about this geological work and his colleagues at Lamont generally knew less. He might have been taught something about continental drift as an undergraduate geology major (as some American students were then), but at Lamont the environment was different. A direct consequence of the postwar expansion of geophysics under military sponsorship was the influx into the field of men with training in physics or engineering. An entire generation of earth scientists was working on geological and geophysical problems, knowing little geology and even less about the history of geological thinking—these men did not even know Alfred Wegener's name. Those who knew about continental drift were generally dismissive of it, particularly because leading geophysicists such as Harold Jeffreys opposed it. Few knew that leading geologists in the 1920s and 1930s had taken the idea seriously, and even fewer knew why.[122]

In April 1957, Heezen wrote to a used bookseller in London to find a copy of Alfred Wegener's *The Origins of Continents and Oceans*.[123] He also exchanged reprints with the elderly but still-active Arthur Holmes.[124] Neither engagement seems to have altered his thinking. Moreover, like many Americans, Heezen had a chauvinistic reading of continental drift: he credited the theory to the American geologist Frank Taylor (1860–1938), who in 1926 had published an article in the *American Journal of Science* proposing a general equatorial migration of the continents.[125] Heezen read Taylor's model as implying that the Pacific Basin was experiencing compression, but if the ridges were rifts, then the ocean basins were in extension and Taylor was wrong.[126] But Heezen's reading was wrong: Taylor proposed not that the continental blocks were moving toward the Pacific but that they were moving toward the equator, which indirectly compressed the margins of the Pacific.

By 1956, it was clear that the central Atlantic was in tension—splitting down the middle and spreading apart—yet its margins showed little or no evidence of compression, suggesting that perhaps the ocean as a whole was growing. From there, it was a simple step to the conclusion that the Earth as a whole was growing. Indeed, an expanding Earth made sense if one considered the Atlantic Ocean only. If other oceans were similar, then Earth had to be growing. Moreover, it was possible not to be overly focused on the Atlantic and still come to the conclusion that the planet was expanding; this was the position of Australian geologist S. Warren Carey (1911–2002), who championed the idea in the mid-1950s (and beyond).[127] Carey accepted Wegener's arguments that the continents had once been united; the question was how they split apart. He concluded that rifts were key to the answer. The East African Rift (and its extensions into the Red Sea and the Gulf of Aden) was clearly a place where the crust was pulling part, as were the Mid-Atlantic Ridge and the medial ridges of the Indian and Southern Oceans. These ridges, with their axial seismicity, had been long known; only the discovery of the central trough was recent.[128] But where was the crust being pushed together? Although he claimed to have come to it independently and his articulation of it was more succinct than Carey's, Heezen adopted the view that there was no zone of compression comparable to the zones of extension.[129] Therefore, the Earth must be expanding.

Most geologists flinched at a theory whose mechanism was even more obscure than continental drift, but Heezen found support in the work of Hungarian geophysicist László Egyed (1914–1970), who himself drew on the work of British physicist Paul Dirac (1902–1984). In the 1930s, Dirac had proposed that the force of gravity might decrease over time, which would permit the planet to expand. A time-varying gravitational constant was examined in 1958 by Princeton physicist R. H. Dicke (1916–1997), who published an extensive discussion in *American Scientist*.[130] The theory was elegant. It was a simple explanation that accounted for the interconnected quality of the ridge-rift system: the worldwide network of mid-ocean ridges composed Earth's broken seams. This network explained the jigsaw fit of the continents, as well as the paleontological and stratigraphic evidence from the geological record for the progressive breakup of Pangaea.

Heezen presented his arguments for Earth expansion in 1957 at the American Geophysical Union, and from 1958 onward he became a vocal and persistent advocate of the expansion theory.[131] He was insistent that "strong and persistent patterns in the topography of the deep-sea floor ... impose important restrictions on the directions of any projected continental displacements"—in other words, the continents were separating from the mid-ocean ridges but in no case appeared to be moving toward them.[132] He made his case both in scientific and in popular venues. In 1959, he was quoted in the *New York Herald Tribune* while attending the first International Oceanographic Congress at the United Nations: "Whatever the mechanism, the expanding Earth explains the facts."[133]

The congress was the first of its type and attracted considerable media attention. *Time* ran a story in September 1959 entitled "How Oceans Grew." Tapping into the Cold War spirit of US-Soviet competition and unabashedly promoting the hometown team, the article claimed (absurdly) that the Lamont work had been done "on a shoestring," whereas allegedly better-funded Russian compatriots "had comparatively little to say." It also effaced Marie Tharp, crediting the discovery of the Mid-Ocean Ridge rift to "Lamont Men Maurice Ewing and Bruce Heezen."[134]

In 1960, Heezen brought these ideas together in *Scientific American*, in a piece entitled "The Rift in the Sea Floor." The article rehearsed the history of seafloor exploration—how he and Tharp discovered the rift—and offered three possible theoretical interpretations: continental drift, Earth expansion, and convection current theory. Heezen acknowledged that the paleomagnetic work—the apparent polar-wandering paths—indicated that the continents had changed their positions relative to the poles and to one another. However, his response was not to address these data but to raise difficulties with continental drift, of which there were three. The first was an old one: if the continents drifted through the seafloor, where was the evidence of compressive deformation? Second, if the ocean were subject to such deformation, then "one would expect to find new crust forming somewhere on the [ocean] floor" (as opposed to only on the ridges), but such new crust was nowhere to be found. Third, if the oceans were opening up along the ridge axes, they must be closing up somewhere else, but "there is no evidence for such a reciprocal action."[135]

Each of these was a well-known complaint, but rather than engaging with the responses they had provoked, Heezen simply moved on, claiming that Earth expansion explained the paleomagnetic evidence as well as drift did: "Expansion of the Earth would change the relative positions of the continents in a way that would satisfy the different polar wandering curves— much as inflation of a balloon changes the orientation of points drawn upon it."[136] Heezen might have ended his argument there, except for the fact that Ewing favored the third alternative: convection currents without continental drift. Convection currents were the mechanism for continental drift offered by Arthur Holmes and that would soon be incorporated into and accepted as part of plate tectonics. Ewing, however, had adopted a variant in which the drift component was dropped. In this view, there was no need to displace the continents. The geological evidence that had led Holmes to accept drift in the first place was just ignored.

Heezen claimed that whole Earth expansion was fully consistent with the paleomagnetic data, but that was not true. Few geophysicists involved in the apparent polar-wandering work agreed that whole Earth expansion could explain it: the changes in orientation were too great. In fact, a colleague at the University of British Columbia felt that "expansion would not change the relative positions of the continents at all," it would only change their separation.[137] Heezen replied that he should have written that "the expansion of the Earth could change," but this qualification would hardly satisfy the factual constraints emerging from continental paleomagnetism.[138] So now there were two bodies of data—geological and continental geomagnetic—that Heezen simply dismissed. Heezen was aware of the paleomagnetic data, but his files contain no exchanges with Keith Runcorn, P. M. S. Blackett, Kenneth Creer, or any of the other geophysicists doing this work. Nor did he reach out to any geologists—other than the aging Holmes—who understood the geological case for drift. Evidently, he had developed his views almost entirely in isolation, without attempting to discuss the crucial continental paleomagnetic data with anyone outside Lamont.[139]

Ironically, in 1959 Holmes had raised the issue of the apparent polar-wandering path evidence with Heezen, whose answer was a non sequitur: with Earth expansion, "the continents are not required to float through the oceanic crust like icebergs in the sea. The oceanic crust has formed as the continents were displaced by expansion, thus avoiding the difficult mechanical problems of continental drift."[140] Heezen was not addressing the paleomagnetic data; he was changing the subject. He was also ignoring the evidence of the deep-sea trenches, which he interpreted (following the views of colleagues at Lamont) as extensional features.[141] Most egregiously, he was ignoring (or dismissing) Holmes's own work and contesting an earlier version of continental drift from which Holmes (and others) had already moved on.

Most people outside Lamont thought the trenches were either compressional or strike-slip features, but inside Lamont, the extensional view still held and fit with Heezen's expansionist views. He wrote: "Marginal trenches are now interpreted as extensional features in no way attributable to crustal compression (Talwani, Sutton and Worzel, 1959). Thus the continental margins seem to be dominated by faulting and subsidence (Ewing and Heezen)."[142] These Lamont interpretations were shared by few elsewhere, but Heezen presented them as if established fact, and that seems to be how they were treated there.

How was it possible for a man as talented as Heezen to ignore so much evidence? These arguments and interpretations give the impression that he (and many of his Lamont colleagues) had been living in a closed world—innocent (or dismissive) both of the views of geologists in other places and of those who had considered these problems before. When he belatedly learned about them, he paid insufficient attention to understand them. Indeed, in a comprehensive account of the basis for his views presented in a symposium, "Concepts of the Expanding Earth and Continental Drift," held at Columbia in December 1959, he asserted bizarrely: "We have all been taught that the earth is shrinking and for a century few have offered any serious objections to this unfounded assumption. The ... tentative and speculative conclusion that extensional deformation dominates the tectonic fabric of the sea floor seriously challenges the assumption of a shrinking earth."[143]

This was completely wrong. The postulate of a cooling Earth had been refuted by the discovery of radiogenic heat in the early twentieth century and by the demonstrations by Lord Rayleigh, John Joly, Arthur Holmes, and others that radioactive elements were widespread in common rock-forming minerals. Indeed, the *failure* of the contracting Earth assumption was central to Wegener's argument that a new theory was needed. Scarcely a geologist in America still thought the Earth was shrinking.[144] Heezen was arguing with a view of the Earth that had been refuted half a century before.

In contrast, tensional forces had been a focus of attention at Lamont for some time. While Heezen's and Tharp's discovery of the rift at the center of the Mid-Ocean Ridge clearly demonstrated that the ridge was being pulled apart—that the ocean was in tension at their center—other work at Lamont had wrongly suggested that the trenches were in tension, too. Since 1947, Worzel had worked on gravity measurements over the ridge, work supported jointly by the Geological Society of America and the ONR. While gravity measurements had not been considered particularly useful in the 1930s, the development of long-range rocketry had changed that. Gravity measurements had come to be considered of great "value for the precise navigation of rockets and jet-propelled missiles." In 1959, Worzel explained: "The shape of the Earth is of military value for the precise navigation of rockets and jet-propelled missiles. Gravitational effects near islands, mountains, and continental coasts often cause deflections from the vertical. These make the precise location by astronomic means of ... long-distance navigational stations subject to errors greater than a mile. The only known means of evaluating these errors is by a suitable gravitational survey of the region surrounding each station."[145] Gravity measurements were most valuable close to military navigational stations; much of the Lamont work focused on the region of the Navy Caribbean

SOFAR station, adjacent to the Bermuda field station that Gordon Hamilton ran. The gravity work in Bermuda led Ewing and Worzel to revisit Hess's (and Ewing's own) earlier work in the Caribbean, but they rejected Hess's interpretation that the gravity anomalies there revealed compressive forces. Instead, they argued that the negative anomalies were best explained by a thin crust overlain by water or sediments, and a thin crust would be produced not by compression but by extension.[146] If so, the Atlantic was being pulled apart at both the center and the margins. It was perhaps a short step from an expanding Atlantic Ocean to an expanding Earth, at least for those at Lamont.

One might wonder whether the problem here was simply Bruce Heezen, who, as we've already seen, was in many ways a troubled person. But Ewing and Worzel shared his views, which were not merely idiosyncratic but also ill informed. The closed world of Lamont science, reinforced by Navy protocols and habits of secrecy and reluctant release of data, seems to have contributed not only to idiosyncratic intellectual interpretations but also to an astonishing degree of ignorance.

Recall Ewing's discussion in the 1947 proposal for the mapping of the Mid-Atlantic Ridge, in which he asserted that the rugged topography there indicated "the strong probability that submarine canyons similar to those found off the Atlantic coast of the United States have been carved into the flank of the Middle Atlantic Ridge."[147] According to Daly's theory, the canyons along the Atlantic (or other coasts) were carved by submarine mudslides developing at the toes of deltas. Because no rivers ran on the mid-ocean ridges, it was *impossible* for canyons to develop by this means on the ridges.

No one ever imagined Doc Ewing as a geologist, but one might have thought that a man writing a proposal about submarine canyons would have made some effort to understand the theory he was proposing to test. Similarly, Heezen rejected continental drift in part because he was thinking about Taylor's version of it—a version he misunderstood—and in part because he had not assimilated that geologists had moved passed the original version of the theory in which continents plowed through the seafloor. But the problem was not simply a lack of understanding of specific geological theories and claims. The problem was that thinking at Lamont was entirely inwardly focused. Prevented by military secrecy from discussing their data and ideas with others, Lamont scientists had developed idiosyncratic interpretations of the world, ignored important counterevidence, and argued against views that no one actually held.[148] Secrecy contributed to intellectual isolation, and this isolation had epistemic consequences.

Heezen and Ewing were by no means the only scientists ever to hold idiosyncratic theoretical views, but their work sits in a larger historical context in which no one at Lamont played a significant role in reinvigorating the debate over continental drift. In the early stages of the plate tectonics revolution, no institution in the world had as much relevant data and did as little with it. Once it became clear that seafloor spreading was gaining wide and serious acceptance elsewhere—including at Princeton and Cambridge—Lamont scientists jumped on the bandwagon and made major contributions to confirming and developing the theory.[149] A similar pattern prevailed at Scripps. Despite the huge amount of relevant data compiled and stockpiled in the United States, the initiative for reopening the debate over continental drift came from outside the Navy-dominated institutions and mostly outside the United States altogether. To underscore this point, consider how the theory of seafloor spreading was confirmed.

The Confirmation of Seafloor Spreading

Until 1963 or so, Earth expansion was a live hypothesis: it accounted for some significant empirical data, some prominent scientists supported it, and it had not been refuted.[150] But by 1964 the situation had changed. The British paleomagnetic work had led Hess to revisit the question of global tectonics, and in 1962 he published his landmark paper on the splitting of the ocean at median ridges as a mechanism for moving continents. Meanwhile in 1961, Robert Dietz had published an almost identical argument and given the theory a name: seafloor spreading.[151]

The crux of the idea was that the seafloor behaved like a giant conveyor belt: the crust and upper mantle split apart along the mid-ocean ridges, riding passively on mantle convection currents and driving continental drift. Hess famously glossed his hypothesis as "geopoetry" to deflect potential antagonists who would have immediately seen the connection to the discredited idea of drift. But it was the same as the version of drift proposed by Arthur Holmes (as a modification to Wegener's original formulation) thirty years before. The intellectual issue at stake, of course, was not whether the idea was new or old, but whether it was right. This would be answered by evidence provided by seafloor magnetism.[152]

It was well known that igneous rocks recorded the magnetism of the field prevailing at the time the rocks solidified from magma.[153] On this basis, in 1963, British geologist Drummond Matthews (1931–1997) and his graduate student at Cambridge University, Fred Vine (b. 1939), proposed a "rider" to Hess's hypothesis: if the seafloor spread, with the crust and upper mantle moving symmetrically away from the ridges, then the oceanic crust should record polarity reversals in the Earth's magnetic field in the form of remanent magnetism in the newly created crust. This magnetism would appear as symmetrical bands or stripes of normal and reversed magnetism, detectable by towed magnetometers. In 1955, magnetic stripes had been observed in the Pacific off the western coast of California and then mapped in detail in 1956; the results were published in a widely read 1961 paper in the *Geological Society of America Bulletin*. At the time, no one had known what they meant, but in light of the Vine-Matthews hypothesis, they might mean a lot. Cambridge was abuzz.[154]

Another source of buzz was the paleomagnetic work of P. M. S. Blackett (1897–1974) and colleagues at Imperial College, London, and of Keith Runcorn (1922–1955), first at Cambridge and later at Newcastle-upon-Tyne. By the early 1960s, these scientists had shown that the continents had changed their positions relative to both the magnetic poles and to each other. Hess summarized this work: "Paleomagnetic data ... strongly suggest that the continents have moved by large amounts in geologically comparatively recent times.... This migration of the poles as measured in Europe, North America, Australia, India, etc., has not been the same for each of these land masses. This strongly indicates independent movement in direction and amount of large portions of the Earth's surface."[155] The time had come to bring these data, ideas, and people together. In late 1965, Blackett, Bullard, and Runcorn organized the "Symposium on Continental Drift" for the Royal Society. The meeting was a who's-who of marine geology and geophysics at the time: Ken Creer, Victor Vacquier, Bill Menard, Ron Girdler, Joe Worzel, Robert Fisher, Tuzo Wilson, Harold Jeffreys, and Gordon MacDonald, as well as the elder statesman of mantle convection, Felix Vening Meinesz.

It was also a turning point of scientific opinion in Great Britain. Particularly important was Teddy Bullard's now-unqualified support, expressed in his quantitative demonstration of the jigsaw puzzle fit of the continents— later dubbed the "Bullard fit."[156] Runcorn presented the details of the argument from terrestrial paleomagnetism, showing how the apparent polarwandering paths of Europe and North America diverged, consistent with drift but inexplicable on either a static or an expanding Earth.[157] Ron Girdler (1930–2001), from Newcastle, proposed that the Red Sea and the Gulf of California were incipient ocean basins, where "new oceanic crust is evolving in regions of continental break-up." Noting the shallow seismicity in these regions, he suggested a conceptual framework that would soon be accepted: "Regions of shallow seismicity help to locate regions of rising convection currents in the mantle [and] regions of deep ... focus earthquakes locate regions of descending convection currents."[158] Tuzo Wilson (1908–1993) presented a visionary paper in which he imagined how geological interpretation might look if one accepted continental drift as a general principle and worked from there.[159]

Conspicuously absent were any significant contributions from Lamont. Indeed, when the organizers first composed their list of whom to invite, no one from Lamont was on it. Worzel happened to be on sabbatical at Cambridge; when he learned about the plans, he spoke up and got Heezen and Ewing invited. Explaining to Heezen in early March: "I guess you will be over for the Continental Drift Symposium. Lamont was completely overlooked in its formation. My remarks resulted in you and Doc receiving invitations."[160] Ewing did not attend but Heezen did; Worzel participated as a discussant.

This historic meeting—which in hindsight has been viewed as a turning point—left both Heezen and Worzel unmoved.[161] When the symposium finished, Worzel penned a full-page aerogram to Ewing (something he did regularly) on various topics but scarcely mentioned the meeting: "They had held the symposium on continental drift at the Royal Society, asking me to be a discussant at a late hour" was all he had to say.[162] At Cambridge, however, the buzz was loud and getting louder. The following day Worzel wrote again, describing the Vine-Matthews hypothesis but still dismissing the whole thing as a fad: "In the continental drift symposium the only new considerations over those discussed the last time it had its fad, were the paleomagnetic data and the extensive mid-ocean ridge. This latter is now held to be a factory making new crust in the oceans, which then spreads out to form the nice uniform thin layers which form the new ocean crusts exactly in thickness and velocity the same as the much more ancient ocean crusts. The factory for chewing up old ocean crusts of the size of the whole Atlantic is ignored."[163] Worzel's contribution was to insist that, because one important element remained unexplained, the whole idea could be dismissed. He had listened to the greatest minds in twentieth-century earth science make the case for a paradigm shift and seen the data taken by most of them to be incontrovertible. In hindsight, he was a witness to history. But he was unable to see that history was being made.

The symposium had no more effect on Heezen, who presented a paper entitled "Tectonic Fabric of the Atlantic and Indian Oceans and Continental Drift." The talk was abundantly illustrated by lantern slides, and its published version was accompanied by full-page reproductions. Yet despite the fact that no one had better data on the structure of the mid-ocean ridges—where seafloor spreading was initiated—Heezen rejected the Vine-Matthews hypothesis and clung to whole Earth expansion. Although the paper was ostensibly coauthored by Tharp, Heezen concluded in the first-person singular: "The writer believes that a general expansion of the earth better explains the sea floor tectonic fabric than the recently popular convection current hypothesis."[164] Like Worzel, Heezen dismissed seafloor spreading as something that was just "popular."

When the organizers wrote to Heezen after the meeting to thank him for his contribution, Teddy Bullard made a point of scrawling at the bottom: "Don't forget to congratulate Marie on the splendid Indian Ocean sheet."[165] No doubt his praise for the map was sincere, but two weeks later, writing to Walter Munk at Scripps, he commented on the intellectual isolation of the Lamont scientists: "We had a most interesting discussion on continental drift at the Royal Society recently, 80 per cent of people here seem 95 per cent convinced, but various Americans (Joe Worzel, etc.) seem not to recognize a band wagon when they see one."[166]

By this point Heezen had been committed to the expanding Earth explanation for nearly a decade. The theory, of course, fit his Atlantic data; if one knew only about the Atlantic Ocean—an ocean that splits down its middle but has no "factory for chewing up old ocean crusts"—one might well conclude that Earth expansion was required. But while Heezen and Tharp had been detailing the structure of the Atlantic, Bill Menard had been doing the same for the Pacific—which was very different.[167] Indeed, when Keith Runcorn first invited Heezen to England for the Royal Society meeting, Heezen accepted in a four-line reply. He had received Runcorn's invitation, he was happy to come, he would talk about the Indian Ocean, and he couldn't "wait to hear what Bill Menard will say."[168]

Bill Menard and the Marine Geology of the Pacific

Henry William Menard (1920–1986) was one of the leading earth scientists of his generation, a geologist who rose to a position of prominence as a researcher, educator, and scientific administrator. Born in Fresno, California, he received his BS in geology from the California Institute of Technology in 1942 and then served in the Navy. Like Hess, he put his earth-scientific training to work in the war, serving as an interpreter of aerial photographs in "selecting targets, assessing damage, identifying military and industrial installments," mostly in the Pacific theater. He also prepared charts and maps in the Solomon Islands and "corrected admiralty charts of the French Channel coast."[169] He was promoted from ensign to lieutenant, after the war keeping the latter rank as a reserve officer.

Menard resumed his education to earn his master's degree in geology at Caltech in 1947 and his PhD at Harvard in 1949, working on sediment transport.[170] From there he moved to the newly formed Oceanographic Studies

Branch of the US Navy Electronics Laboratory (NEL) at Point Loma, California, where he began the work that would occupy him for two decades: the "bathymetry and geomorphology of the deep sea [and] the theory of turbidity currents and the transportation of marine sediments."[171] The latter was a direct extension of his PhD work; the former was a specific interest of the Navy. With Navy support, Menard would do in the Pacific what Heezen would do in the Atlantic.

In 1955 Menard moved up the road from Point Loma to La Jolla to become associate professor of geology at Scripps.[172] He focused his attention on the marine geology of the Pacific, participating in over twenty expeditions and more than one thousand Aqua-Lung dives, collecting and compiling reams of bathymetric data and publishing on the structure and sedimentology of the deep sea. Most of his publications had descriptive titles—geology and bathymetry of this and that region—and did not indicate the theoretical issues on which they might bear. But that did not mean that Menard was not thinking about theory; he understood that his work was relevant to tectonic questions.[173]

In 1964 he published *Marine Geology of the Pacific*, a textbook that was translated into Russian and Chinese and might have become the definitive text on the subject had it not been for developments in tectonics. *Marine Geology of the Pacific* was born obsolete, as its theoretical framework was overshadowed by the confirmation of seafloor spreading. The following year, Menard accepted a position as technical officer in the White House Office of Science and Technology (1965–1966) and subsequently focused his energies on scientific administration. From 1978 to 1981, he served as director of the US Geological Survey.[174]

Today Scripps describes Menard as a man who "is perhaps best known for his promotion of the theory of plate tectonics before it was widely accepted in the scientific community."[175] But just as Tharp misrepresented Heezen's contribution to plate tectonics, Scripps has misrepresented Menard's. Like Bruce Heezen, Bill Menard is historically more conspicuous for what he didn't contribute to plate tectonics than for what he did.

Menard was well placed to have played a major role. Maintaining high-level clearances throughout his career, he had access to all the relevant bathymetric, geological, magnetic, and other data potentially available to interpret the Pacific structure. He was professionally and financially secure, having been promoted to full professor in 1961, and he earned additional income from an independent diving business consulting to oil companies working offshore in Southern California.[176] He spent 1962 on sabbatical at Cambridge, where he participated in the great swirl of conversation. He had both the epistemic and social resources at his disposal to make a major contribution, the kind of opportunity most scientists only dream about.

Yet like Heezen, he missed his opportunity. It was not until 1966—after Vine and Matthew's explanation of seafloor spreading had been confirmed through analysis of seafloor magnetics (in part by Lamont scientists, who finally did recognize the bandwagon and scrambled to climb on) that Menard changed his views. He explained to Teddy Bullard in October of that year: "You will find me converted to hanging on the Vine. I finally took the time to look at the magnetic data in detail. The symmetry seems decisive."[177] The reams of restricted bathymetric data to which Menard had access brought him to seafloor spreading no sooner than most of his colleagues and later than some.

Most of Menard's published work focused on the "fracture zones" of the Pacific: long escarpments, generally trending east-west, that rose well above the abyssal seafloor. The type example was the Mendocino Escarpment, which Menard first mapped in 1950 on the Scripps Mid-Pac expedition, co-sponsored by the Navy Electronics Laboratory (NEL). This was a gargantuan feature: a 1,200-mile-long escarpment trending westward from Cape Mendocino in northern California (where the San Andreas fault heads northward off to sea) and rising as much as ten thousand feet above the surrounding seafloor (fig. 5.8).

MENDOCINO ESCARPMENT is traced for 1,200 miles by this topographic map of the bottom off Cape Mendocino in northern California. The Escarpment is represented by the narrow band of contours running from left to right near the bottom of the map. The contours are numbered in fathoms below the surface of the sea. One fathom is six feet.

Figure 5.8 Fracture zones of the Pacific. The Mendocino Escarpment is traced for 1,200 miles by this topographic map of the bottom off Cape Mendocino in Northern California. The escarpment is represented by the narrow band of contours running from left to right near the bottom of the map, numbered in fathoms below the surface of the sea.

From Henry W. Menard, "Fractures in the Pacific Floor," *Scientific American* 193 (1955): 36–41. Illustrations by Irving Geis, reproduced with permission from the estate of Irving Geis and HHMI.

Between 1949 and 1955, Scripps and NEL collaborated on eight major Pacific expeditions. Extensive bathymetric measurements revealed that the Pacific floor was highly irregular, characterized by seamounts, swells, and steep escarpments. Menard focused particularly on the latter and by 1955 had identified four major, parallel escarpments off the western coast of North America: Mendocino, Murray, Clarion, and Clipperton. The Murray was 1,900 miles long and seven thousand feet high, characterized by rift features at its western end and deep narrow troughs with mountainous peaks and submerged volcanoes on either side. The Clipperton had less relief—only two thousand to three thousand feet—but was more than 3,000 miles long.

How did these features form? Their scale suggested massive crustal deformation, but of what kind? With the possible exception of the San Andreas Fault, few features on land came close to matching these, and they were remarkably *straight*, following great circles across the ocean floor. On land, such prominent features were invariably associated with earthquakes, yet these fracture zones were aseismic.

Menard vacillated. In his first article on the subject, published with Robert Dietz in the *Journal of Geology* in 1952, he considered three hypotheses for the origins of the Mendocino Escarpment: a submerged continental slope, a reverse fault, or a strike-slip fault. In a 1955 article in *Scientific American*, he focused his sights deeper to the fundamental causes of the features and considered two options. One was that the fracture zones and the San Andreas Fault were complementary, produced as a by-product of regional, plastic deformation. Following a suggestion by Vening Meinesz, Menard proposed that they were produced by migration of the Earth's crust over the mantle early in Earth history. In this account, the escarpments were relict tectonic features, perhaps as old as three billion years. Menard took pains to say, "This has nothing to do with the hypothesis that continents 'drifted' in comparatively recent geological time."[178] But if the escarpments were billions of years old, why had they not been worn down?

Alternatively, the fracture zones might be caused by active convection currents in the mantle, which, "if they do exist ... might exert a drag on the underside of the crust that would cause fractures."[179] In the mid-1950s, Teddy Bullard, Roger Revelle, and geophysicist Arthur Maxwell had revived the discussion of mantle convection to account for higher-than-expected heat flow over the oceans, suggesting that convection brought heat into the upper mantle under the oceans and compensated for the relative dearth of heat-producing radioactive elements in the oceanic crust.[180] Perhaps deformation associated with convection caused the fractures, but how?

In 1956, Menard concluded that an ancient age for the escarpments made sense only if there had been "subsequent rejuvenation by other processes." He leaned toward "youthful" convection, tentatively concluding that "an annular convection current rising near the Hawaiian Islands and sinking near North America stressed the crust and produced the fracture zones by plastic deformation."[181] Menard looked to contiguous features on land to confirm the idea that these were global-scale responses to mantle deformation and found them in the Channel Islands and transverse ranges of California, which appeared to be extensions of the Murray Fracture Zone (fig. 5.9).[182]

Meanwhile, as Menard was mapping the Pacific bathymetrically, others were mapping it magnetically. In 1958, Menard and his Scripps colleague Victor Vacquier (1907–2009), the inventor of the fluxgate magnetometer (designed to detect antitank land mines in World War II and later used in antisubmarine warfare), published preliminary results of a magnetic survey of the region surrounding the Murray Escarpment.[183] Further detailed work on the US Coast and Geodetic Survey ship *Pioneer* by Ronald G. Mason (1916–2009) and Arthur Raff (1917–1999) revealed a pronounced north-south trend to the magnetic intensity pattern, with the entire pattern apparently offset about eighty nautical miles across the escarpment (with the northern side displaced to the right) (fig. 5.10). The authors interpreted the region as a lava field, with flows filling linear depressions in an older topography. The varying

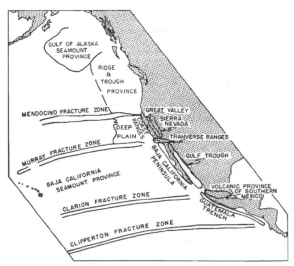

Figure 5.9 The relationship between oceanic fracture zones and contiguous features on land, 1956.

From Henry Menard, "Deformation of the Northeastern Pacific Basin and the West Coast of North America," *Bulletin of the Geological Society of America* 66 (1955): 1149–98 (fig. 14). Reprinted with permission of the Geological Society of America; permission conveyed through Copyright Clearance Center, Inc.

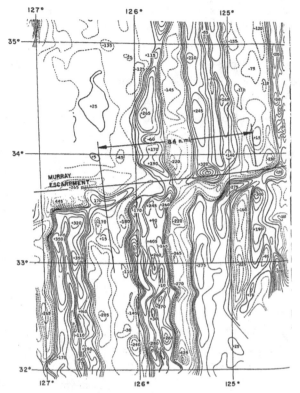

Figure 5.10 The distribution of magnetic anomalies in the Pacific, relative to the fracture zones, 1958. This map shows "the total magnetic intensity of the area outlined in [their] Figure 1. The contour interval is 50 gammas (1 gamma equals 10^{-5} gauss). The north-south trending features are interpreted as being lava flows that fill the linear-depressions of old topography. Note the horizontal shift along the Murray escarpment." From Henry Menard and Victor Vacquier, "Magnetic Survey of Part of the Deep Sea Floor off the Coast of California," *Research Reviews, Office of Naval Research*, (1958) 1–5.

magnetic intensities were interpreted as different flows, with the offsets in the anomalies indicating that the escarpment was a transcurrent fault: "The Murray escarpment is reflected clearly on the magnetic map.... It interrupts the great north-south trending features. The continuity of the latter is re-established by slipping the area north of the Murray escarpment 84 nautical miles to the west along the line of the escarpment. The magnetic survey thus shows that, in addition to the vertical displacements of 1,200 feet, there is a right lateral displacement along the Murray escarpment which occurred after the pouring out of lavas on the ocean floor."[184] In other words, the "fracture zones" were giant strike-slip faults; Menard and Vacquier held to this interpretation for the next eight years.

While Heezen saw the entire Atlantic splitting apart at its seams in one great rent, Menard pictured the Pacific being shredded along multiple parallel fracture zones. This was not necessarily a problem—the Atlantic and Pacific might be different—but it did raise a question: Were two entirely different processes at work under these two great oceans?

Menard could not make up his mind. Having rejected Vening Meinesz's convection theory in 1956, he returned to it in 1958 in a paper with Scripps colleague Robert Fisher (b. 1925). He called Vening Meinesz's fossil convection theory "the most general" explanation for the observed features—despite the evidence presented with Vacquier that the displacements along the fracture zones had continued into geologically recent times. He rejected Hess's suggestion from his 1938 work on gravity anomalies and tectogenes that "faults mark the lateral boundaries of convection currents that formed the associated [ocean] trenches" on the grounds that it applied only to "special situations," but left unexplained what those special situations might be.[185]

By 1960, further mapping in the eastern Pacific Ocean had revealed another significant feature: a topographic high, now called the East Pacific Rise, running roughly north-south along the eastern Pacific, southward from the Gulf of California. Unlike the fracture zones, the East Pacific Rise was seismically and volcanically active and exhibited high heat flow. Now Menard had a feature in the Pacific comparable to Heezen's Mid-Atlantic Ridge, although its position could hardly be described as medial, as its northern end merged into continental North America. Menard once more changed his mind. Writing in *Science* in 1960, he embraced the heat-flow work of Bullard, Maxwell, and Revelle, invoking active convection for the origins of the East Pacific Rise: "A hypothesis of youthful convection currents in the mantle, suggested by Bullard, Maxwell and Revelle, to explain high oceanic heat flow offers a simple qualitative explanation of all the facts given above.... The horizontal limb of the convection cell moves the crust outward and thins it at the crest of the rise by normal faulting along the tension cracks. Blocks are displaced different distances by wrench faulting on fracture zones because of variations in intensity of convection along the rise."[186] The East Pacific Rise lay atop a rising convection cell, and the fracture zones were wrench faults along which oceanic crust was displaced varying amounts in response to varying intensity of convection. How or where the displacements originated was left unanswered, although a suggestion had been offered by Harry Hess, whose preprint was widely circulating. But Menard stopped short of embracing either seafloor spreading or continental drift. On the contrary, like Ewing, he seems to have been looking for a means to accommodate convection without displacing the continents. His discussion is worth quoting at length, as it shows him struggling to make sense of data that he found confusing and contradictory:

> Continental drift, as suggested by the parallelism of the Atlantic Coast lines and the crest of the mid-Atlantic Ridge, has been a very attractive concept for continental geologists [*sic*], particularly since it was revitalized by the paleomagnetic evidence for polar shifts and possible drift. Marine geologists, on the other hand, have been reluctant to accept the concept ... because they find no evidence for it in the geology of the sea floor.... However, if a random distribution of relatively short lived "oceanic" rises is accepted, the picture is entirely different. If all rises were in the center of ocean basins it would not be clear whether the convection current, or another agent, which produced the rise centered itself relative to the margins of the basin or created the basin. With rises bordering the Pacific and penetrating Africa, it appears more probable that most rises are centered because the margins of the basin have been adjusted by convection currents moving out from the center. If so, the African and East Pacific rises may mark relatively young or rejuvenated currents which have not yet had time to produce much continental displacement.[187]

Menard acknowledged the British paleomagnetic data and how strongly they suggested drift, but his arguments about the positions of the rises suggests that (like Heezen) his greatest concern was to accommodate *his* data—particularly, the fact that the East Pacific Rise was not located in the middle of the Pacific Ocean. This challenged the idea that the ocean basins were created when a continent split: if the ridges and rises were remnants of that process, they should occur in the center of the basin—which was true in the Atlantic but not in the Pacific.[188]

In later years, Menard would characterize his views at that time as the "sequential hypothesis"—a theory by which "ridges are of different ages and convection acts at different times." In hindsight it is certainly possible to extract this conclusion from his writings.[189] But what stands out more clearly is not so much what Menard did believe as what he didn't: he rejected the hypothesis of seafloor spreading because of its implication that the entire ocean floor was young.

Menard had read Hess's preprint, but he argued that the structure of the oceans was inconsistent with any theory that required the oceans to have been entirely remade in relatively recent geological times. True, the fracture zones had recently displaced the surrounding crust, but the crust itself—with its guyots and volcanic islands, rises, ridges, and fracture zones—had too much topography to have been recently "swept clean."

In 1960, Menard joined forces with Heezen to write the chapter "Topography and Structure" for Maurice Hill's multivolume compendium *The Sea*. The chapter did not actually appear until 1963, giving the two men ample opportunity to revise it if they chose to, but they did not: the published version cites neither Hess's nor Dietz's paper on seafloor spreading. On the contrary, the heavily descriptive paper avoids taking a theoretical stand except for the issue of the age of the ocean basins, and they dismiss the mounting evidence for young oceans on the grounds that sampling so far may simply have failed to have found older rocks: "If present sampling is considered to be adequate, the ocean basin floor is no older than Cretaceous. However, further sampling may indicate greater age and it seems more logical to assume that the basin is very old."[190] Why more logical, they did not say.

The years 1962 and 1963 passed with no apparent change in Menard's views. Then came 1964—an important year for Bill Menard as for other earth scientists. He attended the Royal Society meeting on continental drift, as well as a major symposium on the upper mantle organized by the International Union of Geological Sciences in New Delhi. The year 1964 also saw the release of *Marine Geology of the Pacific*, published by McGraw-Hill in a series of texts intended to become standard references. Many did: De Sitter's *Structural Geology*, Grim's *Clay Mineralogy*, Heiskanen and Vening Meinesz's *The Earth and Its Gravity Field*, and Turner and Verhoogen's *Igneous and Metamorphic Petrology* are all familiar titles to a generation of (now-aging) earth scientists. But as Menard himself admitted, the book was obsolete upon arrival.[191] The descriptive information was still valid, but the discussion of the competing interpretive hypotheses was not.

At the Royal Society, Menard had reiterated his view that the fracture zones were transcurrent faults that offset the worldwide rift system but were not part of it, that rises were not necessarily located in the middle of ocean basins, and that their locations were controlled by continental rather than oceanic processes: "The location of the continents exerts a primary control over the position of rises, even though rises shape the ocean basins in which they lie."[192] Menard emphasized Vacquier's work showing that the magnetic anomalies on either side of the Pacific fracture zones were displaced anywhere from one hundred to one thousand kilometers. The persistent picture was one of segments of the oceanic crust moving independently, in some cases over great distances. Menard put forward a complex hypothesis involving mantle convection triggered by continued crustal differentiation, explaining vaguely, "The chief conclusion to be derived from the distribution of rises is that they roughly circle continental nuclei which suggests that some phenomenon related to continents exerts a primary control over their location."[193]

Like Heezen, Menard was at pains to find a theoretical explanation that would account for *his* ocean, and, like Heezen, he settled on an idiosyncratic view. At the New Delhi symposium on the upper mantle, he pressed the point. The seafloor-spreading hypothesis had an appealing simplicity, but it did not seem to explain why "Africa and South America are almost completely surrounded by oceanic rises and ridges.... It is not enough to move Africa and South America apart from a single ridge; many rises and ridges enter into the problem."[194]

Nor could Menard see how seafloor spreading would explain the displacement of the magnetic anomalies along the fracture zones, displacement that "appears to be almost completely horizontal."[195] He preferred the idea that transient convection currents were centered on the continents and that the oceanic effects—including offsets along the fracture zones—were a by-product.

In a book chapter "Sea Floor Relief and Mantle Convection," published in 1966 in the multivolume *Physics and Chemistry of the Earth*, Menard continued to press a position that was increasingly obsolete: "Neither the facts of marine geophysics nor the hypothesis of convection, of some sort, in the mantle require acceptance of the bolder hypothesis that the sea floor is periodically swept clean by convection or that the oceanic crust is easily created or destroyed."[196] Frank Press once suggested that every scientist ought to be permitted to expunge one contribution from the scientific literature.[197] If Bill Menard could do so, no doubt this would be it.

The Culture of Navy-Sponsored Scientific Work

Like Bruce Heezen, Bill Menard bet on the wrong horse. The two men were in similar positions: they both had access to an enormous amount of data but lacked perspective on it. We might conclude that they were too close to the data, unable to step back to see the larger picture. But why? Why was it that the men who did make sense of the seafloor data were mostly outside the United States and almost entirely outside the institutions that were so well supported by the Navy?

Consider the culture of scientific work that Navy patronage fostered. Throughout the 1950s, the general attitude at Lamont and Scripps, as well as to some extent at Woods Hole, was to collect as much data as possible.

Maurice Ewing is said to have believed that as long as Navy money was flow-ing, he should keep his ships at sea, collecting as much data of as many types as possible; when the money ran out, there would be time enough to analyze it.[198] A similar attitude prevailed at Scripps, driven in part by financial con-siderations but also by the belief—no doubt fostered by the Navy's seafaring culture—that going to sea was what *counted*.

In her history of the institution, Elizabeth Shor noted that Scripps stal-warts, including her husband George Shor and Russell Raitt (1907–1995), were "usually planning the next expedition before they [had] worked up the results of the previous one(s)—and have often departed without even a re-morseful glance at the unfinished data. In this they were not alone among oceanographers."[199] These men considered it glamorous and pleasurable to go to sea, far more so than staying at home to analyze it. (This is one reason data analysis was often left to women).[200] Seafaring has always been a part of what it means to be an oceanographer: Roger Revelle often referred to himself as a sailor, and Columbus Iselin prided himself on actually sailing to his job at Woods Hole.[201] When Harald Sverdrup was trying to determine what might unite his disunited scientific staff—and to determine what made a scientist an oceanographer rather than a chemist or physicist who studied the ocean or a biologist who studied the life in it—he settled on the act of going to sea (chapter 1).

Navy patronage reinforced this impulse: when Raitt and Shor collected data and left it for others to analyze, they were doing what their patrons wanted. Oceanographers would not lose their Navy funding if they neglected theoretical questions, but they might lose it if the stream of useful data dried up. This is not to suggest that anyone ever told oceanographers not to analyze their data or to leave it to women to analyze. Nor is it to suggest that men like Shor and Raitt were not doing what they wanted to do. Rather, it is to empha-size that in science, as in life, you generally get what you pay for, and the Navy was paying for bathymetric data, not the geological interpretation of it. The Navy did not care why there were ridges and escarpments; it simply needed to know, for navigational and other purposes, where they were.

For the scientists involved, this was a potentially risky situation, be-cause when data gathering outruns its interpretation, it makes a scien-tist vulnerable to being scooped; others may interpret the scientist's data before he or she has a chance to. Raitt and Shor evidently did not worry about this, but Heezen, Tharp, and Menard did. It would certainly explain why they were protective of their data: that knew that it would be easy (if security restrictions were lifted) for someone else to skim the cream. In-deed, in his book *Science, Growth and Change*, written in 1965 during a year in the White House Office of Science and Technology, Menard suggested that this was what smart scientists *should* do. Fields grew fastest when they were new, and "good scientists ... shift subfields when they consider the cream skimmed."[202] In hindsight, Menard probably regretted that he did not do that himself.[203]

When Worzel was in Cambridge in 1964, Scripps's Bob Fisher was visiting Cambridge at the same time. Worzel wrote to Ewing, dismissing Scripps and Cambridge in one breath: "Bob Fisher attacks you on the world wide fissure in the ridge, claiming it a zone rather than one continuous feature.... I suspect it is partially a protective mechanism because he feels this Scripps inferiority about leading in geophysics. [But we] are united at least in our scorn for the British attitude of their 'obvious superiority.'"[204] In fact, the British had a reason to feel superior, for despite all the data that Scripps and Lamont had collected, it was British scientists who realized what it meant. British scientists reopened the debate over continental drift through their work on continental paleomagnetism, British scientists proposed the critical test of seafloor spreading, and British scientists presented the first data indicating that it was likely true. Many things were different in Great Britain; one of them was a more open climate for discussion and data sharing.

We cannot prove that Heezen or Menard would have made better sense of their data had they discussed it with more people, worked under different conditions, or had access to related information that could have aided their interpretations, but the circumstantial evidence supports that view. The issue is not one of delay, a word that historians dislike as it seems to imply a "proper" time frame for events that reality failed to follow. The issue is the way patronage can shape a scientific culture.

The development of scientific knowledge is a process by which certain problems are set aside while others are pursued and resolved. If the issues that are set aside are publicly known and discussed, then they can and likely will be revisited. But if these issues are known only to select individuals—or the pertinent data are sequestered—important questions may be left not merely unanswered but also unaddressed. As Shils had argued in the 1950s, if they are lost from sight, it may take years or decades before they are rediscovered, if at all.

The Impact of Secrecy on Earth Science

In the development of plate tectonics, many classified documents were destroyed and historical information has been lost. Was scientific information lost, as well? Given the aggressive document destruction pursued by Ewing at

Lamont and the inconsistent record keeping elsewhere, it seems likely.[205] As Raymond Siever later put it: "Those of us not part of those groups knew nothing of the classified research. Until this material, if ever, becomes declassified, we will not know exactly how much we could have known that would have materially advanced the science generally."[206]

It was not just a particular theory that was affected by data sequestration but earth science as a discipline, as well. This may be a lasting impact of Cold War patronage. As several historians have noted—and this story underscores—during the Cold War some areas of earth science were handsomely funded while others were left to tread water or wither.[207] The manner in which plate tectonics developed, based on geophysical work that was largely secret and then suddenly revealed, left most geologists out in the cold. This created a widespread perception that traditional geology was obsolescent, if not actually obsolete.[208] Moreover, the financial largesse bestowed on geophysics had a synergistic effect: as geophysics grew and prospered, smart young people were attracted to it and it came to be seen not just as a more vibrant science than geology but as a more effectual one. As a result, more funds, departments, and faculty positions were dedicated to it, and geophysics did become a better, stronger, more effectual science.

When lecturing on the history of continental drift and plate tectonics, I am often asked why geologists expressed so little interest in the ocean floors; some historians have criticized geologists for being "landlocked" and narrow-minded.[209] But geologists were not so much uninterested in studying the oceans as unable to do so. As Joe Worzel stressed, oceanographers wanted to share their data with geologists, who were better qualified to interpret it and would no doubt have been interested, but security restrictions prevented them. Geologists almost certainly would have contributed more had they been permitted to, and had that happened, their reputations—and that of their science—would have been stronger and the history of earth science very different. The story of the *Pioneer* survey—and its interpretation by historians and philosophers of science—will prove the point.

The Impact of Secrecy on History and Philosophy of Science

In the 1980s, when critical sociological analyses of science were au courant, some observers noted that nearly all the evidence used in the 1920s to argue for continental drift had been valid all along. Some geologists had accepted this evidence at the time, but others had not. This seemed to suggest a degree of intellectual incoherence: if the evidence of drift was valid, were scientists acting irrationally when they rejected it? Or was the evidence somehow insufficient, and the irrational ones were the mavericks who accepted continental drift prematurely?

Philosophers of science Larry Laudan and Rachel Laudan took up the defense of scientific rationality by arguing that different groups of scientists may (rationally) subscribe to different epistemic standards. The early adopters in this case were convinced by the consilience of field evidence and the theory's explanatory power—which was rational—but the recalcitrant scientists were waiting for the gold standard of scientific demonstration: a hypothesis that predicts a novel fact that proves to be true. The novel fact was the existence of seafloor magnetic stripes. In the 1960s, their standard was met, so they were rational, too.

The Laudans' analysis was appealing insofar as it offered a reminder that intellectual diversity is not anarchy. However, it suffered from a fatal flaw: it wasn't true. The Vine-Matthews hypothesis did not predict a novel fact that turned out to be true. Seafloor stripes were already known; Ronald Mason and Arthur Raff had discovered them in the 1950s. What was true was that none of the men involved in their discovery had been able to explain them.

Ronald Mason studied physics as an undergraduate at Imperial College and then earned a masters in geophysics, hoping it would provide more diverse opportunities than life in the lab. After his war service, Mason was offered a position as a lecturer in geophysics at Imperial, and in 1951 he spent a sabbatical leave at Caltech. In 1952, while listening to presentations on marine seismology, he began to ponder what other geophysical techniques could be applied at sea. He had heard about Project Magnet, a joint effort between the ONR, the US Geological Survey, and the Naval Ordnance Lab to use airborne magnetometry to map magnetic patterns associated with volcanoes and atolls in the Pacific.[210] Industry had developed magnetometers for use in petroleum and mineral exploration and by the Navy for minesweeping and submarine detection; Mason realized that one could get more detailed and accurate data towing a magnetometer behind a ship than by setting it on an unstable airplane. In this way, one could glean information not just about islands and atolls but also about the structure of the entire seafloor.

Mason asked Russ Raitt if anyone at Scripps had thought of doing this; the story goes that Roger Revelle overheard the conversation and invited Mason to do the job.[211] In fact, Heezen and Ewing and others at Lamont had already done what Mason envisaged—a combined magnetometry-bathymetry survey over the mid-Atlantic using a fluxgate magnetometer towed behind a ship—

but the results had not yet been published. The Lamont scientists generously agreed to loan a magnetometer, which was towed for six months and eight thousand miles on the *Capricorn* expedition in late 1952 and early 1953.

Capricorn was an important expedition for several reasons, not least of which was the discovery by Revelle and Arthur Maxwell of anomalously high heat flow under the Albatross Plateau (later recognized as part of the East Pacific Rise), work that formed the basis of their paper with Teddy Bullard suggesting that high heat-flow values in the oceans reflected mantle convection.[212] The trip also demonstrated the feasibility of Mason's idea: it turned out to be easy to tow a magnetometer behind a ship and also cheap, because most of the components were readily available as military surplus.[213]

The *Capricorn* data were not of much scientific use, however, because single lines across a wide expanse of ocean were hard to interpret. Mason called them "long unrelated profiles" that "give no indication of the lateral extent or direction of anomalous magnetic trends, and, therefore, provide no basis for quantitative geological interpretation."[214] What they needed were closely spaced lines over a specific region. They needed to make a a magnetic map of the seafloor.

In 1955, Menard told Mason of a survey planned by the US Coast and Geodetic Survey to obtain detailed bathymetry of the entire western coast of North America. The purpose—highly classified at the time—was to map the submarine topography to determine where to put SOSUS hydrophones.[215] The plan was to cover an enormous area: from 32° to 52°N latitude, and approximately 1,250 miles from north to south and 300 miles out to sea. With Scripps technician Arthur D. Raff, Mason asked the Hydrographic Office for permission to join the cruise and tow a magnetometer. The initial answer was no: these were the early years of the SOSUS program and officials were afraid Mason and Raff would slow their operations. (They were also likely nervous about a foreigner on a mission so highly classified.) Revelle intervened and permission was granted.

The SOSUS system was designed to locate Soviet submarines with a high degree of precision, a task that would succeed only if the detectors themselves were accurately located. Seamounts were an excellent place to locate hydrophones—but only if you knew where they were. The Navy was applying state-of-the art navigation to the survey, and this proved crucial for the science, too: Mason and Raff were able to position their magnetic measurement to within three hundred feet. This was a remarkable degree of accuracy for a survey of its kind, and it yielded a distinct pattern of alternating zones of normally and reversely magnetized rocks running subparallel to western coast of North America. Plotted on a map, they looked like zebra stripes: irregular in fine detail but startlingly regular overall. Fortuitously, the ship's east-west tracks, designed to produce coherent bathymetric profiles, turned out to be ideal for mapping the north-south linear anomalies. Preliminary data from the magnetic survey were published in 1958 in the *Geophysical Journal of the Royal Society*, and the full analysis appeared in 1961 in the *Bulletin of the Geological Society of America* (figs. 5.11 and 5.12). Mason later recalled, "Nothing like this had ever been observed before, either on land or on sea."[216]

Scientists often exaggerate the novelty of their findings, but nothing like this *had* ever been seen and the paper triggered considerable discussion. Canadian geophysicist Lawrence Morley (1920–2003), who independently conceived the explanation of seafloor spreading (but had his paper rejected by both *Nature* and *the Journal of Geophysical Research*), has called the Mason and Raff map "the trigger that set off the escalation of investigations and ideas that culminated in the theory of plate tectonics."[217] Morley dropped what he was doing to work on an explanation, because Mason and Raff admitted they had none. Nor did anyone else: the map of the stripes remained

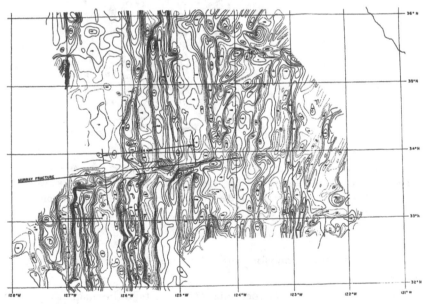

Figure 5.11 Initial results from the *Pioneer* survey of magnetic anomalies, 1961. Note the "distinctly linear pattern in the south-west corner of the map area," which was their starting point and persuaded them to continue with the survey.

From Arthur Raff and Ronald Mason, "A Magnetic Survey off the West Coast of the United States between Latitudes 32 degrees n. and 36 degree n. longitudes," *Bulletin of the Geological Society* 72, no. 8 (1961): 1259–65 (fig. 2). Reprinted with permission of the Geological Society of America; permission conveyed through Copyright Clearance Center, Inc.

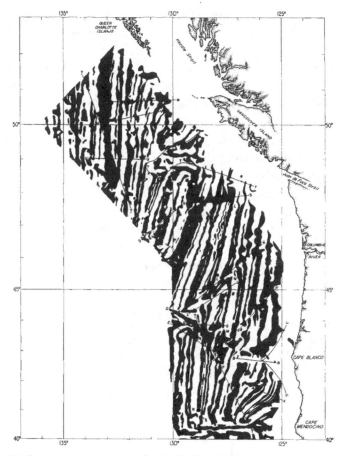

Figure 5.12 The magnetic stripes on the Pacific floor, 1961.

From Arthur Raff and Ronald Mason, "Magnetic Survey of the West Coast of North America, 40 Degrees n. Latitude to 52 Degrees n. Latitude," *Bulletin of the Geological Society* 72, no. 8 (1961): 1267–70 (fig. 1). Reprinted with permission of the Geological Society of America; permission conveyed through Copyright Clearance Center, Inc.

in the literature for "about four years with no plausible explanation as to their cause."[218]

Menard and Vacquier had focused on the offsets in the magnetic pattern rather than the magnetics themselves, stressing that across the Murray Fracture Zone it was possible to match the patterns by assuming a right lateral displacement of about one hundred miles. It was this observation that led them to interpret the fracture zones as transcurrent faults. However, across the Mendocino and Pioneer Fracture Zones, there was no match. Vacquier concluded that the displacements must extend beyond the distance of the mapped region, so he extended the survey using a proton precession magnetometer supplied by Navy (the technology was new and the Navy wanted to know how well it worked), running east-west lines along the fracture zones until he found what appeared to be matches: left-lateral displacements of 710 miles and 160 miles along the Mendocino and Pioneer, respectively, and right lateral displacements ranging between 95 miles and 425 miles along the strands of the Murray fault zone.[219] Revelle called these discoveries "the most important geophysical discovery of the last ten years."[220]

Today, these fracture zones are understood as seismically inactive extensions of ridge-to-ridge transform faults: they are created along offsets in the mid-ocean ridges and then displaced laterally as the seafloor spreads. Magnetic patterns are displaced along these zones because the ridges along which they originally formed have offsets, but the amount of the offset is nowhere near as great as Vacquier imagined and the direction of offset is the opposite. (They are not *transcurrent* faults but *transform* faults; see fig. 5.13) But whether the magnetic patterns were offset to the right or the left, a little or a lot, there was a bigger issue at stake in 1961: what were the stripes? The insight of Vine and Matthews was to link the stripes both to seafloor spreading and to magnetic-field reversals.[221]

The recognition of reversely magnetized seamounts was critical to Vine's and Matthews' idea. From Hess's work on guyots, seamounts were accepted as volcanic structures younger than their surrounding crust, so if there were field reversals, then some seamounts might display reversed polarity. Conversely, if there were seamounts with reversed polarity, this would be strong evidence that field reversals occurred. The data from the *Pioneer* survey included a number of reversely magnetized seamounts.

Did Mason realize he had reversely magnetized seamounts in his survey? If he did, why did he not pay attention to them as proof of field reversals? In 2001, Dan McKenzie posed these questions to Mason, who replied:

> By the time of our survey I had accepted field reversals and … realized that some of the seamounts in the area … were reversely magnetized. But I do not recall any serious discussion of the fact, and I did not seek to exploit it. The major reason was the lack of suitable bathymetry. The ship obtained continuous east-west profiles of bathymetry 8 km apart in the north-south direction, but these were highly classified and I was denied access. However, I was given spot depths at approximately 15 km intervals along the profiles, but I considered these inadequate for serious work. With better bathymetry we might have tried modeling the magnetization of some of the seamounts, as we later did … elsewhere.[222]

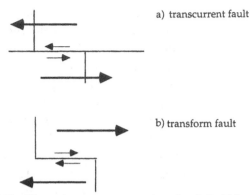

a) transcurrent fault

b) transform fault

The difference between transcurrent and transform faults. (a) In a transcurrent (or strike-slip) fault, the direction of movement can be determined from the offset of a feature intersecting the fault. If the feature is moved to the left, it is a left-lateral fault, as shown here. The north side of the fault has moved to the left (west), the south side of the fault has moved to the right (east), and the fault may continue indefinitely. (b) In a ridge-to-ridge transform fault, a section of the mid-ocean ridge is fractured perpendicular to its length. In this case, the right side of the ridge is moving to the right (east), the left side is moving to the left (west), and the sense of motion is opposite of that illustrated in (a). Note also that the fault does not extend indefinitely, but terminates against the north-south running ridge segments.

Figure 5.13 The difference between transcurrent and transform faults.

From Naomi Oreskes, *Plate Tectonics: An Insider's History of the Modern Theory of the Earth* (Boca Raton, FL: Westview Press, 2001).

To analyze the seamounts properly—indeed, even to know for sure that they *were* seamounts—one needed bathymetry. Harry Hess could have obtained the necessary data, or Bruce Heezen or Bill Menard, but Mason was British and denied access. Fred Vine confirmed this point: "Everyone was mystified by the fact that despite the relatively strong magnetization of the volcanic rocks of the ocean floor there was no systematic correlation between topography and the magnetic anomalies developed over it. In many ways, this was the crux of the problem."[223] But this was not quite right; it was not that the magnetism and topography were not correlated but that the correlation had not been recognized because Mason lacked the bathymetric data.

Seamounts and volcanic islands had large anomalies, but their sources had not been worked out—how much of it was remanent, how much was induced by present field? To analyze this accurately, one needed to calculate the induced component. This required knowing the volume of the seamount, which in turn required having high-resolution bathymetric data. Vine and Matthews had that. Mason and Lawrence Morley did not. Indeed, this is what distinguished their landmark 1963 paper from the conceptually identical one that Morley submitted. Vine and Matthews did not simply propose that mid-ocean ridge spreading might produce seafloor magnetic stripes; they showed that it had done so in a specific setting. That setting was neither the Atlantic nor the Pacific—the oceans Americans knew so well—but the Indian.

In late 1962, Matthews had undertaken detailed magnetic and bathymetric work in the Indian Ocean as coordinator of the UK contribution to the International Indian Ocean Expedition. He returned from the Indian Ocean with a large quantity of magnetic data, including a detailed survey of the crest of the Carlsberg Ridge in the northwestern Indian Ocean at 5°N. It was the largest and most well-known mid-ocean ridge at that time.[224] With detailed bathymetry, he and Vine were able to calculate the three-dimensional parameters of the anomalous areas, correct for induced magnetization, and determine the thermo-remanent component. They used a computer program developed by one of Mason's colleagues at Imperial College that had been designed specifically to interpret magnetic anomalies from a three-dimensional source region. They explained: "Work on this survey led us to suggest that some 50 per cent of the oceanic crust might be reversely magnetized and this in turn has suggested a new model to account for the pattern of magnetic anomalies over the ridges. The theory is consistent with, and in fact a virtual corollary of, current ideas on ocean floor spreading and periodic reversals in the Earth's magnetic field." The key insight was that as oceanic crust formed, it would be magnetized in the direction of the prevailing Earth's field, and each new increment of crust would be similarly magnetized. But if a field reversal occurred, then the next increment would be reversely magnetized: "Thus, if spreading of the ocean floor occurs, blocks of alternately normal and reversely magnetized material would drift away from the centre of the ridge and parallel to the crest of it."[225]

This was the crux of their paper, but the full details of the analysis seemed too long for a letter to *Nature*. Without them, however, the paper was "rather long on interpretation and speculation and short on original data."[226] Maurice Hill provided the solution, Vine explained, by giving "us permission to include two long, unpublished, magnetic profiles across the crests and flanks of the North Atlantic and northwest Indian Ocean ridges acquired by the [Cambridge] group in 1960 and 1962 respectively. The title [of the paper] was changed from 'Magnetic Anomalies over the Oceans' to 'Magnetic Anomalies over Oceanic Ridges' and the introductory paragraphs were added to set the science and incorporate the ridge profiles."[227] By luck, Hill had magnetic profiles across the Mid-Atlantic Ridge, whose topography Tolstoy and Ewing had published in 1949, and across the Indian Ocean where Matthews had just mapped. This allowed them to produce a simple and compelling picture that

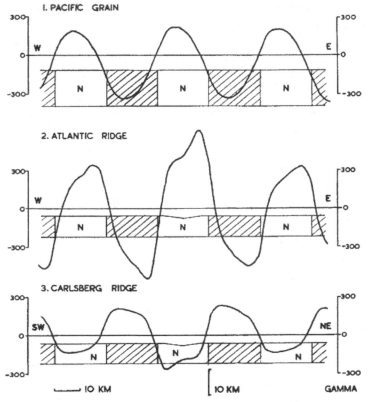

Figure 5.14 The magnetic profiles over the Pacific, Atlantic, and Indian Oceans, 1963. From Frederick John Vine and Drummond Hoyle Matthews, "Magnetic Anomalies over Oceanic Ridges," *Nature* 199, no. 4897 (1963): 947–49. Reprinted with permission from Springer Nature.

clearly illustrated the relevant pattern, as well as its evident resemblance to the one mapped by Mason and Raff off the coast of California (fig. 5.14).

In later years, Teddy Bullard would dismiss Morley's claim to co-discovery, arguing that if he had really believed in his idea, he would have fought harder to get it into print: "He can't really have believed he had something important or he would have got it published. It is said to have been rejected by *Science* and *Nature*. Why? Who were the referees? Why didn't Morley ... insist? I certainly would not have allowed Vine and Matthews' paper to languish unpublished."[228] Cambridge scientists may feel entitled to insist that their papers be published in *Nature*, but even if Morley had had a powerful mentor like Bullard to argue on his behalf, he lacked the data that Vine and Matthews had. He had the same idea but not the same data with which to support it. What he did not know—and could not have known—was how much data supported his views but was hidden from sight. Access to bathymetric data made the difference for Vine and Matthews, and lack of it may explain why Morley failed where they succeeded.[229]

In 1967 Vine and Hess joined forces to do additional work on magnetics in relation to seafloor spreading; Hess wrote to the Naval Oceanographic Office to request the necessary bathymetric data: "Fred Vine and I have oriented basalt cores from last year's drilling operations at Mid-way Island. Fred has completed a laboratory magnetic study of the cores. To finish off a manuscript, we need to know a little more about the topography around Midway and the positions of nearby seamounts. Could you send us a copy of BC1805N and the present master collection sheets of soundings for the 1805N area? I have the data here only up to 1951."[230] In that year—1951—Hess had written his first memo urging declassification of the sounding data.

Hess was also asking for magnetic data from the Mediterranean, explaining "I know [the Naval Oceanographic Office] has a great deal of magnetic data on the Mediterranean area.... We are also interested in magnetic data in the Caribbean and east of the Lesser Antilles.... Data from Puerto Rico south to Venezuela would be very interesting to us."[231] Hess knew what to ask for, and fourteen days later he received it.[232] Others were not so fortunate, lacking the connections, knowledge, or wherewithal to run the Navy gauntlet. The issue was not simply knowing what to ask for and being willing to ask; it was also being willing to spend the time and energy to extract information from the system, to navigate the "iron curtain." As one senior oceanographer put it, "People get discouraged just by the prospect of it."[233]

In 1968 the US Navy finally released the classified bathymetric information for most of the world's oceans. Oceanographer of the Navy Instructions 3130.6C "effected declassification of a large quantity of ocean bathymetry data and thus facilitates a significantly broader world-wide dissemination of this important and useful information."[234] National security was evidently not compromised, but for the scientists in this story it was too late. Plate tectonics was established. Harry Hess died the following year.

6 *Why the Navy Built* Alvin

Plate tectonics was a powerful explanatory theory, but it also made testable predictions. One of these was that the eruption of mid-ocean ridge basalts would generate hydrothermal circulation and seafloor hot springs. Throughout the 1950s and early 1960s, heat-flow measurements near mid-ocean ridge crests revealed much lower values than expected for the crests to be the sites of ascending mantle convection currents (chapters 4 and 5). However, the low values would make sense if the ridges were sites of hydrothermal activity, as seawater circulating through the fractured ridge crests would cool them. This would also generate hot springs: cold ocean water would be heated as it percolated through hot basalt and would vent on the seafloor in underwater geysers.

By the early 1970s, scientists were making plans to search for ridge-crest hydrothermal vents. What happened was not merely a confirmation of plate tectonics—by the mid-1970s already widely accepted—but one of the scientific finds of the century. Diving in a novel submersible vessel named *Alvin*, oceanographers witnessed spectacular submarine geysers erupting black fluids. Subsequent measurements and calculations would indicate that the color was caused by tiny mineral particles precipitating from fluids venting at temperatures as high as 400°C.[1] Crowded around the vents were biological communities of startling abundance and beauty. In the months and years that followed, scientists would identify hundreds of new of species in complex vent communities that thrived in darkness, rooted in chemosynthesis rather than photosynthesis.

The discovery of seafloor-vent ecosystems has been described as having come "as a complete shock."[2] Many scientists believed that the deep sea was devoid of life, for without light how could there be life? As Robert Kessler and Victoria Kaharl recalled in the mid-1990s, "The discovery of the first hydrothermal vent communities in the deep sea ... astonished scientists around the world."[3] Oceanographers called it "one of the most exciting developments since the beginnings of the study of oceanography."[4] In a world where many people thought there was not much new under the sun, scientists had found something new under the sea, with wide-ranging implications.

The vent discovery led to considerable interest in its practical applications, from exploration for economic mineral deposits to the development of heat-resistant organic molecules.[5] Told this way, the discovery fits the narrative that has dominated public rhetoric about science in the twentieth century: that basic science yields unanticipated practical benefits as progress flows from discovery to application. In the standard telling, the vent discovery is a triumph of curiosity-driven research.[6] But this is at best half the story, because the discovery of seafloor hydrothermal vents was also an engineering accomplishment, made possible by a novel technology designed to allow humans to explore and work in the deep-ocean environment. And the history of this technology reverses conventional wisdom, because *Alvin* was not invented to do basic research but to satisfy the military demand for deep-submergence capability.[7]

World War II, Allyn Vine, and the Problems of Underwater Sound

Alvin was the brainchild of Woods Hole oceanographer Allyn C. Vine (1914–1994), who dedicated his career to Navy oceanographic problems. In the late 1930s, Vine was working on a master's degree in physics at Lehigh University, where Maurice Ewing (1906–1974) was teaching. During the summers from 1937 to 1939, Vine followed Ewing to Woods Hole, where they worked alongside Joe Worzel, Brackett Hersey, Columbus Iselin, and Athelstan Spilhaus developing marine geophysical techniques (chapters 2 and 3).

Vine joined Woods Hole full-time in 1940. That same year Iselin succeeded Henry Bigelow (1879–1967) as director and was assiduously cultivating military contacts, proactively suggesting ways his scientists might aid the anticipated war effort.[8] Vine became involved in several war-related projects. Most important from the military perspective was the development and installation of bathythermographs on Navy ships.[9]

The crucial importance of ocean temperature to underwater sound transmission had been recognized in the early 1930s, when sonar operators working on routine training exercises found that their equipment worked well in the morning but poorly in the afternoon, particularly on fine days; this came to be known as the "afternoon effect."[10] Some observers wondered whether sonar operators were dozing off after lunch, but certain consistent patterns of variability—from area to area and at different times of day—were sufficient to suggest that the effect was caused by the ocean itself. In 1934, a group was created at the Naval Research Laboratory to investigate further. Collaborating with Woods Hole scientists, they soon discovered it was caused by the effect of changing temperature on the velocity of sound as surface waters warmed in the afternoon sun. Suddenly, it became clear that accurate temperature measurements were crucial to effective sonar operations. Moreover, because temperature affects density, which in turn affects buoyancy, accurate temperature profiles would be useful to many aspects of submarine operations. As Scripps oceanographer Fred Spiess (1919–2006) later put it, it was becoming clear that oceanographers understood things relevant to both anti- and prosubmarine activities.[11]

But there was more to it than routine "relevance." In typical midlatitude-ocean surface waters with normal salinities, near 35 percent, the velocity of sound is approximately 1,500 meters per second.[12] With decreasing temperature the velocity decreases and acoustic waves are refracted as they travel though water layers of varying temperature, just as light rays are refracted through media of varying density. Because ocean temperatures generally decrease with depth, sound waves are generally increasingly refracted as they travel downward into the ocean's colder layers. The effect is particularly pronounced as sound travels through the thermocline—the zone of rapid temperature decline below the surface layer that was central to Henry Stommel's analysis of ocean circulation (chapter 2, see fig. 2.13).[13] The downward refraction of sound waves as they traveled through the thermocline could cause a beam to pass beneath a shallow target and miss it entirely. The implications of this were not difficult for submariners to grasp (fig. 6.1).

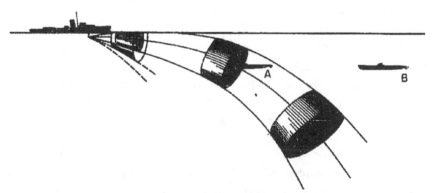

Figure 6.1 The downward refraction of a surface sound wave in relation to a typical temperature-depth profile. At A, the submarine is detected; at B it is not.

From NDRC, *Physics of Sound in the Sea, National Defense Research Committee Summary Technical Reports,* originally issued as Division 6, vol. 8, ed. Lyman Spitzer Jr. (1946; Research Analysis Group, Committee on Undersea Warfare, National Research Council, 1969).

To exploit this understanding required precise and accurate temperature-depth profiles, so in the summer of 1940 the newly created National Defense Research Committee awarded one of its first grants to Woods Hole for research on the effect of temperature on underwater sound transmission.[14] Woods Hole's first-ever year-round staff was assembled to work on the project; its principals were Ewing, Worzel, and Vine.

For the BT to be useful in combat, it had to be quick and easy to use, so Ewing and Vine spent the summer of 1940 refining the instrument so it could be readily raised and lowered from a moving ship. Vine also worked on a submarine version; in doing so, he began to develop a particular empathy for his military colleagues. On one occasion he opposed his scientific colleagues' preference for the metric system, arguing to keep BT measurements in imperial units on the grounds that "the poor kids from Kansas that were dragged into the Navy shouldn't have to learn the metric system while they were seasick."[15] By the end of 1943, Vine, Ewing, and Worzel had manufactured seventy-five BTs in the Woods Hole machine shop (all in imperial units) before a contract for mass production was signed with the Submarine Signal Company of Boston.[16]

The Discovery of the Sound Channel

The BT measured and recorded temperature—the primary control on sound transmission—but sound is also influenced by pressure and salinity. In most areas of the open ocean, salinity variations are minor, but pressure variations with depth are not, and while decreasing temperature decreases the velocity of sound in water, increasing pressure increases it (fig. 6.2). Thus, submariners diving into the deep ocean would encounter opposing trends: falling temperature would reduce the sound velocity, but increasing pressure would increase it. The net effect would create a velocity minimum zone

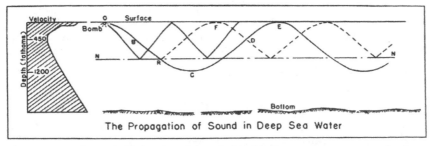

Figure 6.2 A schematic illustration of the relationship between a temperature-depth profile and the creation of a sound channel. Note the bomb detonated at the surface.

From "Long Range Sound Transmission," Interim Report No. 1, Contract Nobs-2083, August 25, 1945, declassified March 12, 1946 (fig. 6).

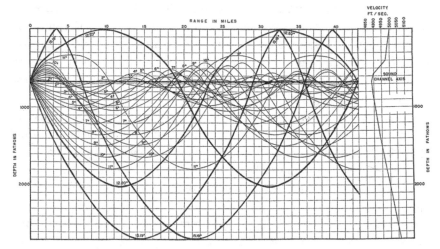

Figure 6.3 Maurice Ewing and Joe Worzel's ray tracings in the sound channel.
From M. Ewing and J. Worzel, "Long Range Sound Transmission," *GSA Memoir* 27 (1948): 1–40 (fig. 5).
Reprinted with permission of the Geological Society of America; permission conveyed through
Copyright Clearance Center, Inc.

whose position would depend on the exact temperature-pressure profile encountered.[17]

Ewing and Worzel made a crucial realization. Within the velocity minimum zone, sound waves would be effectively trapped: waves moving upward
would be refracted down, and waves moving downward, refracted up. Ewing
and Worzel called this the "sound channel" to indicate that acoustic transmissions would be confined within a narrow zone rather than bouncing up
and down between the ocean's surface and the seafloor, as would otherwise
be the case (fig. 6.3).

In a confidential report written in August 1945, they explained:

> If a non-directional sound source is placed at the depth of minimum ve-
> locity, hereafter called the axis of the sound channel, the sounds that start
> at angles above the horizontal are bent downwards due to sound refrac-
> tion [and] those which start at angles below the horizontal are bent up-
> wards.... Thus, a sound starting [in the channel] will cross and recross the
> axis indefinitely until it is absorbed, scattered, or intercepted by some ob-
> stacle. Sounds traveling in this manner are channeled: hence, the name
> sound channel.... These sound channel sounds have been heard at far
> greater distances than any other man-made sounds.[18]

Ewing and Worzel predicted that charges exploded within the sound channel could be positively identified at distances as great as ten thousand miles.
In 1944, they confirmed it: a charge released off the coast of West Africa was
heard in the Bahamas.[19] In 1947, the capacity to send or detect sound signals
over great distances became the basis for sound fixing and ranging, or SO-
FAR, a new Navy system for search and rescue. In later years, the deep sound
channel would often be referred to as the SOFAR channel.[20]

The sound channel had another critical implication. The bouncing of
acoustic rays around the axis of the channel produces zones in which rays
converge, creating both particularly intense acoustic signals and zones where
the rays are refracted away, creating shadow zones. These shadow zones are
"insonified"—that is, insulated from sonar transmissions. Sound is unable
to penetrate. As a later report put it with characteristic dryness, these regions have "great practical importance for Naval sonars."[21] But the pattern
of convergence and shadow zones is complex, with their locations depending on the distribution of water temperature and pressure, and changing
not just through the course of a day (as the scientists had discovered) but
also with the seasons. The net effect is not simply sound or silence, but regions of greater and lesser sound transmission whose positions were contin-
ually changing. Detailed, regularly updated information on ocean conditions
would be crucial to exploiting this effect.[22]

Vine spent much of 1943 and 1944 installing BTs on Navy vessels and
training Navy crews in their use. By the end of the war, many thousands of
BT profiles had been compiled, mostly by women "computers."[23] In 1972 Vine
received the Navy Oceanographers' Commendation for his work, said to have
contributed to the "savings of untold numbers of lives, and millions of dollars
in ships and equipment."[24]

From Underwater Sound to Deep Submergence

Naval historian Gary Weir has said that Woods Hole work on underwater
sound transmission "touched every aspect of naval warfare between 1940
and 1945."[25] It was the sort of work that helped to justify the creation of the
Office of Naval Research (ONR) to foster research in the postwar period.[26] In
1976, retired rear admiral Robert K. Geiger (1924–2013) recalled: "At the end
of the war, the Navy was convinced that the continuation of this fundamental research was an indispensable part of a far-sighted naval defense program.
The impenetrability of the sea to electromagnetic radiation and the relative
difficulty to locate underwater objects by sound indicated that submarines

would occupy a prominent position in [all future] naval warfare. In this respect, it was important to continue research in the fundamental properties of underwater acoustics, which was to remain the principle detection means of pro- and anti-submarine warfare."[27] Similar work had been done during the war on the West Coast at the University of California Division of War Research (UCDWR), which in 1946 became Scripps's Marine Physical Laboratory (MPL). In 1965, MPL's director Fred Spiess summarized his lab's activities as "a steady effort to increase our knowledge of the generation, propagation and detection of energy in the ocean."[28] In later years he would describe their conception of the ocean as a "medium of sound transition."[29] But Navy support for oceanography went beyond matters that were directly associated with underwater sound. Scientists like Vine had demonstrated that basic oceanographic knowledge could have material value in combat operations, and many in the Navy were sympathetic to furthering oceanography, as a science, on this account.

The most important early ONR contract at Woods Hole was the "basic task contract" to increase "knowledge of oceanographic phenomena" (chapter 2).[30] Looking back on the contract after its first five years, ONR program manager Gordon Lill summarized: "Under the present contract … the Woods Hole Oceanographic Institution has continued to act as a pilot plant of new ideas and new methods in the military aspects of oceanography, and concurrently has carried on progressive research in theoretical, basic oceanography."[31]

If Woods Hole was a pilot plant, the foreman was Al Vine. A man of wide-ranging and restless intellect, Vine was constantly thinking of ideas with military relevance. He once proposed hiring blind people to work as sonar operators, noting that "most of us are acoustic amateurs compared to those whose principal contact with the world is through sound."[32] On another occasion he envisaged a highly equipped submarine to serve as a refuge for "key government officials and a few others" in the event that a massive nuclear exchange left the surface temporarily uninhabitable. (He did not say where they would go or what they would do when they emerged.)[33] Vine did not hesitate to convey his thoughts to high-level government officials, as in 1969 when he penned a six-page letter to Henry Kissinger on the importance of maintaining the internationalism of the high seas.[34] Often he was ahead of his contemporaries, as in 1956 when he urged more attention to the problem of nuclear waste disposal, noting that it not been attacked with "anything like the vigor with which the production [of nuclear materials] was attacked [and] in the long run … may be theoretically, technically, and economically more difficult."[35]

Vine was particularly interested in vessels; he called himself a "Navy and vehicle-oriented oceanographer."[36] In 1945, he presciently envisaged the day when a submarine, perhaps carrying nuclear weapons, would be lost on the seafloor. When that occurred, the Navy would be left helpless without deep-submergence capability. In a confidential "Memorandum on Submarine Rescue," he noted that normal submarines are limited in their depth capacities by their "primary offensive purposes," which require them to have numerous hull openings and to be able to carry heavy equipment. But a submersible vehicle whose primary purpose was simply to go deep could have a heavier, stronger hull and reach much greater depths—perhaps twenty thousand feet. Vine thought it made sense to begin to develop a deep-submergence vehicle for rescue and salvage purposes.[37]

There was also a more immediate justification for a deep-submergence vehicle: the problem of the "deep scattering layer." As the name suggests, this is a layer within the ocean, beginning at a depth of approximately six hundred feet, in which sound transmissions are highly scattered and therefore not effectively propagated. First recognized by Russell Raitt (chapter 5) and colleagues at UCDWR, it appeared as a phantom layer that reflected sound as if it were the seafloor.[38] However, it was clear that it was far too shallow to *be* the seafloor; Martin Johnson (1893–1984), the pioneering biological oceanographer who coauthored *The Oceans*, suggested that marine organisms were perhaps reflecting the sound. If so, then the phantom "bottom" should rise and fall with time of day as organisms moved toward the surface in the evening; this prediction was confirmed shortly before war's end. Soon after, Brackett Hersey and colleagues found evidence that the sound was being reflected off gas bubbles released by fish.[39] A vehicle that could travel into the scattering layer would be a powerful means to study the effect and determine possible remedies.

By the mid-1950s, there was a third justification: SOSUS. The sound channel had been discovered during World War II, but its full exploitation in antisubmarine warfare was a Cold War development, as underwater listening came to be seen as imperative in the face of prowling Soviet submarines. In 1949, the Navy had announced its intention to develop SOSUS, in 1952 a test array had been installed at Eleuthera in the Bahamas, and by the mid-1950s the system was in place.[40] (SOSUS was the classified name for the system; the unclassified name CAESAR was given to the production and installation of the array.)[41] By the late 1950s, the system involved more than one thousand hydrophones and thirty thousand miles of undersea cables; eighteen land-based listening stations were able to detect acoustic signals of less than one watt at ranges of several hundred kilometers (and perhaps more).[42]

With the laying of hydrophones and cables on the seafloor, a new issue arose: inspecting, maintaining, and repairing them.[43] As Vine wrote in a memo to Paul Fye in October 1960, "Manned deep submersibles are badly needed ... to carry out the job of survey, supervision of equipment, and trouble shooting."[44] Elsewhere he continued, "Such a craft could be used to supervise, check, and do routine inspection of apparatus and perhaps be used to retrieve equipment or ordnance."[45] The Navy's commitment to installing instrumentation on the seafloor led to the demand for access to that instrumentation. Although it might seem obvious in retrospect that the technical capacity to visit the depths of the ocean would be of spectacular scientific interest, early memoranda say next to nothing about that.

A False Start: The *Aluminaut* Program

There had already been several attempts to reach great depths in the oceans, but none involved what could be called an operational vehicle.[46] The most famous vessel was the steel-hulled bathyscaphe *Trieste*, built by the Swiss physicist Auguste Piccard (1884–1962) and completed in 1952. A two-meter sphere attached at its top to a large gasoline-filled float; buoyancy was created by the gasoline, ballast by iron pellets. The pellets were held in place by an electromagnet: if power were lost, the ballast would drop and the sphere would return to the surface.

Piccard had received support for his projects from the Belgian, Swiss, and Italian governments, but in 1955 *Trieste* was sitting in dry dock for lack of operating funds. Robert Dietz, the marine geologist who along with Harry Hess would develop the theory of seafloor spreading, was at that time working with the Office of Naval Research in London and thinking about using *Trieste* for scientific research.[47] Dietz arranged for Jacques Piccard (1922–2008)—Auguste's son and *Trieste's* pilot—to present his work at a National Academy of Sciences workshop on deep-sea research.[48]

At the workshop Dietz emphasized that the Navy would be an appropriate agency to fund a scientific research program based on *Trieste*, but it would do so only if oceanographers expressed real enthusiasm. After the academy passed a unanimous resolution of support, the Navy contracted with Piccard for a series of dives in the Tyrrhenian Sea, off Naples.[49] The successful program brought Navy officials and civilian scientists to a depth of ten thousand feet and convinced the Navy to purchase *Trieste* and house it in San Diego for scientific and military purposes. In January 1960, piloted by Piccard and accompanied by Navy Lieutenant Don Walsh, *Trieste* made a historic dive to the bottom of the Challenger Deep in the Marianas Trench—35,805 feet. Humans had reached the deepest place on Earth.[50]

But *Trieste* was not a substitute for what Vine had in mind. For one thing it was difficult to maneuver. The sphere could not hover silently, or with precision, because maintaining neutral buoyancy required alternately jettisoning gasoline and iron ballast; for an experiment in underwater listening during the 1957 season, Piccard stabilized the craft at one thousand feet by attaching a long line of gasoline-filled plastic bottles.[51] Moreover, to refill the ballast after a dive, it was generally necessary to return to port. By Piccard's own admission, *Trieste* was a prototype, built on meager funds and very hard to use.[52]

A more attractive proposition to Vine would be something more like the *Soucoupe Sous-Marin*—in English called the *Diving Saucer*—a small maneuverable submersible built by an ambitious Frenchman named Jacques Cousteau (1910–1997). *Soucoupe* was an impressive little vessel but had a depth capacity of only 980 feet. Vine wanted the maneuverability of *Soucoupe* with the depth capacity of *Trieste*, and he wanted it at Woods Hole. He found it in a plan for a vessel called *Aluminaut*.

In 1957, naval engineer Edward Wenk Jr. (1920–2012) had led a National Academy of Sciences Committee on Undersea Warfare study on pressure hulls for deep-diving submarines. The committee was one of many established by the academy at the end of World War II to maintain connections between academics and military established during the war, and the focus of the study was whether submarines could go deeper through hull improvements. The shipbuilding industry was tradition bound and rarely considered using any material but steel, but alternative materials, such as fiberglass-reinforced plastic, titanium, and aluminum alloys, had excellent strength-to-weight ratios and might permit submarines to reach greater depths. When factors such as availability and ease of fabrication were included, high-strength aluminum alloys showed the greatest promise.[53]

In 1958, Louis Reynolds (1910–1983), chairman of Reynolds Metals Corporation, commissioned Wenk to undertake further work at the Southwest Research Institute (SRI) and develop a design for a deep-diving aluminum submarine.[54] Wenk envisaged a sub that could withstand depths to fifteen thousand feet—more than 60 percent of the ocean's expanse—and Reynolds envisaged an expanded market for aluminum. Reynolds and Wenk approached Captain Charles B. Momsen Jr. (1920–2002), the director of undersea programs at the ONR, with a proposal to build *Aluminaut* for the US Navy.[55]

As a research unit, the ONR could not build a submarine, but it could rent one if it had a research purpose; Vine prepared a technical report formally proposing the *Aluminaut* program to support scientific research at Woods Hole. According to Vine, eleven Woods Hole scientists had expressed interest in using the vessel in their research; six of these were "intensely interested in the deep scattering layer." In particular, the acousticians wanted to know "to what extent the reverberation is divided between many small creatures and a few large ones," perhaps including whales, sharks, porpoises, and giant squid. Vine particularly singled out the giant squid, saying that it was "hoped that large deep squid can be found and their living habits can become better known."[56]

Other proposed areas of investigation included geological studies, including bottom features such as ripple marks and tracks of marine organisms, currents, submarine troughs, and seamounts; physical and chemical oceanography, especially studies of bottom currents; gravity; and sound transmission. The latter was redundant, because in some sense the *entire* subject was sound transmission. Although the Woods Hole biologists might learn about marine life *qua* marine life, the focus on the scattering layer was motivated and framed by its salience to sound transmission. Indeed, all the topics were relevant to submarine warfare. Currents, though listed under "geological studies," were not normally thought of as a geological topic, but they were of interest to any Navy-oriented oceanographer as well as, obviously, to the Navy itself.

The same may be said of seamounts. Vine described them as "of particular interest," but to most geologists they were no more interesting than other features of the ocean basins and less interesting than many.[57] As we saw in chapters 4 and 5, there was far more reason to be concerned with the mid-ocean ridges, which had been proposed as the sites of continental rifting and drifting; or the trenches, which were associated with major and somewhat mysterious gravity anomalies; or the fracture zones of the Pacific, which were utterly unexplained. Yet seamounts figured prominently in Vine's proposal and in many other reports to the ONR at the time, because of their salience to SOSUS. Crucially, you could put a hydrophone on a seamount in areas of the ocean that were otherwise too deep.[58]

In the spring of 1960, Vine sent a draft of his report to the Reynolds Metals Corporation; Reynolds wondered whether Vine had been insufficiently explicit about the military purposes. Vine explained to Paul Fye that "Reynolds felt that maybe our proposal was too research-minded and not enough Navy-oriented to please the Navy higher-ups and perhaps we should add a classified naval warfare addendum to our initial proposal."[59] Vine disagreed because overemphasizing the operational aspects ran the risk of stepping on Navy toes, and it was the ONR's job to negotiate the appropriate balance. It was also unwise to compromise the scientists' interests too readily. Vine expected a certain amount of haggling, but he did not intend to give away too much too soon: "I explained [to Reynolds] the running context [within the Navy] between the combat oriented groups and the research groups, [and] explained that as middle man it was the job of the ONR to negotiate any re-orientation of our proposal. Woods Hole normally cooperates with the Navy on applied problems and would do so again but we would not sign our basic interests away before the fact. If only a highly ASW [antisubmarine warfare] short-range program is envisaged then we would lose interest in regards to *Aluminaut*. Reynolds concurred that our approach was probably right but risky."[60] This was a motif that ran through many Woods Hole communications: the distinction between short- and long-term goals, and between the "highly ASW" from the generally "Navy-oriented." These distinctions were central to how Vine and Fye understood their relationship with the Navy. They were eager to work on matters that were broadly "Navy-oriented," but they also considered it important to take the long view and avoid projects that were too "highly ASW."

This approach protected Woods Hole's interests, they believed, because it allowed for creativity and created space for basic research. It assigned the program organization and structure to scientists—even if the general orientation was toward military concerns—and prevented the Navy from micromanaging things. In short, so long as a program was Navy-oriented but not too highly ASW, scientists would be able to remain in charge. Reynolds feared that if the project were not explicitly framed in military terms, the Navy would reject it; Vine and Fye feared that if it were too strongly framed that way, scientists would lose control. The challenge was to strike the right balance.

In February 1961, in a meeting at the ONR in Washington, DC, the parties found common cause.[61] Reynolds Metals would design and build *Aluminaut* based on Wenk's SRI research; the Bureau of Ships would vet the vehicle design and specifications; and the construction would be subcontracted to the Electric Boat Division of General Dynamics.[62] Once completed, the vessel would be rented by Woods Hole for three years; ownership would then transfer to the Navy.[63] And the ONR would foot the bill.

These terms were laid out in a statement of agreement among the ONR, Woods Hole, and the Reynolds Corporation. The plan was to work from the general terms discussed in February to develop and sign a formal agreement between Woods Hole and Reynolds for the lease of the vessel once it was ready; Reynolds would negotiate a separate construction contract with Electric Boat. The effort immediately became snarled. Vine envisaged a three-year program costing $3.2 million, covering "a year of preparation (during construction) and two years of operation." The ONR was prepared to support something more like $1 million for the first year of the project, with a promised "1,000,000 or slightly more for each of the next two years."[64] Vine was happy with the ONR's commitment, but not with Reynolds's demand for a leasing fee of $500,000 per year, which he feared would leave little for the technical work. Moreover, Reynolds had yet to guarantee that the vessel would be fully insured and would "operate about as outlined in the specification."[65]

Woods Hole and the ONR thought they had agreed in February that Reynolds would transfer title of the vessel at the end of the three-year lease term, but Louis Reynolds did not agree that he had agreed.[66] In early correspondence, Dwight Day, the retired Navy admiral who was directing the *Aluminaut* program for Reynolds, had made it clear that Reynolds would not want to sell the vessel. Louis Reynolds envisaged *Aluminaut* in grand historic terms and intended at the end of its useful life to present it to the Smithsonian. Neither the ONR nor Woods Hole accepted that. They expected that Reynolds would build the vessel and then hand it over to them, and that they, not Reynolds, would decide its use and fate.[67]

Vine was particularly unhappy about the vessel design. Mimicking a sperm whale in size and shape, the initial design contained only two small portholes at the front of the vessel and no visibility at the back or sides (fig. 6.4). This would severely constrain its usefulness for scientific observations—typical for a conventional submarine and one reason conventional submarines were rarely used for scientific research.[68] In early discussions with Day, Vine had underscored the urgency of high standards of vision and maneuverability, which he called the "two prime requirements that need improvement [with respect to Reynolds's early designs] and cannot stand degradation."[69] Eight months later no progress had been made; Woods Hole engineer William Schevill complained that "at this rate I think I'd rather tow a camera from a surface ship" (on towed instrumentation, see chapter 7).[70]

The visibility issue came to a head when the *New York Times* ran a front-page article about *Aluminaut*. Using information supplied by Reynolds, the newspaper reported that "the pilot will sit amidships and will have no view of the outside." Vine clipped the article and twice underscored this sentence, which confirmed his suspicions that Reynolds was barreling forward with insufficient regard to scientific concerns.[71]

Reynolds finally agreed to add two more windows to the vessel design and to transfer title after three years; in November 1961 a contract was signed between Reynolds and the Electric Boat Company for the design and construction of *Aluminaut*. The vessel would be forty-four feet long with an eight-foot hull diameter and weigh 150,000 pounds. It would carry a crew of three—two scientists and a pilot—to depths of fifteen thousand feet. With an average

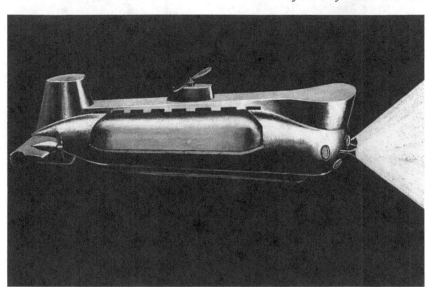

Figure 6.4 The Design of *Aluminaut* 1960. Note the lack of visibility except for the small portholes at the front. Produced by the Reynolds Metal Corporation for Woods Hole Oceanographic Institution.

From *Tomorrow through Research* 15, nos. 1–2, Southwest Research Institute, p. 2, in Woods Hole Office of the Director Records (Paul McDonald Fye), 1942–1979, AC-09.5, "Aluminaut: Publicity, 1961," Data Library and Archives, Woods Hole Oceanographic Institution. Reprinted with permission from the Woods Hole Oceanographic Institution.

speed of two knots, it would have a submerged range of one hundred miles and sufficient air to dive for seventy-two hours. While its depth capacity was considerably less than *Trieste*, at fifteen thousand feet *Aluminaut* would nevertheless have access to more than half of the known depths of the sea and all the depths at which the Navy was likely to send or receive sound. While the Navy would not reveal the depth capacity of its existing fleet, press releases allowed that fifteen thousand feet was "many times deeper than modern subs can dive."[72] The launch was scheduled for 1963 at a projected cost of $2 million.

With each step in the project, the parties issued press releases playing on the exploratory and heroic images invoked by *Aluminaut*'s name. Both Woods Hole and Reynolds laid claim to the conquest of "inner space" (or sometimes "liquid space"); Fye frequently noted that, although the ocean floor was only four miles away, we knew scarcely more about it than the far side of the moon. *Aluminaut* represented the forefront of technology, but it was also friendly: Reynolds's publicity images presented an anthropomorphic vessel reminiscent of the Little Engine That Could. Publicity images suggested that *Aluminaut* would make the deep sea not only accessible but also hospitable (fig. 6.5).

Figure 6.5 Reynolds company publicity images: (a) artist's rendition of *Aluminaut* exploring the deep sea; (b) a smiling Louis Reynolds holding a scale model of *Aluminaut*, produced by the Reynolds Metal Corporation for the Woods Hole Oceanographic Institution.

(a) From WHOI Image collection, 1930–ongoing, AC-44, "Alvin Construction—Cetacean, Aluminaut, Trieste: Three photo albums, annotated: Alvin testing and construction; Lukens Steel, MIT, Hahn and Clay, Litton Industries; Rainnie and McCamis; negatives, April 1963, August–September 1963," Data Library and Archives, Woods Hole Oceanographic Institution; (b) From *Tomorrow through Research* 15, nos. 1–2, Southwest Research Institute, p. 2. In WHOI Office of the Director Records (Paul McDonald Fye), 1942–1979, AC-09.5, "*Aluminaut*: Publicity, 1961," Data Library and Archives, Woods Hole Oceanographic Institution. Reprinted with permission from the Woods Hole Oceanographic Institution.

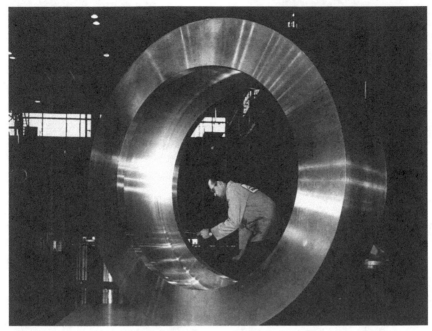

Figure 6.6 The aluminum ingot for the *Aluminaut* hull, produced by Reynolds for Woods Hole.

From WHOI Image collection, 1930–ongoing, AC-44, "*Aluminaut*: incl. mock-up; underwater, published in *Oceanus* v.31:4, #6057," Data Library and Archives, Woods Hole Oceanographic Institution. Reprinted with permission from the Woods Hole Oceanographic Institution.

Not surprisingly, each party emphasized the aspect of the project most resonant with its mission. Fye focused on the scientific and exploratory aspects, stressing that *Aluminaut* would extend "our capabilities for a wide variety of geological, biological, and physical research at all depths, such as studies of the submarine canyons, the edge of the continental shelf and the daily vertical migration of marine animals."[73] Reynolds emphasized the demonstration of aluminum's capabilities, claiming that the metal's unique combination of high strength and low weight made it "the only practical material strong enough to resist the external pressures at super depths."[74] In a heavily publicized move—perhaps a publicity stunt—Reynolds Metals cast "the largest ... aluminum ingot ever made" (31,750 pounds) for use in *Aluminaut* (fig. 6.6).

Louis Reynolds suggested that his company's entry into deep submergence was driven by patriotism, particularly the desire to enable the nation to explore new frontiers. One press release claimed, "We felt that the *Aluminaut* project was so urgent and so much in the nation's interest that we voluntarily went ahead with the program" even before the Navy had expressed direct interest.[75] While television producer Gene Roddenberry would soon immortalize (outer) space as the final frontier, Reynolds reminded Americans that the deep sea was Earth's final frontier. Just as railroad technology had made the West accessible, deep-submersible technology would make the ocean accessible. "Private enterprise has an obligation to expand the world's frontiers, and undersea exploration is one of the last great challenges left on earth,"[76] Reynolds claimed. *Aluminaut* would allow man to fulfill the Old Testament promise of "dominion over the sea."[77]

Frontier narratives implied resource exploitation, and Reynolds publicity materials promoted the idea of humans mining and farming the ocean floor.[78] Brochures offering artist's renditions of human colonies on the continental shelf suggested that the seafloor was rich in mineral resources and undeveloped land available to feed the world's burgeoning populations. Mining the resources of the ocean floor was not, and never had been, part of the scientists' or the Navy's motivation for deep submergence, but it satisfied the futurist image that Reynolds was weaving around *Aluminaut*. "The ocean floor is a vast, relatively unknown treasure house of minerals, oil, chemicals, and food sources," Reynolds told the *Wall Street Journal*. "You can expect soon an international race for rights, claims and physical possession of potentially rich underwater mineral deposits, seaweed beds, and fishing grounds."[79]

Of the three partners in the project, the Navy presented its interests most matter-of-factly. "The interest of the Navy," the ONR noted flatly in its September press release, "embraces both oceanography and the obvious implications for submarine developments of the future, including the possible use of aluminum for the hulls of future deep-diving submarines. The Navy also expects to gain a better understanding of problems associated with anti-submarine warfare."[80] Behind these vague generalizations for public consumption, however, the Navy knew exactly what it wanted *Aluminaut* to do. As Woods Hole engineer William Rainnie Jr. (1924–1985) explained to physicist (and later NASA administrator) Robert Frosch (b. 1928) in the summer of 1962, "The initial use for this craft will be the inspection of sonar arrays off Bermuda next summer."[81]

Bermuda was the site of the Artemis project, a Navy experiment in long-range active listening; Frosch was the project's director. In 1950, the Navy had established a study group, Project Hartwell, to explore "the technical limits to effective anti-submarine and anti-mine warfare."[82] One of the group's recommendations was the establishment of a civilian scientific laboratory dedicated to underwater sound. This became the Hudson Laboratories of Columbia University, in Dobbs Ferry, New York, heavily staffed with Columbia graduates who had worked on the Manhattan Project.

SOSUS was a passive system—one listened for noise—and that worked because Soviet submarines were very noisy. But sooner or later the Soviets would make their ships quieter; then one would have to seek them actively by sending signals and listening for their return.[83] In 1957, the ONR asked scientists to initiate a project to determine the technical and economic feasibility of a long-range, low-frequency active sonar system able to detect Soviet submarines at up to five hundred nautical miles.[84] The Artemis project began in 1958, first under the direction of Hudson's Robert Frosch and then the lab's associate director, physicist Alan Berman (b. 1926).[85] Unlike most Navy projects and research groups, the code name was not an acronym; it referred to the Hartwell Project scientist who first suggested the project's feasibility, Harvard physicist Frederick Hunt (1905–1972).[86]

The Artemis work was largely done at the Hudson Labs but included scientists from MPL, Woods Hole, the Naval Electronics Laboratory, the Naval Research Laboratory, and Bell Laboratories. The task was extremely challenging, primarily for lack of equipment capable of producing signals of sufficient strength to overcome propagation losses and background noise. When Berman took over in 1963, he confessed that he had initially been "appalled by the obvious difficulties ... [and] had grave doubts as to the wisdom of pursuing the project."[87] Nevertheless, the project moved forward, driven by pressure from the Navy and the willingness of project scientists to be "brave" in the "battle for decibels."[88] The first modules had been installed in 1961, near Bermuda, and a second, larger installation was scheduled for the summer of 1963.[89]

The success of the battle for decibels would depend in part on the conditions of battlefield, so a major concern was to "uncover the environmental factors that would limit the specifications and performance of the system."[90] This led to a ten-year research effort, led by Woods Hole's Brackett Hersey, on environmental factors affecting the Artemis system. This included scattering by bottom reflection, temperature-dependent velocity variation, and the impact of deep currents on seafloor instrumentation. There was also the issue of deep-sea access. How would you get the hydrophones to the seafloor, emplace them, and retrieve them if you needed to repair, upgrade, or replace them? As described in an early MPL proposal: "The receiving hydrophone system for Project Artemis will require installation of a large, complicated set of acoustic elements on the bottom of the ocean in deep water. Array elements must be designed and evaluated and techniques must be developed for placing these elements on the deep sea floor."[91] Whatever scientific work *Aluminaut* might do in the future there was a task waiting right then: supporting the technology that was sustaining the Navy's ability to detect Soviet submarines.

From *Aluminaut* to *Alvin*

By early 1962, Electric Boat was moving forward on tests and design work for *Aluminaut,* but things were not going well. Particular worry arose over the yield strength of the prototype cylinder. Tests on samples taken "from the center of the worked metal in the axial and radial directions fell below the design yield (60,000 psi) by several thousand psi," and these problems worsened as the months went by. Forging of hull components was several weeks behind schedule "due to delays cause by surface rupturing on ingots and hemihead tooling modifications.... Forgings have experienced repeated surface cracking, and extensive conditioning has been required between operations, delaying the early phases of the forging process.... As a result of the repeated cracking, Reynolds has ordered all remaining unworked ingots back ... for re-homogenizing."[92]

The engineering problems convinced Paul Fye that the project needed closer supervision. In March, he appointed physicist Earl Hays (1918–1985) in charge. A Navy officer in the Pacific during World War II, Hays had worked at the Brookhaven and Los Alamos National Laboratories before joining Woods Hole as chief assistant to Bracket Hersey on underwater sound. (He would later become chair of Woods Hole's Department of Ocean Engineering.) Hays would be assisted by James W. Mavor Jr. (1923–2006), who would oversee the metallurgical work, and by William O. Rainnie Jr., the graduate of the Navy submarine service who would become *Alvin*'s first pilot. Vine would serve as a consultant to the project on oceanographic use and applications.

Mavor believed that the forging problems meant that construction of *Aluminaut* was still some time away, and it was unclear how long the problems would take to fix. There were significant unanswered technical questions and insufficient coordination between the various parts of the project responsible for testing, verification, analysis, and design. Although *Aluminaut* was experimental, it still needed to be designed and tested as a "future operating vehicle"; no one wanted an experiment once men were in it underwater. Mavor also noted that Woods Hole was the only prospective operator and suspected that, for that reason, the project was not receiving sufficient scrutiny.[93]

Things unraveled fast. In April 1962, only seven months after the glowing press releases of the previous September, Fye wrote to the ONR describing what he called "the last chapter in the sad story of negotiations with Louis

Reynolds."[94] It turned out the parties had never actually agreed on terms. Woods Hole and the ONR had acceded to Reynolds's demand of $500,000 a year for three years on the condition that the Navy would then assume title, and respecting Reynolds's desires for recognition of its historic significance, *Aluminaut* would be offered to the Smithsonian at the end of its useful life. But between January and April, Reynolds had incrementally increased his monetary demands; by April he was asking for an annual lease fee of $712,455 for the first three years and $100,000 per year after that. Backing off his initial position of refusing to sell, he offered three alternatives. One, Woods Hole could purchase the vessel on completion for $3 million. Two, the Navy could take over the construction contract with Electric Boat by paying his costs incurred, which he estimated to be $650,000 so far. Three, if the Navy had lost interest, he had other customers to whom he could sell *Aluminaut* at a profit. Either way, Reynolds claimed, he intended to build a second *Aluminaut* right away.[95]

Now it was Woods Hole and the ONR's turn to balk. Rear Admiral Leonidas D. Coates (1907–1989), chief of naval research, reiterated unambiguously that his group was unwilling to enter any agreement that did not transfer title after three years.[96] Because the technology was new and untested, the Navy needed the flexibility to modify the vessel, which could be assured only if the Navy owned it:

> The retention of title to the *Aluminaut* beyond the three years previously agreed is completely and irrevocably unacceptable. The tenuous technology upon which the *Aluminaut* is based gives no assurance of its continued usefulness beyond any stipulated period. The undefined character of the research program which will be coupled to a deep research vehicle makes complete flexibility in the use of the vehicle, in its testing, and in the decisions to modify it in whatever way imperatively the prerogative of the scientific group responsible for the program. Only by obtaining title and consequently full responsibility for the *Aluminaut* can the Navy assure the attainment of the objectives indicated above and the concomitant safety of all personnel involved in the research program.[97]

Coates emphasized that all parties involved had taken risks in exchange for anticipated rewards; for Reynolds, this was the development of new applications of aluminum technology. The ONR "would never have considered a deep research vehicle of the size, complexity, and tenuous technological merit of the *Aluminaut* in the first place were it not for the implicit and expressed interest of the Reynolds Metals Company in this venture. No encouragement was ever afforded that group except on a shared cost basis in view of their expressed interest in the advancement of aluminum technology, the aluminum market, and [the] public image of the Reynolds Metals Company through the resulting publicity." The terms offered by Reynolds were "wholly unacceptable."[98]

Fye was offended at Reynolds's implicit threat to sell *Aluminaut* to a third party (which, given Mavor's observations, was certainly a bluff) and at the fact that Reynolds was treating *Aluminaut* as an ordinary commercial venture. That was not how the project had been conceived. Rather, it had been a partnership in which each party had an interest: "The Navy is interested in deep-sea vehicles and deep-sea research, Reynolds is interested in aluminum technology and fabrication, and Woods Hole in oceanography. Each partner was to make a contribution towards the successful prosecution of the project."[99] Now Reynolds was behaving like an ordinary contractor, as if this were not an experimental technology and his firm should not be expected to share the risks.

The offer to sell at costs incurred was unacceptable because there had been no agreement on accounting procedures, so no means to determine what the costs to date had been and no way to control future costs given that the vessel was unfinished. Throughout the project, the estimated costs had risen steadily, raising the specter of injury to other programs. Indeed, there was already grumbling in some Navy quarters. Fye felt that it would be "impossible to have additional funds earmarked to *Aluminaut*.... The mere attempt to do so would solidify the already considerable opposition to the project that existed in certain segments of the US Navy."[100]

There was grumbling in the scientific ranks as well. Hays was a good choice to direct the project, given the engineering difficulties being encountered, but it did not satisfy the promise of Vine's original proposal that the chief scientist on the project be a "well-rounded oceanographer" with a "keen personal interest in the scientific use" of the submersible.[101] This may have contributed to skepticism about the project among Woods Hole scientists, but there was a deeper issue. Earlier that year Fye had confided in a colleague that there "had been some opposition on the part of our senior people to having this project within the Institution."[102] The issue was *Aluminaut*'s purpose, which most Woods Hole staff concluded was not scientific. Shortly before the project began to unravel, Mavor complained to Fye that a lack of scientific purpose was contributing to staff skepticism. "The *Aluminaut* has been spoken of an instrument which may be used for science. Various general uses have been proposed by the staff, but I have seen nothing specific tying *Alumi-*

naut characteristics with the scientific uses. Some senior staff members have condemned the project for this reason."[103]

Mavor suggested that it might help to develop some specific projects "directly related to *Aluminaut* which may help to justify" it, such as an "experimental study of crystal structure (in the Al)."[104] This suggestion was hardly likely to warm the hearts of Woods Hole oceanographers: a project on the crystal structure of aluminum might interest engineers, metallurgists, or even perhaps mineralogists, but it would not be an *oceanographic* project. It also reveals the divide growing at Woods Hole between science and engineering—the divide that Henry Stommel had noted when he suggested that it was fine for Woods Hole engineers, but not Woods Hole scientists, to work on "need-to-know science" (chapter 3). Meanwhile Vine's earlier suggestion that *Aluminaut* might be used to study marine life seems to have disappeared entirely.

At the ONR, officials were even less enthusiastic about taking over the construction contract with Electric Boat, lest it release Reynolds from a stake in the project's success. Reynolds had an interest and "presumably ... a competence in aluminum technology," and therefore an interest in seeing the project succeed. Were the Navy to take over, "failure would always be construed as Navy incompetence, but a failure with Reynolds in charge would make it clear that the fault was in design or materials."[105] Left to their own devices, the Bureau of Ships would not have chosen an untested material like aluminum, the ONR now insisted; the Navy would have "specified the hull materials, and it would have been one they knew well, i.e., one they have already invested a lot of time and money in to determine its characteristics and fabrication techniques."[106] As for the alternative of buying the vessel outright for $3 million, neither Woods Hole nor the ONR had $3 million to spend. Fye concluded it was time to cut their losses. Coates concurred, feeling that the course of events over the previous six months "made clear the intentions of Reynolds ... to ignore, abrogate, or amend the agreement reached in my office in February 1961."[107] Woods Hole staff continued to follow the progress of *Aluminaut*, but they no longer anticipated using it, and Coates invited Fye to submit a proposal for a second attempt to procure a deep-submergence vehicle.

Things would be different the second time around. Whereas previously the project had moved forward informally on the basis of Reynolds's prior research, Woods Hole now put the project out to competitive bid.[108] In May 1962, a formal solicitation was mailed to seven companies inviting bids for the design and construction of a research submarine operating to a depth of six thousand feet.[109] Every element of the project was specified: hull materials

(HY-100 steel, titanium, or aluminum), safety factors, material and fabrication inspection, testing, and a penalty of $500 per day for delay beyond a specified delivery date of ten months after the award of the contract.[110] Assuming a contract was signed in June, this meant April 1963.

Autonetics, Electric Boat, General Mills Electronics, General Motors Defense Research Laboratory, Lockheed Aircraft, Philco, and United Aircraft were invited to a bidders' conference in Boston at the end of May. Vine and Hays explained their goals: "We intend to work near the bottom in areas that may be rough with currents of several knots. Other work will involve midwater depths (not near the bottom) hovering, runs along tracks at various depths, ascents and descents. Good control and maneuverability required, especially for bottom work. For example, we shall want to pick up specific bottom samples, remain in a position to take a core sample, take pictures of special outcrops, etc."[111] While describing the projects in broadly scientific terms, they made clear that time was of the essence and implied that if the contractors could not satisfy the April 1963 deadline, Woods Hole might yet purchase *Aluminaut*. "*The urgency of the short bid time and 10 month delivery is controlled by research needs next spring*. If we cannot meet this schedule, we may drop the project as we are still negotiating for ALUMINAUT."[112] When pressed at the bidders' conference about the penalty for delays, Vine noted the "urgent need for deadline, and possible losses to Government if not met."[113] The urgency was not, of course, created by basic science. Questions that had taken decades (or even centuries) to answer could stand to wait a few more months. The urgency was created by Artemis.

Only two firms responded, Autonetics and General Mills, and neither bid was conforming. After a second meeting in Washington, Autonetics came back with a conforming bid at $595,000 and General Mills at $498,500. (Recall that Reynolds was demanding $3 million!) The Bureau of Ships was enlisted to assess the safety features of the proposed vessels, to evaluate the designs, and to prepare "a set of Safety Provisions and Specifications to guide us in our overseeing of the craft's construction."[114] In July, the ONR approved Woods Hole's plan to proceed with contract negotiations with General Mills.

Since 1960, the Electronics Group of General Mills had been working on the design for a nineteen-foot, two-man, steel-hulled submersible for scientific exploration called *Seapup*; in 1961 they had proposed to build it for as little as $100,000.[115] The new submersible was tentatively named *Seapup VI-A*. Over the next few months, construction proceeded with none of the tension associated with *Aluminaut* and only a small fraction of comparable correspondence. General Mills did the job they were hired to do.

But while construction was going smoothly, a good deal of time had been lost over the *Aluminaut* debacle; they were behind schedule with respect to Artemis. As Fye explained to the chief of the Bureau of Ships in June 1962, "We have initiated procedures for the procurement of a vehicle to meet our research needs as well as special problems related to the *Artemis* project which must be solved during the summer of 1963."[116] While science and exploration were providing the public justification for the program—in effect, the cover story as well as internal justification for Woods Hole's involvement—it was Artemis that was setting the schedule and creating the pressure to meet deadlines.

Woods Hole staff were still monitoring the progress of *Aluminaut*, but it was increasingly clear that, even if it were completed, they would not want it. In September, Mavor replied to a query from Fye: "Regarding your question 'Does this caliber of work indicate that we will not trust the final structure enough to use the *Aluminaut* if it is ever completed?,' my answer is no [we will not].... Most of the many unanswered technical questions existing a year ago remain unanswered.... As things are going now, I think there is little chance that we will know enough about *Aluminaut* when it is completed to trust it."[117] Woods Hole had "not entirely divorced" from *Aluminaut*, but Mavor and the others were restricting their efforts to "keeping abreast with the fundamental metallurgical and technical hull problems."[118] Their focus was on the vessel being built by General Mills.

The troubles with *Aluminaut*—particularly the safety concerns raised by the engineering problems—and the pressure to get *Seapup* done by April 1963 raised the question of whether a manned submersible made sense: perhaps hydrophones could be serviced by remotely operated vehicles? This idea was being pursued at Scripps (chapter 7); the desire to have a crewed submersible at Woods Hole clearly contained an element of competition between the two institutions. But institutional rivalry was scarcely sufficient justification for a $3 million research program. Given the historical association of deep submergence with spectacle and the difficulty that the details of the underwater acoustics program could not be discussed, Fye began to worry that some people might consider the whole venture "just a stunt."[119]

In July, Rainnie had written to solicit advice from Robert Frosch (soon to become deputy director of ARPA, and later assistant secretary of the Navy for research and development); Ralph Kissinger Jr. (1914–2000), a retired naval engineer who had worked at SRI on the *Aluminaut* design; and Fred Spiess. "This venture is a unique and inherently uncertain one, for we are pushing the 'state-of-of-the-art' reasonably hard," Rainnie wrote. "In order to assure that this is in the best interest of marine science and the National Program in Oceanography, we have requested you to study and advise us as to whether this is, in fact, a reasonable and sensible way to proceed, considering, but not limited to, the following areas: 1. Scientific needs, 2. Technical Feasibility, 3. Adequacy of design, 4. Cost, 5. Reasonable Probability of Accomplishing the Objectives."[120] The letter was placed in a folder labeled "doubt letter."[121]

The three recipients met in Washington and penned a reply supporting the Woods Hole effort. The scientific need, they wrote, had "been the subject of many committee meetings and reports and is generally well documented. The primary mission ... will be in bottom investigation in geology, geophysics, biology, observations of bottom installations [i.e., hydrophone arrays] and aid in work on such installations. The secondary mission will be in midwater tasks in physical oceanography, underwater acoustics and biology [i.e., studies of the deep scattering layer]." The project did not require "any engineering know how which is not available today"; it simply required "first class engineering and workmanship."[122]

The only slight note of discord was struck over the "probability of accomplishing the goal," because of the extremely tight schedule, which did "not leave any room for mistakes or delays whether engineering or administrative." Rainnie had repeatedly underscored the "required delivery date of April 30, 1963 of a fully tested vehicle," but the advisers suggested that Woods Hole was pressing too hard on the delivery issue.[123] "While the delivery date is important for accomplishment of a particular mission, failure to meet the time requirement will not prejudice the general usefulness of the vehicle," they concluded. Frosch of course knew about Artemis, so it seems they were trying to encourage Woods Hole to slow down, do the job right, and build a vessel that would be useful for research even if they missed the deadline for the upcoming Artemis installation: "This is a needed and overdue attack on a vehicle for investigation of many problems.... The deep research vehicle has been conferenced to death and at this stage the right thing to do is to build something."[124]

Before the research vehicle could be completed—indeed, just three weeks before the April deadline that Rainnie had stressed—a tragic event would underscore the imperative of deep submergence. It was one of the major US losses of the Cold War: the sinking of the submarine *Thresher*.

The *Thresher* Disaster

In 1963, the US nuclear submarine *Thresher* (SSN-593) was lost at sea while undertaking a test dive 220 miles off the coast of Cape Cod. Completed in 1960, the boat had spent 1961 in an extended shakedown cruise, followed by a year of repairs, adjustments, and retrofits. On April 9, 1963, it was finally ready to go out again. On April 10, it was lost, along with the entire crew of ninety-six men and twelve officers, plus twenty-one civilians who were on board.[125] It had cost $45 million dollars to build. It was the worst submarine disaster in naval history.[126]

The tragedy was heavily publicized in front-page newspaper articles and leading stories on radio and television; it reverberated throughout American society. The enormity was felt not only in terms of the human and monetary losses but also in terms of American aspirations for global power. *Thresher* was the fifteenth nuclear submarine in the US fleet and the first in its class of fast-attack submarines coming on line in the early 1960s. Together with the companion ballistic missile submarine, the fast-attack submarine was a crucial part of the nuclear triad, designed to protect the United States from the ultimate Cold War threat—the preemptive first strike—by ensuring that any Soviet attack would be met by an immediate response from submarine-launched ballistic missiles. Although the *Thresher* did not carry nuclear warheads, it was nuclear powered, and the whereabouts and condition of its reactor were unknown.[127]

The loss of any submarine full of men would have been horrible, but this disaster was imbued with added poignancy because *Thresher* was state-of-the-art: the fastest and deepest-diving submarine in naval history. It was considered a beautiful boat; management and workers at the Portsmouth Naval Shipyard where the boat was built "looked upon *Thresher* as their finest creation. They were proud of her."[128] Even the radical left-wing folk singer Phil Ochs was inspired to write about it, so great was the sense of loss even among those who disagreed with the politics that drove the nuclear Navy. "For she'll never run silent," he gently sang, "she'll never run deep / for the ocean has no pity, and the waves they never weep, they never weep."[129]

Thresher had departed the Portsmouth Naval Yard the previous morning to conduct sea trials escorted by the USS *Skylark*. On the morning of the second day at sea, the crew commenced the first deep dive, which "appeared to *Skylark* personnel to proceed satisfactorily until about 9:13 am, when *Thresher* reported to *Skylark*, 'Experiencing minor difficulties. Have positive up angle. Am attempting to blow. Will keep you informed.'" At 9:16 a second message was received that seemed to include the words "test depth" and then at 9:17 the words "nine hundred north."[130] That was the last message received, although later the first lieutenant on *Skylark* would testify to hearing one more thing:

A. [A] sound that registered with me as being familiar because of the fact that I heard a lot of ships breaking up during World War II after having being torpedoed at depths. It sounded as though there was a compartment collapsing or something similar to that nature.

Q. Did you hear anything in addition to the sound which you identified as similar to a compartment breaking?

A. No, sir.

Q. Can you describe that sound to the court?

A. It is a rather muted, dull thud.[131]

Five minutes after *Skylark* first received communications indicating that something was wrong, *Thresher* was lost. The sea was calm; there were no other ships in the area. The summary report of the Navy Court of Inquiry dryly noted that "there was no evidence of sabotage or enemy action in connection with the loss of the *Thresher*."[132] What had happened, and where was the boat?

Within hours, the chief of naval operations ordered a Court of Inquiry, which was underway by the next day.[133] Because *Thresher* was a nuclear submarine, it also fell under the jurisdiction of the US Congressional Joint Committee on Atomic Energy.[134] While the US Congress and Navy Court of Inquiry spent May and June trying to determine what went wrong, the Navy was desperately trying to locate the lost hull. The *Skylark* had commenced a sonar search immediately and additional vessels and aircraft were quickly brought to the area.[135] As it happened, Woods Hole's *Atlantis II* was on a cruise about a hundred miles away and immediately joined the search team.[136] Nothing was found in the initial hours except a small oil slick and two rubber gloves, both right handed.

As it became clear that *Thresher* was lost, the Navy mobilized scientists to help in the search: from the Naval Research Laboratory, the Naval Oceanographic Office, the Naval Ordnance Laboratory, the Hudson Laboratories, Scripps, Lamont, and Woods Hole. The technical effort would be led by the ONR's senior oceanographer, Arthur Maxwell (the man who worked on heat flow with Roger Revelle in the 1950s; chapter 5); the civilians deployed included Vine and Hersey from Woods Hole, Spiess from Scripps, Ewing and Worzel from Lamont, and Frosch and Berman from the Hudson Labs. The analysis group would be based at Woods Hole; their task was to "analyze and interpret the results of the search effort and to act as a day-to-day scientific staff for the on-scene commander."[137] That man was Navy Captain Frank A. Andrews (1921–2014), commander of Submarine Development Group II, who would lead a flotilla of twenty-eight Navy warships, several other specialized Navy vessels, and civilian research ships including *Atlantis II* and Lamont's research vessel *Conrad*.

The *Atlantis II* had recently finished sea trials and was equipped with the latest instruments, including an improved underwater towed camera that was an extension of one of Ewing's and Vine's earliest projects at Woods Hole. Photographs of twisted metal and perhaps paper, later recognized as part of a large debris field, provided the first tangible evidence of the lost hull. This was followed by further underwater photography by the *Conrad*, which provided the first evidence of heavy objects definitely attributable to *Thresher*. This was a start, but small bits of metal or paper might travel far from their source—a problem that worsened with each passing day—so the mere presence of debris did little more than confirm what was already known. The *Thresher* had sunk in 8,500 feet of water. Given the strength of ocean currents and the time it would have taken for the hull to reach bottom, there was a very large area in which it could have come to rest.

Looking for a mangled hull in 8,500 feet of water over an area of hundreds of square miles was worse than the proverbial needle in the haystack; at least in a haystack there is light. Underwater photography relied on a supplied light source, which illuminated only a very small area. It could take months, or longer, to search the ocean floor this way. Moreover, the bottom sediment in the region was very soft: might the hull have struck bottom with sufficient force to impel itself entirely into the mud? If so, there would be nothing visible from the surface and underwater cameras would be useless. If the boat were completely (or even mostly) buried, there might be no sonar signal either.

The scientists responded to this problem by applying all available means of detection—photography, echo sounding, magnetometry, radiation—and analyzing the data as it came in.[138] A great deal of hope was placed in sonar; Vine tried his hand at a series of calculations to produce profiles of the likely sonar signal of various objects, among them a plank of wood and a school of tuna.[139] Scientists at the Hudson Labs built a physical model of the seafloor topography in the region. Yet at root the problem was indeterminate, for there was no example of a sonar signal from a submarine on the seafloor. Moreover, if the hull were broken into several pieces or buried in the marine mud, the signals might be very weak. (At one point it was suggested that an old submarine be towed and sunk in the search area for comparison.[140]) The search team also undertook a radiation survey. Detectable radiation would be the clearest evidence they were close to the target, but it would also mean that the sub was leaking radioactive materials—potentially a worst-case disaster. Fortunately, the radiation survey found nothing.[141] *Trieste* was available to go down to the seafloor, but without a target there was no use sending it.

While sonar and radiation proved inconclusive, other efforts paid off. Most

important were the estimates of the effects of ocean currents on the disposition of the hull. Vine and his colleagues calculated the likely direction of the
vessel when the crew lost control, based on information on vessel speed and
direction at the time of the accident and the prevailing ocean currents. They
were able to calculate a vector of travel as the boat went down and narrow
the search area to a ten-square-mile region. This became the basis for a more
focused search using towed camera, dredging, echo sounding, and magnetometers.[142] On May 14, photographs taken from the *Atlantis II* revealed a large
debris field, replete with paper, wire, and bits of twisted metal. Dredging the
site soon produced a battery plate positively identified as belonging to a submarine of the same class as *Thresher*, as well as several O-rings. A few weeks
later, using magnetic anomalies as guide, *Conrad* photographed an oxygen
bottle, a hydrophone, and a ten-foot piece of sheet metal.[143]

In August, *Trieste* was sent from the West Coast in an attempt to find and
photograph the hull. But it lacked a precise target and in any case had little
capability in underwater navigation. *Trieste* found a few more bits and pieces
but still no hull. In September, the North Atlantic weather put a stop to the
operations. During the winter of 1964, *Trieste* was fitted with a new float and
mechanical arm and renamed *Trieste II*. In the spring, *Trieste II* was towed
to Boston while the task force prepared to resume its work. In March, *Conrad* photographed a twenty-foot square section from the bow of the hull.[144]
On June 23, 1964—almost fifteen months after the disaster—the task force
identified the location of the hull, broken into five pieces, at 42°N and 65°E,
almost due east of Woods Hole in 8,250 feet of water.

The task force was now joined by the USNS *Mizar*, probably the most sophisticated oceanographic ship at the time. A recently converted Antarctic
supply ship that had been assigned to the Naval Research Laboratory, *Mizar*
was equipped with state-of-the art towed underwater cameras, magnetometer, radiation monitoring equipment, side-looking sonar, and a sophisticated underwater tracking system to enable the ship to locate both itself and
the towed equipment with respect to transponders placed on the seafloor.[145]
Three days after the hull was located, *Mizar* sailed from Boston on June 26
and proceeded to the area identified by the *Conrad* photographs and magnetic
anomalies. By the following day, *Mizar* had photographed the five major sections of the hull and much of the debris field around it (fig. 6.7). Over the next
two weeks *Mizar* cameras took over twenty thousand photographs of the hull
and surrounding terrain.[146]

The pictures were unmistakable. Navy officials hoped that *Trieste II* might
be able to get closer to the wreck, stay longer in its vicinity, and perhaps dis-

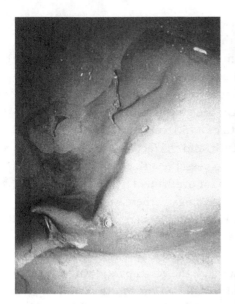
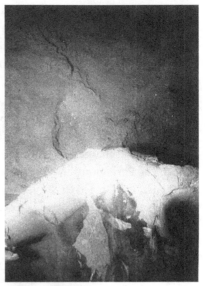
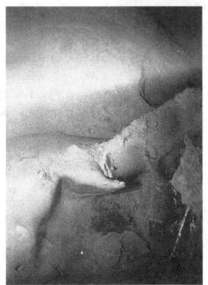

Figure 6.7 Mizar photos of the *Thresher* hull at the bottom of the sea, partly buried in the marine mud.

Images courtesy of the late David van Keuren, Naval Research Laboratory.

cover crucial information on the cause of the disaster. On June 21, *Trieste II* began the first of a series of dives to the hull. A series of markers, called "fortune cookies," had been dropped from the surface to help the pilot navigate the bottom, but the effort was still fraught with difficulties, particularly because bottom currents made it difficult to lower and maintain the bathyscaphe in the appropriate position. On the very first dive, a fire broke out in the main propulsion system at a depth of 8,250 feet; catastrophe was averted only by the lack of oxygen at that depth, and the fact that the batteries had completely discharged by the time the vessel resurfaced.[147] The vessel was sent to the Portsmouth Naval Shipyard for overhaul of its electrical systems. In August, it returned to the search area.[148] Again, there were difficulties, especially since the repairs had left the vessel with less horsepower overall and it could manage a bottom speed of only 0.6 knots in the face of bottom currents of 0.3 knots.[149]

Two dives failed completely. On a third, the vessel passed within twenty feet of the hull but the camera failed. Finally, on the fourth dive, using the *Mizar* locating system, *Trieste II* located and sat atop a portion of the wrecked hull. A representative of the Naval Research Laboratory described it as "one of the most amazing operations we could have imagined."[150] As the pilot slowly maneuvered the vessel to its target position, the crew at first saw nothing. After checking their coordinates, Frank Andrews ordered the pilot to set down on the seafloor, which he did, but again, nothing. Finally, the pilot turned the vessel around, only to discover that the bottom they had been resting on was in fact the *Thresher* hull. Now they were able to take detailed pictures. The hull was not buried but had plowed a considerable way into the bottom sediment, pushing up large quantities of mud deposited as a one- to two-inch crust on its rear portions (fig. 6.8).

Oceanographic knowledge and technology had brought the task force to the right place, but it had taken more than a year and was fraught with difficulties. The *Trieste*'s work underscored the value of being able to get down to the seafloor, but Navy personnel were dissatisfied with how long it had taken, how difficult it had been, and how close to a second disaster they had come.

In 1965, Captain Andrews summarized the *Thresher* search experience in the *Naval Engineers Journal*. Despite its success, the efforts of the task force "demonstrated only too clearly the degree of ignorance and inability which surrounded the entire business." *Trieste II* had obtained invaluable information but "lacked the maneuverability and ground speed to exploit effectively her other capabilities."[151] Considerable time had been lost towing the vessel to the location, which even under good conditions could be moved at no more

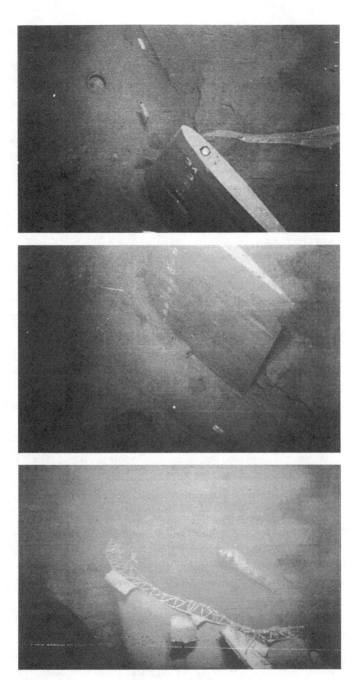

Figure 6.8 *Trieste II* photos of the *Thresher* hull at the bottom of the sea, partly buried in the marine mud.

Images courtesy of the late David van Keuren, Naval Research Laboratory.

than four knots. Heavy seas slowed the process further and damaged the overlying float and externally mounted equipment, and it was difficult and dangerous to get men in and out when seas were greater than three to four feet high. The *Trieste* personnel had managed some rather daring maneuvers in heavy seas, but "this type of performance could not have been carried out as a daily diet or soon someone or something would have been injured badly."[152] Even when all went well, every dive of a few hours was followed by at least an equal number of hours of maintenance, repair, and battery recharging, often lasting well into the night. And since *Trieste II* was virtually a new vehicle compared to the original, the crew was still on a steep learning curve. Overall, the vessel had fared poorly under the intense operational conditions.

Andrews concluded that the Navy "recognized the need for a long range study and development program in deep search, salvage, and rescue techniques."[153] This recognition went all the way to the top, as the secretary of the Navy announced the creation of the Deep Submergence Systems Review Group, to be headed by Rear Admiral Edward C. Stephan, commander of the US Naval Oceanographic Office. The group's charge was to "examine the Navy's plans for the development and procurement of components and systems related to the location, identification, rescue from and recovery of deep submerged large bodies, including submarines, from the ocean floor."[154] The review group was divided into ten subdivisions, including science, industry, administration, operations, engineering, and research. Heading "research" was Fred Spiess; heading "science" was Al Vine.

The group recommended the establishment of a permanent Navy Office in Deep Submergence—the Deep Submergence Systems Project (DSSP)—whose primary goal would be development of "a small manned-submersible operating from a mother submarine to provide an all-weather capability for rescuing submarine personnel down to the disabled submarine's crush depth."[155] Fye communicated with Admiral Stephan about the role that Woods Hole's new vessel—soon to come on line—might play.[156] Stephan noted that having such a vehicle would necessarily entail finding uses for it when no immediate disaster was at hand. "Realizing that whatever capability will be developed will not be used frequently," he wrote, "we feel that such equipment should be conceived against its probable background usefulness in other phases of science and engineering."[157] Vine was right: the Navy did need *Alvin*. Woods Hole's efforts to create and operate a deep-submergence vessel were in line with the Naval demand to have effective deep-submergence capacity in place the next time something like the *Thresher* loss occurred. They did not have long to wait.

From Hydrophones to the Hydrogen Bomb

On June 5, 1964, the long-awaited submersible was finally commissioned at Woods Hole. The Electronics Division of General Mills had meanwhile been purchased by Litton Industries, which completed the contract and delivered the promised vessel, albeit thirteen months late.[158] (Meanwhile Reynolds had almost completed *Aluminaut*, at a cost of $3.5 million; it would be commissioned in September.)[159] The completed boat was twenty-two feet long with an eight-foot beam and displaced thirteen long tons. Design specifications called for a submerged range of twenty to twenty-five miles traveling at a normal speed of 2.5 knots, with a maximum of 6–8 knots. Its depth capacity was six thousand feet, not enough to reach the mid-ocean ridges, much less the abyssal plains, but enough to cover the continental shelves and upper slopes.[160] In keeping with Vine's long-standing concerns about visibility, there were four ports permitting the pilot and scientist to see in front of and below the vehicle. Its name would be deep-submergence research vessel (DSRV) *Alvin*. Why Alvin? It was a contraction of the name of the man who had long been waiting for it: Allyn Vine.[161]

Alvin was immediately put to work. A memo to the file, dated June 15, 1965, articulated its first job:

> The DSRV *Alvin* will be utilized as a means of transportation to permit visual confirmation of the conditions of the Artemis modules and associated cables heretofore not possible. Additionally, modules that may be shielded due to their location behind a cliff or in a canyon will be noted. As a corollary to visual inspection, it is strongly desirable to retrieve a representative module from the 1963 implantment in order to have some means for prediction of the useful life of the field. Retrieval of the earlier modules (1961–1963) would provide additional data in regard to the effect of the undersea environment as related to the array.[162]

This was the urgent work referred to in previous memos, the reason *Alvin* was needed "by 1963." The first Artemis modules had been installed in the summer of 1960 near Plantagenet Bank, Bermuda. A few more had been added in 1961, and in the summer of 1963, the major installation had gone forward: 210 modules containing thirty-two hydrophones each.[163] (*Alvin* had not been available for the 1963 installation, but it had proceeded anyway: the modules had been laid from a surface ship.) Initial results were good. In a meeting in September 1963, just three weeks after the completion of the installation, Berman reported that its success "affirmatively answers the question of the feasibility of large array installations in the ocean."[164] But how was it doing two years later? Apparently not so well. A "large number of modules have faults including some which have no useful output. In addition, apparent failures have occurred in some buoyancy spheres causing the respective modules to assume a horizontal position."[165] Program scientists knew which modules had failed, but they did not know why.[166] Were they broken, entangled, or hidden behind geological or topographical obstructions? What was the condition of the functional hydrophones after up to four years in salt water? How much longer were they likely to last? Visual inspection, photography, and selective module retrieval would help to answer these questions.[167]

Project scientists had also installed thermistors, inclinometers, and current meters to track environmental conditions that could affect sound transmission and array performance, but some of these were not working properly. In a 1963 memo discussing the problem, J. C. Munson of the Naval Ordnance Laboratory concluded that the temperature measurements needed to be independently verified; he proposed "mount[ing] a thermistor on Alvin [to] determine the temperature structure over as much of the array field as possible."[168] *Alvin* was finally available to do this.[169]

Alvin would fulfill its obligations to the Artemis project in the summers of 1964 and 1965 and would undergo repairs and adjustments during the intervening winter. Then, in the summer of 1966, it would be available to undertake its first season of scientific work. But before that could happen, disaster struck again. On the night of January 22, 1966, William Rainnie received a call from officials at the ONR that he later optimistically recalled as "the suspenseful beginning of an episode that would bring international recognition to the Woods Hole Oceanographic Institution and *Alvin*, its small deep diving (6000 foot) research vessel."[170] At the time, however, the main emotion he felt was not suspense but horror. Rainnie learned on the telephone that the US Air Force had lost a hydrogen bomb.

The Palomares Disaster

While the US Navy maintained its readiness during the Cold War primarily through underwater listening, the US Air Force achieved readiness by maintaining aircraft in the skies. Instituted in 1961, the policy of "airborne alert" kept a portion of the Strategic Air Command aloft at all times to minimize vulnerability to a preemptive first strike.[171] As part of this policy, and to ensure that planes were ready to reach distant targets at any time, the Air Force maintained a fleet of tanker planes capable of midair refueling. Under Operation Chrome Dome, planes routinely engaged in midair refueling one or more times during lengthy missions.[172]

On January 17, 1966, routine turned to wreckage as an Air Force B-52 bomber carrying four one-megaton hydrogen bombs collided during midair refueling with a KC-135 tanker plane. The altitude was 30,500 feet, the location was southeast Spain, and the town below was the coastal village of Palomares. Seven men were killed: all four crew of the tanker and three of the seven on the B-52. The four bombs fell to Earth, along with the mostly intact hull of the tanker, the smashed fuselage of the B-52, and tons of debris and shrapnel. Miraculously, no one on land was hurt, and the bodies of the seven men were quickly recovered. But where were the bombs?

One was found within a few hours of the accident, nearly undamaged, and two more were found the next day. The latter had scattered radioactive debris over a wide area but fortunately had not detonated their "physics packets."[173] The fourth bomb could not be found. At the end of the week, the US Air Force acknowledged that it was lost.

Press coverage was extensive and for the Air Force humiliating. *New York Times* humorist Russell Baker offered advice on how to find a missing thermonuclear device: act nonchalant and the bomb—like a pair of lost pliers hiding themselves until they realize you don't care—will surely turn up. The Air Force's mistake was to act as if the bomb mattered; the Air Force should have said it had plenty more and that anybody who found it could keep it, and "within 12 hours of this announcement, the missing bomb would have turned up peering sheepishly out of some farmer's haystack."[174] Herblock, the irreverent *Washington Post* cartoonist, drew a peasant on a burro being questioned by a stiff, saluting American officer: "Perdóneme. Ha usted visto un ... Uh ... H-bomb?"[175] Europeans did not see the humor: the French newspaper *Paris Jour* published a map illustrating a worst-case scenario—a thousand-kilometer radius of devastation and contamination that covered all of Spain and Portugal, most of North Africa, and extended well into France.[176]

The US military insisted in public that the bomb was safe, as its triggering mechanisms had not been activated and the chance of an unintended explosion was "essentially negligible."[177] As the Air Force put it, "In general, weapons are designed so that a positive event or sequence of events peculiar to its planned mode of delivery or attack must occur before a weapon will produce a significant nuclear yield."[178] This sequence would be initiated only on orders for an actual attack. It was stressed that "built-in safeguards had been perfected through years of extensive safety testing" and that the United States had never inadvertently detonated a nuclear device. That was the public position; in private Air Force officials were gravely concerned. After it was all over, the Air Force acknowledged that that the bomb could have detonated and called it "reassuring" that it had not.[179]

Scientists from the Sandia National Laboratory, who had designed the ordnance system for the bombs, were recruited to determine where it might have landed. In addition to making calculations based on the weapon's shape, location at the time of the accident, and prevailing wind directions, they also interviewed eyewitnesses. One Spanish fisherman regretted that he had been unable to save the airman he had seen go down in the Mediterranean, and this proved a critical piece of intelligence: an aerodynamics expert realized that what the fisherman had seen was not an airman, but the bomb, still attached to its parachute, which was designed to deploy in an accident just like this. The bomb was somewhere in the Mediterranean Sea. The US ambassador to Spain took a swim to reassure the public there was no risk, but his bravado had little impact. The bomb had to be found. As one report put it, recovery was "mandatory."[180] And the military had an additional problem: the bomb belonged to the Air Force, but marine salvage belonged to the Navy.[181]

The recovery operation would be enormous, involving thirty-four vessels and more than 3,400 military and civilian personnel.[182] The *Mizar* was again recruited for underwater photography, along with three submersibles, *Cubmarine*—a shallow submersible built by the Perry Corporation—*Aluminaut*, now fully functional, and *Alvin*. Seven days after the Navy request for help, *Alvin* was air lifted from Otis Air Force Base to Rota, Spain, with a crew of three pilots—William Rainnie, Marvin McCamis (1923–2004), and Valentine Wilson (1924–2001)—and a scientific staff of eight headed by Earl Hays.[183]

The operation was long and arduous. The bulk of the effort was concentrated on sonar scans, but these proved difficult to interpret. The bottom topography was extremely rough, with steep cliffs and numerous narrow submarine canyons that generated abundant reflections and reverberations. Again, the operators were stymied by inexperience: no one knew what the sonar signal of an H-bomb on the ocean floor would look like. The Navy also mobilized an underwater sonar system: Westinghouse's Ocean Bottom Scanning Sonar (OBSS), a towed vehicle that in principle could operate down to thirteen thousand feet but in practice proved ineffective because the sonar signals could not be positively identified and the rough bottom topography and inaccurate bathymetric charts led to collisions that damaged the instrument package.[184]

When sonar targets were identified, the Navy followed up with visual inspection. In shallow waters, divers were sent to investigate, but navigational uncertainties meant it was often difficult for the divers to find the specified targets. They did find all kinds of shrapnel and debris, including a urinal from the KC-135, but no bomb. *Cubmarine* successfully identified eighteen pieces of aircraft debris, but it was limited to six hundred feet. *Mizar's* towed-camera system could in principle be used at any depth, but like OBSS it was unable to maintain a constant height above the rough bottom and suffered several collisions.[185] This left *Alvin* and *Aluminaut*. Both had depth capacities more than adequate to cover the search area. Although the water depth involved was not as great as in the *Thresher* search, the submarine topography was far more treacherous. Yet the very difficulty of the terrain made it suitable for *Alvin*, which could go deep and maneuver through small spaces.

Alvin made its first dive in February and from then on maintained a grueling schedule of dives nearly every day, with each day lasting ten to twelve hours.[186] *Aluminaut* dove nearly as often. On one occasion *Aluminaut* hit bottom in an area where charts claimed a greater depth; the pilot was forced to slide along the bottom until reaching a water depth appropriate for the ballast he was carrying. Even then, the sub wouldn't move. The near-panicked crew shook it free by running back and forth in the cabin. Later, they found that gobs of mud had stuck to the sub's bottom, weighing it down nearly permanently.

While *Alvin* and *Aluminaut* pilots searched their assigned areas, *Mizar* did the same with its underwater camera system. Some photographs revealed tracks on the seafloor—probably mostly from bottom dwellers—leading the *Alvin* team to wonder whether the bomb had left a track, too. On March 1, *Alvin* divers saw a furrow that they believed could have been caused by the bomb sliding down hill. They attempted to follow it, but lost the track and had to resurface before they could find it again. Ten days passed in which they searched fruitlessly, then were reassigned to a new area. The pilots protested, expressing their frustration in letters home to Paul Fye. "The admiral here is no great shakes," Earl Hays wrote, "sort of a scream and holler man. I think we have convinced him that we are reasonable people, and are interested in doing all we can, but that we are not in the Navy and not likely to be."[187]

On March 12, the *Alvin* team was given permission to return to the area where the track was first observed. On March 13 they found it again but were unable to follow it to its end before running out of battery power. March 14 saw seas too rough to dive. Finally, on March 15, two months after the accident, they found the track for a third time. Now they followed it backward downslope, keeping the track in clear view through the bow ports and the propeller clear of the sloping bottom. It worked: they could see the bomb's parachute and part of its fin. The Navy designated the object "contact #261."

The water depth was 2,550 feet, the bottom slope was 70°, and the site was very near cliffs that plunged more than one thousand feet. A misstep could send the bomb over the cliffs, perhaps permanently. *Alvin's* vaunted maneuverability would be put to the test. *Aluminaut* was sent down to hold the position while *Alvin* resurfaced, and *Mizar's* precise navigational system was used to determine the exact coordinates.[188] A buoy was placed at the surface three hundred yards away.

Alvin was not strong enough to retrieve the heavy weapon, so a plan was developed to attach a hook and cable and pull the bomb up from a surface ship. On March 24, *Alvin* successfully managed to hook up to the weapon only to have it slip away, slide farther down slope, and become lost again. Meanwhile the Air Force had been tracking the headlines and their tone was increasingly critical. "'Safe' A-bomb missing in Spain Plane Crash," had given way to "Forty Days and Still No Bomb: US Leaders Silent," "US May Never Find Lost Bomb," and "Sun of Death Nearly Sets Coast Ablaze."[189] The *New York Times* ran the headline, "H-Bomb Searchers Fail Again."[190] The Soviets accused the United States of violating the Limited Test Ban Treaty on the grounds that its intent was to prevent contamination of the oceans and atmosphere. International pressure was mounting to allow the United Nations or the International Atomic Energy Commission to participate in the search.[191]

An admiral would later describe the next nine days as "agonizing," as *Alvin* and *Aluminaut* searched and re-searched.[192] Finally, on April 2, *Alvin* found the bomb again, 120 yards down slope and three hundred feet deeper than before. Three attempts to attach cables to the weapon failed.[193] In one attempt, the bomb was pulled about three hundred feet, but when pilot Valentine Wilson followed it, *Alvin* was nearly entangled in the billowing parachute.[194] Had that happened, there would have been no means to get the crew out.

The bomb was sitting in soft mud at the edge of a deep ravine; the situation was very tense. At this point, the Navy made a decision to stop further manned dives and attempt the recovery with a cable-controlled underwater vehicle (CURV), a towed device designed to recover test torpedoes at depths to two thousand feet. Over the previous weeks, CURV had been hurriedly modified to reach three thousand feet and its claw enlarged to fit the diameter of the missing bomb. On April 4, CURV was lowered into the water and successfully attached the first of three nylon cables to the parachute. A second cable was also successfully attached, but in the process of connecting the third one, CURV became entangled in the bomb's parachute. It was unclear whether the cables could stand the weight of both objects, but the only option was to try to haul them up together. At 7:40 a.m., April 7, the bomb and CURV appeared at the water's surface.[195] Scuba divers immediately plunged in to wrap wire straps around the bomb for hoisting onto the deck of the

Figure 6.9 The US military official photograph of the recovered H-bomb.
From Place et al., *Palomares Summary Report*, 1975.

waiting Navy ship, the USS *Petrel* (fig. 6.9). Security was extremely tight. Few people had ever actually seen a hydrogen bomb.[196]

Eighty days and hundreds of press releases after the accident, the bomb had been retrieved six miles off the Spanish coast from a depth of 2,850 feet. The total cost of the operation was $10 million, of which $1 million, paid by the ONR, covered the efforts of *Alvin* and *Mizar*.[197] According to the Navy summary report issued a year to the date of the recovery, the Palomares operation was "the largest concentrated underwater search in history"—a "milestone in oceanic history."[198] It was the first time a large object had been salvaged from the deep sea. *Alvin* returned to the States to do the work in Bermuda "interrupted by the higher priority mission in Spain."[199]

The retrieval was highly publicized. Returning again to the trope of the Little Engine That Could, "Little Alvin and the Bomb" presented the story of the "little sub that could."[200] While Woods Hole and Litton had an obvious self-interest in promoting Alvin's role, the Navy agreed that *Alvin* was one of "the most valuable units" of the task force.[201] While CURV had pulled the bomb up, *Alvin* had found it. Moreover, it was the *Alvin* pilots who realized that the bomb might have left a track in the mud, putting the operation on its eventual road to success. In Swanson's words: "The weapon was actually found as the result of deduction by operational personnel (the crew of *Alvin*) that it might have slid down a slope and left a track."[202]

The successful retrieval of the hydrogen bomb changed the way the Navy felt about lost objects. There were many objects lost in the deep seas and previously presumed beyond recovery, but they created little anxiety, because if the United States couldn't recover a lost object, then neither could the Soviets. Now the possibility of recovery had been transformed into the necessity of it. As the Navy summary report on the operation put it, the "historical security of the deep seas has been ... radically breached." The success of the Palomares operation "has now generated a requirement for an equally favorable outcome in any subsequent similar situation."[203]

Working beyond the Mission Profile

Alvin had clearly paid back the Navy investment; now it was time to do some science. In the summer of 1966, *Alvin* finally began to do what had long been planned. After completing further Artemis inspections in July, *Alvin* undertook a series of seven scientific dives. Most involved working out the system, but on July 18, 1966, "*Alvin* [completed] the first truly scientific dive to a depth of greater than 100 feet."[204] Now the pace of scientific work began to accelerate. Woods Hole marine geologist K. O. "Ken" Emery (1914–1998) began a series of investigations of submarine canyons off the US Atlantic coast. Marine biologists began the long-promised study of mid-depth marine fauna related to the scattering layer, and in 1967, Emery and colleagues found evidence of human habitation on the floor of the Chesapeake Bay in an area that they could now prove had been exposed during Pleistocene low-sea-level stands eight thousand to ten thousand years earlier.[205]

Still, the overall scientific return by the end of 1967—the period for which the ONR had promised funding—was modest. The summer diving season was short, and *Alvin* was busy with military projects for much of it. Each set of dives led to another set of repairs and modifications, and scheduling priorities always put Navy work first. As Fye put it in the summer of 1966, *Alvin* would work in Bermuda inspecting "underwater structures" and then, "time and schedules permitting, certain Woods Hole investigators will join the expedition for [scientific] dives."[206] In the first three years, time and schedules did not permit very much. But scientists at Woods Hole were not exactly beating down the doors to use *Alvin*, because, in fact, few scientists at Woods Hole saw how it could be of use. In retrospect, we might imagine that everyone would have wanted to use *Alvin*, but this was not the case. Many scientists simply did not see how to use it in their research programs.

As we have seen, there had long been a question about how scientifically useful *Alvin* really was. In the early months of 1962, when negotiations over *Aluminaut's* construction were unraveling, the staff had addressed the question of what science might be done on the vessel, assuming it became operational. About "a dozen" Woods Hole staff members had had input into Vine's original proposal for *Aluminaut* and Vine felt it was now "time to: a) see if any additional people wish to use *Aluminaut* or if there are new kinds of experiments. b) discuss scientific programs sufficiently to insure the planned instruments are adequate. c) be sure those who want to can help decide the requirements and design of specific instruments."[207] Vine prepared a questionnaire entitled "Desired Work on *Aluminaut*," which was sent to the entire Woods Hole staff. Their replies were not preserved in the director's files, but the types of questions posed speak to the undeveloped quality of scientific thinking about deep submergence. The first question: "What are you most interested in finding out or doing?"[208]

The open-endedness of Vine's questionnaire could be viewed as a virtue—in principle, any kind of scientific work could be done—but it also reflected a lack of attention to the purported scientific goals of the deep-submergence project. Both Vine and Fye's files contain reams of discussions about programs costs and timetables, metallurgical specifications, and military applications, but conspicuously missing is any sustained discussion of the scientific work that might be done.[209] In fact, there is scant discussion of science at all. No wonder the staff were skeptical that *Alvin* would do scientific work; science was nearly absent from its planning.

When Paul Fye defended himself to his staff in his 1961 winter of discontent (chapter 3), one of the specific charges he addressed was that "the Scientific Policy Committee unanimously voted against the *Aluminaut* and that I, in turn ... went ahead with it anyway." Fye allowed that it was "true that this group of senior staff members met several times to consider this project [and] invited the key designer, Dr. Edward Wenk, to Woods Hole to discuss it," but he insisted that, despite their reservations, "after very careful consideration [they] did vote to go ahead with the project."[210] Fye also claimed that "about 15 of our own scientists had developed a program and scientific need for the craft which would keep it busy for one to two years." But in the copious documentary record of the *Aluminaut-Alvin* program, there is not a single document to support this claim or even one like it. So far as the historical record shows, a scientific program had never been developed in the planning for *Aluminaut*. In any case, the very fact that the charge was made and seemed plausible to Woods Hole staff—and that the scientific policy committee was against *Aluminaut*—is evidence enough. The military value of a deep-submergence vessel was never seriously challenged, but its scientific value was doubted more than once.

Vine's career was built on the development of vessels, instruments, and equipment, so his scant attention to scientific questions is not surprising. Moreover, in the early stages of the project detailed discussion of science was not required, and as Louis Reynolds noted, too much science might have scared the Navy away. What is surprising is the almost complete lack of correspondence from Woods Hole scientists to Vine, since any scientist aspiring to use the new vessel would have had to solicit his support.

The lack of such correspondence suggests that few, if any, scientists expected *Aluminaut* or *Alvin* to be used for science. It is also consistent with the accusation of many Woods Hole scientists that Fye simply wasn't interested in science (chapter 3). The Woods Hole trustees had recruited him because of his Navy affiliations, which they hoped would enable him to strengthen and expand the institution's financial base, and he ultimately earned the respect of his staff—and survived the Palace Revolt—to become Woods Hole's longest-serving director. But he was not a person to whom his scientists looked for intellectual guidance and inspiration. His staff did not come to him to vet ideas. Nevertheless, we might still expect to see evidence that they tried, at least on pragmatic grounds, to gain his support for their projects. But the files are silent on this account. As far as the records show, there simply wasn't much discussion of the scientific purposes of the deep-submergence vessel in its early years. *Alvin* was proposed by men who had built their careers applying technical knowledge to military missions. Science, they assumed, would follow—as it had with magnetometry, seismic refraction, and bathymetry—but they didn't think deeply about what kind of science would follow, how it would follow, or what would happen if it failed to follow.

Once *Alvin* was operational and became known to the larger scientific community, however, Fye began to receive inquiries from scientists eager to use it for their research. Interestingly, most of these were from *outside* Woods Hole. Now there was a new obstacle: cost. If scientists wanted to use *Alvin*, they had to find the money to pay for it, and *Alvin* was very expensive. One interesting request came from Canadian marine biologist Frederick Aldrich (1927–1991), a professor at St. John's University in Newfoundland. Aldrich proposed an investigation of the enigmatic giant squid—one of the very ideas Vine had mentioned in his 1960 report proposing the *Aluminaut* program. The giant squid is the largest known marine invertebrate—made famous a century before by Jules Verne—but it was very poorly known scientifically.[211] Aldrich included a startling photograph of the giant beast, which could not have failed to capture Fye's attention (fig. 6.10). Fye replied

Figure 6.10 Image of giant squid found at Ranheim in Trondheim, October 2, 1954.
Courtesy of the NTNU University Museum of Natural History and Archaeology in accordance with
their open-access policies.

to Aldrich, but his answer was negative. He explained, regretfully, that *Alvin* was booked for the entire first year for work near Bermuda and requests had already begun to pile up for the following year. But more to the point was money. Fye did not doubt the merits of Aldrich's project; there simply were no funds to pay for it. He explained:

[Besides scheduling] another problem which we would face in using *Alvin* for your work in observing giant squid is that of funding for her. All of her operations to date have been paid for from the contract which provided funds for her construction. At some future date she will of course become just another research vehicle of the institution and each of the research projects for which she is used would pay for their fair share of the operating costs of that year.... I hope I have not sounded too discouraging. I did want you to know the facts of the situation as we see it now and not have your hopes set too highly.[212]

In his original proposal, Vine had emphasized that "any deep submersible should be available to all oceanographers who want to use the device for worthwhile programs in the same manner that *Atlantis* has been used by both WHOI and non-WHOI oceanographers."[213] That was an admirable aspiration, but it was not realized, at least not in *Alvin*'s first decade and not while *Alvin* was funded by the US Navy. It would be some time before *Alvin* became just another research vessel, in part because its potential uses were extremely specialized and in part because operating costs were higher than the already-high costs of ordinary research vessels.[214] Despite all the explicit promises and implicit expectations, there was no provision for basic scientific research, and there never had been. It would take a great deal of work to create that provision, and it would not come from the Navy. It would come from civilian funding agencies, and then the history of *Alvin* would be rewritten as if it had been intended to do basic research all along.

7 Painting Projects White: The Discovery of Deep-Sea Hydrothermal Vents

While Woods Hole was building *Alvin*, the laboratory that began life as the University of California Division of War Research (UCDWR) was taking a different approach to studying the deep sea, focusing on remotely operated vessels. After the war, UCDWR had moved to Scripps, where it had grown and prospered as its Marine Physical Laboratory (MPL) under the direction of physicist Fred Spiess.[1] With extensive funding from the ONR, the Bureau of Ships, and the Bureau of Naval Weapons, MPL scientists had developed a sophisticated instrument named Deep Tow, short for "deeply towed sounding system." As its name indicated, this device—really a package of devices—could be towed behind a surface ship on a long tether that enabled it to sail close to the seafloor to collect topographical, magnetic, and seismic data.[2] But like *Alvin*, it was not developed to satisfy scientific curiosity. In fact, even more so than *Alvin*, it was tied to a specific Navy program: an antisubmarine missile system, Subroc.

Imagine a missile launched from a submarine into the atmosphere to re-enter the ocean at some distance away. It could travel much faster and farther than a torpedo and potentially devastate an unsuspecting target. The Navy called the idea Subroc—for submerged, launched rocket. But just as intercontinental ballistic missiles could be effective only if submariners had very precise control of the launching site, submerged launched rockets could be effective only if submariners knew exactly the position of the target—for which they needed very precise sonar information. Because sound waves from distant targets are bounced off the seafloor at angles that depend on bottom topography, they would also require accurate and detailed information on that topography—more accurate than could be obtained by narrow-beam echo sounding from a surface ship. As Spiess explained in a summary report to the ONR, "Our broad objective is determination of the ways in which sound interacts with the seafloor and the description of the sea floor in terms relevant to these interactions." As usual, a broad objective framed a specific problem: "Motivation came initially from studies of use of bottom reflected paths for fire control purposes."[3] In other words, Subroc.

Deep Tow was initially funded by of the Bureau of Naval Weapons, but it also functioned through the ONR's general support of MPL. Like Woods Hole, Scripps had broad-ranging ONR contracts supporting mission-relevant research. One of these was Nonr-2216, which began in 1958 under the title "Marine Physics."[4] Nonr-2216 was MPL's equivalent to Woods Hole's "basic task" contract, providing roughly half of the laboratory's annual operating budget and supporting "all phases of research at MPL ... as they apply to the solution of Navy problems in the ocean, or in the ocean-air or ocean-earth interfaces."[5] This included underwater acoustics and communications, geomagnetics, gravity measurement, seismic profiling, signal processing, and "solutions to problems involving underwater structures and cabling from underwater sensors to shore or ship based processing stations"—in other words, SOSUS. As the scientists' annual report for 1965 put it, "All elements of the effort at MPL is [sic] aimed toward the direct solution of problems peculiar to the Navy."[6]

A particularly peculiar problem involved the seafloor. MPL had been involved in work on the physical properties of the seafloor since its early days as the UCDWR, primarily for its relevance to underwater sound propagation. This continued after the war with funding from the Bureau of Ships, and later from the ONR. A major motivation for this work had been the recognition of sonar bearing errors due to bottom slope.[7] By the mid-1960s, the emphasis had shifted from broad surveys to "detailed examination of particular important areas," arising "from such incidents as [the] THRESHER and H-bomb searches."[8] In 1965 Deep Tow was "substantially accelerated" by funding from the Deep Submergence Systems Project.[9] Spiess would later say that Deep Tow was a system designed "to search for things on the sea floor" (and he did not mean fish), but this was not entirely true: it was a system designed to study the seafloor, and later that would include things lost on it.[10] Later still, it would include basic science.

In 1967, graduate student Tanya Atwater (b. 1942) and her adviser, Scripps assistant professor John Mudie (b. 1938), used Deep Tow to produce the first detailed topographic map of a spreading center: the Gorda Rift at 41°N off the California coast.[11] (Their work was published as the lead article in *Science* in February 1968, making Atwater possibly the best-known graduate student in the United States at the time.) Interinstitutional rivalry led Woods Hole scientists to envisage dives to the Mid-Atlantic Ridge in *Alvin*.[12] In March 1968, J. D. Phillips (b. 1938), a Woods Hole geophysicist, wrote to colleagues suggesting that *Alvin* could be used to surpass Atwater's work.[13] There was a snag, however: the Navy would not necessarily pay for it.[14]

When Paul Fye had defended himself against the "Palace Revolt" by insisting on his commitment to basic science (chapter 3), the top two scientific topics in which he claimed to be interested were the origins of ocean basins and the stability of features of the crust.[15] These subjects had now come to the fore. But Phillips was right about the snag: there was no money to pay for it. The ONR was not prepared to pay for a scientific study of the Mid-Atlantic Ridge, and Woods Hole had no others funds to cover *Alvin*'s activities. In fact, the increase in scientific activity that *Alvin* enjoyed in 1967–1968 coincided with signals that the ONR might not be prepared to continue funding *Alvin* at all.[16] The three-year initial commitment had been fulfilled, and there was talk of the Navy buying its own submersibles. In 1963, the company Hahn and Clay, subcontractors to General Mills, had built two six-foot-diameter steel spheres, one of which became *Alvin*. The other, plus a third, would soon become the Navy submersibles *Sea Cliff* and *Turtle*.[17]

Alvin's success at Palomares had actually worsened its prospects, because the Navy concluded from the episode that submersible technology was too important to leave in civilian hands. As Admiral Swanson noted, a historic milestone the first time around would be an operational requirement the next, and so a principal recommendation of his 1967 report, "Lessons and Implications for the Navy," was that the DSSP accelerate its efforts to procure its own salvage technology capable of operating down to twenty thousand feet.[18] Moreover, the emphasis should be on remotely operated vehicles, because of "an order of magnitude increase of bottom time and immunity from concern over operator safety."[19] (This was a reference to the Palomares rescue, which had coming alarmingly close to disaster with the near loss of *Alvin* and the men in it.[20]) Moreover, if the Navy ever found itself in need of a crewed submersible, there were several available in the private sector.[21]

Indeed, in just a few years the world of deep submergence had been transformed. General Dynamics had finally completed *Aluminaut*, along with a set of shallow-diving submersibles, including *Star I*, launched in 1964 with a depth capacity of 200 feet, *Asherah*, built in 1964 for the University of Pennsylvania for archeological investigations down to 600 feet, and *Star II* and *III*, launched in 1966 with depth capacities of 1,200 and 2,000 feet, respectively.[22] These were not direct competitors with *Alvin*, but Westinghouse had completed *Deep Star*, a saucer-shaped vehicle modeled after Cousteau's *Soucoupe* with a depth capacity of 12,000 feet.[23] The Lockheed Missile and Space Company had produced *Deep Quest*, which in February 1968 set a record for a true submarine (no external float) with a dive to 8,310 feet.[24] Lockheed had also been commissioned by the DSSP to produce *DSRV-1*, which could attach to the hatch of a sunken submarine to provide deep-sea rescue capability; this was due to come on line in 1970.[25] Other private contractors, including Grumman Aerospace, North American Rockwell, Union Carbide, and General Oceanographics had submersibles either on board or on the drawing board.[26] And then there was the Perry *Cubmarine*, which had not found the H-bomb but had worked efficaciously at Palomares.

According to an article in *Product Engineering* in March 1966, there were fifty-seven submersibles worldwide, the majority built by American contractors, and at least ten of which had depth capacities over ten thousand feet.[27] Moreover, in 1968, the Navy commissioned *Dolphin*, a diesel-electric research submarine capable of operating at depths of four thousand feet.[28] *Alvin* was no longer special; it was just one vessel among many. If naval needs were intermittent, it might be better and cheaper to subcontract as they arose.[29] If they were continuous, then the Navy would be better off with submersibles of its own. Either way, *Alvin* had become neither unique nor essential. Fye's suggestion to the ONR that they provide a block grant for *Alvin* flew in the face of this reality. Not surprisingly, the ONR rejected it.[30]

While the Navy looked to distance itself from *Alvin*, the National Science Foundation (NSF) resisted taking it on. The foundation had many other projects in need of support, and funding *Alvin* would have consumed a significant fraction of the entire ocean sciences budget. The NSF did support ship costs, but program officers felt that their highest priority was to fund projects that the Navy did not fund, and *Alvin* was the Navy's baby.[31] Moreover, the ONR was cutting back all around, so many oceanographers were looking to the NSF, including Fred Spiess and the Deep Tow group. In a memo to Scripps director William Nierenberg, Spiess explained: "Basic ONR and DSSP interest [in Deep Tow] still remains high, but the latter group is strongly hardware oriented and the former is operating under tight budget restrictions."[32] He suggested that NSF block grants help cover the costs of ship time associated with Deep Tow.

Funding was not *Alvin*'s only problem. Since the early 1960s, the Navy had been under pressure from Secretary of Defense Robert McNamara to reorganize and reform. This led to the 1966 reorganization of Navy bureaus into system commands, with a heightened sensitivity to who was responsible for what; *Alvin* did not fit clearly into any command.[33] Worse still, *Alvin*'s troubles rested against a backdrop of increasingly sharp questions about military funding of (ostensibly) civilian research. On university campuses across the country, students and faculty were challenging US involvement in Vietnam and questioning the military presence on campus, both in the laboratory through funding of scientific research and in the classroom through Reserve Officers' Training Corps programs (chapter 5).[34]

Perhaps for these reasons — or because the strong ties forged during World War II between military officers and civilian scientists had by this point weakened — the Navy was turning away from the academic community and toward program officers and for-profit consultants.[35] How these larger events contributed to funding pressures at the ONR is beyond the scope of this chapter (or this book); what we know is that by 1968 tensions over the costs of operating *Alvin* were escalating. William Rainnie wrote numerous letters and memos to ONR officials over matters such as who paid for downtime and transit time. *Alvin* was frequently traveling to Bermuda to help with tests of underwater vehicles and weapons systems at a facility known as the Atlantic Undersea Test and Evaluation Center (AUTEC), and Rainnie suggested that the Navy pay for the travel time.[36] But AUTEC officials felt that since *Alvin* was funded by the ONR, the costs were already covered. Rainnie complained to Fye that they needed to "seriously try to conclude an equitable agreement with AUTEC for near future survival and that we press ONR hard to supply the ultimate answer."[37]

Rainnie also felt pressed to reconcile AUTEC demands with the augmented scientific aspirations of Woods Hole staff. He had an optimistic opinion of ONR willingness to support basic scientific research, but increasingly events were not in not accord with that view. "There are strong indications that ONR prefers our course of independent scientific exploration," Rainnie claimed, "but CNR [Chief of Naval Research] has to be prodded into putting his money where his mouth is."[38] The CNR was prodded, but not as Rainnie had hoped. In a letter to the chief of naval operations, copied to Fye, he reminded all parties of the terms of the program: "The deep submersible *Alvin* is owned by the Navy and operated by the Woods Hole Oceanographic Institution under contract to the Office of Naval Research. The mission objectives are to conduct deep ocean research in support of the Navy's goals and operate within the mission profile."[39] But what was "the mission profile," and who defined it? These questions had been raised at Scripps in the 1930s and at Woods Hole in the late 1950s and early 1960s. They now arose again as scientists fought to use *Alvin* for basic scientific investigations.

Working beyond the Mission Profile

The question of the mission profile — and who defined it — came to a head in late June with a request from the US Navy Underwater Sound Laboratory for

Alvin to spend six weeks on an urgent project near the Azores.[40] While the un-classified letters did not specify the mission, the timing and location reveal that it was the search for another lost submarine, *Scorpion*.

The *Scorpion* disaster was less publicized than *Thresher*, but in some ways substantially worse. The latter had gone down at a known time and place, accompanied by an escort who heard the crew's last communications. The *Scorpion* had disappeared without a trace. Officials only realized the boat was lost when it failed to return to port on May 27; it had last been heard one week earlier, south of the Azores and on track for home. Ninety-nine men were missing, presumed dead, and the Navy had no idea where they were. Acoustic signals—possibly associated with the sub's implosion—enabled the Navy to narrow the search to a twenty-mile radius, but finding it was still going to be a very hard job.

Alvin's summer schedule had already been set, including a number of scientific projects that had been waiting for some time, and the Azores work would take weeks or longer. After considerable soul searching, Fye decided to decline the request. In response, the lab dramatically increased the amount they were willing to pay—from $30,000 to $150,000—and the ONR instructed Fye to accept the money and do the job. Perhaps scarred from the Palace Revolt, or perhaps feeling that the ONR was pushing too hard, Fye held firm. After all, he was not an officer, and this was his decision to make. Trying to be diplomatic, he summarized his position this way: "Woods Hole is desirous to be responsive to the needs of Navy laboratories but must do this in harmony with the scientific commitments."[41]

On July 3, Fye telephoned ONR Captain Van Ness to see if the Navy work could perhaps be delayed until the following year. It was clear that the ship was lost—this would be a salvage mission, not a rescue—so time was not of the essence. Moreover, Fye argued, the Navy's inflexible attitude was undermining scientific support for the deep-submergence program, support that he had worked hard to build. "We have been slowly building the support of scientists in the *Alvin* program and concept—[the] proposed operation (USL) THIS YEAR could well destroy the confidence & support & hence the potential utility of the DSRV as a research tool."[42] Van Ness was unmoved and insisted the work be done right away.[43] Fye finally agreed on the condition that any scientist who felt he could make use of the trip be allowed to come along. (Although the issue did not come up, it could only have been a "he," because the Navy would not permit women to participate.)[44] Sedimentologist Charles Hollister took up the offer (on Hollister, see chapter 8).

The constraints on *Alvin* were undeniable. *Alvin* operated, as the chief of naval research crisply put it, "within the mission profile." Priority went to tasks requested by the Navy and scheduled research projects would be pushed aside when Navy needs dictated, and because the diving season was short, it was easy for military demands to push science off the schedule entirely. Those who thought otherwise had reality placed before them. "Independent scientific exploration" could take place as time and money permitted, but it was clear that time rarely permitted and money never did: there was no money earmarked for science. Despite what anyone said or hoped, *Alvin* had not been built to do basic research, and—except insofar as vehicle development was considered research—research was not what the ONR had undertaken to finance.

None of this was unreasonable from the Navy perspective: the ONR had supported *Alvin* to satisfy Navy needs. There was never any promise of funding for science. Like Deep Tow at MPL, *Alvin* was developed with the aim of solving "problems peculiar to the Navy."[45] In contrast, Fye wanted *Alvin* to be available for scientific purposes, and it was certainly implied in many exchanges that it would be. Recall when Rear Admiral Stephan noted that a deep-submergence vessel "should be conceived against its probable background usefulness in … science and engineering."[46] *Alvin*'s designation—DSRV—implied that its purpose was research, and the 1962 call for bids invited designs for a "Research Submarine."[47] The numerous press releases describing *Alvin* as a vehicle developed for "military and scientific purposes" also implied that it would be a dual-use vehicle, with distinct and independent scientific purposes. ONR officials were apparently comfortable with this implication; there is no evidence that the ONR ever attempted to disabuse anyone of the idea that *Alvin* would be a research vessel, serving research goals. But as we have seen, whether that was an accurate characterization depended on how one interpreted the word *research*.

Among both scientists and historians it has been widely accepted that, in its early years, the ONR effectively functioned as "the federal government's only general science agency."[48] And in the 1950s that may have been true. By the 1960s, however, the federal government had a defined general scientific research agency—the NSF—yet ONR officials still continued to insist that they supported both basic and applied research and that the creation of NSF had not made them obsolete. The chatter around *Alvin* was consistent with this. An example is the ONR's 1961 press release in which it noted that "the interest of the Navy" in deep submergence "embraces both oceanography and the obvious implications for submarine developments of the future."[49] At *Alvin*'s commissioning, Admiral James Wakelin Jr., assistant secretary of the Navy for research and development, expressed his pleasure at being present as they commissioned "this important research submarine into the service of

science" and his hope that *Alvin* would have a major impact on "scientific research, particularly basic research important to the national interests."[50]

Woods Hole and ONR press releases used the language of parallel paths:
military and scientific purposes, submarine developments and oceanography,
applied and basic science. These related but distinct activities would coexist
and thrive in symbiosis. While the Navy and Woods Hole might interpret the
word *research* differently, the generally understood implication was scarcely
disputable: *research* was different from *operations*.[51] But therein lay the rub. To
a ship commander, Artemis was research, because it was not operational.[52]
For some physicists and acoustic engineers, it was research, too. When the
first Artemis installations were put in place in 1963, project director Alan Berman reminded his team that this was not (yet) an operational system: "Those
among you who have been keeping score will observe that we may be as much
as 20 dB below what would be required for an operational *Artemis* system. You
would, of course, be correct. What we possess is research equipment."[53] Until
the system worked, work on it was (in his view) research.

If Woods Hole scientists were disappointed that what they considered
research was getting short shrift, they had in part their own colleagues to
blame. Allyn Vine often (and perhaps deliberately) blurred the distinction
between oceanographic and operational research, as when he insisted at the
Alvin bidders' conference that the vessel had to be ready by the summer of
1963 for "research projects." By this he meant the Artemis module inspections. Was he being disingenuous, or was he using "research" as a cover story?
Perhaps, but most likely not. Vine understood that Artemis was still under
development; his notion of research overlapped with Berman and others who
were developing new techniques, new instruments, and expanded capabilities. To them, this was research, because it was not operational. In any case,
Artemis was classified, so Vine could not discuss it. Research was sometimes
at least in part a cover story.

For most Woods Hole scientists, however, research was emphatically not
Artemis. Research, as William Rainnie put it, was "independent scientific exploration," work in which scientists chose the methods of investigation and
the time and place of study. It was what Henry Stommel had stood up for in
the Palace Revolt: science driven by the desire to understand a portion of the
world. It was science driven by curiosity rather than by a mission—military
or otherwise. It was science in which the choice of topics was determined by
scientists not bureaucrats, individuals not bureaus. Such research might turn
out to be useful, but utility was not its motivation.

From this perspective, what would be the point of Woods Hole's involvement with *Alvin* if there were no promise of independent science? Woods

Hole would become just another government laboratory—as Stommel, Jo-anne Malkus, and their partners in the Palace Revolt had feared and Ray Siever had concluded Woods Hole already was. If the Navy had not made an explicit promise, there had certainly been an implicit promise that *Alvin* would do science, at least in the context of "background usefulness." But that was not happening. This highlighted a second rub: if the vessel lost its fore-ground usefulness, then the Navy's justification for supporting it was also lost. *Alvin* was (metaphorically) sunk either way.

If time and money had been available, the summer of 1968 would have been the perfect time to visit the mid-ocean ridges. With the theory of plate tectonics crystallizing around the idea of plate motions as rotations on a sphere, 1968 was arguably the single most exciting year in the history of earth science. With attention focused on mid-ocean ridges as divergent plate boundaries, Woods Hole was perhaps the best-placed institution in the world to answer the question of how seafloor spreading really worked. But at the end of the 1968 diving season, *Alvin* (literally) sunk. While being launched on a dive, the cables connecting it to the mother ship broke and the vessel spun out of control. Fortunately, the pilot and two passengers were able to escape, but the vessel landed on the seafloor at 5,198 feet, where it sat until the fol-lowing diving season, when it was salvaged and returned to Woods Hole for repairs.[54]

The accident brought the funding crisis into sharp relief. The salvage oper-ation was difficult and costly, and repairs would be, too. Moreover, scientists recognized that to answer big questions about plate tectonics, they needed to reach the seafloor, and for the Mid-Atlantic Ridge this meant a depth ca-pacity approaching ten thousand feet. Talk of retrofitting with a titanium-alloy hull to accommodate such depths inevitably led to the question of who would pay for it.

In an interview in 2001, Robert Frosch suggested that, with the excep-tion of rescue and salvage capability, the Navy had never been especially in-terested in very deep submergence for the simple reason that there was no fighting down there.[55] This issue had come up over Project Glaucus—a joint Naval Electronics Laboratory and Bureau of Ships effort to create an experi-mental submarine, *Dolphin*, commissioned in 1968 and capable of diving to four thousand feet. From the Navy perspective, this was very deep diving, and there was a fair difference of opinion over its value. "Most people accepted the value of deep submergence for "evasion of ASW attacks," a 1963 NEL re-port noted, "[but] the possible advantages of deep submergence for offensive actions is more controversial. It can be argued that almost all of its potential targets today would be at a shallower depth or on the surface, and present methods would require it to come to shallow depths for an attack.... However, deep diving submarines in possession of the enemy could swiftly alter this outlook."[56] If the Soviets were developing deep-attack capability, however, there was no evidence of it.

Fye may not have been a great scientist, but he was astute enough to realize that a historic scientific opportunity was about to slip through his institution's fingers, and to his credit, he spent the next twenty-four months trying to establish a system for financing and operating *Alvin*. At one point he pinned his hopes on the Defense Advanced Research Projects Agency (DARPA), which operated AUTEC and whose program directors were prepared to come forward with some funds. But meanwhile the ONR was cutting back. In 1971, the ONR cut *Alvin*'s projected budget for fiscal year 1972 by a third. Somehow Fye persuaded the ONR of the value of the retrofit, which went forward that year, funded by the ONR, NSF, and the US National Oceanographic and Atmospheric Administration (NOAA).[57] But in his report for 1973, William Rainnie noted that "one of the major problems facing the section this year was finding adequate support in the face of reduced federal budgets for submersibles."[58] At the end of the year, the retrofitted *Alvin* had a depth capacity of twelve thousand feet and the prospect of no budget with which to use it.[59]

The FAMOUS Expedition

The idea of exploring the mid-ocean ridges with submersibles had also occurred to French geophysicist Xavier Le Pichon (b. 1937), who had recently returned to France from Lamont, where together with colleagues Jack Oliver, Lynn Sykes, and Bryan Isacks he had helped put together (at the eleventh hour) the global picture of plate tectonics.[60] The new theory made mid-ocean ridges the center of seafloor spreading (chapters 4 and 5), but there was still considerable uncertainty about what exactly was happening there.

Geologist Harry Hess had first proposed the theory of seafloor spreading (chapter 4), but his version was not the one that would ultimately be accepted. He had proposed that the seafloor was composed of peridotite—the dense iron- and magnesium-rich rock that comprises Earth's upper mantle but is rare in the continental crust—and theorized that uplift at the ridges was caused by thermal expansion associated with regions of high heat flow, combined with volume changes accompanying the hydration of peridotites exposed to seawater. In essence, he was arguing that the seafloor *is* the mantle; what we call oceanic crust is simply mantle that has been hydrated by interaction with seawater. In contrast, Scripps marine geologist Robert

Dietz had argued that the seafloor was basalt and the ridges built up by submarine volcanic eruptions.

By the early 1970s, most scientists favored Dietz's account—samples dredged from the ridges were almost invariably basalt, not hydrated peridotite—but the number of samples was modest and the ridge structure known primarily from low-precision echo sounding.[61] Moreover, Hess, who died in 1969, had been a very influential figure. His convictions were not to be lightly dismissed, particularly since he had almost single-handedly reopened the continental drift debate in the United States after most of his geological and geophysical colleagues had closed it.[62]

There was also the problem of terrestrial heat flow. It had long been recognized that the amount of heat emanating from the Earth's interior was higher over the Mid-Atlantic Ridge than in other parts of the ocean. In the 1920s, British geologist Arthur Holmes had used this as evidence of mantle convection currents rising beneath the ridges, which he proposed as the driving force of continental drift. In the 1930s, Dutch geodesist Felix Vening Meinesz, working with Hess, had suggested that mantle convection currents could explain certain observed gravity anomalies in the ocean basins: negative anomalies would occur where the crust was downbuckled above descending convection currents (chapter 4).[63] After the war, these ideas stimulated British geophysicist Teddy Bullard and Americans Roger Revelle and Arthur Maxwell to study terrestrial heat flow in more detail (chapter 5).

Before the war, Bullard had undertaken land-based heat-flow measurements and greatly improved the available instrumentation for accurate measurements. In 1949, he visited Scripps, where he began to develop the first practical device for measuring heat flow through the soft mud of the seafloor.[64] In the 1950 Mid-Pac expedition, Revelle and Maxwell used Bullard's device in the first comprehensive study of heat flow through the oceanic crust (chapter 5). They expected oceanic heat flow to be lower than continental heat flow, because the latter contains more heat-producing radioactive elements; Bullard had predicted it to be three times lower.[65] Instead, they found that it was nearly the same. Bullard suggested that the anomaly could be explained if mantle convection currents brought interior heat to the surface beneath the ocean floor, compensating for the deficit of radiogenic heat, but he acknowledged that the idea was speculation. There was no direct evidence of convection currents, so to invoke the oceanic heat flow as evidence of them was to risk circularity.[66]

When convection currents began to gain credence in 1961 and 1962 with the ideas of Hess and Dietz on seafloor spreading, geophysicists recalled that Bullard had suggested the idea in 1952. Yet as late as 1963, he was still hedging his bets: "It has been suggested that the pattern of heat flow might indicate the locus of an ascending convection current in the mantle, though there is no compelling reason to suppose this is so." Scientists still needed a "more thorough survey of many features of the ridges."[67]

Another problem was the East Pacific Rise. A detailed heat-flow survey by Richard Von Herzen (1930–2016) and colleagues on the Scripps Rise-Pac expedition revealed a pattern of generally very high heat flow over the rise but scattered very low values near its crest. The high values fit the convection model, but the low values did not.[68] Heat flow in the Gulf of Aden and the Gulf of California displayed a similar pattern, suggesting that (ignoring the few low values) the gulfs were crustal rifts above rising mantle convection currents.[69] By the early 1970s, similar patterns of mostly high but occasionally very low heat flow had been measured over the Mid-Atlantic Ridge as well.[70]

The young Cambridge geophysicist Dan McKenzie (b. 1942) raised more questions. In 1967, McKenzie had calculated the expected heat flow though the ocean crust, assuming seafloor spreading and conductive cooling of basalts intruded at the mid-ocean ridge. His calculations were successful at predicting the overall heat flow from the ocean crust, but near the ridge axes, the measured values were lower than predicted.[71] Without the hypothesis of seafloor spreading, the oceanic crust appeared to be too hot. With it, it appeared to be too cold.

That same year, McKenzie (with Scripps geophysicist Robert Parker) formulated the first quantitative statement of plate tectonic theory. Scientists at Scripps and Lamont now recognized the bandwagon, and in hindsight many participants have recalled the exhilarating feeling of pieces clicking sharply into place.[72] But the heat-flow data did not click. In fact, they were a major stumbling block. Some scientists objected in general: why should mantle convection happen to neatly compensate for the deficit of radioactive elements in the oceanic crust? Others were troubled by McKenzie's calculations: why should the heat flow at the ridge crests be lower than predicted by the seafloor-spreading model?[73] Still others accepted the overall pattern but remained bothered by the anomalously low values around the ridges. As John Sclater (b. 1940), a graduate student who worked on heat-flow data, later recalled: "We expected to observe a relatively smooth increase from near normal [heat flow] on the flanks to a factor of four higher than normal over the crests of the ridges. The apparently random occurrence of low values completely confused us."[74]

There was a possible explanation: hydrothermal circulation through the fractured basalts of the ridge crest. If seawater percolated through the ridge crest, it would cool the hot rock, depressing the overall heat flow and accounting for the discrepancy between theory and observation. It would mean that conductive cooling was not the only mechanism in play. Convective circulation would account for the sporadic very low values in areas of otherwise generally high heat flow: these were the zones of fluid flow.[75] By 1972, the idea was gaining credence. As one geophysicist wrote, "Once it is established that a substantial part of the surface heat loss at the ridge crest is non-conductive, no great imagination is needed to select the most likely mechanism: hydrothermal circulation in the oceanic crust."[76] Iceland sat atop the Mid-Atlantic Ridge and was laced with geysers and hot springs; this demonstrated that hydrothermal circulation was a feature of at least one mid-ocean ridge where it was visible above sea level, so perhaps there were similar features where the mid-ocean ridge was hidden underwater.[77]

Other evidence was consistent with hydrothermal activity. It had long been known that waters at the base of the Red Sea were anomalously hot and salty.[78] If the Red Sea was an incipient ocean—where the continental crust had just broken apart—then the brines could be hydrothermal fluids generated by seafloor volcanism.[79] At MPL, John Mudie suggested to Fred Spiess in July 1968 that one good reason to buy a new temperature probe for the lab would be "to detect signs of hydrothermal activity or recent volcanism of the seafloor in oceanic ridges. Hydrothermal activity is responsible for the 'hot brines' of the seafloor in the Red Sea, and it is not unlikely that they may also exist in other regions of present tectonic activity."[80]

Still more evidence came from rock magnetism. In the late 1950s, geophysicist Ted Irving had helped reopen the continental drift debate with a clever analysis of rock paleomagnetism. Studies of rock paleomagnetism had revealed *relative* motion between the continents and the poles, but was it the continents that had moved or the poles? If the latter, then all the continents should show the same apparent polar-wandering paths, but if the former, they would have different paths. By comparing apparent polar-wandering paths from different continents, Irving had demonstrated that the continents had indeed moved independently of one another; this was the work that Harry Hess cited to persuade his American colleagues that it was time to look again at continental drift. In subsequent work, Irving showed that ridgecrest basalts were heavily oxidized, with a dramatic decrease in their magnetic remanence on either side of the ridge crest.[81] This could be explained by hydrothermal alteration: circulation of seawater would oxidize the iron-bearing magnetic minerals and decrease the intensity of their magnetic signatures.

Geochemists had something to say as well. In the mid-1960s, scientists on the Rise-Pac expedition had found distinctive metal-rich sediments on the crest of the East Pacific Rise. In a paper published in *Economic Geology*—a journal dedicated to understanding ore-forming processes—they argued that the enrichments were caused by venting of hot fluids "probably related to magmatic processes at depth," and that the process might "serve as the basic enrichment mechanism and source for the elements in some ore formations."[82] The rise was an extension of the Gulf of California, where, as in the Red Sea, metalliferous brines were probably the result of magmatic-hydrothermal activity—and they suggested a worldwide association of mid-ocean rift systems, magmatism, hydrothermal processes, and ore deposits.

On the 1972–1973 Trans-Atlantic Geotraverse (TAG) expedition, scientists found manganese crusts near 26°N on the Mid-Atlantic Ridge closely associated with a water-temperature anomaly of 0.11°C–0.14°C. They interpreted these crusts as the products of hydrothermal circulation, in which cold water seeped into fractures, was heated at depth, and vented "as submarine springs along fracture systems."[83] The water-temperature anomalies occurred in a region of magnetic lows, and the scientists suggested that hydrothermal alteration was responsible for the decreased magnetic intensity (because hydrothermal alteration can destroy magnetic minerals such as magnetite, converting it to nonmagnetic iron oxides and hydroxides). Indeed, they noted that this reasoning could be flipped: regions of low magnetic intensity could be used as guides to recent or active hydrothermal sites.

Scientists were beginning to wonder whether the frequently observed concentrations of iron and manganese on the ridge crests were the result of widespread, even pervasive, hydrothermal activity in response to seafloor spreading.[84] Among the scientists who put forward this idea was Scripps graduate student, J. B. "Jack" Corliss (b. 1936), who had completed his PhD in 1970 on the subject of mid-ocean-ridge basalts. (He had also briefly studied philosophy of science with Paul Feyerabend at Berkeley before switching to geology.) Supervised by geologist Tjeerd "Jerry" van Andel (1923–2010), Corliss had analyzed the mineralogy and geochemistry of Atlantic mid-ocean-ridge basalts toward an understanding of their genesis, in a project funded by the NSF and the American Chemical Society. He found that the basalts were geochemically diverse, suggesting compositional inhomogeneities in the upper mantle where they originated or in the conditions of partial melting when they formed. But to interpret primary geochemical variability, he needed to sort out the effects of secondary alteration. Corliss found that his basalts were depleted of many of the same elements that were enriched in the adjacent sediments.[85]

These effects were consistent with the long-standing idea of hydrothermal exhalations or volcanic emanations accompanying submarine volcanic activity and transporting elements from the magma into seawater. Economic geologists working with ore deposits had long believed that submarine "exhalations" were responsible for the formation of some sulfide ore deposits— sometimes called exhalative deposits—but exactly how this happened was never specified. Indeed, the vaporous exhalations had always seemed a bit mysterious—if not occult—and economic geologists were sometimes viewed by their colleagues as less rigorous scientists than they ought to be because they invoked causes, like exhalations, that had never been observed. Corliss, though, had observational evidence.[86] Indeed, the various line of evidence— heat flow, geomagnetism, geochemistry—were all consistent with the effects of hydrothermal alteration, which leaches metals from rocks while cooling them, altering their magnetic signature and producing metalliferous fluids. Hydrothermal activity might also generate hydrothermal vents, as seawater heated by interaction with fresh basalt became buoyant and erupted in underwater geysers.

Against this backdrop of interest and expectation, Le Pichon, by then head of the Centre Océanologique de Bretagne in Brest, France, proposed a joint French-US collaboration to study the mid-ocean ridges. The Centre National pour l'Exploitation des Oceans (CNEXO) had recently purchased its own submersible, *Cyana*, a "diving saucer" with a 1.2-inch steel hull and a depth capacity of ten thousand feet. They were also still operating Piccard's *Archimède*, with its six-inch steel hull and awesome depth capacity of thirty-six thousand feet. Le Pichon recruited American colleagues to organize the Mid-Atlantic Ridge Workshop in an attempt to garner the support of the US National Academy and the NSF, who soon agreed to support the program.

The French-American Mid-Ocean Undersea Study—FAMOUS—was launched in 1973 as a comprehensive study of a section of "typical" mid-ocean ridge: the length between 36°N and 37°N latitude, four hundred miles west of the Azores.[87] Submersible dives coupled with detailed bathymetry, magnetometry, and other techniques would permit a comprehensive investigation of a section of mid-ocean ridge to determine the nature of the geological and geophysical processes taking place there. The project would be led by co-chief scientists Le Pichon and Jim Heirtzler (b. 1925), and it would live up to its ambitious name.[88]

The project began with seven French dives in *Archimède*, achieving the first direct look at the seafloor. The following summer, the two nations joined forces to complete an unprecedented forty-four dives: twenty in *Cyana*, seventeen in *Alvin*, and seven in *Archimède*. A dozen scientists participated, among them on the US team Heirtzler and Robert Ballard (b. 1942) (later of *Titanic* fame) from Woods Hole and Jerry van Andel (at Oregon State University), and on the French team Jean Francheteau (1943–2010).

The area of investigation was chosen because of its proximity to an adequate staging area in the Azores and in view of existing data. Thirty previous cruises across the mid-Atlantic at this latitude had produced a plethora of magnetic and bathymetric information with which to plan specific dives.[89] In addition, the divers were blessed with a good map: a photomosaic of the area produced by the US Naval Research Laboratory (NRL) using its light-behind camera system (LIBEC). Cofunded by the NRL and the NSF, LIBEC had been developed in response to the *Thresher* experience and consisted of a towed camera in which a high-intensity electronic flash was mounted several feet behind the camera. With the camera towed third feet above the seafloor, LIBEC could illuminate an area up to forty-five feet across; the USNS *Mizar* obtained 4,882 usable pictures of the FAMOUS area. When compiled into a photomosaic, the pictures provided a good sense of the overall orientation and morphology of the axial rift zone, its fault scarps, and its lava flows.[90] The scientists also prepared by visiting areas of active volcanism on land. The French traveled to Iceland and the Afar Triangle in East Africa; the Americans went to Hawaii. When the time came to dive, they had a good idea of what to expect.

The forty-four dives produced over three thousand pounds of rock samples—including the first abundant samples of fresh basalt ever recovered from seafloor—and approximately one hundred thousand photographs. More important than the numerical yield, the expedition produced a vivid new image of the seafloor. The study demonstrated conclusively that the mid-ocean ridges were the result of submarine basalt eruptions, cut apart by extensive faulting. (There was no peridotite.) The main rift valley, located approximately nine thousand feet below the ocean surface, was twenty miles across and five thousand feet deep: a seafloor Grand Canyon. Volcanism was confined to a narrow zone about a mile or two wide at the center of the axial rift valley. The volcanic rocks in this axial zone were very fresh; away from the axis, the basalts were older and more weathered. Still, they were no older than about one hundred thousand years, confirming average spreading rates of about one inch per year.[91] Perhaps more interesting than the average spreading rate, however, was the discovery that spreading was variable and asymmetric: over the past seven hundred thousand years, the ridge had spread about one-third of an inch per year to the west and one-half of an inch per year to the east.[92]

The volcanic rocks were cut by huge, prominent faults that created very complex relief, including underwater scarps as high as one thousand feet, with talus slopes along their bases.[93] These topographic details had not been resolved by earlier seismic refraction and echo-sounding studies, proving the value of direct human observation: important features had been undetected by even the Navy's most sophisticated technology.[94] Moreover, there was an intangible value to direct human observation, made manifest by the enormous popular interest that the discoveries provoked. Within a few months, the expeditions were being featured prominently in the pages of *National Geographic*.[95]

The vivid pictures of submarine pillow lavas cut by giant fault scarps left little doubt about the role of the mid-ocean ridges in global tectonics. The ridges were the sites of submarine basalt eruptions, and the faults displacing them were not the transcurrent faults that Bill Menard and Victor Vacquier had imagined (chapter 5). They were transform faults—segments that connected pieces of the seafloor that were moving apart. They were not offsetting the plate boundary; they *were* the plate boundary. Although no hydrothermal activity was seen, one dive recovered samples of metal-rich sediments, indicating geologically recent hydrothermal activity and confirming that the anomalously low heat flow over the ridges was likely the result of dissipation by convective circulation.[96] The long-standing tranquil image of the deep sea was shattered, replaced by direct observation of a tectonically active region littered with the debris of underwater volcanic eruptions and landslides, and topography as rugged as anything seen on land. Seafloor spreading was no longer a hypothesis, at least not to these scientists. It was an established fact.

Saving *Alvin*

The FAMOUS expedition had been funded by the NSF, NOAA, and CNEXO and promoted by the cooperation of prominent scientists, several of whom had recently themselves become famous for their contributions to plate tectonics. Its success was attributable to a high degree of cooperation and coordination among individuals and agencies. However, it was also attributable to the availability of good data from earlier Navy work, tangible evidence of the value of military-scientific cooperation. With the success of the expedition, one might have concluded that *Alvin* had finally lived up its promise and that its future would no longer be in doubt.

In fact, *Alvin* was nearly bankrupt. In September 1974, Fye wrote to Van Andel: "We are in considerable jeopardy with funding for the *Alvin* project.... NSF, NOAA, and the Navy agreed to get together and let us know soon whether or not they can collectively save the program. It would be a sad day for Woods Hole and oceanography if this program is terminated just at the time of its greatest pay off."[97] Fye had managed to convince DARPA and NOAA to help keep *Alvin* operational while he tried to negotiate its future, but the sums he was working with were very small and the amount he needed very large. For fiscal year 1974, he had obtained commitments of $75,000 from NOAA, $28,000 from the ONR, and $21,500 from DARPA; he needed about $1 million.[98] The NSF program officers were sympathetic but adamant that the agency could not be responsible for such a large, ongoing expense. Making matters worse, the ONR had communicated its intention to end its support for the program entirely. In what it euphemistically called a "no fund equipment loan contract," the ONR offered to hand over *Alvin* without further financial support. If Woods Hole could not or did not want to take it, *Alvin* would be mothballed.[99]

Fye began speaking to anyone who would listen, including Atomic Energy Commission chairwoman and marine biologist Dixie Lee Ray, the Massachusetts senator Edward Kennedy, and the Massachusetts representative and speaker of the House Thomas "Tip" O'Neill. With the support of the Massachusetts congressional delegation, Fye was able to convince the NSF to create a partnership to save *Alvin*. The solution, ironed out toward the end of 1974, was to make *Alvin* a UNOLS facility.[100]

UNOLS—University National Oceanographic Laboratory System—was a network established in 1971 to make oceanographic facilities more readily available to scientific investigators.[101] It solved two problems: one, that ships placed heavy financial burdens on the institutions that kept them, and two, that scientists in institutions without ships might have good ideas. Under the UNOLS program, the costs of maintaining a ship at one institution was subsidized by the NSF or other agencies in exchange for making it available to investigators from other institutions; the total commitment would be $900,000 per year. On November 11, 1974, the NSF, the ONR, and NOAA signed a letter of agreement for the continued support of *Alvin* as a UNOLS platform.[102] UNOLS solved Woods Hole's funding problem and made *Alvin* available to a much wider swath of the scientific community, paving the way for continued work on mid-ocean ridges.

The Discovery of Deep-Sea Hydrothermal Systems

In 1967, Tanya Atwater had arrived as a graduate student at Scripps to find the place "in chaos."[103] In 1967–1968 the pieces of the plate tectonic puzzle came together (although the term *plate tectonics* was not yet in wide use), and scientists were scrambling to reinterpret their data and make their mark in the revolution. After completing her work with John Mudie on the Gorda Rift, she undertook a piece of work that would change geological thinking. Using previously collected paleomagnetic data from the eastern Pacific, Atwater demonstrated that the San Andreas Fault system of coastal California had been created when the eastern portion of the Pacific plate was completely subducted under the western continental margin of North America. The remains of the Pacific—what had previously been its western half—began to move northward, creating the San Andreas Fault. In the language of the new theory, a convergent plate margin became a transcurrent boundary. Previous interpretations based on the geology of California had placed the origins of the San Andreas in the late Cretaceous period, some seventy-five million years ago, yielding a very small offset rate: about one-third of an inch per year. Atwater's analysis of the paleomagnetic data proved that the actual age was no more than twenty million to thirty million years, meaning the northward rate of motion for the Pacific plate was far greater: about two inches per year. Not only was this much faster than most geologists had thought possible, it was also much faster than the spreading rates measured in the Atlantic. These results were published in the *Geological Society of America Bulletin* in 1970.[104]

Atwater realized that if the Pacific was spreading faster than the Atlantic, then the frequency of volcanic eruptions there must be greater and the rocks hotter. Therefore, the odds of finding hot springs were greater along Pacific spreading centers than Atlantic ones. This was good news for the Scripps group, as it implied that, if the goal had been to find vents, the FAMOUS scientists had not looked in the best place.

But the Pacific is a very large ocean, so the search area was huge. However, Atwater had another useful insight: that vent fluids would have to mix with surrounding seawater and therefore produce substantial regions of warmer-than-average water. One could therefore narrow the search by making water-temperature measurements near ridge crests, and this could readily be done with Deep Tow.

Temperature Anomalies at the Galápagos Rift

From December 1971 through February 1973, Mudie led a thirty-five-thousand-mile expedition to the Pacific with funding from the NSF, the ONR, and the DSSP. Named South Tow for its emphasis on regions south of the equator, it was a diverse and ambitious scientific data-gathering effort, including studies of ocean biology and circulation as well as seafloor magnetics, seismic reflection, and heat flow. It also had a military purpose. As explained in the funding proposal to the ONR, the cruise was to include "operations related to the sea floor search problem."[105] By this the funding proposal meant the Deep Submergence Salvage Program: Leg 5 would work to "sharpen up and shake down the deep tow and related ship's equipment," particularly Deep Tow's side-looking sonar, which could be used in a future search for a lost submarine.[106] But the technology to search for lost submarines could also be used to understand deep-sea geophysical processes, and Legs 6 and 7 would be a Deep Tow survey of the Galápagos Rift. Leg 6 would be a general reconnaissance, and Leg 7, led by John Sclater, a detailed heat-flow study over a region of anomalously low values near ridge crest.[107]

Sclater had been of two minds on the heat-flow question. On two earlier expeditions to the Galápagos, in 1969 and 1970, a modest amount of magnetic, heat-flow, and bottom water-temperature measurements had been collected; Sclater had interpreted heat-flow lows as a sampling artifact.[108] However, in 1973 Sclater and MPL graduate student Kim Klitgord revisited these data, paying particular attention to water-temperature anomalies near the spreading center. They interpreted them as evidence of hydrothermal activity near the ridge crest, which could "account for the variability of heat flow across the spreading center and the difference between observed and theoretical heat flow" models.[109] But colleagues at Lamont had given similar data a different interpretation: they attributed local areas of lower-than-average heat flow to "downward propagating thermal waves induced by temperature variations at the sediment-water interface"—that is, transient thermal effects caused by annual variations in bottom water temperatures.[110] The Lamont scientists noted the possibility of some circulation of fluids into porous layers within the ocean-floor basalts but placed little emphasis on it. The problem of heat flow at the ridge axes remained unresolved.

Sclater hoped to resolve it. Joining him were John Mudie, Robert Detrick, and Kim Klitgord from MPL, Richard Von Herzen from Woods Hole and his graduate student David Williams, and several other scientists and technicians. They focused on a portion of the Galápagos Rift, a mid-ocean spreading center located near 86°W latitude along the equator, about two hundred miles northeast of the Galápagos Islands. The ocean floor there was nearly new (less than four million to six million years old), the spreading rate was fast (about four to six centimeters per year), and detailed bathymetry and sediment distribution maps were available from earlier work.[111] Most important, the cruise would cross the Pacific spreading center in a position essentially equivalent to a particularly good magnetic profile, Lamont's *Eltanin-19*, which had helped to persuade many people (particularly at Lamont) of the reality of seafloor spreading.[112] Correlation of detailed heat-flow measurements with good-quality magnetic data might enable the scientists "to use heat flow averages as an age-predictive tool, and possibly lead to some insight on the thickness of the lithosphere."[113] If heat flow decreased systematically away from the ridge crest, then heat-flow measurements could be used to date ocean floor in areas where the magnetic anomalies were not as clear as *Eltanin-19*.

The scientists were able to make accurate measurements within five kilometers of the spreading axis on crust as young as 150,000 years.[114] The average heat flow over the crest was far less than predicted by theoretical models of a conductively cooling lithosphere but consistent with additional convective or advective heat loss—that is, with hydrothermal circulation. Moreover, the overall heat-flow pattern was extremely complicated and variable near the ridge crest. The highest values were found near fault scarps and topographic highs and on sediment mounds up to twenty-five meters in radius and rising five to ten meters above the surrounding seafloor.[115]

The scientists also completed over one hundred thousand individual water-temperature measurements, including two near-bottom horizontal water-temperature profiles.[116] Echo sounding from an acoustic data telemeter placed just above the middle thermistor ensured accurate elevations (critical, because ocean water temperature varies with depth): The high sensitivity temperature profiles at bottom revealed spikes up to several hundredths of a degree centigrade. While tiny by surface standards, from the perspective of the frigid waters of the deep sea, they were substantial (fig. 7.1).[117]

The heat-flow data were consistent with the variability Von Herzen had observed in his earlier work, and the water-temperature measurements consistent with the venting of hot waters into the deep ocean. The scientists made the link to hydrothermal systems explicit. A press release issued by the Scripps Public Affairs Office on the completion of the cruise in February 1973 explained: "The high variability of heat flow temperatures recorded is believed to be caused by hydrothermal circulation of ocean water among the rocks, causing a condition resembling the hot springs and geysers of Yellowstone Park or those found in Iceland."[118] The scientists put it this way: the results provided "reasonably clear evidence for hydrothermal vents ... from the bottom water temperature anomalies."[119] The only other plausible explanation was heat transfer from a submarine volcanic eruption, but that would be transitory and would not explain the metalliferous sediments: "The observed temperature anomalies are more probably caused by thermal

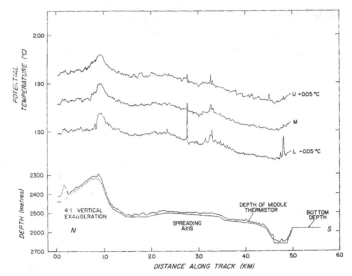

Figure 7.1 Bottom water horizontal temperature profile, showing anomalies over the Galápagos Ridge Crest.

From D. L. Williams, R. P. Von Herzen, J. G. Sclater, and R. G. Anderson, "The Galapagos Spreading Centre: Lithospheric Cooling and Hydrothermal Circulation," *Geophysical Journal of the Royal Astronomical Society* 38, no. 3 (1974): 587–608 (fig. 8). Reprinted with permission of Oxford University Press.

plumes rising from hydrothermal vents. The correlation between heat-flow maxima and the occurrence of small sediment mounds ... suggests that these mounds are hydrothermal vents."[120]

One could argue that the scientists had discovered hydrothermal vents on the seafloor. One could also argue that the TAG scientists had discovered hydrothermal vents in the mid-Atlantic. Both groups had detected hot waters associated with metalliferous sediments, and both interpreted their observations as seafloor hydrothermal vents. They had "seen" the vents with their instruments, but they had not seen them with their eyes.

The Rise Workshop and Pleiades Expedition

In April 1976, MPL sponsored a workshop to coordinate submersible studies of fast-spreading centers in the Pacific. The FAMOUS expedition had clarified the character of the slow-spreading Mid-Atlantic Ridge, but the Pacific was younger, hotter, and spreading faster. Moreover, the East Pacific Rise was observably different in two additional ways. One, it displayed a central axial block, twenty to forty kilometers wide and two hundred to four hundred meters above the surrounding seafloor, but no central rift valley. Two, it was located, as its name indicated, in the eastern Pacific; it was not a mid-ocean ridge.[121] To the north, at least, it did not connect to the worldwide system of mid-ocean ridges but merged into the San Andreas system, raising important questions about ocean-continent transitions.

The "Rise" workshop, as it came to be called, included most of the leading scientists involved in ridge studies—Tanya Atwater, Robert Ballard, Jack Corliss, Xavier Le Pichon, John Mudie, Fred Spiess, and Jerry van Andel—as well as Navy representatives from the DSSP and the Office of Naval Research and Development Activity. The outcome was a proposal for a series of dives to the East Pacific Rise and the Galápagos Rift to explore fast-spreading centers and continent-ocean plate boundaries. There would be six specific topics of investigation: the structure and origins of the axial blocks, the nature of the volcanic processes, the nature of hydrothermal processes, geodetic measurements, magnetic measurements, and the study of transform faults and transform-rise intersections.

Each of the topics had been the subject of earlier investigations using instruments carried on or towed behind surface ships; the emphasis here was on the additional information that could be gleaned by direct observation, measurement, and sampling from a submersible. For example, published profiles of the East Pacific Rise were invariably plotted with great vertical exaggeration (often twenty to one hundred times), making the margins of the axial blocks look like vertical cliffs, but in fact the slopes were all less than 20°, less than the resolution of most echo sounding. Similarly, numerous samples of volcanic rocks had been dredged from surface ships, but to evaluate primary igneous and hydrothermal processes, one needed spatially accurate sampling; to evaluate paleomagnetism, one need oriented measurements in situ.[122] And to answer the plate tectonic question—what is the rate of plate motion?—scientists needed to place geodetic instruments on the seafloor.[123]

The existence of seafloor hydrothermal systems was no longer really in question, for the indirect evidence was abundant and (to most of the scientists involved) compelling. So why was a submersible program needed? The answer, to scientists, was that it would enable them to make "direct observations ... to establish the true nature of the phenomena involved."[124] For the scientists on the FAMOUS expedition, seafloor spreading had been established by seeing its effects with their own eyes. The Rise group wanted to do the same for vents: to discern them not just indirectly but also through the direct observation that would remove any doubt.

The NSF was committed to funding research on mid-ocean spreading centers, but scientists still needed the Navy for ships.[125] Moreover, the success of the FAMOUS expedition had hinged in no small part on the availability of a precise, detailed, bathymetric map. Deep Tow could make precise measurements over small areas, but the best tool for detailed mapping of large areas was the multibeam sonar system of the Naval Oceanographic Office (NOO). "Since [NOO] is the only organization in the United States which has this system, it is critical that early discussions be held with the Navy to insure that these surveys are conducted," the workshop report noted. The scientists needed both Navy ships and Navy data.[126]

Meanwhile Deep Tow work was continuing, as Scripps scientists launched the Pleiades cruise back to the Galápagos Rift. Like many cruises, this one supported several different research projects, but the first two legs were specifically designed to investigate hydrothermal processes at the spreading center. Leg 1 consisted of two projects relevant to seafloor vents: "fine scale aspects of the sea floor" and "hydrothermal processes on the Galapagos Rift." The principal investigator on the first was MPL geologist Peter F. Lonsdale (b. 1948); on the second it was Ray F. Weiss (b. 1943), a geochemist in the Isotope Laboratory of Scripps's Geological Research Division. Both Weiss and Lonsdale were recent graduates: Weiss had completed his PhD in 1970, Lonsdale in 1974, and both had stayed on as researchers at Scripps.[127]

Lonsdale's dissertation, "The Abyssal Geomorphology of a Depositional Environment at the Exit of the Samoan Passage," had focused on topography and sedimentation. These were among the features affecting underwater sound, and one publication from Lonsdale's thesis, "Near Bottom Acoustic Observations of Abyssal Topography and Reflectivity," was published in 1974 in the proceedings of an ONR symposium, "Physics of Sound in Marine Sediment."[128] This was a classic MPL synergy: sound could be used to examine sediments and sediments affected sound.

Lonsdale's thesis committee included Woods Hole's Charles Hollister, and his project followed from Hollister's work on the role of deep currents and abyssal sedimentation (chapter 8). Until the mid-twentieth century, most people had thought that ocean-floor topography was mainly structural in origin—controlled by faults and submarine volcanism and blanketed by an even layer of sediment, quietly deposited as planktonic debris and fine-grained materials slowly sank through the water column to the deep seafloor. But as we saw in chapter 5, in the 1930s Dutch geologist Philip Kuenen had demonstrated experimentally that turbid currents could transport large quantities of coarse sediments from coastal areas to the deep sea, and in the 1950s Lamont scientists led by Maurice Ewing and Bruce Heezen had demonstrated the existence of turbidity currents in the Hudson Canyon off the coast of New York.

In the 1960s, Hollister and others had shown that the sediments deposited by turbidity currents could be reworked by strong currents that scoured the seafloor and sometimes generated significant topography. Consistent with the Stommel-Arons model (chapter 2), these deep currents were strongest along the western boundaries of the ocean basin. They were also strong in some seafloor channels, such as the Samoan Passage (north-northeast of the island of Samoa), the main channel through which Antarctic Bottom Water travels northward into the Central Pacific. Lonsdale chose this for his dissertation study. Using echo sounding, air-gun reflection profiling, and Deep Tow, as well as coring and dredging of sediments, Lonsdale showed that the patterns of sedimentation and bedforms were consistent with active, directionally consistent deep currents.[129] It was becoming increasingly apparent that the seafloor was active not just tectonically but sedimentologically as well.

Weiss had studied dissolved gases in seawater, and he now focused on the use of helium to detect mantle fluid sources. Argon, helium, and neon are "conservative quantities" in seawater: they are not normally added or subtracted by geochemical or sedimentary processes, so they can be used as tracers of ocean circulation. However, helium isotopes (He-3 and He-4) may be introduced into the deep ocean from Earth's interior or by interaction of ocean water with rocks derived from the interior. Therefore, if one were to find them in anomalous concentrations, it would be evidence of material introduced into the ocean from Earth's interior. But water samples are easily contaminated by atmospheric gases, and without accurate knowledge of the gas solubilities, it would be difficult to recognize contamination.[130] For his thesis, Weiss had analyzed the solubility of argon, helium, and neon in seawater and developed a sampling device for collecting gas-tight sealed water samples. He would apply this sampler on Pleiades to look for evidence of elements derived from Earth's interior.[131] Together, Weiss and Lonsdale's projects would make up the first leg of Pleiades.

The second leg would be co-led by Jack Corliss, an assistant professor at Oregon State University, and Woods Hole's Richard von Herzen. The title of their project was the same as Weiss's: "Hydrothermal Processes on the Galapagos Rift." Corliss had proposed that heavy-metal depletion in ridge basalts was a result of hydrothermal alteration; this was his chance to test the hypothesis. His leg would attempt to find direct evidence of hydrothermal processes on the Galápagos Rift.[132]

Leg 1 ran from April 29 to June 12. The participant list records twenty-four persons, including three women: graduate student Kathleen Crane (in charge of transponder navigation), technician Kathy Poole, and Weiss's wife, Portia, who signed on as a volunteer. The goal was "deep towing, coring, dredging [and] hydrocasts," but this description fails to convey the intellectual import.[133] Weiss would be taking physical samples along with temperature-depth-salinity measurements. This was crucial, because ocean temperature varies with depth, so accurate depth records would be essential to differentiate between a thermal anomaly and a fluctuation in the elevation of the instrumentation. Salinity also varies with temperature (warmer waters can dissolve more salt), so accurate salinities would be needed to rule out anomalies associated with water-mass movement. When and if a real anomaly was detected, they would be able to fill a sample bottle and, with luck, capture vent fluids.

The scientists encountered an extremely complicated temperature field with numerous high-temperature spikes, but most were accompanied by salinity variations suggesting mixing of turbulent waters over the ridge crest. However, immediately over the fissures at the spreading center, they detected strong temperature anomalies that departed from regionally prevailing temperature-salinity characteristics, all located directly over the central axial fissure of the rift. The most extreme of these showed an increase of about 0.2°C, an order of magnitude greater than previously found. They managed to collect water samples from three of these; the samples contained anomalous concentrations of He-3. This was strong evidence that the components of this water had interacted with mantle-derived materials, either rocks or fluids beneath the sediment-water interface.[134] The scientists took extensive bottom photos of the area, and in some of them they saw red and orange sediments, perhaps hydrothermal iron oxide deposits. They also saw clamshells on the seafloor, near the temperature spikes. They left transponders to mark the sites.[135]

Leg 2 ran from June 13 to July 22. There were thirty-one participants. Crane and Legg were the only holdovers from the first leg; the new members included Jack R. Dymond, an associate professor at Oregon State, and David Williams, working for US Geological Survey. In the cruise report, Corliss wrote that "several observations of significance [were] reported to us by Peter Lonsdale and Kathy Crane." These included "bottom photographs at the ridge crest of fissures associated with thermal anomalies, [that] reveal yellow to orange surficial deposits which appear to be hydrothermal deposits."[136]

One might say that seafloor hot springs had been discovered, as the scientists worked to identify sites for *Alvin* dives to sample the fluids. The goals were "to determine if a program of dives with deep submergence research vehicle *DSRV Alvin* in the survey area have [*sic*] a reasonable chance of sampling hydrothermal fluids and contributing significantly in other ways to our understanding of submarine hydrothermal systems, [and] to measure background levels of various physical and chemical parameters in the sediment and water so as to understand the geological, geophysical, and geochemical setting of the hydrothermal activity and to allow us to place the observations from the submarine in a well-defined context."[137] In their minds, the existence of ridge-crest hydrothermal systems was no longer in doubt. The goal was to learn more about them.

The scientists confirmed and expanded on the results reported by Lonsdale and Crane. Eleven additional measurements revealed temperature anomalies of 0.1°C – 0.2°C, and several samples dredged manganese-rich crusts "which appear to be hydrothermal deposits."[138] At the site of the largest temperature anomaly, bottom photos showed "extensive yellow and orange deposits." They cored the sediment and recovered muds with "orange to brown lumps and streaks interpreted as hydrothermally deposited Fe and Mn." They placed additional transponders, sonar reflectors, and sediment traps "within 100 meters of two prime dive sites."[139]

Returning to the Galápagos in *Alvin*

Clearly there was hydrothermal activity at the Galápagos Rift. Water temperatures were elevated, geochemical signatures indicated mantle input, and there were sedimentary mounds rich in the metal oxides characteristic of terrestrial hot springs. The Pleiades scientists presented these results at the annual meeting of the American Geophysical Union: a set of three multi-authored presentations, with Corliss, Crane, and Weiss as first authors, detailing the results with respect to the sediment mounds, the structure of the rift, and the thermal and geochemical evidence for hot plumes.[140] Meanwhile Weiss and Lonsdale prepared two papers for submission to *Nature*.[141]

Lonsdale also prepared a separate paper discussing the biological evidence obtained from the Deep Tow photographs.[142] In a paper submitted to *Deep Sea Research* in April 1977, he described a dense and "remarkable community of large and abundant benthic organisms," dominated by "unusually large" bivalves associated with zones of hydrothermal discharge.[143] At another site, at 3°S on the East Pacific Rise, where warm water was also detected, photographs revealed an abundant community of sea anemones and sea pens. These clusters of biological activity were highly localized; in six hundred bottom photographs taken at 3°S, not a single sea anemone was observed away from the one observed cluster. Lonsdale concluded that the dense biological communities, both on the East Pacific Rise and on the Galápagos Rift, were associated with hydrothermal vents. Just as the TAG scientists had suggested that zones of low magnetic intensity could be guides to hydrothermal sites, Lonsdale suggested that zones of dense clusters of benthic filter feeders might similarly serve as guides.

Meanwhile, Corliss, van Andel, and Von Herzen were on their way back to the Galápagos. In 1976, they had submitted a proposal to the NSF for a series of *Alvin* dives—justified by the earlier heat-flow work—and were on their way.[144] Since Corliss had no diving experience, van Andel recruited Robert Ballard, with whom he had worked on FAMOUS. Other participants included Jack R. Dymond, David Williams, and Tanya Atwater from the earlier cruises; Louis I. Gordon and Deborah Stakes from Oregon State University; Robert Hessler from Scripps; and MIT geochemist John Edmond.[145] Crane, who had meanwhile compiled a map of the rift based on the Deep Tow work, was invited to help navigate back to the site.[146] Thirteen scientists would make dives, including the three women: Atwater, Stakes, and Crane.[147] The team reached the area in mid-February 1977.

The transponders left behind the previous year were designed to remain silent until reactivated by a designated acoustic signal. On February 13, 1977, Crane was able to reactivate the transponder they had left at the site they called Clambake I—but only after a tense five of hours trying.[148] The National Geographic Society had provided the means to set up an onboard photography lab so the scientists could develop their photographs immediately to guide their explorations. On February 19, they developed a set of pictures in a region of temperature spikes and saw blankets of clams covering the seafloor. This was no clambake; these clams were alive. It was time to dive.

What they found has been described as "one of the most exciting developments since the beginning of the study of oceanography."[149] Yet the initial observations of these communities read less like a eureka moment and more like a page out of Henry David Thoreau. Van Andel kept a personal diary, and just before turning in on the evening of February 17, 1977, he wrote:

> Today the first dive and a glorious one it was ... The landscape was both gorgeous and geologically fascinating. It was totally unlike FAMOUS; no gigantic vertical crags rising hundreds of feet into the gloom, no steep flow fronts of bizarre lava forms, no white snowfields amongst black rocks, no faults no uplifts.... Instead, a gently undulating terrain of smooth broad lava domes, glittering in the lights.... And in the middle of this stark and barren vastness, hard, prickly, new and untamed, a small oasis, perhaps an acre, sharply defined, with coral gardens, pink and gold anemones, white crabs in great variety and profusion, yellow, brown, liver-colored fish, medusoid large clumps of some kind of mussel ten inches long, crevices filled

with their huge bleached shells looking from afar like rims of snow in a boulder field. It was like an aquarium, huge but carefully arranged with elegance and taste and grace and enough to see to suit the public every step.... What produced this little paradise in the ... sea floor desert? I have no idea. Warm springs perhaps, which we suspect in this area? Time for bed; it was a lovely day and tomorrow will be my second dive.[150]

In a series of twenty-four dives in February and March 1977 along the axis of the Galápagos Rift, thirteen scientists observed and photographed active hydrothermal springs surrounded by thriving biotic communities.[151] Although by this point the vents were not a surprise, the profusion of life crowded around them was.

All the sites contained abundant signs of life, but the dominant organisms were distinct at each. The site they designated the Garden of Eden was characterized by spaghetti-like acorn worms, giant tube worms, limpets, and fish, and an unusual animal that resembled a dandelion, later identified as a new species of siphonophore. (Distantly related to the Portuguese man-of-war, siphonophores resemble jellyfish but are attached to the bottom by thin filaments; they were first discovered on the *Challenger* expedition one hundred years before.) Most remarkable was the size of individual organisms: clams a foot across, tube worms many feet long. While the overall biodiversity at each site was low, the total biomass was great. At Clambake II, the clams were dead; this appeared to be a fossil hot spring. The site they called Oyster Bed was actually covered by mussels (revealing geophysicists' ignorance of biology—perhaps a trivial point, but perhaps not, as we shall see in chapter 9).[152]

The press coverage that followed was laced with superlatives—*amazing, astounding, incredible, fantastic,* and *profound*—and the inevitable comparisons with Jules Verne.[153] Articles appeared in *National Geographic,* the *New York Times,* the *San Francisco Chronicle, Time, Newsweek,* and many other venues.[154] Many participants were interviewed extensively (although conspicuously not the women); even before his work on *Titanic,* Ballard was becoming a celebrity.[155] Jack Corliss told the *San Francisco Chronicle* "that what we are finding will prove to be the greatest discovery in the history of benthic biology since the discovery that life was even possible in the deep sea." This claim may have seemed exaggerated to biologists who read it in the newspaper, but the discoveries did raise fundamental questions as to how vent environments sustained abundant life.[156] This question would be taken up by the Rise workshop participants, who would soon be on their way the East Pacific.

From Warm Springs to Black Smokers

The scientists on the Galápagos dives had observed warm springs venting at a pleasant 10°C–20°C.[157] But some people had expected more. In a paper published in May 1976—while the Pleiades scientists were taking samples over the Galápagos—Northwestern University scientists Tom Wolery and Norman Sleep had calculated the rate of fluid circulation required to explain the divergence between the measured and predicted heat flows and concluded that fluids should be venting at temperatures as high as 300°C.[158] In other words, there should be true hot springs, and so the search continued. In 1978, Jean Francheteau led a multinational team to search for vents along the East Pacific Rise. With funding from CNEXO, the NSF, and the Mexican government, the project was dubbed RITA—for the Rivera and Tamayo Fractures Zones near the southern tip of Baja California.

The Tamayo Fracture Zone—really a ridge-to-ridge transform fault but named before the existence of such faults was recognized—is the southernmost extension of the San Andreas Fault. North of it, the Sea of Cortez separates mainland Mexico from Baja California, and the tectonics there are dominated by short spreading centers connected by long segments of ridge-to-ridge transforms. Although the Sea of Cortez is an incipient ocean basin, the main displacements there are transcurrent, not extensional. The Rivera Fracture Zone, to the southwest, marks the transition into a more typical mid-ocean-ridge setting dominated by extensional tectonics. From here southward, the East Pacific Rise consists of long stretches of spreading centers displaced by short sections of ridge-to-ridge transform faulting (fig. 7.2). The region between the two fracture zones—21°N on the East Pacific Rise— would be the focus of the study. With a spreading rate more than twice that of the Mid-Atlantic Ridge, the closely spaced ridges and fractures zones along the East Pacific Rise were a good place to look for hydrothermal activity.[159]

Diving in *Cyana*, Francheteau and his colleagues discovered massive sulfide mounds exposed on the seafloor: hills, cones, and tubular structures up to ten meters high and averaging five meters across. They were colorful—red, white, brown, black, yellow, and gray—and dominated by the iron sulfide mineral pyrite (or its polymorph, marcasite)—the dominant mineral in most economic sulfide ore deposits. The mounds also contained iron oxides and hydroxides, mostly goethite and limonite, minerals associated with the weathered surfaces of terrestrial sulfide deposits on land, and amorphous silica, commonly found in hot springs and geysers. Later analysis of the sulfides revealed ore-grade concentrations of zinc, copper, silver, and other metals—up to 29 percent zinc and 6 percent copper.[160] The scientists also found what ap-

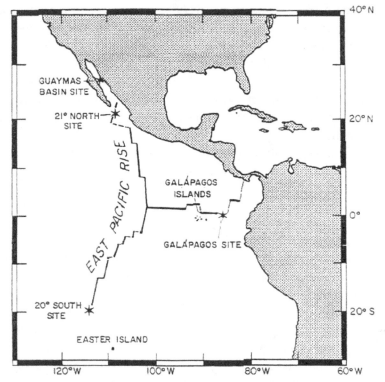

Figure 7.2 Map of the Galápagos Ridge Crest region where hydrothermal fluids were first detected.

From MacDonald and Spiess, "Workshop Report," press release, n.d., Box 63, Folder 11, "Rise and Gala-pagos Rift File," Scripps Office of Communication Records, SAC 13, Special Collections & Archives, UC San Diego. Reprinted with permission from the Scripps Institution of Oceanography.

peared to be a fossil-vent community: an area approximately fifteen by one hundred meters along a fissure covered with dead clams the size of "dinner plates."[161] Some of the clams were partially dissolved, indicating that they had been dead for some time and suggesting that the loci of hydrothermal activity had shifted as new fracture zones opened in response to tectonic activity.

Between February and March, the team completed twenty-one dives to depths of three thousand meters with traverses covering fifty-seven kilometers.[162] The mineral assemblages strongly suggested an association with hydrothermal systems, and the tubular structures—alternating layers of crystalline pyrite, marcasite, and/or sphalerite, amorphous silica, and iron hydroxides—could be vents in which minerals had deposited on walls of the tubes as hot fluids moved through them. Still, they had not found the fluids.

The following year, another group finally did, including some members of the original Deep Tow team. In April 1979, Fred Spiess and Ken Macdonald returned with Deep Tow, *Alvin*, and a large scientific team. The key paper from their discovery would have twenty-two authors, including four Mexican geologists from the Universidad Nacional Autónoma de México—Diego Córdoba, José Guerrero, Arturo Carranza, and Víctor Díaz García—plus Ballard, Francheteau, Hessler, and Atwater (then an assistant professor at MIT), Scripps colleagues Miriam Kastner and John Orcutt, and Scripps graduate student Rachel Haymon.[163]

The goal was to obtain a higher degree of detail and resolution in mapping an ocean spreading center than previously achieved by bringing instruments down to the seafloor in *Alvin*. Detailed gravity measurements might help to detect a magma chamber below the ridge axis. Magnetic measurements might permit detections of the regions of freshest volcanism (and therefore hottest rocks) or, alternatively, of most intensely altered rocks (and therefore most recent or active hydrothermal circulation). They also planned to make direct observations of the massive sulfide deposits and to look for active hot springs near the field of dead clams found the previous year.[164]

Smart, lucky, persistent, or all of the above, they found what they sought: true hot springs, venting at temperatures as high as 380°C.[165] There were about twenty vents in all, mostly along a parallel line, presumably marking a major fissure. As in the Galápagos, the springs were surrounding by flourishing biotic communities, but there was something new as well: they were discharging black fluids from discrete mounds up to ten meters above the seafloor. The fluids were venting from the mounds at rates of one to five meters per second, creating buoyant plumes tens to hundreds of meters above the seafloor. The effect was like smoke billowing from a blast furnace, so they dubbed the mounds hydrothermal "chimneys" and the springs "black smokers."[166]

The scientists were able to take samples both directly from the chimneys and from the sediments surrounding them. The dominant chimney minerals were pyrite (FeS_2), anhydrite ($CaSO_4$), and chalcopyrite ($CuFeS_2$), plus several other minerals typically associated with ore deposits: sphalerite, gypsum, limonite, goethite, and pyrolusite. In effect, they were seeing a mineral deposit forming before their eyes. As they recorded in the cruise report, "This is the first direct observation of hydrothermal sulfide minerals being deposited on the sea floor."[167] They might have gone further: it was arguably the first direct observation of economically valuable minerals being formed in any natural environment.

Analysis showed that the fluids contained extremely high concentrations of dissolved metals, sulfide, and silica—precisely the constituents expected in an ore-forming fluid. The black color was caused by tiny mineral particles, precipitated when the metal-rich fluids mixed with cold waters and became saturated with respect to metal sulfides and oxides. In some cases, the fluids were dominated by amorphous silica, producing "white smokers." At least as significant as the chimneys themselves, albeit less photogenic, were the sediments surrounding them, which contained up to 45 percent iron, 34 percent copper, and 15 percent zinc, plus significant traces of lead, cobalt, silver, and cadmium. If sustained across time and space, concentrations like this would form the sort of ore deposit over which speculators would be scrambling and smart investors would be double-checking to ensure the samples weren't salted.[168]

Press reports focused on the potential economic aspect, and media attention contributed to some implausible claims, such as that the vents might be an exploitable source of geothermal energy.[169] *Newsweek* quoted Ballard as describing the vents as containing a "bonanza of minerals"; the *San Francisco Chronicle* wrote of "mineral deposits of incalculable value." The *New York Times* suggested that "vast deposits of much needed minerals" might be present beneath the world's mid-ocean ridges.[170] In retrospect these claims were grossly exaggerated; perhaps they were motivated by late 1970's anxieties about shortages of oil, minerals, and other natural resources. In reality, none of the scientists involved (or, likely, the journalists reporting on their findings) understood the economics of mineral exploitation.[171] If they had, they would have recognized that neither the scale of the deposits nor the economics of deep-sea mining would justify exploitation of this resource in the foreseeable future. What the deposits *did* represent was a watershed in the history of geological science.

Geologists were used to seeing effects and imagining their causes, viewing matters after the event and trying to reconstruct what had happened. In the mid-nineteenth century, William Whewell (the man who coined the term *scientist*) famously declared that "we only know causes by their effects," and in geology this was certainly true.[172] His French counterpart and positivist Auguste Daubrée complained that geology suffered as a science, because geologists were merely witnesses after the fact, but positive knowledge required scientists to be "witnesses in the moment."[173] (For this reason he advocated an expanded role for experiments and physical models in geology.) Earth scientists were now witnessing a major geochemical process in the moment. They did not have to speculate about what they were seeing; they were observing it. They were witnessing one of the key phenomena that had justified the establishment of their science in the first place: the creation of economically valuable mineral deposits.

In 1980, a third major discovery was made. In a joint US-Mexico expedition on the research vessel *Melville*, scientists led by Peter Lonsdale investigated thermal anomalies at 27°N in the Guaymas Basin, part of the Gulf of California, southeast of San Diego. Lonsdale had been to the area in 1977 in *Sea Cliff* (under Navy command) and had retrieved a single sample of a sulfide-bearing, ferromanganese-rich talc, interpreted as a hydrothermal sinter.[174] Ore deposits were known from adjacent land areas, and metalliferous brines had been found in the nearby Salton Sea, so the scientists were optimistic that something significant would be found.

The Guaymas Basin is an active spreading center very close to land—Baja California to the west, mainland Mexico to the east—and it is very different from both the Galápagos Rift and the East Pacific Rise. Only three million years old, and as little as one hundred kilometers wide, the gulf fills with sediment eroded off adjacent landmasses, which are transported by turbidity currents to its deepest portions. Abundant plankton blooms supply further sedimentary debris. As a result, submarine lavas do not erupt onto the seafloor but intrude into a pile of soft, wet sediments. Lonsdale and his colleagues called this "sediment-smothered sea floor spreading."[175]

The scientists used Deep Tow to survey the site, take photographs, and detect temperature anomalies, in this case up to 0.4°C above ambient seawater. Then they sampled the seafloor sediments. They found evidence of hydrothermal fluids at temperatures up to 315°C, high geothermal gradients (up to 10°C per meter), and world-record heat flows. Moreover, they found that the interaction of hot water with sediments rich in organic matter was generating hydrocarbons, which reacted with metal-rich pore fluids to deposit metal sulfides in the sediment pile.[176] While a great deal of attention was paid to the hydrothermal chimneys and metalliferous sediments of the East Pacific Rise, the sediment-hosted deposits of the Guaymas Basin resembled actual sulfide ore deposits far more closely, particularly the type found on land in sedimentary rocks. They also had a much greater likelihood of being preserved over the course of geological time than the beautiful but fragile chimneys of the bare oceanic ridges, suggesting that this environment was a better model for how economically valuable ore deposits actually formed.[177]

Media coverage culminated in 1980 with a National Geographic Society special featuring Robert Ballard, *Dive to the Edge of Creation*, which aired on nationwide television on January 8, 1980.[178] The original question—what controls heat flow over the ridges?—had receded in the face of the more popular biological discoveries. Ironically, scientists had answered a question they had not set out to answer—how do sulfide ore deposits form? The biggest biological question—how does life thrive where there is no light?—remained unanswered.[179]

Life on the Seafloor and Other Matters

Two hypotheses emerged to explain vent life. One was that rising thermal plumes set up countercurrents that carried nutrients into the region. The other was that the vent communities were self-sustaining, based on the chemosynthetic activity of autotrophic bacteria, specifically, the oxidation of hydrogen sulfide.[180] Hydrogen sulfide (H_2S) is common on the Earth's surface under anoxic conditions; it is the source of the rotten-egg smell in swamps and marshes and in surface hot springs. In the ocean, it occurs mostly in stagnant basins such as the Black Sea. In deep-sea hot springs, it is generated by reaction of seawater with sulfur-bearing minerals in seafloor basalts. Hydrogen sulfide is toxic to most forms of life, but some bacteria can metabolize it, and the vent fluids contained large amounts of such bacteria.[181] Studies soon suggested that these bacteria were key to the story.

Most terrestrial ecosystems are rooted in photosynthesis, but in vents the base of the food chain is chemosynthetic: bacteria use the oxidation of hydrogen sulfide to hydrogen sulfate (H_2SO_4), and the energy released by that reaction, as the basis for life.[182] Chemoautotrophic bacteria had been discovered in the late nineteenth century by Russian microbiologist Sergei Winogradsky (1856–1953), who subsequently demonstrated their role in nitrogen fixation by leguminous plants. He also found them in terrestrial hot springs, but most of his colleagues dismissed this as ecologically unimportant.[183] In any case, no one had identified a complex ecosystem based on autotrophic bacteria as a primary food source. However, the high concentrations of hydrogen sulfide and large numbers of bacteria in vent fluids suggested that these communities were chemoautotrophic. The evidence that confirmed it was the discovery that the bacteria were found living symbiotically inside the tissues of several of the major faunal groups.

Scientists soon recognized that the diets of many vent organisms are composed predominantly or exclusively of these bacteria. One of these is the giant tube worm, *Riftia pachyptila*, whose taxonomy has taken decades to resolve.[184] Lacking a digestive track, *Riftia* lives by absorbing carbohydrates metabolized by bacteria that live inside it. Several of the major faunal groups that make up the vent communities, including *Calyptogena*, the giant clams, and *Bathymodiolus*, the dominant mussel, are now known to thrive symbiotically on the bacteria that make their homes within their tissues.[185] Autotrophic bacteria in the vent environments also live on the surfaces of organisms, rocks, muds, and sulfide mounds.[186] Some invertebrate species exploit these bacteria by grazing on them, while others extract organic molecules released into the seawater when the bacteria die.[187]

Figure 7.3 Dr. Meredith L. Jones holding a preserved sample of a giant vent worm, 1981. Smithsonian Institution Archives. Image No. SIA2010–1520.

Besides *Riftia,* scientist have recognized numerous previously unknown organisms at vents, including *Alvinella pompejana* (fig. 7.3), sometimes referred to as the Pompeii worm, which thrives at exceptionally high temperatures—at least 40°C and perhaps as high as 110°C—and has been found attached to the walls of chimneys.[188] More than 95 percent of animals discovered at vents to date are previously unknown species, and altogether more than twenty new families, ninety new genera, and nearly three hundred new species have been identified.[189]

Perhaps even more important than the discovery of new species was the appreciation of the extreme environments in which they live. By 1983, it had been demonstrated that some of these bacteria were hyperthermophilic, thriving at temperatures over 100°C.[190] By the mid-1990s, vents had been found in all the world's oceans, at temperatures as high as 400°C.[191] These discoveries suggested that life might perhaps thrive in other environments previously thought sterile, including the interior of the Earth and the Jovian moons.[192] The word *extremeophile* has been coined to describe life that thrives in extreme environments, and the question of how organic molecules survive extreme heat without degradation remains an active topic of study, one that also has practical applications.[193]

Many thermophilic microorganisms recovered from vents belong to group Archaea—as the name suggests, thought to be among most ancient of extant organisms. This has led to speculation about a possible connection among hot springs, autotrophic bacteria, and the origins of life.[194] Geologists believe that Earth was hotter and more volcanically active in the early part of its history than today, and hydrothermal venting may have been ubiquitous in the early oceans. While Earth's surface was bombarded by meteorites during the first four hundred million years of its history, submarine environments might have been relatively placid places where life could have begun to evolve unmolested.[195] Vents are extreme microenvironments, with strong chemical and physical gradients: temperatures ranges of 400°C can occur across spaces of just a few centimeters. While we now know that most on-ridge vents are short lived, some off-axis vents have persisted for hundreds of thousands of years, creating environments where life could have gained a toehold. A dense plethora of microchemical and microphysical environments, sustaining far-from-equilibrium conditions, around long-lived off-ridge vents, might have been just the thing to permit the amalgamation of amino acids, perhaps adsorbed onto charged mineral surfaces, into early forms of life.[196] As one group of researchers has put it, "hydrothermal vents truly appear to be special among the various settings that have been considered for the origins of life."[197]

Seafloor Mineral Deposits

While the discovery of hydrothermal vents—with their possible relevance to the origin of life—was presented as *Alvin's* great "basic science" discovery, the relation of vents to economic mineral deposits was presented as its great "applied science" one. Very quickly, the promise of economically recoverable metal sulfide deposits was being fit into a narrative of a basic science investigation that serendipitously turned out to have useful applications. Within a few years, there were numerous government-sponsored studies and workshops on deep-sea mining, and even a US Department of Interior Draft Environmental Impact statement for proposed mining leases on the Gorda Ridge.[198]

Geologists working in the mining industry viewed these claims with a jaundiced eye, as there was no shortage of sulfide deposits on land that could be mined more easily and cheaply than minerals under two kilometers of ocean water. In hindsight their skepticism proved correct: to date no economic metal sulfide deposits have been recovered from the deep sea. Although the grades of these deposits are locally impressive, the overall tonnage is small compared to most minable deposits and the costs of recovery far higher.[199]

These deposits were useful, however, for understanding terrestrial ore deposits. It is clear that massive sulfides in basalt host rocks, such as the famous Troodos deposits in Cyprus—mined since the Bronze Age—were formed in seafloor environments similar to those scientists observed from *Alvin*. In 1985, Rachel Haymon and colleagues reported fossil tube worms in a massive sulfide deposit in Oman—a discovery that geologists around the world were discussing over campfires as proof the deposits formed on the seafloor.[200] Today, geologists accept that the Guaymas Basin offers an explanation for how sediment-hosted ore deposits form, firmly linking ore deposits to their tectonic settings. All this helped to cement the notion that *Alvin* was a "basic science" program that proved to have useful applications.

As for heat flow—the problem that led geophysicists to the ridges in the first place—by now the answer is anticlimactic: detailed flow measurements confirmed the role of convective cooling at the ridge axes. The heat-flow pattern at ridges is not a monotonic drop away from the axis but an oscillating pattern consistent with overall high heat flow punctuated by local zones of strong convective cooling.[201] More than two-thirds of Earth's heat flux is lost through ocean basins, and about a third of this is released through convective circulation at vents.[202] Ironically, however, this convection is no longer thought to be the driving force of plate tectonics but the side effect of moving slabs whose motions are dominantly controlled by subduction of their dense, cold, old ends. The places where oceanic crust forms—the ridges— are now generally viewed as less interesting than the subduction zones in which it is consumed.

Painting Projects White

From the perspective of natural science, the discovery of seafloor hydrothermal vents was by far the most important thing *Alvin* ever did. The results transformed scientific thinking about heat flow and mantle convection, ocean chemistry, the origins of ore deposits, and life itself. They helped to transform our vision of the deep sea from a barren wasteland to a fertile eco-system, perhaps the ultimate incubator of life.[203] But these profound discoveries began with prosaic problems: the transmission of underwater sound in submarine warfare and the salvage of materials lost at sea. The scientists involved were curious people, but the context in which they made these discoveries was not that of curiosity-driven research. Their discoveries were the end product of a two-decade-long interaction that fostered the pursuit of scientific questions that could be amalgamated with and yoked to the concerns and needs of the US Navy, during a time of political anxiety over what might be lurking or lost in the sea.

For most scientists, the important part of the story is the ending: that scientists made major discoveries about the natural world. This is reflected in the way *Alvin* is presented by scientists as a vessel dedicated to civilian science, its success a triumph of curiosity-driven research. Some have gone so far as to expunge *Alvin*'s military pedigree. When Robert Ballard wrote a retrospective on *Alvin* in 2000, he began with its commissioning "at Woods Hole in 1964," ignoring the first three years when *Alvin* did mostly military work.[204] An article published in *Science* in 2002 on an occasion described as *Alvin*'s thirty-fifth birthday referred to its "3700 dives … since 1967."[205] But *Alvin* was *not* thirty-five in 2002; it was thirty-eight. Its first three years of its life and the dives it did on behalf of the US Navy had been expunged.

In *Water Baby*, a popular history of *Alvin*, science writer Victoria Kaharl acknowledged the Navy's role in funding and supporting the project but maintained the myth that the military connection was incidental, an excuse used to "justify the sub's existence."[206] This is consistent with the stories scientists have told of "painting projects blue"—putting a veneer of military relevance on basic science projects.[207] The historical facts bely these myths. In fact, they indicate the inverse: that scientists working on military-sponsored projects painted their projects *white*—pretending that the military projects were motivated by basic science when patently they were not.

As we saw in chapter 6, at the start of the *Alvin* project many oceanographers doubted that a deep-submergence vessel would be of much scientific use. Those who believed it could be of use had to fight for a vessel that actually *would* be of use, and when it was built, they had to fight for the opportunity to use it. Then, when the Navy had achieved its ends, scientists had to fight for *Alvin*'s survival. Later, scientists would imply, suggest, or even claim outright that both *Alvin*'s creation and its research program were driven by scientific considerations, but this was simply untrue. The reality was that it was driven first by the anticipated and then by the actual need for military access to the deep sea environment. Both the idea of *Alvin* and its technical realization were rooted in military needs, as were the instrument specifications and delivery deadlines. When scientists got to use *Alvin* for nonmilitary activities, they still had to fit "mission profile." Projects that did not fit were not fostered.

Science did not lead the *Aluminaut* and *Alvin* program; it followed. Recall Jim Mavor's complaint to Paul Fye in the early 1960s that Woods Hole scientists were skeptical about *Aluminaut* because of the gap between its design features and any plausible scientific use. Mavor's complaint is reinforced by the recollections of Fred Spiess, who noted that, from the perspective of basic science, "*Alvin* was a solution in search of a problem."[208] In hindsight it may seem hard to believe that Woods Hole scientists would not have been thrilled by the prospect of going to the deep sea, but the reality is that most were not, and the context in which *Alvin* was developed explains why.

This is not to say that the scientific questions that were highlighted as part of the deep-submergence program—bottom currents, the daily vertical migrations of animals, the distribution of seamounts—were *fabricated*. Not at all. They were phenomena in the natural world that scientists might have wished to investigate. But they came into focus via military salience, because they were questions that these particular vessels—whose specifications were determined by their military mission—were able to address. Topics that had no military salience—such as the giant squid—were shunted aside, even when the scientific value was acknowledged to be high. Most tellingly, when conflict arose and Paul Fye tried to protect his scientists' interests, the ONR insisted that he enforce the "mission profile"—and he did. When that mission was fulfilled, the ONR sought to discontinue its support.

Moreover, there is something absurd about the idea that scientists pulled the wool over Navy eyes. To suggest that scientists could paint projects blue—that they could fool their military patrons into supporting irrelevant research—is to suggest that military officers were acutely naïve. Such a view seems implausible on its face. After all, if we reject as simplistic the idea that scientists were duped or used, we should equally reject the idea that their military patrons were. Paul Fye, Maurice Ewing, Roger Revelle, and Fred Spiess had enormous respect and even admiration for the military men with whom they worked; they did not consider them at all unsophisticated. Indeed, on the occasion of a presentation in 1960 to the Ad Hoc Committee on Basic Research of the President's Science Advisory Committee, Maurice Ewing wrote in his notes, "Navy: A general remark about the navy's support of basic research may be made. In general, their administrators are well informed and not easily misled."[209]

Conclusion

There has been a deep and critical connection between military programs—particularly the development of new technologies—and the growth of natural knowledge about the ocean. That much seems uncontroversial.[210] However, the story in this chapter supports the potentially more controversial claim that military patronage not only empowered and enabled scientists by supporting them to develop and use these techniques and technologies, but also disempowered them to the extent that military concerns came to dominate the scientific agenda in fields whose funding came primarily from military sources. The historical evidence demonstrates that the agenda of American oceanography after World War II was structured by the needs—real and perceived—of the US Navy, and that scientific priorities at Woods Hole and elsewhere were set to a large degree by those needs.[211]

This is not to say that scientists were duped or used, or that the relationship was unidirectional. The US Navy did not merely call on scientists; scientists also called on the Navy.[212] Wenk, Vine, Fye, and Spiess were motivated to develop tools and understandings that would both help them to do their job and help the Navy do its job. But the constitution of "their job" was itself subject to interpretation. Some scientists, like Henry Stommel, Joanne Malkus, and William von Arx, believed unequivocally that their job was to increase understanding of the natural world, and only that. As Stommel put it, the scientist has "a deep-seated drive to wrestle personally with the unknown and to create understanding ... where before there were not even questions." This was contrasted—and in his mind incompatible—with an agenda structured by the Navy's "need to know."[213] Others, however, including Fye and trustee E. Bright Wilson at Woods Hole and Maurice Ewing and Charles Drake at Lamont, were equally unequivocal in their conviction that part of the job of their community—if not necessarily each individual in it—was to serve the nation's military needs.[214] Moreover, many scientists relished their roles as experts serving the nation. The report that inspired *Aluminaut* emerged from a US National Academy of Sciences committee, one of many venues where scientists proffered their expertise to potential military patrons.[215]

One way to account for the military-scientific relationship is through the concept of mutual orientation, developed by historian Paul Edwards in his study of Jay Forrester and the scientists on Project Whirlwind. Edwards notes that scientists and engineers who worked closely with US Air Force sponsors came to understand and internalize their sponsors' concerns; they frequently found themselves thinking about solutions to Air Force problems. They also came to understand which sorts of projects would be likely to find support, so after a while would not bother to propose projects they knew would gain no traction. Conversely, Edwards argues, Air Force officials were influenced by their interactions with computer scientists, who made them aware of new possibilities (recall the "salutary influence" that James Dorman at Lamont insisted that scientists had). In Edwards's words, "The source of funding, the political climate, and their personal experiences oriented Forrester's group toward military applications, while the group's research eventually oriented the military toward new concepts of command and control."[216]

It is not clear how much oceanographers and geophysicists oriented Navy patrons toward new concepts, but it is clear that oceanographic information helped to make certain systems—like SOSUS and Artemis—conceivable and possible. Like the computer scientists who worked on Project Whirlwind, the oceanographers who developed the *Alvin* program were not exploited by the Navy, if by that we mean used to their detriment or against their will. Rather, they actively sought opportunities for Navy sponsorship and attempted to forge a symbiotic relationship, one that sometimes proved genuinely mutual and sometimes did not. Clearly, the interactions between scientists and Navy officers were diverse; causal arrows can be found in both directions. But it is also clear that Navy officials were not naïve. As one memo from the deputy assistant oceanographer of the Navy put it in 1969, any proposed science plan had to "demonstrate relevancy and responsiveness of programs with Navy needs," and these needs were to be specified in terms of "specific functions" rather than "broad mission categories."[217] As marine chemist Andrew Dickson has put it, scientists were not free to do just anything on the Navy dime.[218]

A different account has been offered by sociologist Chandra Mukerji, who concluded that oceanographers served primarily as a reserve labor force on which the state could call as needed. When not needed, she concluded, they were more or less free to do as they wished. There are elements of this story that fit that picture, particularly if we consider technologies as part of that force.[219] The work of oceanographers in the *Thresher* search, Admiral Stephan's belief that science could provide the background use for *Alvin* at times when the military had no foreground use, and the calling up of *Alvin* and *Aluminaut* to find the lost H-bomb at Palomares certainly fit the bill.

Mukerji's thesis, though, fails to account for the changes that occurred over the course of the Cold War. Oceanographers may have in part served as a reserve labor force—and, as we shall see in our final chapters, ended up vulnerable for just that reason—but their science was not expanded and strengthened during the Cold War to be that. Scientists were funded to answer specific questions, work on specific projects, develop specific instruments, and take on specific tasks. Not all these tasks had direct relations to particular weapons systems or operational activities, but many did. These included crucial systems like SOFAR and SOSUS and weapons programs like Polaris. Crucially, when projects no longer seemed useful to the Navy, they were terminated, even when scientists insisted on their continued intellectual value.

However, to say that oceanographers were not a reserve labor force — or at least not primarily a reserve labor force — is not to claim that they were autonomous, either. This point may seem so obvious that one is compelled to ask, why have scientists resisted this seemingly obvious conclusion? Why not just admit that Navy funding was beneficial but not unalloyed? Why *bother* painting projects white? I will return to this question in the conclusion of this book.

8 *From Expertise to Advocacy: The Seabed Disposal of Radioactive Waste*

In March 1961, the US Navy articulated a ten-year plan for its "entire ocean-ographic effort ... including basic and applied research, ocean surveying, undersea research vehicles, instrumentation, ship construction, facilities, and the training of oceanographic manpower." The document, known as TENOC, for the *Ten Year Program in Oceanography 1961–1970*, summarized the Navy's aspirations for oceanography: "Advancing our knowledge of the oceans and thereby increasing the effectiveness, within the oceanic environment, of naval operations, weapon systems, and ship and equipment design."[1] The argument was simple: because the Navy operated in the sea, it needed to understand the sea: "The successful design, construction, and operation of a powerful modern Navy depends ... on how well the environment is understood."[2]

In 1945, this argument might have been controversial, but by the 1960s it was conventional wisdom. Past investments in oceanography had produced "significant" results, and "future investment can be anticipated to return manifold benefits in weapon and equipment improvements and more efficient and safe operations."[3] Research funds (mostly from the ONR) would triple over the following decade, from just over $10 million dollars in 1961 to more than $33 million in 1970 (fig. 8.1). Including funding for ships, facilities, surveys, and information dissemination, the total research and development budget would grow to over $91 million. Roughly one-third of that flowed to just three institutions: $12.6 million to Woods Hole, $10.5 million to Columbia and Lamont (including the Hudson Laboratories), and $9.6 million to Scripps.

By 1971—the end of the TENOC decade—there was no question that scientific understanding of the ocean environment had greatly advanced. Long-standing questions about ocean circulation and chemistry, about the structure and origins of ocean basins, and about the character of the seafloor had

Activity or Lab	Budget Line Item	1961	1962	1963	1964	1965	1966	1967	1968	1969	1970	
Surveys		9311	9321	10271	11483	11895	11720	13625	16500	16500	17245	
Forecasting		755	1562	2044	1887	1624	1537	1537	2327	2327	2327	
Info Dissemin.												
Info Dissemin.		832	1195	1213	1228	1236	1251	1258	1258	1268	1270	
MIL Ocean		14403	13870	20215	23195	24310	25365	28425	30020	31775	32885	
Ships		4200	13600	35400	30200	28800	30200	29200	29200	28400	-	
Inst		1850	3075	6508	5827	4576	5790	4660	4560	5100	3890	
Facil												
Basic Research		10045	9722	15105	16950	19740	21650	24525	26820	29785	33415	
Summary	TOTAL	41396	52345	90756	90770	92181	97513	103230	110685	115155	91032	
	Inst (Aug)	-	1000	-	-	-	-	-	-	-	-	
	(Aug) Basic Research	935	1412	-	-	-	-	-	-	-	-	
GRAND TOTAL		42331	54757	90756	90770	92181	97513	103230	110685	115155	91032	

TABLE 18   GROSS SUMMARY OF TENOC FUNDING (Funds in Thousands)

Figure 8.1 Gross summary of TENOC funding, table 18.

From Navy Department of the United States, *Department of the Navy Ten Year Program in Oceanography: TENOC, 1961–1970* (Washington, DC: Department of the Navy, Office of the Chief of Naval Operations, 1961).

been answered and Navy support was a crucial component of that success. Without the funds, equipment and logistical support provided by the US Navy, scientists would simply not have been able to collect the data, explore the realms, or complete the analyses that they did. Yet by the early 1970s, Navy-academic bonds were fraying.

The personal ties forged during World War II that had linked Columbus Iselin, Maurice Ewing, Roger Revelle, Paul Fye, Bruce Heezen, Harry Hess, and Bill Menard to admirals and other officers had become largely the stuff of legend.[4] By the late 1970s, Iselin, Ewing, Hess, and Heezen were dead; Revelle and Fye had retired; and Menard had moved to scientific administration. A new generation of scientific leaders had emerged, who generally lacked the military experience that had helped their predecessors forge alliances with active-duty officers. The "golden age" of Cold War Navy-sponsored oceanographic work was tied to a particular generation of men who had either fought in World War II or worked closely with those who did. It was, in this sense, at least as much the result of World War II as the Cold War, a product of the affinities and affiliations that had been forged during the former and remained robust into the latter, but only for two decades.

Navy support had never been a blank check, but in the 1950s and early 1960s it sometimes seemed as if it were: funds had flowed freely within broad mandates and with only modest oversight. By the 1970s, this was no longer the case. With the 1970 passage of the Mansfield Amendment, Congress insisted that military research and development funds be assigned to projects whose military character was clear; no doubt this caused some military officials to scrutinize their budgets and allocations more carefully. Among oceanographers, however, the amendment received little mention; no project discussed here was explicitly reduced or curtailed with reference to the Mansfield Amendment. However, a concern that appeared with increasing frequency in the late 1970s and early 1980s was the problem of too many oceanographers competing for too few funds.

The TENOC document was based on the recommendations of the scientists on the US National Academy of Sciences Committee on Oceanography, who wanted to breed more of their kind. In 1945, Harald Sverdrup had considered it a top priority to increase the number of trained oceanographers; fifteen years later this theme was repeated in TENOC. One of ten specified priorities was to "significantly ameliorate the manpower shortage in the field."[5] The 1960s and 1970s saw rapid growth in both the size and the number of academic programs in oceanography. By 1985, Woods Hole's dean of graduate studies Charles Hollister estimated that US graduate programs were producing sixty-five to seventy new physical oceanographers every year, and that did not include marine geologists or geophysicists.[6]

As the number of oceanographers swelled, so did competition for funds. By the mid-1970s, anxieties over funding were cropping up in the notes, memos, and internal reports of academic oceanographers and institution directors with an urgency not felt since the 1930s. Paul Fye described his situation in July 1974: "We find ourselves in rather severe financial crisis ... [in particular] the lack of money for ship operations."[7] Funding for scientific research continued to grow throughout the 1970s, but the number of oceanographers competing for those monies and the number of ships that had to be maintained had increased even more.[8]

By the late 1980s, the Navy's commitment to academic oceanographic science had decreased substantially, not just in relative terms but in absolute ones, too. Budgets for Navy-sponsored research were mostly flat from 1970 to 1983, and ONR-sponsored research had decreased significantly: in constant dollars from $6.4 million to only $4.1 million (fig. 8.2).[9] The next generation of oceanographers found themselves working much harder for Navy support than their teachers had, and also searching for alternative funding sources.[10]

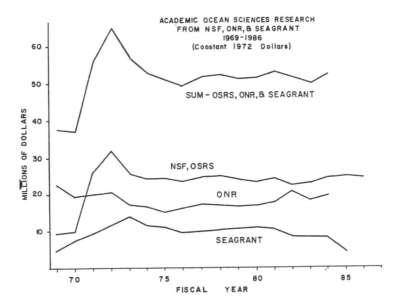

Figure 8.2 Funding for oceanographic research, 1970s to 1980s, based on materials prepared by Charles Hollister for the House Armed Services Committee, Subcommittee on Research and Development, Hearings on "Health of Basic Research in Oceanography," March 1985. (a) Total Navy research support, 1969–1985, in constant 1972 dollars. Sponsored research peaked in 1972 with a sudden increase in NSF support in that year, and then leveled off. (b) Total ONR support for oceanography compared with other disciplines. ONR support for oceanographic research declined dramatically from the mid-1970s to the mid-1980s but was more or less compensated by increased NSF support. (c) Total dollars and constant 1971 dollars awarded and man-months [sic] of support provided at Woods Hole Oceanographic Institution, 1971–1984. As costs rose, similar budgets were able to sustain less research. (d) NSF marine science budget in nominal and real dollars, 1975–1985. Ostensible budget increases were outpaced by inflation.

From Statement of Charles Davis Hollister, PhD, FAAAS, FGSA, Senior Scientist, Dean and Member of Directorate of the Woods Hole Oceanographic Institution, before the Research and Development Sub-Committee of the House Armed Services Committee on Health of Basic Research in Oceanography (SE 31) in the Office of Naval Research, pp. 5, 7, in Charles Davis Hollister Papers, 1967–1998, MC-31, "Congressional Testimony House Armed Services Committee, March 28, 1985," Data Library and Archives, Woods Hole Oceanographic Institution.

This chapter and the next examine the pitfalls and problems that befell oceanographers as they strove to adapt to the changing circumstances of oceanography. In this chapter, we explore how Woods Hole dean and marine geologist Charles Hollister struggled to maintain support for his scientific program while the political exigencies that had inspired it were waning. In the final chapter, we examine how a group of physical oceanographers tried

OFFICE OF NAVAL RESEARCH

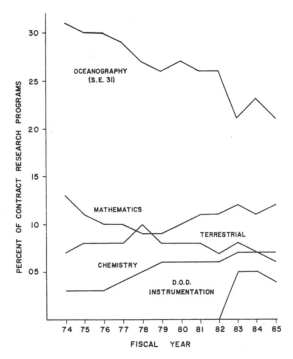

Figure 8.2 (continued)

to adjust their research priorities in response to changing societal needs, only to find that a goodly portion of society did not want their help—at least not in the way they were offering it. We turn first to Charles Hollister.

Charles Hollister and Radioactive Waste Disposal at Sea

Had he not died in a hiking accident at the age of sixty-three, Charles Hollister (1936–1999) could have been said to have lived a charmed life. Smart, self-confident, and rich, he was born into California's famous Hollister family, who arrived in the United States in the 1630s and in California in 1866, when they purchased a huge tract of pristine California coastline just west of Santa Barbara.[11] Raised on the family ranch and educated at the exclusive Chadwick School on the Palos Verdes Peninsula, Hollister then studied at Oregon State University and earned his PhD at Lamont, working with Bruce Heezen.

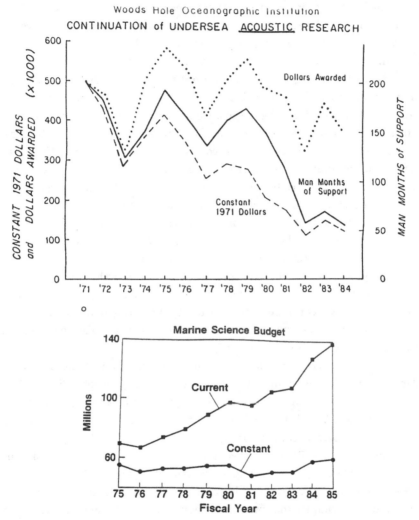

Figure 8.2 (continued)

Hollister and Heezen were among the first marine geologists to document the sedimentologically active nature of the seafloor. This was a matter of considerable interest to both the Navy and the AT&T corporation for the potential impact of deep currents on underwater hydrophone arrays and telecommunication cables (chapters 4 and 5). Heezen and Hollister had demonstrated that currents sweep the abyssal depth, scouring the seafloor in some areas and depositing large amounts of sediments in others. Hollister's 1966 article "Shaping the Continental Rise by Deep Geostrophic Contour Currents," published in *Science* and coauthored with Heezen and Lamont colleague W. F. Ruddiman, helped establish the existence of "contour currents"—powerful currents that run along topographic contours of the continental rise—and gave the name *contourites* to the deposits created by them.[12]

This work challenged the conventional wisdom that deep marine sediments accumulated slowly in geosynclines, gradually subsiding basins or troughs along continental margins. Geosynclines were conceptualized as quiescent, with no significant source of sedimentary input but for the occasional turbidity current that carried coarse clastic materials from the continental margin into the deep marine environment and, in particular, no lateral sediment source. Hollister's work showed that this picture was wrong. First, the dominant materials in contourites were very fine-grained, clay-rich sediments, inconsistent with turbidity currents as their source. Second, the turbidity-current model suggested a dominant transport direction from continent to ocean, yet ripple marks and current lineations seen in underwater photographs showed that the dominant transport direction in the Atlantic was parallel to the coastline rather than perpendicular to it. This suggested that transport was controlled not by turbidity currents generated on the edge of the continental shelf but by the deep density-driven circulation posited by the Stommel-Arons model (chapter 2).

Besides providing a brilliant empirical confirmation of the Stommel-Arons model, the data pointed to a radically different view of sedimentological processes and established Hollister's reputation as a rising young star.[13] He was hired at Woods Hole, where he spent the rest of his career, working primarily on abyssal sedimentary dynamics. In the 1970s, he organized the High Energy Benthic Boundary Layer Experiment, or HEBBLE, to evaluate the mechanisms leading to large-scale movement of abyssal marine sediments.[14] He also pioneered the design and use of a large-diameter-piston coring device to obtain samples that preserved their internal sedimentary structure. Only four years after completing his PhD, Hollister coauthored with Heezen a gorgeous coffee-table book, *The Face of the Deep*, summarizing a decade of deep-sea discoveries.[15] By the late 1970s, the ONR considered him "a principal leader in the field of sediment dynamics" and a man who delivered on his research promises.[16]

Hollister also had administrative skills, and by 1989 he had risen to the rank of senior scientist, vice president of the corporation, and associate director of external affairs at Woods Hole. From 1979 to 1989 he also served as dean of graduate studies of the joint Woods Hole-MIT graduate program in oceanography. Proud of his western US heritage—and flaunting a showy masculinity—he continued to hike, climb, ski, hunt, and fly-fish throughout his life. He served as president of the American Alpine Club, became a fellow of the elite Explorer's Club, and served as deputy sheriff of Santa Barbara County. In biographical sketches, he noted among his accomplishments the ascent of the Vinson Massif, the highest peak in Antarctica, which had earned the team the John Oliver La Gorce Medal of the National Geographic Society. When Hollister organized a scientific conference in winter, the schedule was arranged to reserve afternoons for skiing.[17]

Hollister's assertive outdoor life was matched by an equally assertive approach to science: he pursued ideas with vigor, dedication, and a hefty dose of self-confidence. He was an energetic man who liked to organize initiatives and see his visions realized. But there were forces affecting his science that he could not control. As a student in the 1960s, Hollister was a beneficiary of Navy largesse. But by the 1970s, Navy priorities were changing, and Hollister needed to figure out who would support his work and with what justification.

At first, he had little trouble adjusting. As the ONR decreased its support for deep-ocean research, Hollister moved several of his projects to the National Science Foundation (NSF). But in 1976, the NSF declined to fund a project to differentiate the effects of turbidity currents and contour currents on the Icelandic Rise on the grounds that the proposal lacked novelty. One reviewer wrote: "My major objection pertains to the overall objectives of the investigation, which seem to duplicate … Hollister's own PhD thesis."[18] So Hollister shifted again, this time finding a new mission in the disposal of radioactive waste at sea and a new patron in the Sandia National Laboratory's Subseabed Waste Disposal program.[19] For the next fifteen years, Hollister directed a study group at Woods Hole known as the Deep-Sea Geology Study Program, served as the coordinator for the Department of Energy's (DOE) US Seabed Disposal Program Site Suitability and Selection programs, and was task-force leader of site selection for the DOE's International Seabed Working Group, which examined potential waste disposal sites outside US jurisdiction.

But there was another snag. By the mid-1980s, most people had concluded that waste disposal at sea was a bad idea, and political and scientific attention turned to land-based options. In 1986, at the direction of the US Congress, the DOE ended its seabed disposal research programs to focus on evaluating a potential repository site at Yucca Mountain, Nevada. This time, Hollister did not simply shift to find another patron. Instead, he decided to fight, challenging the wisdom of this decision in the same assertive manner that he hunted, fished, and summited mountains. In doing so, Hollister's science and politics became deeply entangled, as he defended subseabed disposal not merely on scientific grounds in scientific venues, but on political grounds in the halls of Congress and in the mass media.

The Seabed Disposal of Radioactive Waste

The idea of using the deep sea as a final resting place for high-level radioactive waste had been considered as early as the 1950s (chapter 2). Low-level wastes had been dumped into the oceans for some time: between 1946 and 1970 the US Atomic Energy Commission (AEC) had approved the dumping of eighty-six thousand containers of low-level wastes at four licensed sites, three in the Atlantic and one in the Pacific.[20] European countries— particularly the United Kingdom and France—had also dumped wastes at sea, and in 1967 the European Nuclear Energy Agency organized a cooperative dumping effort.[21]

In public, the AEC had generally made upbeat claims about the problem of waste disposal, downplaying the challenge and at times even implying there was no challenge. In its 1959 annual report to Congress, for example, the AEC wrote: "Waste problems have proved completely manageable in the operations of the Commission and of its predecessor war-time agency, the Manhattan Engineering District.... There is no reason to believe that proliferation of wastes will become a limiting factor on future development of atomic energy for peaceful purposes."[22] In private, however, the AEC was concerned about the lack of a suitable means to dispose nuclear waste and, among other possibilities, was considering dumping it in the deep sea.[23] But if waste were dumped in the deep sea, would it stay there?[24]

In 1955, the Atomic Energy Commission convened a meeting at Woods Hole on marine nuclear waste disposal. Arnold Arons later recalled that the "AEC people pointed with concern to the mounting volume of wastes being stored in situ [at power plants] and correctly predicted the rather frightening situation that would develop toward the end of the century.... They sought advice about the possibility of oceanic disposal."[25] Another participant was William von Arx, who concluded that, given the present state of knowledge, any attempt to use the oceans to dispose of raw fission products "must be regarded as involving serious risk."[26] Others agreed. The report of the meeting was largely pessimistic: not enough was known about the deep ocean to say whether wastes put there would remain there, whether deep circulation would distribute radioactive materials throughout the ocean or even rapidly return them to the surface environment, or whether there would be adverse effects on benthic life.

Arons recalled, "no one was prepared to make categorical predictions, [as] too little was known about abyssal circulation."[27] Having discerned that "AEC people might be interested in supporting some fundamental research despite the pessimistic atmosphere," he approached them about funding: "The AEC people were very receptive to this approach and accorded some support."[28] Some of that funding was used to make calculations of the rate at which materials might circulate between the surface and the abyssal environment. Oceanographers and geochemists had considered the question and obtained very different answers. Revisiting the *Meteor* data, German oceanographer Georg Wüst suggested a decadal time scale for midlayer circulation and no more than three hundred years for deep waters to be out of contact with the surface environment. These numbers were reinforced by calculations of dissolved oxygen, which suggested as little as 140 years, hardly conducive to long-term waste isolation. In contrast, Lamont geochemist Wallace Broecker had used carbon 14 tracers to calculate a mean residence time for North Atlantic deep water of 650 years.[29] His colleague Lawrence Kulp suggested an even more sluggish environment, in which "tropical Atlantic bottom water descended from the surface at a high altitude as long as 1800 years ago."[30] None of these numbers was entirely reassuring, but Broecker and Kulp's offered some promise.[31]

Ten years later, the Stommel-Arons model would show that deepwater circulation was significant and could therefore perhaps redistribute waste deposited in the deep sea; Hollister and Heezen's work showing that the deep marine environment was sedimentologically and oceanographically active validated that view. These discoveries seemed to dash any hopes that nuclear wastes deposited on the continental shelf, slope, or rise would rest in peace.

At the same time, the establishment of plate tectonics indicated that the interior regions of oceanic plates—the abyssal plains—might be tectonically stable for many millions of years. If this were so—and by the 1970s few doubted that it was—then the interior portions of tectonic plates might be just the place to dispose of waste, if one could get it there, past the technical, logistic, and political obstacles. As a man who had ascended some of the world's most challenging peaks, Charles Hollister was certain that you could.[32]

Hollister began to develop his views on the deep seabed as a suitable site for waste disposal in the mid-1970s, focusing on the analysis of the deep-sea sediments in the midplate environments to understand both the nature of the sediments themselves and the environments in which they had been deposited. His large-diameter-piston coring device was designed to permit collection of samples large enough to preserve sedimentary structures, and therefore the evidence of the nature, extent, and strength of deep currents or other disturbances.[33]

Knowing that radioactive waste would never really go away, Hollister entitled his proposal "A Logic and Decision Model for the Seabed Geologic Seclusion of Radioactive Waste."[34] The basic concept was this: plate tectonics provided a theoretical framework for identifying the stable portions of the ocean whose tectonic behavior was highly predictable, and where the sedimentary processes were primarily depositional rather than erosional, so materials put there would stay there. They would not "go away," but they would be effectively isolated.

Hollister paused little over the question of benthic life (see chapter 9 for a similar pattern in a different context): "We do not attempt to discuss the complex biological communities in the ocean provinces, but merely call attention to the overall biological productivity of surficial waters as an important indicator of mid-water and bottom biological productivity."[35] He was referring to the principal scientific argument against seabed disposal—the risk of contamination of the marine food chain—and suggesting that this risk was negligible. Because biological productivity is greatest in the photic zone (near the sea surface) and declines with depth (he argued), there was little life to worry about in the abyssal environment, and biological contamination could be ignored.[36]

This was consistent with the view he had presented in a proposal to the Atomic Energy Commission in 1973, in which he had referred to the north-central Pacific as a "Marine Desert."[37] In the nineteenth century, the view of the ocean as a watery desert was perhaps defensible, but in the twentieth, far less so. Perhaps Hollister thought that the regions with abundant life, such as the ridge-crest vent environments, were exceptions that proved the rule (chapter 7). Perhaps because the vents were specifically associated with plate boundaries, that could be taken to mean that away from those boundaries, benthic life would be scant. Either way, he set aside the biological concerns and became the leading US scientific advocate for seabed waste disposal.[38]

He called his idea the MPG option, for midplate, midgyre (fig. 8.3). One could deposit wastes in a well-understood portion of the abyssal ocean, he argued, in locations that were central with respect to two critical variables: plate boundaries and oceanic gyres. Plate boundaries were tectonically unstable, and the edges of ocean gyres were oceanographically unstable. At either boundary, materials placed there might be redistributed. But the centers of plates and gyres were like the eye of a hurricane: the calm in the middle of tectonic and sedimentological storms.[39]

Hollister built directly on the bathymetric work of Heezen and Tharp

A portion of the Pacific Ocean Floor, showing the mid-plate, mid-gyre (MPGs) study areas for the disposal of high-level nuclear wastes. These locations are purely research stations and do not represent any official citing decisions. The symbol :— represents the depth in feet below sea level. (© 1969 National Geographic Society)

Figure 8.3 Hollister's proposed "MPG"—midplate, midgyre—solution for deep disposal of nuclear waste. Hollister illustrates his proposed sites against the backdrop of the physiographic map of the Pacific seafloor, developed by his graduate adviser Bruce Heezen and Marie Tharp.

From Hollister, "The Seabed Option," *Oceanus* 20 (1977): 18–25. Image credit: Heinrich Berann/ National Geographic Creative; reprinted with permission of NG Image Collection.

(chapter 5)—defining suitable locations against the backdrop of the physiographic map of the seafloor—and on recent advances in physical oceanography, which helped to identify areas of the seafloor least subject to significant deep circulation. It was a state-of-the art solution to a thorny practical problem, at least from the scientific standpoint.

By the mid-1970s, Hollister had assembled a team of colleagues to aid in preliminary site selection, getting advice and support particularly from Woods Hole chemist Vaughan Bowen, who had for many years received support from the AEC for research on the fate of radionuclides in marine environments and was advising Sandia on site-selection criteria for waste disposal. Unlike Hollister, Bowen did not believe that radionuclides delivered to sediments could be assumed to stay put. On the contrary, he thought that "prediction of the movement of any released waste would be uncertain," so a better understanding of the sediment column as a natural barrier was needed.[40] Hollister identified two potential sites in the north-central Pacific for more detailed study.

In 1979, he wrote a long, detailed paper for a symposium on Marine Sciences and Ocean Policy at University of California, Santa Barbara. By this time, nuclear generating capacity in the United States was growing more slowly than predicted, mostly because of high costs, although to some extent also because of public opposition, which increased after the 1979 Three Mile Island accident. Hollister nevertheless argued that safe waste disposal was the principal impediment to expanded capacity, and there was a simple, effective solution waiting to be pursued: the MPG solution. To this he now added an additional component: find a place where the abyssal muds were deep and emplace the wastes within them. With their low permeability and high adsorption capacity, the muds would ensure that any radioactive materials that leaked from their containers would be adsorbed onto the muds.[41] Hence the emphasis on subseabed disposal: waste would not be dumped on the seafloor but encased in the abyssal mud.

The Nuclear Waste Policy Act of 1982

Hollister's scientific research was part of a broad program by the US government to evaluate waste disposal options, which accelerated in 1982 with the passage of the Nuclear Waste Policy Act. This act charged the Department of Energy with developing a permanent repository for high-level wastes by 1998. Mined geological repositories would be pursued as the primary disposal mechanism, but alternatives would be researched as well.

The DOE formalized its efforts to evaluate subseabed disposal. A 1981 overview suggested that feasibility studies could be completed by 1988 and an engineered system "designed for demonstration in the 1990s."[42] The work would involve basic scientific research into the abyssal environment, but the goal was to identify a suitable site for waste disposal consistent with that time frame. Hollister, however, had already been working on exactly this. A contract from 1978 gives an idea of its tenor: First, Hollister would serve as the coordinator of the US Seabed Program Site Suitability and Selection activities and task-force leader of site selection for the International Seabed Working Group. This would be the "principal" task in fiscal year 1979 and would include site investigations in the eastern North Atlantic and the South Pacific and South Atlantic. By the end of the year, DOE would receive preliminary site-suitability assessments based on both historical and new data. Second, Hollister would develop a "paleo-environmental/predictability model" of the sites. Using information obtained from core samples, he would create a model of environmental conditions over the previous ten thousand to one hundred thousand years and use that to assess probable future conditions at the sites. This would "provide a framework for ultimate model spanning 1,000,000 to 10,000,000 years."[43] Third and fourth, Hollister would complete a report on "generic site suitability criteria" for low-level waste disposal from naval reactors, working in conjunction with Sandia scientists and engineers and other federal personnel, and identify generic study regions in the western North Atlantic on the basis of these criteria. The sites would be at least "100 km on a side and no more than 200 miles [sic]" from the US coastline. Fifth, he would develop a "research and development plan for describing the present environment" of the sites, with a focus on characteristics that would allow their ranking or exclusion.[44] He would also provide predictions of "possible processes that might affect the site suitability for the next 100 to 1000 years," based on creating "a complete record of the depositional environment over the past 10,000 to 1,000,000 or more years" that would offer "the basis for assessing a site's usefulness as a repository."[45]

Matters did not, however, proceed as he expected. Hollister's Sandia first-year funding had amounted to nearly $100,000 ($99,877 to be exact), but in the years to follow, the amounts would fall off dramatically. In 1979–1980, his funding was cut in half, and in 1981 to a mere $8,000 (fig. 8.4).

Many people had serious reservations about ocean disposal, not just for the reasons raised in the 1950s. Hollister's idea, endorsed by many at Sandia, was that by burying the wastes in the sediments, rather than simply placing or dumping canisters on the seafloor surface, disruptions associated with deep currents and chemical corrosion by salt water could be avoided. In this respect, he was revisiting an idea proposed by Allyn Vine in 1949 to load wastes into torpedoes and drop them from surface ships; the acceleration of their free fall would propel them into the soft muds of the seafloor, where they would remain (fig. 8.5). Were the capsule to leak, crack, or corrode, the low permeability and high adsorptive capacity of the muds would prevent radionuclides from dispersing.

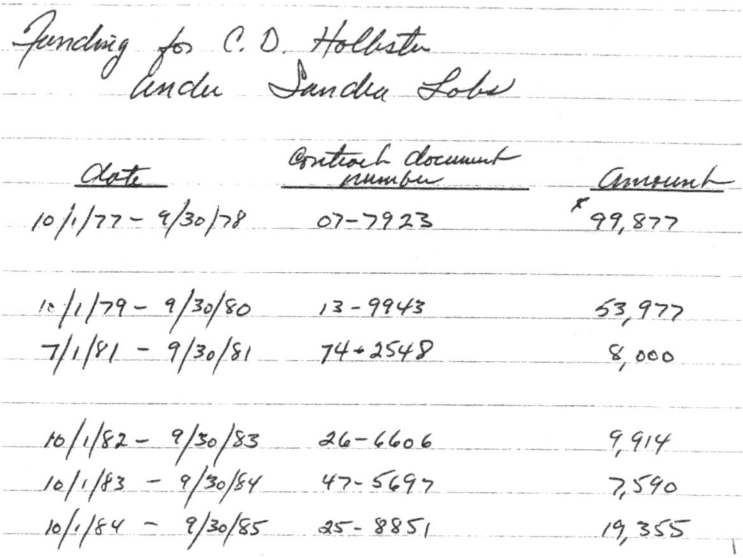

Funding for C. D. Hollister
Under Sandia Labs

Date	Contract document number	Amount
10/1/77 – 9/30/78	07-7923	99,877
10/1/79 – 9/30/80	13-9943	53,977
7/1/81 – 9/30/81	74-2548	8,000
10/1/82 – 9/30/83	26-6606	9,914
10/1/83 – 9/30/84	47-5697	7,590
10/1/84 – 9/30/85	25-8851	19,355

Figure 8.4 Charles Hollister's summary of research funding under Sandia Laboratory Contracts, 1977–1985.

From Charles Davis Hollister Papers, 1967–1998, MC-31, "Proposals, 1960s–1970s," Data Library and Archives, Woods Hole Oceanographic Institution. Reprinted with permission of the Woods Hole Oceanographic Institution.

The snag was that marine disposal of high-level radioactive waste was prohibited by the London Dumping Convention on the Prevention of Marine Pollution by Dumping of Wastes and Other Matter, adopted in 1972. The US Marine Protection, Research, and Sanctuaries Act of the same year, informally known as the Ocean Dumping Act, also prohibited it.[46] In 1983, a moratorium would prohibit even low-level waste disposal at sea, pending further scientific, technical, and social-scientific studies.[47] Although subseabed disposal might be technically sweet, it was also almost certainly illegal.[48]

Hollister and other advocates of subseabed disposal tried to argue that their proposal was not dumping: it was insertion, as the waste materials would be emplaced into the seabed, not dumped onto it. Hollister suggested that the use of sites beyond national borders would in fact simplify the legal issues, because it "minimizes participation by states and localities." Hollister had a point: state and local opposition was emerging as an issue in land-based disposal. His implication, however, that the open ocean was beyond regulatory reach was silly. Whether the proposal was to place waste on or in the seabed, it would fall within the jurisdiction of the London convention, as well as international maritime law, which nearly all experts thought it would violate.

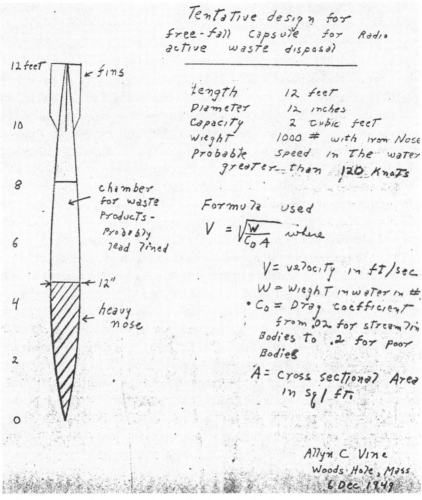

Figure 8.5 Al Vine's 1949 drawing for the disposal of radioactive wastes in free-fall capsules.
From Charles Davis Hollister Papers, 1967–1998, MC-31, "A. Vine design for active waste disposal capsule, December 1949," Data Library and Archives, Woods Hole Oceanographic Institution. Reprinted with permission of the Woods Hole Oceanographic Institution.

The 1982 Nuclear Waste Policy Act required the DOE to "develop and manage a system" for the permanent disposal of nuclear waste. It also established the office of Civilian Radioactive Waste Management, under the DOE, to evaluate at least three potential sites for a permanent geological repository. These activities would be funded through a fee on commercial nuclear power generation—the Nuclear Waste Fund.[49] The DOE chose nine sites in six states for evaluation, but work fell behind schedule and budget as the agency encountered substantive public opposition. In 1987 the Congress directed the DOE to focus on only one site: Yucca Mountain, on the edge of the Nevada Test Site, a location that fell under federal jurisdiction.[50] The subseabed disposal option was not eliminated by this move—in 1983 the DOE had presented to the Office of Management and Budget a five-year budget for subseabed research, and Hollister's funding situation improved (fig. 8.4). But in each of the two following years, allocated funds were less than originally budgeted, and for 1986 the DOE allocated no funds at all.[51] In February 1986, the Office of Civilian Radioactive Waste Management officially terminated the subseabed disposal project.[52] The reason for DOE's decision was explained to Hollister by Edward Boland, congressman from Massachusetts' second district. It was simple and straightforward. Politically, seabed disposal was going nowhere, so there was no point in further research. In his words, there "is no constituency for, and a lot of opposition to, disposing of high-level wastes at sea."[53]

While some geologists wondered about the wisdom of choosing a tectonically active site with a history of recent volcanism, the Nevada option was consistent with what many scientists had argued all along: that the ocean was too poorly understood to dispose high-level wastes there.[54] It was also consistent with the engineering argument that it would be easier to engineer and monitor a terrestrial environment. In any case, the government had made its decision and scientists would have to accept it. In a memo to the project scientists, the program director at Sandia Laboratories wrote, "I request that all participants accept the fact that [the Office of Civilian Radioactive Waste Management] will not continue to fund the project. Furthermore, I hope that no one will do anything to question the wisdom of this decision."[55] But Hollister did question it. More than that, he actively fought it, not simply as a scientific decision but as a political one.

From the mid-1980s until his death in 1999, Hollister wrote numerous popular articles and letters to editors defending seabed disposal and attacking the proposed repository at Yucca Mountain. He made television appearances, including on *Firing Line* with William F. Buckley. He wrote letters to editors of leading magazines and newspapers. He enlisted Woods Hole trustees to support his efforts. He cultivated members of Congress, and when that did not work, he organized a lobbying effort, under the guise of "education."

Hollister Goes to Washington

Hollister's earliest notes on the matter date from 1984, when he corresponded with the staff of New Jersey congressman William J. Hughes, who had taken up the question of whether the London convention prohibited subseabed disposal of nuclear waste. In November 1983 (on the same day that the Hon. Bill Richardson regaled House colleagues with news of the Espanola, New Mexico, High School marching band, which was preparing to represent the state in the Rose Bowl parade), Hughes had introduced a resolution in Congress that the United States recognize that subseabed disposal fell within the legal framework of the London Dumping Convention and was prohibited by both the convention and by US law.[56] Cosponsored by Gerald Studds of Massachusetts, Claudine Schneider of Rhode Island, Barbara Boxer of California, and eight others, the resolution urged that the United States recognize "that subseabed emplacement falls within the framework of the London Dumping Convention, which defines 'dumping' as any deliberate disposal at sea of wastes," and affirm that "the Convention not be eroded by carving out special exceptions from its provisions as new technologies are developed."[57]

Hughes's resolution was motivated by an impending consultative session of parties to the convention.[58] The US State Department was hedging on whether subseabed disposal was or was not prohibited; Hughes wanted the department to state unequivocally both that the London Dumping Convention was the appropriate regulatory framework and that it prohibited high-level nuclear waste disposal at sea, whether dumped, dropped, inserted, encased, or emplaced. Until such time as it was changed by international agreement, the convention was the relevant rule on the matter and the United States should not act unilaterally to reinterpret it.[59]

The State Department responded to Hughes by saying that it was not prepared to take a position on whether subseabed disposal was or was not dumping—insisting that international legal opinion was divided—but that it would agree that the London Dumping Convention was the appropriate forum in which to decide the matter. The department proposed the "consensus position" that "no such disposal should take place unless and until it is proved to be technically feasible and environmentally acceptable, including a determination that such waste can be effectively isolated from the marine environment, and a regulatory mechanism is elaborated under the London Dumping Convention to govern the disposal into the seabed of such radioactive waste."[60]

The State Department's position might to a politician have seemed plausible, but it created a double bind: no one could demonstrate the technical feasibility of the subseabed disposal option without pursuing it, but that seemed to be prohibited by the convention, and the department would not advocate modifying it when the technical feasibility remained in doubt. Hollister decided to try to undo this double bind. In 1987, he and a group of colleagues organized the Seabed Association. Its purpose would not be to ask for money for scientific research—which, for better or worse, scientists routinely do.[61] Nor would it be to offer technical testimony when asked—something scientists also routinely do. It would be to advocate for subseabed disposal, including the possibility of modifying the London Dumping Convention.

The approach was overtly political, its primary focus Congress. Donations were solicited from both individuals and institutions; the latter included Lamont, the University of Washington and the University of Hawaii, and Woods Hole agreed to "match the average contribution (up to $5000)." A memo accompanying a draft charter and position paper for the Seabed Association declared, "WE NEED MORE MEMBERS. PLEASE SOLICIT THE SUPPORT OF YOUR COLLEAGUES."[62] With these funds, Hollister and his colleagues began to collect information on members of Congress who might be sympathetic to their views or whose support would be essential for success, as well as their staff.[63] They especially focused on the Nevada delegation, whom they hoped for obvious reasons to enlist as allies. One file, labeled "Congressional Liaison," contained the business cards of the staff to Nevada senator (and later Senate majority leader) Harry Reid and Nevada governor Richard Bryan, whom Hollister had already met, with notes on when he had met them. Where politicians were up for reelection, the notes mentioned whom they were running against.

Hollister's closet ally in the project was a man named John Kelly, the founder of Austin-based JK Research Associates. Kelly was named chairman of the Seabed Association, and he took the lead on making contacts in Washington, cultivating staffers, and arranging meetings. To help get the project started, he lent the group money out of personal funds; later he billed $5,389 in "invoices payable" for his firm's efforts.[64] In a diagram outlining the "key players," Kelly was identified as "John Kelly, lobbyist" (fig. 8.6). Charles Hollister had made the decision to lobby Congress on behalf of nuclear waste disposal at sea and to pay for a professional lobbyist.[65]

Hollister already had extensive contacts in Congress, in part through Woods Hole associations and in part through earlier congressional testimony on science funding. One of these contacts was Massachusetts senator Edward M. Kennedy.[66] In 1984, Kennedy had visited Woods Hole, where Hollister and others had "made a compelling case . . . for discretionary funding for innovative research" separate from the NSF peer-review process—a position Kennedy later supported in congressional hearings with NSF director Erich Bloch.[67] Hollister had argued that too much money was controlled by the "conservative peer-review process," which (he felt) encouraged "low-risk, tried-and-true approaches." In his view, more money should be placed at the discretion of laboratory directors; this would help to foster bolder, more cre-

KEY PLAYERS

SCIENTISTS

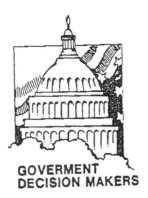

Seabed Association
    Dr. John Kelly - Lobbyist

Woods Hole Oceanographic Institute
    Dr. Charles Hollister

Other Oceanographic Institutions

GOVERMENT
DECISION MAKERS

CONGRESS

Pro -

    Senator James McClure (R-Idaho)

Con -

    Senator Bennett Johnston (D-Louisiana)

INDUSTRY

Edison Electric Institute

American Committee on Radwaste Disposal
(ACORD)

Figure 8.6 Diagram outlining the "key players" in the subseabed disposal program. Note the description at the top: "John Kelly, lobbyist." This was prepared for a DOE report on the Office of Subseabed Disposal and sent to Charles Hollister with the notation "FYI Here is a briefing package I would use if I were called upon to brief senior management."

From Walter Warnick, Office of Subseabed Disposal Research, ER-80, Briefing package, in Charles Davis Hollister Papers, 1967–1998, MC-31, "Congressional Contacts, 1988," Data Library and Archives, Woods Hole Oceanographic Institution. Reprinted with permission of the Woods Hole Oceanographic Institution.

ative initiatives. Innovation, he asserted, was the product of "individuals with vision, genius, and the courage to take risks."[68] No doubt Hollister considered himself such an individual; his approaches to Congress were certainly risky.

At first things went well. Hollister was invited to testify in July 1987 before the Senate Committee on Energy and Natural Resources, where he spoke at length. The committee provided a three-page list of follow-up questions, mostly focusing on legal and political aspects. Hollister was asked, for example, to outline the relevant international legal framework; to explain which international laws would have to be amended to allow subseabed disposal; and to describe the positions of Japan, the Soviet Union, and other governments with Pacific coastlines. One question posed was, "Do all the governments with Atlantic coastline support sub-seabed disposal in the Atlantic?" Other questions addressed engineering and sociological aspects: What kind of ship would bring fuel to the site? What would happen if there were an accident? How much would it cost if the United States adopted subseabed disposal as its primary means of disposing of spent fuel?

In addressing these questions, Hollister was well outside his wheelhouse, but he nevertheless expounded at length. Many of his answers could have been challenged by experts in relevant fields. He asserted, for example, that because the UN Convention on the Law of the Sea affirmed the freedom of the seas for peaceful purposes, it therefore "sanction[ed] subseabed R&D activities"—an artful interpretation at best but also irrelevant because the United States was not a party to the convention. As for the impact of a ship, Hollister acknowledged that the port facility would require "emergency response capability" but asserted that, even if materials escaped from a seabed repository, it would cause no harm to humans. He left the question of harm to other species unasked and unanswered.[69]

While Hollister was not lobbying for the passage of a specific bill pending in Congress, he and his Seabed Association Colleagues were lobbying to convince members to support specific appropriations. In a letter to oceanographer John Byrne, president of Oregon State University, Kelly summarized the state of affairs and suggested how to address them through the appropriations process: They faced several major obstacles. Although the Nuclear Waste Policy Act of 1982 did not prohibit the investigation of alternative disposal methods, the earmarked funds derived from a fee on the nuclear power industry could be used only to establish an actual repository—a provision that the nuclear power industry strongly supported. However, the House Appropriations Committee opposed funding for subseabed disposal from any source, viewing it as a distraction from repository development. Kelly believed that the House position was deliberately designed to eliminate alternatives—that members believed the only way to site a repository was to "eliminate all alternatives and force a state to accept [it]." He therefore suggested that the group stress that the success of the land-based proposal was by no means certain. Nevada could not be forced to accept Yucca Mountain, and the political issues facing seabed disposal were therefore no worse than those facing land-based disposal.

Kelly might have been right about the House strategy, and he was certainly right that the success of Yucca Mountain was not assured. But he was wrong to claim that the political issues facing seabed disposal were no worse than those facing Yucca Mountain, because the subseabed required not only agreement by the relevant parties in the United States, but also broad international agreement.

In suggesting that Yucca Mountain would fail, moreover, they were attempting to construct a self-fulfilling prophecy, and this was tricky territory in opposing Yucca Mountain, Hollister and his colleagues were opposing their own former (and potentially continued) patron, the DOE, which was on record as opposing the continuation of the subseabed program. To attempt to discredit land-based disposal was to cast doubt on the patron whose support they needed and on whom they had previously depended. How could they win by alienating DOE? Their answer was that funds should be found that did not compete with the Yucca Mountain program. Kelly explained: "We have tried to make peace by presenting subseabed ... as a complement to an overall waste management strategy."[70] But this was disingenuous, because in other contexts they were actively opposing Yucca Mountain, working to *ensure* that it would fail.

Kelly insisted that if only people knew more oceanography, they would view the matter as he did: "Many people simply do not believe that subseabed disposal will work. This is symptomatic of a broader lack of knowledge and understanding of oceanography."[71] Kelly urged Byrne to talk with Oregon senator Mark Hatfield to try to "revitalize his support," and suggesting that the appropriations bill "would be the appropriate vehicle."[72] But it was not just general support they sought; Kelly and Hollister had a specific ask. That was to add language to the appropriations bill to establish an office for subseabed research within the Civilian Radioactive Waste Management Program and to create a university-based consortium to conduct that research. The consortium would include Woods Hole, Scripps, and Lamont, as well as Oregon State University, the University of Washington, the University of Rhode Island, and Texas A&M—all schools with strong oceanography programs.[73]

Kelly also developed a memo in which specific language for the appropriations bill was suggested, and he prepared a briefing for Warren Rudman of New Hampshire and several other senators whose states were potential targets for a second repository. By supporting the marine alternative, he suggested, they could avoid the political risk of a land-based repository in their home states.[74] While the scientists publicly insisted on their nonpartisan commitment to apolitical scientific research, Kelly's notes for the organization of the proposed consortium show otherwise. Under qualifications for the chairman of the consortium and director of the Office of Subseabed Disposal, they wrote "Republican or clearly nonpartisan."[75]

In September 1987, Kelly testified to the House Committee on Interior and Insular Affairs Subcommittee on Energy and Environment, representing the Seabed Association. His eight-page statement covered what was by then familiar ground—how seabed disposal was safe, how it was not dumping, and why it should be pursued as a backup to Yucca Mountain by providing an "inexpensive hedge against possible failure of the very costly land repository program." The termination of the scientific research had had nothing to do with its technical merits (or lack thereof), he insisted, but was based on the desire on the part of DOE to close off alternatives: "The subseabed disposal project is perceived as a potential threat to the land repository program. There is a strong feeling that continued research on subseabed disposal would divert the effort to site a land repository and that the only way to force a people to accept a land repository is to cut off all alternatives."[76]

Kelly argued that it was still possible to identify and characterize three marine sites (one in the North Atlantic, one in the North Pacific, and one in the US Exclusive Economic Zone), develop designs for a seabed disposal system, assess the potential environmental impacts, and educate the public about the option of subseabed disposal by 1995—the target date under the law for identification of the first repository.[77] Kelly asked that monies be authorized from the Nuclear Waste Fund, because "the ratepayers and utility industry will be the primary beneficiaries and because it will be politically impossible to secure adequate funds from general revenues."[78] On this last point, at least, Kelly was right.

In October, Kelly compiled a status report for the members of the Seabed Association. From February to September, they had expenses of $22,356, including Hollister's and Kelly's testimonies to Congress, bank charges, legal fees, office supplies, printing, telephone, postage, and the biggest item: "management labor," or his salary. The report noted with satisfaction that every piece of pending legislation on radioactive waste had a subseabed component, and the signs were good for both the House and the Senate to pass bills with a provision for the program. On a separate page, he listed all the potentially relevant pending legislation—seven different bills in all (fig. 8.7).[79]

LEGISLATION

"High-Level Radioactive Waste Storage Act of 1987," (S.1266), introduced by Senators Evans (R-WA), Murkowski (R-AK), and Hecht (R-NV).
> Calls for regional storage facilities and has a strong subseabed section that establishes the university consortium, sets objectives, and authorizes funding for the consortium ($150 million over 8 years).

"Subseabed Nuclear Waste Disposal Research Act of 1987," (S.1428), introduced by Senator Hecht (R-NV).
> Establishes the Office of Subseabed Disposal Research within the Office of Energy Research of DOE, establishes the university consortium, and sets objectives.

"Nuclear Waste Policy Act Amendments of 1987," (S.1668), introduced by Senators Johnston (D-LA) and McClure (R-ID).
> Reduces the number of land sites to be characterized for a mined repository from 3 to 1, provides incentive payments of $100 million per year to the host state, and calls for a report on subseabed disposal within 9 months. S.1668 is the primary legislative vehicle.

"Nuclear Waste Policy Act Amendments of 1987," (H.R.2888), introduced by Mr. Udall (D-AR) and others.
> Establishes a Nuclear Waste Policy Commission to review the program, including implementation of section 222 (alternative methods such as subseabed), and make recommendations to Congress in 12 months.

"Nuclear Waste Policy Act Amendments of 1987," (H.R.2967), introduced by Mr. Udall (D-AR).
> Similar to H.R.2888, and also creates a super Negotiator to facilitate selection of repository sites. H.R.2967 is the primary legislative vehicle.

"High-Level Radioactive Waste Storage Act of 1987," (H.R.3077), introduced by Mr. Morrison (R-WA).
> Regional storage facilities and a strong subseabed section that establishes the university consortium, sets objectives, and authorizes funding for the consortium ($150 million over 8 years). Similar to S.1266.

"Subseabed Nuclear Waste Disposal Research Act of 1987," (H.R.3499), introduced by Mr. Jones (D-NC), Chairman of House Committee on Merchant Marine and Fisheries.
> Establishes the Office of Alternative Disposal Methods in the DOE Office of Civilian Radioactive Waste Management, establishes the university consortium, sets forth objectives, authorizes funding for the consortium ($150 million over 8 years) and for NOAA ($50 million over 8 years) to provide technical support. This is the best bill for subseabed.

Figure 8.7 John Kelly's list of pending legislation in Congress relevant to the future of subseabed research. Note the comment on the last bill listed, H.R. 3499: "This is the best bill for subseabed." Hollister, Kelly, and their colleagues were not only tracking the issue in Congress and "communicating" with members about the scientific background but also expressing preferences for specific legislative action.

From Status Report, 21 October 1987, The Seabed Association, in Charles Davis Hollister Papers, 1967–1998, MC-31, "Congressional Contacts, 1988," Data Library and Archives, Woods Hole Oceanographic Institution. Reprinted with permission of the Woods Hole Oceanographic Institution.

The Nuclear Waste Policy Amendments Act of 1987

At the end of 1987, Congress passed the Nuclear Waste Policy Amendments Act, which established the Nuclear Waste Technical Review Board as an independent agency within the Executive Branch to review the technical and scientific validity of Department of Energy efforts to find a permanent repository for radioactive waste.[80] Buried in this bill was the provision for which Hollister and Kelly had lobbied: the establishment of an office for subseabed research within the Civilian Radioactive Waste Management Program and the creation of a university-based consortium to conduct the research. Section 224(b) allowed for the funding of research through the university consortium: "the Secretary may make grants to, or enter into contracts with, the Subseabed Consortium ... and other persons.[81] The scientists had got what they wanted—or so it seemed.

In January 1988, Woods Hole chemist Derek W. Spencer (1934–2006) wrote thank-you letters to Edward Kennedy, John Kerry, Gerry Studds, and several others who had supported the effort. The consortium would have two goals: first, to build scientific consensus on the feasibility of subseabed disposal, and second, "the task of educating the public about subseabed disposal." One might argue that research is research only if it is open ended—that is, if the results are not known in advance—so any effort to "build consensus" on a technical question is, ipso facto, epistemically inappropriate. But Spencer made clear that his goal was not epistemic but political: "With the decision to characterize only the Nevada site, subseabed disposal becomes the primary back-up if the land program falters, as it has to date."[82]

Spencer's comment—"as it has to date"—betrayed their view, or perhaps their hope, that the Nevada site would fail and the US government would be forced to turn to the subseabed option. But even their supporters were not sure they actually wanted seabed disposal to go ahead. One of them was the junior senator from Massachusetts, John Kerry, whose staff indicated that the senator would support them, but only if they were "very careful to state that it is a research issue and that the possibility of nuclear waste disposal in the ocean is very small."[83] In any case, approving something in law is one thing, funding it is another, and had the scientists had been ever so slightly more alert (or less self-confident), they might have noticed a problem: the Budget Reconciliation Act of 1987 allocated no funds for the subseabed research program. Section 224(b) permitted the secretary to make grants and enter into contracts, but it did not require him to. In contrast, Section 224(a) required the secretary to report to Congress within 270 days on the state of knowledge regarding seabed disposal, including the legal framework and likely costs, and to make a recommendation regarding its future. But 270 days was hardly adequate time for the kind of research program the scientists had in mind.

When they realized what had happened, Hollister and Kelly hastily contacted DOE officials. Someone at DOE was sympathetic enough to request congressional permission to transfer funds from other DOE science activities, but it was denied. Robert Roe, chairman of the House Committee on Science, Space, and Technology, suggested that if the DOE were anxious to maintain this program, it should take the funds from "other program funds within the Department ... such as, for example, the Departmental Administration account."[84]

That idea was not taken up and Hollister continued to try new angles. Hoping to entice Louisiana senator Bennett Johnston into his camp, he invited scientists from Louisiana State University to join the consortium. He reached out to Guy Nichols, chairman of Woods Hole's board of trustees, who knew Johnston.[85] On Woods Hole letterhead, Nichols wrote: "My friends in Washington tell me you have an excellent chance (and are the obvious choice) to be the next Senate Majority Leader. I expect a sizable war chest available to assist other Democratic hopefuls in their campaign efforts, and in your own next election, will help achieve this goal. This leads me to suggest a Bennett Johnston fundraiser here in Massachusetts is very much in order."[86] While Nichols worked the legislative angle, Hollister worked other contacts, including the godfather of all things nuclear, Edward Teller. Hollister asked if Teller might not approach his contacts in the nuclear power industry to persuade them that it would make sense to divert a small portion of the Nuclear Waste Fund to the subseabed option. Teller declined, explaining his own view was that the waste problem should be solved through reprocessing.[87]

Hollister Goes Public

Hollister blamed the turn of events on the nuclear power industry. Committed to bringing a repository on line as fast as possible, they had opposed any option that might be interpreted to cast doubts of the safety, feasibility, or desirability of Yucca Mountain. Still, based on conversations he and Kelly had had with industry executives, Hollister had thought that they would accept his argument that seabed disposal could be a complement, not an alternative, to Nevada. He was enraged to discover that was not so.

Hollister penned an intemperate letter to Charlie McNeer, an executive at the Wisconsin Electric Power Company and representative of the Edison Electric Institute, with whom he and Kelly had talked at length and on whose support they had mistakenly thought they could count. The utility group's position had been that it would allow subseabed disposal to go forward if "it could not in any way be seen to divert attention of funds from aggressive implementation of the current nuclear waste disposal program."[88] Hollister had read this as permissive support; he felt duped.[89]

"I feel as though I have been torpedoed," he wrote in a draft letter to McNeer. "The utility industry ... has effectively blocked implementation of Section 224. This will, I think, turn out to have been a grave mistake," he threatened. He continued:

> If, by cutting off the subseabed option, the utility industry hopes to foster a "license at any cost" mentality that will force [the Nuclear Regulatory Commission] and [Environmental Protection Agency] to approve the Nevada site even if there are significant technical uncertainties, they are mistaken.... With no back-up disposal method, failure to license the Nevada site will have severe consequences for the energy security of the United States. With no solution to the waste problem in sight, the nuclear power option will be foreclosed, and the international market in nuclear technology will be handed to France, Japan, and other nations. As the price of oil goes up like a rocket in the next decade and concern about the greenhouse effect restricts the use of coal, the United States (and the utility industry) will face a crippling crisis. Is this risk worth the dubious advantage of killing subseabed disposal?[90]

Hollister sent a draft of the letter to John Kelly for comments, who scrawled on it: "Don't send. Let's talk."[91]

Hollister was right about the Nuclear Regulatory Commission and the Environmental Protection Agency—who in the end would not approve the Yucca Mountain site—but he had willfully misread the industry position. In September 1987, Kelly had given a presentation to the Utility Nuclear Waste Management Group (UNWMG), part of the Edison Electric Institute, attempting to persuade the group to support subseabed disposal. In response, the group published a position paper whose conclusions and recommendations were unambiguous:

> The UNWMG has concluded that:
>
> • Any activity, funded by the Nuclear Waste Fund, that is not directly aimed at furthering the geologic disposal program undoubtedly will delay the geologic program and result in corresponding program cost increases. It is crucial to maintain funding and focus for activities related to the work at Yucca Mountain....

- ... For any fiscal year, Congress will appropriate only so much money from the Nuclear Waste Fund. Funding of subseabed disposal will reduce funding for the geologic disposal system.
- Congress has wisely prohibited the use of funds from the Nuclear Waste Fund for activities unrelated to geologic waste disposal ... Subseabed disposal is not geologic disposal.... Based on review of the subseabed disposal concept, the UNWMG recommends that:
- Subseabed disposal research should not be funded from the electricity consumer-funded Nuclear Waste Fund.
- Congress should be urged now and in the future to resist requests to fund subseabed disposal research from the electricity-funded Nuclear Waste Fund.
- Should Congress decide to pursue such research, funds should be provided from general appropriations.[92]

Congress had followed these recommendations precisely.

Given the consensus in favor of Yucca Mountain among Congress, the nuclear power industry, and the DOE, as well as opposition to marine disposal by environmental groups and legislators from many coastal states, Hollister might have conceded that, technically attractive or not, seabed disposal was simply not going to happen. Instead, his position became even more entrenched and his arguments more overtly political. The Seabed Association claimed that it wished to preserve the subseabed option should the terrestrial repository program fail, but in fact Charles Hollister was *trying* to make the terrestrial option fail.

Teaming Up with William F. Buckley

In June 1991, Hollister appeared on the television show *Firing Line* with William F. Buckley, the well-known provocative conservative commentator. In the episode "Is There a Problem of Nuclear Waste?," Hollister answered there was not, so long as the United States embraced seabed disposal. In the press release for the show, to be moderated by Michael Kinsley of the *New Republic*, the promoters effused: "Probably no other public affairs program devoted to nuclear waste has ever been fun.... Mr. Hollister is a relaxed and genial guest."[93] But if he appeared relaxed on television, he was anything but in private. Hollister was angry and determined, and he and Buckley had been working behind the scenes to get his message across.

Hollister had helped Buckley with an editorial that appeared on June 7, 1991: "Nuclear Waste: A Solution." It began by introducing Hollister, a "cowboy"-turned-scientist who knew a heck of a lot about mud—more, in fact, than just about anyone. When one listened to Hollister, Buckley gushed, one realized that all the fuss about nuclear waste was just a load of "caterwauling." The answer was literally as simple as dirt: bury the waste in mud, ocean mud. However, this simple answer had yet to be heard because "liberals" preferred to get their answers from Jane Fonda rather than the Woods Hole Oceanographic Institution. If Americans listened instead to a real scientist, a "true-blue oceanographer," this is what they would learn: "About half the Earth is covered by vast underwater fields of clayey mud resembling creamy peanut butter. Miles thick, these muds carpet vast areas of deep-sea basins.... Negatively charged ions on the edges of these extremely fine mud particles are attracted to the positively charged ions of such heavy metals as cadmium, zinc, mercury, iron, magnesium, lead, cesium, and plutonium. This attraction causes the heavy metal ions to stick the mud particles."[94] Americans would all know this were it not for the misinformation propagated by naïve liberal environmentalists. He sent a copy to Hollister for his comments, instructing him not to worry about the "vernacular": "it's the way I talk to The Enemy."[95]

On television three weeks later, Hollister also used the vernacular but eschewed the details of surface chemistry in favor of the politics of waste disposal. Introduced as senior scientist and vice president of Woods Hole Oceanographic Institution, Hollister emphasized that the abyssal ocean was "geologically the dullest place on Earth"—where nothing had happened and nothing would happen for millions of years. If the wastes were put into missile-shaped canisters and dropped appropriately, they would embed themselves in the sticky mud.[96] Of course, no one wanted nuclear waste in his or her backyard, but the abyssal ocean floor was no one's backyard.

Hollister acknowledged that funding for his own research had been ended by the government's decision not to pursue subseabed disposal, but he insisted that this had nothing to with his advocacy for this option.[97] He also insisted that the government decision had nothing to do with either technical feasibility or international law, but solely with the desire of the US Department of Energy to ram forward Yucca Mountain by denying attention to alternatives. One ex–assistant secretary at the Department of Energy, he claimed, had said to him, "Nothing, Charlie, will erode public confidence in our ongoing efforts [more] than if we studied a viable alternative." The DOE, Hollister suggested, was deliberately stifling scientific research for political ends. "The political reality," he concluded, "is that it is not in the interest of the Department of Energy to pursue such an option.... The political reality is that we are hamstrung."[98]

Hollister was right that politics—particularly international politics—played a role in the conclusion that seabed disposal was not feasible, but no one denied that. He was constructing a straw man when he suggested that they did. Moreover, Hollister's defense of the seabed disposal was also political. Although money was originally a factor leading him to try to preserve the seabed disposal program, the issue had gone well beyond that; his continued pursuit amounted to a political campaign. That campaign had a technical foundation, to be sure; Hollister clearly believed that seabed disposal was a technically sound alternative. But on what grounds did he attack Yucca Mountain as unsound?

Yucca Mountain was designed as a "high and dry" repository—above the water table in the Nevada desert—and its reliability hinged on accurate understanding and modeling of the hydrology of the unsaturated zone. What did Hollister know about the hydrology of the unsaturated zone? Not much.[99] His expertise was in marine sedimentology, not hydrology.

The case for Yucca Mountain also relied to a certain extent on the argument that local tectonic activity would not significantly affect the hydrologic conditions; Hollister's background as a geologist gave him insight into general questions of tectonism, but not into the specifics of how tectonics might affect the hydraulic conditions there. Hollister blurred this distinction. In a letter to the Agency for Nuclear Projects in Carson City, Nevada—the state agency charged with overseeing the federal high-level waste disposal program—he wrote, "As a geologist I see absolutely no defense for putting a high level waste repository in the tectonically active basin and range province of Nevada."[100]

Hollister stressed the tectonic unpredictability of Yucca Mountain, a point with which other geologists might have concurred. But in making this claim, he implied that the marine environment was predictable, or at least sufficiently predictable that one could say with confidence that nothing undesirable would happen if wastes were put there. Was he right? Was the ocean environment that well understood, or the technical basis for subseabed disposal so undisputed that it was reasonable for Woods Hole, as an institution, to advocate it? Hollister thought so, but others did not. In a response to the Buckley editorial, and perhaps under pressure from Woods Hole staff who disagreed with Hollister, the Director's Office issued the following disclaimer: "The Institution's duty is to foster the pursuit of objective research and education, not to endorse a particular position.... Most of WHOI's scientists perform basic research, investigations of the fundamental processes in the marine environment. These processes must be understood before we can predict the fate of materials in the ocean. We are still a long way from understanding all the fundamental processes in the ocean."[101] Hollister's actions contradicted this directive: he had endorsed a particular position—for seabed disposal, against Yucca Mountain—and he had done so not merely as Charles Hollister, American citizen, or even Charles Hollister, PhD, but as Charles Hollister, senior scientist at Woods Hole Oceanographic Institution, dean of graduate studies and vice president of the corporation, titles he never failed to invoke in media statements and congressional testimony. After all, as Buckley had argued, from whom did you want to get your answers: Jane Fonda or Woods Hole?

Who Speaks for Woods Hole?

The issue of ocean dumping came to the fore in 1991 with respect not only to radioactive waste but to waste in general. In January, Woods Hole had hosted an international conference supported by the Sloan Foundation, "The Abyssal Ocean Option for Future Waste Management," to examine the feasibility of sewage disposal in the abyssal ocean. Participants included leading oceanographers, geochemists, sedimentologists, and engineers from the United States and Europe. A report of the conference on the front page of the *New York Times* described considerable support for abyssal waste disposal among Woods Hole scientists, including some who dismissed opposition to it as environmental hysteria.[102]

The article described the abyssal ocean as "the planet's most useless real estate"; elsewhere, Scripps geochemist Edward Goldberg would call it "waste space."[103] It noted that scientists were "pushing ahead with an unpopular proposal"—specifically, a request to Congress to amend existing law banning ocean dumping to allow for one million tons of sewage sludge to be shipped to two sites that were three hundred miles offshore, about halfway to Bermuda. The point, they insisted, was to track and monitor the effects to better understand the fate of toxics in the deep ocean, but environmentalists viewed this as a loophole that could surely be used for other purposes once opened.

Environmentalists erupted. The matter had been settled, they argued, when Congress enacted the Ocean Dumping Ban Act in 1988, so why were Woods Hole scientists trying to unsettle it?[104] Why was Woods Hole advocating dumping hazardous wastes at sea? The reaction created considerable consternation in the Director's Office, which quickly disclaimed the charge. The scientists were "not advocating ocean disposal," Woods Hole's director Craig Dorman insisted in a letter to the editor of the *New York Times*; "they are concerned that an option for a particular and very problematic part of our total waste stream not be closed without adequate knowledge." This was a research conference, he insisted, to investigate the issue from scientific and technical perspectives. He closed his letter by underscoring the distinction between research and advocacy, inquiry and lobbying: "It is important to distinguish between a research plan and advocacy.... It is also important to distinguish between scientific inquiry and lobbying. While some scientists may be asking the government to amend the anti-dumping law to permit controlled research, the Woods Hole Oceanographic Institution, as a private, non-profit research institution, is not an advocacy organization and has not sought such a change."[105] This was true: Woods Hole had not sought such a change. But Hollister had, and he had done so using the institutional imprimatur. So where was the line? By law, Woods Hole scientists could seek to influence legislators through educational activities, but the institution could not seek to influence legislation without risking its nonprofit status.[106] According to Internal Revenue Service rules, "An organization will be regarded as attempting to influence legislation if it contacts, or urges the public to contact, members or employees of a legislative body for the purpose of proposing, supporting, or opposing legislation."[107]

Hollister as a citizen had the right to reach out to elected officials in any way he saw fit, but he had not acted merely as a citizen: he had advocated for seabed disposal in letters written on Woods Hole stationery, in phone calls made from his Woods Hole office, and with money supplied in part by the institution and often in respect of specific pending legislation. Trustee Guy Nichols had written on Woods Hole stationery offering to raise funds for Bennett Johnston with a barely veiled suggestion of quid pro quo.[108] Derek Spencer had made it perfectly clear where he stood. Their actions may have been technically or pragmatically justified—they may even have been right— but they were not apolitical. They were not just asking for money for "controlled research"; they were advocating for radioactive waste disposal at sea and against disposal at Yucca Mountain.

It was not just opponents of ocean dumping who thought so; supporters did, too. They seized on what looked to them to be the institution's advocacy and tried to use it to bolster their position. An editorial in *USA Today*, for example, praised the Woods Hole scientists, naming their request for an exception to the legal ban on ocean dumping as "the Woods Hole plan." In an invited reply to the editorial, staff scientists at the Environmental Defense Fund (EDF) also took the view that the scientists were acting as advocates, stressing that the initiative to create a loophole to the ban on ocean dumping was driven by scientists who wanted not merely to study the problem in the abstract but also to amend the law.[109] Dorman's reassurance that their work was only "research" was hollow, the EDF scientists argued, for what was the point of studying something unless you thought there was a chance of doing it? No one who thought it was a flat-out bad idea advocated further study of it.[110] And Woods Hole scientists were clearly not asking to study it to show how bad it was.

Ecologist George Woodwell (b. 1930) was the director of the Woods Hole Research Center, an organization dedicated to scientific research relevant to environmental questions located in the town of Woods Hole but independent of the Woods Hole Oceanographic Institution. He was deeply disturbed that the Oceanographic Institution seemed to be endorsing deep-ocean disposal, and he wrote to Dorman to register his dismay. Woodwell rejected Dorman's argument that the sponsored conference did not represent advocacy. Clearly, at least some Woods Hole scientists were actively advocating deep-ocean disposal, and the point of the conference was to move things in that direction. Moreover, both the conference and Dorman's defense of it ignored the very things that Woods Hole scientists had discovered about the deep ocean over the previous two decades. True, the quantity of life in the abyss was low, but its diversity was high, and it included many unique life forms. It was shocking to Woodwell that Dorman would be cavalier on this point.

Woodwell told Dorman that he had "tried to see this project in the most constructive and generous light possible" but had found himself "firmly placed among your critics." Woods Hole Oceanographic Institution scientists were using an alleged ignorance about the ocean to justify their call for research, but scientists knew a lot about the oceans, including the life within it: "It is, after all, your own Oceanographic Institution's research … that led to the recognition of the extraordinarily high biotic diversity and low density of life in the abyssal benthos. And the continued revelations of the *Alvin* and its successors in exploring the benthic fauna … constitute one of the most exciting and provocative frontiers of science.… The proposal [for ocean waste disposal suggests] that we do not take seriously our own discoveries about the oceans and flies in the face of what virtually every citizen has learned from us to accept as elementary reality."[111]

This was not the first time that the question of the line between research and advocacy on seabed disposal had been raised. Already in 1976, Vaughan Bowen had taken exception to the tone of a Sandia report that sought to "demonstrate" the viability of deep-sea disposal: "I can't emphasize— evidently—too loudly or too often that we are not involved, and must not represent ourselves as being involved in demonstrating (or in selling) anything! What we are trying to do is to evaluate the suitability and feasibility of the use of the deep ocean for waste emplacement. If we let 'selling' terminology creep into this sort of paper then our scientific credibility disappears just as if we peddled bottles of snake oil!"[112] Bowen criticized an earlier report that had suggested the site was safe for waste disposal without actually examining or monitoring the site. He challenged a number of the report's premises, including the assumption that "all the [radio]available activity is released continuously over a ten-year period after disposal," as well as the assertion that this was a conservative view. Bowen thought it neither true nor conservative. It was more probable, he thought, that in areas of current activity the drums would be leached most quickly in the initial period after dumping. Moreover, the possibility that the drums would be "tumbled about" by currents "could not be dismissed without more detailed study."[113]

Bowen also criticized the assumption that all relevant radionuclide transport would be by diffusion. Again the issue was currents. He wrote: "Such an assumption is untenable in considering an area of bottom current activity as strong as is shown by our samples. It becomes very important to ascertain whether the dump site seafloor is being swept clean by a continuous systematic bottom current."[114] He concluded that it was a mistake to plan disposal of any hazardous materials without detailed oceanographic study of the specific area under consideration and a mistake to dispose of any hazardous materials without monitoring and "careful period reassessment of the suitability of the area and of the behavior of the materials disposed."[115]

What did scientists know about the deep ocean at that time? In 1981, the question was addressed in a special issue of *Oceanus*, "The Oceans as Waste Space?" Pun intended, the idea of the oceanic expanse as either a space for waste or wasted spaced was clearly designed to provoke, or at least capture attention. Lead author and Scripps geochemist Edward Goldberg (1921–2008) wanted people to stop assuming that the oceans were "sacred" and "sacrosanct" and "that any entry of polluting substances is undesirable." He wanted to consider the "profane" position "that the oceans do have a finite [nonzero] capacity to receive some societal wastes."[116]

Goldberg held a view that was commonplace among chemists and engineers in the 1960s and even many ecologists: the solution to pollution was dilution.[117] This was also the position taken by Goldberg's boss, Scripps director William Nierenberg. In 1966, Nierenberg had argued for the ocean disposal of radioactive wastes on the same grounds. In a speech to the American College of Physicians, he insisted that the oceans could absorb "about 100 tons of mixed fission products" while giving rise to "acceptable levels of radioactivity in marine foods."[118]

"Acceptable" was of course a value judgment; Nierenberg's judgment was that the incremental exposure to radioactivity that would be caused by ocean disposal was small compared to what Americans were exposed to from other sources. Perhaps he was right, but the word *acceptable* was often used as a means to discount harm, not by claiming that there would be none—because no one could honestly say that they knew that—but by implying that any harms that might occur would be *inconsequential*. (Later Nierenberg would make the same argument about the atmosphere, first with respect to acid rain and ozone-depleting chemicals and then greenhouse gases, and as the science documenting the harms of these phenomena grew, his arguments became more tortured.)[119] But how "acceptable" was to be defined—and by whom—was never made clear, and by the early 1980s many scientists, particularly biologists, argued that deep-sea disposal was not acceptable.[120]

Goldberg's argument was rebutted by biologist Kenneth S. Kamlet (b. 1946), who took exception to Goldberg's assertion that "we can predict the fate of wastes entering the oceans better that we can predict the fate of wastes introduced on land" and that "effective schemes can be devised for the introduction of highly toxic substances to the marine environment without endangering public health or ecosystems."[121] The science simply wasn't that robust. Indeed, Kamlet argued, it was scarcely developed: "Our ignorance so far exceeds our understanding in the areas of predicting the fate and effects of marine pollutants and of successfully remedying problems once they arise, that it would be irresponsible to presumptively permit ocean disposal of persistent toxic pollutants based on available crude assimilative capacity arguments.... It is foolhardy and wrong to assume that what we do not know cannot hurt us.... What we don't know is almost everything."[122] Ultimately, the argument came down to confidence and the burden of proof: whether one was confident enough about our scientific knowledge to calculate how much waste could be safely absorbed by marine systems—or not—and, in the face of disagreement, who should bear the burden of proof?[123]

Ten years later, oceanographers were still arguing the point, and Dorman was still refusing to concede that the advocates of ocean disposal were advocates—at least in the political sense. One could advocate a position scientifically, he insisted, at a scientific conference, for example, but that was different from advocating it politically. That, of course, was true. In some sense, all scientists are advocates of a sort: they advocate for their data and theories in the halls of scientific conference and the pages of scientific journals. But Dorman was blurring a distinction that most scientists consider quite important: the distinction between arguing in the halls of science and advocating in the public sphere.[124] Hollister and his colleagues were not just arguing with their scientific colleagues; they were attempting to reach into the halls of Congress and, via television, in the homes of the American people. They were not simply defending their understanding of the natural world, or even offering policy advice based on that understanding. They had hired a lobbyist, John Kelly, and created an organization, the Seabed Association, to oppose an existing government program and advocate for an alternative, and they were claiming that their views were warranted by a scientific consensus that did not in fact exist.

As 1991 came to a close, the issue of the line between advice and advocacy was still causing discomfort at Woods Hole. In December, Dorman reiterated the official Woods Hole position in an institutional statement over which his office had labored considerably. The various drafts tried different vocabulary and approaches; the consistent theme was the distinction between inquiry and advocacy, between sharing knowledge and lobbying, and between speaking as an individual and on behalf of the institution.[125] Woods Hole scientists, Dorman asserted, were inquiring, sharing knowledge, and speaking as individuals. They were not advocating, lobbying, or speaking on behalf of Woods Hole.

But where was the line between inquiry and advocacy, or between sharing knowledge and lobbying, or between speaking as an individual and speaking on behalf of the institution in which one worked? Dorman claimed that Woods Hole scientists involved had not "lobbied" to have the laws regarding ocean dumping changed. That was disingenuous at best, for they had hired a lobbyist to try to change appropriations bills to ensure continued funding for the seabed disposal program, and they had made it clear that they wanted to see the law changed. Changing the law, in fact, was a central focus of their efforts.

Consider a December 1990 trip that Hollister had made to Washington with Derek Spencer. They had scheduled meetings with officials at the Office of Technology Assessment, the White House Office of Science and Technology Policy, the US Environmental Protection Agency, the Office of Management and Budget, and the National Oceanic and Atmospheric Administration. Hollister's notes record that one goal of the trip was to collect the names of relevant members of Congress, and their staff, whose support would be needed for any change in law or policy. After meeting with Robert Niblock of the Office of Technology Assessment, he wrote the following note for his file:

> Involve key congressional staffers into any discussion about waste in the ocean [including the new] Chief of Staff for George Brown of House Science and Technology [committee].[126]
> Be sure NOPS (National Ocean Policy Study Subcommittee) of the Senate Committee on Commerce, Science and Transportation is informed and

preferably involved (majority chair is Ernest Hollings, vice chair is John F. Kerry).

Should have same treatment for Senate Committee on Environment and Public Works which has jurisdiction over environment, ocean protection, and toxic substances. Majority chair is Quentin Burdick, Ranking Minority Chair is John Chafee, Chief Counsel for the minority is Steven Shimberg (key contact).

Advice from Niblock is to work these authorization committees first and be sure to stress that the US should maintain a lead in the technology of deep ocean disposal ... Jack White of Energy R& D office for State of New York at Albany should be involved. He is a very savvy and powerful constituent."[127]

In February 1991, Hollister made a return trip to Washington to meet with an assistant to energy secretary Hazel O'Leary. In a memo to Dorman, Hollister explained how the contact with O'Leary had been arranged by a DC contact and "Washington liaison hopeful." (Woods Hole had been considering hiring a professional Washington liaison.) The objective of the meeting was to brief the aide "on the subseabed disposal concept and to obtain an audience with the Secretary on the same subject." The meeting was very successful; the aide "agreed to set up a brief in about 6 weeks' time."[128]

Hollister and Spencer meanwhile prepared a confidential memo for Dorman on the deep disposal concept. It was an explicit plan for deep-ocean disposal—by then renamed deep-ocean isolation—complete with a discussion of a private firm that stood ready to implement it.[129] Denying any scientific or technical uncertainties, the memo asserted that deep-ocean isolation "uses established technology in a cost-effective, low-risk manner and produces no harmful environmental consequences. [It] is a relatively inexpensive and reliable method of directly depositing environmentally isolated processed sewerage sludges onto and into the deep ocean floor where its environmental effect is minimal."[130] This was deceitful. True, the technology to dispose of sewage waste was established, but it had never been used for nuclear waste. Sewage was just not the same.

The man proposed to take on the technical task was one Christian Kongsli of Oslo, Norway, the managing director of Energy Drilling Company of Liberia.[131] "He is an international financier in the offshore energy field, arranging and participating in consortiums to purchase and operate offshore oil-drilling rigs in Europe, the United States and Canada," the memo explained. It detailed Kongsli's proposal to develop deep-ocean isolation based on existing offshore drilling technologies, how much it would cost, and why it would be different than dumping. Derek Spencer had separately been in communication with Kongsli, writing that he was "firmly convinced of the potential of the abyssal ocean for future waste management." His conviction was also that they "would be able to convince the regulatory agencies to adopt [it] as a pilot program."[132] They were proposing a private, for-profit venture in collaboration with a drilling company incorporated in Liberia, a country well known for permitting dubious ships to register under its "flag of convenience."

Hollister's and Spencer's consideration of environmental concerns was more dismissive than ever, insisting that any concerns were based on the "naïve yet prevalent view that all ocean dumping is inherently evil. This monolithic conception of the ocean ignores its vast complexity."[133] Moreover, having in one breath used nearshore sewage disposal to insist on the safety of their proposition, in the next breath they insisted that disposal of nuclear waste in the deep sea was both different and safer: "There is substantial environmental difference between injecting sewage sludge or sewage waste water in the shallow coastal waters of the continental shelf, where the vast majority of sea life resides, and depositing encapsulated sludge onto the deep ocean floor, the abyssal plains, where negligible current, cold temperatures (approximately 2°C), and the absence of sunlight and protein reduce the development of plant and fish life to a fraction of what exists in shallow coastal waters. The placement of large amounts of encapsulated sludge into these deep sedimentary basins would have no effect on the environment."[134]

Some environmentalists in the 1990s may well have been naïve, and many people, including scientists, probably did not appreciate the ocean's vast complexity, but the idea that anyone who expressed reservations about the seabed disposal concept—including fellow scientists such Vaughan Bowen and George Woodwell—held that all ocean disposal was inherently evil was an absurd caricature. To assert that an untested technology would have no environmental effect was at best arrogant, and at worst dishonest. To characterize the deep-ocean environment as devoid of currents that might disturb materials placed there flew in the face of not only the discoveries that Woods Hole scientists had made but also Hollister's own work, the very basis of his claim to be an expert on the deep marine environment.

Hollister's Position Worsens: The Nuclear Waste Policy Act of 1995

In 1995, the US Congress proposed an amendment to the Nuclear Waste Policy Act forbidding subseabed disposal and "any activity related to it." Not surprisingly, the Seabed Association had tried to prevent the act's passage. In a memo regarding communications with the staff of Alaska Senator Frank Murkowski, Kelly had suggested they propose to enlist the University of Alaska in a public education program: "We understand that subseabed disposal would evoke public concern, and we are prepared to undertake a public education effort in conjunction with the scientific research program, which would involve the University of Alaska."[135] Colleagues in Alaska could help "educate" the public to accept seabed disposal. Hollister and Kelly used the term *education* advisedly, because their efforts to influence the positions of members of Congress would be illegal if they involved the use of federal funds.

The distinction between education and lobbying had come up before when Hollister organized a series of seminars for congressional staffers. In the autumn of 1990, he had initiated an outreach effort designed to "increase people's understanding of the oceans." The three-day intensive seminar would focus on marine science in relation to environmental issues, national defense, and economic competitiveness. While Hollister spoke broadly of "people," including "industrial leaders, Congressional staffers, educators, and other national decision makers," Hollister's "Congressional Education Week" would primarily target congressional staff.[136]

The first workshop was scheduled for October 18–21, 1990, and Hollister sought support from the DOE, NOAA, and the ONR—all federal agencies. In a letter to ONR director Fred Saalfeld, Hollister wrote, "We are concerned that important decision makers do not have an understanding of how important a knowledge of ocean science is in informed decision-making about global change."[137] Saalfeld agreed, offering $20,000 to support the seminar and suggesting that Admiral Paul Gaffney, assistant chief of naval research, be invited. He also reminded Hollister of the legal limits on the use of federal funds: "Since you intend to invite Congressional staffers to the seminar, please be especially sensitive to the law prohibiting ONR from using its funds to lobby. Having received my $20K, please insure the seminar only informs its attendees and in no sense attempts to influence or suggests policies for the Congress."[138]

Saalfeld had reason to be concerned. The Byrd Amendment—enacted just the year before—specifically prohibited the use of federal funds "to pay any person for influencing or attempting to influence an officer or employee of any agency, a Member of Congress, an officer or employee of Congress, or an employee of a Member of Congress" in connection with federal grants, contracts, loans, cooperative agreements, and so on.[139] The Seabed Association had done precisely those things with private money. If Hollister continued any of those activities using federal funds, it would be illegal.

The seminar went ahead, followed by two more. The second, "Ocean Science and National Security," held in February 1991, included members of the staff of Hawaii senator Daniel Inouye, Massachusetts senator John Kerry, South Carolina senator Ernest Hollings, and several others.[140] The third, "The Ocean, Climate Change and the Environment," in October 1992, included eighteen staff members and legal counsel to the House Committees on Science, Space, and Technology; Merchant Marine and Fisheries; Interior and Insular Affairs; Energy and Commerce; and Appropriations—and the Senate Committees on Environment and Public Works, Energy and Natural Resources, and Armed Services.[141] The ONR supported these additional seminars, and Saalfeld continued to remind Hollister not to cross the line from education to advocacy. "Please keep constantly in mind that you are prohibited from using public funds to lobby Congress," he wrote again in 1991. "All your efforts in a seminar that uses my funds must be devoted solely to informing the attendees, and cannot in any way suggest policies."[142]

No doubt these seminars did educate congressional staffers about scientific matters, but read in the context of Hollister's other activities, there is little doubt they were intended to influence Congress and to suggest policies. What was Hollister doing if not trying to influence employees of Congress to be sympathetic to the seabed program? It hardly seems coincidental that the "educational" seminars came right on the heels of Hollister's failed attempts to obtain funding for the seabed research provision. Hollister was not merely interested in informing Congress; he was determined to restore authorizations for the seabed program, determined to get the necessary allocations into appropriations bills. Indeed, when Hollister did not like a person's position, his answer was often to "educate" them. On the occasion of a proposed cutback in federal NSF funds in 1981, Hollister wrote an urgent message to the Woods Hole directorate arguing it needed to contact the staff director of the House Science, Research, and Technology Committee, who was "no friend" of the relevant programs. Hollister's suggestion? "He should be educated ASAP."[143]

Despite these "educational" efforts, by the mid-1990s, seabed waste disposal—radioactive or otherwise—was a moribund, if not deceased, idea. The DOE was moving forward with the site evaluation of Yucca Mountain, and in 1993 two annexes to the London Dumping Convention banned all dumping of radioactive waste at sea. Whether technically feasible or not, marine disposal was not politically feasible. In 1994, the US National Academy of Sciences' Committee on the Management and Disposition of Excess Weapons Plutonium concluded as much: "Subseabed disposal would face intense political opposition from many quarters, and a complex web of national and international legal hurdles and regulation.... Any proposal for disposal in or below the oceans is likely to provoke intense public and political opposition, both within the United States and internationally.... Overcoming the political, legal, and regulatory hurdles (including providing experimental data that do not yet exist) would be difficult, uncertain of success, time-consuming, and expensive.... The committee does not believe such an approach should be pursued."[144]

Hollister nevertheless continued to assert that the reasons for the death of the subseabed program were illegitimate. He compiled newspaper reports on problems with Yucca Mountain and closely followed a dispute among DOE scientists over whether disposed warhead wastes might go critical in a subsurface environment. Reported on the front page of the Sunday *New York Times* under the sensational headline "Scientists Fear Atomic Explosion of Buried Waste," the claim had been dismissed by a Los Alamos peer-review committee as lacking technical merit—and in any case, it was irrelevant to Yucca Mountain, as the DOE had no plans to dispose weapons-grade waste there—but Hollister carefully filed it with his other anti–Yucca Mountain clippings.[145] He also continued to correspond with the staff of Senator Murkowski. Above all, he insisted it was not too late to consider the seabed alternative. On the contrary, "the need for a back-up to land disposal at Yucca Mountain is greater than ever."[146]

In 1995, something happened that gave him renewed encouragement. Site-suitability studies at Yucca Mountain were far behind schedule and Energy Secretary O'Leary had been quoted as saying, "I feel very uncomfortable on the path we are pursuing with no alternative to the geologic disposal option."[147] This was the opening Hollister needed. In March, he and Kelly organized a conference call with several stalwart colleagues to discuss the next round of action. One idea was to approach the National Academy of Sciences with a request for funds to study "ocean options for high level and low level waste," suggesting that "Yucca Mountain is finished." The academy declined. Another was to approach the Central Intelligence Agency with ideas to research "plutonium sequestration from the anti-terrorist perspective." They declined as well.[148]

The participants on the call agreed that they would continue to argue that the federal government should fund the office that had been created by section 224(b) of the Nuclear Waste Policy Amendments Act and allow the Subseabed Consortium to get to work.[149] With a budget of $2.5 million, they could demonstrate concept feasibility within six years, and the energy secretary could decide whether to pursue international cooperation. They also added a new angle: emphasize subseabed as a supplemental plan in the event that Yucca Mountain was not volumetrically sufficient, as the capacity of the seabed, they repeatedly pointed out, was "virtually unlimited."[150]

Again they worked their contacts with members of Congress—this time including Senators John Glenn of Ohio, Ted Stevens of Alaska, and Pete Domenici of New Mexico. They also persuaded Woods Hole trustee Guy Nichols to contact Senator Johnston again. Johnston replied that he had "little hope that Congress would fund" their research program, because the DOE was already over budget on Yucca Mountain and had been asked to take on the responsibility of interim storage of spent nuclear fuel. Under such circumstances he saw "no chance of initiating a new research program," concluding vaguely: "My staff has talked extensively with Charlie Hollister and John Kelly about subseabed disposal in the past. I will see that they keep in touch and remained informed on the subject."[151]

Hollister and Kelly prepared briefing notes (including "quotable quotes" to use in letters and enumerating counterarguments to anticipated criticisms), and again they contacted industry representatives and the energy secretary's office (although her staff, according to one memo, were "skeptical.")[152] Actually, the secretary was more than "skeptical," as one memo admitted: "The general consensus within DOE was that any proposal for restarting the research on subseabed disposal would be 'dead on arrival.'"[153]

With few options left, Hollister wrote letters to a number of wealthy private patrons who he had reason to believe might lend their support. In one letter, he dismissed the opposition as "extremists" hoping to "predetermine the outcome," noting that "if we ... work the legal and political issues, and build public knowledge and understanding, the perception of subseabed disposal could be quite different several years from now."[154] Momentum, however, was running in the other direction. In September, Republican senator Larry Craig of Idaho and five cosponsors introduced a bill to amend the Nuclear Waste Policy Act explicitly to prohibit subseabed or ocean-water disposal of any spent nuclear fuel or high-level waste, or to use funds for "any activity relating to the subseabed or ocean water disposal of spent fuel or high-level radioactive waste."[155]

The politics here might seem counterintuitive: congressional Republicans were not generally known as environmental allies. But the explanation is easily found in the utility industry's position: they continued to oppose any action that might be seen as diluting, deferring, or in any way interfering with the repository at Yucca Mountain.[156] Because Hollister and his colleagues had repeatedly criticized the Nevada option, support for the subseabed alternative was interpreted as interfering with Yucca Mountain. Kelly referred to Craig's bill as an "industry bill," because the nuclear power utilities were determined that Yucca Mountain be approved and opened as soon as possible.

Hollister had been working the issue for two decades, and one sympathetic colleague tried to convince him to let it go. Bud Ris, the president and executive director of the Union of Concerned Scientists, wrote Hollister a sympathetic but frank letter suggesting that it was time to lay down arms.[157] Regardless of the technical merits, subseabed disposal just wasn't going to happen:

> I believe the entire [environmental] community would oppose any initiative in this area, no matter how sound the technical merits. Common sense just makes it impossible for any of them to believe that seabed disposal of high level waste will be designed, managed, and operated flawlessly, without any risk to the public or environment. No study or endorsement by UCS [Union of Concerned Scientists], WHOI, the NAS or anyone else will convince them otherwise. They fear that seabed disposal of high level waste—no matter how safe—would create a precedent for using the oceans as a "dump" again, rolling back twenty years of environmental progress.... Although I realize that you ... are only proposing that the seabed option be 'studied' more thoroughly than in the past ... few others will see it that way.[158]

Hollister was again unpersuaded, and his efforts increasingly looked just a little bit desperate. In 1996, for example, he and John Kelly considered approaching the EPA's Office of Indoor Air and Radiation, to see if perhaps they would fund the seabed provision.[159] They would not: radioactive waste was DOE's problem.

Having failed repeatedly with Congress, Hollister concluded that his only option was to try again with the public. In the fall of 1996, he teamed up with science writer Steven Nadis to tell the story of how the DOE had squashed a technically feasible option for nuclear waste disposal. "It was a clear case of 'not invented here,'" Hollister was quoted as saying, in a piece published in the October issue of *Atlantic Monthly*. The DOE had rejected the idea not based on science nor on the formidable legal impediments, but simply "because the sub-seabed researchers never really fit in with mainstream DOE culture." Kelly concurred, calling the DOE's position "extremely superficial."[160]

The Nadis piece provoked a flurry of angry letters. Like Buckley before him, Nadis had based his views almost entirely on discussions with Hollister, who was well outside the mainstream of both environmental and scientific opinion. Clifton Curtis of Greenpeace International referred to a "stack of documents" that he had sent Nadis, "all of which made it clear that the dominant view of the scientists and others is that seabed burial is a bad idea." Hollister insisted that his own early work for the DOE demonstrated the technical feasibility of seabed disposal, but Curtis noted that the work had never been subject to public discussion or external peer review. Eugene Roseboom Jr., a geologist with the US Geological Survey, agreed that some antinuclear groups "exaggerate the risks and play upon the public's excessive fears of radiation," but Nadis had effectively done the same by neglecting to mention the "favorable features" of Yucca Mountain and focusing instead on "dramatic but technically minor objections[, all of which] have been thoroughly studied and publicly reviewed."[161]

Perhaps the most telling criticisms of the article came from marine biologist Boyce Thorne-Miller, who raised the issue Hollister had for twenty years blithely dismissed: the risk to ocean ecosystems and the potential for contamination of the marine food chain.[162] Miller wrote: "Hollister, who is not a biologist, clings to the belief that deep sea life is insignificant. Marine biology has advanced beyond that view many years ago, but Nadis did not interview marine biologists. Those who worked on the Sandia project would, of course, agree with Hollister that they should have been able to do more research— scientists always believe that more research is needed, because research is their livelihood. However, they would also tell you that one of the discoveries of that project was an immense, hitherto unknown, diversity of animal life on and in the deep sea floor."[163] Miller stressed that Hollister stood to gain personally from research on ocean dumping: he was not a neutral witness, not an objective scientific adviser, not a disinterested expert. On the contrary, he would benefit financially, scientifically, and professionally from the research for which he was such a passionate advocate. Miller concluded, "I am always fascinated that scientists are considered objective sources of information while environmentalists are assumed to have devious personal agendas."[164] Hollister's agenda might not have been devious, but it was personal.

Still Hollister did not back down. He took up the drumroll when invited to deliver the 1997 Doherty Lecture on Oceans Policy, sponsored by the Center for Oceans Law and Policy and delivered in the Congressional office buildings on Capitol Hill, and again in 1998 when he and Nadis coauthored a piece in *Scientific American.*[165]

The *Scientific American* piece would turn out to be Hollister's last major comment on the subject. More than twenty years had passed since he first addressed the idea, but Hollister's views were unchanged: the seabed was stable and predictable, the muds were as thick as peanut butter, and the land-based alternative was going nowhere. The only departure was that Hollister no longer left it open whether the wastes should be dropped from a surface ship or be placed into holes drilled into the seafloor: Hollister dropped the passive emplacement options and insisted that, using technology that had been perfected in the deep-sea drilling program, waste-bearing canisters would be sedulously emplaced into drill holes and carefully backfilled (fig. 8.8). With this, he argued more forcefully than ever that seabed disposal was not dumping, "a wholly inappropriate label." To call it dumping "makes as much sense as calling the burial of nuclear wastes in Yucca Mountain 'roadside littering.'"[166]

Ideas, Information, Advice, and Advocacy

Many years ago, the sociologist of science Robert Merton argued that a central norm of scientific communities is disinterestedness.[167] Scientists can be objective in the evaluation of knowledge, Merton insisted, because there are no outside interests—financial, political, egotistical—driving their judgments. Since then, various sociologists have highlighted the ways in which Merton's view of science was unduly idealistic, a point that in recent years has become increasingly obvious as molecular biologists and medical school faculty have flocked to create high-tech start-up firms or to take on lucrative consulting relationships with pharmaceutical companies.[168] These relationships have stirred controversy precisely because they involve a potential conflict of interest—or, more properly, a conflict between interest and disinterest.[169] Can one judge a drug objectively when one has a financial stake in bringing it to market? Or properly inform patients in a clinical trial when failure to enroll them might threaten the viability of the trial? Don't climate scientists have a stake in emphasizing the threat of anthropogenic climate change? And is it possible for anyone involved in a debate to be truly disinterested? After all, the very fact of one's involvement implies some kind of interest.

For decades, tobacco industry funding imprinted the results of biomedical research, and similar problems arise today with research on fracking funded by the gas industry or pesticides funded by the chemical industry.[170] But the problem transcends obvious commercial conflicts of interest. As a society, we need scientists to alert us to problems, threats, and opportunities of which we might not otherwise understand or even be aware. Society would not have been able to protect itself from stratospheric ozone depletion had leading scientists not acted as sentinels, stepping out of the lab—both metaphorically and literally—to alert us to the threat and suggest how to address it.[171] Nei-

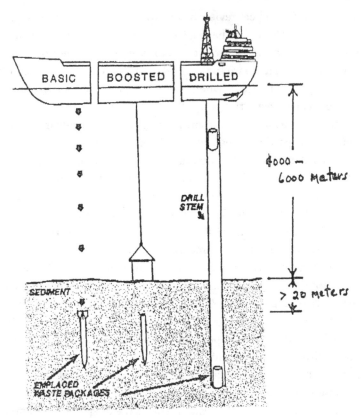

Figure 8.8 "Subseabed Concepts" illustrating three ways waste packages could be emplaced in the seabed, to thereby persuade an audience that this was not "dumping." From "The Subseabed Disposal Project: Briefing Book 1985," JK Associates, 1985, p. 7, in Charles Davis Hollister Papers, 1967–1998, MC-31, "Sub Sea Bed Disposal Project Briefing Book, 1985," Data Library and Archives, Woods Hole Oceanographic Institution. Reprinted with permission of the Woods Hole Oceanographic Institution.

ther would we know today that the changes we are witnessing in our climate are caused by greenhouse gases, had scientists not done the work to demonstrate it. In this regard, scientists serve an essential sentinel function.[172]

It was not wrong, therefore, for Charles Hollister to try to make people aware of the potential to dispose of radioactive waste in the seabed, and up to a point it was not wrong for him to try to make the case. But where to draw the line between ideas, information, advice, and advocacy? How do we decide where one ends and another begins? How can we distinguish between reasonable interest and unreasonable self-interest?

Within scientific communities, competing claims and counterclaims are sorted through scientific discussion and peer review: only claims that with-

stand the scrutiny of peer review have earned the right to be considered scientific, and only those that have garnered consensus have earned the right
to be considered facts.[173] But policy recommendations are not facts, so how
to judge them? What are we to think when an individual scientist advocates
a particular public policy or when scientists promote policies outside their
arena of proximate expertise?

Scientists have an interest in promoting both their research programs in
general and the particular research results on which they have staked a prior
professional claim, even when no finances are involved. As historian Spencer
Weart has emphasized, disputatious scientists can be relied on to agree on
one thing: that there should be more money for research, particularly their
own.[174] Yet societal trust in scientists as experts hinges at least to some degree
on an implicit assumption of independence: that scientists can be trusted to
tell the truth because they are free of conflicts of interest—or at least freer
than lawyers paid to represent a client, politicians with a district to represent, or corporate executives with a product to sell. But what happens when
scientists are selling a product? Surely this was Vaughan Bowen's point when
he argued against the language of selling: that it was at odds with the epistemology of inquiry that is supposed to be a hallmark of scientific objectivity
and undergirds scientists' credibility.[175]

Our dominant models of science insist that scientists can and will change
their mind in the face of evidence—and we have many fine examples where
scientists have done just that.[176] Call it the virtue of epistemic openness. Was
Hollister open? Was he prepared to change his mind about seabed disposal?
Was there any evidence that would have led him to modify his views? The
historical evidence suggests not.

We have seen that Hollister dismissed the concerns of biologists that radionuclides escaping into the seawater might work their way up the food chain,
because he was convinced that the marine muds would capture any stray components that might leak from the waste canisters. Were the environment of
the abyssal plains entirely stagnant, this argument might have been robust.
But even by 1980 there was substantial evidence that this wasn't the case.

In 1979 Sandia had commissioned a series of papers from experts in physical oceanography to address "the current state of knowledge with regard
to the physical oceanographic questions that must be answered generally
if high level nuclear waste is to be disposed of on or under the seabed."[177]
The experts met several times and gathered to present their final results at a
workshop in Big Sky, Montana, in January 1980.[178] The results were highly discouraging for the seabed concept. One paper, by Scripps oceanographer Larry
Armi (who was funded by both the NSF and the ONR) documented evidence

of significant abyssal mixing caused by boundary-layer turbulence. This mixing was revealed by visibility measurements, which showed that abyssal mud particles were entrained into the water column for a distance of ten to one hundred meters above the bottom—the so-called bottom boundary layer—and then dispersed into the open ocean from there.[179] Even if clay particles adsorbed leaked radionuclides, those clay particles could themselves be dispersed through the bottom boundary layer into the rest of the ocean.[180]

Other studies offered evidence of exchange between the bottom boundary layer and the rest of the abyssal ocean. Christopher Garrett of Dalhousie University discussed phenomena that contribute to mixing of the whole ocean, including wind shear at the surface, sinking of cold waters in the polar regions as part of the thermohaline circulation, internal waves leading to shear instability or overturning of density stratified layers, upwelling, and mesoscale eddies.[181] Vaughan Bowen and colleagues presented data showing that fallout plutonium, entrained on downward-sinking particles, could be remobilized and flushed "into the general circulation pattern of the North Pacific bottom water."[182] The correlation of plutonium with oxygen levels was weak, showing that "this is not a front moving out from its source, but the result of rapid incorporation of this tracer into the deep water gyre circulation." This meant that if plutonium escaped from a disposal site, it could readily be dispersed. Biologist Michael Mullin noted that placing canisters in abyssal mud did not protect them from organic activity, as many benthic creatures are burrowers whose activities disturb the mud.[183] In short, setting aside legal and ethical questions, there were plenty of scientific reasons to suspect that subseabed disposal might not work; Hollister's advocacy did not represent a consensus of scientific opinion.

Throughout the debate, Hollister ignored substantial technical evidence that subseabed disposal might not be a good idea, insisting that the obstacles were only political. Nor did he acknowledge his own self-interest in pushing the seabed option. Hollister left a long paper trail, but nowhere is there evidence that he ever considered how self-interest might be conditioning his scientific views, or how it was leading him to discount not only public opposition but also the concerns of own colleagues. In hindsight, one colleague put it diplomatically: "I have mixed views regarding Charley's receptiveness to accepting potential problems with the seabed burial concept."[184]

All parties to debates have interests that condition their responses to evidence and arguments, and all scientists have views conditioned by their technical expertise and experience.[185] And there are situations in which technical expertise leads more or less directly to recommendations, as when epidemiologists advise us to stop smoking and oncologists tell us to wear sunscreen.[186]

But Hollister's actions were not merely a result of his technical expertise; they were also driven by considerations that went beyond the question of technical viability. Initially, he was motivated to maintain a productive research program, one that provided a significant portion of his funding for deep-sea investigations in a climate of shrinking financial support. But as time went on, the motivations changed. In the end, they were probably not financial at all.

By the 1990s, Hollister was busy with Woods Hole administration; financial support for his deep-sea research would no longer have been much of a concern. Why continue to make the case? One reason might have been pride. Hollister had staked a claim on the desirability of subseabed disposal, he was confident that his views were correct, and he was determined to convince others that those views remained correct.

But perhaps there was something even deeper. Recall Hollister's congressional education seminars and his determination to "educate" members of Congress and their staff. Hollister had spent a great deal of time trying to get people to listen to his views. Had the US government pursued the seabed option, Hollister's audience would have grown enormously. He was one of the world's leading experts on abyssal sedimentation. A decision to dispose of America's nuclear waste in the deep sea would have placed him at the center of an important policy question, making him a man on whom the US government would need to rely. It would place him in the limelight—and for a long time.

Conclusion

The desire for influence is not rare among scientists, and scientists certainly know many things to which politicians, business leaders, and citizens would do well to listen. Scientists also have an obvious interest in asking for research support, and they must seek support if they are to do their work. But we can usefully distinguish between interests—which all people have—and conflicts of interest. Scientists' need for research funding is an open agenda, not a hidden one. However, Charles Hollister did not merely ask for research funding in intellectual domains he considered interesting and important. He advocated a specific public-policy position—one in his personal self-interest—under the cover of scientific expertise and objectivity.

Hollister's choices undermined Dorman's claims that Woods Hole as an institution took no advocacy position except that of advocating basic research. When Hollister spoke, when he testified, when he wrote letters to the editor and appeared on television, it was never merely as Mr. Charles Hollister, resident of Falmouth, Massachusetts, but always as Dr. Charles Hollister, senior scientist, dean of the Woods Hole graduate program, and vice president of the Woods Hole corporation. He used his scientific credentials and credibility in an explicitly political way and drew on the prestige of his institution to promote a particular policy position. His scientific expertise, his political advocacy, and his desire to continue his research on abyssal environments were entwined in a cat's cradle that in principle might have been disentangled but in practice never was.

In a 1991 report deeply critical of the US nuclear waste program, *High-Level Dollars, Low-Level Sense*, released by the Institute for Energy and Environmental Research, authors Arjun Makhijani and Scott Saleska noted that the entire US radioactive-waste management program was plagued by a conflict of interest at its core: the DOE and the Nuclear Regulatory Commission regulate and manage nuclear waste disposal while promoting nuclear power and building nuclear weapons.[187] The DOE in particular has had the conflicting agenda of ensuring the safety of waste disposed at any repository while continuing (until quite recently) to produce more of those wastes.

These conflicts have been recognized by environmental and citizens' groups, but less recognized have been the conflicts of interest among the scientists who have worked on those programs. Scientists who worked for the DOE on Yucca Mountain, either directly as employees of the national laboratories or indirectly as academics receiving DOE research support, took seriously the need to properly evaluate the site, but the desired answer was always clear. Hollister was not wrong to notice that and be troubled by it. In principle, DOE scientists could have said that Yucca Mountain was unsuitable, but in practice the weights were heavily on one side of the evaluative scales. With each passing year, as millions more dollars and thousands more hours of scientific labor were invested, the scales became weighted further. For the scientists involved, evaluating Yucca Mountain was like an election in the Soviet Union: there was only one candidate.[188] Hollister was right to object to that.

He may also have been right when he complained that the seabed option had been poisoned because the "politics got too far out in front of the science."[189] On some level, that is true of the entire civilian nuclear waste disposal program. Many scientists would argue that there are several places in the United States where radioactive wastes could be safely stored but have never been properly evaluated for social and political reasons. But in any policy decision, politics is part of the story. Hollister knew that, of course, or why else would he have dived so deeply into them?

Since World War II, American scientists have played an increasing role as purveyors of information and advice to the federal government.[190] Nowadays there is scarcely a prominent scientist in a policy-relevant field who has not offered some kind of information or advice to some branch of government. If the scientist is an expert in the area under discussion, there is always the possibility that his or her funding could be affected by the decisions being made.[191] This creates the potential for conflict of interest. When an individual—as opposed to an institutional authority, such as the US National Research Council or the Intergovernmental Panel on Climate Change—gives advice, this also raises the question of the boundary between scientific knowledge and individual opinion. This is one reason the category of scientific consensus has in recent years come to the fore: to distinguish between scientific conclusions, broadly accepted by the relevant technical experts, and personal opinions or idiosyncratic positions.[192]

The problem, I suggest, is not that Hollister became involved in politics per se; scientists do not lose their rights of citizenship when they earn a PhD or even enter into a contested public debate. The problem is that he exaggerated the technical basis for his position and ignored evidence that did not comport with it. He also drew on the authority of his institution to present his opinions as if they were matters of fact, and he was never forthright about his personal stake—not even with himself.[193]

9 *Changing the Mission: From the Cold War to Climate Change*

In February 2000, the director of the Scripps Institution of Oceanography, physicist Ed Frieman (1926–2013), looked back on the final years of the Cold War. One might have thought it a time for celebration—the Cold War was ending and the United States would win—but he recalled it as a "troubling" time both for the nation and for Scripps: "Tensions had reached new highs [and] threatened to boil over."[1] The speech that followed, however, revealed that he was not talking about the late 1990s but about the entire period from 1980, when Ronald Reagan became president, to 1996, when Frieman retired.

The "golden age" of Navy largesse really lasted only one generation. From World War II through the early 1960s, abundant funding enabled oceanographic institutions to expand their activities in ways that they might have previously imagined but would never have been able to achieve. As we have seen, that expansion led to anxieties about freedom, autonomy, and the direction of oceanographic science. In time it also led to new anxieties about money, as directors needed enlarged budgets to keep their enlarged institutions humming, and already by the late 1960s, oceanographers were worried once again about money. By the 1970s, Navy support of oceanographic research had begun to wane under the influence of the Mansfield Amendment—which required the US military to focus its research dollars on its military mission—and the Navy's decision to fund more internal research and to build its own research vessels.

Whatever the challenges of working with the Navy, it was hard to imagine another organization that could support oceanographic science on a comparable level or another motivation that would be as compelling as the Soviet threat. The Cold War placed the intellectual interests of oceanographers into a broad and compelling—even urgent—global political context. How would their science thrive without a big new question to address, a new central motivating concern? As Frieman put it, "The vacuum left by the Cold War's end left us unsure and uncertain about the future. What would the next big issue be"?[2] Charles Hollister tried but failed to persuade the US government that the answer was the disposal of radioactive waste (chapter 8). What finally did turn out to be persuasive was the emerging issue of anthropogenic climate change.

A New Opportunity in Global Climate Change

There was a strong argument to be made for oceanographers to focus on climate change.[3] Some oceanographers, famously Scripps director Roger Revelle and geochemist Charles David Keeling, had a long-standing interest in the issue. As part of the International Geophysical Year, Revelle helped to obtain the funding that enabled Keeling to begin the detailed measurements of atmospheric carbon dioxide that would later bear his name—the Keeling Curve—and win him the National Medal of Science. Revelle and others had also begun to publish serious scientific papers on carbon dioxide and climate, and to talk to the media about its potential implications.[4]

By the late 1950s, many oceanographers and geophysicists were aware of the scientific work suggesting that increased carbon dioxide from burning fossil fuels could alter the climate, and they were pondering how they might contribute to scientific understanding. In 1957, for example, Woods Hole director Columbus Iselin discussed the matter in a speech to the Newcomen Society for the History of Engineering and Technology. He mentioned it only in passing, in a section of the speech addressing radioactive waste, but his manner suggested that the basic facts were scarcely controversial: "In the case of carbon dioxide in the atmosphere, man went ahead and began burning fossil fuels at a prodigious rate before he considered the possibility that he might be altering the climate on Earth in an unfavorable manner."[5] Soon after Iselin's speech, Woods Hole's William von Arx began to argue not only that the question of carbon dioxide, fossil fuels, and climate deserved oceanographers' attention, but also that weather and climate could offer oceanography a better motivating framework than the military.

One of the leaders of the Woods Hole Palace Revolt (chapter 3), von Arx felt it had been a mistake to link the fortunes of his science to the US Navy, and he continued to make the case even after the revolt failed. Von Arx pulled no punches: oceanography was "in bondage," tethered to the needs of the Navy. He was not ungrateful for the support he received, nor unmindful of the practical considerations that had linked the Navy to American oceanographic science, and vice versa. Chief among these was the extreme expense of doing serious oceanographic work, which suggested to him that oceanographers would always be compelled to tie their science to some extrinsic concern. The challenge was to find the particular extrinsic concern that was most logically and naturally tied to their substantive intellectual interests, the domain in which the interests of scientists and their patrons were most fully aligned. That, in von Arx's opinion, was not warfare, but weather and climate. "If, because of operational expense, it is to be the eternal fate of oceanography that its pursuit must be tied to some compelling cause, it would seem that the problems of weather and climate would provide its most appropriate justification," he wrote.[6]

Oceanography had a well-established link to meteorology. Dynamic oceanography had developed in the early twentieth century from the Bergen school of dynamic meteorology, and during World War II oceanographers and meteorologists had worked together on weather prediction and surf forecasting (chapter 1).[7] By the 1950s, the importance of air-sea interactions for both disciplines was widely recognized.[8] Oceanographers like Walter Munk worked to understand the role of wind and weather in generating waves, swell, and currents, while meteorologists like Joanne Malkus were recognizing the role of the sea surface in generating hurricanes and other weather events.

It was not simply that weather and climate were plausible justifications for basic oceanographic research, von Arx argued. It was that they were better justifications than military concerns. The central difficulty with Navy patronage was not that the Navy did not share oceanographers' interest in the ocean; the difficulty was that the Navy mission focused too much attention on technology at the expense of science. His views in this regard presaged the argument made in the 1980s by historian Paul Forman: that military patronage in physics fostered a science of technical mastery and gadgeteering.[9] Von Arx believed that military patronage was problematic because it fostered a culture of technological bravado at the expense of conceptual understanding. A focus on weather and climate, however, could change that: "This refreshing change of 'mission' in ocean research would draw a different sort of people into marine science. There would be more thought-centered effort and less thing-centered preoccupation as with deep submersibles … and other elements of technological derring-do which "big science" tends to encourage."[10] Foreshadowing conclusions that would soon become commonplace, von Arx noted that human effects on the environment were increasingly evident, including early signs of what scientists would come to call global warming. Indeed, well before the emergence and popularization of the concept of the Anthropocene, von Arx noted that humans were changing the planet on a geophysical scale. A particular concern was the impact of fossil-fuel combustion on the planetary radiation balance: "Man has become a *force of nature.*

He ... is altering the radiation balance ... and by his vigorous consumption of fossil fuels the concentration of carbon dioxide."[11]

The Long History of Concern about Global Climate Change

It had been known since the nineteenth century that carbon dioxide was a greenhouse gas—highly transparent to visible light, fairly opaque to infrared—and that its presence in our atmosphere makes Earth a comfortably warm planet.[12] By the mid-twentieth century it had become broadly accepted among earth scientists that changing concentrations of atmospheric carbon dioxide (due to natural causes) had the potential to affect global climate by altering Earth's radiative balance. This concept had been developed in the mid-nineteenth century by Irish physicist John Tyndall (1820–1893), who demonstrated that carbon dioxide and water vapor are greenhouse gases, and it was invoked at the end of the nineteenth century by America's most famous geologist, T. C. Chamberlin (1843–1928), to explain the ice ages.[13] Around the same time, Swedish chemist Svante Arrhenius (1859–1927) suggested that a changing concentration of atmospheric carbon dioxide might also occur from unnatural causes, specifically burning fossil fuels.

In 1923, British geologist R. L Sherlock (1875–1948) published a book entitled *Man as a Geological Agent*. Its central thesis was what would later become the Anthropocene concept: that the scale and extent of human activities were now so great as to compete with natural forces. Sherlock discussed a wide variety of geological and geophysical impacts, including forestation, farming, erosion, mining and quarrying, and construction of dams and harbors; he drew in particular on the work of American paleontologist George Perkins Marsh (1801–1882), who in 1874 had published *The Earth as Modified by Human Action*, which considered potential meteorological effects of deforestation. He also drew on Chamberlin and Arrhenius, concluding the book with a chapter on climate. Chamberlin, Sherlock noted, had argued that carbon dioxide removal from the atmosphere, as "a consequence ... of the vast mass of carbon locked up by animals and plants ... during the Carboniferous period," could have caused the ensuing Permian glaciation. If he was right, then a "reversal of the process," whether by natural causes or human activities, could lead to global warming.[14] This was also Arrhenius's conclusion: "Arrhenius thought that if the amount of carbon dioxide in the air were increased three-fold, the temperature of the Arctic regions would rise by 8 or 9°C."[15]

It is not clear how widely read the book was, but it is clear that by the 1930s more than a few scientists were taking up the question.[16] In 1931, E. O. Hulburt (1890–1982), a physicist at the US Naval Research Laboratory, took up the carbon dioxide question in the context of the ice ages, calculating that doubling or tripling atmospheric carbon dioxide would increase surface temperatures by 4 and 7 Kelvin, respectively, and those temperatures were "about the same as those that occur when the earth passes from an ice age to a warm age, or vice versa." He therefore concluded (tautologically) that "the carbon dioxide theory of ice ages ... is a possible theory."[17] One objection, however, was that the specific infrared radiation absorbed by carbon dioxide was also absorbed by water, so adding or subtracting carbon dioxide in a water-bearing atmosphere would have little effect. Hulburt showed data to suggest that this was probably incorrect because the spectral overlap was only partial, but he also suggested that more work was needed on this important point. Meanwhile, also in the 1930s, British engineer G. S. Callendar (1897–1964) began to argue that the atmospheric carbon dioxide concentration was rising from burning fossil fuels and was already having detectable effects in northern latitudes. Callendar returned to the question in the 1950s, compiling existing global measurements to suggest that carbon dioxide had risen from a baseline value of approximately 290 parts per million at the turn of the century to a value of 320 and rising just fifty years later (fig. 9.1).[18]

Callendar published these results in geophysical journals—*Tellus, Weather,* and the *Quarterly Journal of the Royal Meteorological Society*—and his work, along with Hulburt's, was taken up by American physicist Gilbert Plass (1920–2004) in an important article published in 1956 addressing the infrared absorption question.[19] As a research scientist at the Lockheed Corporation, Plass had been studying atmospheric infrared absorption. Using an early digital computer and revised laboratory measurements of the carbon dioxide absorption band by the aptly named W. H. Cloud, he showed that

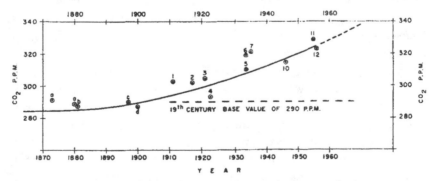

Figure 9.1 Measurement of atmospheric CO_2, showing rise in atmospheric concentration in the first half of the twentieth century.

From G. S. Callendar, "On the Amount of Carbon Dioxide in the Atmosphere," *Tellus* 10 (1958): 243–48 (fig. 1). Reprinted in accordance with Taylor and Francis open-access policies.

the negative critique was incorrect.[20] Better resolution of the spectral lines showed that the overlap was much less than generally supposed; increasing atmospheric carbon dioxide could have a significant impact on climate.[21]

Plass calculated the radiative flux upward and downward from the Earth's surface over one-kilometer increments for three different carbon dioxide concentrations—current, doubled, and halved—and he found that a doubling of atmospheric carbon dioxide could lead to a 3.6°C increase in average global temperature. Halving carbon dioxide produced an equivalent decrease. With the calculations modified to include a "reasonable average cloud distribution over the Earth," the result for carbon dioxide doubling was 2.7°C—less than without clouds but still enough to cause major climate impacts—which confirmed that changing carbon dioxide concentrations could have caused the climatic changes observed in geological history.[22] Tyndall had argued that the effect of carbon dioxide on climate was a *vera causae*—a true and demonstrable cause; Hulburt had concluded that it was "a possible theory."[23] What Plass added and underscored was the very significant point that the climate was very sensitive to even small changes in carbon dioxide: "The radiation calculations predict a definite temperature change for every variation in CO_2 amount in the atmosphere. These temperature changes are sufficiently large to have an appreciable influence on the climate [because] a relatively small change in the average temperature can have a large effect on the climate."[24]

Plass's paper dealt mainly with the long-standing debate over carbon dioxide fluctuation as an explanation for the ice ages, but he also addressed the impact of human activities. Noting that humans were adding more than 6 billion tons of carbon dioxide per year to the atmosphere by burning fossil fuels—several orders of magnitude greater than natural inorganic sources—he concluded that human activities had greatly "disturbed the CO_2 balance." Plants might take up some of the excess carbon dioxide through increased photosynthesis, but that would be a transient effect, as the carbon dioxide would be rapidly returned to the atmosphere as the plants died and decayed. The oceans might also remove some carbon dioxide, but at a rate that was very much slower than the rate of human addition. He thus affirmed Callendar's result and went further, predicting that human activities were leading toward an increase in atmospheric carbon dioxide of 30 percent by the end of the century and a corresponding increase in average global temperature. "The influence of the extra CO_2 on climate will become increasingly important in the near future as continuously greater amounts of CO_2 are released into the atmosphere by man's activities," he wrote.[25] It was a matter not of if but of when.[26]

The link between carbon dioxide, climate, and fossil-fuel combustion was reinforced a year later, in a now-famous paper (also in *Tellus*) by Roger Revelle and his Scripps colleague, geochemist Hans Suess (1909–1993). Revelle and Suess emphasized the historically unprecedented character of human activities in the mid-twentieth century. Humans had become geological agents, returning to the biosphere in just a few centuries the accumulated organic carbon stored in rocks over the course of hundreds of millions of years. In hindsight, this paper is often cited as a clarion call to the dangers of global warming, but that is not quite right. Revelle and Suess argued their point opportunistically, in the literal (and nonpejorative) sense of that word: they saw an opportunity to use human actions as a probe to greater understanding of the controls on the Earth's radiative balance: "Human beings are now carrying out a large scale geophysical experiment of a kind that could not have happened in the past nor be reproduced in the future. Within a few centuries we are returning to the atmosphere and oceans the concentrated organic carbon stored in sedimentary rocks over hundreds of millions of years. This experiment, if adequately documented, may yield a far-reaching insight into the processes determining weather and climate."[27]

In the coming years, as world population and resource use continued to grow, the impact of increased carbon dioxide would likely be large enough to produce detectable climatic effects. Therefore, they suggested, scientists should monitor carbon dioxide and document any effects that might occur: "In contemplating the probably large increase in CO_2 production by fossil fuel combustion in coming decades, we conclude that a total increase of 20–40% in atmospheric CO_2 can be anticipated. This should certainly be adequate to allow a determination of the effects, if any, of changes in atmospheric carbon dioxide on weather and climate throughout the Earth."[28] As Plass had already noted, an important uncertainty was just how much carbon dioxide accumulated in the atmosphere as compared with the fraction taken up by the biosphere and absorbed by the oceans. If a good deal were taken up by the oceans or biosphere, then the effect on the climate system might be modest, but if most remained in the atmosphere, then the effect would be substantial. This question was taken up by Scripps geochemist Charles David "Dave" Keeling (1928–2005), who, as part of the International Geophysical Year, began meticulous measurements of atmospheric carbon dioxide at Mauna Loa, Hawaii.[29] Keeling's time-series data soon showed that about half of released carbon dioxide was "missing" and presumed to be absorbed into the oceans or taken up by plants. The remainder was in the atmosphere, where its concentration was on an upward march. By the mid-1960s, Keeling's work had been noticed in science policy circles in Washington, DC.[30]

The US government had invested considerable resources in weather mod-

ification programs—primarily cloud seeding for rain making, but also attempts to control or create tropical storms and hurricanes and other severe
weather events; there had also been substantial private-sector investment
in cloud-seeding technology.[31] In 1953 the US Congress had established an
Advisory Committee on Weather Control, charged with evaluating the various public and private initiatives in cloud seeding. In 1957 the committee
reported to President Eisenhower that weather modification did work—at
least somewhat and in some cases—and recommended further study.[32] But
this raised the possibility that, if weather and climate could be deliberately
modified by human activity, were they not also capable of being modified
inadvertently?

The issue of man-made climate change earned the moniker *inadvertent
weather modification*. The term was coined by geophysicist Gordon MacDonald
(1929–2002), who in 1963 chaired the Panel on Weather and Climate Modification, charged with "a deliberate and thoughtful review of the present
status of activities in this field and of its potential and limitations for the
future."[33] A wunderkind with a BA and PhD from Harvard, MacDonald had
coauthored with oceanographer Walter Munk the acclaimed monograph *The
Rotation of the Earth* when he was only thirty years old. Soon he was serving on
numerous US government science advisory panels, including President Lyn-
don Johnson's Science Advisory Committee (1965–1969) and later Richard
Nixon's Council on Environmental Quality (1970–1972).[34] If Revelle was the
first scientist to capture widespread scientific attention for the issue of global
warming, MacDonald was the first to capture significant political attention
for it.[35]

Two reports were issued, one in 1964 and a second in 1966. The panel concluded that there was no in-principle scientific objection to weather modification. The obstacles were statistical—it was difficult to prove that any given
intervention had had the intended effect—and ethical-weather modification
experiments might go awry. But this stimulated another thought: If humans
could modify the weather intentionally, might they not also do so uninten-
tionally? If adding silver iodide could seed clouds, then what might adding
other components do? By the time of the second report, the panel had concluded that inadvertent weather modification was a real possibility, and the
most important potential cause was the steadily increasing atmospheric carbon dioxide from the burning of fossil fuels.[36]

MacDonald was influenced in part by conversations he had had with
Revelle, who had first alerted him to Keeling's emerging results.[37] In 1965, a
report by the Environmental Pollution Board of the President's Science Advisory Committee had noted that "by the year 2000 there will be about 25%

more CO_2 in our atmosphere than at present [and] this will modify the heat balance of the atmosphere to such an extent that marked changes in climate, not controllable through local or even national efforts, could occur."[38] An appendix written by Revelle and Keeling, coauthored by geochemists Harmon Craig and Wallace Broecker and meteorologist Joseph Smagorinsky (one of the founders of numerical weather prediction), explained why this mattered.[39] Plants absorb carbon dioxide, but it is quickly returned to the atmosphere when the plants die. Burning wood only slightly accelerates the process by which extracted carbon dioxide is returned, since the wood would soon decay anyway. But burning fossil fuels accelerates the carbon dioxide return nearly infinitely. Carbon that has been stored in rocks over the course of the Phanerozoic time period—more than five hundred million years of Earth history—would, without human intervention, mostly stay there. However, because of human activities, it was being returned, from a geological standpoint, nearly all at once.

Comparison of Keeling's first five years of measurements with the known volume of carbon released from fossil-fuel combustion gave a preliminary answer to Revelle and Suess's question: "Almost exactly half of the fossil fuel CO_2 apparently remained in the atmosphere."[40] This was far more than Revelle had expected and meant potentially serious long-term consequences such as melting of the Antarctic ice cap and thermal expansion of seawater, which would yield significant global sea level rise. "By the year 2000," they concluded, "the increase in atmospheric CO_2 will be close to 25%. This may be sufficient to produce measurable and perhaps marked change in climate, and will almost certainly cause significant changes in the temperature and other properties of the stratosphere."[41]

These comments about carbon dioxide were admittedly made in an appendix—the main report dealt with climate change only in passing. In the mid-1960s, most earth scientists—focused on geological rather than human time scales—still believed that the planet was heading (naturally) toward the next ice age. If they considered human impacts (and most did not, considering them to be outside the domain of their science), they perhaps expected accelerated cooling caused by sulfate aerosols and other particulate emissions. With coal still the dominant source of fossil-fuel energy, the effects looked to be larger than any possible warming effect. As MacDonald later summarized, "In 1969, it seemed plausible that our activities could either lead to a disastrous ice age or to an equally disastrous melting of the polar ice caps."[42] But the majority view leaned toward the former.

Things began to change as more scientists learned of Keeling's carbon dioxide measurements at Mauna Loa, which showed that the concentration of atmospheric carbon dioxide was continuing to climb.[43] By the 1970s, a number of scientists were building models to try to predict when a detectable climate signal might occur. In 1978, *Oceanus*, the official journal of the Woods Hole Oceanographic Institution, released a special issue, "Oceans and Climate," which began with an introduction by Robert M. White (1923–2015), who had just stepped down as the first NOAA administrator and was serving as chairman of the National Research Council's (NRC) Climate Research Board:

> We now understand that industrial wastes, such as carbon dioxide released during the burning of fossil fuels, can have consequences for climate that pose a considerable threat to future society. The Geophysics Research Board of the National Research Council in its recent report, "Energy and Climate," foresees the possibility of a quadrupling of the CO_2 content of the atmosphere in the next two centuries with a possible increase of 6 degrees Celsius in global surface temperatures. Changes of such magnitude accompany climatic shifts from glacial to interglacial epochs.... Experiences of the past decade have demonstrated the consequences of even modest fluctuations in climatic conditions [and] lent a new urgency to the study of climate.... The scientific problems are formidable, the technological problems, unprecedented, and the potential economic and social impacts, ominous.[44]

Several of the papers in the *Oceanus* volume cited the climate modeling work of National Center for Atmospheric Research scientist Stephen Schneider, and Syukuro "Suki" Manabe and R. T. Wetherald at Princeton, which predicted that doubling atmospheric carbon dioxide could increase average global temperature by 1.5°C to 3°C.[45] These numbers were low compared to Hulburt and Plass's figures, but they were still concerning.

The following year, scientists began to address the issue in earnest. The World Meteorological Organization (WMO) held the first World Climate Conference to address the question of anthropogenic climate change; White served as chair. Recognizing the "all-pervading influence of climate on society," the conference issued an "appeal to nations" to "foresee and prevent potential man-made changes in climate that might be adverse to the well-being of humanity."[46] The WMO acknowledged that the causes of climate change were diverse, but one stood out as particularly concerning: "the burning of fossil fuels, deforestation, and changes in land use [that] have increased the amount of carbon dioxide in the atmosphere."[47] It was likely that further increases would contribute to a gradual warming of the planet as a whole, which would in turn affect the distribution of temperature and precipitation. The WMO called on world governments to increase efforts to collect high-quality climate data and to support research to better understand climate change and its impacts. (They also called for world peace.)

Meanwhile in the United States, the US Department of Energy had asked the JASONs—the reclusive group of scientific advisers established in the early 1960s to advise on scientific questions relevant to national security—to consider the long-term impact of atmospheric carbon dioxide on climate. MacDonald led the study, which concluded that at current rates of fossil-fuel use, atmospheric carbon dioxide was likely to double by 2035 and significantly perturb the climate system. The study group built two models, one numerical and one analytic, and both suggested that doubling atmospheric carbon dioxide would most likely produce a mean global temperature rise of 2°C–3°C. This was comparable to the "altithermal" period of four thousand to eight thousand years ago, when sea level rose dramatically around the globe, leading to the various cultural accounts of major floods and drowned civilizations that developed at that time.

Such seemingly small changes could produce large social and economic impacts. "The Sahelian drought [1968] and the Soviet grain failure [1972]," they noted, "illustrate the fragility of the world's crop producing capacity, particularly in those marginal areas where small alterations in temperature and precipitation can bring about major changes in total productivity." Changes of this magnitude "would ... have significant demographic effect." Moreover, 2°C–3°C was the expected average change. The warming could be as much as 10°C–12°C at the poles due to polar amplification. This is the feedback loop in which melting ice exposes bare land and water, which then absorbs more heat (because they are darker and less reflective than ice), which leads to more warming, and so on. Manabe had identified this effect in climate models, and the JASON committee agreed that it would occur. The net results of change on this scale were unlikely to be good: "Any change would produce stress and possibly disaster in some parts of the world since so many aspects of society have adapted, with very large investments in their infrastructure, to the [present] climate."[48]

The JASON report reached the White House, where President Carter's science adviser Frank Press wanted to ask the National Academy of Sciences for a second opinion. The academy enlisted renowned MIT meteorologist Jule Charney (1917–1981). The Charney committee affirmed that carbon dioxide was rising, that it would likely continue to rise, and that it would likely have significant physical, social, and economic impacts. In their words: "If carbon dioxide continues to increase, [we] find no reason to doubt that climate changes will result, and no reason to believe that these changes will be negligible."[49]

In their proposal for the Charney Report, the academy had stressed what the JASONs had also noted, that "the close linkage between man's welfare and the climatic regime within which his society has evolved suggests that such climatic changes would have profound impacts on human society."[50] They encapsulated the state of the science: "A plethora of studies from diverse sources indicates a consensus that climate changes will result from man's combustion of fossil fuels and changes in land use."[51] It was 1979, and scientific experts had a consensus that increasing atmospheric carbon dioxide was going to change the climate.

But when would these changes occur? That was a trickier question. While there was a consensus that changes would occur—and, with a few dissenting voices, that they would be largely adverse—there was not as yet consensus on when.[52] Many scientists at the time talked of effects becoming evident by the end of the century, but one person suggested that changes were perhaps already under way: meteorologist John Perry, the chief staff officer for the academy's Climate Research Board. In an article in *Climatic Change*, he suggested that scientists might be underestimating the problem. In their models and analyses, most scientists focused on doubling carbon dioxide, but Perry pointed out that this was just a convenient point of comparison. "Physically, a doubling of carbon dioxide is no magic threshold ... If we have good reason to believe that a 100 percent increase in carbon dioxide will produce significant impacts on climate, then we must have equally good reason to suspect that even the small increase we have already produced may have subtly altered our climate," he concluded. "Climate change is not a matter for the next century; we are most probably doing it right now."[53]

Edward L. Miles, a professor at the School of Marine Affairs at the University of Washington, agreed. He ventured that not only that people would underestimate how rapidly changes might occur but also would overestimate our ability to deal with them. History showed that human societies were highly sensitive to climate and that small changes could produce large social effects. Yet governments were slow to respond, because the "direct incremental costs of regulation occur in the present, while the benefits and total societal costs are in the distant future."[54]

But it soon became clear that the costs were not so distant. When the 1980s produced the warmest decade on record and the US midwestern states experienced searing drought, some scientists became convinced that climate change was under way and the costs were beginning to be felt.[55] In the summer of 1988, the US Senate held hearings on the question, where James E.

Hansen, the director of the NASA Goddard Institute for Space Studies, testified that the human fingerprint on the climate system had been detected, although not that any specific drought could be attributed to it. Hansen's testimony was reported on the front page of the *New York Times* under the headline "Global Warming Has Begun, Expert Tells Senate."[56]

The *Times'* revered science writer, John Noble Wilford, characterized the testimony as transformative, quoting Environmental Defense Fund scientist Michael Oppenheimer saying that Hansen had added "an air of scientific respectability and immediacy to the environmental movement's longstanding warnings about the build-up of carbon dioxide in the atmosphere."[57] But others—including many of Hansen's colleagues—were not yet convinced. Both basic physical theory and climate models predicted that global warming should occur; scientists agreed on that. The National Academy of Sciences had already said so. But was it already occurring? Climate is naturally variable and weather records are messy and inconsistent; could one say with certainty that the observed effects were something other than measurement error, instrumental noise, or natural variability?[58] Had Hansen really observed the signal above the noise? Blame it on scientific conservatism, reticence, erring on the side of least drama, or plain competitiveness, but most of Hansen's colleagues felt that more work needed to be done.[59]

Taking the Ocean's Temperature

Throughout the history of the earth sciences, effects have often been observed before scientists could account for their causes. Geologists knew there were ice ages before they could say why, they recognized giant thrust sheets in the Alps before they could explain how they got there, and arguments for continental drift were built on empirical evidence before anyone understood mantle dynamics.[60] Global warming was different. It was the converse of what earth scientists were used to. The cause was established—no one questioned that carbon dioxide was a greenhouse gas, responsible for the balmy state of our planet—and even the most ardent skeptics accepted that its atmospheric concentration was on the rise. The question was whether that cause was yielding an effect. Oceanographers began to think that they might have the means to answer that question.[61]

In 1971, the dean of the Graduate School of Oceanography at the University of Rhode Island, John Knauss (1925–2015), had come to the same conclusion as von Arx: that the future of oceanography lay in addressing global environmental questions. Writing to a colleague at the National Oceanic and Atmospheric Administration, he suggested global monitoring as a future direction in oceanographic science: "Concern for the environment is not a passing fad. The problems are not going to go away."[62]

Revelle agreed, and in 1982 summarized the issue in *Scientific American*. The carbon dioxide question was really a set of questions. One was how much carbon dioxide would be added to the atmosphere in the coming years. This was a question of human activity, primarily clearing forests and burning fossil fuels, and there was little reason to suspect that those activities would end any time soon. A second was whether this increase would change the global climate; the consensus of scientific opinion was that it would. The third was whether changes were already occurring. This, he argued, was the question of the hour: "Mathematical models of the world's climate indicate that the answer is probably yes, but an unambiguous climate signal has not yet been detected."[63] To find that unambiguous climate signal, oceanographers proposed an ambitious project: acoustic tomography of ocean climate, or ATOC.

Scientists today accept that there has been an increase in global average temperatures of about 1°C since the Industrial Revolution; in 2014 the Intergovernmental Panel on Climate Change (IPCC) declared that the warming of the climate system was "unequivocal."[64] But why did it take so long—more than fifty years if we count from Revelle and Suess and Plass, more than a century if we count from Arrhenius—to reach this conclusion? Part of the answer is the difficulty of interpreting historical temperature records, which are noisy, of variable quality, and geographically clustered. To obtain global averages from historical records involves numerous inferences and assumptions; there is no thermometer that permits direct measurement of the average temperature of Earth.[65] But what if you could measure the Earth's average temperature? In the early 1980s, a group of oceanographers at Scripps and Woods Hole posed this question, and their answer was that you could, by using the speed of sound in the oceans.

The idea was derived directly from Cold War work. As we have seen in previous chapters, sound transmission was a major motivation for Navy support of oceanographic research—a centerpiece of the SOFAR navigation system and the SOSUS machinery of underwater acoustic surveillance—and the link between sound velocity and temperature was very well studied. (At Scripps, some scientists had jested that their institution should be renamed the Scripps Institution of Underwater Listening and Location.[66]) At the center of all this work was the recognition that the speed of sound in water is temperature dependent, and variations in temperature (and density) alter sound transmission.

This central fact—that the speed of sound in water is temperature dependent—could be applied to the question of global warming, because if the planet as a whole were warming, then necessarily the oceans were as well. If the oceans were warming, then the speed of sound through them should be increasing. Since the oceans cover three-fifths of the globe, an average ocean temperature would be a measure of the average global temperature. Moreover, it would be a better measure of global warming than atmospheric measurements, because the oceans are less temporally and spatially variable. As Walter Munk liked to put it, the oceans have less "weather" and more "climate."[67]

But how, exactly, do you take the temperature of the ocean? One can no more place a mercury thermometer into the ocean to get a global average than into the air. But there is a way to measure the ocean temperature. Because sound in the ocean can travel for very long distances—this was the insight that led to SOFAR—its speed over those distances integrates the average temperature of the water along the way. Release sound from a high-intensity source and record the travel time to a receiver, and in effect you measure the average temperature of the water mass through which the sound has traveled. A long-range transmission, say from Honolulu to Half Moon Bay, California, would integrate the thermal conditions of the water between those two points. Acoustic velocity actually *is* a thermometer.[68]

Ocean acoustic tomography was developed in the 1970s and 1980s by Munk and MIT oceanographer Carl Wunsch (b. 1941) based on an analogy with medical imaging of the interior of the human body and seismic imaging of the interior of the Earth.[69] The technique is acoustic because it relies on sound waves; it is "tomographic" because it creates an image using vertical slices through the water column (CT scans are computer-assisted tomography, from *tomos*, Greek for "cut" or "slice," as in *atom*, the unit of matter that cannot be cut). The measurements from the numerous pathways are integrated to create a picture of the ocean through which the sound has traveled.[70] The technique relies on acquiring and integrating sound waves that have traveled over various possible "ray paths," and the sound channel permits the propagation of low-frequency sound with minimal attenuation (fig. 9.2; see also chapter 2).[71] If scientists took measurements across the world's oceans, collected the ray paths, and analyzed the travel times, then they would in effect be taking the ocean's temperature. If they repeated this over a decade, they could determine whether the oceans were warming and the result would be entirely independent of terrestrial or atmospheric records.[72] Sweet.

In 1974, Woods Hole had hosted the first International Workshop on Low-Frequency Sound, sponsored by the antisubmarine warfare programs of the

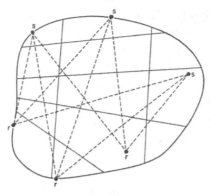

Fig. 1. Perturbations in acoustic travel time from any source *s* to any receiver *r* (dashed paths) are used to estimate sound speed perturbations in an arbitrary grid area (solid lines).

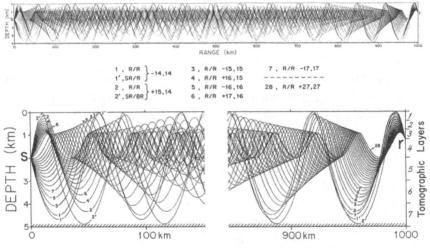

1 , R/R	}-14,14	3 , R/R -15,15	7 , R/R -17,17
1' , SR/R		4 , R/R +16,15	- - - - - - - - -
2 , R/R	}+15,14	5 , R/R -16,16	28 , R/R +27,27
2' , SR/BR		6 , R/R +17,16	

Fig. 2. Ray diagram for a 1000-km range, using the mean sound speed profile along a path between Eleuthera and Bermuda. The source is at 2000-m depth, the receiver is axial at 1280 m. For simplification only those rays that travel downward towards the receiver are plotted. The first and last 150 km are shown on an enlarged scale, with the earliest nine rays labeled in order of their arrival; the associated ray identifiers [equation (3)] are tabulated. The scale 1 to 7 on the bottom right identifies the layers used in the vertical tomographic slice (Table 2, Section 5). We are indebted to John Clark of the Institute for Acoustic Research for this figure.

Figure 9.2 Multipaths from source to receiver.

From Walter Munk and Carl Wunsch, "Ocean Acoustic Tomography: A Scheme for Large Scale Monitoring," *Deep Sea Research Part A: Oceanographic Research Papers* 26, no. 2 (1979): 123–61 (figs. 1–2). Reprinted with permission from Elsevier.

Office of the Chief of Naval Operations. Introductory remarks were given by Brackett Hersey, a longtime Woods Hole scientist with naval connections dating to the 1930s (chapters 3 and 6). Hersey linked the Navy's interest in underwater sound studies to the need to "forecast the performance of existing sonars."[73] Sound had been used to identify seamounts and underwater volcanoes, to reveal currents, gyres, and eddies, and to locate missile impacts and the position from which the lost nuclear submarine *Scorpion* had last been heard just before it disappeared in the Atlantic Ocean in June 1968 (chapter 7). There were thus many excellent opportunities to advance understanding in the context of "the potential enemy's capabilities and characteristics."[74] Underwater sound was as important in the 1970s as it had been in the 1940s; the ONR supported acoustic research at Woods Hole at the level of $1.5 million per year, second only to its omnibus grant in dollar value.[75]

ATOC built on this history of Navy largesse and interest; it also drew on well-tested hardware. The SOSUS network would provide the equipment needed to detect the sound transmissions, which were the same as the transmissions used for ocean acoustic surveillance of Soviet submarines. The scientists would rely on a technology that was well tested, well maintained, and well funded and that had global reach.[76]

The Conceptual Groundwork

In 1979, Munk and Wunsch laid the conceptual ground for ATOC in a forty-page paper published in *Deep Sea Research*, "Ocean acoustic tomography: a scheme for large-scale monitoring." They built on studies carried out on behalf of the Navy in the 1960s, when Gordon Hamilton, working at the Columbia University Geophysical Research station in Bermuda, in conjunction with the Navy SOFAR station, had used the Atlantic Missile Range hydrophone array to detect temperature variation on a scale of ten to one hundred kilometers (fig. 9.3; on Hamilton, see chapter 5). The concept also rested on geophysical inverse theory—a mathematical technique used in seismology to determine Earth structure from seismic wave arrival times. John L. Spiesberger, Munk's former graduate student and postdoctoral fellow, fleshed out the details.[77] Writing in the *Journal of the Acoustic Society of America*, Spiesberger and colleagues demonstrated that the expected multipaths could indeed be resolved and identified using ray theory. While transmitter drift posed some challenges, overall the differences in travel times along reciprocal transmissions revealed clear differences in propagation speed along different paths, demonstrating that the project was feasible not just in principle but in practice. Most important, they had demonstrated that they could separate the noise of short-term and moderate scale ocean "weather" from the global signal of "climate."[78]

Munk and Wunsch's paper had originated in a JASON committee summer project supported by the Defense Advanced Research Projects Agency (DARPA); both their work and Spiesberger's had also been supported by the

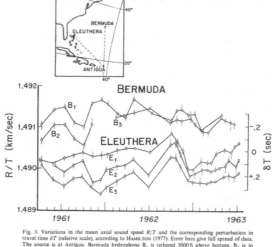

Fig. 3. Variations in the mean axial sound speed R/T and the corresponding perturbation in travel time δT (relative scale), according to Hamilton (1977). Error bars give full spread of data. The source is at Antigua. Bermuda hydrophone B_1 is tethered 5000 ft above bottom, B_2 is in shallower water on bottom, both are at 875 m beneath the surface. Eleuthera hydrophones E_1, E_2, and E_3 are all on bottom at about 1000-m depth.

Figure 9.3 The idea of ocean acoustic tomography, showing "variations in the mean axial sound speed R/T and the corresponding perturbation in travel time δT (relative scale), according to Hamilton (1979)."

From Walter Munk and Carl Wunsch, "Ocean Acoustic Tomography: A Scheme for Large Scale Monitoring," *Deep Sea Research Part A: Oceanographic Research Papers* 26, no. 2 (1979): 123–61 (fig. 3). Reprinted with permission from Elsevier. The original data come from Gordon Hamilton.

ONR.[79] In 1981, they joined forces with ten other colleagues to form the Ocean Tomography Group, with support from the ONR and the National Science Foundation. In a paper published in *Nature* in September 1982, they reported on the results of a proof-of-concept tomography experiment in a three-hundred-kilometer square southwest of Bermuda; the results were consistent with direct measurements from ship- and airborne surveys.

Ocean temperature could of course be measured locally in various ways; the power of the ATOC concept came from the large volume of ocean water that tomography sampled. As we saw in chapter 2, sampling had long been the bane of oceanographers; traditional methods—Nansen bottles, bathythermographs, CTDs—were drops in the bucket, sometimes literally.[80] How did one know if samples were representative? Acoustic tomography solved that problem, as well as the problem of short-term temporal variation, allowing scientists to see through the ocean weather to the ocean climate. Hence, the name: acoustic tomography of ocean climate.[81]

While the *Nature* article portrayed acoustic tomography primarily as a new sampling technique—a form of remote sensing—Spiesberger and colleagues were strengthening the link from ocean to global climate. Spiesberger had moved from Scripps to Woods Hole and wrote a draft proposal to the ONR for a feasibility study of an ocean acoustic thermometer to measure global climate change. Initial studies had been done in a small area of the Atlantic where the Navy had long done acoustic work; the next step would be a test over a larger area, using a transmitter in Hawaii and a receiver off the California coast: "The purpose of this measurement is to determine if the received signal to noise ratio is sufficient to permit tomographic inversion of the multipath arrival times which would provide information on the average heat content of the ocean. If this measurement is successful, a future proposal will be made to establish a series of acoustic monitoring stations in order to measure variations in the ocean's thermal structure. These measurements would be made over long periods of time perhaps spanning decades or more."[82]

Spiesberger and his colleagues explicitly linked their proposal to anthropogenic climate change: the point of pursuing measurements for "decades or more" was to track the effects of global warming as carbon dioxide continued to accumulate in the atmosphere, much as Revelle had proposed twenty-five years earlier. Because the thermal effects should become increasingly evident over time, the longer the timeframe the clearer the results would be: "One can imagine measurements extending for 100 years or more where perhaps the gradual heating of the oceans due to the increase of carbon dioxide could be detected. Just as astronomers have established observatories where measurements have been taken for hundreds of years, the oceanographers might establish an acoustic observatory of the type described herein."[83] As Spiesberger recollected to Henry Stommel in 1989, "our intention [was] to set up acoustic observations to detect hypothetical greenhouse effects on climate change" over the long term.[84]

Contemporary scientists frequently lament the difficulty of undertaking long-term monitoring programs, in part because the reward structure of science favors projects that can be brought rapidly to publication. (Dave Keeling famously recounted late in his life how hard it had been to sustain support for the carbon dioxide monitoring program for which he ultimately won the National Medal of Science.[85]) Despite the challenge, these scientists were prepared to undertake a long-term ocean climate-monitoring program because they recognized the significance of the question, as well as the rewards that would come to any scientist or group of scientists who could provide a conclusive answer.

Von Arx had thought that changing the mission to weather and climate would enable oceanographers to escape the Navy yoke, but there is scant evidence that the scientists who developed ATOC had that in mind. On the contrary, they relied initially on their existing Navy patrons and designed the project to build on Navy technology and facilities. Rather than construct a new purpose-built observatory, for example, the ONR would support the use of data-processing and recording equipment at the Cooley Electronics Laboratory at the University of Michigan and the Navy communications facility in Centerville Beach, California.

Meanwhile, at Scripps, Ed Frieman (1926–2013) had embraced both the general idea that oceanographic institutions should pursue climate change research and the specific idea of ATOC. A distinguished physicist and science administrator whose career had included a stint as director of the Office of Energy Research and assistant secretary of energy in the Carter administration, Frieman had a long-standing interest in energy and environmental issues. In the aftermath of the Arab Oil embargo of 1973–1974, he had been involved in discussions of alternative energy possibilities, including nuclear fusion and synfuels, as well as DOE support of atmospheric science research. Like John Knauss, Frieman had come to believe that global change was the next most important scientific question for oceanographers to address. In 1987, he wrote to the dean of graduate studies at the University of California, San Diego, discussing research priorities. "Global change" would be the number-one priority for the future, as "momentum was building in Washington," and Scripps was "well-positioned to assume a substantial leadership role."[86] ATOC fit that framework.

Two years later, Munk published two articles proposing a full-scale test of acoustic tomography to detect global warming, based at Heard Island in the southern Indian Ocean. It was a remote location—logistically very difficult—but technically it was ideal, because there were unimpeded ray paths from the island into all five world oceans, and these paths could reach research stations in Brazil, India, South Africa, Australia, and the east and west coasts of the United States (fig. 9.4). The specific choice of Heard Island was inspired by an accident: in 1960, three hundred pounds of high explosives were inadvertently detonated in the sound channel near Perth, Australia, and the sound waves had been detected in Bermuda. If the world oceans were warming, they would detect it.[87]

Meanwhile Spiesberger and his colleagues had preliminary results from the northeastern Pacific. In three sets of measurements, using a set of moored arrays in the northeastern Pacific, they had collected arrival times for five to ten sound paths covering a distance of approximately three thousand meters and demonstrated that the paths were similar from year to year at each receiver location (fig. 9.5). They summarized these results in 1991 in the *Journal*

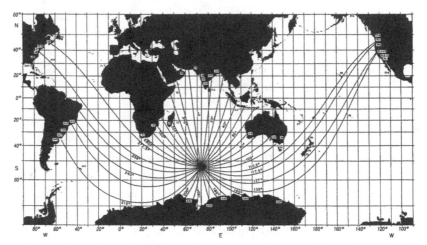

Figure 9.4 Location of Heard Island, showing ray paths to all the world's oceans and oceanographic research institutions on five continents.

From Walter Munk and A. M. G. Forbes. "Global Ocean Warming: An Acoustic Measure?" *Journal of Physical Oceanography* 13 (1983): 1765–77 (fig. 5). Reprinted with permission of the American Meteorological Society; permission conveyed through Copyright Clearance Center, Inc.

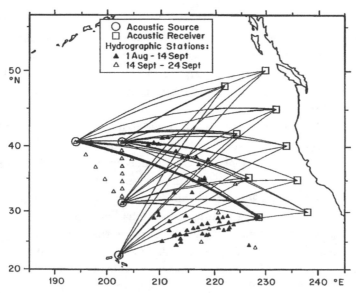

Figure 9.5 The location of the acoustic tomography experiment, with locations of four sources and nine receivers in the northeastern Pacific.

From J. L. Spiesberger and Kurt Metzger, "Basin Scale Tomography: A New Tool for Studying Weather and Climate," *Journal of Geophysical Research* 96, no. C3 (1991): 4869–89 (fig. 1). Reprinted with permission of John Wiley & Sons, Inc.

of Geophysical Research, as well as in a Woods Hole internal report, concluding that the results indicated "that it is possible to accumulate long records of arrival times that might be used to characterize the thermal state of the ocean," and thereby address "the problem of measuring oceanic thermal changes associated with global warming brought on by increases in greenhouse gases."[88]

A Small Glitch

The Heard Island Feasibility Test was scheduled to begin in January 1991. Low-frequency sound from a single source near Heard Island would be monitored at sixteen sites around the globe. Climate models suggested an expected anthropogenic warming signal of .020°C per year at the ocean-atmosphere boundary, decreasing to 0.005°C per year at a depth of one kilometer. This was equivalent to a sound speed change of about 0.1 second per year over a thousand-kilometer path, a very small amount. To achieve this level of sensitivity, the test would have to be able to identify and measure ray-path travel times to an accuracy of ten to fifty milliseconds. From a technical standpoint, the key question was whether a coherent signal could be identified over such distances and the travel times measured to the required degree of accuracy; Spiesberger's data suggested they could. Everything was going according to plan. Everything, that is, until the concern arose that the sound transmissions might adversely affect marine life.

Munk explained a few years later: "The issues in [the Heard Island Feasibility Test] were: can signals generated by currently available acoustic sources be detected at ranges of order 10 Mm [megameters, i.e., thousands of kilometers], can coded signals be "matched filtered" to measure travel time to better than 0.1 s, and can this be done without harm to local marine life?"[89] This, however, was a bit of revisionist history, because in the early ATOC proposals there is no discussion of potential harm to marine life. Tomography had been conceptualized and developed by physical oceanographers and engineers, and the prospect of interfering with marine life had evidently not occurred to them (or at least not strongly enough to discuss in their papers or proposals). Perhaps this was because they were making use of established technology, whose safe operation they took for granted. Perhaps because no biologists were involved in developing the project, the issue never came up. Or perhaps because, as Robert Oppenheimer famously said about the hydrogen bomb, it was so technically sweet that it just drew you in.

However, quite a few marine animals hear in low-frequency sound ranges, most notably whales, and in the United States the Marine Mammal Protection Act of 1972 and the Endangered Species Act of 1973 prohibit the "taking" of any federally listed endangered or threatened species without legal authorization. "Taking" is defined very broadly, to include harassing, harming, pursuing, shooting, wounding, killing, trapping, capturing, collecting, or attempting to do any such things. Because many species of whales are endangered or threatened, the project came to the attention of the National Marine Fisheries Service (NMFS). Officials there concluded that legal authorization was required.

But the project was set to go, and given its technical complexity, as well as the involvement of colleagues from around the globe standing by to monitor the received transmissions, it would be no small task to reschedule, so project participants drew on their contacts in Washington to expedite the approval process.[90] Meanwhile, they set sail from Fremantle, Australia, hoping the permits would arrive in time. They did, but only just: eight days before the scheduled start of transmissions. They also came with stipulations. Four marine mammal observers were to be aboard a dedicated survey vessel, including a bioacoustician to conduct acoustic monitoring. Three additional observers traveling with the sound source would "monitor the effects of the transmission on marine mammals close to the source," with comparative observations made before and after each transmission. If "marine mammals are sighted or heard ... within 5 km of the source," or "in the event of injury or mortality of one animal," the transmissions were to be delayed or suspended.[91] The biological observations were to be submitted to the NMFS within ninety days of completion of the experiment for independent evaluation. With little time to spare, the oceanographers recruited a local biologist, Ann E. Bowles, who studied animal bioacoustics at Hubbs-SeaWorld in San Diego, to perform the biological monitoring.[92] In 1994, Bowles would earn a PhD at Scripps, but at the time her highest degree was a BA in linguistics.

The transmissions were completed in five days, and the experiment was over by the end of the month.[93] The scientists noted that the sonar clicks of sperm whales were notably absent during the transmission periods but concluded that the whale silences were not "associated with long-term effects."[94]

The referent of the acronym ATOC had by now evolved into acoustic *thermometry* of ocean climate—perhaps to better communicate its intent—and the scientists applied for and received funding from a new US federal government program, the Strategic Environmental Research and Development Program, or SERDP, which was established by the US Congress in 1990 as a program within DARPA to make military systems available for civilian scientific research and to "harness some of the resources of the defense establishment ... to confront the massive environmental problems facing our nation and the world today."[95]

Between the time that the scientists had first developed the idea of an ocean acoustic thermometer and when they first began to use it, the Cold War had ended. Despite Frieman's anxieties, this was good news, as SERDP would support research that either made use of existing military technologies and databases to address environmental issues, or developed new knowledge and technologies that helped US agencies to address their environmental obligations. It was a swords-to-plowshares program, except that the swords were actually still swords. The topic of study had shifted from Soviet submarine surveillance to global climate surveillance—but the hardware, the scientific know-how, and the patrons were the same.

From Heard Island to ATOC

In February 1993, Scripps was awarded a $35 million contract to run ATOC; phase 1 would run for two years from February 1994 through January 1996. Given the logistical difficulties of working in a location as remote as Heard Island, they would move closer to home, using Navy facilities in Hawaii and California and placing acoustic sources in the nearby ocean. For receivers, they would rely on the existing bottom-mounted, horizontal hydrophone arrays maintained by the US Navy as part of SOSUS System, as well as several additional vertical mounted arrays.[96] During the proposed initial two-year project period, a twenty-minute signal would be released every four hours, up to six times per day.

To manage such a large project with complex logistics, Scripps contracted with a private science and engineering firm with extensive military contracts, Science Applications Inc. (SAIC). Among other things, SAIC helped prepare the necessary permit applications. The scientists understood that they would need permits, particularly as the California source would be located on Sur Ridge (and linked by a twenty-two-mile cable to the Point Sur Naval Facility on the California coast), at the southern end of the Monterey Bay. This was an area of California famous for its spectacular coastline with abundant marine life, and with a human population highly attuned to environmental issues. Moreover, the source itself would be within the boundaries of the Monterey Bay National Marine Sanctuary, a federally protected marine area (fig. 9.6). Before a full-scale program could go forward, environmental concerns would have to be addressed. The scientists began to gather information on pertinent species, particularly whales and sea turtles, and to plan a marine mammal study program to accompany ATOC.[97]

A major question was whether they would have to obtain a "small-take exemption"—a formal legal release from the laws prohibiting activities that might adversely affect threatened or endangered marine species—or whether

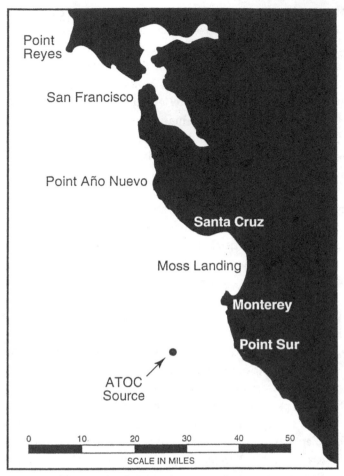

Figure 9.6 Location map of proposed ATOC source in the Monterey Bay National Marine Sanctuary.

From Box 12, Folder "Marine Mammal Correspondence Pre-September 1993," Acoustic Thermometry of Ocean Climate Project Records, 1991–1996, Scripps Institution of Oceanography Archives, SAC 32, Special Collections & Archives, UC San Diego. Reprinted with permission of the Scripps Institution of Oceanography.

they could proceed without it. According to an official with the Fish and Wildlife Service who reviewed the ATOC proposal, the term *harass* in the Marine Mammal Protection Act would cover any "intentional or negligent act or omission that creates the likelihood of injury to wildlife by annoying it to such an extent as to significantly disrupt normal behavioral patterns which includes breeding, feeding, or sheltering."[98] Because the acoustic transmission would penetrate a large portion of the Pacific, the potential harassment was considerable.

In March 1992, Bowles wrote to the Office of Protected Resources at the NMFS with a request to proceed without a formal exemption. While discussing plans for the Hawaii source to be located at Kepuhi Point on the island of Kauai (at eight hundred meters depth, with a source transmitting at 70 Hertz and 195 decibels) she argued that there would be no impact and therefore no take: "We do not anticipate any deleterious effects to the hearing, migration, communication or reproduction of marine mammals off the island of Kauai, although some species may avoid the transmission site during the first few months of exposure."[99] What was the basis for this conclusion? Small cetaceans and monk seals have poor hearing in the relevant frequency range, and humpback whales, which do hear well in that range, swam through the region only for a few months. Therefore, she doubted that there would be major effects. "Within the zone of influence, humpback whales may alter swim direction and exhibit subtle changes in behavior until they habituate," she speculated, but this was too minor to constitute a take.[100]

NMFS representatives saw the matter differently. They concluded that under the law any behavioral changes would represent a take, and so the ATOC team would have to apply for a small-take exemption. This involved formal public hearings, followed by a 60-day comment period, a 120-day period for agency response, and whatever additional time it took for the Office of Management and Budget to approve the exemption. NMFS officials did not suggest that the scientists would not, in the end, get the exemption, but they did suggest that the process would likely take a year.[101]

Bowles and her colleagues were shocked at the prospect of such a long delay for what seemed to them purely bureaucratic reasons, and they tried to find a way around it. Exceptions to the rule were permitted for a small number of specific reasons falling into two main categories. The first was commercial activity, which ATOC clearly was not.[102] The second was for scientific activity in aid of the purposes of the Marine Mammal Protection Act, including survival and recovery efforts such as captive breeding and bona fide scientific research that might aid such efforts.[103] You could do scientific research that might disrupt marine mammals in the short run if the goal were to understand and better protect them in the long run.[104] That was not the purpose of ATOC, although the idea of pretending that it was occurred to them, as Bowles explained to David W. Hyde, a physical oceanographer at SAIC who was serving as ATOC project manager.[105] She wrote, "You can't get a scientific research permit because ... NMFS already knows you are doing the work for other reasons."[106]

Bowles lamented that none of the required steps—public hearings, public comments, agency responses—were (in her opinion) "really necessary to protect the marine mammals; in fact they foster resistance among agencies and commercial operations to approach the NMFS at all. Other sources of noise, such as shipping, most tomographic experiments, and private vessels are completely unregulated. Therefore, as usual, the scientific community is getting picked on." She noted that the Marine Mammal Protection Act was up for reauthorization and suggested that the Department of Defense "put some lobbyist to work to try to get the regulation changed."[107]

Bowles recognized, however, that the law was the law (at least for the time being), and so she focused her attention on getting the small-take exemption. If she succeeded, then at the very least the project scientists would have "no problems with stupid and useless monitoring requirements above and beyond what we have already agreed to, if we can just get through the regulatory paperwork."[108] Bowles was not alone in feeling that Marine Mammal Protection Act regulations were "stupid and useless"—many of her colleagues felt the same way. They also shared the view that scientists involved in the project were the most qualified to judge its impacts. Again, the NMFS viewed the matter differently.

Estimating the Take

The NMFS guidelines required the scientists to calculate the numbers of animals likely to be "taken," based on numbers of marine mammals and other species that inhabited the affected waters, swam at depths where the transmissions could be heard, and were able to hear in the relevant frequency range. These were not trivial estimates to make, particularly because there were limited data on marine mammal numbers and habits. Moreover, because the ATOC signal was purposefully designed to penetrate the entire ocean, the affected area was the entire ocean—or a great deal of it—and on some interpretations one might end up "taking" practically every whale in the sea. Even Bowles admitted, "It looks like we're going to get a permit to take an astronomical number of whales."[109]

Bowles thought that this would not be a problem—"apparently no one raised an eyebrow about the 386,000 marine mammals we were supposed to disturb at Heard Island"—but she was concerned that an early draft application made it seem as if ATOC was a done deal. This, she worried, could cause problems. "I'd like to suggest a couple of minor changes that may save us some trouble later," she wrote to David Hyde. "You should emphasize that you are developing this Acoustic Observatory rather than treating it as a *fait accompli*. This covers your backside and emphasizes the fact that this project is really research."[110]

Meanwhile, Bowles had solicited advice from Christopher W. Clark, a biologist-engineer at Cornell specializing in the acoustic response of marine mammals, and Daniel P. Costa, professor of ecology and evolutionary biology at the University of California, Santa Cruz. Clark had worked with the Navy on Whales 93, an initiative to use SOSUS—by then renamed the Integrated Undersea Surveillance System (IUSS)—to locate and track whales on ocean-basin scales; Costa was working with the ONR on the announcement of a program in low-frequency sound and in the process of developing his own grant proposal to study sound effects on elephant seals.[111] Bowles wanted their input on a draft plan, to be included in the permit application, for a biological component to the ATOC research program.

Both Clark and Costa were disturbed by what Bowles had sent them. They found the plan poorly crafted and thought that it read very much as if ATOC were a *fait accompli*. The biological work at Heard Island was too short and too limited to demonstrate anything, and it was misleading for the ATOC scientists to suggest otherwise. What was needed, both felt, was to draw a clear distinction between monitoring and research, and to propose and execute a properly designed research program. The proposal as it stood was formulated as a monitoring program—to observe and detect any adverse effects; both Clark and Costa argued this was a bad approach, scientifically and politically. Scientifically, it did not constitute research for the simple reason that it would not answer any scientific questions. If you did observe effects, you would not be able to explain why, and if you didn't, you'd be stuck with the problem of equating absence of evidence with evidence of absence. Either way, you were trying to do something logically impossible: to prove a negative. Politically it looked precisely like they were covering their derrières, trying to avoid the biological issues more than to address them. They agreed with Bowles that the biological program needed to be "really research," but so far it wasn't.

What ATOC needed, Clark argued, was a well-structured scientific research program, capable of determining whether systematic behavioral changes were occurring and, if they were, whether those changes were comparable to other changes associated with oceanographic or meteorological variables. The studies should include analysis of species-specific vocal rates and repertoires, location and tracking of individual whales, characterization of whale migration tracks and corridors, and evaluation of species-specific spatial and temporal distribution; this degree of detail was necessary to produce meaningful results. Solid data on these matters would enable scientists to determine whether ATOC changed anything and also contribute to basic understanding of marine mammals. In short it would be real research. It would also help to address another concern. Cetacean biologists were following the project closely, and a properly structured research program would help to persuade them that ATOC would be beneficial rather than harmful. As Clark put it, "Whales are not going to go away."[112]

A month later Clark was frustrated that the proposed program was still vague and structured only to passively monitor the mammals, which was unlikely to satisfy the cetacean biologists. In a letter to Hyde, Clark warned that the marine mammal community was very aware and "very sensitive to the ATOC program," and that "we will receive an unusually large amount of attention and scrutiny." He concluded emphatically: "WE MUST HAVE A PLAN AND CLEARLY DEFINED, COHERENT OBJECTIVES OR WE'RE GOING TO [GET] RELATIVELY POOR RETURNS ON THIS FANTASTIC OPPORTUNITY. WE MUST BE VERY SUCCESSFUL IN THIS RESEARCH PROGRAM, NOT JUST MILDLY SUCCESSFUL."[113]

Costa agreed, blaming their problems on Bowles's inexperience: "After reading over the permit [application] as it stands now, I'm not very optimistic. It is way too vague.... NMFS is very concerned with actual numbers, not the vague statements made by Ann. After review of her draft, it is apparent that she has never written an NMFS marine mammal permit before.... I have written far too many of these damn things in the past and know only too well the kinds of comments NMFS is likely to ask.... You can put it in as it is, but I guarantee you will get it back."[114] Later that day, after speaking with several folks at NMFS and the Marine Mammal Commission, he sent a second email: "NMFS is concerned that Ann doesn't have a clue as to what the issues are and I agree."[115]

Costa also agreed that it would not be enough merely to get *some* data; they needed enough data of sufficient quality to answer the key biological questions in a convincing manner. Their goal should not be to get "past the paperwork" but to do the job right. "The goal is to have a well-designed scientific program that is defensible and stands on its own.... The future after [the] ATOC pilot program is contingent on quantitative data on the impact or lack thereof of ATOC emissions [*sic*] over the next 30 months. Just getting a permit is half the problem. Getting solid data, so that 2 years from now you can go to the table and say, 'look here are data that show.... This is what I am trying to help you achieve.'"[116]

To their credit, Bowles and Hyde took this advice to heart and worked to improve the specificity and focus of the proposed marine mammal research program as part and parcel of the first phase of ATOC.[117] In the end, $2.9 million dollars would be earmarked for a scientific research program that would involve acoustic tracking of animal response to the ATOC transmissions, direct observations through aerial surveys, and monitoring of specific individual marine mammals equipped with tracking devices. The revised application emphasized that this would not be just a monitoring program, designed to detect adverse impacts of ATOC, but a research program to improve scientific knowledge and understanding. They further stressed that full-scale acoustic tomography would not proceed until phase 1 resolved the question of biological impacts. In September, the scientists submitted a draft application to the NMFS for a Scientific Research Permit, with a small-take exemption, under the Marine Mammal Protection Act.[118] After a further round of revisions based on comments from the NMFS, a revised application was submitted on December 8, 1993.[119]

The application and Hyde's cover letter emphasized the features intended to minimize the impacts on marine mammals, although in most cases these features had technical justification as well. Transmitting over a somewhat smaller area than the Heard Island test, the scientists would be able to use a somewhat less intense source (195 decibels instead of 206–21) at a slightly higher frequency (70 Hertz versus 57 Hertz), which should lessen the environmental impacts (and was also less expensive). The 70 Hertz frequency was suggested as a better range in which to work because it was similar to noises to which marine life was already exposed. Rather than using a surface source on a boat, they would locate the Point Sur transmitter on a sea mount at a depth of 880 meters. This would place the source below the highly productive surface layers of the sea; it would also maximize long-range transmission in the sound channel. (The Kauai source would be similarly located.) Finally, they added a new element to their protocol: the transmissions would begin with a five-minute ramp-up period intended to permit sensitive animals to swim away before receiving the full impact.

A critical eye might have noticed that both the cover letter and the proposal made no reference to Heard Island and therefore no mention that certain program specifications were changes over the feasibility study design. One might think that Hyde would have wanted to emphasize those changes, to stress the project scientists' willingness to modify protocols in response to biological concerns. Instead, he downplayed any suggestion of early missteps, insisting that "from the outset, the design of ATOC sound transmissions has been particularly sensitive to the regulations of the Marine Mammal Protection Act."[120]

If by "ATOC" he meant the project as it developed after Heard Island, then this statement was true. But if one viewed Heard Island as the first step of ATOC, it was not. Nor was it true if one considered the scientists' attempt to avoid having to apply for a small-take exemption. In the special issue of the

Journal of the Acoustical Society of America presenting the Heard Island results, seventeen papers dealt with acoustic transmission; only one dealt with biological effects.[121] Chris Clark had already noted this, writing to Hyde in July 1993: "We suffer from the history of the Heard Island Experiment. No matter how you try, people see ATOC as an extension of HI [and they] do not like the way HI was handled."[122] The permit request also referred to the proposed first phase as "experimental research ... to establish the feasibility of a future global ocean climate monitoring program." Again, this was true insofar as the Heard Island work had established only the feasibility of detecting sound waves over the required distances and adequately identifying acoustic arrival times, not the feasibility of a large-scale, ocean-monitoring program. It was also a reasonable characterization, given that the permit under consideration was only for the first two-year phase of the study. This characterization, though, could be read as downplaying the scientists' intent, clearly stated elsewhere, to run an ocean climate-monitoring program over the course of a decade or longer.

Bowles was certainly right about one thing: the permitting process produced voluminous paperwork. And this paperwork makes clear that the physical oceanographers and engineers involved with ATOC presumed that the transmissions would be benign without much scientific evidence to support or refute that view. Repeatedly they argued that transmissions of this type were acceptable because the Navy had already used them for decades. This was dubious on two counts. Most Navy sonar operations involved passive listening; it was not clear how much the Navy had yet operationalized active sonar systems. But even if it had, it was possible—and indeed, there was at least some evidence to suggest it was probable—that past Navy work had done harm but had gone undetected because no one had paid adequate attention. Scientists and Navy officials had assumed the transmissions were benign, but they did not know that to be the case.[123]

The marine mammal research program now proposed was an improvement over what had gone before, but it was not research in the sense of having a primary purpose to produce new knowledge about marine life. Perhaps more to the point, the ATOC scientists faced a clear challenge of overcoming their structural bias. It was not the case that either outcome—harm or no harm—would be equally acceptable to them; for ATOC to proceed, they needed the outcome to be no harm. Biologists might have been equally interested in knowing whether low-frequency transmissions did or did not affect marine mammals, but not the oceanographers. A discovery that the ATOC transmissions had measurable impacts on marine life would be damaging, even fatal, to the project.

Hyde did not manage to transcend this optimistic bias. On the contrary, his overt presumption that there would be no harm was evident everywhere. In his cover letter to NMFS, for example, he wrote: "We believe that the ATOC transmissions will have no significant impact on marine mammals.... As designed, a key program objective *will be to demonstrate that the planned global network can be operated without any adverse effects on marine mammal populations.*"[124]

This was the sort of language pharmaceutical companies might be tempted to use in planning a clinical trial of a promising new drug. It was the sort of language that the US Department of Energy used to describe its studies designed to demonstrate the safety of its proposed nuclear waste repository at Yucca Mountain, Nevada.[125] It did not convey epistemic neutrality. Moreover, Hyde's assertion in his cover letter that significant impacts were unlikely was at odds with information found in the proposal itself, which noted that "no studies of marine mammal reactions have been conducted to acoustic sources confined to this part of the acoustic spectrum [70 Hertz]."[126] Most previous studies of response to noise were "short-term, surface-based measurements ... of startle response or transitory avoidance ... a poor method for estimating long-term changes in behavior that could have significant biological consequences, such as increased activity, avoidance of critical habitat, interruption of feeding or failure to find a mate." The bottom line: "Long-term health effects of noise have never been examined in marine mammals."[127]

Bowles had claimed that the behavioral changes observed at Heard Island were "not associated with any long-term effects," but the entire experiment had lasted less than a month. In fact, there were almost no data to support or refute the presumption of no long-term effects. The ATOC scientists would insist they were working to get those data, but the NMFS regulations stipulated that to be eligible for a scientific research permit a project could not be a cover for others activities and must "ultimately benefit the marine mammals studied." This was a sticky wicket, for it was one thing to claim that ATOC would do no harm, another to assert a positive benefit. This was no doubt why Clark had stressed the importance of a bona fide research program that would benefit understanding of marine mammals overall, but ATOC scientists seemed still to be struggling with the point.

In December, Walter Munk became involved, and with Hyde he penned a letter to the NMFS stressing that what they were now calling the Marine Mammal Research Program (MMRP) was an "ambitious effort," for which they had "assembled an impressive group of renowned marine mammal scientists."[128] Bowles had been taken off the case: Chris Clark would direct the program, Dan Costa would be the principal investigator for the California team, and their work would be facilitated by a program manager, Clayton Spikes, chief engineer with a firm known as Marine Acoustics Inc. Spikes had served as technical coordinator for IUSS applications in cetacean research, marine mammal tracking, and fishing policy enforcement.[129] The scientists would also establish an independent expert advisory board.

The biological team would complement "the exceptional group of under-water acoustic scientists involved with the ATOC project." The latter would carefully describe the physical environment of the transmissions to make sure that the conditions under which observations were made were well doc-umented, enabling "verifiable testing for any possible significant changes in their behavior or distributions as a function of measurable data (source range, received level, signal-to-noise ratio, etc.)." The overall goal was a "re-search data collection program that is sensitive enough to detect any po-tential changes, and [will] be able to associate those changes with the char-acteristics of the physical acoustic environment in which the animals are exposed."[130]

It still remained unclear if establishing whether ATOC would hurt marine mammals constituted a "benefit," and the scientists were in a double bind. No one knew—and there was no way to know—whether these studies were likely to have adverse effects. The only way to find out was to do the studies, which, if they did have adverse effects, would be disallowed. As the permit ap-plication noted, "We do not have a basis for estimating the nature or level of these influences without an experimental program that exposes significant populations of potentially affected species over a time span of one or more seasons."[131]

This would be a key issue in the public discussion that ensued. There was no reason to suppose that ATOC would be of major benefit to marine life and a great deal of concern that it could do harm. This concern was felt not just by ordinary citizens but also by scientists at other universities, in federal agencies, and on the staff of environmental organizations.

Public Outcry over ATOC

Public hearings on ATOC were announced in the *Federal Register* on Febru-ary 3, 1994, but word was already spreading in the marine mammal commu-nity. By Scripps' own estimate, the ATOC take could be up to 670,000 animals per year, encompassing ten different species of whales, eight species of dol-phin and porpoise, and various seals, sea lions, otters, and turtles, including several threatened or endangered species.[132] While the word *take* in this context meant any effect, no matter how small, biologists began to dispute the claim that the effects would be small.

As word of the project spread, opposition grew among marine biologists, conservationists, and, especially, aficionados of whales. As one conservationist put it, "whale lovers went wild."[133] Led by Dalhousie University biologist Hal Whitehead, a sperm whale expert, and Linda S. "Lindy" Weilgart, a postdoctoral fellow at Cornell University's Bioacoustics Research Program, opponents of the project took to the internet, creating a Listserv (marmam@uvvm.uvic.ca) based at the University of Victoria.[134] They soon had over 1,500 subscribers. Postings warned of severe damage to marine mammals and suggested that the rushed character of the original permitting process was a deliberate attempt to avoid public discussion.[135] While many of the emails were quite emotional, the people posting them were not necessarily ignorant of science. Many of the postings came from graduate students, offering well-informed discussion of acoustics, tomography, and even Walter Munk's status in the oceanographic community.[136]

Senior members of the scientific community also weighed in. Louis Herman, director of the Kewalo Basin Marine Mammal Laboratory, Honolulu, publicly emphasized a point noted by Bowles in her 1992 correspondence with the NMFS—that the ATOC signal fell within the frequency band of the humpback whale song. He did not agree that the effects would be minor until the whales "habituated." Rather, he thought ATOC might obscure the whales' sounds, rendering them less detectable or even unrecognizable.[137] The ATOC permit acknowledged that long-term effects would be difficult to detect, he noted, yet it was precisely such long-term effects that were "of the greatest concern."[138]

As it became clear that the hearing would be contentious, Scripps director Frieman sent a letter defending the project to a long list of senators and representatives, urging them to send faxes to the NMFS in support of the project; he supplied both the NMFS fax number and a sample letter. Frieman sought to steer the conversation away from whales and back to global warming, emphasizing that ATOC could provide concrete data to assess whether climate models were correct in predicting measurable effects from the observed increases in atmospheric greenhouse gases: "The current projections of global warming are largely based on computer modeling [and] there are no measurements of ocean temperature which can be used to assess the modeling predictions. ATOC's ability to measure annual change in ocean temperature of as little as 0.005 degrees Celsius over very large ocean areas will fill in a critical missing piece in the global warming puzzle."[139]

Meanwhile, the mass media had picked up the story. Many Scripps scientists would later blame the negative public reaction on an article in the *Los Angeles Times*, published on March 22 to coincide with the first public hearings, with an alarming headline asserting that ATOC could make whales deaf.[140] The widely syndicated article probably did help mobilize opposition, but the project had already been placed in the public eye in a feature story in the *San Diego Union Tribune*, where the possibility of deafness had been raised not by hysterical environmentalists or sentimental whale lovers but by Navy insider Sam Ridgway.

Ridgway was chief veterinarian for the Navy's marine mammal research program, and he had worked for over thirty years at the Naval Control and Ocean Surveillance Center at Point Loma, San Diego. The article explained that the Navy had been using marine mammals for some time, training dolphins, whales, and sea lions to one day "help fight a war." The Point Loma facility housed dozens of marine mammals, more than fifty of which were considered "surplus—some of them retired from active duty stretching back to patrolling Vietnam's Cam Ranh Bay." Three beluga whales, captured in 1977, had been used in experiments related to submarine warfare beneath the polar ice caps; dolphins had been used in sonar research; whales had wires attached to their heads, recording their brain waves. The article suggested that Ridgway's long experience made him uniquely suited to evaluate the impact of the ATOC transmissions, and he felt that it was possible for whales' hearing to become overloaded by the ATOC sounds: "Continued exposure to this degree of sound could result in some degree of deafness," he concluded.[141]

Juxtaposed in the article with Ridgway's conclusion were Walter Munk and David Hyde insisting otherwise. Munk was quoted saying that there was "a great deal of scientific literature that would suggest that the sound levels we're generating do not do any damage." Hyde was quoted as flatly asserting that the ATOC transmissions "cannot cause long-term hearing damage."[142] The next day—March 22—the *Los Angeles Times* ran its front-page story, "Undersea Noise Test Could Risk Making Whales Deaf." The framing was the clash of the titans: on one side oceanographers promising to solve the problem of global warming (and arguing that the potential harm to all marine life from global warming was far greater than the potential harm to a few species from ATOC) and on the other side, marine biologists wanting to save the whales.

The whales were represented by the Cornell postdoc Lindy Weilgart. The ATOC sounds could cause deafness in nearby whales, she asserted, "leaving them unable to navigate or find food." She noted that the ATOC broadcasts at 195 decibels were "10 million times as loud as the 120-decibel levels that were known to disturb some whales," concluding: "We are invading an ocean habitat that so far has been untouched by man. It's an experiment of tremendous implications and we are doing it without a clue of what it would do."[143] In a line that was widely quoted, she concluded that a "deaf whale is a dead whale."[144]

ATOC scientists would seize on these comments as proof that the public had become inflamed on the basis of a misunderstanding. Weilgart's comments about the logarithmic decibel scale would have been accurate had they referred to sound transmission in air, but transmission through water is different, and 195 decibels in water does not have the same effect on an eardrum as 195 decibels in air does.[145] Moreover, anyone who knew anything about the Navy's extensive undersea programs—or the history of fishing or telegraphy—knew that to say that the deep-ocean habitat was "untouched by man" was just plain wrong.[146] In contrast, by ATOC scientists' own account, the ATOC signal was equivalent to 110 decibels in air—a level of noise comparable to a rock band—so even on a technically accurate comparison, the ATOC noise was very loud.[147]

The *Times* article was syndicated in local papers across California, including the Santa Rosa *Press Democrat*, the *Paso Robles Press*, the San Luis Obispo County *Telegram-Tribune*, the Pleasanton *Valley Times*, the Hayward *Daily Review*, the *San Ramon Valley Times*, the Watsonville *Register-Pajaronian*, the Alameda *Times Star*, the Pinole *West County Times*, the San Jose *Mercury News*, and many more. Almost anyone who read a newspaper in California that day would likely have seen it, as well as many people beyond the state, as the article ran in the *Orlando Sentinel*, *Detroit News*, *Denver Post*, and elsewhere. Many of the syndicated headlines were even more inflammatory than the original: several referred to the acoustic source as a "boombox," and the Portland, Maine, *Press Herald* changed risk to fact: "Sound-Blast Proposal Imperils Sea Creatures: The High-Decibel Experiments, Part of Global-Warming Research, Would Harass and Kill Whales and Dolphins."

Perhaps the fairest article was run by the *San Francisco Chronicle* under the headline, "Undersea Plan to Study Global Warming Inflames Foes." This time the framing was biologist versus biologist: Hal Whitehead versus Dan Costa. Whitehead argued that if whales were not actually deafened, they might nevertheless suffer hearing loss and disturbances in feeding and socializing. Costa argued that most scientists were convinced that there was "no evidence" that marine mammals were damaged by the loud, low-frequency noises already part of the underwater environment. "Annoyed," perhaps, but "certainly not endangered."[148]

The US National Academy of Sciences Ocean Studies Board had meanwhile been reviewing the relevant evidence. Their 1994 report offered no clear guidance; the panelists found "almost no quantitative information" on which to base a conclusion. What data were available suggested a wide range of sensitivity among marine mammals.[149] Most marine mammals probably barely noticed low-frequency sound, but others clearly did. Discussing baleen whales (which include the gray, right, fin, and humpback whales), the committee allowed that the effects could "conceivably range from potential hearing loss and gradual deafness for the entire species—and eventual extinction" to "practically no discernible impact."[150]

One might say that this conclusion was singularly unhelpful, except for the fact that the law was clear that the burden of proof was on the group seeking to be exempted from the law. Their obligation was to demonstrate that no harm would be done (so that the law was not violated), or, if some adverse effects were unavoidable, that the overall scientific justification and benefits were clear (so the violation was warranted). This was a standard that the ATOC scientists were challenged to meet, and the hearings did not go well for them.

NMFS Hearings, March 1994

Initials hearings were held at the NMFS headquarters in Maryland. Walter Munk testified on behalf of the project, emphasizing that the oceans were not only an important reservoir of global heat and carbon dioxide but also a "reservoir of ignorance." ATOC could diminish that ignorance and ensure that policy decisions were made on the basis of complete information. Christopher Clark, emphasizing his past work studying the impact of explosives used in oil and gas exploration, supported Munk's conclusion that biological effects would likely be small, and he assured the assembled group that, given his expertise, any harm to marine mammals would be "a particularly acute concern" to him.[151]

Various written statements were submitted in support of the project, including a seven-page letter from one of America's most distinguished scientists, Robert White. White had edited the 1978 special edition of *Oceanus* on oceans and climate, and he had been tracking the issue for a long time. He been among the first American scientists to warn of the dangers of global warming, highlighting the problem in the 1960s in his capacity as first (and only) administrator of the Environmental Services Administration (a precursor to the EPA) and the first administrator of NOAA.[152] In 1979 he served as chairman of the first World Climate Conference, and in the 1980s he continued to track the issue as president of the University Corporation for Atmospheric

Research, president of the National Academy of Engineering, and vice chair of the National Research Council. Probably no scientist in the United States, except possibly Roger Revelle, knew and cared more about the issue of climate change than Robert White.[153]

White argued that information on global ocean temperatures was essential to determine "whether climate warming is unequivocally occurring."[154] While marine mammal protection was important, the need for information about global climate change was urgent, and land-based and satellite measurements were unlikely to be as conclusive as ATOC. The decision was a difficult one, he allowed, but on balance, the threat of global warming outweighed the potential risk to marine mammals.[155]

Biologists did not agree. Besides Whitehead and Weilgart, others testifying against ATOC included Robbins Barstow, a past president of the Cetacean Society. Oceanographers seemed to be taking offense at the questions being posed—particularly that anyone should doubt their good intentions—but Barstow argued that the public and other scientists had the right to "question and debate the merits of this request and its implications for marine mammals and ocean ecology."[156] Inadvertently proving the point, *Ocean Science News* contemptuously dismissed Barstow's views: "The hysteria of some in the marine mammal community has finally reached the point where the love of 'Flipper' is endangering the good and important work of all ocean scientists. It is time to call a halt to this craziness."[157]

Crazy or not, newspapers across the country continued to run the story. The day after the hearings, the *Los Angeles Times* article was picked up by more West Coast papers, including the *Seattle Times* and the *Oregonian*, and the Associated Press released a similar article that traveled to many additional papers. The mantra "a deaf whale is a dead whale" was spreading around the country. The *Los Angeles Times* also published a follow-up article describing how activists were mobilizing to stop the project. A marketing director in Los Angeles was quoted: "This is a nightmare. I've been calling everyone I know. I've been calling senators and the governor. It would be criminal to do this."[158] This article was also widely syndicated; in the San Jose *Mercury News*, the marketing director's quotation was printed under the headline in large bold font.[159]

ATOC was no longer a scientific project being evaluated by scientists on scientific terms. It was a matter of public affairs, and the NMFS had given the public until Friday to submit additional comments. Submit they did. An NMFS spokesman described the public response as "unprecedented," as letters, faxes, phone calls, and emails began to pour into the NMFS offices, not only from ordinary citizens but also from members of Congress.[160]

On March 23, a group of representatives from Pacific Rim congressional districts—Patsy Mink of Hawaii and George Miller, Ron Dellums, and Sam Farr of California—wrote to Commerce Secretary Ronald Brown requesting an extension of the public comment period on the permit applications. (The NMFS is part of NOAA, which is part of the Department of Commerce.)[161] Senator Barbara Boxer asked for the public hearings to be held in California so that affected parties could participate; she was seconded by George Miller of California, chairman of the House Natural Resources Committee; by Gerry Studds of Massachusetts, the chairman of the House Merchant Marine and Fisheries Committee who had also been involved in the discussion of seabed radioactive waste disposal; and by California's senior senator, Dianne Feinstein.[162] Both Boxer and Feinstein also wrote to the NMFS; Feinstein queried whether the experiment could be done at another site with less impact on marine mammals.[163]

Adverse media coverage continued, as editorials against the project began to appear across the country. The San Francisco *Examiner* took a particularly derisive position: "Imagine what it would be like if aliens from space decided—in the name of science—to target the Earth's inhabitants from their orbiting ships with megadecibel blasts of noise that could frighten or deafen many of the people below. Substitute humans for aliens, and you pretty much have the scenario for an experiment proposed by the Scripps Institution of Oceanography."[164] The Ventura *Star Free-Press* called the project "frightful"; the *Seattle Post-Intelligencer* called it "goofy." Alluding to claims that many marine mammals were deaf in the 70 Hertz frequency range, the *Los Angeles Times* suggested that it was the Scripps scientists who were deaf, while the *San Francisco Chronicle* concluded that they were both deaf and dumb: "Whales and dolphins, which are known to have a high degree of intelligence, must be wondering just how lethally dumb their terrestrial mammalian cousins can get."[165] An op-ed writer in the *Santa Barbara News-Press* asked: "These people are all supposed to have college degrees, aren't they? The only rational explanation for this scheme is that the Scrippsites have already run this experiment on themselves, scrambling their brains beyond recognition."[166]

In their attempts to reassure the public, some Scripps scientists further inflamed them. David Hyde was quoted saying, "We are not out to harm a single whale with these underwater sounds," insisting that he and the other ATOC scientists were prepared to stop the experiment "on evidence that it is harming a single animal."[167] But the question was never whether ATOC scientists were out to harm whales—no one had ever alleged that. The question was whether they would do so negligently—stupidly—and Hyde's cavalier comment underscored the point: it was hardly credible that they would give up ATOC on the death of a lone fish or barnacle.[168] Hyde also accused Congress of being in a "reaction mode," yet to many a "reaction mode" seemed appropriate. Some Scripps scientists acknowledged that perhaps they had not adequately informed the public, but most took the approach that the best defense was offense, accusing the public of being emotional, hysterical, and ill informed. Others blamed the marine mammal experts for "irresponsibly alarming the public with misconceptions."[169]

As faxes and phone calls poured in and press coverage spread to the *Washington Post*, the NMFS agreed to the congressional request to extend the public comment period and hold additional hearings in Hawaii and California. Meanwhile, resistance grew. On March 31, the Advisory Council of the Monterey Bay National Marine Sanctuary called for delay until more information was gathered.[170] On April 5, the *New York Times* quoted the distinguished marine biologist and former chief scientist of NOAA Sylvia Earle as saying, "If you further damage the patient, the earth, while you try to take its temperature, then maybe the method is flawed."[171]

NMFS had scheduled California hearings for April 18 but agreed to a delay. Director Frieman wrote another letter, sent to over sixty-eight members of Congress, in which he tried to counter the "deeply disturb[ing]" media coverage and stressed the various steps that had been taken both to minimize impacts and detect any that might occur.[172] He also wrote to the Scripps associates—prominent individuals who supported the institution financially—to reassure them that the Scrippsites had not had their brains scrambled. On the contrary, ATOC was both necessary and well conceived: "The most significant effort to date to determine if greenhouse gases are indeed causing a heat transfer to the oceans as part of global warming." Revealingly, he insisted that "until global warming is better understood, governments will not be able to take effective steps to counteract its negative impact."[173] Like many ATOC scientists, Frieman assumed that the chief obstacle to policy action was a lack of scientific proof of the problem; thus, modest harms could be justified by the greater good that would come from obtaining proof. Among the letter's recipients were Jonas Salk, Walter Cronkite, philanthropists David Packard and Cecil Green, actor Ted Danson, and many others.

Scripps moved into damage-control mode. Alarmed that press coverage and editorials were almost uniformly negative, and ignoring that much of the adverse commentary was coming from *scientists*—particularly cetacean biologists—Scripps staff insisted that the negative views were due to misinformation and misunderstanding. They concluded that a concerted public relations effort was needed. In April, the communications office prepared a package of materials seeking support from counterparts at Woods Hole and other oceanographic institutions: "ATOC scientists were unable to convince this audience [at the NMFS hearings] that the project would do no real harm to the marine ecosystem and have been fighting a major public relations battle since then.... The media attack has resulted in public hysteria, and has created political pressure for NMFS and ATOC. At this point, we need your help to serve as the point of contact for your local media and government offices."[174] The package included copies of model letters to send to relevant officials and suggestions for specific actions, such as calling local science writers; it also included a model letter from a "concerned citizen." Scripps also solicited feedback on a proposed "fact sheet" on ATOC, which included a set of "frequently asked questions."[175] One portion read:

Q. How are ATOC's acoustic sources designed to minimize impact on marine mammals?

A. The sources will radiate about 200 watts of acoustic energy, much less than many sonars, communications, and geophysical research sound sources which have been in use for many years. The ATOC signal is about the same level as radiated by an individual large ship traveling at 20 knots speed. [The] signals ... travel along the ocean's "deep sound channel," buried for most of their path within the ocean's background noise....

Q. Are the sounds emitted from the acoustic sources "blasts" ... ?

A. No. The ATOC source transmits a very low frequency sound, spread from 60 to 90 Hz, which sounds like a distant rumbling to the human ear. Its energy is in the frequency band below the range most animals hear.

Q. Will the ATOC underwater sounds deafen whales, dolphins, seals or sea lions?

A: Absolutely not. No physiological damage will occur to marine life as a result of ATOC sounds, even if they dive deep. Ships pass by animals hundreds of times a day without their sounds harming them. Scientific data and years of observations support this finding.[176]

And:

Q. Did the scientists consider marine mammals when they designed ATOC?

A. Yes. Since its inception, ATOC has worked with leading marine biologists to design ATOC signal characteristics, transmission schedules, and source locations.[177]

This "fact sheet" was circulated among Woods Hole scientists, one of whom was behavioral ecologist Peter Tyack, an expert on whales and dolphin communication.[178] Tyack sent it to Chris Clark, who wrote a long, detailed critique.[179] Responding to the question of how ATOC's acoustic sources were "designed" to minimize impacts on marine mammals, he wrote emphatically: "Wrong question. I think it is unwise and slightly untrue to say that, except for the 5 min. ramp up, the source characteristics were designed to minimize impact on marine mammals. The reasons were based on [the] oceanographic experiment's needs." Responding to the question of whether the signal would deafen mammals, he wrote, "This is scientifically not a true response." Responding to the statement "Ships pass by animals hundreds of times a day without their sounds harming them," he wrote "(*we don't know this*) ... Scientific data (*What data?*)."[180]

Clark offered the following alternative:

> Although there are no specific scientific experimental data relating low-frequency underwater sound levels with auditory damage, we believe that there is little chance that animals will suffer physiological damage as a result of ATOC sounds. The animals (i.e. blue, finback, humpback whales, etc.) believed to have the most sensitive hearing in the frequency ranges of the ATOC sounds are not known to dive deeply enough to come with a range (~500–600 feet) of the loudspeaker that might cause temporary loss of hearing. The animals (i.e. toothed whales, sea lions, turtles, seals) that are known to dive to great depths probably have poor hearing in the frequency range of the ATOC sound. However, scientific knowledge on this subject of the effect of loud, low-frequency sounds and marine mammals is extremely limited, and it is for this reason that we are supporting marine mammal research.[181]

On the question of whether scientists had considered marine mammals when they first designed ATOC, he wrote: "A8: Since its inception *(Is this really true?)* ATOC has worked with marine biologists to design ATOC signal characteristics, transmission schedules, and source locations. *I would take issue with this statement. ATOC is designed as an oceanographic experiment not a [marine mammal] experiment.*"[182] Clark also noted a comment to the effect that scuba divers would not be able to hear the sources either, to which he replied, "This diver statement is not a documented statement at all."[183]

The fact sheet was indeed misleading. Besides the problems that Clark noted, there was also this: what mattered was not what the ATOC signals would sound like to a human ear or even to most animals, but what they would sound like to particular marine mammals whose echolocation system depended on their sensitive ability to detect low-frequency sound. Some cetacean biologists had suggested that man-made background noise generated by ships, sonar, and other sources might contribute to whale stranding by masking echolocation signals, so if the ATOC signal were similar to these sources, that was not reassuring. Nor was there scientific evidence to support the claim that the ATOC signal would be "buried for most of their path within the ocean's background noise"; this was one of the questions at stake.

Moreover, by posing the question in terms of "deafening," Scripps scientists were perpetuating the very mistake for which they had criticized others. Deafening was the extreme of potential harm, but lesser yet still serious harms had also been flagged as concerns. Ships did pass animals hundreds of times per day, but there was no comprehensive study of the impacts. As for military sonar projects, they were generally secret, so most biologists would have no way of examining, much less demonstrating, whether they had contributed to past harms. As the NRC had already concluded, scientists didn't know, and if the Navy knew, it was not saying.[184]

Clark's response was forwarded to Director Robert Gagosian, who concluded that Woods Hole should not join forces with Scripps on the press release and instructed the Woods Hole communications office to keep some distance: "We still want a low profile. If [our] people ... don't agree with the SIO [Scripps Institution of Oceanography] answers, then we don't want SIO as our spokesperson. I know that Peter [Tyack] does not agree with the 'absolutely not' answer to the whale deafening question. I don't want us to be associated with what we consider incorrect answers."[185]

Spring 1994: The Debate Continues

Things were going badly for the ATOC team. The hearings in Hawaii were contentious, and the NMFS decided to delay the California hearings until May. Meanwhile, the scientists had decided to revise the marine mammal research program design under the advice of an advisory board assembled the previous December.[186] In February the board had submitted an extremely detailed, seven-page, single-spaced summary of recommendations. Noting that ocean acoustics research had been going on for many years "with little or no consideration of its potential effects on marine mammals," they suggested that the present crisis offered an opportunity to redress that imbalance.[187]

Echoing Clark's advice, the advisory board suggested that scientists should not merely monitor the effects of ATOC's initial eighteen to twenty-four months phase but should "develop the basis for predicting the effects of the long-term ATOC program."[188] This would not be easy, because the kind of broad objectives that would satisfy marine biologists "would be difficult to meet under any conditions, and especially within the next two years." As if to underscore the point that the oceanographers' sanguine view had not been grounded in science, the board noted, "At present, little is known about the natural history, demography, and normal behavior of many potentially affected species. Although short-term behavioral and distributional effects of ATOC are amenable to study with the proposed approach, it will not be practical to answer all the basic questions about potential long-term and population effects within the scope of the ATOC MMRP as presently proposed." It was no use claiming that the project would be stopped if adverse effects occurred if you hadn't specified in advance—and, worse, didn't actually know—what constituted an adverse effect. The board recommended that "the MMRP and ATOC as a whole should develop a list of potential marine mammal responses that, if they occur, would be considered sufficiently severe and adverse to warrant a change in the experimental procedure and/or suspension of ATOC transmissions."[189]

Two months later, the board suggested additional adjustments to the protocol. Originally, the Marine Mammal Research Program was slated to run concurrently with the first phase of climate studies, but this meant that the experiment lacked a control. "This represents a poor experimental design," they noted, suggesting that the first five months of the program be dedicated exclusively to the MMRP, in the following controlled manner: a one-week period with twenty-minute ATOC broadcasts every four hours, followed by a two-week period with no broadcasts, repeated for five months. This would allow for comparison of mammalian behavior with and without ATOC signals. In the first two experimental periods, the signal would be run at 180 decibels, "the level at which 50% of gray whales indicated an avoidance response" in previous studies. Then in the following five periods, the signal would be increased to 195 decibels, the level at which ATOC proposed to operate.[190] The source should be controlled by biologists (not oceanographers or technicians) with authority to suspend operations immediately if any "significant disturbance" was observed. At the end of the five-month period, a two-day workshop should be convened to permit open discussion with "colleagues, interested parties, and the media"—and the results reviewed by the MMRP advisory board, the NMFS, and the Marine Mammal Commission.[191] The key point, as Dan Costa emphasized, was to design the protocol in a manner that was optimal for observing marine mammals—rather than optimal for ATOC—and to seek objective, independent peer review. This was good science. It was also the only way to regain credibility and "seize the high ground."[192]

ATOC scientists were not quite ready to seize the high ground, but they did agree to two modifications. First, they would prepare environmental impact statements for both the California and the Hawaii sites. Second, they agreed to accept the recommendation not to proceed until after the Marine Mammal Research Program had submitted its results and had those results independently reviewed. The NMFS announced that no permits would be issued until after the draft environmental impact statement had been submitted and assessed.[193]

Despite these changes, the May 16 hearings in Santa Cruz went much the same way as the previous ones. Then the conflict escalated as a consortium of environmental groups—the Natural Resources Defense Council, the Sierra Club Legal Defense Fund, the Environmental Defense Fund, Earth Island Institute, the American Oceans Campaign, the League for Coastal Protection, and the Humane Society of the United States—filed suit in federal court to stop the project. The plaintiffs accused the researchers of violating the National Environmental Policy Act, the Marine Mammal Protection Act, and the Endangered Species Act.[194] Scientists who saw themselves as environmental saviors found themselves cast in the role of environmental villains.

More Stakeholders Join In

Christopher Clark was not the only expert who noted that certain ATOC claims were not supported by evidence and tried to persuade the ATOC team to adjust their approach. Robert J. Hofman, the scientific program director for the Marine Mammal Commission in Washington, did so, too. He had been invited to comment on an early draft of the environmental impact statement; in June—just a few weeks after Gagosian had decided that Woods Hole would not cosign the Scripps fact sheet—and he had this to say: "Parts of the draft as presently written appear to make claims or statements which are not supported by data or analysis.... One example is the sentence in the Abstract which reads: 'all available data from research on marine mammal hearing indicate there will be negligible, if any, impact from ATOC transmissions.' It seems to me that it would be more accurate to say something like—'Available data are insufficient to conclude that ATOC transmissions will have negligible effects on marine mammals and the Marine Mammal Research Program described herein has been designed to resolve the uncertainties.'"[195] The ATOC scientists were either unwilling or unable to take this advice.

The NMFS was legally required to notify other pertinent agencies and to invite their comments on the permit application, among them the US Environmental Protection Agency, the US Fish and Wildlife Service, and the California Coastal Commission. Their responses demonstrated to the NMFS that worries over ATOC were not restricted to hysterical environmentalists and ignorant citizens. The principal federal agencies responsible for environmental protection and species management shared several of their concerns, and they said so unequivocally.

John Turner, chief of environmental services at Fish and Wildlife, expressed his department's belief that "the proposed research has the potential to harm several species of marine mammals." They were particularly concerned with the "potential to interfere with long-distance communication by the great whales," noting that "gray whales have been documented to abandon a breeding lagoon for several years which was ensonified by dredging noises. The whales did not return to the lagoon until after the dredging terminated."[196] Given that the ATOC ramp-up period was only five minutes, a gray whale (even if swimming at top speed) would not be able to escape the affected area unless, by luck, it had happened to be close to its borders. If the ocean were already filled with disruptive noise, as the oceanographers insisted, that was no justification for adding more. Finally, there was the question about whether the project was really needed and whether it would produce the promised results: "Documentation is lacking supporting the assertion that the magnitude and duration of the proposed experiment sound level is required to accurately measure ocean temperature changes."[197] Turner concluded on behalf of his agency: "The [Fish and Wildlife Service] recognizes that ATOC is an innovative method to measure global warming but believes that potential effects to living marine resources have not adequately been addressed.... The [Fish and Wildlife Service] recommends that approval of their permit be postponed until enough data and evidence is available to conclude with confidence that marine mammals will not be harmed by the experiment."[198] The Environmental Protection Agency shared similar concerns and reminded the NMFS that by law any environmental impact statement must include evaluation of alternatives, including the alternative of "no action." All things considered, in this case that might be the right choice.[199]

The California Coastal Commission was aggrieved that Scripps had tried to evade its authority by asserting that there would be no impact on the coastal zone.[200] The commission begged to differ. The California Coastal Zone Management Act and the California Environmental Quality Act prohibited any development where there were feasible alternatives of lesser impact or required mitigation measures to lessen the impact, and it was up to the commission, not Scripps, to make that determination. By July, the commission had concluded that "the proposed ATOC program would affect resources of the California coastal zone," and therefore potentially would violate not only federal law but state law as well.[201]

Legal counsel advised the Scripps scientists to prepare a "show and tell" to help smooth relations with the California Coastal Commission.[202] Scripps was hoping to get the commission to grant a "no-impact" waiver, but the external relations department at the University of California, San Diego, advised them that no-impact waivers were generally understood to apply to "routine types of projects which they have seen over and over again." In contrast, "a project that has engendered considerable public attention and which is undergoing a full [environmental impact report] is [not] a common occurrence."[203] Scripps staff met with several commissioners and exchanged letters and faxes, but nevertheless continued to argue that the commission lacked authority to review the project. Not surprisingly, the commission disagreed.

The Draft Environmental Impact Statement

In November 1994, the draft environmental impact statement for the Point Sur site was released, followed in December by the one for the Kauai site.[204] The documents were extensive and detailed, and considered the option of moving to alternative sites, but in the end they remained committed to the original sites for technical reasons: they were optimal for transmitting in the sound channel and capturing sound over a large expanse of ocean, and they were proximate to existing underwater and land-based faculties. The environmental impact statement also held to the claim that there would be "no significant impacts."[205] From the permitting perspective, the most important part was the biological research program, which had been revised in several ways to resolve the uncertainties and produce bona fide scientific research.[206]

The revised program followed many of the key suggestions of the MMRP advisory board, above all addressing the double bind that the only way to determine whether ATOC would do harm was to undertake the very activities that were alleged to cause it. It did this by dividing the research into three phases. Phase 1, lasting from April to October 1995, was a biology-only program consisting of baseline data collection on marine mammals, with no ATOC transmissions. In phase 2, transmissions would begin, but the source would remain under biologists' control, and the transmission cycle would be greatly reduced as compared to the original concept: only one twenty-minute transmission every four days (versus twenty-minute transmissions every four hours, around the clock, every period). This would allow any affected animals at least three days respite. After one month, preliminary results would be compared with the phase 1 baseline and the project would be modified or aborted if the results provided evidence of disturbance or harm. Assuming this was not the case, phase 3 would be a two-year ATOC experiment, in 1996 and 1997, more or less as originally planned (75 Hertz, 195 decibels, and a duty cycle consisting of twenty minutes full intensity, every four hours). Still, biologists would continue the monitoring activities, and the experiment could be halted at any time if unacceptable impacts were detected. Finally, the Scripps scientists committed themselves to operating "at the minimum power level necessary to support MMRP objectives and feasibility operations."[207] Crucially, ATOC had been made dependent on the MMRP rather than the other way around.

The ATOC scientists had come a long way from their earlier insistence that no harm would be done. Opponents of the project appreciated the change, as well as that the new report included helpful bibliographies and summaries of existing knowledge about the effects of low-frequency sound, including a "crash course in marine bioacoustics."[208] But many still remained skeptical. As Canadian biologist Paul K. Anderson put it, the draft environmental impact statement underscored "the extreme difficulty faced by the MMRP in attempting to assess risks to marine animals."[209]

A persistent question was whether the time frame was sufficient to detect adverse effects. Fish and Wildlife Service field supervisor Craig Faanes noted that the proposed baseline—April to October—"may not be of sufficient duration to determine whether a species is negatively affected," particularly if it did not include a mating season. He also raised the question Chris Clark had noted earlier: which "methods and criteria" would be used to determine whether a species had been affected? How would they determine effects on organisms that "are difficult to observe, not present during the time of year the data are gathered, or those for which little information is known on their behavior patterns prior to the project?"[210]

Costa had already noted this problem: How can you prove that an absence of observed effects means an absence of actual effects? Although provisions were emplaced for shutdown if "an unacceptably significant disruption of the behavioral patterns of a marine animal" was observed, who would decide what was unacceptably significant?[211] Paul Anderson noted that the problem of optimistic bias remained: "Although the objectives are formulated in classic null-hypothesis format ... the introductory paragraphs suggest a philosophical bias. The stated objective is to 'validate' the assumptions that "reactions from marine mammals are unlikely at ATOC received levels <120 dB at distances of >20 km."[212]

Just as ATOC scientists had trouble abandoning the claim that ATOC would do no harm, they had trouble abandoning the presumption that the MMRP would demonstrate that it would do no harm. Indeed, when the MMRP advisory board had first been appointed, its charge emphasized that

a "key program objective will be to demonstrate that the planned global network can be operated without any adverse effects on marine mammal populations."[213] That was in 1993; two years later the optimistic bias was still in
place.

Regulations developed by the US Council on Environmental Quality stipulated that comparable presentation of alternatives was the key to "sharply defining the issues and providing a clear basis for choice among options by the
decision-maker and the public."[214] EPA officials reviewing ATOC had strongly
recommended the "development of a range of reasonable alternatives that ...
satisfy the basic project purpose (collecting data on global warming) but have
fewer potential adverse effects on marine biological resources."[215] The draft
environmental impact statement did not do that. It evaluated alternatives
within the basic ATOC framework in terms of geographic sites, duty cycles,
and other technical details and components, but it did not comparably evaluate the "no ATOC" option.

Meanwhile, another change was taking place: ATOC scientists adjusted
their views about what ATOC was intended or could be expected to achieve.
Walter Munk explained: "With regard to climate, my views of what we should
focus on have been modified over the last two years. A stand-alone detection
and mapping of the greenhouse-induced changes over and above the ambient
changes will take a long time, a few decades if [climate model] predictions are
correct. I now think our emphasis should be to test, and help improve, current climate models."[216] Munk made a similar qualification when he wrote to
the California Coastal Commission in June, in anticipation of their approval.
Rather than suggesting that ATOC would detect, confirm, or prove global
warming, he argued that the point was to test climate models:

ATOC is intended to observe the ocean on the large space scales that characterize climate—3,000 to 10,000 kilometers—so that modelers will be
able to:

A. test their models against the changes seen by ATOC over a few
   years
B. and, if, and when, the models prove adequate at hindcasting, use
   those same models to make climate predictions.

By testing and improving climate models now, ATOC can make progress toward greenhouse predictions later.[217]

Munk's modified position was reasonable, and certainly more defensible
than the earlier suggestion that ATOC would prove global warming. It was

also consistent with the position that John Spiesberger and his colleagues had argued back in 1983: that meaningful results would take decades. But it was a very different framing than the one ATOC advocates had started with, and it substantially lessened the justification for the project. After all, why risk harm to whales if the experiment wouldn't prove global warming, anyway?

California Coastal Commission Hearings: January 1995

On January 6, 1995, the California Coastal Commission held public hearings in Santa Cruz and announced that written comments would be accepted through January 31. By this point, many people were engaged with the issue, and hundreds of cards, letters, and faxes flowed in from scientists, conservation organizations, and ordinary citizens.[218]

Some of these comments were positive: Scripps scientists had persuaded a number of political leaders to endorse the project, including the San Diego congressional delegation. A few leading scientists also offered endorsements, including David Green, chairman of the NRC committee on low-frequency sound; Kenneth Norris, founding president of the Society for Marine Mammalogy; and James Hansen, who hoped (albeit perhaps ambiguously) that the commission's "considerations of these proposed measurements will give appropriate weight to their potential value for understanding global climate change."[219]

However, even where support was expressed, it was often in a qualified manner. The members of the Monterey Bay Chapter of the American Cetacean Society were divided, so they had decided to focus their comments solely on the MMRP component. If certain concerns were addressed—outlined in a four-page letter about the research protocols for whales and how the decision to terminate the project would be made—then they would support the project.[220] John Pearse, professor emeritus of biology at UC Santa Cruz who had been a member of the NRC's Committee on Low-Frequency Sound and Marine Mammals, supported the program, not so much because of what it would tell us about global warming but because it "represents a tremendous opportunity to obtain much of the information needed to better protect both marine mammals and their food sources.... Without such information, we have no basis for protecting marine mammals, and other forms of marine life, from anthropogenic sound."[221]

Two rare expressions of unqualified support on the biological side came from the Monterey Bay Aquarium. The first was from Stephen K. Webster, PhD, director of education and member-at-large of the Monterey Bay National Marine Sanctuary Advisory Council. Webster felt that ATOC had been misrepresented by the press and misunderstood by the public, and that it "never deserved the play it received in the public press and resulting debate." The combination of "irresponsible 'tabloid' journalism, misinformed self-proclaimed 'experts,' and issue-hungry, fundraising fringe environmentalists has elevated this issue to a level of controversy it never deserved, and still doesn't." The MMRP would provide "accurate, objective, and scientific supportable conclusions," and permit "an informed decision about the future of ATOC."[222]

The second was from James Barry, an assistant scientist at the Aquarium Research Institute, who found most of the public reaction illegitimate. It is "ludicrous to think that uneducated, but well-meaning, citizens unfamiliar with ocean physics, sound transmission in water, behavior and physiology of marine mammals, fishes and invertebrates, and other information relevant to the ATOC proposal, should outweigh the highly informed and carefully considered views of experts in these fields." Certain that anyone opposing the project was letting their emotions interfere with their reasoning—and equating ATOC physical oceanographers with "informed conservationists"—he insisted that "the opinions and advise [*sic*] of highly informed conservationists, like the scientists proposing this study, are based on fact rather than emotion, and should be heeded. Although public opinion is important, it should carry little weight for [the] advisory committee."[223]

Barry had worked himself into a state of high dudgeon, but the point that he, Webster, and many other ATOC supporters seemed to miss (or ignore) was that a good deal of the "public" criticism was actually quite well informed. Some were scientists not associated with ATOC; others were conservationists and environmental groups that had done their homework; some were educated citizens who had taken the time to learn about the issue and express their views. In any case, whether they were based on fact, emotion, or a mixture of both, the NMFS was legally required to hear them.

The most obvious well-informed critics were the cetacean biologists, who recapitulated their arguments with greater precision and scientific detail. Lindy Weilgart and Hal Whitehead each submitted line-by-line critiques of the draft environmental impact statement; Weilgart's ran to nineteen pages. ATOC scientists had accused her of a poor grasp of physics; she in turn accused them of a poor grasp of biology. Among other things, their discussions of marine mammal hearing were based on fallacious analogies with human hearing. Some animals might become acclimated to the noise, as the environmental impact report suggested, but it was equally possible that they might become hearing impaired, and in the short term it would be "impossible to determine which is happening."[224] The ramp-up period was supposed to enable sensitive creatures to move away, but that presupposed that they could swim fast enough and knew which way to go, which might be expecting too much even from intelligent marine mammals, much less invertebrates. In any case the presupposition was unverified.

ATOC's own advisory board had doubted that the time allotted for initial scientific studies would be adequate to collect data on long-term effects. In fact, Weilgart noted, the scientists had not even allotted enough time to analyze the results they would have. It was "unrealistic to expect the MMRP to complete a substantive analysis of all types of behavioral reactions, and to prepare a comprehensive report suitable for external review, within 1 mo. after the end of data collection."[225]

ATOC scientists claimed to be sensitive to the biological concerns and willing to stop the program if necessary, but their commitment to the project led to a persistent and obvious optimistic bias. Where scientific data were available that suggested adverse effects, the draft environmental impact statement minimized or dismissed them; where there were no data, it assumed no effects. ATOC's timetable showed them proceeding with phase two for six months before the pilot study final report was expected, belying the claim that phase two would hinge on the MMRP results. Clearly, ATOC scientists had still not fully embraced the biologists' concerns. Weilgart summarized:

> The DEIS [Draft Environmental Impact Statement] is unconscionably dismissive of likely adverse impacts on marine life.... Conclusions of "minimal impact" are repeatedly made, even when these conclusions are based on completely unsubstantiated assumptions.... We are asked to accept complete guesses at the auditory sensitivities (thresholds) of the vast majority of species in the study area, particularly the endangered large whales.... ATOC's own independent scientific advisory board states that "ATOC documents assume hearing damage ... will not occur if received levels of ATOC sounds are below 150dB. The Advisory Board notes that this assumption may or may not be true, but there are no supporting data from marine mammals" (MMRP AB report, June 12, 1994).... [The] document consistently attempts to downplay the very real risks that this project presents, concluding minimal impact when effects are unknown or even with evidence to the contrary.[226]

Whitehead also underscored the optimistic bias. In areas where he had expertise, he found that the document was "often seriously wrong, invariably in the direction of minimizing the potential impacts." It asserted that refinements to the protocols for the MMRP were agreed to at a meeting in July 1994, but Whitehead was at that meeting and denied this characterization of it. "There was no agreement whatsoever about refinements to the project protocols. In fact, the whole project, from its overall goals and structure, to the details of the research program, was severely criticized by non-ATOC participants."[227]

Despite certain improvements, Whitehead believed the MMRP was unlikely to provide the data needed to make an informed decision. Reiterating the point raised by Chris Clark months earlier, he maintained that ATOC had been "designed to fit the needs of the physical oceanographers," and the proposed modifications changes were inadequate to resolve the biological questions. "The MMRP may produce some interesting information, but, as it stands, it has no hope of examining the effects of the ATOC source on cetaceans." The time frame was too short, the geographic scale too small, and the "chances of detecting any of the four 'unacceptable effects' [that would trigger project termination] are virtually zero."[228]

This last point was crucial, for the assurance that no harm would be done hinged on the promise to abandon work if "unacceptable" effects occurred. According to the proposal, the four "unacceptable" effects were avoidance or abandonment of previous high-use areas; an increase in at-sea observations of dead animals or strandings; an increased incidence of stressed, emaciated, or diseased animals; and a decrease in calving and pupping rates or total population size. But Monterey Bay—where the study would take place—was not a high-use area, so it was logically impossible for avoidance or abandonment of a high-use area to be observed. Dead animals and strandings are rare events, so even if one or more were observed, it would be impossible in the time allowed to show a statistically significant change. To measure stress, emaciation, or disease in cetaceans, one would have to collect individuals, and there were no plans for that. As for calving and pupping rates, with the exception of gray whales, there were no generally accepted figures, so there would be no way to know if the rates were changing. As for total population, natural variability appeared to be very high, so even if significant changes were observed, what would that mean? For cetaceans, Whitehead dismissed the MMRP as "almost completely useless." His complaints were reiterated by a research associate in Clark's laboratory at Cornell, who observed that the draft environmental impact statement was characterized by "an underemphasis on those potentially negative impacts on marine life that cannot and will not be measured."[229]

Why Environmentalists Opposed ATOC

The purpose of ATOC was to confirm the reality of global warming, and so scientists involved had expected environmental organizations to be natural allies. This expectation was thoroughly confounded. Virtually all the comments submitted by environmental groups were thoughtful, detailed, and well informed. Virtually all acknowledged that global warming was a serious problem and that additional information could be useful. Yet not one environmental group was prepared to support ATOC. They all agreed with Weilgart and Whitehead that the analysis was flawed, particularly in assuming no impacts or minimal impacts where data were lacking. ATOC was presented as an environmental project, yet its biases showed a disregard for one of the major concerns of modern environmentalism: protection of threatened and endangered species and their habitats.

This argument formed the core of a detailed critique by the Center for Marine Conservation (CMC), based in Washington, DC. The purpose of an environmental impact statement, they noted, was to "assist decision makers in choosing a course of action that minimizes adverse effects," but because this statement never really considered alternatives, it failed in that role: "NOAA, the National Marine Sanctuary Program, the independent scientific community, and the general public were expecting a report that would realistically portray the uncertainties related to the feasibility and safety of the project, that would describe a range of alternatives, and present a preferred alternative that would pose negligible risks to the federally recognized ecological values of the Sanctuary. Instead, the document's ambiguities, inaccuracies, and treatment of uncertainties has intensified rather than quelled concern over ATOC. Ironically, most of these critics support the overall goal of ATOC to improve our understanding of global climate change."[230] The CMC complaint was not in the least bit hysterical; on the contrary, it quite rationally underscored the central illogic that the draft environmental impact statement repeatedly defaulted "to a conclusion of no expected significant impact," when the raison d'être of the MMRP was the dearth of scientific basis for predicting the impact. The repeated use of terms including *nonexistent*, *negligible*, and *minimal* to describe potential impacts undermined the report's credibility. "This tendency to dismiss uncertainty exacerbates rather than alleviates questions regarding impacts," they concluded.[231]

All data evaluation involved judgment calls, the CMC acknowledged, but in their enthusiasm and haste to assert "no impacts," Scripps scientists had made errors of fact as well as judgment. It was said, for example, that "no information exists on noise impacts to salmon," but this was incorrect; there was "abundant evidence that salmonids hear and behaviorally respond to low-frequency sounds." In fact, "repetitive low-frequency sounds are now being used to deflect juvenile salmonids from sloughs leading to water export pumps in the Sacramento–San Joaquin Delta"—which could hardly be efficacious if the salmon were deaf! And at the same time that the draft environmental impact statement claimed that the ATOC program was being designed in a manner to minimize adverse effects, it also admitted that the choice of the source location was based on economics and pragmatics, namely, a "minimum cable run to shore" and "close logistical support." The authors' "vested interest in the project" had undermined their capacity to produce an objective report.[232]

While public and media attention had focused on whales, other marine life could be affected, too, yet almost no effort had been made to address this. This theme was taken up by the Pacific Fishery Management Council, whose members noted that the monitoring program was focused on marine mammals, not marine life, even though extensive scientific literature documented the effects of sound on fish. They, too, found the draft environmental impact statement dismissive, as in the suggestion that injury to fish was insignificant because any injured fish would just be more easily caught by predators—after all, all fish get eaten sooner or later—as if that were not an ecosystem disruption![233]

Nearly all the environmental groups stressed what the EPA had already noted months before: the law required genuine consideration of the "no action" alternative. Kauai Friends of the Earth argued that this should have included not only alternative means of measuring the speed of sound in water but also alternative means of evaluating global warming. Scripps had dismissed scientific alternatives to ATOC with what the Kauai group called the "astounding proclamation" that all other scientific methods of addressing global warming are "either included in the project as proposed, or would not meet project objectives." This hardly seemed a meaningful evaluation. The Kauai Friends noted that, consistent with its "presumptive" tone and "hubris," the report ignored "hundreds of comments offered by the public" at the hearings in California and Hawaii in 1994.[234] They, too, made the observation that in the absence of evidence of harm, Scripps had assumed or asserted there would be no harm. This was patently unscientific: a classic error of logic that undergraduates are taught not to make.

Kauai Friends of the Earth also put their finger on an essential question: ATOC had been presented as a crucial experiment, which would answer the question of global warming. But how definitive would it be, really? How definitive is any scientific experiment? The Kauai Friends had obtained a copy of the original Scripps proposal for ATOC (submitted to ARPA in May 1992) and found ample evidence that, even by the scientists' own reckoning, ATOC was no sure thing. Scientists had provided detailed discussions of uncertainties about spatial resolution and the analysis of ray paths, and sources of error regarding the interface between ATOC measurements and global circulation models, satellite data, and sea-surface temperatures. This was normal scientific practice, but it underscored that all experiments are subject to interpretation. It also underscored a crucial fact often elided by ATOC advocates: detecting global warming was not the same as confirming the human role in it. Actually, Walter Munk had admitted as much: "The definitive statement on this issue would appear to be contained in the lead article in the special issue of the *Journal of the Acoustical Society of America* on the Heard Island Experiment wherein the authors (Munk et al.) state the following: '. . . Finally it is important to emphasize that acoustic thermometry addresses the issue of measuring climate change (ambient or otherwise) in the oceans; it does not tell us anything about the underlying causes and about the effects on the atmosphere.'"[235] The United Nations Framework Convention of Climate Change, adopted in 1992, committed its signatories to preventing "dangerous *anthropogenic* interference with the climate system," but if ATOC could not distinguish between anthropogenic and natural change, then what was its point?[236] The Kauai group concluded that it had no point, other than advancing the careers of the scientists involved.

Would ATOC Be Scientifically Definitive?

In their public comment, the Kauai group hit upon an issue that scientists had been debating privately for some time: Would ATOC be scientifically definitive? How strong *was* the link between the speed of sound and ocean temperature? Could it be measured precisely enough?

In 1990, when John Spiesberger and colleagues submitted the paper "Basin-Scale Tomography" to the *Journal of Geophysical Research: Oceans*, it was initially rejected, with reviewers questioning whether the technique was sufficiently sensitive and accurate to detect the small changes expected.[237] At the time, the total mean global temperature change from anthropogenic warming was only about half a degree Celsius, so the expected annual change was only a tiny fraction of a degree, perhaps one or two hundredths. One reviewer claimed that monthly variability in ocean thermal structure was as great as variability across years, which, if true, would undermine ATOC's ability to separate the signal from the noise.[238]

Technical uncertainties included the accuracy of the estimated ray paths, the accuracy of the clocks and recording devices used, errors in the measured distance between source and receiver, and even errors introduced by the curvature of the Earth.[239] The positions of the moored sources were tracked acoustically by bottom transponders, whose positions were in turn determined using the Global Positioning System (which at that time was very new), and the positions of the bottom-mounted receivers were located using transit satellite-navigation systems; these positioning systems potentially introduced error. Errors might also be introduced by the assumption that eddies and internal waves were not confounding the results.[240] If these errors were consistent they would not matter, because the goal was to discern progressive changes rather than absolute values. But given ocean complexity and variability, could anyone ever prove that the errors were consistent?

A significant cause for concern among reviewers was whether the sound-wave travel paths were adequately understood. One reviewer noted that travel time along steep rays is "highly sensitive to properties at turning points," such as the upper boundary of the sound channel, and it would be easy to misidentify rays with similar travel times: "I'd like to be convinced that the interpretation of travel times is as straightforward as the text suggests."[241] Another reviewer thought the authors were promising more than they could deliver and repeating the "tomography sales pitch that can be found in a dozen or so previous publications":

> The central theme, that travel time measurements can be interpreted as temperature measurements, is an axiom of ocean acoustic tomography and certainly not new or under scrutiny. The more important and much more difficult question to answer is: Along what path through the ocean is the travel time being measured? The determination of path is glossed over with ... idealized examples where models and measurements match perfectly. But, for many of the long path[s] in the Pacific ... structure is ambiguous and possibly even chaotic and unknowable. There is a big step missing between the examples and most of the real transmission channels.[242]

Another reviewer suggested that the basic principles of ocean acoustic tomography were not universally accepted, an interesting state of affairs considering how much money had already been spent: "Ocean Acoustic Tomography ... is still regarded as a rather controversial subject in Oceanography, because of the inverse relationship between science and funding that seems to apply: high investment for over 10 years and as yet very little return, a rather ironic twist to the meaning of inverse method.... I am not trying to be sarcastic, but [it is] important that an article on a subject that is still regarded with considerable suspicion be damn good.... By and large, oceanographers have yet to be convinced."[243]

Spiesberger and colleagues revised and resubmitted their paper with a new title, "Listening for Climatic Temperature Change in the Northeast Pacific," but again they met objections. The reviewers acknowledged the authors' reputations for high-quality work but continued to worry about the theoretical foundations: "Is ray theory valid or not? . . . The authors need to determine whether ray theory applies or not—or more properly what aspects of ray theory may apply—and then either use it or not as the case may be..... If I had reviewed [an earlier paper] for the *Journal of the Acoustical Society* I would have rejected it."[244]

Public presentations inevitably have a different character and tone from expert presentations and peer reviews—reviewers are expected to be skeptical and scrutinize details—but these comments show that acoustic tomography was not nearly the cut-and-dried scientific technique that it supporters sometimes implied. ATOC was a highly complex scientific endeavor with myriad possible confounding factors, and it was difficult even for experts to judge the odds of success.

When one looks closely, one finds this point acknowledged in the draft environmental impact statement: "Whether the ATOC technique will provide useful climatic information depends on [a number of factors. ATOC] ... is subject to fundamental uncertainties about the extent to which acoustic means can detect ocean climate changes."[245] If it failed to detect a warming signal, it could be because there was no warming, but it could also be that the experiment had failed to detect it. That is true of any negative result, but the additional component was that the accuracy required was extremely great, and the accuracy of the system was unknown. It was one thing to say that ocean acoustics was a well-established science, which it was, but another to say that ATOC was an established technique, which it was not.

One peer-reviewed paper evaluating the feasibility of the ATOC approach in 1993 had concluded that, when all the uncertainties were considered, the chance of detecting the expected greenhouse-induced warming was "realistic."[246] One could conclude from this language that there was also a realistic chance of *not* detecting the expected signal. David H. Rind, a climate modeler at the Goddard Institute for Space Studies (where James Hansen was director), suggested that even if one did detect a signal, it would still be hard to know what it meant, in part because of the ocean's complexity and heterogeneity: "It's conceivable that [even in the face of global warming] the water could warm in one portion and cool in another, and that would confuse the issue as to what is happening. It's not as if this data is going to smack you in the face and say, 'This is happening.'"[247]

In fairness to the ATOC scientists, they had not claimed that their data would smack you in the face. Walter Munk had stressed in 1994 that the results would not provide "a stand-alone methodology for monitoring ocean climate variability."[248] Rather, he suggested that, together with satellite altimetry and other sources of information, the data would provide a means to monitor the changing conditions of the ocean, complementing climate models and enabling scientists to test their predictions. That was a fair claim. But as the debate became public and polarized, at least some ATOC's champions—including Scripps director Frieman—did assert that ATOC would reveal whether anthropogenic warming was underway, and that would make a crucial difference to policy. For this reason, they suggested, the value of ATOC would compensate for any harm to ocean life. Biologist Paul Anderson thus concluded that how one viewed ATOC ultimately depended on four things. As he put it:

A—Believe ANY risk to marine mammals or other organisms is unacceptable.

B—Believe that it is POSSIBLE to adequately address the risk in this case.

C—Trust that ATOC will modify procedures satisfactorily, or abort, if MMRP results so indicate.

D—Believe that our world society is capable of taking corrective action should unassailable greenhouse data become available.[249]

Believing in the capability to take corrective action was clearly the most important. It was more than plausible to argue that there was no such thing as unassailable greenhouse data and that the ATOC data could (and likely would) be challenged in the same manner that other evidence of global warming was being challenged.[250] It was also plausible to argue that, even in the face of unassailable evidence, society was simply incapable of taking the necessary corrective action.[251]

Would ATOC Be Politically Definitive?

No scientific experiment is perfect, but some experiments do produce results that nearly all scientists find compelling and persuasive, and it is certainly possible that ATOC could have done that. But even if ATOC proved scientif-ically persuasive, this did not mean that it would have the impact on policy that Frieman and others had suggested.

The person who made this point most cogently was Rodney Fujita of the Environmental Defense Fund (EDF). Fujita had a PhD in marine ecology and had worked with the IPCC on the effect of elevated sea-surface temperatures on coral reefs; no one could accuse him of being ill informed. Moreover, if any environmental organization would have been thought likely to support ATOC, it would have been EDF; global warming was one of the organization's central concerns. But Fujita insisted that ATOC scientists had not been forth-coming about the insufficiency of necessary evidence: "The fact that there exists virtually no evidence bearing on the question of how marine organ-isms might respond to the ATOC sound source should be squarely acknowl-edged.... The key to good policy making on this issue is to freely acknowledge the great uncertainties surrounding the potential impacts of ATOC and work to reduce them, rather than attempting to paint a rosy picture that shows that the impacts are likely to be insignificant. Unfortunately, the draft [envi-ronmental impact statement] ... consistently makes the error of concluding that if no evidence for a significant impact exists, the impact must be non-existent."[252]

He allowed that EDF could support the project, but only under certain conditions. These included moving the source outside the sanctuary; estab-lishing "better defined, objective thresholds for adverse impacts that would result in [program] termination"; involving environmentalists and citizens in establishing the termination criteria and determining when they had been met; and recruiting independent, outside experts to assess the results of the pilot study. But the real issue—the one that struck at the heart of the justi-fication for taking the risks that ATOC entailed—was whether ATOC would actually answer key questions about global warming and, if it did, whether that would make any political difference.

Fujita urged the NMFS to recognize that the "potential for sweeping changes in global warming policies resulting from the ATOC data is low" and to weigh the risks and benefits of the project in this light; to acknowledge the dearth of scientific data pertinent to predicting the risks; and to "come to grips with the limitations of science [and] recognize that uncertainty about the impacts of ATOC will always remain, because the habits of marine mam-mals, the complexity of the marine environment, and the difficulty of do-ing controlled experiments that isolate cause and effects relationships in the ocean will often prevent the drawing of strong inferences. Above all, we hu-mans need to avoid hubris."[253] Even if all these conditions were met, however, there was another point to consider. While ATOC scientists accused their opponents of irrationality, they had fallen into an illogic of their own, or at least a counterfactual: the presumption that knowing the scientific facts would lead to a solution to global climate change. All the available evidence suggested that this was not so. Even if ATOC produced evidence that was compelling to scientists, that did not mean it would be compelling politically:

> None of us should be overly optimistic that data generated by ATOC, no matter how accurate or precise, will result in a dramatic improvement in climate change *policies*. ATOC could reduce key uncertainties about ocean heat uptake, [and] while a reduction in this uncertainty, better climate models, and a more definitive indication that global warming is occurring—all potential benefits of ATOC—would definitely be helpful, they are probably not the most important factors limiting progress toward taking action to prevent global climate change. Vast economic and political interests continue to resist significant changes in the current patterns of fossil fuel use and deforestation that are driving climate change, and they are not expected to disappear in the foreseeable future.[254]

Events over the following two decades would prove Fujita sadly correct.[255]

Comments from Citizens: We Don't Need It and We Don't Trust It

ATOC scientists often asserted that public opposition was based on ignorance, but the public commentary from the California Coastal Commission hearings belies that claim. Citizens in all walks of life—from retirees to schoolchildren—weighed in, and some of them had clearly done their homework. A dive master wrote a six-page letter describing the effect of low-frequency sound transmissions on divers, including several graphs; a teacher of deaf students wrote to emphasize the "well-known adverse effects of long-term exposure to very loud sounds."[256] Others were less fact focused but not necessarily irrational: one letter began clearly: "Why I think California Acoustic Thermometry of Ocean Climate Project (ATOC) is a REALLY BAD idea."[257] Other comments overlapped those of the environmental groups: that the oceanographers were arrogant and dismissive of public concerns, that ATOC might do harm, and that scientists' claims that ATOC would be definitive and therefore move policy were naïve about political reality.

Numerous citizens noted the irony of oceanographers asking the public to respect their expertise while they disrespected the expertise of marine biologists. Citizens reminded the commission that the biological team for the

Heard Island work had been cobbled together at the last minute and that its
leader, Ann Bowles, did not even have a PhD. Others suggested that the scientists seemed to think they were above the law, as they had originally not
even bothered to apply for permits for Heard Island, had attempted to claim
exemption from the Marine Mammal Protection Act, and had tried to deny
the legal authority of the California Coastal Commission. Others took offense
at the "patronizing" and "dismissive" tone of the draft environmental impact
statement, which failed to answer many of the issues raised at public hearing.
This made it seem as if they considered the hearings just another aspect of
the law to be skirted.[258]

A retired radar and sonar engineer wondered how oceanographers could
be so sure that animals would not be harmed. "I have yet to encounter a scientist that can communicate with a whale," he wrote, "yet we purport to
know what they hear and how they interpret it."[259] Others objected to the
irony of undertaking the experiment in a marine sanctuary; several took particular offense at the argument, offered in the draft environmental impact
statement, that the sanctuary was a particularly good place to study the potential adverse effects because of the large number of marine mammals there.
Most respondents considered that logic perverse, if not insane; one respondent compared it to "nineteenth century science, where you had to kill ...
an animal in order to study it."[260] Another asked, "Is this how we go about
protecting our endangered species and marine mammals, by turning them
into experimental guinea pigs, potentially stressed and disrupted, violated
in their own habitat? . . . The name 'sanctuary' should speak for itself."[261] And
another: "What is a sanctuary for, if not to protect the marine life within
it?"[262] The oceanographers' alleged concern for marine life was belied by the
project's history: "We all know that the marine mammal component of this
project only came about because of public concern and outcry. If knowledge
about marine mammals is so critical, then cancel ATOC and initiate the marine mammal research free from the pressure" of ATOC.[263]

The cultural position of whales as an icon of animal intelligence, loyalty,
and musical ability clearly contributed to citizens' sense of outrage.[264] It was
common knowledge, many argued, that whales used low-frequency sound
to navigate and communicate with other whales, so it just wasn't plausible
that ATOC would have no effect. Arguing that ATOC would add only a small
increment to the background hum of existing noise pollution was akin to
justifying more air or water pollution on the grounds of what we had pol-
luted already.[265] If folks had not previously known the extent of ocean noise
pollution, they did now, and it only proved that "there is an immediate need
for noise reductions to make the oceans quieter."[266] Perhaps the most compas-

sionate respondent was a woman in Santa Cruz who wrote, "I know it's hard to let go."[267]

One interesting argument was that it didn't actually matter whether ATOC harmed marine life, because the whole project was superfluous. As the Kauai Friends of the Earth had noted, the United States was a signatory to the UN Framework Convention on Climate Change, which existed because leading climate scientists—most famously James Hansen—had already concluded that global warming was under way.[268] While not all citizens were necessarily tracking the progress of climate science, those who were paying attention to ATOC were probably more likely than the average American to know that scientific evidence of climate change was already mounting and that Hansen had testified in Congress to that effect.

Several respondents argued that the $35 million ATOC price tag would be better spent on developing solar, wind, or tidal power. One characteristic letter read: "We already have plenty of studies and data showing that there is global warming. That is a given. We do not need another experiment that proposes to tell us what we already know."[269] The retired radar engineer concluded sagely by noting that ATOC could actually delay action on the grounds that the science was not yet complete: "The decade or more needed to collect data may be used as an excuse for not taking positive action ... now."[270] Another put it this way: "Whether ATOC helps or hurts efforts to correct global warming is a political judgment call—it might work either way."[271]

Similar views were expressed by environmentalists, including Naomi Rose, a biologist for the Humane Society of the United States who was interviewed by the *Orange County Register* in April 1994. "We already know that pumping man-made greenhouse gases into the atmosphere is probably bad news. Instead of studying things to death, we should act to reduce the amount of greenhouse gases that get released," she said.[272] This was also the view of the *San Francisco Examiner*. An editorial in March concluded: "The money could be far better spent on finding better and alternative energy sources—wind, solar, and even tidal—that would reduce humanity's dependence on fuels that could contribute to global warming."[273]

For many citizens, ATOC had the wrong affiliations to be credible. While scientists were proud of the swords-to-plowshares aspect of the project, for onlookers the military association was sometimes grounds for suspicion. Bill Dietrich, a reporter for the *Seattle Times*, had already noted that "rather than taking this as an example of post–Cold War conversion, critics already unhappy with Navy experiments with captive dolphins regarded it with suspicion."[274] UC Santa Cruz physics professor Stanley Flatté put it more bluntly: "Folks thought it was some kind of secret Navy project."[275] He was right: many people read ATOC's military associations and its Department of Defense funding as evidence that ATOC was actually a cover story for a secret military project.

Various respondents wondered why ATOC was funded by the Department of Defense rather than the Department of the Interior or the EPA. One respondent demanded sarcastically: "What other projects have been funded by the Advanced Research Projects Agency?.... Do they have a track record in environmental science they would care to share with us?"[276] Others questioned why portions of the project were classified and why the project relied on Navy facilities. A woman in Hawaii objected to the installation of "the underwater bomber"; presumably she had confused the underwater "boomer," as several newspapers described it, with a bomber.[277] Other letters were more sophisticated but expressed the same idea: that the public was not being told the whole truth—perhaps not even any part of the truth. A man named David N. Seielstad, in the upscale community of Princeville, Hawaii, had read the draft environmental impact statement in detail and wrote at length (with citations to specific pages in the report) about the manifold reasons the program seemed suspicious: "From the beginning the ATOC proposal has had the aroma of a military research project. It is funded by DOD monies. It is administered by the US Navy. The originators of the project seem to be going to great lengths to disguise and conceal the true nature and the purpose of the project. In the proposal (p. 62), provision is made to 'manage classified aspects of the project....' The Johns Hopkins University Applied Physics Laboratory (a major Navy research and development contractor) is to use its clearance and store [any classified] data."[278] Seielstad was right about the associations—the Applied Physics Laboratory was a major Navy contractor, and various aspects of ATOC were classified. Other military associations had not been well explained or were covered up, such as links to the Pacific Missile Range Facility, a fact that was first denied and then affirmed. (The facility's website touts it as the "world's largest instrumented multi-environment range capable of supporting surface, subsurface, air, and space operations simultaneously."[279]) Seielstad noted that when scientists are "less than candid," the public gets suspicious. After a close reading of the draft environmental impact statement, he had concluded that ATOC was a cover story: "What is ATOC really? It is being promoted as a study of global warming.... Who could be opposed to that? [But] ATOC is ... only masquerading as environmental research."[280]

Similar sentiments were widely expressed. One letter asked the organizers to "stop insulting the intelligence of the human race" with their "global warming greenwash."[281] Another insisted that "if global warming was the true priority, then the use of tax dollars would be more wisely spent in the areas of clean energy and … efficiency" and that the classified nature of ATOC "implies that this has nothing to do with global warming, rather it is a military operation intended to improve submarine detection."[282] Still another demanded: "You should be honest with the American public about the true nature of these experiments. If the purpose is to learn more about global warming, why the classified designation? Please respond."[283] And: "If the Navy wants to sell us defense research cloaked as environmental concern, they should have gone to the CIA or NSA and kept their mouth shut."[284] And finally: "It's amazing to me that the public is viewed as being so stupid that we would believe that the Navy is suddenly concerned about global warming."[285]

One argument in defense of the project was that the Navy had been using this sort of acoustic transmission for decades; this was a major reason the scientists involved considered it unlikely that ATOC would do harm. But to many critics, it proved the Navy was used to operating without environmental oversight. For them, the US military as steward of the environment was simply not plausible. In an interview some years later, Naomi Rose put it this way: "The oceanographers asked: 'Why would you even think we would hurt the environment?' and environmentalists responded, 'Why would we think you wouldn't?'"[286]

Perhaps the best evidence of the widespread presumption that ATOC was a military project is that ATOC's defenders thought it was a military project, too. Among the rare expressions of public support, several were based on the corollary presumption that ATOC was necessary to defend the United States from its enemies. One man, commenting on the Kauai draft environmental impact statement, wrote: "I am not opposed to this research project, and I encourage you to press on with it. I believe in our military, and the importance of being defensively prepared. I do not agree with the efforts of Greenpeace and others of the liberal left to cripple our ability to defend our country. I am a conservative American, so if they are against your project, I am for it." Similarly, from a couple in Seaside, California: "We are adamantly FOR the ATOC project. Do not let the Santa Cruz Marxists stop this important work" (fig. 9.7).[287]

Conditional Approval

Amid all the claims and counterclaims—oceanographers and climate scientists mostly on one side, and biologists, conservation associations, and citizens mostly on the other—one person attempted to affirm to what was right on both sides of the argument. That person was Sylvia Earle, scientist,

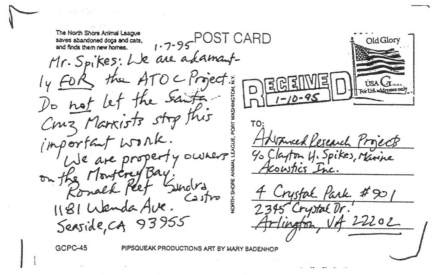

The North Shore Animal League
saves abandoned dogs and cats,
and finds them new homes.

1·7-95 POST CARD

Old Glory

USA G.....
For U.S. addresses only.

RECEIVED
1-10-95

Mr. Spikes: We are adamantly FOR the ATOC Project.
Do not let the Santa
Cruz Marxists stop this
important work.
    We are property owners
on the Monterey Bay.
    Ronald Reef   Sandra
1181 Wanda Ave.   Castro
Seaside, CA 93955

TO:
Advanced Research Project
C/o Clayton H. Spikes, Marine
Acoustics Inc.

4 Crystal Park #901
2345 Crystal Dr.
Arlington, VA 22202

GCPC-45   PIPSQUEAK PRODUCTIONS ART BY MARY BADENHOP

Figure 9.7 Postcard comment from supporters of ATOC.

From Box 11, Folder "MMRP draft EIS/EIR vol. 1 comments received and responses, Jan. 1995 C1–C37," Acoustic Thermometry of Ocean Climate Project Records, 1991–1996, Scripps Institution of Oceanography Archives, SAC 32, Special Collections & Archives, UC San Diego. Reprinted with permission of the Scripps Institution of Oceanography.

engineer, deep diver, and grande dame of American marine science, whose diverse accomplishments ranged from setting a world women's depth-diving record in 1979 to serving as NOAA chief scientist from 1990 to 1992. Smart, articulate, and beautiful, Earle was well known for her unwavering dedication to protecting the oceans as the "blue heart" of our planet. She did not accept the argument that ATOC would do no harm, but she did believe that with a properly constructed marine mammal research program, it could help advance much-needed scientific understanding and produce a net benefit. In making their final decision, the California Coastal Commission quoted her at length:

It seems obvious that the proposed research will, in fact, have some impact on the behavior of marine organisms.... I share with many others deep concerns about adding additional stresses to ocean ecosystems already modified by recent human activities ranging from overfishing to various kinds of pollution including high levels of "noise pollution." ... However, I am convinced that the greatest threat to the health of the oceans and to the planet as a whole is lack of knowledge and the profound mistakes in judgment that result from ignorance. Therefore, I believe it is important to try to resolve the problems associated with ATOC, if possible, and find ways to fill the enormous gaps in understanding the nature of the ocean and the effects of human activity on marine life.

Earle acknowledge the debate about "the protocols and the likelihood of success of the ATOC research and the MMRP," but she stressed the high caliber of the scientists involved and felt that if anyone could extract meaningful results from the proposed research, it would be them. She wrote: "Whatever is decided ... there should be protocols in place for discontinuing activities that appear to be causing problems—even without definitive proof of damage. [Yet it] may well be that more damaging than the effects of the ATOC project and the MMRP is the ignorance that will continue if such research is not conducted. With or without these projects the volume of noise in the sea is likely to increase significantly. It is vital that better understanding be gained of what this means to marine life, to the health of the oceans, and to the human future."[288]

The commission agreed and gave its approval for a two-year project, conditional on the "combination of the monitoring and protective measures incorporated into the project, the up-front commencement of the MMRP, and the relatively short (two-year) duration of the project prior to seeking any further permanent authorization."[289] Project scientists were required to return to the commission for approval in the event of any significant project modifications; if evidence developed "documenting adverse effects on marine resources" different from those already anticipated; and for any extension beyond the two-year period. They also would be required to add a fisheries biologist to the MMRP advisory board and to include effects on fish as part of their analysis.[290]

The commission still had concerns, noting that the likely impacts to important marine species were unquantifiable and that "determining the least environmentally damaging alternative prior to receiving monitoring results is similarly difficult."[291] They continued: "This is not an easy issue to resolve, given how little is currently known about marine mammal responses to sound, combined with the difficulty of monitoring these responses.... Even with monitoring, assessing whether the project will be beneficial or detrimental may be difficult."[292] They thus concluded, clearly with some reluctance, "Since the only way to determine the project's impacts is to proceed in the short term and study its impacts, the authorization of a two-year initial ATOC project is warranted."[293]

The oceanographers made two additional concessions: one, to agree to the commission's insistence that they establish a clear set of "termination criteria" for curtailing sound transmissions; and the other, to move the sound source out of the marine sanctuary. They would place it instead on the Pioneer Seamount, forty-eight nautical miles west of the Pillar Point Air Force Tracking Station near Half Moon Bay, twice as far from land as the original site.[294] These changes formed the basis for negotiation with the parties that had sued them. On June 2, 1995, after months of difficult negotiations and a few additional concessions to biological concerns—including that the scientists would make no claim regarding the MMRP proving or disproving long-term impacts—the parties agreed to an out-of-court settlement.[295] With the agreements in place, the NMFS issued permits for the initial phase of ATOC MMRP to begin.

What Did They Find?

Engineering tests began on October 28, 1995, and the project was in full swing early in the new year. An interim review in mid-1996 revealed no significant problems, and following the agreed-on terms, the project was allowed to continue through early 1998. Meanwhile, the NRC was preparing to review the MMRP results, taking as its charge the review of both the specific MMRP results and any independent advances in understanding the impacts of low-frequency sound since the last review in 1994. But even before the MMRP was over, scientific articles had begun to appear, particularly in the *Journal of the Acoustical Society of America*, in which the (widely critiqued) optimistic bias was clearly still intact. A paper published in May 1997, for example, concluded that whales would not hear the signal unless they dived to a depth of four hundred meters, and therefore, "the ATOC signal will probably have minimal physical and physiological effects on cetaceans," although no direct evidence was offered to support that claim.[296] Another paper, published in 1998 in the same journal, subjected three species of rockfish in an enclosure in Bodega Bay, California, to an "ATOC-like" signal; the fish showed no evidence of response. In fact, the alleged "ATOC-like" signal peaked at 153 decibels, well below than the actual 175 to 195 decibels of ATOC transmissions.[297]

When the NRC issued its report, it became clear that this optimistic bias was unjustified. The conclusions of the hundred-page report, released in 2000, were summarized in the first paragraph of the executive summary:

> Some of the MMRP observations ... showed no statistically significant effects of ATOC transmission. For these observations, the Committee could not distinguish among true lack of effect and insufficient observations, small sample sizes, and incorrect statistical treatment of data. A somewhat clearer lack of significant effects of the ATOC transmissions was

demonstrated in observations of elephant seals' diving behavior near the Pioneer Seamount source. Some statistically significant differences between control and exposure were found for other species, including (1) an increase in average distance of humpback whales from the California source and (2) increased dive duration for humpback whales off Hawaii. The MMRP found no obvious catastrophic short-term effects as a result of transmissions from either source, such as mass strandings or mass desertions of source areas.[298]

The worry that the MMRP science would be insufficient to resolve the crucial questions was validated. Some statistically significant effects had been observed, but it was not clear whether they were biologically significant. In other cases, no statistically significant effects were observed, but it was not possible to say whether this was because there were no effects or because there were insufficient data to detect any effects. The result was predictable— and, indeed, had been predicted.

While the executive summary took the understated tone characteristic of NRC reports, the detailed report was sharply critical. The central problem was the one that ATOC's critics had long pointed out, and that Chris Clark had tried to address: the design of the project was inadequate to answer the key biological questions. MMRP was essentially a biological "retrofit" onto a program designed for another purpose. Its design was based on neither maximizing the relevance to marine mammals nor minimizing the impacts, but on the project's primary goal of measuring the speed of sound. "As a consequence," the NRC concluded, "the results of the MMRP do not conclusively demonstrate that the ATOC signal *either* has an effect *or* has no effect on marine mammals in the short or the long-term."[299] After five years, months of hearings, hundreds of comments, and millions of dollars, the question of whether ATOC would or would not harm marine life remained unanswered— and oceanographers had not demonstrated global warming, either.[300]

In a footnote, the committee quoted at length Peter Worcester, an ATOC program scientist, who had explained the basis of the program design in written correspondence to the committee: it was based on maximizing the chances of obtaining a meaningful signal over the noise. The ocean is a very variable place, so the observed signals would be affected by many different things; the scientists needed to design the sampling system to differentiate between these sources of short-term variability and the temperature signal of interest. Worcester explained, "The duty cycle is actually set by the need to avoid aliasing [i.e., confusion or conflation] of rapidly changing oceanographic phenomena.... The combination of ocean phenomena led to a [sampling] cycle ... consisting of 1 day with six 20-minute transmissions at 4-hour intervals to adequately sample tidal variability, occurring every fourth day to adequately sample ocean mesoscale variability."[301] This, of course, was precisely the complaint that environmentalists had leveled from the start, and the claim to which Clark had taken exception in the Scripps press release: ATOC was not designed to minimize impacts on marine life. It was designed to detect global warming. All the subsequent attempts to adjust that design amounted to a patch-up job, with the unsatisfactory results that Clark had predicted.

It was beyond the committee's charge to recommend whether ATOC continue, but the results suggested that ATOC scientists had been at best premature in their assertions that the project would do no harm. The committee stressed that even now there were inadequate grounds to make such judgments, and that developing grounds would take "a more sustained and integrated approach than has been the case in previous research."[302]

The committee might have left the matter there, but another issue concerned its members: how and why a group of physical oceanographers had found themselves in the position of designing a marine mammal research program. In an implicit rebuke of the quality of science supported by mission-driven agencies, the committee suggested that the research that was needed should not be sponsored by ARPA or the Navy, but by the scientific agencies that funded biological research. ATOC aspired to be basic science, but it had been pursued in the spirit of mission-driven science, without independent peer review and with insufficient attention to corollary scientific issues. When the scientists had tried to address those corollaries, they had created a project that was so poorly designed that it was unable to answer the questions it needed to answer. The committee offered this recommendation: "Mission-oriented agencies should ensure that the research they sponsor will not only contribute to their immediate missions but also answer basic scientific questions.... Most importantly, all of these projects should receive strict peer review and be evaluated on the quality of the science proposed."[303] By "peer review," they meant review by independent and acknowledged experts in the field. If the question was the impact of sound on cetaceans, the relevant reviewers should be cetacean biologists. If the question was about fish, the reviewers should be ichthyologists. And so on.

This point was reiterated in a findings and recommendations section, in a conclusion that could well have been applied to the whole US Navy oceanographic research program. It was a conclusion with which William von Arx, Henry Stommel, and all the members of the Woods Hole Palace Revolt (as well as the Canadian squid biologist Frederick Aldrich) would have concurred. It was the problem of mission-driven science: "Most marine mammal studies are funded from mission-oriented sources. At this time the greatest source of funding for marine mammal research is the ONR. However, by its nature, ONR-funded research tends to be focused on questions of practical importance to the Navy, and is not necessarily responsive to the broad interests of scientists seeking to learn more about the basic biology of marine mammals. Scientist-driven fundamental research could significantly improve our understanding of hearing and the effects of low-frequency sound on marine mammals, as well as our overall understanding of the acoustic behavior of these animals."[304] The Navy focused on matters of practical import to the Navy. At times those matters aligned with issues of scientific import, creating important domains of knowledge, but at times they did not, leaving significant domains of ignorance.

The impact of underwater sound on marine life was a domain of ignorance. The Navy had spent untold sums to understand the propagation of underwater sound and had studied animals to help understand that propagation, but the impact of the sound on those animals was essentially unknown. In more than half a century of scientific investigation, the basic science that might have answered the key questions about ATOC's impact on marine life had never been done. The fact that the Navy had left this area so understudied proved the point: it was naïve to expect Navy-funded research to answer questions peripheral to Navy interests (or, for that matter, to claim that it had ever done so except inadvertently). To understand marine mammals, a program had to be designed to do that. The same might well have been said of any area of oceanographic research. Indeed, of any scientific research at all.

The NRC report could hardly be read as validation, but it had stopped short of saying that ATOC should be stopped, and so the ATOC scientists did not abandon hope. On the contrary, anticipating a positive recommendation from the NRC, they had been preparing a new draft environmental impact statement and small-take permit application for the project's continuation, which they released in the spring of 2000 at the same time as the NRC report. Once again, the oceanographers tried to put an optimistic spin on the situation, and once again biologists opposed them. But this time the biologists were fortified with the knowledge that many of their predictions had come true. In September 2000, Paul Anderson posted a scathing denunciation: "Both the [new] DEIS and the small take permit application pretend that the Acoustic Thermometry of Ocean Climate Marine Mammal Research Project effectively dispelled any concerns as to the effect of these sounds on marine mammals. The ATOC MMRP not only did not demonstrate long-term effects, but ... it failed to adequately investigate short-term responses. The proposal for continuation of ATOC is based on false [premises]."[305] NMFS agreed: the MMRP had failed to demonstrate no harm and so the permits would not be renewed. The scientists were instructed to remove their instrumentation. The project ended on a tragic note in August 2000 when the twelve-thousand-pound transmitter was retrieved from the Pioneer Seamount: a winch operator was struck in the head by a piece of equipment and killed.[306]

Coda

In October 2002, a federal judge halted a National Science Foundation–funded project using intense compressed-air blasts to study the structure of the seafloor in the Gulf of California after two beaked whales were found dead on the nearby Mexican coast with evidence of hearing damage. US District Magistrate James Larson based his ruling in part of evidence that the Navy considered sounds above 180 dB to be "potentially harmful to marine mammals."[307]

That particular incident involved compressed air—not low-frequency active sonar—but it brought into high relief the entire issue of sound in the ocean, including the diverse scientific and military activities that made use of underwater sound. Two years later, when the Navy introduced SURTASS-LFA, a towed low-frequency active sonar system for submarine surveillance (in effect, a towed version of Artemis), environmentalists were poised to object.[308] This time, they were armed with substantial scientific evidence that supported their claim that at least some adverse effects could be expected. In response, in 2004, Congress amended the Marine Mammal Protection Act to add a provision addressing "incidental takes" that occur in the context of a "military readiness activity."[309]

Meanwhile, the National Research Council continued to track the science, issuing new reports in 2003 and 2005, while scientists continued to try to assess the cumulative effects of anthropogenic stressors on marine life.[310] None of this work definitively answered the question of the impact on marine life of either noise in general, or low-frequency sonar in particular, but it did prove that the ATOC scientists had been premature in insisting their project would do no harm. The 2003 NRC report, for example, noted that "use of one of the Navy … mid range active sonar [systems] was found to contribute to a stranding incident in the Bahamas."[311]

Overall, the dominant thrust of the NRC reports was the conclusion that sufficient data to determine the impact of underwater sound on marine life—particularly long-term impact—were lacking. By the NRC's judgment, direct evidence of marine mammal death has not been found, but there are sufficient data to support the conclusion that there are impacts, both on marine mammals and on fish.[312] The question is how significant the effects are.[313]

The 2005 NRC report took as its charge the attempt to clarify whether observed effects, such as changes in vocalization patterns, are biologically significant, explaining, "an action or activity becomes biologically significant to an individual animal when it affects the ability of the animal to grow, survive, and reproduce."[314] These scientists noted that short-term disruptions to behavior may or may not be biologically significant—more research was needed—but they also allowed "that wild animals rarely engage in activities that are not biologically significant (even play is not frivolous)."[315] An observed effect was therefore likely to have at least some biological effect and it would be mistaken to assume that effects that appear minor to us are minor to the organisms that experience them.

Recall the ATOC "fact sheet" that described how the ATOC noise would sound to a human ear: the relevant issue was never what those sounds would sound like to a human ear, but what they would sound like to—and what impact they might have on—marine life. That was what ATOC scientists did not know when they began their project, and still did not know when the project ended. Indeed, after more than two decades, three NRC reports, and several lawsuits, the science of underwater sound and its impacts on marine life was still unresolved. But so far as both the scientific community and the courts were able to judge, it was not correct to say that underwater sound has no adverse effect.

In 2007, a coalition of environmental groups (including the California Coastal Commission) sued to stop the Navy use of middle-frequency sonar off the California coast on the grounds that it could harm whales, dolphins, and other marine life. A US district court judge in Los Angeles agreed and ordered the program halted. The case settled in 2008, but only after the judge (who was described by the *Los Angeles Times* as having "spent years poring over studies about whale deaths and injuries after Navy exercises") issued a set of rules that the Navy would have to follow. Previous guidelines that the Navy had developed—such as slowly reducing and then shutting off sonar power when whales or dolphins came with two hundred yards—were, in her words, "grossly inadequate."[316] However, the judge stopped short of banning sonar activities entirely, and another lawsuit, filed in 2012, argued that the exemptions given to the Navy were too broad. In 2016, this case reached the US Ninth Circuit Court of Appeals, which found in favor of the plaintiffs.[317]

Since the conflict over ATOC, the Navy has worked with the NMFS to develop guidelines to minimize impacts of its sonar activities. According to the

US Department of Justice, "The Navy and NMFS have developed an extensive set of mitigation measures to reduce potential impacts [of sound] to marine mammals [which are] expected to consist of short-term behavioral modifications that the [Marine Mammal Protection Act] classifies as 'Level B' harassment."[318] According to the law, level B harassment is defined as "any act of pursuit, torment, or annoyance which has the potential to disturb a marine mammal or marine mammal stock in the wild by causing disruption of behavioral patterns, including, but not limited to, migration, breathing, nursing, breeding, feeding, or sheltering."[319] In other words, independent judges have found, and the Navy has acknowledged, that harassment caused by sonar operations does, in fact, occur. And we have not solved the problem of global warming.

Conclusion
The Context of Motivation

During the Cold War, oceanographers and marine geologists and geophysicists worked productively with Navy patrons to learn a great deal about the deep sea; scientists justly look back on that period with pleasure and describe it as a golden age. Yet Navy priorities were inevitably different from scientific ones, and this sometimes led to conflict.

Some readers might be tempted to dismiss those conflicts as personal, or perhaps sociological, but the analysis offered here shows that they had epistemic consequences. Navy sponsorship at times impeded the advancement of knowledge rather than promoting it. Potentially significant projects were delayed or rejected on the grounds that they did not fit the mission profile, and many questions were neglected for lack of military salience. The US Navy never told oceanographers what to think and only rarely told them what to do, but military interests framed oceanographic studies in impactful ways, and inevitably some matters received more attention than others. It was never true, as one historian has recently claimed, that scientists during the Cold War "received virtually unlimited research funds to investigate whatever they wanted."[1] On the contrary, it was clear even at the time that there were definite limits to what the Navy would fund and both benefits and costs to Navy support. Is it possible in hindsight to evaluate both the benefits and the costs, the gains and the losses?

Most of the oceanographers in this story would have rejected the notion that Navy money was tainted, but some did worry that the Navy was pulling them in directions that were not altogether good. Yet at the same time, and perhaps somewhat perplexingly, nearly all of them insisted that the Navy was funding "basic research." This conclusion attempts to make sense of the complex issue of the impacts of Navy patronage in terms both of what was gained and what was lost—what was done and what was left undone—when the context of motivation was military defense oceanography.[2]

The Pure Science Ideal

In insisting that their work was "basic" science, the oceanographers in this story were invoking what historians have called the pure science ideal.[3] Like many slogans and catch phrases, this one has meant different things to different people, but we cannot dismiss it as too malleable to be meaningful because it has been highly persistent as an actor's category—a term with meaning to the people who used it. In the late nineteenth century, American scientists used the pure science ideal to argue for increased federal government support for scientific research; in the twentieth century they invoked it to justify the creation of the US National Science Foundation as a largely self-governing agency. The foundation, established in 1950 and heralded by its first historian as a "patron for pure science," has been viewed as exemplary for its commitment to investigator-driven scientific research.[4] It has been, and remains for many scientists, the best source of "untainted" money.[5]

Yet ironically (and probably not coincidentally), the ideal of pure science— in the past variously glossed as scientific research that is free, undirected, curiosity driven, investigator initiated, fundamental or, most commonly, "basic"—was most vehemently promoted at the very moment when American science was becoming tightly coupled to the geopolitical ambitions of the national security state. Moreover, this was a change from what many had argued before. American earth scientists, in particular, had argued before World War II that a sharp separation between pure and applied science was neither to be found nor desired. Pure and applied science were, in the words of one scientific leader, two sides of the same coin.[6] After the war, however, as money poured into earth science to support costly investigations in oceanography, meteorology, seismology, isotope geochemistry, glaciology, and more, scientists insisted that they were pursuing basic science even when their work was funded by the US Department of Defense, and even when their projects were manifestly connected to national-security concerns and, in some cases, even to specific military missions.[7]

In this story, *Alvin* is the clearest example. Why did oceanographers insist that the submersible was designed and built for basic research when the facts were otherwise? More broadly, why did so many oceanographers insist that the Navy supported basic research when its projects were understood as military defense oceanography, framed and constrained by the Navy need to know?

One reason these oceanographers had a positive view of Navy patronage is clear: the Cold War was a good time for them. Oceanographers and marine geophysicists working under Navy patronage answered many important scientific questions and resolved long-standing scientific debates. They were able to do these things because there was a confluence of interest between what many oceanographers (particularly physical oceanographers) wanted to learn and what the Navy needed to know. Henry Stommel and Arnold Arons wanted to study ocean circulation. Harry Hess was interested in seamounts and guyots. Bill Menard was utterly taken by seafloor topography and the Pacific Ocean fracture zones. Engineers at Woods Hole were excited to build *Alvin*. The Navy was interested in these matters and shared many of these goals, albeit for different reasons. For these men, Navy support did not create a conflict, because they were not being asked to do anything they did not want to do. On the contrary, they were being enabled to do what they wanted to do. The Navy was supporting *their* basic research. But then why not just admit that this was done under the auspices of military defense oceanography and defend that as a fine thing? Why downplay the military connections?

Many of these military projects were classified and so could not be discussed except in the most general terms, but scientists could have explained that their work was broadly relevant to antisubmarine warfare or ocean acoustic surveillance or deep-sea rescue and salvage without revealing classified details. Some of them did. Scripps's Fred Spiess took pride in the usefulness of his work and made no attempt to hide or downplay his military connections. But he was an exception. Moreover, classification does not explain why, later, scientists and popular writers expunged *Alvin*'s early history from their accounts of it, or why many oceanographers continued to deny the military applications of their work long after projects had been declassified.[8]

I have come to believe that the answer to this question is that these scientists were working to advance knowledge under circumstances that they worried—and their leaders had often publicly insisted—were at odds with that advance. Because their beliefs about the freedom of science were so deeply held—and often tied to their politics—the option to question and revise those beliefs did not appear to be available.[9] Their scientific conclusions were subject to revision in light of new evidence, but their philosophy of science proved less epistemically open, and so they resisted the truth about their circumstances and insisted that they remained fundamentally free, even when that was manifestly not the case. This, I suggest, is why scientists painted their projects white: to admit that their work was driven by extrinsic considerations would have been to suggest that it was—or at least might have been—tainted. Because, after all, some science is tainted.[10]

Tainted Science

In recent decades, revelations that the US tobacco industry funded research intended to obscure and conceal the causal links between its products and adverse health effects has led to questions about the propriety of accepting further research funds from that industry; many universities no longer accept tobacco industry funding.[11] Some scientific journals will not accept articles even for consideration if the research was tobacco industry funded.[12] Research shows that these journals are right to do this: peer-reviewed studies of the impact of funding sources on investigations, with respect to not just the health effects of tobacco use but also the safety and efficacy (or lack thereof) of pharmaceuticals and other medical interventions, has demonstrated that the interests of sponsors can affect research results.[13] These impacts are so substantial that they have a moniker: "the funding effect."[14] Thus, increasingly journals require funding disclosures, to illuminate for editors, reviewers, and readers potential conflicts of interest.

The rationale is clear and compelling. When a particular outcome is desired (and investigators know or suspect what that outcome is), it can affect research in myriad ways, both minor and profound.[15] Not all patrons want accurate information. Misinformation and disinformation can be of value— and when it is, some groups will seek to create, sustain, and promulgate it.[16]

The facts of funding bias are consistent with the intuitions of many scientists, as well as nearly all historians, sociologists, and ordinary people, that *of course* it matters who pays for science. Funders want particular questions answered, and sometimes they also want particular answers—whether or not those answers are true. In theory, peer review should sort out biased research and faulty findings, but in practice peer review is a weak filter.[17] Research has demonstrated that a good deal of problematic science has passed peer review and been published in respected journals. Not all of this bad science is industry funded, and not all industry-funded science is bad, but quite a lot of it is.[18]

This helps to explain why so many scientists believe that "pure" or "basic" science is superior to that which is "applied," "directed," or "mission driven"; they believe that the latter is subject to potentially distorting external pressures to which the former is not.[19] If so, it would make sense that scientists would insist on the purity of their research—on its basic and fundamental character—even when their funders have specific goals and concrete purposes in mind. To allow that their research was funded by an outside organization with an interest in something other than the advancement of learning would be to raise the specter of taint. It would also raise the specter of loss of independence and freedom, the very issues that Denis Fox and Claude ZoBell back in the 1930s, and Henry Stommel and his colleagues in the 1960s, did raise.

Unalloyed Science for Unalloyed Knowledge

The freedom of scientists under Navy patronage to be driven by intellectual interests rather than geopolitical tensions was at the heart of the questions raised at Scripps in the late 1930s, and again at Woods Hole in the early 1960s, about the desirability of military funding. It was what Denis Fox and Claude ZoBell called the "freedom of science," what Henry Stommel called the "drive to wrestle personally with the unknown." Navy support created the opportunity for Stommel to spend his life wrestling with the unknown, but it also involved constraints that, as William von Arx had noted, were always there "in a ghostly way."[20]

To say that scientists working with the Navy had both opportunities and constraints is not, however, much of a finding: we can say that about anyone working in any system. Moreover, there is nothing in principle wrong with being paid to solve a problem: doctors, lawyers, and engineers do this every day. So do exploration geologists and geophysicists. Spiess and Vine thought it was a good thing to have an operational problem to solve. But that is not how most of their colleagues saw it.

Consider again the idea of "pure" science. The opposite of *pure* may be *impure*, but it may also be mixed. Another opposite of *pure* is *alloyed*, and alloys are created because they are in some characteristic better suited than pure metals for a particular purpose. The conditions under which oceanographers worked during the Cold War were clearly mixed: a mix of freedom and constraint, a mix of generous funding and fiscal limitations, a mix of autonomy and external direction. We might therefore conclude that Cold War oceanography was an alloyed science—suited to the purposes of the period—a blend of different but compatible needs and interests. However, the notion of an alloyed science does not quite explain why so many oceanographers fussed about freedom and autonomy and insisted that they were being funded to do work they wanted to do.

I suggest that a large part of the answer is that these oceanographers believed that scientific integrity depended on epistemic independence. They believed that, to find the truth about the world, they needed to be personally independent, their science driven by curiosity and curiosity alone.[21] And it wasn't just oceanographers; throughout the second half of the twentieth century, American scientists clung to the myth of unfettered research and insisted on the imperative of intellectual autonomy and the centrality of curiosity as the primary driving force of science.[22]

Sociologist Thomas Gieryn has focused attention on the sociological dimensions of "boundary work" in helping to establish and maintain professional identity and independence. No doubt when scientists insist on the imperative of autonomy, it contains elements of this.[23] But scientists' insistence on their own autonomy also reflects their beliefs about the nature of knowledge, about what is required to learn about the world, about what it takes to find truth. It reflects an epistemology that presumes that it takes unalloyed science to produce unalloyed knowledge.

Consider Stommel's ideal scientific agent: a man or woman wrestling with nature, unimpeded by potentially interfering considerations. He or she is free to pursue leads wherever they go, and thereby discover diverse and sometimes unexpected truths about the natural world. Stommel's vision is an epistemological one. It is a belief about what it takes to produce reliable knowledge. But it is also a moral claim, because truth is good and falsehood is not.

Consider the antonyms for the terms *pure, basic,* and *fundamental.* The dictionary gives a long list of antonyms for *pure* and nearly all of them carry negative ethical connotations: *impure, sullied, dishonest, fake, counterfeit, invalid, ingenuine, unreal, contaminated, corrupt, dirty, tainted, uncertain,* and even *vulgar.* The antonyms of *basic* are not quite so emotionally charged, but they do suggest lesser value: *auxiliary, inessential, insignificant, minor, nonessential, secondary, unimportant, unnecessary, extra, peripheral.* The antonyms of *fundamental* include *trivial, minor, subordinate,* and *needless.* What scientist would want to do work that is unnecessary, peripheral, or needless, much less counterfeit? And while an alloy is a mixture that may be good or bad depending on your purposes, something that is alloyed is debased. (The very word *debased*— suggesting the presence of base metals—implies a loss of value.) The implication is clear: scientists want to do work that is central and not peripheral, reliable and not counterfeit, dignified and not debased. They want to do work that is unalloyed, and they want their work to be *good.*

Steven Shapin has suggested that the early modern ideal of scientific knowledge was rooted in gentlemanly ideals of personal and fiscal autonomy: scientists had to be gentlemen because only they had the requisite autonomy to be intellectually trustworthy.[24] Whether or not one can trace an unbroken thread from the seventeenth century to the twentieth, one can say that the ideal of scientific research as an activity of independent individuals has been particularly resonant in American culture, where it overlaps with prevailing beliefs about the power and virtues of individualism and competition. Historian Paul Lucier has suggested that the trope of "unfettered research" was an essential element in the professionalization of American science as "no-strings-attached cash became the holy grail of the professionalization process." Aspiring late nineteenth-century men of science earned their keep in diverse ways, but "whatever occupations [they] might actually have engaged in, they were treated as short-term steps toward this long-term objective."[25] Yet even as these men aspired to remove themselves from the commercial economy, they valorized the principles of autonomy, individualism, and unfettered competition that in the late nineteenth century were thought to characterize it.

David Hollinger has noted that the ideal of the autonomous scientific agent competing in a free market of ideas closely parallels the autonomous economic agent essential to classical formulations of enterprise capitalism.[26] In this sense, laissez-faire individualism is enshrined in the foundations of both Anglo-American political-economic and Anglo-American scientific thought. Vannevar Bush expressed this sentiment when he wrote *Science: The Endless Frontier*, the brief not only for what would become the US National Science Foundation (albeit in a different form) but also that provided the culturally accepted justification for spending taxpayer dollars on activities that previous generations had rejected as elitist and arcane.[27] Bush insisted that "scientific progress on a broad front results from the free play of free intellects, in the manner dictated by their curiosity for exploration of the unknown."[28]

In essence, American scientists in the second half of the twentieth century subscribed to a theory of knowledge (sometimes explicit, more often not) in which personal freedom was requisite for scientific progress and curiosity provided the appropriate context of motivation. Yet the reality of their scientific lives was that the lion's share of their work—and nearly all the monetary and logistical support for it—was motivated and justified by Cold War tensions and perceived political imperatives. It is therefore perhaps not surprising that so many of them insisted that context was irrelevant and that extrinsic considerations drove their patrons but not them.

Epistemologies have histories, and Vannevar Bush's views were rooted in an already-established discourse about the role of science in American democracy and political economy. At the nation's inception, Thomas Jefferson's colleagues rejected his call for a Department of Science as inappropriate for a republic; they would accept the US Coast Survey as the nation's first scientific agency only because it was clearly requisite to national defense and trade. Throughout the nineteenth century, American scientists would find support and motivation for their work in the context of the expansion of American economy and territory, as well as national defense, but they continued to struggle to overcome arguments that pure science was elitist, undemocratic, and, especially, *European*.[29]

In the late nineteenth century, American physicist Henry Rowland pled for pure science as an alternative to the crass materialism he saw dominating American society in the Gilded Age; he argued against a perceived hostile environment.[30] Some decades later, sociologist Robert Merton defended the ideal of disinterested science in response to totalitarian threats to both scientific and political freedom. His most famous essay—"The Normative Structure of Science," published in 1942—stressed the importance of disinterestedness in the context of an ideal scientific community that is open, sharing, and fully communicative. The additional element carried by disinterestedness is the preservation of the credibility of the scientific expert in contradistinction to the shill, the snake-oil salesman, and the charlatan.[31]

Merton's most full-throated defense of scientific autonomy was published earlier—in 1938—just as military support for American science was ramping up in anticipation of entry into World War II, which is to say, just as the story told in this book begins. In *Science and the Social Order,* Merton noted the enormity of the pressures under which scientists in Nazi Germany had found themselves, suggesting that the natural autonomy of science made it threatening to totalitarian regimes. While Merton recognized that "pure science" was an idealization that could be interrogated on a number of grounds, he argued that its central function as a norm was to protect scientific integrity by ensuring, or at least attempting to ensure, its autonomy:

> One sentiment which is assimilated by the scientist from the very outset of his training pertains to the purity of science. Science must not [in this view] suffer itself to become the handmaiden of theology or economy or state. The function of this sentiment is to preserve the autonomy of science. For if such extrascientific criteria of the value of science as presumable consonance with religious doctrines or economic utility or political appropriateness are adopted, science becomes acceptable only insofar as it meets these criteria. In other words, as the pure science sentiment is eliminated, science becomes subject to the direct control of other institutional agencies and its place in society becomes increasingly uncertain.... The exaltation of pure science is thus seen to be a defense against the invasion of norms that limit directions of potential advance and threaten the stability and continuance of scientific research.[32]

Thus, what might at first seem hypocritical or even dishonest—that American scientists were insisting that they were doing pure science (and merely painting their projects blue for credulous military men) at the very moment when they were trading autonomy and openness for a greater measure of material support—is not hypocritical at all.[33] It is what we should expect. These scientists were guarding their integrity by insisting that they had not sacrificed core values. They were defending their research as worthy of societal trust and support. They were insisting that their science remained untainted.

Moreover, an argument that in hindsight we find unconvincing may nevertheless have been sincerely held. Historian Loren Graham (1998) has shown that many scientists in the Soviet Union believed in dialectical materialism, which often informed their science in productive ways. Sigrid Schmalzer has suggested that Chinese scientists generally shared Mao Zedong's goal of making Chinese science "self-reliant."[34] And we have seen that many American scientists in the Cold War accepted the goal of containing communism, even if that was only occasionally made explicit.

Scientists often share the politics of their patrons; they may also find ways to adjust their views to accommodate practicalities. Robert Kargon and Elizabeth Hodes have described how conservative American scientists in the 1930s who believed in laissez-faire capitalism struggled when these views came into conflict with the opportunities presented by expanded federal funding of scientific research.[35] Federal funding went against these men's stated principles, but in the end they softened their principles and accepted the money.

In chapter 1 we saw how some scientists who benefited from military largesse had previously opposed state funding of science as a threat to scientific integrity. As followers of Michael Polanyi and members of the Society for Freedom in Science, they feared that state funding would mean state control, raising the specters of suppression and oppression.[36] But when war broke out and government funds began to flow, most of them shifted their positions and joined their colleagues in accepting state support. Some justified their change of heart by the exigencies of World War II (and then the Cold War); others denied that it was a change of heart by insisting that the work they did was pure and that they had state support without state control. In chapter 4 we saw Roger Revelle and Maurice Ewing invoking the ideal of "independent scientific exploration" as a lever to try to wrest some measure of autonomy from their patrons and to argue that excessive control by patrons would undermine the science that the patronage was intended to foster. And throughout this story (with the notable exception of the Woods Hole Palace Revolt), oceanographers mostly focused on the positive: the ways in which Navy support facilitated work that would otherwise have been difficult or impossible and empowered them to achieve goals that might otherwise have remained unachievable. After all, money is a form of freedom, and if smart scientists have money to spend, they are bound to learn something. Conversely, as Frederick Aldrich learned to his dismay when he sought to use *Alvin* to study the giant squid, ideas with no money go nowhere.[37]

Given this, we can appreciate oceanographers' positive view of the US Navy as their sponsor. The Navy was a good patron for science because it was rich, because its funding and logistical support enabled scientists to do work they wanted to do, and because a confluence of interest meant that a good deal of the work that the Navy supported was intended to answer questions about the natural world. What the Navy needed to know overlapped considerably with what scientists wanted to know. Moreover, as we have seen throughout this book, and Navy officers themselves stressed, Navy interest in the ocean ranged widely. As the TENOC report put it, "the successful design, construction, and operation of a powerful modern Navy" depended on how well the ocean environment was understood.[38] A wide range of scientists could pursue their interests under Navy sponsorship.

The net result of Cold War Navy-scientific cooperation was that knowledge about the ocean was expanded, and not just in a minor way. Questions that had been debated for centuries—about the nature of the deep-sea circulation, the structure and origins of the seafloor, and whether or not continents moved—were addressed and answered to scientists' satisfaction. In science, as elsewhere, money is power: in this case, the power to build new instruments, to travel to new places, to investigate, to study, to support graduate students, and ultimately to learn. Navy support was powerfully enabling.

But the Navy was a good patron for another, more epistemological reason: because it had no prior interest in the outcome of the investigations it sponsored, no expectation of what the right answer should be. Scientists funded by the Navy did not have the autonomy to study anything they wanted, but they did have something that was equally or even more important: the autonomy to design their studies as they saw fit and pursue leads where they led. They had this freedom because the Navy shared their desire for accurate information. This is relevant not only for our understanding of Cold War oceanography, but also for our understanding of contemporary science.

To make this point clear, we have to be willing to use a word that makes some science studies scholars flinch: the US Navy was a good patron because it wanted to know the *truth*. We have to be able to speak about truth without scare quotes.

Whether ocean circulation was shallow or deep, whether it was wind driven or density driven, whether guyots were rare or widespread: in all these cases and many more, it was in the Navy's self-interest to know the answer. The Navy's interest was concrete and immediate and what it needed was knowledge. Accurate knowledge. Comprehensive knowledge. Good information would help its men and officers to perform their tasks, whereas incorrect or incomplete information could spell disaster. A missed shadow zone could mean the difference between escaping an enemy and being torpedoed. An undiscovered or wrongly located seamount could mean a devastating submarine crash. We might say that the Navy's epistemic stance was one of neutrality: it had no interest in the factual outcome of the work it sponsored, apart from a compelling interest in it being factual. This was no small thing, because, as we've noted, not all patrons are interested in facts. But during the Cold War the US Navy was.[39]

The Navy did put pressure on scientists to put operational needs before intellectual ones, and it did restrict what scientists could publish.[40] But nowhere have we seen evidence that Navy officers pressured scientists to find a particular result, to design a study or interpret their data in a particular way, or to prefer one theoretical interpretation over another. From an intellectual perspective, I would argue, this is the critical feature that accounts for scientists' overall positive view of their military patrons despite the various difficulties they encountered along the way. The key element that makes sense of the stories told in this book is that the Navy did not interfere with scientists' epistemic flexibility and interpretive independence.

The Navy interest in good information (along with its ample purses) enabled it to support science in a manner congenial to scientists and productive of authentic, robust understanding. And this is key feature that distinguishes the Navy during the Cold War from other sorts of patrons—such as the tobacco or chemical industries—who have sometimes supported science not to advance knowledge but to cloud, confuse, or even obstruct it.[41] This is not an argument for "pure" science in any absolute sense; we have seen throughout this book that the Navy was not supporting pure science. It was supporting useful science, but what made that science useful was its veracity. To be useful it had to be true.

Returning to Steven Shapin, he has also stressed that, gentlemanly origins or not, science has never been pure.[42] Scientists have always needed funds to pay for experimental apparatus, laboratory supplies, the costs of fieldwork, and the salaries of laboratory and observatory assistants. Once science became a professional activity, scientists needed salaries, too. In the twentieth century, these costs mushroomed as scientific research increasingly relied on sophisticated instrumentation and "research platforms," such as particle accelerators, gigantic telescopes, space satellites, and, as I have stressed here, oceanographic vessels. With the rare exceptions of some independently wealthy individuals who funded their own investigations, scientists have always needed patrons.[43] To suggest that patronage, in some general way, is problematic is to ignore history and beg the question at hand. That question is not, Does patronage affect science? Inevitably, it must. Rather, the question is: How does patronage affect science? Or, better, how did this particular form of patronage affect this science at this time? How did it structure what scientists found themselves working on and, more important, found themselves *wanting* to work on?[44]

To return to our story, we can now ask: How did the Cold War context motivate some lines of research and discourage or even block others? How did it affect what scientists thought their goals should be? How did it influence their self-perception, how they understood their relationships with colleagues in other fields and academic departments and even the world at large? Above all, how did it help to determine what they cared about?

Incompetence or Hubris?

One of this story's more perplexing aspects is why brilliant scientists — committed to knowledge based on evidence — dismissed or discounted evidence from other fields, in some cases even other branches of their own field. We saw this in chapter 8, when Charles Hollister refused to accept the evidence of turbulence in the deep benthic boundary layer, and even more clearly in chapter 9, when Walter Munk and colleagues insisted that ATOC would have no significant biological impacts in the face of clearly articulated objections and evidence from biologists.

The physical oceanographers who conceptualized and developed ATOC did not initially consider potential impacts on marine life, at least not in any serious way. When they were forced to do so, their initial reaction was to assert that any impacts would be insignificant, despite their lack of relevant evidence, expertise, or knowledge. Then, when it became clear that potential biological impacts would have to be addressed, they continued to discount the concern. They insisted that the outcry was the result of public scientific illiteracy and emotional overreaction, even when the "public" comments were offered by qualified experts in marine biology.

Recall the comments of Robert J. Hofman, the scientific program director for the Marine Mammal Commission in Washington, DC, when asked to comment on an early draft of the ATOC environmental impact statement. The ATOC scientists had made numerous claims that were "not supported by data or analysis," he observed, including the claim that "all available data from research on marine mammal hearing indicate there will be negligible, if any, impact from ATOC transmissions." This was simply false: the reality was that the available data were "insufficient to conclude that ATOC transmissions will have negligible effects on marine mammals."[45] Even after relevant data were collected under the auspices of the ATOC Marine Mammal Research Program and evaluated by the US National Research Council—and the latter concluded that the assertion of "no harm" was not defensible given available scientific evidence—ATOC scientists continued to insist that they were right and everyone else was wrong. It might seem that the oceanographers were committing the logical fallacy of conflating absence of evidence with evidence of absence, and indeed, they did do that. But it was worse than that. They were making claims that even their own studies suggested might not be true.

In 2003, the NRC noted that marine mammals were known to change their vocalization patters in the presence of anthropogenic noise. How did they know this? Among other things, it had been documented by Ann Bowles in an article published in 1994 based on the original Heard Island experiment. In a special volume of the *Journal of the Acoustical Society of America*, Bowles and colleagues reported that sperm whale "clicks, clangs and a few codas" were detected during 24 percent of the experiment baseline period but fell to zero during their transmissions. They also found that several individual marine mammals changed course to avoid the direction from which the transmissions were coming. "The results," they wrote, "suggest that ... whales could have altered their distribution in the immediate vicinity of the ... transmissions."

Bowles stressed that these results were very limited, which they were. But rather than inquire further as to what they might signify, she tried her best to explain them away, emphasizing that the whales "returned or were replaced by new individuals quickly when transmissions stopped." All well and good, but if the full-scale ATOC program were implemented, the transmissions would have been continual for a decade. How would this affect them? She speculated that "in the long run animals might have habituated well to the transmissions," but the operative word here was *might*. What she actually knew—what her own evidence showed—was that marine mammals had been affected: whales went silent and changed their course. In her own words: "Changes in behavior of pilot whales and sperm whales provided unequivocal evidence of behavioral effects of the transmissions."[46] The ATOC team did not just dismiss data other people had collected; they dismissed data they had collected!

How could they have done this? When asked, most participants referred back to their long history of Navy-sponsored work, suggesting that none of the earlier projects had done any harm. But none of those earlier projects had been subject to comparable scrutiny, not by the Navy, not by independent biologists, and certainly not by the public. It was only because ATOC was a civilian project—outside the norms under which they had previously operated—that scientists faced public scrutiny, and that was something for which they were unprepared.

All experts operate to some degree in communities that are segregated from other experts, not to mention from the lay publics, but military security and classification meant that Cold War scientists had even less interaction outside their domains than would otherwise have been the case. Even members of Congress frequently lacked information about classified scientific work; Harry Truman famously had to be informed about the Manhattan Project when he suddenly became president of the United States upon the death of Franklin Roosevelt. As historian Paul Edwards has stressed, on many different levels the Cold War created a closed world, and that closure had both social and intellectual consequences.[47]

US Navy patrons accepted that for projects to succeed, they needed to give scientists some intellectual latitude.[48] They also accepted that it was appropriate as patrons of science to permit publication to the extent compatible with Navy interests. But they did not encourage scientists working on classified projects to interact with other experts, much less broader communities. As Joe Worzel complained, classification made it impossible to interact with scientific colleagues in related disciplines who lacked security clearance, even when you had good reason to believe that such interactions were necessary for success. Scientists working on classified projects certainly did not feel that it would be a good idea to talk freely about them to the public or the press, to write popular accounts, or to explain the meaning and significance of their work in broader venues and larger terms.

To a historian of twentieth-century science, this stands out as a significant shift: American earth scientists before World War II routinely wrote popular books and articles even on seemingly arcane topics. William Bowie, chief of the Geodesy Division of the US Coast and Geodetic Survey, wrote pieces for the *New York Times* and *Popular Science* and even had a radio program on geodesy, hardly the sexiest of scientific fields. Throughout the nineteenth and early twentieth centuries, American scientists presented public lectures, particularly if their work involved expeditions to glamorous, exotic, or dangerous places. Historian Paul Lucier notes that scientific lectures to the public were so common in the nineteenth century that *Scientific American* called them an "established institution."[49] Some scientists did this for money; fees for lectures and newspaper articles could be important sources of financial support, particularly for research that had an expeditionary component. (Before World War II, when funds for scientific research were scant, payments for popular accounts and public lectures were an important means of supporting scientific work. Popular accounts might also elicit interest from private patrons and clubs, such as the elite New York–based Explorers' Club, of which Charles Hollister would later be a member.[50]) Other scientists, like Bowie, who worked at federal agencies felt that the public who paid their salaries deserved to know how their monies were being spent.

After World War II, this changed. When funding for scientific research increased, outward communication from the scientific community decreased. Although it is a challenge to read historical silences, one cannot help but notice this one: earth scientists in the postwar years simply did not engage the general public in the manner that they previously had. Before the war, scientists lived in a world where support for research was hard to come by and outreach might yield useful connections. During the Cold War, outreach was at best tricky: classification sometimes made it impossible and a steady stream of funding made it unnecessary. In the Cold War, scientists lived in a world where they were accountable to one another and to the patrons who paid the bills, which in oceanography mostly meant the Navy. Scientists who had worked with the Navy for decades were used to proceeding without public scrutiny, and the notion of the public as ultimate patron was rarely if ever even raised. It is hard not to conclude that their inability to communicate with colleagues in other disciplines, much less the general public, was a consequence of this state of affairs. When conditions changed, not only did they not know how to communicate, they did not even realize that they needed to. What is worse, when the need became apparent, many of them resented it.

This inability to communicate was exacerbated by the hubris that led these scientists to believe so strongly in their projects—so strongly in the premise that they were necessary for national defense and perhaps even world peace—that they brushed aside not only any suggestion that they might do harm but also any suggestion that the harms might be unjustified. In oral histories and interviews, Cold War oceanographers frequently noted the important role they had played in containing communism. Some felt that by virtue of the weapons they had helped to build and the delivery systems they had helped to perfect, that they had saved (or at least helped to save) American democracy.[51]

When the Cold War ended, these men looked for a new mission, and they found it in climate change: as they had saved the world from communism, they would save it from climate change. The problem, however, was that the world didn't want to be saved—or at least, not in the way that these oceanographers were proposing. As many citizens noted and ecologist Rodney Fujita stressed, there was already substantive evidence that anthropogenic climate change was under way; the obstacle to action was not lack of knowledge but lack of (political) power. To Maurice Ewing it was obvious that helping the Navy do its job was meritorious and scarcely needed explaining, much less defending, and he resented it when Columbia students and colleagues suggested otherwise. When Columbia voted to eliminate classified research, he left the institution in disgust. To Charles Hollister, it was obvious not only that his expertise on the seafloor should be respected by politicians and citizens but also that his conclusions should be heeded. When that did not occur, he did everything in his power to try to sustain the seabed research program and also to discredit the Yucca Mountain alternative. To the scientists involved in ATOC, it was obvious that ATOC should be pursued.

The ATOC scientists were unprepared for the public outcry over ATOC; they had never experienced such press interest or scrutiny in anything they had done before.[52] In particular, scientists working with the Navy had not had to concern themselves with environmental impacts. It was not so much that they did not care about marine life but that they had never had to care. As the chief of naval research had put it so succinctly when rejecting using *Alvin* to study giant squid, it wasn't part of the mission profile.

Scientists supported by taxpayers might reasonably have expected to have to explain their work—its significance, impacts, meaning, and hazards (both moral and physical)—but these scientists had no habit of doing so, no experience, and for that reason little ability. When asked to explain themselves, their response essentially amounted to "trust us, we're experts."[53] But as we have seen, the public did not trust them. Physical oceanographers who had worked for decades on problems related to submarine warfare were excited by the prospect of applying their knowledge to a new, important peacetime problem, but what they failed to realize is that those decades of work had cast them in a certain light, as men with certain affiliations and affinities. Throughout the Cold War, many things had been hidden, so it was not unreasonable for citizens to wonder what they were hiding now. The cost of their military associations in this case was not truth, but trust. Scientists who had worked for decades studying the ocean as a theater of warfare were simply not credible when they presented themselves as guardians of the ocean as an abode of life.[54]

Accountability and Mode 2 Science

The history recounted here fits several aspects of what Helga Novotny, Michael Gibbons, and their colleagues have called "mode 2" science.[55] These scholars identify two primary types, or modes, of scientific research. Mode 1 is characterized by knowledge produced by expert practitioners of a discipline, judged by fellow experts and driven by disciplinary priorities. Mode 2 is characterized by diverse practices of knowledge production, motivated by society's needs, and judged by broader communities with a stake in the outcome. One may query whether mode 1 science—accountable only to its practitioners and motivated by purely disciplinary concerns—has ever actually existed; scientists have always had patrons, who in turn have always had expectations. History belies the suggestion that there was ever a time or a kind of science in which scientists were accountable only to themselves.[56] But the idea that there may be different modes of accountability, shifting with political, social, and cultural context—that contexts of motivation may bring with them shifting frameworks of accountability—does illuminate a key aspect of the ATOC story and the end of the Cold War broadly.

When physical oceanographers first developed the ATOC concept, it was based on work they had done before. These scientists saw an opportunity to use existing knowledge and technologies in a novel way—moving from warfare to weather, as von Arx had put it—but within the same basic structures of work and patronage. They published their early papers in traditional venues and sought funding from their traditional patrons. When they turned to the Strategic Environmental Research and Development Program (SERDP) as a new source of funding, it was knowing that it had been created by the Department of Defense to forge a link from Cold War to post–Cold War research. SERDP was designed to create continuity, not disjunction.

In fact, the oceanographers faced several disjunctions. It was not just that the social and fiscal contexts had changed; the law had changed, too. The Marine Mammal Protection Act (1972) and the Endangered Species Act (1973) made illegal certain activities that in the 1950s and 1960s had been perfectly permissible. By the early 1990s, these laws had been in place for two decades; one might think this was sufficient time for the scientific community to adjust. But consider this: Walter Munk began his career at Scripps in the 1930s; by the time he led ATOC, he had been doing oceanographic research with Navy support for nearly sixty years. That was a long time. Long enough, perhaps, to have become accustomed to a certain way of doing things and to breed a certain confidence that those ways were good ones. Likewise, for Maurice Ewing at Lamont: when he faced questions from Columbia colleagues in

1968, he had been working under Navy auspices and conditions of secrecy for thirty years. Times had changed, but these men had not.

It was not that oceanographers had not been accountable before, but the range of people to whom they had become accountable—the range and diversity of people who were aware of what they were doing and felt warranted to comment and even to intervene—was much greater. While oceanographers had obviously not always agreed with their Navy patrons on all issues, they had shared a basic belief in the value—even the necessity—of the projects that they were undertaking. The public, however, did not necessarily share that belief, and when oceanographers insisted that they were right and their critics were wrong, both their colleagues and the general public began to question the presumption that these were people in whom they should place their trust.

Opportunities and Opportunism

The triumph of earth sciences in the second half of the twentieth century is that questions that had been long considered moot were conclusively answered; the triumph of earth scientists is that they effectively used the opportunities presented by the Navy's interest in their science. In this sense, we can say that the scientists in this story were opportunists in the positive sense of the word: oceanographers took advantage of the opportunities created by postwar circumstances in a flexible and creative manner.

Cognitive scientist Donald A. Norman notes that most human action is opportunistic; we take advantages of circumstances and act as opportunities arise. For Norman, this is a good thing: opportunistic action is flexible and involves "less mental effort [and] less inconvenience" than premeditated action. It may even be "of more interest," because it is more likely to be innovative.[57] If Norman is right, then scientific work may be at its best when it is opportunistic. In that case, the success of oceanography in the Cold War is no mere coincidence or contingency. On the contrary, the opportunism of men like Harald Sverdrup, Henry Stommel, Harry Hess, Maurice Ewing, Bruce Heezen, and Paul Fye, and women like Joanne Malkus and Tanya Atwater, was precisely why they were successful.

But there is a negative side to opportunism, the pejorative sense with which we are all familiar: an opportunist is unprincipled and therefore not to be trusted. If scientists are opportunists and the public has discerned this, albeit vaguely, perhaps this has implications for public trust? The scientists involved in ATOC were opportunistic in applying their Cold War knowledge of ocean acoustics to the study of global climate change. Charles Hollister was opportunistic when he championed the seabed radioactive waste disposal to advance his own research programs. Maurice Ewing built an entire career on the opportunities offered to him by the US Navy. These men all expected society to welcome their work, and for a long time, society did. But they were surprised, disappointed, irritated, angry, resentful, and even hurt when that no longer occurred. Ewing left Columbia in a huff. Hollister turned to lobbying and public relations, neither of which worked out as he hoped. The ATOC scientists attempted to convince their adversaries with reassurances, but these reassurances rang hollow, in part because environmentalists saw them not as newfound allies but as opportunists pursuing fame (if not fortune) at the expense of marine life.

Trust in Science?

In 1975, sociologists Todd La Porte and Daniel Metlay found that most Americans at the time felt that the effect of science and technology on their lives was largely beneficial, and they trusted scientists more than politicians or business leaders. (This is still largely true, even as overall trust has declined.[58]) The authors suggested that this trust was founded in a distinction between factual and valuational premises: scientists are trusted to the extent that they are perceived to make decisions based on facts, not values. They also noted that most citizens distinguish between science and technology, feeling largely favorable about the former but wary about the latter.[59] "Should this distinction be lost," they concluded, "[negative] attitudes now mainly associated with technology could spill over to scientific research as well." Their conclusion echoed Merton's earlier argument about the risks to science of being "subject to the direct control of other institutional agencies."[60] If oceanographers were primarily serving the military, why should the public have trusted their pronouncements? If the tiger can change his stripes, what does that say about the tiger?

Scripps in the 1990s was indeed trying to change its stripes. Consider once more the press release on ATOC, which opened with the following claim: "Scripps Institution of Oceanography began nearly 100 years ago as a marine biological laboratory seeking knowledge about the world's oceans. Studying the life that inhabits the oceans and the environment in which they live has been the primary concern of Scripps scientists. Today that commitment continues."[61] This was simply untrue. Scripps was founded in 1903 as a marine biological laboratory, but that identity had been rejected in the 1920s when Scripps refashioned itself as an oceanographic institution. Harald Sverdrup had been brought to La Jolla in the 1930s to make the oceanographic aspiration a reality, and he did so—as we saw in chapter 1—by downgrading biological studies in favor of physical and chemical ones. These choices antagonized the staff biologists, Denis Fox and Claude ZoBell, and Sverdrup's long saga of troubles in the late 1930s and early 1940s had as much to do with their perception that he was disrespectful of biological work as it did with his political commitments, patriotism, or anything else.

When World War II ended, Sverdrup's personal inclination toward physical over biological science was reinforced by the Cold War context of motivation. In May 1945, Sverdrup and Richard Fleming composed a document entitled "Memorandum on Post-war Studies of Oceanography of the Surface Layers" in which they offered what would become accepted wisdom in the years to come: that oceanographic knowledge was essential for modern naval operations.[62] To underscore the point, they compiled a chart enumerating specific military problems and the fields of oceanography relevant to them. On the left were the scientific topics: surface temperature distributions, currents, waves, swell and surf, bottom sediments, tides and tidal currents, marine meteorology, transparency of seawater, and—the final item—biology and chemistry. On the right were the military matters that could justify efforts in that domain: vessel design, amphibious landings, weather prediction. What stands out in their analysis is not only the dominance of physical over biological questions but also the far greater specificity offered for the former. Physical oceanography was divided into eight categories related to eight different sets of military problems, while all of biology and chemistry were lumped into a single category motivated by only one Navy concern (and a very old-fashioned one at that): fouling and corrosion (fig. C.1).

After the war, when Roger Revelle was proposed as the next Scripps director, the biologists on staff rebelled. They wanted the next director to redress the imbalance created by wartime concerns, not to reinforce and institutionalize it. Their preferred candidate was Daniel Merriam, a fisheries biologist who ran the Bingham Oceanographical laboratory at Yale in close collaboration with the Department of Zoology. They argued that he could help diversify the institution's patrons and restore equity for biological work.[63] Instead, they got one who solidified the links to the US Navy and for whom biological matters were rarely of top concern.

Revelle's focus on the physical sciences was driven to a large extent by his conviction that the military would be the guarantor of scientific research in oceanography in the years to come. This was expressed concretely in revisions to the graduate curriculum de-emphasizing biology—in fact, eliminating marine biology altogether—and a conscious decision to reject a major role for fisheries investigations and focus instead on what he called the "ba-

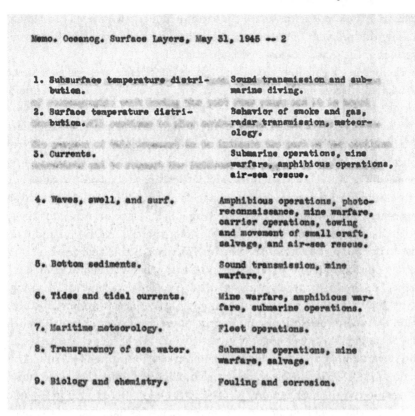

Figure 10.1 "Memorandum on Post War Studies of the Oceanography of the Surface Layers," prepared by Harald Sverdrup and Richard Fleming, May 31, 1945.
From Office of the Director Records, 1920–2013, Scripps Institution of Oceanography Archives, SAC 1, Special Collections & Archives, UC San Diego. Reprinted with permission of the Scripps Institution of Oceanography.

sic problems of the sea."[64] Consider this comment on the work of the Scripps Marine Life Research program, which had a long-standing relationship with the California fishing industry and had of late been studying the California sardine. Revelle wished to refocus the program on the "dynamic analyses ... of the processes in the sea," and he suggested that the problems of the sardine industry might be best addressed by fundamental studies "treating the sardines *en masse* as if they were particles of a fluid somewhat similar to seawater."[65]

One can take Revelle's point that advances in understanding fish migrations might ensue from a better understanding of the currents in which they swam and sardines could be used to track those currents, but was life itself not a basic problem of the sea? Was it not one of the *most* basic problems of science? To be sure, the travails of the sardine industry were not a fundamental scientific problem, but neither was the long-range detection of Soviet submarines or the salvage of lost hydrogen bombs.

The balance between physical and biological science arose as well at the Woods Hole Oceanographic Institution. In 1955, biologist Alfred Redfield, one of the first scientists recruited by Henry Bigelow in the 1930s and who later served many years as associate director, raised the matter of the place of biology in the institution. Like his counterparts at Scripps, Redfield was concerned that biology was being relegated to second-class status. The response of the Woods Hole scientific advisory committee suggests that he was right. Biology, they suggested, should have a place at Woods Hole, but physical oceanography would set the research agenda: "The committee considered Dr. Redfield's discussion of the place of biology in the Institution and feels that further efforts to stimulate support for work in this field are desirable. Perhaps the best kind of activity would be biological work which demanded the use of the Institution's resources and knowledge in physical oceanography, but which also pointed toward providing ... new information which could be used in physical oceanography."[66] Just as one can accept Revelle's point that physical oceanographic investigations might help the fisheries industry, one can accept the Woods Hole advisory committee's point that it made sense for Woods Hole to prioritize biological studies that could interact with physical oceanography in productive ways. But their framing and phrasing make clear that biology was being viewed as handmaiden to physical oceanography. To the extent that choices had to be made, biology was expected to support physical oceanography, rather than the other way around.

Let me not overstate the case: biological questions were not abandoned at oceanographic institutions. Research on sardines continued, oceanographers studied life in the context of the deep scattering layer, and dolphins attracted a certain amount of attention. But overall, one particular set of questions was preferred, and the history told here explains why. Oceanographers after World War II built a science consistent with creating the kind of oceanography that the US Navy needed.

The Social Construction of Ignorance

Historians Robert Proctor and Londa Schiebinger have promulgated the term *agnotology* to describe the study of ignorance, and in recent years a significant literature has developed on this theme.[67] One point of this literature is to show that it is not unreasonable to pose questions about scientists' failures to explore certain domains or to learn certain things.[68] Knowledge has a history and ignorance does, too.[69] When we permit this framing, we can pose an important question that might otherwise get lost: what were the intellectual costs of letting the US Navy define the mission of American oceanography in the Cold War?

One obvious answer involves climate change. As we have seen, in the late 1950s oceanographers were well aware of climate change as an issue of scientific and social import. Roger Revelle was one of the first American scientists to pay sustained attention to the issue; his 1957 paper with Hans Suess describing increased atmospheric carbon dioxide as a "great geophysical experiment" is now considered a classic.[70] The paper was a call to arms, suggesting that scientists should pay attention to this "great experiment" and design investigations to discern its effects, if any, on weather and climate. William von Arx made a similar argument a few years later in his 1965 memo on "science in bondage." These ideas built on a long tradition in oceanography that recognized the links between oceans and climate. Harald Sverdrup had focused on air-sea interaction as an important component of oceanographic research in the 1930s, and when NOAA administrator Robert M. White took up the question in 1978 in a special issue of *Tellus*—Woods Hole's in-house journal—the issue was entitled "Oceans and Climate" (chapter 9).

Oceanographers could have taken up climate change as a central research focus at any point in the 1950s, 1960s, or 1970s, and a handful did. The most important was Charles David Keeling, who (with Revelle's support) began the systematic measurement of atmospheric carbon dioxide in 1958 and dedicated the rest of his career to it. Over the following five decades, Keeling produced the graph that now bears his name and is engraved on the wall of the National Academy of Sciences headquarters in Washington, DC: the Keeling Curve. In 2001, he was awarded the National Medal of Science for this work, and his son Ralph carries it on today. One might have thought that students would have been clamoring to work with Keeling *père* on such a significant topic, that monies would have flowed freely toward it, and that the subject would have generated a significant, expanding research program. But before his death Keeling wrote poignantly of the difficulties he had faced in obtaining and sustaining funding, and colleagues have noted that he had very few graduate students.[71] There are no doubt several reasons this was so—Keeling suggested that his work was viewed as "monitoring" rather than "research," and colleagues have suggested that his sedulous approach to laboratory measurement made him a tough taskmaster. But one important dimension of his story is indisputable: the Navy was just not much interested in measuring carbon dioxide. Indeed, if von Arx's judgment was correct, the Navy was not interested in climate at all—which is (as we saw) why von Arx wanted to focus on it![72]

Oceanographers at Scripps did not become seriously interested in carbon dioxide and climate change until the late 1980s, more than a decade after colleagues at the Goddard Institute of Space Studies (James Hansen), the National Center for Atmospheric Research (Stephen Schneider), or the Geophysical Fluid Dynamics Laboratory (Kirk Bryan, Syukuro Manabe, Richard Wetherald). It was also nearly a decade after the National Research Council produced its first major report on the topic, chaired by MIT meteorologist Jule Charney.[73] Most of the scientists who took up climate change in the 1970s and 1980s were meteorologists or atmospheric physicists rather than oceanographers. Revelle, an oceanographer, and Hans Suess, a marine geochemist, recognized the significance of carbon dioxide and climate in the 1950s and called on colleagues to pay attention. Von Arx did the same in the 1960s, White in the 1970s. But for the most part, the call went unheeded.

As an institution, Scripps took up climate research with vigor only after Navy funds were on the decline. In 1987, director Edward Frieman proposed that the institution make a conscious decision to develop climate change as a research focus; his proposal to the University of California was framed explicitly in the context of the need for a new, motivating focus in the face of declining Navy interest and support.[74] In later years, as climate change came to the fore as both a scientific and a social issue, the university would make much of the legacy of Roger Revelle, emphasizing his leadership and vision on climate change.[75] But this is revisionist history. Revelle had wanted to lead on climate—he had tried to lead—but the reality is that few of his colleagues followed.

Physical oceanographers were not wrong to shift their attention to climate science in the late 1980s, and they were not wrong in the early 1990s to propose ATOC. As Rodney Fujita and others noted, a better understanding of ocean heat uptake would definitely have been a useful thing. But their representation of ATOC was not merely that it was a project that would help us learn about important aspects of global warming, which their colleagues at NASA had already concluded was under way. They implied that it was the crucial experiment that would prove global warming and thereby break the political logjam. But as several commentators noted, by 1995 there was already a good deal of scientific data showing that climate change was under way, as well as an emerging political commitment to doing something about it. In 1992, world leaders had signed the UN Framework Convention on Climate Change and President George H. W. Bush had pledged to act on that commitment. In 1994, atmospheric physicist Benjamin Santer and his colleagues demonstrated that greenhouse gases were driving tropospheric warming, a finding that led the IPCC the following year to declare that the "balance of evidence" suggested that anthropogenic climate change was under way.[76] Perhaps this is why James Hansen was lukewarm in his endorsement of ATOC. Like so many members of the public, he understood that oceanographers were proposing to prove something that he and many of his colleagues in climate modeling felt they already knew.[77]

If physical oceanographers had a more balanced appraisal of the role of science in the climate change debate, they might have understood that their project was not as crucial as they made it out to be and perhaps walked away from ATOC before they had invested so much in it. They might have understood, as Fujita stressed, that what the world most needed at that juncture was not more science, but more political will. They might have realized that ATOC came too late.

We can tell a similar story about Woods Hole, which nowadays has an entire institute dedicated to understanding the role of the oceans in climate change. It, too, was late to the climate table.[78] The scientists who might have earlier led such an effort—George Veronis, William von Arx, Melvin Stern, and especially Henry Stommel—all left in the aftermath of the Palace Revolt.[79] Oceanographers could have been leaders in understanding climate change, but for the most part they were not. Navy patronage helps make sense of why.

Moreover, to the extent that the world needed more ocean science in the 1990s, ATOC made clear that what was needed most was not physical oceanography, but marine biology. Recall that a principle argument used in favor of ATOC—and the default assumption that it would do little or no harm—was that the Navy had been using this sort of underwater sound in its operations for decades. This was true. So why was there so little information about its effects?

The answer by now is clear: the Navy concentrated its patronage on matters that were pertinent to its mission, particularly submarine warfare. It focused—as Paul Fye aptly put it—on what it needed to know. In marine science, this led to a heavy emphasis on physical oceanography and geophysical studies of bottom topography. Research on biological matters was pursued to the extent that it was deemed pertinent (think fouling organisms and the sonar profile of tuna, as well as Sam Ridgway's work with the dolphins and whales at Point Loma).[80] But one of the realizations that emerged from the ATOC debacle was just how little biology had been deemed pertinent. The Navy had been using low-frequency sound for decades—even using marine mammals in its work—yet it had never funded the scientific research that would have permitted the National Research Council to evaluate how that sound affected marine mammals.[81] The Navy had paid even less attention to fish. The ONR did fund basic science—so did other Navy bureaus—but it was basic science that "fit the mission profile." Biology sometimes fit the bill, but mostly it did not. This becomes acutely clear when we consider fish.

Forgetting Fish

In his 1965 annual report to the ONR, Fred Spiess encapsulated the focus of his research group as "knowledge of the generation, propagation and detection of energy in the ocean."[82] That was a fair characterization, and not just of the Marine Physical Laboratory, but of a goodly portion of Cold War physical oceanography. Fair enough: the Navy needed to understand the ocean as a medium through which energy was propagated. But there was no comparable Navy-funded laboratory at Scripps dedicated to understanding the ocean as a place in which fish and other organisms lived. Navy support led to an expansive understanding of the ocean as a theater of warfare, but not so much as an abode of life, and that—even more than climate change—is the road not taken of this story.

Nineteenth- and early twentieth-century oceanographers had been deeply interested in marine life. Biological sampling was a central activity of the great *Challenger* expedition, and its chief scientist, Wyville Thomson, was a zoologist. In the early twentieth century, the dominant justification for expanding oceanography as a modern, international science was its relevance to fish. In 1902, when Swedish oceanographer Otto Pettersson spearheaded the formation of the International Council for the Exploration of the Seas, fish were a major part of its context of motivation. A decline in European fish stocks was already attracting scientific attention and social concern, and Pettersson and his colleagues felt that coordinated oceanographic research could determine the cause and suggest remedies.[83] Some investigators, for example, considered that an observed abrupt decline in a fish stock in a particular region might simply indicate that the fish had gone somewhere else; others suggested that the impacts of ocean currents on nutrients might be involved. These hypotheses could be tested through appropriate oceanographic work.

For the next three decades, diverse oceanographic investigations were undertaken, driven by the belief that a better understanding of currents, of the distribution of temperature and salinity, and of nutrients could help explain fluctuations in fish populations. These fluctuations had serious consequences for European countries that relied on fish as a source of food and economy. Even Harald Sverdrup thought that understanding the distribution of nutrients and plankton was an important element of oceanographic work because of its relevance to fish.[84]

Fast-forward to 1998. The School of Marine Affairs at the University of Washington has organized a symposium to celebrate its twenty-fifth anniversary. Among the presentations is a paper by Dayton L. Alverson and Karma J. Dunlop on the status of world marine fish stocks.[85] Everyone at the meeting is well aware that fisheries around the globe are under duress. The loss of Newfoundland cod has made newspaper headlines; comparable collapses have occurred in Russia, Germany, France, and the United Kingdom. And it is not just cod: the populations of several species that people like to eat, including orange roughy, Chilean sea bass, and bluefin tuna have collapsed. Leading Canadian biologist Daniel Pauly and his colleagues have introduced the alarming idea of "fishing down the food chain"—that as stocks of the largest and most desirable fish are depleted, fishermen begin to target lower trophic levels; their theory is strongly supported by data from the UN Food and Agricultural Organization.[86] (A few years later, Dalhousie University scientists Ransom Myers and Boris Worm would publish results in *Nature* suggesting that ocean ecosystems have been so systematically harvested that only 10 percent of wild fish populations remain.)[87] In this context, Alverson and Dunlop raised a shocking point: despite the diverse lines of evidence that suggested severe problems in global fisheries, there was no general agreement as to how to evaluate fish stocks, so it remained unclear whether current fishing levels were unsustainable for most species. Even more startling, in a panel discussion of their paper, it was noted that a majority of marine species remained unidentified.[88]

At the end of the twentieth century, many (perhaps most) American biologists would have said that natural history—with its emphasis on taxonomy and classification of species—was a relic of the past, a very nineteenth-century activity, redolent of Wyville Thomson, if they even knew who he was. The exciting work in biology was in molecular biology, genetics, and especially the cutting-edge field of synthetic biology.[89] But in fact, the basic natural historical work of identifying, naming, and classifying the world's marine life had never been completed. Scientists at the end of the twentieth century not only did not know the size of most fish populations; they did not even agree on how to determine their size. Nor did they know how many different species of fish there were. Even among identified species, life histories were poorly known: scientists did not know, for example, how long many species of fish lived or whether females were fertile throughout their lives, which in turn made it virtually impossible to make accurate analyses of populations.[90]

For all the spectacular advances in oceanography in the twentieth century, basic questions about fish and fisheries remained unanswered.[91] Scientists literally did not know how many fish there were in the sea.[92]

As the world entered the twenty-first century, it became clear that the collapse of the world's fisheries was a major environmental catastrophe—as severe as terrestrial biodiversity loss and perhaps more so. Of particular concern has been the impact of fishing on deepwater species. An estimated 40 percent of the world's trawling grounds are now in waters deeper than the continental shelves, where fish grow slowly and can live to a great age.[93] For the remaining 60 percent in shallow waters, the issue is no longer one of "risk," but of documented damage.

In 2000, the US secretary of commerce declared a disaster in the West Coast groundfish fishery. In 2003 the Canadian government announced the complete closure of the Newfoundland cod fishery after a moratorium implemented in 1992 had not produced sufficient recovery to justify reopening.[94] This closure was not merely a matter of environmental impact, but of social and economic impact as well. As Canadian minister Yvonne Jones explained: "The closure of northern … cod fisheries indefinitely is devastating news to Newfoundland and Labrador. It will have an enormous economic impact…. The life blood of many communities … has been taken away."[95] Jones could have gone further: the end of the Newfoundland cod fishery marked the end of a source of food and livelihood that previously sustained communities in the northeastern Atlantic region for more than three centuries.[96]

One might suppose that the Newfoundland cod fishery was mismanaged (and on some level that would be inarguable), but similar patterns were being recognized around the globe. In 2003, the European Commission reported that closures or very severe catch limits had been enacted or were recommended for "all cod stocks outside the Baltic Sea," and as well as for whiting in the Irish Sea, sole in the western English Channel and Bay of Biscay, hake and lobster off the Iberian Peninsula, and many more.[97] Around that same time, the Pew Oceans Commission released the first of a series of preliminary reports on the state of the world's oceans. Their report on fish and fisheries concluded that the combined effects of overfishing, bycatch, habitat degradation, and fishing-induced food web changes have fundamentally altered the composition of ecological communities and the structure, function, productivity, and resilience of marine ecosystems.[98] In 2008, one study would predict the collapse of all wild fish populations by 2050.[99] And in 2012, United Nations Secretary General Ban Ki-moon announced that more than half of global fisheries were exhausted and a third of critical marine environments—including seagrasses, mangroves, and coral reefs—destroyed.[100]

In the years since the University of Washington conference at which Alverson and Dunlap stressed what we didn't know about fish, it has become clear that we do know something deeply troubling. The collapse of the world's ocean ecosystems is an environmental calamity of enormous proportion, arguably as significant as anthropogenic global warming, if not more so (and intricately interconnected to it). While this global environmental catastrophe was unfolding, where were the oceanographers? If Pettersson and his colleagues recognized the developing problem in 1903 and thought that oceanographers could help to solve it, what happened?

Framed this way, the answer is clear: social and political context had shifted the priorities of oceanographers by altering the problems that were understood to justify government investment in their science—in short, by changing the context of motivation. European oceanography—in Scandinavia, in Germany, in the United Kingdom—was badly damaged during World War II and took a long time to recover. During that time, the United States—as in so many areas of science—became the leader in oceanographic research, where, as we have now seen in detail, attention focused on military affairs. (The Soviet Union continued to sustain a focus on fisheries research, but US scientists seem not to have paid much attention.)[101] With the US dominating global scientific oceanography, and the needs of the US Navy dominating US oceanography, oceanographers in the late twentieth century constructed reliable knowledge about the ocean as a physical medium through which sound and submarines might travel, but they also constructed substantial ignorance about the ocean as an abode of life. There is no reason in principle why attention to physical oceanography should have crowded out good biological work, but in practice, it did. It is not just in hindsight that we can see this. As the conflicts at Scripps in the 1930s and the Palace Revolt at Woods Hole in the 1960s revealed, scientists at the time saw it, too. (At Lamont, which was always Navy funded, there was never any biological work to crowd out.)

The forces driving the collapse of ocean ecosystems, like the forces driving climate change, are not primarily the result of ignorance, but to the extent that ignorance plays a role, we might suppose that there is still time for scientists to pick up the dropped intellectual stitches. Fisheries scientists have certainly tried. In 2010, the Pew Charitable Trusts and Alfred P. Sloan Foundation announced the culmination of their "landmark" Census of Marine Life, a ten-year study involving 2,700 scientists from eighty nations.[102] The census generated countless scientific papers, numerous books and conference proceedings, and even inspired a song.[103] But like ATOC, it came too late. In this case, much too late. Scientists were trying to count something that to a large extent was already gone.[104]

In theory we can always go back and study the things that we have previously neglected. In practice, however, we cannot, because the world is not the same place. As Heraclitus put it long ago, we cannot step twice in the same river; he might also have said the same ocean. Many of the fish that we have valued—as well as many others that we should have valued but did not—are gone, and those that remain live in ecosystems that are drastically altered. It has even been argued that the entire field of marine conservation is based on a fallacy: designing management plans for populations that have already been decimated and building those plans on baselines that are far below what they were before the onslaught of global human predation.[105] We cannot go back and pick up where we left off because the ocean is now a different place. Much of what we would want to study is no longer there. The world has been changed; we cannot rewind the tape. We can understand what we have done and why we did it, but we cannot go back and do something else instead.

The Giant Squid, Once More

At the beginning of this narrative, we noted both historians' interest in whether military support had "distorted" American science in the Cold War and their discomfort with that term. In the chapters that followed, we saw that some of our historical actors felt that military funding was distorting— using that very word. We may argue whether the word *distorted* is the best one to describe what we have documented; perhaps it is better to say that throughout the history recounted here scientists encountered both *epistogenic*, or knowledge-producing, and *agnotogenic*, or ignorance-producing, forces, tendencies, and trends.

We might expect scientists who are not well funded to complain that they lack sufficient funds, but even among those whose work was amply supported—Henry Stommel, William von Arx, Joanne Malkus, Fred Spiess, and even Paul Fye—the concern arose that there might be a time and place when an important question might arise that would be neglected because it lacked Navy relevance.[106] Stommel understood this in terms of a romantic (pastoral!) ideal of scientific freedom, in which each man or woman sets his or her own scientific course. Fye expressed it in terms of "an essential degree of freedom" needed to pursue research.[107] Spiess recognized that the Navy was "strongly hardware oriented," with a tendency to support machines rather than men, building rather than thinking.[108]

We have seen how Scripps scientists who focused on biological questions, already marginalized in the late 1930s, were further marginalized as military concerns became central, and how Frederick Aldrich was turned away from using *Alvin*. We have also seen how *Alvin* almost didn't make its most famous and important scientific contribution, as the Navy declined to fund a scientifically exciting but militarily irrelevant investigation of hydrothermal systems at the mid-ocean ridges. And we have seen how geologists who might have made good use of Navy data were unable to do so because of security restrictions. We will never know what these men and women might have done had they been as well supported as their colleagues whose interests did mesh with the Navy's need to know, but we can surmise that they would have done something, and it might have been significant. Consider, once more, Frederick Aldrich and what he didn't get to learn about the giant squid.

Aldrich proposed his investigation in 1965. Fast-forward to 2005, when Japanese scientists for the first time observe and photograph a giant squid in its native habitat.[109] The results are published in the *Proceedings of the Royal Society* and widely reported in the popular press. The *New York Times* heralds the discovery as "a long-awaited breakthrough," quoting one scientist: "This has been a mystery for a thousand years."[110] Compared to a thousand years, a delay of forty is perhaps no big deal. There is of course no way to know what Aldrich would have found, or what anyone else might have done in those intervening forty years. But it is clear that Aldrich's project was not rejected because it was uninteresting or scientifically insignificant. If in using *Alvin* Aldrich had been able to jump-start a serious study of the giant squid, maybe—just maybe—we would today have a better understanding of the evolution of vision and sight, and maybe even a better answer to the creationists favorite counterargument against natural selection.[111] Or maybe we wouldn't. But we would certainly have learned something that, forty years later, people definitely thought mattered.

Epilogue: Motivation Matters

Navy sponsorship created men who called themselves "Navy- and vehicle-oriented oceanographers" and the geopolitical context of the Cold War—particularly the imperative of elucidating underwater sound propagation for submarine and antisubmarine warfare—made it possible for them to obtain funds for a variety of projects, such as *Alvin*, which otherwise would have been too expensive and arcane to undertake.[112] But the Cold War not only led to financial support and political justification; it also provided psychic motivation, by creating a sense of importance, a sense of being needed, a sense of being part of something larger than oneself.

When Charles Bates, a geophysicist in the Hydrographic Office (and later an ARPA program director) wanted to recruit young scientists from Princeton in 1948, he wrote to his academic colleagues in the geology department asking them to identify promising candidates and enumerating the benefits of a job at Hydro. Besides the good pay, ample vacation, and intrinsic interest of the work, there was a major advantage that was often overlooked: "Since the work is of a military nature, there is also a certain satisfaction in conducting work for the Armed Services during the current Cold War."[113]

Al Vine believed that meaningful motivation was requisite for most men to do good science. Speaking at a conference on deep-sea research in 1956, he thought that money without clear motivation was hurting his science: "Most of these things [being proposed for the future] are copied from work done in the past when people had far less money than we have now. I am not worried about money as much as about the zeal of people to identify the problem and the particular vehicle that will help them solve it."[114] A decade later, when the Office of the Oceanographer of the Navy sought to define its agenda, one memo noted that defense and nondefense programs could thrive in tandem, because there was "no conflict in the[ir] motivation."[115]

In 2001, Fred Spiess prepared a historical retrospective for the Acoustical Society of America. He began his paper by talking about "the personal relationships that made those times productive and enjoyable." The Navy context fostered a tremendous sense of teamwork, he recalled, as civilian scientists, Navy scientists, and Navy officers worked side by side to define problems and find solutions. Spiess therefore focused his paper neither on data or theories, nor instruments or experimental practices, but — of all things! — on *committees*. These included the Hartwell committee that helped push forward SOSUS, and Project NOBSKA, which first met at Woods Hole in 1956 and laid the foundations for what would become the Polaris submarine program. These committees generally met in the summer, often for a month or more, and members came to know one another well. Each time they reconvened they renewed their sense of communal purpose: "Each of us returned home having made new friends and with an expanded view of where our own ideas might fit into the wider scheme of things." Spiess thus entitled his paper "Motivating the Underwater Acoustics Community."[116]

William von Arx wanted his science to be "inner-directed," but Spiess found it satisfying to have an extrinsic motivation connected to the "wider scheme of things." He also found it satisfying to be part of a collegial group of like-minded men, working together on significant matters. While von Arx and his colleagues defended the virtues of basic science — which they interpreted as fundamentally individualistic and nonutilitarian — Spiess defended the virtues of collegiality and usefulness. The distinction between basic and applied science could be parsed in terms of intellectual centrality, but it could also be parsed in terms of ambitions and aspirations, which is to say, in terms of motivation.[117]

At one extreme, the *Thresher* and *Scorpion* searches and the H-bomb recovery at Palomares were military projects; they had clearly defined objectives related to military hardware and their time frame was immediate. The purpose of scientists' involvement was not to answer questions about the natural world, but to help the Navy solve the problem at hand as quickly as possible. At the other extreme, the nature of vent ecology is a scientific problem: no one expected that understanding tube worms would have military salience and it mattered little how long the questions took to answer. Not unexpectedly, the Navy paid for all of the former but little of the latter.[118] In this sense, we can say that (with some exceptions) the Navy did not pay for basic science.

Yet a distinction that is clear on its edges can be blurred in the middle, and much of the story told in this book lies in that middle. To Henry Stommel, Joanne Malkus, and William von Arx, Artemis was applied science (or engineering), but to Alan Berman it was research because it was not an operational system. The boundaries of research and application were necessarily subjective, depending on one's goals. Edward Frieman, the physicist who took the reins of Scripps in the mid-1980s, once commented that "to a physicist, all oceanography is applied science."[119] To a historian, the crucial question is not whether oceanography in the Cold War was basic or applied, but how the Cold War context and military patronage shaped what was learned—as well as what was not learned—about the natural world.

Philosopher of science Susan Haack has aptly noted that "the external environment in which scientific work is conducted can affect which questions are thought worth investigating ... and even, sometimes, what results are reached ... and the environment may be more, or less, hospitable to thorough honest investigation."[120] Quite so. But in this case to think of military considerations as "external" is to miss the point: oceanography grew and prospered in the Cold War because oceanographers shared or internalized the values of their Navy patrons.[121] Oceanographers chose to work on subjects that interested them, but those interests were bred, developed, and sustained in a context in which some lines of inquiry were well funded and rewarded—materially, intellectually, and emotionally—while others were not. Fields that were nourished flourished, and those that were not did not.

Moreover, as scientists trained students, the interests of the next generation were weighted toward issues originally driven by Cold War concerns, even after military funding had decreased or ceased, and even after the political context that inspired them had changed.[122] Military concerns were naturalized, so the extrinsically motivated became the intrinsically interesting. One scientist at the Naval Electronics Laboratory encapsulated this experience in 1967, when he wrote to Bruce Heezen describing work he was doing on sound velocity in marine sediments on behalf of a particular Navy project. "This sounds like a drag," he confessed, "but, surprisingly, I find myself really interested in it!"[123] And when I was in graduate school at Stanford at the tail end of the Cold War, the dean of the School of Earth Sciences admonished us not to do just anything that struck our fancies, but to focus on "the next most important question."[124] But the next most important question depends on what was answered last, and that, in turn, depends to a substantial extent upon who has been footing the bill.

Scientists were not naïve about the way in which Navy patronage fostered and sustained some areas of their science and not others, and periodically they tried to diversify their funding sources for just this reason. In 1968, Fred Spiess suggested to Scripps's director William Nierenberg that they solicit more NSF support for Deep Tow so that his group might address a wider range of earth science topics: "Development and use of our deeply towed geophysical instrumentation and navigation system has ... opened up a wide range of new geological-geophysical research opportunities.... It seems appropriate, therefore, to request partial NSF support for this work in order to insure our ability to attack problems of basic geological importance without necessity of weighing Naval relevance."[125]

Paul Fye had earlier come to the same conclusion, particularly with respect to the cost of vessel operations. In a letter to NSF earth science director William Benson in 1959, he summarized his situation:

Since the war, the cost of ship operations, other than those related to applied research projects, has been borne mainly by the Office of Naval Research. Although this support has been generous, and in the case of physical oceanography has been unrestrictive in terms of specific research, these funds have been inadequate to provide an essential degree of freedom in the planning of field operations and the development of research programs. Such programs have always to a certain extent been dependent upon associations with military application-type research projects and this has invariably influenced to some degree the direction of the research itself.[126]

Of course it did. How could it not?

Acknowledgments

This project has taken a long time to come to fruition, and along the way I have incurred innumerable debts. If I were to attempt to list all the people and institutions to whom I am grateful, I would not only kill more trees than I have already but I would surely forget and offend someone about whom I care greatly. I hope, therefore, that it will suffice to collectively thank all the generous colleagues at the University of California, San Diego—in the History Department, in the Science Studies program, at the Scripps Institution of Oceanography, at Sixth College, in the UCSD Provosts Council, and in the Scripps Institution of Oceanography archives and UCSD special collections—who aided me along the way, as well as my generous colleagues in the Department of History of Science at Harvard University who did not abandon hope that this book would, indeed, get finished. I am also indebted to the invisible college of historians of earth science with whom I have had the good fortune to be associated with these past thirty years as we collectively transformed the history of earth sciences into a "legitimate" academic field; I could not have wished for a wiser or more generous group of colleagues. I am also deeply grateful to the scores of earth scientists who gave generously of their time to discuss their work. In highlighting their contributions, I hope this book in some way repays them.

Some of the episodes discussed in this book were fraught, and all of them were complex. To the extent that I have done justice to any of them, I am grateful to the many oceanographers and geophysicists who were willing to speak with me about them, sometimes on the record, oftentimes off. Readers may notice that I did not undertake formal oral history interviews for this book. There were several reasons for this choice: because many of the scientists involved had already been formally interviewed and often on more than one occasion; because, for obvious reasons, I preferred to rely on contemporaneous documentary records rather than post hoc recollections; and because a number of scientists were willing to talk to me—over lunch, over coffee, at the Scripps snack bar, and the like—but did not wish to participate in formal interviews. Some diligent readers may note that I occasionally invoke "personal communications." These refer to such informal conversations and are used to fill in details that were not otherwise available.

I would be remiss if I did not singularly acknowledge former Scripps director Charles Kennel, who provided the seed money that enabled me to start this project, and Scripps archivists Deborah Day and Carolyn Rainey, whose amazing work over several decades literally made my work possible. I am also very grateful to Lynda Claassen, director and chief curator of the Mandeville Special Collections Library at UCSD.

I would also like to acknowledge the support of the Woods Hole Oceanographic Institution. I owe a particular debt of gratitude to former director Robert Gagosian for granting me great archival latitude, to the late Arnold Arons for speaking freely about difficult things, and to archivists Lisa Raymond and Dave Sherman.

I also owe a shout-out to my "gang of three" at Scripps—Lynn Russell, Richard Somerville, and Larry Armi—with whom I discussed all the questions and problems of this project over innumerable lunches at the UCSD faculty club and rather a few too many glasses of wine in other places. I also owe a great debt of gratitude to the late David van Keuren, who helped me identify important sources at the Naval Research Laboratory and located the underwater photographs of the USS *Thresher* hull, and to Matthew Crawford, David Spanagel, Keith Benson, Mott Greene, Paul Farber, and Elena Aronova, who read the entire manuscript with a generous spirit and eagle's eye. A huge thank you, as well, to my smart and patient editors at the University of Chicago Press, first Christie Henry and then Karen Darling, and my remarkable research assistants, Hannah Conway, Erik Baker, and Aaron van Neste, who helped me fact-check and correct the final manuscript; their work has made it possible for me to do mine. I am also grateful to my graduate students Colleen Lanier-Christensen, Dani Hallett Inkpen, and Gustave Lester, whose astute thinking helps to sharpen mine. And then there is Ron Rainger, with whom I shared so many hours in the archives, and who I wish had lived to see this work realized.

For financial support, I am deeply grateful to the National Science Foundation, the American Philosophical Society, the University of California, and Harvard University Faculty Development Funds. The publication of this book was supported by a subvention from the FAS Tenured Publication Fund, Division of Social Sciences, Harvard University Faculty of Arts and Sciences.

Above all, I am grateful to my husband, Kenneth Belitz, who has never doubted the value of my work (even when I have), and to my daughters, Hannah Belitz and Clara Belitz, who may well have doubted it but stuck by me anyway.

Portions of chapter 1 were previously published as Naomi Oreskes and Ronald Rainger, "Science and Security before the Atomic Bomb: The Loyalty Case of Harald U. Sverdrup," *Studies in the History and Philosophy of Modern Physics* 31B (2000): 309–69, copyright 2000 by Elsevier. Portions of chapters 6 and 7 were previously published as Naomi Oreskes, "A Context of Motivation: US Navy Oceanographic Research and the Discovery of Sea-Floor Hydrothermal Vents," *Social Studies of Science* 33, no. 5 (2003): 697–742, copyright 2003 by Sage Publications.

Sources and Abbreviations

The research for this manuscript was done over the course of more than a decade, and as such, some of the archives have been processed for the first time or reprocessed since the time the research was conducted and archival locations were recorded. We know this is the case for the Harry Hess Papers at Princeton, the Operational Scientific Service Records at Woods Hole (previously just the papers of the Submersible *Alvin*), the SIO-ODR collections, and the SIO-Revelle papers. I have done my best to ensure that locations provided in citations are accurate, but I encourage researchers wanting to follow up in these collections to consult archivists at those institutions. There are also a number of citations to documents, mostly in chapter 1, that were collected by Ronald Rainger at the National Archives and Record Administration prior to the year 2000. In the course of completing this volume, I realized that a number of these citations were incomplete. Before these citations could be verified in person, the National Archives and Record Administration closed all locations to the public due to the COVID-19 global pandemic. These include the records for NACP-HO, NACP-NRDC, NACP-OCNO, NACP-ONR. I have provided as much location information as I could based on the late Professor Rainger's notes, previous publications, and the assistance of NARA archivist Nathanial Patch.

ABBREVIATIONS USED IN NOTES

BCHP Bruce C. Heezen Papers, Accession 10-018, Smithsonian Institution Archives, Washington, DC

EBWP Papers of Edgar Bright Wilson Jr., courtesy of Harvard University Archives, Cambridge, MA

ECBP Papers and correspondence of Sir Edward Crisp Bullard, 1907–1980, 1916–1984, Churchill Archives Centre, University of Cambridge

FOIA Freedom of Information Act

HHP Harry Hess Papers, Special Collections, Princeton University Library, Princeton, NJ

HUSFBI　FBI file on Harald U. Sverdrup, FOIA 93-788.

HUSN　US Navy file on Harald U. Sverdrup, FOIA 321-195.

MEP　Maurice Ewing Papers, 1912, 1925–1974, Dolph Briscoe Center for American History, University of Texas at Austin

NARA　National Archives and Records Administration

NACP　National Archives at College Park, MD

NACP-HO　Hydrographic Office Security Classified General Correspondence, Records of the Hydrographic Office, RG 37, National Archives at College Park, MD

NACP-NDRC　Records of the National Defense Research Committee, Records of the Office of Scientific Research and Development, RG 227, National Archives at College Park, MD.

NACP-OCNO　Aerology- General Correspondence, 1919-1946, Records of the Deputy Chief of Naval Operations (Air), Record of the Office of the Chief of Naval Operations, RG 38, National Archives at College Park, MD

NACP-ONR　General Correspondence of the Office of the Coordinator of Research and Development, Records of the Office of Naval Research, RG 298, National Archives at College Park, MD.

NRLT　*Thresher* search files, US Navy Research Laboratory Archives

SBF　Scripps Institution of Oceanography Biographical Files, 1850–2013, Scripps Institution of Oceanography Archives, SAC 5, Special Collections & Archives, UC San Diego, La Jolla, CA

SIO-ATOC　Acoustic Thermometry of Ocean Climate Project Records, 1991–1996, Scripps Institution of Oceanography Archives, SAC 32, Special Collections & Archives, UC San Diego, La Jolla, CA

SIO-FOX　Denis L. Fox Papers, 1924–1983, Oceanography and Marine Science, SMC 10, Special Collections & Archives, UC San Diego, La Jolla, CA

SIO-FRIEMAN　Edward A. Frieman Papers, Oceanography and Marine Science, SMC 77, Special Collections & Archives, UC San Diego, La Jolla, CA

SIO-HUBBS　Carl Hubbs Papers, 1912–1979, Oceanography and Marine Science, SMC 5, Special Collections & Archives, UC San Diego, La Jolla, CA

SIO-MENARD　Henry William Menard Papers, Oceanography and Marine Science, SMC 18, Special Collections & Archives, UC San Diego, La Jolla, CA

SIO-MPL　Marine Physical Laboratory Records, 1962–1990, Scripps Institution of Oceanography Archives, SAC 15, Special Collections & Archives, UC San Diego, La Jolla, CA

SIO-NIERENBERG　William Nierenberg Papers, 1932–2000, Oceanography and Marine Science, SMC 13, Special Collections & Archives, UC San Diego, La Jolla, CA

SIO-OCR   Scripps Institution of Oceanography, Office of Communication Records, SAC 13, Special Collections & Archives, UC San Diego, La Jolla, CA

SIO-ODR   Office of the Director Records, 1920–2013, Scripps Institution of Oceanography Archives, SAC 1, Special Collections & Archives, UC San Diego, La Jolla, CA (director name in parentheses)

SIO-REVELLE   Roger Revelle Papers, 1933–1991, Scripps Institution of Oceanography Archives, SMC 6, Special Collections & Archives, UC San Diego, La Jolla, CA

SIO-SVERDRUP   Harald U. Sverdrup Manuscripts, 1937–1977, Scripps Institution of Oceanography Archives, SMC 121, Special Collections & Archives, UC San Diego, La Jolla, CA

SIO-ZOBELL   Claude E. ZoBell Papers, 1931–1989, Scripps Institution of Oceanography Archives, SMC 131, Special Collections & Archives, UC San Diego, La Jolla, CA

SSF   Scripps Subject Files, 1903–1979, Scripps Institution of Oceanography Archives, SMC 121, Special Collections & Archives, UC San Diego, La Jolla, CA

UCLACF   Office of the Chancellor. Administrative files (University Archives Record Series 359), UCLA Library Special Collections, University Archives, Los Angeles

UCPR   University of California, Office of the President Records. CU-5, Series 2, Bancroft Library, University of California, Berkeley

WHOI-HOLLISTER   Charles D. Hollister Papers, Data Library and Archives, Woods Hole Oceanographic Institution, Woods Hole, MA

WHOI-ODR   Records of the Office of the Director, Data Library and Archives, Woods Hole Oceanographic Institution, Woods Hole, MA

WHOI-OSSR   Operational Scientific Service Records, Data Library and Archives, Woods Hole Oceanographic Institution, Woods Hole, MA

WHOI-STOMMEL   Henry Melson Stommel Papers, Data Library and Archives, Woods Hole Oceanographic Institution, Woods Hole, MA

WHOI-VINE   Allyn Vine Papers, Data Library and Archives, Woods Hole Oceanographic Institution, Woods Hole, MA

WHOI-VON ARX   William von Arx Papers, Data Library and Archives, Woods Hole Oceanographic Institution Woods Hole, MA

Notes

INTRODUCTION

1. Krimsky, *Science in the Private Interest*, attributes a version of this to Tufts University president Jean Mayer; it is also discussed in Haack, "Scientific Secrecy," 67. I first heard it in the early 1990s.
2. Biagioli, *Galileo Courtier*; Miller, *Dollars for Research*; Dupree, *Science in the Federal Government*; Kohler, *Partners in Science*; Crosland, *Science under Control*.
3. Epstein, *Impure Science*.
4. See, e.g., Hounshell and Smith, *Science and Corporate Strategy*; see also Crow and Bozeman, *Limited by Design*.
5. Richter et al., "In Vivo Effects." See also vom Saal and Hughes, "An Extensive New Literature"; vom Saal and Welshons, "Large Effects from Small Exposures"; Vandenberg et al., "Regulatory Decisions"; Bekelman et al., "Scope and Impact." For a summary of this research, see Oreskes et al., "Why Disclosure Matters."
6. Joravsky, *Lysenko Affair*; Roll-Hansen, *Lysenko Effect*; Graham, *Lysenko's Ghost*.
7. Joravsky, *Lysenko Affair*.
8. Marine research was also done at Stanford's Hopkins Marine Station in Pacific Grove and at the University of Washington's Oceanographic Laboratories at Friday Harbor, which was supported from the 1930s onward by the Rockefeller Foundation. On the history of Friday Harbor, see Benson, "Summer Camp." On marine biological laboratories more generally, see the other contributions to that volume, as well as Benson, *Laboratories*, and Benson, *Why American Marine Stations?*; Dexter, "History of American Marine Biology"; Rainger et al., *American Development of Biology*.
9. Burstyn, "American Oceanography."
10. Raitt and Moulton, *Scripps Institution*.
11. Harald Sverdrup to Roger Revelle, 5 December 1940 and 13 November 1940, Box 1, Folder 17, SIO-Sverdrup. Revelle, of course, would later become one of the most famous earth scientists of his generation, mentor to Al Gore, and

the scientist who did more than anyone in the United States to call attention to the issue of anthropogenic climate change.

12.   Greene, *Geology in the Nineteenth Century*; Oreskes, *Rejection of Continental Drift*.

13.   For more on Crary and Rutherford's contributions, see Lawrence, *Upheaval*.

14.   Greene, *Geology in the Nineteenth Century*; Oreskes, *Rejection of Continental Drift*.

15.   Ewing et al., "Geophysical Studies."

16.   Miller, "Geophysical Investigations," 811.

17.   Memorandum: "Preliminary Draft for Comment and Criticism: Oceanography in the Navy Post War Program," N428D/Gordon G. Lill to Dr. [Emanuel] Piore et al., 17 October 1947, Box 5, Folder "Operation Crossroads, 1947–47," 1, SIO-Revelle. Gordon Lill signed the cover letter, but Roger Revelle signed the report itself, so it appears that Revelle wrote the letter and Lill forwarded to Piore and other Navy officials.

18.   NDRC, *Physics of Sound*; Rainger, "Constructing a Landscape"; Weir, *Ocean in Common*; Doel, "Constituting the Postwar Earth Sciences"; Doel, "Defending the North American Continent"; Hamblin, "Pursuit of Science."

19.   Fleming, *Meteorology in America*; Fleming, *Fixing the Sky*; Slotten, *American Science*.

20.   Greenberg, *Pure Science*; Sapolsky, *Science and the Navy*; Weir, *Ocean in Common*; Doel, "Constituting the Postwar Earth Sciences"; Hamblin, "Pursuit of Science"; Hamblin, *Oceanographers and the Cold War*; Hamblin, *Arming Mother Nature*.

21.   Hamblin, *Oceanographers and the Cold War*, xv. Hamblin's focus is on international cooperation, so despite our near-identical topic, our treatments are very distinct. For an excellent overview of science and the American state in the Cold War, see Wolfe, *Competing with the Soviets*. Wolfe usefully reminds us that the Cold War was a period of unprecedented growth in the US government generally, so the growth of science, per se, is not as distinctive as is sometimes suggested. Still, its impacts are nevertheless important and worthy of understanding.

22.   Cloud and Clarke, *Shutter Darkly*; Cloud, "World in a Barrel"; Doel, "Constituting the Postwar Earth Sciences"; Hamblin, "Pursuit of Science."

23.   Inman, *Oral History*.

24.   "The Scientists"; see also Pfeiffer, "Naval Research."

25.   See particularly Edwards, *Closed World*; Weir, *Forged in War*; Weir, *Ocean in Common*; Craven, *Silent War*; Westwick, *Secret Science*. For studies on Cold War weapons systems, see Hall, *Lunar Impact*; Mackenzie, *Inventing Accuracy*; DeVorkin, *Science with a Vengeance*; DeVorkin, "Military Origin"; Dennis, *Change of State*; Leslie, *Cold War*; Hevly, "Tools of Science"; Galison, *Image and Logic*; Kevles, "Hot Physics"; Forman, "The Maser."

26.   Sapolsky, *Science and the Navy*.

27.   Chargaff, *Heraclitean Fire*.

28. For the role of some of these other Navy bureaus, see Sontag et al., *Blind Man's Bluff*.

29. Forman, "Behind Quantum Electronics." Ron Rainger argued that the division of basic and applied was used mostly by scientists to try to carve out a protected domain, but for the Navy that distinction had little use. Navy emphasis was on the "instrumentalities of war," and this was true at the ONR as well as in the operational bureau. Ron Rainger, personal communication with author, 9 July 2003. Unfortunately, Ron became ill before he was able to publish this work, but Ron Doel pursues related themes in "Constituting the Postwar Earth Sciences."

30. Forman, "Behind Quantum Electronics"; Kevles, "Hot Physics." See also Hounshell, "Rethinking," for a summary.

31. Wolfe, *Competing with the Soviets*, 3–4, gives a succinct summary of the standard historical critique of the concept of distortion.

32. Ibid.

33. For a fuller discussion of the Kevles-Forman debate and its methodological lessons, see the introduction to Oreskes and Krige, *Global Cold War*. For more on the historical study of ignorance, see Proctor and Schiebinger, *Agnotology*. On the inherent flexibility of science as a category, see Gieryn, *Cultural Boundaries*.

34. See, e.g., Oreskes and Krige, *Global Cold War*; Wolfe, *Competing with the Soviets*; Wolfe, *Freedom's Laboratory*; Solovey and Cravens, *Social Science*; Solovey, "Science and the State"; Hounshell, "Rethinking the Cold War"; Dennis, "Accounting"; Dennis, "Change of State"; Dennis, "Historiography"; Dennis, "First Line"; Hamblin, *Arming Mother Nature*; Hamblin, *Oceanographers*; Hamblin, *Poison in the Well*; Edwards, *Closed World*.

35. DeVorkin, "Who Speaks," engages a similar issue with respect to US astronomy after World War II, showing that astronomers were also concerned with the question of whether government funding would lead to government control, and specifically the question of who speaks for science.

36. For background on the question of patronage in science, see Crosland, *Science under Control*; Mirowski and Sent, *Science Bought and Sold*. With some notable exceptions (Biagioli's *Galileo, Courtier*, a book that engenders considerable disagreement among historians of astronomy), scholars who have studied scientific patronage have rarely been willing to claim that that patronage caused scientists to work in particular ways, much less to conclude certain things about the natural world. For instance, Shapin and Schaffer, *Leviathan*, argued that general political principles and inclinations affected how Hobbes and Boyle interpreted the evidence of a vacuum, but they did not claim that their specific patrons caused these scientists to work on the questions they did. On the ideology of freedom in or of science, see Wolfe, *Freedom's Laboratory*.

37. Confidential memorandum: "The Military Aspects of the [ONR] Geophysics Branch Program," n.d. [c. 1947–1948], Box 26, Folder 36, p. 7, USSSC. This doc-

ument appears to have been coauthored by Roger Revelle, Gordon Lill, and others; the comment about Woods Hole and Scripps is most likely to have come from Revelle.

38. Weir, *Forged in War*. See also Craven, *Silent War*.

39. There is another reason I did not focus on the military side of this story: I could not have furthered that line of inquiry with any degree of specificity without obtaining security clearance, which would have meant that what I wrote would have been subject to military review, vetting, and potential censorship. I made the decision, therefore, to rely on Freedom of Information Act requests, declassified documents, and unclassified archival materials— along with publicly available materials on military politics and programs and scientists' accounts of their interactions with military patrons. The latter may be skewed, but that is part of the story, too.

40. On this point, see Edwards, *Closed World*.

41. Gregg Mitman has discussed the use of dolphins in Navy research, and a potential threat to whales is discussed in chapter 9. See Mitman, *Reel Nature*.

42. The need for a strong military presence in the seas as a basis for economic power was earlier argued by Mahan, *Influence of Sea Power*. During the Cold War, the argument in the public sphere tended to be cast less explicitly in economic and more in political terms, specifically "containing communism" as a means of protecting American freedom. Many historians would nevertheless argue that economic considerations still loomed large. For my views of the relations between broad economic and political arguments and the history of science, see Oreskes, "Origins of the Cold War." For an examination of the relations of science, economics, and territorial imperatives with respect to marine biology and fisheries, see Finley, *All the Fish*. For broader perspectives on Cold War science, see Wolfe, *Competing with the Soviets*.

43. George Woolard to James Webb, 14 May 1966, Box 31, Loose letter, HHP.

44. The story told was this: In 1969, as plate tectonics broke open, Higgs presented a "sanitized" version of the data—that is to say, a degraded version in the form of "low resolution 'zebra charts.'" Korgen, "Higgs Receives." For an example of the Navy advising scientists to "sanitize" charts by removing detail in specified regions in order to permit them to be released, see Memorandum, Joseph E. King to Laboratory Director, 9 May 1963, Box 11, Folder 367, "Henry William Menard," SBF.

45. For an outstanding recent philosophical discussion of these matters by one of my own mentors, see Cartwright and Hardie, *Evidence Based Policy*; and Cartwright, "So, What's Evidence?"

46. Shapin, *Social History*, xviii.

47. For a fuller discussion of the Kevles-Forman debate and its methodological lessons, see the introduction to Oreskes and Krige, *Global Cold War*. For more on the historical study of ignorance, see Proctor and Schiebinger, *Agnotology*. On the inherent flexibility of science as a category, see Gieryn, *Cultural Boundaries*.

48. A note on units: Because this book covers a long time period and diverse

people, the units used are not consistent. American oceanographers typically used imperial units until the 1970s, when they mostly (but not consistently) shifted to metric. On due consideration, I decided to retain whatever units my actors used. I apologize for any confusion or annoyance this may create. *NB*: A fathom is six feet (or just under two meters).

CHAPTER ONE

This chapter is adapted from Oreskes and Rainger, "Science and Security." Professor Rainger passed away in 2016.

1.   On American physics during the Cold War, see Paul Forman, "Behind Quantum Electronics"; Kevles, "Hot Physics"; on other sciences, see Oreskes and Krige, *Science in the Cold War*.

2.   See, e.g., Day, "Navy Support"; Benson and Rehbock, *Oceanographic History*.

3.   One might wonder whether the World War I experience entered into their conversations; so far as the archival record shows, it did not, perhaps because most US military support for scientific research during World War I went to physics and chemistry, or perhaps because these scientists simply knew little about that period. Sverdrup, perhaps for obvious reasons, did not discuss war work in Germany.

4.   Raitt and Moulton, *First Fifty Years*, 119.

5.   Sverdrup, "Dynamics of Tides in the North Siberian Shelf"; Sverdrup, *The Norwegian North Polar Expedition*; "Six Years in the Arctic," typescript prepared for a public lecture, 4 December 1936, Box 1, Folder 2, "Lectures and Manuscripts," SIO-ODR. Sverdrup, "Informal Autobiography."

6.   Fram, "The Maud Expedition, 1918–1925," Polar Exploration Museum, https://frammuseum.no/polar_history/expeditions/the_maud_expedition __1918-1925_/; see also Nansen, *Fridtjof's Nansen's Farthest North*.

7.   Sverdrup, "Dynamics of Tides in the North Siberian Shelf"; Sverdrup, *The Norwegian North Polar Expedition*; Sverdrup, "Six Years"; Sverdrup, "Informal Autobiography."

8.   Memo, 11 August 1936, Box 1, Folder 1, SIO-ODR (Sverdrup).

9.   Sverdrup to Vern O. Knudsen, 1 February 1938, Box 1, Folder 1, SIO-ODR (Sverdrup).

10.   Sverdrup, "Informal Autobiography"; Revelle and Monk, "An Appreciation"; Munk, "In Memoriam"; Friedman, "Expeditions of Harald Ulrik Sverdrup"; Nierenberg, "Harald Ulrik Sverdrup."

11.   Raitt and Moulton, *First Fifty Years*, 101; Eric Mills, *The Scripps Institution*.

12.   The Rockefeller Foundation, "Oceanography." Most of the Rockefeller support was during the period when T. Wayland Vaughan was director.

13.   Sproul to Sverdrup, 20 March 1936, and Sverdrup to Sproul, 11 April 1936, Box 1, Folder 8, SIO-ODR (Sverdrup); Raitt and Moulton, *First Fifty Years*; Shor, *Probing the Oceans*. A further complication was that Scripps scientists did marine biology but not biological oceanography. Sverdrup considered

the distinction partly a matter of motivation and approach and partly a matter of working at sea, or not. For a historical appraisal of the distinction, see
Appel, *Marine Biology/Biological Oceanography*. For a contemporary appraisal,
see the discussions in Sverdrup et al., *The Oceans*.

14.   Raitt and Moulton, *First Fifty Years*, 129.

15.   Sverdrup, 30 April 1938, Box 1, Folder 2, SIO-ODR (Sverdrup).

16.   Various documents, Box 6, Folder 202, SBF.

17.   Of the two men, ZoBell would have the more distinguished career. He published his first paper on microbial life at deep-ocean depths in March 1936
("Vertical Distribution of Bacteria," in *Bulletin of the American Association of
Petroleum Geologists*). From this research, ZoBell also began forming hypotheses about the role of microbial organisms in modifying the Earth's geological
and chemical composition, including asserting that microorganisms played
an important role in the production of hydrocarbons. Overall, ZoBell is
credited with discovering sixty-five species of ocean bacteria and authoring
almost three hundred scientific papers and one textbook, including sixty-six
papers on petroleum microbiology. See "Claude ZoBell, 84, Marine Microbiologist," *New York Times*, 16 March 1989, https://www.nytimes.com/1989/03
/16/obituaries/claude-zobell-84-marine-microbiologist.html; Bass, "ZoBell's
Contribution"; and Conway, "A Lively Stone."

18.   Cullen, *Down to the Sea for Science*.

19.   On the issue of identity in field versus lab science, see Kohler, *Landscapes and
Labscapes*; Kuklick and Kohler, "Science in the Field."

20.   Claude ZoBell, Marine Microbiology Progress Report for Year Ending
15 March 1936, Box 21, Folder 626, SBF.

21.   Box 2, Folder 60, SIO-Fox. Later he would be a strong supporter of the loyalty
oath. His political views were complex. They were strongly anticommunist
but also radically democratic, even libertarian. He was an advocate of faculty governance and tried to use the loyalty oath affair as a springboard to
achieve greater faculty governance in the UC system.

22.   Fox to Sverdrup, 25 February 1937, Box 4, Folder 139, SIO-Fox.

23.   In 1936, his promotion to assistant professor triggered a discussion as to
what his title ought to be. His preference was assistant professor of marine
biochemistry and physiology. Fox to Vaughan, 29 May 1936, Box 6, Folder 204,
SBF. This apparently was not to Vaughan's liking; the title Fox received was
assistant professor of physiology of marine organisms. Sverdrup to Sproul, 31
December 1936, Box 1, Folder 8, SIO-ODR (Sverdrup). See also Fox to Sverdrup, 27 November 1936, Box 4, Folder 139, SIO-Fox; Fox, *Again the Scene*.

24.   Lecture outline, Box 1, Folder 12, SIO-ZoBell.

25.   ZoBell to Sproul, 17 November 1943, Box 21, Folder 627, SBF.

26.   McEwen, 1937, Box 1, SIO-ZoBell.

27.   McEwen, 1937, Box 1, SIO-ZoBell; Mills, "Useful in Many Capacities"; Mills,
"The Oceanography of the Pacific."

28.   Sverdrup et al., *The Oceans*.

29. Sverdrup was even less sympathetic to geology and actively tried to prevent Francis Shepard, who pioneered the study of submarine canyons (chapter 4), from becoming a member of Scripps faculty. In a private memo discussing what would happen to Shepard at the end of his Geological Society of America–funded project on the Gulf of California, Sverdrup was explicit that he did not wish to keep him on: "The work of Shepard is not oceanographic. We shall continue work in sedimentation which is a problem in oceanography, but to us details of the submarine topography are of small interest and the geological character of the rocks forming the bottom is insignificant." Notes from Pacific Grove, Stanford, Berkeley, Davis, 15–22 October 1937, Box 1, Folder 6, SIO-ODR (Sverdrup). In this opinion, Sverdrup was very wrong; see chapters 4 and 5.

30. Norwegian oceanography had been very concerned with fisheries; for some scientists, strengthening dynamic oceanography meant moving away from specific concerns about fish populations. Rozwadowski, *The Sea Knows No Boundaries*.

31. Raitt and Moulton, *First Fifty Years*, 123. The *E. W. Scripps* was formally presented to the institution in December 1937, after considerable work by Sverdrup to obtain the funding, find the boat, and have it refitted. Ironically, only four years later it was handed over to the Navy for work at the UC Division of War Research.

32. Plans for use of … the *E. W. Scripps*, Box 1, Folder 11, SIO-ODR (Sverdrup); T. D. Beckwith to Loye H. Miller, 9 December 1937, Box 1, Folder 11, SIO-ODR (Sverdrup). The issue of what was and was not oceanographic came up as well in communications with the UC Life Sciences Group, whom Sverdrup invited to submit a proposal for work on the *E. W. Scripps*. Evidently they were well aware of Sverdrup's boundaries, for their reply carefully divided their proposed work into two categories: "oceanographic" and "not oceanographic." Among the latter—work they proposed might be done when the vessel was not being used for oceanographic work—they included soil sampling for microorganisms, algae and myxomycetes, and fungi, precisely the kind of work ZoBell might have done.

33. On "fitting in" in science, particularly during a time of transition in science, see Zeller and Reis, "Wild Men."

34. Plans for use of … the *E. W. Scripps*, Box 1, Folder 11, SIO-ODR (Sverdrup); T. D. Beckwith to Loye H. Miller, 9 December 1937, Box 1, Folder 11, SIO-ODR (Sverdrup).

35. Raitt and Moulton, *First Fifty Years*, 129.

36. Maienschein, *Transforming Traditions*; Oreskes, *Rejection of Continental Drift*. In the larger world of biology and earth science, the methodological trend was in favor of laboratory work; Fox and ZoBell were doing work that elsewhere would have been considered "cutting edge." It was a peculiarity of Scripps and Sverdrup's commitment to seagoing oceanography—and just plain bad luck for Fox and ZoBell—that their work was considered problematic there.

37. Discussion with ZoBell, 10 September 1936, Box 1, Folder 6, SIO-ODR (Sverdrup).

38.   Memo, 18 December 1936, and Sverdrup to Sproul, 31 December 1936, Box 1, Folder 8, SIO-ODR (Sverdrup); Robert M. Underhill to ZoBell, 9 January 1937, Box 21, Folder 627, SBF.

39.   ZoBell to Underhill, 14 January 1937, Box 21, Folder 627, SBF.

40.   ZoBell to Sverdrup, 31 May 1938, Box 1, Folder 11, SIO-ODR (Sverdrup).

41.   Sverdrup to Loye H. Miller, 25 March 1938, with attached memos, Box 1, Folder 14, SIO-ODR (Sverdrup).

42.   Sverdrup to Sproul, 12 January 1938, Box 1, Folder 12, SIO-ODR (Sverdrup).

43.   ZoBell to Sverdrup, 5 November 1936, SBF 21.626; see also Sverdrup to Navy Bureau of Construction and Repair, 18 December 1939, Box 1, Folder 15, SIO-ODR (Sverdrup). Some work on the problem was already being done by the 1910s, but I have not determined who funded it.

44.   A. H. van Keuren to ZoBell, 11 May 1935, Box 21, Folder 626, SBF.

45.   ZoBell to Sverdrup, 5 November 1936, Box 21, Folder 626, SBF; ZoBell to Whedon, 8 February 1937, Box 21, Folder 627, SBF; Sverdrup to Navy Bureau of Construction and Repair, 18 December 1939, Box 1, Folder 15, SIO-ODR (Sverdrup); Sverdrup to L. A. Nichols, 11 December 1939, Box 1, Folder 15, SIO-ODR (Sverdrup); and various materials, Box 1, Folder 9, SIO-ODR (Sverdrup).

46.   Sverdrup to L. A. Nichols, 11 December 1939, and Sverdrup to Nichols, 19 December 1939, Box 1, Folder 15, SIO-ODR (Sverdrup).

47.   Sverdrup to George Taylor, 12 June 1940, and Sverdrup to D. G. Maclise, 28 June 1940, Box 1, Folder 16, SIO-ODR (Sverdrup).

48.   Various documents, Box 1, Folder 17 and 18, SIO-ODR (Sverdrup). For comparison, this was one of SIO's four main outside funding sources in 1939–1940, besides the Scripps Endowment. The largest was a $10,000 grant from the Geological Society of America (GSA) for an expedition to the Gulf of California. Another was a $2,000 grant from the California Department of Fish and Wildlife for research on sardines. There was also support for staff salaries through the US Works Progress Administration, which appears to have been very generous—funding a number of technicians, secretaries, and stenographers—but it did not provide direct funds for research. What is perhaps most significant is how hard these other funds generally were to get and how tightly they were guarded. The paperwork surrounding the GSA grant is enormous, yet the amount given was still not enough to cover the project's full costs. Sverdrup to Revelle, 13 November 1940, Box 1, Folder 17, SIO-ODR (Sverdrup). By comparison with the GSA, or with the myriad supplications Sverdrup made to Robert Scripps to raise the money for the *E. W. Scripps*, obtaining Navy money was nearly effortless. The difference was not lost on Sverdrup—or on Revelle.

49.   Sverdrup to Sproul, 25 October 1940, Box 1, Folder 17, SIO-ODR (Sverdrup).

50.   Sverdrup to Underhill, 11 January 1941, Box 1, Folder 18, SIO-ODR (Sverdrup).

51.   Sverdrup to Sproul, 25 October 1940, Box 1, Folder 17, SIO-ODR (Sverdrup).

52.   Ibid.

53.   Notes on meeting, 17 February 1941, Box 3, Folder 96, SIO-Fox.

54. Ibid.
55. Ibid.
56. Memo, 20 February 1941, Box 3, Folder 96, SIO-Fox.
57. ZoBell to Sverdrup, 21 February 1941, Box 3, Folder 96, SIO-Fox.
58. Ibid.
59. Nancy Scola, "Public University Researchers Get Cash for Studying GMOs—And the Shaft for Studying Organic Ag," *Grist*, 21 February 2008, https://grist.org/article/monsanto-u/.
60. Baker and Tansley, 1946, Box 3, Folder 102, SIO-Fox.
61. John R. Baker to Fox, 21 November 1944, Box 3, Folder 102, SIO-Fox ; see also Kuznick, *Beyond the Laboratory*.
62. Polanyi, "Rights and Duties of Science"; Polanyi, "The Growth of Thought in Society"; Polanyi, *Science, Faith, and Society*. Articles by Polanyi that circulated among members included "The Growth of Thought in Society" and "Rights and Duties of Science." Although the 1946 piece was circulated after the events described here, it articulates the ideals of the Society already in place. Ironically, within a few years, Scripps was on its way to being almost entirely funded on government grants and contracts, justified not by Marxist principles but by anticommunist ones.
63. Polanyi, *Science, Faith, and Society*; see also Polanyi, "Growth of Thought"; Polanyi, "The Autonomy of Science."
64. "War Makes US, Britain, Keepers of Science," clipping, Box 3, Folder 93, 23 September 1941, SIO-Fox.
65. Like many conservatives then and now, ZoBell was more concerned about the pernicious influence of big government than of big business. This asymmetry in conservative thinking invites more scholarly attention than it has to date received. Perhaps as an individual ZoBell felt he could more easily walk away from a corporate sponsor than a government one, but here I am just speculating.
66. Fox to Sverdrup, 27 July 1943, Box 6, Folder 205, SBF.
67. Memorandum of studies of fouling growth at Scripps, Box 3, Folder 96, SIO-Fox; Sverdrup to Monroe Deutsch, Box 1, Folder 25, 7 August 1944, SIO-ODR (Sverdrup).
68. Box 6, Folder 205, SBF; Box 4, Folder 139, SIO-Fox. Part of this work was done by Olga Janowitz, an Austrian émigré biologist hired as temporary assistant in the summer of 1943. There appears to have been some tension between Janowitz and Fox. She claimed to have achieved good results by "imitat[ing] nature and us[ing] natural waxy films to inhibit both fouling and corrosion." Fox disagreed, arguing that hydrophobic materials needed to be supplemented by detergents: "If certain wetting agents of very low surface tensions be incorporated into a hydrophobic agent such as Vaseline, some very interesting results are obtained." Fox to Bureau of Aeronautics, 9 September 1943, Box 6, Folder 205, SBF. On Janowitz, see letters in Box 1, Folder 23, SIO-ODR (Sverdrup). I mention this incident mainly so as not to "disappear" a rare woman in this story.

69.   P. Forman, "Behind Quantum Electronics"; Kevles, "Hot Physics"; Kevles, *The Physicists*; Oreskes and Krige, *Global Cold War*. See also Friedman, *Appropriating the Weather*; Everest et al., *World War II Underwater*, 117; Weir, "Selling Bellevue."

70.   Iselin, *National Defense*; Rainger, "Constructing a Landscape." The other was in Woods Hole, Massachusetts, near the Woods Hole Oceanographic Institution, where oceanographers had launched extensive studies on the influence of physical factors on underwater sound transmission, funded by the Naval Research Laboratory and the Bureau of Ships (see chapter 2).

71.   AC-09.4, Folder 6: "Richard Fleming to Columbus O'Donnell Iselin, n.d. [May 1941]," WHOI-ODR; see also UCDWR, "Maximum Echo Ranges: Their Prediction and Use," 1941, entry 29, Box 680, Folder UC, NACP-NDRC. UCDWR, Oceanographic Division, "Effects of Diurnal Variations in Temperature on Sound Ranges," September 1941, entry 29, Box 672, Folder UC, NACP-NDRC. UCDWR, Oceanographic Division, "Accuracy of Echo Ranges Predicted from Bathythermograph Observations," 11 December 1941, entry 29, Box 672, Folder UC, NACP-NDRC. UCDWR, Oceanographic Division, "General Conditions for Echo Ranging in the Western North Pacific Ocean," 30 January 1942, entry 29, Box 672, Folder UC, NACP-NDRC.

72.   Levin, "Norwegians Led the Way."

73.   George R. Eckman to J. Edgar Hoover, 23 September 1941, declassified 9 November 1993, HUSFBI.

74.   Eckman to Hoover, 23 September 1941, Exhibit F, HUSFBI.

75.   Ibid., Exhibit C.

76.   Ibid., Exhibit E.

77.   Ibid., Exhibit A; see also HUSN.

78.   George L. Shea to George R. Eckman, 24 July 1941, Exhibit G in Eckman to J. Edgar Hoover, 23 September 1941, HUSFBI.

79.   Oreskes, *Continental Drift*; Lek, "Die Ergebnisse der Strom." This was a major geological and oceanographic cruise to the Dutch East Indies, whose members included Philip Kuenen, later famous for demonstrating the role of submarine mud flows (turbidity currents) in deepwater sedimentation. Together with the work of Dutch geodesist Felix A. Vening Meinesz, it helped to establish the tectonic framework of the Indonesian archipelago, later invoked in the debate over continental drift (see chapter 4). For Lek's thesis work, see Lek, "Tidal Phenomena."

80.   Lek to Sverdrup, 7 March 1939, Box 10, Folder 327, SBF; see also Sverdrup to Lek, 27 March 1939, Box 1, Folder 14, SIO-ODR (Sverdrup).

81.   According to Walter Munk, who knew him, Lek was homosexual, which might account for the late-night visits, or maybe he was just a night owl.

82.   Shea to Eckman, 24 July 1941, HUSFBI.

83.   Ibid.

84.   Ibid.

85.   Ibid.

86. Sverdrup hired several other foreigners on military-sponsored projects. These included Olga Janowitz, Polish meteorologist Wladyslaw Gorzynski, and C. K. Tseng, a Chinese marine biologist who worked a project to develop agar resources. Sverdrup did not like Gorzynski and later removed him from SIO, but he supported Janowitz and Tseng. On Tseng, see Neushul and Wang, "Between the Devil." On troubles that other foreign-born scientists encountered over security issues, see Chang, *Thread of the Silkworm.*

87. Shea to Eckman, 24 July 1941, HUSFBI.

88. Eckman to J. Edgar Hoover, 23 September 1941, Exhibit J, HUSFBI.

89. Ibid.

90. Ibid.

91. Various documents, Box 1, Folder 12, SIO-ODR (Sverdrup).

92. "Goebbels" to Fox, 15 July 1938, Box 1, Folder 13 SIO-ODR (Sverdrup); Box 2, Folder 61, SIO-Fox. Copies of this letter may be found in both Sverdrup's and Fox's papers; however, the copy in the Sverdrup papers appears to have been placed there at a later date by former Scripps archivist Helen Raitt. A note on the margin of the letter, in Fox's hand, says "Xerox copy of a jocular letter from Harald U. Sverdrup, then director of SIO, given to me by him at a party before we were to leave for a sabbatical year at Cambridge University. Translated by Dr. Ted Ennes, 1-18-77."

93. Sverdrup to Fox, 20 October 1938, Box 4, Folder 136, SIO-Fox.

94. Sverdrup to Fox, 8 May 1939, Box 4, Folder 136, SIO-Fox. What was Sverdrup was thinking? On one level, the "Goebbels" letter may be dismissed as poor taste, bad judgment, or just plain peculiar. Despite the Anschluss, the world was not yet at war, and many thought the Soviets would keep the Nazis at bay; the Hitler-Stalin pact had not yet been signed. However, the Nazis had implemented most of the structure of Nazi racial law, including the notorious Law for the Restoration of the Professional Civil Service, which had driven Jewish professors out of their universities, and there had already been a flood of German-Jewish emigration. In later years, Fox would describe the letter as "jocular," but it is difficult to imagine he would have considered it so at the time. See also various documents in Box 2, Folder 61, SIO-Fox.

95. The initial investigations were done by the Military Intelligence Division and the FBI. Later, Navy investigators became involved as well.

96. Eckman to Hoover, 8 August 1941, HUSFBI; see also Eckman to Hoover, 23 September 1941, HUSFBI.

97. John T. Tate to A. H. van Keuren, 9 September 1941, HUSN.

98. Ibid.

99. On the Navy Radio and Sound Laboratory (NRSL), see Weir, "Fashioning." On NDRC Division 6 and the NRSL, see Weir, *An Ocean in Common*; Lawyer, Bates, Rice, *Geophysics*. For a summary of the work done under NDRC Division 6, see Committee of Underwater Research Analysis Group, Principles of Underwater Sound: Originally Issued as "Division 6, Volume 7 of NDRC Summary

Technical Reports" (National Research Council, 1955). The citations to the wartime work of NDRC Division 6 can be confusing. The work of the entire division was first issued as a classified report in 1946 as volume 7 of John T. Tate, Vannevar Bush and James Bryant Conant, *Summary Technical Report of Division 6*, NDRC Washington, DC, 1946. http://www.dtic.mil/dtic/tr /fulltext/u2/221610.pdf. The volume most relevant to the work discussed in this book is volume 7, *Principles and Applications of Underwater Sound*, written by Carl Eckart. This was declassified and published as a stand-alone report in 1954 and republished in 1968. See Munk and Preisendorfer, "Carl Eckart."

100. Revelle would become the fifth director of Scripps (after Carl Eckart, who chaired NDRC Division 6), a founder of the University of California's San Diego campus, a science adviser in the Kennedy administration, one of the first scientists to warn about anthropogenic climate change, and a mentor to a young Albert Gore Jr. See "Past Scripps Directors," Scripps Institution of Oceanography, https://scripps.ucsd.edu/about/leadership/director/past -scripps-directors.

101. HUSN, 1941.

102. Van Keuren to Navy Secretary, 17 September 1941, HUSN.

103. Knudsen to Navy Secretary, 14 October 1941, HUSN.

104. Vannevar Bush to Jerome Hunsaker, 25 October 1941, HUSN.

105. Sverdrup, "Unofficial Autobiography."

106. Sverdrup to Tate, 10 September 1941, HUSN.

107. J. T. Bissel to Deputy Chief of Staff, 20 September 1941, HUSN.

108. Knudsen to Tate, 1 January 1942, HUSN.

109. Later documents identify them by name, but even without those other documents the details make it obvious.

110. Leif was his half brother, who came to the United States at the age of sixteen and became a US citizen in 1917. After graduating in civil engineering from the University of Minnesota in 1921, he worked for the Minnesota Highway Department, and then opened his own engineering company specializing in bridges. During World War II, he rejoined the Army as a colonel and later became General MacArthur's chief engineer in the Pacific. Sverdrup thought that Leif had also had trouble getting clearance because of the relatives in Norway, but I have found no evidence of this. See report 96-10, 14 January 1942, San Diego, CA, HUSFBI.

111. This report was filed in 1943, but it includes details of interviews from early 1942. I cite it in preference to earlier documents that are less detailed and more heavily censored. See Knudsen to Tate, 1 January 1942, HUSN; Knudsen to OSRD, 26 January 1942, recounted in Wallace S. Wharton to J. Edgar Hoover, 19 April 1943, Department of the Navy, Office of the Chief of Naval Operations, HUSN.

112. In some cases, additional details apparent in different documents may be an artifact of censorship. In other cases, interview dates make it clear that some individuals (e.g., Mason, Revelle, Chambers, Fox, ZoBell) were interviewed

more than once. In general, later reports seem to have more detail, suggesting that informants were pressed harder for details as time went on. Wharton to Hoover, 19 April 1943, HUSN.

113. Ibid.
114. Ibid.
115. Both Fox and Chambers had sons, although Fox's would have been only a
child at the time.
116. Wharton to Hoover, 19 April 1943, HUSN.
117. Ibid.
118. Report 96-10, 14 January 1942, HUSFBI.
119. Ibid.
120. Ibid.
121. It is not clear if the FBI interviewed Chambers a third time, or if declassified
materials are unevenly censored, but the FBI report filed in January 1942 includes this charge emanating from Chambers for the first time.
122. Report 96-10, 14 January 1942, HUSFBI.
123. Ibid.
124. Ibid.
125. Report 96-61, 21 January 1942, Chicago; Report 96-44, 27 January 1942, San
Francisco; Report 96-101, 29 January 1942, Los Angeles; Report 77–205,
6 February 1942, Seattle, HUSFBI.
126. Report 96-107, 11 February 1942, Boston, HUSFBI.
127. This allegation likely regards George McEwen, who was involved in the San
Diego Peace Crusade and investigated by the FBI for it.
128. Report 96-107, 11 February 1942, Boston, HUSFBI.
129. Ibid.
130. Knudsen to Tate, 1 January 1942, HUSN.
131. Ibid.
132. Stefansson to Sverdrup, 11 May 1942, Box 1, Folder 20, SIO-ODR (Sverdrup).
133. After May 1942, there is no further correspondence on the subject in
Sverdrup's files. Possibly the work he did that summer "for the Norwegian
government" was connected with polar problems. Telegram: Stefansson to
Sverdrup, 18 May 1942, Box 1, Folder 20, SIO-ODR (Sverdrup).
134. Sverdrup to Commander L. C. Stevens, Bureau of Aeronautics, 12 May 1942,
Box 1, Folder 20, SIO-ODR (Sverdrup); J. E. Sullivan to Sverdrup, 22 May 1942,
Box 1, Folder 20, SIO-ODR (Sverdrup).
135. Sverdrup to Sproul, 15 June 1942, Box 1, Folder 20, SIO-ODR (Sverdrup).
136. Zimmerman and Ackerman to UC Board of Regents, 4 June 1942, Box 1,
Folder 20, SIO-ODR.
137. Sverdrup, 21 May 1942, "Second Memo on Wave Periods and Wave Heights
at La Jolla in Relation to Meteorological Conditions over the North Pacific
Ocean," *Oceanographic Sections* 40 (San Diego: UCDWR, NRSL); Box 1, Folder
2, SIO-ODR. Sverdrup was already working on the problem in May, before
the project officially began.

138. George Turner to Robert Sproul, 21 May 1942, 139: "SCRIPPS General Matters 1936–1959," UCLACF.

139. Ibid.

140. Ibid.

141. ZoBell to Sproul, 17 November 1943, Box 21, Folder 627, SBF.

142. Sverdrup to Sproul, 12 March 1943, Box 1, Folder 22, SIO-ODR (Sverdrup). See also Neushul and Wang, "Devil and the Deep Blue Sea," 2000.

143. Turner to Sproul, 21 May 1942, 139: "SCRIPPS General Matters 1936–1959," UCLACF.

144. Ibid.

145. According to one Scripps scientist (who, ironically, wishes to remain anonymous), there was only one telephone at Scripps in the early days, downstairs in the Old Scripps building, and this was closely guarded by Sverdrup's secretary, Tillie Genter. Turner's comments suggest that the phone was on a party line, or perhaps Genter was in the habit of eavesdropping.

146. Sverdrup, Oceanography for Meteorologists.

147. Turner to Sproul, 21 May 1942, 139: "SCRIPPS General Matters 1936–1959," UCLACF.

148. In later years, Fox and ZoBell would oppose the directorship of Roger Revelle on similar grounds. Revelle was haughty and insensitive; they would have preferred a more humble person, sensitive to the needs and feelings of the faculty. ZoBell made this point on more than one occasion. See various memos in SIO-Fox and also in the papers of Carl Leavitt Hubbs, SIO-Hubbs. The issue got heated when a number of faculty "blackballed" Revelle in replying to a request for their view from Sverdrup. On 24 September, the faculty wrote to Sproul to apologize; they had only meant to convey that, since the war was over, they hoped for an altered focus to "marine resources," particularly food, and did not consider Revelle ideal because "Mr. Revelle does not give promise of being an effective, successful director of the Institution." They reiterated that their first choice was Daniel Merriman, then Richard Fleming and Carl Eckart, in that order. Interestingly, the letter was signed by Walter Munk, along with marine geologist Francis Shepherd and seven other staff, mostly biologists. Box 33, Folder 87, SIO-Hubbs.

149. ZoBell to Sproul, 20 June 1942, Folder 414, UCPR.

150. Case History Memorandum for the File, 23 July 1942, HUSN.

151. Confidential memo from Branch Office, G-2, WDC & Fourth Army, San Diego, CA, 7 July 1942, HUSFBI.

152. Earl G. Harrison to J. E. Hoover, 15 August 1942, US Department of Justice (DOJ) FOIA CO2.12-C.

153. Confidential memo from Branch Office, G-2, WDC & Fourth Army, San Diego, CA, 7 July 1942, HUSFBI.

154. F. A. Calvert to George Burton, 15 November 1942, HUSFBI.

155. Sverdrup to Sproul, 15 April 1942, Box 1, Folder 20, SIO-ODR (Sverdrup); Zimmerman and Ackerman to UC Board of Regents, 4 June 1942, Box 1,

Folder 20 SIO-ODR (Sverdrup); see also Sverdrup to Revelle, 22 June 1943, Box 1, Folder 22, SIO-ODR (Sverdrup); Levin, "Norwegians."

156. Joint Chiefs of Staff Committee on Meteorology (JMC) meetings, 7 January 1943, Box 62, Folder Oceanographic Data Vol. 1, NACP-ONR. 29 April 1943, Box 209, Folder 3 December 1942 Subcommittees, Records of the Joint Chiefs of Staff, RG 218, NACP.

157. JMC meetings, 28 January 1943, 24 February 1943 and 15 April 1943, Box 62, Folder Oceanographic Data Vol.1, NACP-ONR; Knoll to Orville, 22 February 1943, Box 3, Folder H1, NACP-OCNO.

158. Sverdrup to Revelle, 5 May 1943, Box 16, Folder 38, SSF. See also Sears to Sverdrup, 26 May 1943, Box 16, Folder 28, SSF; Sverdrup to Revelle, 1 June 1943, and Sverdrup to Sears, 5 June 1943, Box 16, Folder 28, SSF.

159. Report 96-40, 7 March 1942, Washington, DC, HUSFBI.

160. Ibid.

161. Ibid.

162. Interview by Stuart W. Mark, First Lieutenant, Field Artillery, Memorandum for the record, 12 January 1943, to MID 201, HUSFBI; see also Wallace S. Wharton to J. Edgar Hoover, 19 April 1943, HUSN.

163. If Sverdrup lacked enthusiasm for the British, he would not have been alone among Norwegians. Many were dying in commando raids organized by the British, and by 1943 there was a growing sentiment that the British were cavalier about Norwegian lives. After the death of Einar, Leif shared this view. See George Sverdrup to Deborah Day, 9 October 1993, Sverdrup Family Papers, SMC 47, Special Collections & Archives, UC San Diego Library.

164. Wallace S. Wharton to J. Edgar Hoover, 19 April 1943, HUSN.

165. Buships to Chief of Naval Operations Op 16-B5, 24 June 1942, HUSN.

166. Ibid.

167. Hydrographer to Commander Eighth Fleet, 7 October 1943, and Commander US Naval Forces, Northwest African Waters to Hydrographer, 19 November 1943, both in entry 49, Box 16, Folder H1 (383545), NACP-HO. For background on this subject, see Bates et al., *Geophysics in the Affairs of Man*, esp. 76–77.

168. Bates and Fuller, *America's Weather Warriors*; Nierenberg, "Harald Ulrik Sverdrup"; Levin, "Norwegians."

169. George Sverdrup to Deborah Day, 9 October 1993, Sverdrup Family Papers, SMC 47, Special Collections & Archives, UC San Diego Library.

170. Sverdrup to Roger Revelle, 20 July 1943, Box 1, Folder 23, SIO-ODR (Sverdrup).

171. Sverdrup to Iselin, 15 May 1945, Box 1, Folder 26, SIO-ODR both in entry 49, Box 16, Folder H1 (383545), NACP-HO.

172. Sverdrup to Rossby, 3 July 1944, Box 1, Folder 25, SIO-ODR both in entry 49, Box 16, Folder H1 (383545), NACP-HO.

173. Sverdrup to Sproul, 18 June 1942, Box 1, Folder 20, SIO-ODR both in entry 49, Box 16, Folder H1 (383545), NACP-HO.

174. Friedman, "Expeditions."

175. Nierenberg, "Harald Ulrik Sverdrup."

176. Ibid., 357.
177. Robert Marc Friedman, personal email communication with author, 31 May 2016.
178. Nierenberg, "Harald Ulrik Sverdrup," 357.
179. Bush to Office of the Director et al., 14 September 1944, entry 78, Box 57, Folder Demobilization I, NACP-NDRC.
180. Memorandum, 10–13 January 1945, Box 16, Folder 42, SSF.
181. Harald U. Sverdrup, "Memorandum on Relationship between the Institute of Geophysics and the Contemplated Sonar Research Group at Point Loma," 12 October 1945, Box 1, SIO-ODR (Sverdrup); Sverdrup and Fleming, "Memorandum on Postwar Studies."
182. Sverdrup to Dodson, 1 February 1945, Box 16, Folder H1 (383545), NACP-HO.
183. Revelle to Colpitts, 2 October 1943, and Revelle to Director, SIO, 2 October 1943, both in entry 49, Box 31, Folder 1, A10-1/S68 428232, NACP-HO. Working in cooperation with Fleming and the oceanographic division at the UCDWR, Sverdrup and his staff produced dozens of these manuals for strategic locations throughout the Pacific.
184. Fleming to Chief Bureau of Ships, 15 September 1944, Box 16, Folder H4-11, NACP-HO.
185. UCDWR, "Summary Technical Report of Division 6," 9–12; NDRC, 1946, *Principles and Applications of Underwater Sound*, 119–43 (see n. 99).
186. Fleming, memo to file, Submarine Supplements Conference with Sverdrup, Revelle, and Armstrong, 21 August 1943, Box 1, Folder 23, SIO-ODR (Sverdrup); Colpitts to Harnwell, 24 May 1944; Fleming to Eckart, 2 May 1944 both in entry 77, Box 6, Folder 01.193, NACP-NDRC.
187. Sverdrup, "Plan for Oceanographic Work"; Harnwell to Eckart, Fleming, and Sverdrup, 1 March 1945, Box 24, Folder 12, SSF.
188. Sverdrup, "Memorandum on Relationship."
189. Revelle to Sverdrup, 1 June 1948, Box 1: Correspondence January–February 1948, SIO-ODR (Sverdrup); Revelle and Eckart to Code 410, 20 May 1948, Box 16, Folder 52, SSF; Eckart to Revelle, 10 June 1948, Box 16, Folder 53, SSF.
190. Hess, "Drowned Ancient Islands," 1946; Memo 013B [Oceanography Section] Joint Task Force 1 to 01D, 11 May 1946, Box 6, Folder 29, SSF; Emmery et al., "Geology of Bikini."
191. Menard, *The Ocean*.
192. Fond, "Oceanography Researches"; Menard, *The Ocean*; Oreskes, *Continental Drift*. In 1946 the Navy created a new Oceanography Studies section within the Navy Radio and Sound Lab, by then renamed the Naval Electronics Laboratory (NEL). The NEL would soon undertake classified investigations of the seafloor, and NEL scientists Robert Dietz and Henry Menard would play a major role in the development of plate tectonics.
193. Weisgall, *Operation Crossroads*; Rainger, "Science at the Crossroads."
194. Organization and Personnel of Oceanographic Section, Box 6, Folder 28, SSF.

Scripps personnel on *Crossroads* included Gifford Ewing, Martin Johnson, E. C. LaFond, Herbert Mann, and Walter Munk. Sverdrup participated in a meeting on 28–29 January 1946 to assess the probable effects of the bomb's disturbance on the lagoon. He also discussed issues on the bomb and diffusion. See Sverdrup telegram to Revelle, 2 February 1946, and Sverdrup to Munk, 21 February 1946, Box 6, Folder 27, SSF. On the Bikini Resurvey, see Sverdrup to Revelle and Sargent, 29 May 1947, Box 6, Folder 31, SSF.

195. Iselin to Sverdrup, 10 April 1946, and Sverdrup to Iselin, 12 April 1946, Box 6, Folder 28, SSF.

196. Sverdrup to Loeb, 2 November 1945, and Sverdrup to Dykstra, 26 December 1945, Box 1, Folder 27, SIO-ODR (Sverdrup).

197. The following estimate is derived from a listing of Scripps contracts dated 5 May 1948, soon after Sverdrup had left for Norway. That document, from UCPR 1948:414, lists only contracts with the Bureau of Ships and the Office of Naval Research, which nonetheless totaled $963,475. At the time SIO's other contracts included $400,000 from the State of California for work on sardines. See Eckart to Sproul, 22 July 1948, "Marine Life Research Program," Box 13, Folder 17. The institution also had a number of small contracts with the Beach Erosion Board ($45,000) and the American Petroleum Institute ($19,500). See Box 17, Folder 5, and Box 16, Folder 28, SSF. On military students in classes, see Sverdrup to Office of the President, 2 September 1946, 1946:414, UCPR. See also Ronald Rainger interviews with Dale F. Leipper and Robert O. Reid, 25–26 February 2000, Texas A&M University, College Station, TX.

198. Levin, Norwegians; Oreskes and Doel, "Physics and Chemistry."

199. For example, the 1950 Scripps Mid Pacific Expedition (Mid-Pac) included assessment of radioactivity on the waters and organisms of the Marshall Islands. See Revelle to Vine and Perkins, 9 January 1950, Box 6, Folder 36, SSF. In later years Scripps scientists participated on several nuclear bomb tests in the Pacific, including *Operations Ivy, Wigwam,* and *Redwing.* Other Scripps scientists began studies of radioactive tracers and organic uptake of radioactive substances. See Goldberg, "Accumulation of Radioiodine"; Goldberg and Inman, "Neutron Induced Quartz"; Revelle et al., "Nuclear Science and Oceanography" (copy found in Box 154, Folder 19 SIO-Revelle). By the mid-1950s Revelle and Goldberg were actively involved in National Academy of Sciences studies of the biological effects of atomic radiation, and Revelle had hired other scientists, including Hans Eduard Suess and Gustav Arrhenius, to examine other aspects of radioactivity. These projects all took place after Sverdrup had left. Goldberg will rejoin our story in chapter 8.

200. Hydrographic Office Project NO 4701-ONR Contract N6-ori-lll, Submarine Geology.

201. Fleming to Sverdrup, 8 August 1947, Box 6, Folder 196, SBF.

202. Fleming to Sverdrup, 22 October 1947, Box 6, Folder 196, SBF.
203. Rainger, "Patronage and Science."
204. See, e.g., Slotten, *Patronage*; Fleming, *Meteorology in America*.
205. Kevles, *The Physicists*; Dupree, *Science in the Federal Government*.
206. On the question of orientation see Edwards, *Closed World*. See also discussion in chapter 7.

CHAPTER TWO

1. Memo on postwar studies of oceanography of the surface layers, 31 May 1945, Box 1, Folder 26, SIO-ODR.
2. Weir, *Ocean in Common*; Shor, *Seeking Signals*; Spiess, "Seeking Signals"; Craven, "Ocean Technology"; Craven, *The Silent War*.
3. Cloud, "World in a Barrel"; Wittje, *Age of Electroacoustics*. See also Lichte, "Über den Einfluss."
4. Bureau of Ships, *Prediction of Sound*; Ewing and Worzel, "Long Range Sound"; Eckart, "Refraction of Sound"; NDRC, *Physics of Sound*.
5. Eckart, "Refraction of Sound." Although the BT was considered hugely important to physical oceanography, its analysis was considered boring and menial, and thus it was left to women. For details on its development, how it was used in the war, and why it became women's work, see Oreskes, "Laissez-tomber."
6. As noted in the introduction, I have left intact the units as I found them. I trust readers can convert to metric as needed and apologize for any inconvenience.
7. The BT had been developed in the 1930s by MIT meteorologist Carl-Gustaf Rossby (1898–1957) and his graduate student Athelstan Spilhaus (1911–1998) to study the distribution of water masses in the context of the question of ocean currents.
8. Memo by Robert Deitz to Bill Menard, "Program to Measure Deep Ocean Currents at NIO [National Institute of Oceanography]," 9 March 1955, AC-09, Folder 22, WHOI-ODR. Although this might seem like looking for a needle in a haystack, if there were deep currents, then the bathyscaphe was the only way to get down to them and measure them directly.
9. Hewlett and Anderson, *The New World*; Bruhèze, *Political Construction*; Bruhèze, "Closing the Ranks"; Rainger, "A Wonderful Oceanographic Tool"; Hamblin, *Poison in the Well*; Hamblin, "Let There Be Light"; Hamblin, *Arming Mother Nature*.
10. Stommel, "The Anatomy of the Atlantic"; Stommel suggested that radioactive waste disposal was the major reason to study currents, despite the fact that steamships had made the sailing issues obsolete. However, one should not make too large a point of this, for radioactive waste disposal was a motivation that could be discussed in public, whereas most Navy interests were classified.

11. While the German *Meteor* results had suggested a decadal time scale for midlayer circulation, and that Antarctic bottom water was no more than three hundred years old, carbon-14 measurements by Lamont geochemist Lawrence Kulp suggested a much more sluggish environment, one in which "tropical Atlantic bottom water descended from the surface at a high altitude as long as 1800 years ago." Quoted in memo by Robert Deitz to Bill Menard, "Program to Measure Deep Ocean Currents at NIO [National Institute of Oceanography]," 9 March 1955, AC-09, Folder 22, "Activities ONR, Washington (2/2) 1954–1957," WHOI-ODR. See also Stommel, "Anatomy of the Atlantic." With attention focused on short-lived fission products, such as cesium-137, with a half-life of thirty-three years, it was plausible that the wastes could decay to stability in the deeps before ocean circulation moved them back toward the surface. See also Joseph, *Report of Meeting*. For historical perspective, see Hamblin, *Poison in the Well*.

12. Heezen et al., "Shaping"; Heezen and Hollister, *The Face*; Nowell et al., "High Energy Benthic." Hollister's discovery was highly ironic in light of his later advocacy of deep-sea radioactive waste disposal. See chapter 8.

13. Munk, "On the Wind-Driven"; Munk and Gossard, "On Gravity Waves."

14. Stommel, "Abyssal Circulation" (1957); Stommel, "A Survey"; Stommel, "The Abyssal Circulation" (1959); Stommel, "Large-Scale Oceanic Circulation"; Stommel and Arons, "On the Abyssal Circulation"; Stommel and Arons, "On the Abyssal Circulation II"; Stommel et al., "Some Examples." Munk and Stommel subsequently improved prediction of ocean circulation by adding friction and detailed wind data. This work explained how wind-driven circulation leads to intensification of currents on the western sides of oceans in the Northern Hemisphere (eastern sides in the Southern Hemisphere), helping to explain phenomena such as upwelling currents and El Niño events.

15. Greatbatch and Lu, "Reconciling the Stommel Box Model." For an update on thermohaline circulation and ocean stratification, see also Vallis, "Large-Scale Circulation." For a technical discussion of thermohaline variability, see Paulo Cessi, "Lecture 5: Thermohaline Variability," 2001, http://www.whoi .edu/fileserver.do?id=21419&pt=10&p=17292.

16. As well as in the ancient world. See Peterson et al., "Early Concepts"; Deacon, *Scientists and the Sea*.

17. Peterson et al., "Early Concepts." On Franklin's mapping of the Gulf Stream, see Chaplin, "Knowing the Ocean."

18. McConnell, *No Sea Too Deep*. See also Deacon, *Scientists and the Sea*, 180–83.

19. Defant, *Physical Oceanography*, 575.

20. Mills, "Discovery," says that Lenz's work was ignored for nearly three decades until unearthed by English geologist Joseph Prestwich, then it became quite influential. Defant credits Humboldt and Lenz as a major influence on German oceanographers. Defant, *Physical Oceanography*, 575–76.

21. Discussed by Defant, *Physical Oceanography*, 575; Maury, *Physical Geography*, 171.

22.   On Rennell, see Deacon, *Scientists and the Sea*; Pollard and Griffiths, "James Rennell"; Bravo, "James Rennell"; Gould, "James Rennell's View"; Mills, "Discovery."

23.   Deacon, *Scientists and the Sea*, 229–31; Peterson et al., "Early Concepts."

24.   Deacon, *Scientists and the Sea*, 229–31.

25.   Deacon is known for her dispassion, but even she calls it "incredible" that repeated critiques of the 4° theory, including by "two of the most influential writers [of the] 19th century, Alexander von Humboldt and … Matthew Fontaine Maury, should have had so little apparent effect." She attributes this to the continued use of unprotected thermometers, which (we now know) overestimated deepwater temperatures by failing to correct for pressure effects. Deacon, *Scientists and the Sea*, 280.

26.   Some thought the deep oceans might contain "living fossils"—creatures thought to be extinct but perhaps surviving in remote depths, an idea later used to great effect by science-fiction writers. On the living fossils debate, see Rudwick, *Bursting the Limits*; Mayor, *Fossil Legends*; Mayor, "Suppression of Indigenous Fossil Knowledge."

27.   Deacon, *Scientists and the Sea*, 306. See also Mills, "Discovery"; Rozwadowski, *Fathoming the Ocean*.

28.   Deacon, *Scientists and the Sea*, 311.

29.   Jeffreys, "Deep-Sea Dredging"; Rozwadowski, "Fathoming."

30.   Deacon, *Scientists and the Sea*, 313.

31.   Ibid., 315–16. Carpenter further tested his theory in the Straits of Gibraltar, where he employed an age-old technique to detect a deep countercurrent: a drag constructed from a wicker basket—with an attached iron cross and sailcloth vanes hanging from the cross—was lowered from a small boat. At a depth of about 100 fathoms, the boat moved slowly with the surface current, but as it approached 250 fathoms, it came to a near standstill. The drag evidently had become anchored in the undercurrent. Still, it was not clear whether the Straits—part of a closed basin subject to intense evaporation—were perhaps an anomaly, not comparable with the open ocean.

32.   See, e.g., Fleming, "Climate Dynamics."

33.   Deacon, *Scientists and the Sea*, chap. 15. See also Deacon, "Wind Power." It seems to be a pattern that scientists reject based on quantitative insufficiency phenomena that are later accepted. Alfred Wegener thought it would help his case to offer a variety of possible mechanisms. In fact, it muddied the waters and undermined his central empirical claim, which was the fact of moving continents (rather than the mechanism by which they moved). I take this as a lesson that if one has good data for a surprising phenomenon, one should make the case and leave others to explain it. On the continental drift debate, see Oreskes, *Rejection of Continental Drift*; Frankel, *Continental Drift*. On trusting numbers more than we should, see Porter, *Trust in Numbers*.

34.   Imagine the ocean as a series of layers. The surface layer is dragged along by the prevailing winds but deflected slightly by the Coriolis effect. The

next layer down is subject to frictional drag by the layer above, and again, deflected by the Coriolis effect, and so on down, until the drag effect wears out. The net effect is the Ekman spiral. In practice, the ocean is not a series of perfect layers—eddy viscosity and friction with the seafloor complicate matters—but Ekman's model nevertheless accounted for observed relations between wind and current directions in open-ocean environments.

35. On Bjerknes and the Bergen school of meteorology, see Friedman, "Appropriating the Weather"; Friedman, "Expeditions," 17–27; Benson and Rehbock, *Oceanographic History*; Fleming, *Inventing Atmospheric Science*.

36. McConnell, *No Sea Too Deep*.

37. Spiess, *The Meteor Expedition*.

38. Mills, "Physische Meereskunde"; see also Lüdecke and Summerhayes, *The Third Reich*.

39. Lenz, "Aspirations"; Mills, "Physische Meereskunde."

40. Wüst, "Major Deep Sea Expeditions."

41. Brennecke, "Die ozeanographischen."

42. Mills, "Physische Meereskunde."

43. Ibid., 56.

44. Merz and Wüst, "Die atlantische vertikalzirkulation."

45. Brennecke responded that he had depended on the German results as more reliable. The Miller-Casella protected thermometer had been a major improvement but was still plagued by the problem of re-equilibration as it was raised. In the 1870s, this had been addressed by the invention of the reversing thermometer, in which a glass tube housing mercury has a narrow constriction above the mercury bulb. At the desired depth, the thermometer is inverted. The mercury thread breaks at the point of constriction and slides down the tube, which is calibrated in the reverse direction. When the thermometer is brought up, it reveals the minimum temperature reached. The new thermometer, used on *Deutschland*, was accurate to 0.01°, an order of magnitude improvement. In any case, Merz and Wüst were giving the British far more credit than was due, because even if their data could have been interpreted as revealing deep circulation they hadn't been interpreted that way. See Mills, "Physische Meereskunde"; McConnell, *No Sea Too Deep*.

46. Mills, "Physische Meereskunde," 56–58.

47. Ibid. Mills thinks that Merz never understood the difference—which, if true, is weird given that Merz had studied with Helland-Hansen—as he notes—see his "Deep Sea Expeditions," 70n12. His view is that Wüst had to relearn the dynamic method in the 1920s, that his earlier exposure to it "did not take." Mills, email communication with author, 28 April 2005, and comments on an early draft of this chapter.

48. Mills, "George Deacon."

49. Lenz, "Aspirations of Alfred Merz," 120. According to historian Walter Lenz, Defant was recommended by Sverdrup to succeed him as director of Scripps

in 1946 but rejected because he (Defant) did not speak English. Defant was also the teacher of Lodewyk Lek (chapter 1), who translated the second volume of his *Physical Oceanography* into English. See also Clowes and Deacon, "Deep Water Circulation"; Deacon, "Hydrology"; Deacon, "Note on the Dynamics"; Sverdrup et al., *The Oceans*; Warren and Wunsch, *Evolution*; Spiess, *The Meteor Expedition*; Deacon, *Scientists and the Sea*; Mills, "Physische Meereskunde"; Mills, "Socializing Solenoids"; Mills, "Oceanography"; Mills, "*De Motu Marium.*"

50.   Defant, "Physical Oceanography," 575.

51.   Ibid., 576.

52.   This is a broad generalization: tropical surface waters can be fresher than subtropical ones because of heavy rains in the tropics, and northern polar surface waters are saltier than they would otherwise be because of the removal of fresh water by sea ice. Globally the freshest waters are generated in the Southern Ocean, around Antarctica.

53.   Defant, "Physical Oceanography," 661.

54.   This is another notable (and illogical) pattern in the history of science: interpreting the absence of evidence as evidence of absence. This comes up again in chapter 9.

55.   Storch and Hasselman, *Seventy Years*, 2010.

56.   Sverdrup et al., *The Oceans*, 395.

57.   Ibid., *The Oceans*, 392.

58.   Henry Melson Stommel (1920–1992) Papers, 1946–1996, MC-06, "Guide to Papers of Henry Melson Stommel," WHOI-Stommel. See also Marine Advisory Service Newsletter, July–August 1979, no. 74, MC-06, Folder 3, "Correspondence 1979," WHOI-Stommel. This was a survey of alums of the Graduate School of Oceanography at the University of Rhode Island, so perhaps there was an East Coast bias. Maurice Ewing came second; Matthew Fontaine Maury tied with Harald Sverdrup for third.

59.   William Stelling von Arx papers, 1942–1977, MC-24, "Correspondence 1958," WHOI-von Arx. See also Arnold Arons, "The Scientific Work of Henry Stommel," in Warren and Wunsch, *Evolution*, xvi; Veronis, "A Theoretical Model, xix.

60.   Arons, "Henry Stommel," xvi.

61.   Arnold Arons, personal communication with author, September 1999.

62.   Stommel to Wunsch, 23 September 1990, MC-06, Box 4, "Correspondence 1990," WHOI-Stommel.

63.   Memo, Stommel to McGilvray and Allen, 20 April 1954, MC-06, Folder 2, "Correspondence 1947–54," WHOI-Stommel.

64.   Wunsch, *Henry Stommel.*

65.   Stommel to Russ Davis at SIO, 27 February 1990, MC-06, Folder "Correspondence 1990," WHOI-Stommel.

66.   Stommel, MC-06, Folder "The Royal Society, Answers to a Questionnaire

on the Occasion of Being Elected a Foreign Member of the Royal Society,"
WHOI-Stommel.

67. Ibid.

68. Oreskes, "Laissez-tomber."

69. On the underwater photography work, see Box 65, MEP.

70. Stommel, MC-06, Folder "The Royal Society, Answers to a Questionnaire
on the Occasion of Being Elected a Foreign Member of the Royal Society,"
WHOI-Stommel. It is frequently stated that Stommel was a conscientious
objector and took the job because it would protect him from military service,
but I have found no documentation to confirm or deny this. See, e.g., Veronis, "A Theoretical Model."

71. Arons, "Henry Stommel"; speech by Arnold Arons, (1981) MC-06, Folder 4,
WHOI-Stommel. See also Swallow and Worthington, "Measurements."

72. MC-06, Folder "Notes for Acceptance of Bowie Medal, 1982," WHOI-
Stommel.

73. Arons, "Henry Stommel," xv; Veronis, "A Theoretical Model," xix. See also
Swallow and Worthington, "Deep Currents."

74. Columbus Iselin to Henry Stommel, 30 April 1950, MC-06, Folder "Correspondence 1947–1954," WHOI-Stommel.

75. Munk had also taken up the question of western intensification. See Munk,
"Solitary Wave Theory"; Munk, "Ocean Circulation." In an elegant and more
complete mathematical treatment, Munk had accurately deduced the velocity of the Gulf Stream (to the correct order of magnitude: 10^6 m^3/sec),
proving that the major features of the North Atlantic circulation could be
explained by wind stress acting on the surface. See Stommel, "Gulf Stream."

76. Columbus Iselin to Henry Stommel, 30 April 1950, MC-06, Folder "Correspondence 1947–1954," WHOI-Stommel. On the rejection, see Stommel MC-06, Folder "The Royal Society, Answers to a Questionnaire on the
Occasion of Being Elected a Foreign Member of the Royal Society," WHOI-
Stommel; Veronis, "A Theoretical Model," which affirms that Stommel believed Sverdrup had rejected him because of his popular book. It *is* peculiar
that he was rejected, and it is also possible that the rejection involved loyalty
issues. In 1949, as a visitor to SIO, Stommel was asked to sign the University of California loyalty oath. He did, but he appended a long letter that he
requested be attached to the oath in his file expressing his misgivings with
"the motives behind" the oath. He had signed it, he stated, in good faith, because as a matter of fact he was not a member of the Communist Party, but
he felt "an unpleasant sensation of having done something unclean." Stommel to Sproul, 3 September 1949, MC-06 Folder 2, "Correspondences Robert
Gordon Sproul, 1947–1954," WHOI-Stommel.

77. Stommel's unfavorable impressions were not dispelled in subsequent years.
Munk, he agreed, was brilliant, but Stommel felt that the balance between
theoretical and empirical work at Scripps was skewed and that Munk's highly

theoretical book *The Rotation of the Earth* (1960), written with geophysicist Gordon MacDonald, would be of little help to "people on ships." Stommel to Paul Fye, 10 February 1959, AC-09.5, Folder 24, "Personnel, Stommel, Henry 1953–1957," WHOI-Stommel. (The tuna commission quotation is from the same letter.)

78. AC-09, Folder 54, "Contract Reports N6onr-27701 (Basic Task) 1950," WHOI-ODR. The date was 1 July 1946.

79. AC-09, Folder 80, "Summary Report, Contract N6onr-277, Task Order I, May 2, 1951." WHOI-ODR. See also AC-09, Folder 54, "Contract Reports N6 Nonr 27701 (Basic task) 1950," WHOI-ODR; Letter, June 15, 1951, AC-09, Folder "Activities, Naval, 1959," WHOI-ODR.

80. Schlee, *On the Edge of an Unfamiliar World*; Weir, *Ocean in Common*.

81. Weir, *Forged in War*. Perhaps the question of the balance between physical and biological science was less fraught at Woods Hole than it was at Scripps, because biological science had a secure home at the Marine Biological Laboratory just across the road.

82. "WHOI 58-10 Military Defense Oceanography Conducted during the Period January 1, 1957–August 31, 1957, WHOI, February 1958," AC-09.4, Folder 37, "Grants and Contracts, 1951–1956," and Folder "Grants and Contracts Nonr-769, 1954–1957," WHOI-ODR. For additional discussions of ONR Contracts, AC-09.4, Various Folders on ONR Contracts, WHOI-ODR.

83. Stommel, "The Abyssal Circulation" (1957).

84. Ibid.

85. This is the opposite of what most people presumed then and what many still think even now: that cold polar waters move at depth away from the poles. Stommel holds that in western boundary currents, deep cold water moves away from poles, but everywhere else it diffuses upward and outward toward the poles.

86. Stommel, "The Abyssal Circulation" (1957), 734.

87. Stommel, "A Survey of Ocean Current Theory," 163.

88. Ibid.

89. Ibid.

90. Ibid., 161–63.

91. "WHOI 58-10 Military Defense Oceanography Conducted during the Period January 1, 1957–August 31, 1957, WHOI, February 1958," AC-09.4, Folder 37, "Grants and Contracts, 1951–1956," and Folder "Grants and Contracts Nonr-769, 1954–1957," WHOI-ODR. See also Stommel et al., "Some Examples," 179–80.

92. Stommel et al., "Some Examples," 179–80.

93. In this sense they played a similar role to physical models built by geologists attempting to explain the origins of mountains, See Oreskes, "From Scaling to Simulation."

94. Stommel, "A Survey of Ocean Current Theory."

95. Ibid.

96. Ibid.

97. Eckart, *Principles and Applications of Underwater Sound,* 11 and fig. 2 See also Weir, *Ocean in Common.*

98. Cloud, "World in a Barrel."

99. Weir, *Ocean in Common;* Drew and Sontag, *Blind Man's Bluff;* Craven, *The Silent War.*

100. Oreskes, "Laissez-tomber."

101. Ibid.

102. Stommel, "Anatomy of the Atlantic."

103. Ibid.

104. This was not simply a personal difference between Stommel and Sverdrup; it reflected broader patterns in which Europeans tended to emphasize firm claims and Americans preferred to pursue "multiple working hypotheses" and be antiauthoritarian about them. Oreskes, *Rejection of Continental Drift.*

105. This is an interesting critique in light of the history of plate tectonics, where openness to speculation proved crucial to theory development. Lawrence Morley later recalled that his own work on seafloor magnetized stripes was inspired by the work of Mason and Raff, who described what they had found but admitted they did not have an explanation for it. This enabled Morley to believe that, despite being young and unknown, he might be able to make an important contribution by offering an explanation, even a highly speculative one. See Morley, "The Zebra Pattern."

106. Stommel, MC-06, Folder 6, "Notes for Acceptance of Bowie Medal, 1982," WHOI-Stommel. This was not merely hindsight; Stommel had made a similar point in an article in 1948: "A glance at a treatise of oceanography such as *The Oceans* ... at first gives the impression of great completeness, but upon closer examination the reader will discover how fragmentary and incomplete the theoretical structure of physical oceanography really is." Stommel was not blaming Sverdrup for the state of oceanographic theory, but he was criticizing him for presenting it as more robust than it really was. Stommel, "Theoretical Physical Oceanography," reprint, MC-06, Folder 34: "Originals Not Used 1945–1948," WHOI-Stommel.

107. Stommel, "Abyssal Circulation" (1957), 733–34 (italics added).

108. Ibid.

109. Robinson and Stommel, "Oceanic Thermocline."

110. Stommel, "The Abyssal Circulation," 1959.

111. Ibid.

112. Stommel and Robinson, "Oceanic Thermocline," 296; see also their discussion on 306.

113. Stommel, "The Abyssal Circulation," 1959, 82.

114. Frank Press, personal communication with author, August 2000; Naomi Oreskes to Arnold Arons, 7 September 2000.

115. Fye: Folder "Ryder/Ind: Arons, Arnold, 1950," Arons to G. K. Hartmann and P. M. Fye, Explosives Research Department, NOL (Naval Ordnance Laboratory), 18 April 1950, AC-09.4, Folder 35, WHOI-ODR.

116. Fye: Folder "Ryder/Ind: Arons, Arnold, 1950," Arons to Frank Ryder, 24 April 1950, AC-09.4, Folder 35, WHOI-ODR. On Philip Kuenen, see discussion in Oreskes, *Rejection of Continental Drift*.

117. Smith: Folder "Atomic Bomb Attack, 1950," Arons to Smith, 3 October 1950, AC-09.3, Folder 15, WHOI-ODR.

118. The crucial papers are Stommel and Arons, "On the Abyssal Circulation," 1960; Stommel and Arons, "An Idealized Model—Parts I and II," 1960; Stommel, "Thermohaline Convection," 1961; Bolin and Stommel, "On the Abyssal Circulation—Part IV," 1961; Arons and Stommel, "On the Abyssal Circulation—Part III," 1967; and Stommel and Arons, "On the Abyssal Circulation—Part V," 1972. Why part 4 precedes part 3 is a question whose answer has eluded me.

119. This can be confusing because most of us think of a sink in layperson's terms and imagine water flowing downward. Stommel is using the term here in the technical sense of anything that receives or takes up something, as in the biosphere being a sink for atmospheric carbon. Thus, the abyssal ocean is a sink for cold water, even if that water is moving upward.

120. Stommel and Arons, "On the Abyssal Circulation—Part II."

121. And very close to values now generally accepted. On the issue of radioactive waste disposal in the ocean, see Hamblin, *Poison in the Well*.

122. Stommel and Arons, "On the Abyssal Circulation—Part II," 224. The use of the word *cause* here seems to me a bit confusing, because really they are saying this is the *proof* of abyssal circulation; without it there would be no thermocline. But since Stommel and Arons use the word *cause*, I have retained it.

123. Given current interest in the potential effects of global warming on the thermohaline convection, this a startling paragraph. It suggests that the commonly held view that the thermohaline convection is excruciatingly sensitive to small changes in Arctic cooling may be incorrect. However, considerable work has been done since 1960, so perhaps Stommel and Arons were wrong about this. In any case, the "re-shuffling of the western boundary currents" could have enormous climatological impacts on populations living in areas affected by those currents. See Stommel and Arons, "On the Abyssal Circulation—Part II," 225.

124. Stommel, "Thermohaline Convection."

125. Hansen et al., "Ice Melt."

126. These predictions are discussed by Warren, "Arnold Arons," and in several essays in Warren and Wunsch, *Evolution of Physical Oceanography*.

127. Warren, "Arnold Arons."

128. Stocker, *The Ocean in the Climate System*; Rahmstorf, "The Current Climate"; Wunsch and Ferrari, "Vertical Mixing"; Vallis, *Atmospheric*; Kuhlbrodt et al.,

"On the Driving Processes"; Weaver et al., "Stability of the Atlantic"; Dijkstra, "Ocean Currents"; Stocker, "Past and Future."

129. Roland Emmerich, *The Day after Tomorrow* (20th Century Fox, 2004).

CHAPTER THREE

1. This term is widely used by scientists who know about these events and appears as the label on the relevant file folders in the WHOI archives.
2. Inman, "Oral History."
3. Weir, *Ocean in Common*.
4. Veronis, email communication with author, 21 September 2016.
5. Arons, personal communication with author, September 1999.
6. "North Atlantic Circulation and Water Mass Formation," final report by Valentine Worthington, courtesy of WHOI archives (no report number provided).
7. Weir, "Surviving the Peace."
8. Schlee, *On the Edge*, 273; Raitt and Moulton, *First Fifty Years*, 109. I have seen this quoted many times, although no one I have come across provides a full reference. Raitt and Moulton attribute it to Frank Lillie, the committee chair, in a memorandum to the committee members but give neither a published source nor an archival reference. It was precisely because the government had little interest in oceanography that NAS took it up and sought out private patrons.
9. Rozwadowski, *The Sea Knows*; Rozwadowski, *Fathoming the Ocean*.
10. Rozwadowski, *The Sea Knows*.
11. Miller, *Dollars for Research*; Kohler, *Partners in Science*.
12. Burstyn, "American Oceanography."
13. Schlee, *On the Edge*, 282. See also Weir, *Ocean in Common*.
14. Schlee, *On the Edge*, 282. See also Weir, *Ocean in Common*.
15. Schweber, *In the Shadow*.
16. Quoted without a citation in Greenberg, *Politics of Pure Science*, 136. This book is as good a brief discussion of the question as is available.
17. Von Arx to Hans Panofsky, 12 May 1959, MC-24 Folder "Corr. 1959," WHOI-von Arx. This particular comment was in the context of polar research but seems to encapsulate as well how he viewed the direction of scientific research in general.
18. In this regard, their views were similar to Denis Fox and Claude ZoBell at Scripps (chapter 1). Given the work of Michael Polanyi at this time, and the debates swirling in Europe over socialist planning of science, it is not surprising that at least some scientists would have been pondering this issue.
19. On Malkus, see Fleming, *Joanne Simpson*.
20. Joanne Malkus to Arnold Arons, 24 February 1954, MC-24, Folder "Corp 1958 [*sic*]," WHOI-von Arx. Her insistence on this point might be because others at Woods Hole were, in fact, finding their work being directed. Historian

Eric Mills worked at Woods Hole in the late 1950s with biologist Howard Sanders, who at that time worked on inshore benthic ecology. In the summer of 1959, Sanders told Mills that he was being redirected into deep-sea biology. In 2005 Mills recalled, "Even with Howard's even, optimistic temperament, I got the impression that he was unhappy with this redirection of his work (of course he went on to fame if not fortune as a deep-sea biologist)." Eric Mills, email communication with author, 28 April 2005. This is a good example of how "redirection" did not mean bad science—one might end up doing good and interesting science—but it was something different from what one would otherwise have done.

21.   The other two in the United States were oceanographer Mary Sears and geophysicist Betty Bunce. In Europe, seismologist Inge Lehmann had gained attention for her work demonstrating that the Earth had a liquid outer core and solid inner core.

22.   Gudemann, "Dr. Joanne Malkus." See also Simpson, "Meteorologist." She later divorced Willem and, as Joanne Simpson, became the first woman president of the American Meteorological Society. After she left Woods Hole, she taught for several years at UCLA and then spent the bulk of her later career as chief scientist for meteorology at the Goddard Space Flight Center. Among numerous awards, she received the Carl-Gustaf Rossby Research Medal of the American Meteorological Society in 1983. Some biographical profiles claim that she was the first woman (ever) to receive a PhD in meteorology. I have not been able to verify this; more likely she was the first woman in the United States to do so.

23.   Henry Stommel spent the spring of 1946 at the University of Chicago attending Rossby's lectures. Stommel's notes are a thing of beauty, but he described Rossby in his notebook "vignettes" as "poorly organized" as a lecturer, with "limited material." But we already know that the greatest scientists are not necessarily great classroom teachers. MC-06 30: "Notebook Pages 1946–1957 (1 of 1)," WHOI-Stommel.

24.   Simpson, "Meteorologist," 65. See also Sullivan, "Joanne Malkus Simpson."

25.   Sullivan, "Joanne Malkus Simpson." It will be a fine day when a woman scientist can be lauded for her contributions as a scientist, full stop.

26.   On the history of weather modification research, including cloud formation and seeding, see Fleming, *Fixing the Sky*.

27.   Von Arx would go onto write a widely used textbook, *Physical Oceanography*.

28.   Faller to Von Arx, 11 September 1957, MC-24, Folder "Correspondence 1957," WHOI-von Arx.

29.   Von Arx to Alan Faller, 16 September 1957, MC-24, Folder "Correspondence 1957," WHOI-von Arx.

30.   Faller to von Arx, 5 February 1958, MC-24, Folder "Correspondence 1958," WHOI-von Arx.

31.   William von Arx to Al Faller, 20 February 1958, MC-24, Folder "Correspondence 1958," WHOI-von Arx.

32. Mukerji, *Fragile Power*.

33. Von Arx to Edward Sulkin, 11 November 1958, MC-24, Folder "Correspondence 1958," WHOI-von Arx. The notion of being other-directed was developed in the best-selling book by sociologist David Riesman (with Ruell Denney and Nathan Glazer), *The Lonely Crowd*, first published in 1950. Riesman argued that as American society became more competitive, more institutionalized, and more bureaucratized, other-directed personalities were favored. Von Arx presumably thought that this did not bode well for scientists who wished to be autonomous and "inner-directed." On Riesman, see "David Riesman, Sociologist Whose 'Lonely Crowd' Became a Best Seller, Dies at 92," *New York Times*, 11 May 2002, https://www.nytimes.com/2002/05/11/books /david-riesman-sociologist-whose-lonely-crowd-became-a-best-seller-dies -at-92.html. Sociologist Todd Gitlin argues that Riesman most valorized autonomy; if so, then von Arx and his colleagues would have seen him as an intellectual ally. See Todd Gitlin, "David Riesman, Thoughtful Pragmatist," *Chronicle of Higher Education*, 24 May 2002, https://www.chronicle.com /article/David-Riesman-Thoughtful/1781.

34. Stommel, "Future Prospects."

35. AC-09.4, F: "Administrative History (Paul M. Fye)," WHOI-ODR.

36. In contrast to Coles's statement, at the start of World War II, Iselin was actually the only full-time scientist at Woods Hole. Coles, "An Appreciation," cited in AC-09.4, Folder "Administrative History (Paul M. Fye)," WHOI-ODR.

37. William Von Arx to Al [probably Faller], 20 February 1958, MC-24, Folder "Correspondence 1958," WHOI-von Arx.

38. Joanne Malkus to A. B. Arons, 24 February 1958, MC-24, Folder "Correspondence 1958," WHOI-von Arx.

39. The official Woods Hole history glosses the palace revolt: "Actually there was a fair amount of disagreement over just how large and how business like the Oceanographic Institution should aspire to become, and many employees regretted the loss of the extreme informality and flexibility that had characterized the Institution's first 30 years. When questions of administrative policy came before the Board of Trustees, however, Fye's ideas were supported." "Administrative History (Paul M. Fye)," 75, WHOI-ODR. But the issue at stake was not simply formality versus informality; it was the fundamental purpose of the institution. And the issues did not merely "come before" the Board; rebellious staff insisted on them.

40. Groves, *Manhattan Project*, 140.

41. Memorandum for the Executive Committee, from Paul Fye, 20 January 1961, MC-06, Box 2, Folder "Correspondence 1960–1966," WHOI-Stommel.

42. Ibid.

43. Lewis, *Arrowsmith*; see also Rosenberg, "Martin Arrowsmith."

44. Fission was discovered in 1939; the atomic bomb was tested and then used in combat in the summer of 1945.

45. Memorandum for the Executive Committee, from Paul Fye, 20 January 1961, MC-06, Box 2, Folder "Correspondence 1960–1966," WHOI-Stommel, 3.

46. See, e.g., Bernal, *Social Function of Science*; Polanyi, *Personal Knowledge*. On Polanyi's neoliberal affiliations, see Mirowski and Plehwe, *Road from Mont Pèlerin*.

47. Memorandum for the Executive Committee, from Paul Fye, 20 January 1961, MC-06, Box 2, Folder "Correspondence 1960–1966," WHOI-Stommel.

48. Ibid.

49. Ibid., 3, emphasis added. There is extensive discussion of the Polaris Program in the papers of Maurice Ewing, particularly 154: Polaris. Several letters in this file are to or from Paul Fye.

50. Ibid.

51. Fye to J. Lamar Worzel, 11 April 1960, Box 91, "Polaris, 1960–1964," MEP. Worzel, Bruce Heezen, and others at Lamont embraced this work enthusiastically, and Fye may have been surprised that he did not find similar enthusiasm at Woods Hole.

52. Memorandum for the Executive Committee, from Paul Fye, 20 January 1961, MC-06, Box 2, Folder "Correspondence 1960–1966," WHOI-Stommel.

53. Ibid., 6.

54. Ibid., 7. The full text of his speech on the Miller Bill, HR 4276, can be found in the Wilson papers, Letter to Executive Council, 16 June 1961, and attached speech to the House Sub-Committee on Oceanography. HUG 4878.3, Woods Hole Oceanographic Institution, Box 46, EBWP. In fairness, one should note that this speech make a strong defense of individual initiative and creativity and against excess government control, so one can see why Fye might have taken offense at the staff's characterization of his position. In contrast, the excerpt he chose to use, talking about weapons systems, displayed a lack of resonance with or sensitivity to his staff's concerns.

55. Ibid., 7.

56. Report of the Scientific Advisory Committee of the Trustees, WHOI, 1954–55, AC-09, Folder "Exec: Scientific Advisory Committee 1954–1959," WHOI-ODR.

57. Iselin, 1 July 1958, AC-09.4, Folder 19: "Oceanography and the Navy," WHOI-ODR.

58. Ibid.

59. Ibid.

60. See, e.g., Malkus, *Air and Sea*, 88.

61. The events of the months that followed were so distressing to those involved that more than forty years later, when I approached some of the surviving participants, some found it difficult or even impossible to discuss. At WHOI, the files I found were labeled in such a way as to mark them from other matters—for example, von Arx's files labeled "Palace Revolt."

62. Interview with Elizabeth Stommel by Carl Wunsch, MC-06, Folder "Interviews of Elizabeth Stommel by Carl Wunsch, 1990," 1 of 3, WHOI-Stommel.

63. AC-09.5, Folder "Administrative History (Paul M. Fye)," WHOI-ODR, 75.
64. Stommel to Iselin, Spilhaus, and Wilson, 2 August 1960, MC-06, Folder "Correspondence 1960–1964," WHOI-Stommel, cover letter.
65. Ibid.
66. Ibid., 2.
67. Stommel, notes for the receipt of the Ewing Medal, 1977, MC-06, Folder "National Medal of Science 1989," WHOI-Stommel.
68. Stommel to Iselin, Spilhaus, and Wilson, 2 August 1960, MC-06, Folder "Correspondence 1960–1964," WHOI-Stommel, 3.
69. Ibid.
70. Ibid.
71. Ibid., 5.
72. Ibid.
73. Stommel to Fye, 13 February 1961, MC-06, Folder "Correspondence 1960–1964," WHOI-Stommel, 1–2.
74. Munk and MacDonald, *Rotation of the Earth*.
75. Stommel to Fye, 13 February 1961, MC-06, Folder "Correspondence 1960–1964," WHOI-Stommel, 1–2.
76. Ibid.
77. Ibid., 3.
78. I do not know why Rossby left WHOI; Ewing left to become director of his own institution.
79. Stommel to Fye, 13 February 1961, MC-06, Folder "Correspondence 1960–1964," WHOI-Stommel, 3.
80. Ibid., 3.
81. Although it is also true that the abundant funding and support he received was linked to the usefulness of much of his work to military matters (chapters 1 and 9).
82. Stommel to Fye, 13 February 1961, MC-06, Folder "Correspondence 1960–1964," WHOI-Stommel, 3.
83. Ibid.
84. See Wilson to Malkus, 6 March 1961, MC-06, Folder "Correspondence 1960–1964," WHOI-Stommel. In 1961, Joanne was hoping that she would return to WHOI. See Malkus to Stommel, 8 March 1961, MC-06, Folder "Correspondence 1960–1964," WHOI-Stommel.
85. Malkus to Fye, 28 February 1961, MC-06, Folder "Correspondence 1960–1964," WHOI-Stommel.
86. Malkus, "Large-Scale Interactions." Malkus also argued for the importance of high-quality data, but she argued that knowing which data to collect was the most important thing, and this could be answered only by better theoretical studies. Lack of data was not the principal obstacle to advancing understanding. She made the same claim at the end of the chapter: that the critical area was the fundamental physics of turbulent, heat-driven fluids.

87.   Malkus to Fye, 28 February 1961, MC-06, Folder "Correspondence 1960–1964," WHOI-Stommel.

88.   Ibid., 2. By focusing on the motions of turbulent fluids as the basic foundations of meteorology and oceanography, Malkus was following the program that Rossby (following his own mentor, Bjerknes) had taught her. By claiming that all major advances in nuclear science had been made by academic scientists, she perpetuated a myth fostered by those scientists, a myth that greatly discounted the essential efforts of engineers in taking the atomic bomb from blackboard to reality.

89.   Malkus, *Air and Sea.*

90.   Malkus to Fye, 28 February 1961, MC-06, Folder "Correspondence 1960–1964," WHOI-Stommel, 2.

91.   Sedimentologist Raymond Siever has written that when he was first admitted to the Committee on Oceanography at Harvard, in 1955, he couldn't figure out why its members included a physicist and communications engineer, Frederick Hunt, and a chemist involved in microwave spectroscopy, E. Bright Wilson. He soon learned of their role in World War II; Hunt on underwater sound transmission, Wilson on underwater explosives. Hunt was the man who gave the name to Project Artemis, discussed in chapter 5.

92.   Wilson, *Scientific Research*, 3.

93.   Ibid., 3–4.

94.   Malkus to E. Bright Wilson, 3 March 1961, MC-06, Folder "Correspondence 1960–1964," WHOI-Stommel, 1.

95.   Ibid.

96.   Ibid., 2.

97.   Malkus to Stommel, 8 March 1961, MC-06, Folder "Correspondence 1960–1964," WHOI-Stommel. Malkus copied her letter to Stommel and ten others, including Columbus Iselin and ONR chief scientist Emanuel Piore.

98.   Message from Stommel to Melvin [Stern], handwritten on an internal WHOI memo, 25 January 1961, MC-06, Folder "Correspondence 1960–1964," WHOI-Stommel.

99.   Wilson to Fye, 8 February 1961, HUG 4878.3, Woods Hole Oceanographic Institution, Box 46, EBWP, on 1.

100. Ibid., 1–2 of letter.

101. Ibid.

102. Ibid.

103. Ibid., 2 of letter.

104. Memo to the Chairman of the Board of Trustees, 9 November 1961, HUG 4978.3, Woods Hole Oceanographic Institution, Box 46, EBWP, copies also in WHOI-von Arx papers and WHOI-Stommel.

105. William von Arx, memo, "Institution Policies and Morale," 5–7 November 1961, MC-24, Folder "Palace Revolt, 1961," WHOI-von Arx . Copy also in HUG 4878.3, Woods Hole Oceanographic Institution, Box 46, EBWP.

106. Ibid.
107. Others who wrote included William Schevill, Dayton Carritt, Vaughan Bowen, and William Schevill. See Various Papers, HUG 4878.3, Woods Hole Oceanographic Institution, Box 46, EBWP.
108. George Veronis to Noel McLean, 28 November 1961, MC-06, Folder "Correspondence 1960–1964," WHOI-Stommel. Copy also in HUG 4878.3, Woods Hole Oceanographic Institution, Box 46, EBWP.
109. Bostwick Ketchum to James S. Coles, 22 November 1961, Copy also in HUG 4878.3, Woods Hole Oceanographic Institution, Box 46, EBWP.
110. It is interesting to note that Brackett Hersey was very well known and respected at that time, but a Google Scholar search shows that only a handful of his papers were ever widely cited, and few, if any, are still cited today.
111. Bostwick Ketchum to James S. Coles, 22 November 1961, HUG 4878.3, Woods Hole Oceanographic Institution, Box 46, EBWP.
112. Von Arx to Noel McLean, 5 December 1961, MC-24, Folder "Palace Revolt, 1961," WHOI-von Arx. Copy also in HUG 4878.3, Woods Hole Oceanographic Institution, Box 46, EBWP.
113. Ibid.
114. Ibid.
115. Transcript of notes for staff meeting, 12 December 1961, MC-24, Folder "Palace Revolt, 1961," WHOI-von Arx.
116. Ibid.
117. It's not clear if this is a veiled attack on Malkus; it may well be. Or Fye may have thought that the men were getting overwrought, too.
118. Transcript of notes for staff meeting, 12 December 1961, MC-24, Folder "Palace Revolt, 1961," WHOI-von Arx.
119. James S. Coles to E. Bright Wilson, 22 November 1961, HUG 4878.3, WHOI, Box 46, EBWP.
120. Ibid.
121. Brooks to McLean, 17 December 1961, HUG 4878.3, WHOI, Box 46, EBWP.
122. Wilson to Bostwick Ketchum, 22 November 1961, HUG 4878.3, WHOI, Box 46, EBWP.
123. Various documents in HUG 4878.3, Woods Hole Oceanographic Institution, Box 46, EBWP.
124. Untitled typescript, presentation for meeting with the trustees, 19 December 1961, in MC-24, Folder "Palace Revolt, 1961," WHOI-von Arx. This author of the text is not named, but by a process of elimination it is clear that the prepared text was read by Bill Richardson; see letter to George Veronis, n.d., but replying to his of 19 December, in MC-24, Folder "Palace Revolt, 1961," WHOI-von Arx. A memo from McLean to the trustees says that presentations will be made by Richardson and Veronis, with supplemental remarks by Bostwick Ketchum. The trustees had asked that only two to four faculty attend, but the faculty rejected that suggested and "decided to come en masse" (about

fifteen of them). Memo, Noel McLean to James S. Coles et al., 15 December 1961, HUG 4878.3, Correspondence T-Z, Box 49, EBWP. From von Arx's notes it appears that he also spoke, and perhaps also Stommel.

125.   Untitled notes, stapled to George Veronis to Noel McLean, 28 November 1961, MC-06, Folder "Correspondence 1960–1964," WHOI-Stommel.

126.   Untitled typescript, presentation for meeting with the trustees, 19 December 1961, in MC-24, Folder "Palace Revolt, 1961," WHOI-von Arx. (These are not Von Arx's comments—a separate set with his name on is found in the folder—but those of another colleague who spoke that evening and gave von Arx a copy of his comments.)

127.   Untitled notes, but from content, there are clearly the notes of various faculty for the presentations at the meeting with the trustees, stapled to George Veronis to Noel McLean, 28 November 1961, MC-06, Folder "Correspondence 1960–1964," WHOI-Stommel.

128.   Untitled typescript, presentation for meeting with the trustees, 19 December 1961, in MC-24, Folder "Palace Revolt, 1961," WHOI-von Arx.

129.   Untitled notes, stapled to George Veronis to Noel McLean, 28 November 1961, MC-06, Folder "Correspondence 1960–1964," WHOI-Stommel.

130.   Typescript, from memory W.S. v. A [William S. von Arx], 18 December 1961, MC-24, Folder "Palace Revolt, 1961," WHOI-von Arx.

131.   Von Arx to Veronis, n.d., but replying to his letter of 19 December, MC-24, Folder "Palace Revolt, 1961," WHOI-von Arx.

132.   Typescript for comments to the meeting "Gentleman" (n.d., no name), MC-24, Folder "Palace Revolt, 1961," WHOI-von Arx.

133.   Noel B. McLean, Chairman of the Board of the Trustees, to Resident Research Staff, 19 December 1961, in MC-24, Folder "Palace Revolt, 1961," WHOI-von Arx. Copy also in HUG 4878.3, WHOI, Box 46, EBWP.

134.   Faculty notes for presentations at the trustees meeting, stapled to George Veronis to Noel McLean, 28 November 1961, MC-06, Folder "Correspondence 1960–1964," WHOI-Stommel.

135.   Untitled typescript, presentation for meeting with the trustees, 19 December 1961, in MC-24, Folder "Palace Revolt, 1961," WHOI-von Arx. Around the same time, a trustee argued strongly that they "should look for a director among the actively practicing oceanographers." He mentioned SIO's Walter Munk and Hopkins's Donald Pritchard, as well as Lamont's Maurice Ewing, and for internal candidates, von Arx, Stommel, and Ketchum, with the latter his preference. An active oceanographer would certainly have been more to the liking of the Woods Hole faculty who ran the palace revolt; however, that was not what the trustees decided. Alfred Redfield to "Spike Coles," 17 September 1957, AC-09, Folder "Exec: Selection Committee for Director, 1957–58," WHOI-ODR.

136.   Veronis to Noel McLean, 29 December 1961, HUG 4878.3, WHOI, Box 46, EBWP.

137. Redfield to McLean, 28 December 1961, HUG 4878.3, WHOI, Box 46, EBWP.

138. Ibid.

139. Ibid.

140. Transcript of notes for Staff Meeting, 12 December 1961, MC-24, Folder "Palace Revolt, 1961," WHOI-von Arx.

141. Typescript [Comments, Paul Fye to WHOI Staff], Staff Meeting, 12 December 1961, MC-24, Folder "Palace Revolt, 1961," WHOI-von Arx.

142. Malkus, *Air and Sea*, 88.

143. MC-06, Folder "Interviews of Elizabeth Stommel by Carl Wunsch, 1990," 1 of 3, WHOI-Stommel. Armi also says that several were fired; others deny that there were firings, saying people left on their own accord. The evidence I have found supports a middle ground: that people left "on their own accord" but after it was made clear that they were no longer welcome.

144. This is a fascinating essay, in which von Arx concludes that the high cost of going to sea makes it the inevitable fate of oceanography to be a "science in bondage." Accepting this, oceanographers should choose the applied problem that would best serve the interests of their science, whether fisheries, warfare, or mineral resources. See chapter 6. Von Arx, "A Science in Bondage," February 1965, MC-24, Folder "A Science in Bondage," 2, WHOI-von Arx. See also chapter 9.

145. Stommel to Robinson (unsigned and therefore perhaps unsent?), 26 April 1963, MC-06, Folder "Correspondence 1960–1964," WHOI-Stommel. Dean Ford was Franklin L. Ford; my distinguished colleague, Steven Shapin, was the Franklin L. Ford Research Professor of the History of Science.

146. Veronis to E. Bright Wilson, 1 February 1962, HUG 4878.3, WHOI, Box 46, EBWP. Wilson for his part dismissed the whole thing in hindsight as a kind of inverted Oedipal problem: "I don't really know what the basic trouble is, but I am convinced that there is some fundamental psychological ailment circulating around the Institution.... I know that some of it stems from the good old days when Iselin was the father image down there. I myself found this very comforting during the war. But we all have grown up now and should be able to get along without this kind of guidance." Wilson to George Veronis, 19 February 1962, HUG 4878.3, WHOI, Box 46, EBWP.

147. Von Arx, "Reflections," 15 July 1991, MC-24, Folder "Reflections of War and Postwar Years at WHOI," WHOI-von Arx.

148. Kevles, "Cold War," 263–64.

149. Draft memo, "Establishment of a Scientific Policy Committee of the Woods Hole Oceanographic Institution," AC-09, Folder "Exec: Scientific Policy Committee 1959," WHOI-ODR.

150. Draft memo, "Establishment of a Scientific Policy Committee of the Woods Hole Oceanographic Institution," AC-09, Folder "Exec: Scientific Policy Committee 1959," WHOI-ODR, emphasis added.

151. Ibid.

152. Stommel, "Hank's Introduction to Comments of Oceanographic Wives," n.d., MC-06, Folder "Interviews of Elizabeth Stommel by Carl Wunsch, 1990," 1 of 3, WHOI-Stommel.

153. Note for receipt of Bowie Medal, 1982, MC-06, Folder 6, WHOI-Stommel.

154. Stommel, notes for the receipt of the Ewing Medal, 1977, MC-06, Folder "National Medal of Science 1989," WHOI-Stommel. In an interview with Elizabeth Stommel, Carl Wunsch put the matter more strongly: "CW: Hank never had a kind word to say for him … ES: Yes, he one day came across the [Ewing] medal, which was stuck in the closet somewhere, he said 'Oh, we don't want this.' And he threw it in the waste basket, and you know I let it stay there. I thought, oh well, it's not my business. So I threw it out in the rubbish the next week." Interview with Elizabeth Stommel, MC-06, Folder "Interviews of Elizabeth Stommel by Carl Wunsch, 1990," 1 of 3, WHOI-Stommel.

155. Stommel, notes for the receipt of the Ewing Medal, 1977, MC-06, Folder "National Medal of Science 1989," WHOI-Stommel.

156. Stommel to Wunsch, 23 September 1990, MC-06, Folder "Correspondences, 1990," WHOI-Stommel.

CHAPTER FOUR

1. On the history of tectonics before plate tectonics, see Greene, *Nineteenth Century*. On plate tectonics from the participants' perspectives, see Oreskes and Le Grand, *Plate Tectonics*. A striking feature of many earlier histories of plate tectonics is that they ignored its military context. See, e.g., Frankel, "Continental Drift," and Allwardt, *Global Tectonics*.

2. The other two key lines of empirical evidence were, in chronological order, the terrestrial paleomagnetic evidence collected by European, mostly British, geophysicists—Jan Hospers, Keith Runcorn, P. M. S. Blackett, Ted Irving, and others—which in the 1950s demonstrated mobility of the continents relative to the poles, and the seismological evidence of earthquake foci and slip mechanisms along what were recognized in the 1960s as plate boundaries. See Oreskes and Le Grand, *Plate Tectonics*.

3. Mackenzie, *Inventing Accuracy*.

4. Ewing and Heezen, 1963, Bathymetric Studies Box 5: Hamilton, Gordon, on 1, BCHP.

5. Merton, "Science and Technology."

6. Merton, "Norms of Science."

7. For an account of secrets and science before the rise of institutional mechanisms to create public knowledge, see Eamon, *Science and the Secrets*. On the rise of institutions of public science, see Crosland, *Science under Control*; Morrell and Thackray, *Gentlemen of Science*; Shapin, *A Social History*.

8. Shindell, "Isotope Geochemistry." On scientists accommodating conditions of secrecy, and finding ways to work productively in military boundaries, see Westwick, *Secret Science*.

9. Galison, *Image and Logic*; Schweber, *In the Shadow*; Seidel, "Home for Big Science."

10. Michaels, "Sarbanes-Oxley," 1. For an extensive discussion of this issue in the contemporary context, see Michaels and Vidmar, *Sequestered Science*.

11. Gillispie, *Science and Secret Weapons*. Long and Roland, *Military Secrecy*, conclude that while military secrecy in relation to imminent battles was routine in the ancient and medieval world, secrecy about knowledge relevant to military technologies was the exception rather than the rule.

12. Graham, *Russian Experience*; Schmid, "Defining"; Schmalzer, *Peking Man*; Schmalzer, *Red Revolution*.

13. Westwick, "Secret Science"; Galison, "Secrecy in Three Acts." The use of classified journals is also discussed by Gusterson, *Nuclear Rites*, who argues that the scientists at Lawrence Livermore National Laboratory did not much care for this solution, as it ran "counter to the comfortable orality of knowledge circulation long established among the weapons scientists" (64). In this case, however, the objection is not secrecy per se, since that orality would presumably take place among cleared scientists, but convenience and efficacy.

14. Wright and Wallace, "Varieties of Secrets"; Roosth, *Synthetic*. This is not to say that these conditions of secrecy might not have other adverse effects, such as on collegiality, education, or public trust. See Krimsky, *Science in the Private Interest*.

15. Dennis, "Science and Secrecy Revisited," 1–2, and other papers in Reppy, *Secrecy and Knowledge Production*.

16. Sheila Jasanoff, "Transparency in Public Science," 42. See also Michaels, "Sarbanes-Oxley."

17. Shils, *Torment of Secrecy*, 188.

18. Schweber, *In the Shadow*.

19. Hamblin, "Pursuit of Science."

20. Wang, *American Science."*

21. Oreskes, "Presentist."

22. Proctor and Schiebinger, *Agnotology*.

23. I am cognizant of the fact that I am claiming something that some historians and sociologists of science would dismiss as impossible: to know what someone thought, or did not think. So let me clarify. When Bill Menard declares (chapter 5) in a letter to Teddy Bullard in 1966, "You will find me converted to hanging on the Vine. I finally took the time to look at the magnetic data in detail. The symmetry seems decisive. —B.," this establishes what Menard did *not* think before this point. Similarly, when we can document that Bruce Heezen advocated the expanding Earth hypothesis well into the 1960s, we know that he did not think that seafloor spreading was correct. This is the methodological basis of my argument: that the historical record can reveal the temporality of scientists' views and actions, and therefore what they did not think until certain points. Thoughts are a form of action,

and just as we have historical evidence regarding people's actions, we also have evidence regarding their thoughts.

24. Glen, *Road to Jaramillo*; Oreskes, *Plate Tectonics*. Until doing the work that led to this chapter, I also thought this. Over the years I heard it said many times, and until I did this work, I had no reason to doubt it.

25. James, "Harry Hammond Hess," 111–15.

26. McVay, "In Appreciation," 10–11, 16–17.

27. Hess, "Problem of Serpentinization"; Hess, "Further Discussion"; Hess, "Primary Banding"; Hess, "Gravity at Sea"; Hess and Phillips, "Orthopyroxenes"; Hess, "Geodesy and Geomorphology"; Hess, "Geodesy and Geomorphology"; Hess, "Essay Review"; Hess, "Pyroxenes." The geology of the Skaergaard intrusion was first described in detail in a three-hundred-page monograph by Wager and Deer, *Petrology*.

28. Hess, "Gravity at Sea," 332.

29. Vening Meinesz, "Gravity Expeditions."

30. Oreskes, *Rejection of Continental Drift*.

31. On the relation between topography and gravity, see McKenzie and Bickle, "Volume and Composition; on the assumptions in isostatic models of the 1920s and 1930s and their implications for tectonics interpretation, see Oreskes, *Continental Drift*.

32. Vening Meinesz and Wright, "Gravity Measuring Cruise"; see also Oreskes, "Weighing the Earth"; Oreskes, *Continental Drift*, 236–42.

33. Hess finished his PhD in 1932, taught at Rutgers for a year in 1932–1933, and then was a research associate at the Geophysical Lab of the Carnegie Institution of Washington, in 1933–1934, before being offered a position on the faculty at Princeton in 1934. See James, "Harry Hammond Hess." See also Field et al., *Gravity Expedition*, 1933. Vening Meinesz and Hess were assisted by T. T. Brown of the Naval Research Laboratory, which also provided the base station for the gravity measurements. The work was done aboard the USS *S-48*, commanded by Lieutenant Commander O. R. Bennehoff, and supported by the submarine tender USS *Chewink* under Lieutenant Commander G. A. Miller.

34. Field et al., *Gravity Expedition*, 118.

35. On the origins of the term, see Allwardt, "Tectogene Concept," 489n18.

36. The theoretical expectation was for the crust to return to gravitational equilibrium after being disturbed; the crust would rebound, for example, after being weighed down by a sedimentary or glacial load in much the same way a trampoline rebounds when the person jumping on it flies up into the air. If the crust were not rebounding, then something had to be holding it down. That something could be horizontal compression.

37. Field et al., *Gravity Expedition*, 32, original emphasis.

38. Ibid., original emphasis.

39. F. A. Vening Meinesz to Harry H. Hess, 16 March 1933, Box 29, Folder "1932," HHP.

40. Ibid.
41. Vening Meinesz et al., *Gravity Expeditions*, 133.
42. Ibid.
43. Ibid.
44. Ibid.
45. F. A. Vening Meinesz to Harry H. Hess, 22 September 1935, Box 21, unlabeled folder, HHP.
46. Holmes, "Earth Movements," 590, and cited by Umbgrove, "Short Survey," in Vening Meinesz et al., *Gravity Expeditions*.
47. Allwardt, *Modern Global Tectonics*, argues that Hess was influenced not only by Umbgrove's geology but also by his philosophy of science.
48. Kuenen, in Vening Meinesz et al., *Gravity Expeditions*.
49. Escher, "Volcanic Activity." From present perspectives, Escher's model lacks the asymmetry and depth penetration of the down-going crustal slab, which in theory could have been derived from Wadati's work on the distribution of deep-focus earthquakes but in fact was not widely considered until the work of Hugo Benioff in the 1950s. It also shows continuity between the basaltic oceanic crust and the lower continental crust; today, earth scientists do not believe that there is a basaltic layer in the lower continental crust.
50. Umbgrove, in Vening Meinesz et al., *Gravity Expeditions*, 167–68. See also Molengraaff, "Crustal Movements"; Molengraaff, "East Indian Archipelago"; and Molengraaff, "Continental Drift," 392. Also Molengraaff, *Geologie*.
51. Umbgrove, in Vening Meinesz et al., *Gravity Expeditions*, 172–73.
52. Discussed in Umbgrove, in ibid., 173.
53. Kuenen, in Vening Meinesz et al., *Gravity Expeditions*, 196.
54. This is one way in which tectonics theory has changed since the 1970s; today most scientists believe convection currents are the result, not the cause, of crustal motion: as cold dense oceanic slabs sink into the crust below lower-density continental crust, they drag along the plastic upper mantle below. This is known as "slab pull"; ridge push as accepted in the 1970s is now thought to only represent a small fraction of the driving force. Either way, crustal motions and convection currents are linked in a cause-and-effect relationship, and either way it shows that the argument that continental drift was rejected in the 1920s because of lack of a mechanistic explanation cannot be correct. Both ridge-push and slab-pull models were available at the time; see Oreskes, *Continental Drift*.
55. Kuenen, in Meinesz et al., *Gravity Expeditions*, 195.
56. Ibid.
57. Ibid.
58. Ewing, "Gravity-Measurements."
59. For more on Crary and Rutherford's contributions, see Lawrence, *Upheaval*.
60. Miller, "Geophysical Investigations."
61. Ewing et al., "Geophysical Studies."
62. Miller, "Geophysical Investigations," 811.

63. This is a nice example of how Karl Popper's theory of falsification fails in practice, even if it is appealing in principle. As Thomas Kuhn famously argued in *The Structure of Scientific Revolutions*, scientists are typically loath to abandon existing views without very strong and typically persistent reason. Ewing and colleagues did not view the theory of sunken continents as "falsified," they assumed that the sunken continent was out there somewhere and they had just failed to find it.

64. Oreskes, *Continental Drift*. See also Oreskes, *Why Trust Science?*

65. Maurice Hill, "Geological and Geophysical Investigations of the Floor of the Oceans and of Neighboring Seas Undertaken by the Department of Geodesy and Geophysics, Cambridge," November 1969, Folder B30-33 (1 folder), ECBP.

66. Transcript of a talk presented at Lowell Library, Columbia University, 19 March 1976, in honor of Maurice Ewing (deceased), Folder GT168, ECBP. The reference to sediments as a rubbish dump gives some indication of the great geophysicist's attitude toward geology.

67. Application to the Smithson Fund, 13 July 1937, Folder D344, ECBP.

68. Review of proposal of E. C. Bullard by W. B. Wright, 13 January 1938, Folder D347, ECBP.

69. Ibid.

70. On the empirical culture of American science in the 1920s and 1930s, see Schweber, "Empiricist Temper."

71. Reply by E. C. Bullard to review by W. B. Wright, 13 January 1938, Folder D347, ECBP.

72. Field et al., *Navy-Princeton*, 1933.

73. Ewing, "Gravity-Measurements," 69.

74. Ewing, "Gravimetric Methods," 68.

75. Hess, "Ultramafic Magma," 75.

76. Hess, "Gravity Anomalies"; Hess, "Peridotite Magma," and Griggs, "Mountain Building."

77. Hess generally referred to his Dutch colleague as Meinesz, but the latter always signed his name Vening Meinesz, and that is how Kuenen and Umbgrove referred to him. Hess, "Gravity Anomalies," 71.

78. Ibid. In hindsight, Vening Meinesz's data were largely an artifact of incorrect assumptions about the thickness of the earth's crust and the conditions of the oceanic crust.

79. Ibid., 75.

80. Vening Meinesz et al., *Gravity Expeditions*, 196.

81. Kuenen, "Isostatic Anomalies."

82. Oreskes, "Scaling to Simulation."

83. Hess, "Gravity Anomalies," 76.

84. Kuenen, "Isostatic Anomalies," 170–71.

85. Hess, "Gravity Anomalies," 76.

86. Hess, "Peridotite Magma," 333.

87. Oreskes, "Scaling to Simulation."

88. Earth scientists will recognize this range as consistent with contemporary GPS measurement. See, e.g., Reilinger, "GPS constraints."

89. Griggs, "Creep of Rocks," 639–40.

90. Ibid.

91. Oreskes, "Scaling to Simulation."

92. Griggs, "Creep of Rocks," 640.

93. Ibid.

94. Ibid., 641.

95. Ibid., 643.

96. Ibid., 647.

97. Ibid., 642.

98. Ibid., 616. See also Retrospective given to H. W. Menard, Box 69, Folder 3, SIO-Menard, and discussed in Oreskes, *Continental Drift*, 250–51.

99. For more on this, see Oreskes, *Continental Drift*, 250–51.

100. Hess, "Ultramafic Magma"; "Recent Advances in Interpretation of Gravity Anomalies and Island-Arc Structures," Advanced Report of the Commission on Continental and Oceanic Structure, HHP.

101. Griggs, "Creep of Rocks," 648.

102. Oliver, "Earthquake Seismology."

103. Professor Eldridge Moores says that Hess was already involved in antisubmarine warfare work, during which he met the commander under whom he would later serve in active duty. I have not found any historical documentation to affirm or refute this claim.

104. On why the Navy supported early gravity work, see Oreskes, "Weighing the Earth"; Weir, *Forged in War*. In the 1920s, Navy Secretary Curtis Wilbur was generally pro-science, particularly because he believed that the Navy would soon be run on petroleum, much as Rickover would later believe in the nuclear Navy. Below the level of secretary, the attitude of Navy officers involved appears to have been one of "letting" the scientists work on Navy vessels rather than imagining this had any practical value for the Navy. This began to change in the late 1930s, particularly with Ewing's work on sonar transmissions; see Weir, *Forged in War*; see also chapter 2. The use of gravity data for oil exploration at sea would come later.

105. On the early Navy interest in echo sounding of deep waters and the development of the sonic depth finder, see Weir, "Selling Bellevue."

106. Ibid.

107. In the early 1930s, Navy echo sounding had proved the presence of deeply incised submarine canyons off the coast of California and elsewhere. Such canyons had been suspected in the late nineteenth century but the idea largely ignored for lack of adequate data. When new, high-quality data proved the canyons were real—and dramatic in their aspect ratios—acrimonious debate broke out about their origins. The main interpreter of the California canyons was Scripps marine geologist Francis Shepard, who believed the canyons were drowned river valleys created during Pleistocene low-sea-level

stands and submerged when sea level rose due to melting glacial ice. His principal antagonist was Harvard professor Reginald Daly, who believed they were scoured by submarine mudflows. In 1936, Hess entered the fray with a paper in which he agreed with Shepard that the canyons were incised during a global change in sea level but disagreed about the cause. The low-sea-level stands were not caused by water tied up in glacial ice: the volumes were insufficient to produce the required sea-level drop. Hess suggested instead a change in the ellipticity of sea level, caused by a decrease in the rate of the Earth's rotation, perhaps due to collision with a "small extra-terrestrial body." Thinking as well about the evidence that had motivated Alfred Wegener, he argued that such changes in sea level could account for the intercontinental migration of flora and fauna without "the necessity of drifting continents or uplifting land bridges." Hess later admitted that this was an "outrageous" hypothesis, but he defended it at the time on the grounds that "all possible hypotheses must be kept in mind and the critical data bearing on all of them collected." See Shepard, "Underlying Causes"; Daly, "Origin"; Hess, "Submerged Valleys," 333; Hess, "Further Discussion," 583. Shortly after these papers were published, Philip Kuenen designed the experiment that demonstrated that turbid mudflows (i.e., turbidity currents) could in fact excise submarine canyons, thus supporting Daly's view. Subsequently, the disruption of submarine telegraph cables by turbidity currents proved that they existed in nature, and Daly's view has come to be the accepted one. See chapters 2 and 8.

108. Burnett, *Masters of All*.
109. Hess to the Princeton Department Faculty (Erling et al.), 24 October 1944, Box 25, HHP.
110. Ibid.
111. Hess, "Drowned Ancient Islands," 785.
112. The contemporary explanation is that they formed in high latitudes.
113. Hess to the Princeton Department Faculty (Erling et al.), 24 October 1944, Box 25, HHP.
114. In 1947, for example, he presented "Major Structural Trends of the Western North Pacific," published in 1948 in the Bulletin of the Geological Society of America, a descriptive paper accompanying the publication of a new Hydrographic Office bathymetric chart on the area from Korea to New Guinea. Hess, "Major Structural Trends," 245. See also Hess, "Major Structural Features."
115. Hess, "Major Structural Trends," 245.
116. Hess, "Report of Committee for Geological and Geophysical Study of Ocean Basins, 1943–1944," Box 25, HHP.
117. Ibid.
118. James, "Hess."
119. The report was released in September 1951. I have misplaced the citation to

this, but it is most likely from Box 5, NRC 1952–1953, or Box 31, NRC Committee Reports, HHP.

120. It is difficult to get a precise handle on this issue, because, as historian Jacob Hamblin has noted (Hamblin, "Pursuit of Science"), there appears not to have been a uniform and consistently applied policy for classification or release of Navy data. This is consistent with the comments of many oceanographers, who note that there was never any penalty or even risk to a Navy official for classifying documents that others thought should have been released, but there was considerable risk involved in releasing materials that others thought should have remained secret. Not surprisingly, Navy officials erred on the side of classification, and the burden of proof always rested with those who wanted release. According to Hess's notes, all bathymetric data was routinely classified, with the exception of near US-coastal regions out to the hundred-fathom line, because the latter was already largely in the public domain.

121. "Classification of Factual Material of the Oceans" (Confidential, n.d., c. 1951), Box 17, Folder "Works," downgraded 12 March 2001, HHP.

122. Ibid.

123. On the engineering aspects of the Manhattan Project, see esp. Hughes, *American Genesis*, 381–442.

124. They might have had to make corrections for the variable velocity of sound in water (all of Hess's published work was based on uncorrected data, assuming a constant velocity of 4,800 feet per second), but these corrections were small and significant only in very deep water. In any case, if the Soviets had their own T/depth profiles, they could readily make the required corrections.

125. "Classification of Factual Material of the Oceans" (Confidential, n.d., c. 1951), Box 17, Folder "Works," downgraded 12 March 2001, HHP.

126. Here he is invoking the conventional use of *need-to-know*, not the version that Paul Fye promoted at Woods Hole in the 1960s, discussed in chapter 3. "Classification of Factual Material of the Oceans" (Confidential, n.d., c. 1951), Box 17, Folder "Works," downgraded 12 March 2001, HHP.

127. Siever, "Earth Science Research."

128. Weir, *Forged in War*.

129. Greene, *Geology in the Nineteenth Century*; Oreskes, *Continental Drift*.

130. "Classification of Factual Material of the Oceans" (Confidential, n.d., c. 1951), Box 17, Folder "Works," downgraded 12 March 2001, HHP.

131. "Security Information Report," n.d. but likely March 1952, and "Security Information Report," 20 February 1953, Box 178, Folder 6, Research and Development Board Committee on Geophysics and Geography, Panel on Oceanography, RG 330, NACP. See also Revelle to Ewing, 26 March 1952, Box 245, Folder Revelle, MEP. This incident, as well as the larger issue of Navy data classification, is also discussed in Hamblin, *Oceanographers*, 50–58. Hamblin

suggests that the International Geophysical Year was in part a response to the problem of military data restriction; see his chapter 3.

132. Revelle to Ewing, 26 March 1952, Box 245, Folder Revelle, MEP. See also "Security Information Report," 20 February 1953, Folder 6, Research and Development Board Committee on Geophysics and geography, Panel on Oceanography, RG 330, Declassified 9 JUL 1998, NACP.

133. Revelle to Droessler, memorandum on Classification of Oceanic Soundings, 19 September 1952, Box 2, SIO-Revelle.

134. Ibid.

135. Ibid. Some scientists have pointed out a difficulty with this proposal: if you classify only in areas of importance, you implicitly reveal where those areas are. In contrast, given how many unclassified Navy documents refer to activities around Bermuda, the Caribbean, and the Azores, it is hard to believe that the Soviets did not know the locations of major naval installations.

136. Earl Droessler to Revelle, 26 March 1952, Box 245, Folder Revelle, MEP.

137. Revelle to Ewing, 26 March 1952, Box 245, Folder Revelle, MEP.

138. Worzel to Revelle, 25 April 1952, in Box 16, Folder Revelle-Worzel, BCHP copy also found in Box 245, Folder Revelle, MEP. For additional discussion of civilian scientists who objected to Navy classification of bathymetric data, see Doel et al., "Extending Modern Cartography," 613.

139. Richard Fleming to Maurice Ewing, 5 February 1949, Box 133, Folder Fleming, MEP. His emphasis.

140. Tolstoy did not become a big player in later oceanographic work, despite his early successes. Perhaps there were other reasons, but perhaps Worzel's fears were confirmed.

141. Worzel to Revelle, 25 April 1952, in Box 16, Folder Revelle-Worzel, BCHP. Copy also found in Box 245, Folder Revelle, MEP.

142. Ibid.

143. Ibid.

144. Ibid.

145. Ibid.

146. Ibid.

147. Ibid.

148. Walter Pitman, personal communication with the author, 9 May 2016, New York City; also Pitman, unpublished memoir notes, shared with the author by Tanya Atwater with permission of Walter Pitman, 2016.

149. Worzel to Revelle, 25 April 1952, in Box 16, Folder Revelle-Worzel, BCHP. Copy also found in Box 245, Folder Revelle, MEP.

150. Ibid.

151. "Security Information Report," 20 February 1953, Folder 6, Research and Development Board Committee on Geophysics and Geography, Panel on Oceanography, RG 330, declassified 9 July 1998, NACP.

152. Ibid.

153. Ibid.

154. "Classification of Deep Sea Sounding," Confidential Security Information Memorandum, 17 March 1953, declassified 12 March 2001, Box 31, Folder "Ocean Data Classification," HHP.

155. Ibid.

156. Edward H. Smith to Commanding Officer, Branch Office ONR Boston, 12 February 1951, AC-09, Folder "Nonr–27701 (Basic Task)," WHOI-ODR.

157. "Classification of Deep Sea Sounding," Confidential Security Information Memorandum, 17 March 1953, declassified 12 March 2001, Box 31, Folder "Ocean Data Classification," HHP.

158. Hess to Joseph Cochrane, Hydrographic Office, 9 December 1953, Box 25, Folder US Naval Oceanographic Office (pre-1964), HHP.

159. Ibid.

160. Ibid. This was proposed in Revelle's memo to Droessler, "Memorandum on Classification of Oceanic Soundings," 19 September 1952, Box 2, SIO-Revelle.

161. Ewing to E. R. Piore, 2 December 1954, Box 94, Folder US Dept. of Navy ONR, MEP.

162. Ibid.

163. Ibid.

164. Piore to Ewing, 13 October 1954, Box 94, Folder US Dept. of Navy ONR, MEP.

165. Ibid.

166. Ibid.

167. Ibid.

168. Moynihan, *Secrecy*, 170–71.

169. Luskin to Ewing, 8 November 1957, Box 94, Folder US Dept. of Navy ONR, MEP. This explains why bathymetry remained classified but magnetics did not.

170. Heezen to Kenneth Emery, 5 January 1959, Box 9, Folder "1959," BCHP.

171. Heezen to Adrian Richards at the Naval Electronics Laboratory, 3 February 1961, Box 9, Folder "1960," BCHP. Heezen was quite sloppy in this record keeping, so memos and letters are often found in folders whose dates do not match.

172. Hess, "Gravity Anomalies," 71.

173. Hess, "Major Structural Features," 443.

174. Fisher and Hess, "Trenches," 430.

175. Ibid.

176. Alan Allwardt, *Arthur Holmes*, makes this point in a brief footnote, but seems not to have placed much importance on it.

177. J. I. Merritt, "Hess's Geological Revolution: How an Essay in Geopoetry Led to the New Science of Plate Tectonics," 24 September 1979.

178. Hess, "Ocean Basins," 599.

179. Ibid., 608.

180. Ibid., 603.

181. Latour and Woolgar, *Laboratory Life*; Latour, *Science in Action*; see also

Longino, *Science as Social Knowledge*; Solomon, "Web of Belief"; Knorr Cetina, *Manufacture of Knowledge*.

182. Revelle to Droessler, memorandum on classification of oceanic soundings, 19 September 1952, Box 2, SIO-Revelle.

183. Lovell, *P. M. S. Blackett*; Nye, *Blackett*.

184. Drew, "Submarine Crash."

185. Drew, "Under the Sea."

CHAPTER FIVE

1. Heezen et al., *Floors of the Oceans*. On the significance of their achievement, see Barton, "Marie Tharp"; Doel et al., "Extending Modern Cartography." For Tharp's perspective, see Tharp, "Mapping the Ocean Floor"; Tharp and Frankel, "Mappers of the Deep."

2. On the ways Tharp was often rendered invisible, see Doel et al., "Extending Modern Cartography." There is no question that Tharp, like most women of her time, was denied the opportunities available to comparable men and relegated to secondary and supportive position. However, in the 1970s, as the social climate changed, her contributions became quite celebrated.

3. Heezen became nationally and internationally known, his work featured in the *New York Times*, the *Sydney Morning Herald*, *Time* magazine, and CBS Television. Over the course of his career, he published over three hundred scientific articles and a lavishly illustrated book, *The Face of the Deep* (1971), coauthored with former student Charles Hollister. He was awarded the Henry Bryant Bigelow medal from Woods Hole in 1964, the Cullum Geographic Medal of American Geographical Society in 1973, the Francis Shepard Medal of the American Association of Petroleum Geologists in 1975, and, posthumously, the Walter Bucher Medal of the American Geophysical Union in 1977. See Scrutton and Talwani, *The Ocean Floor*.

4. Note how Tharp was again rendered invisible. Rhodes Fairbridge to Henri Besairie, 8 May 1957, Box 4, Folder "Fairbridge," BCHP.

5. Stommel to Heezen, 28 April 1959, Box 16, Folder "Stommel, Henry," BCHP.

6. Memo, M.R. to W. Arnold Finch, 17 May 1963, Box 9, Folder "Jan.–June 1963," BCHP. On the importance of AT&T support, see Doel et al., "Extending Modern Cartography."

7. Undated and unsigned account of the 1947 Atlantis expedition, Box 69, Folder 9, "Heezen," SIO-Menard.

8. Iselin to Adrian Lane, 13 July 1947, Box 38, Folder "Atlantis 150 Logbook," BCHP.

9. Ewing, "Proposed Survey of the Middle Atlantic Ridge," Box 38, Folder "Atlantis 150 Logbook, 2," BCHP. See also Heezen et al., *Floors of the Oceans*, acknowledgments. On this work and its historical context and significance, see Theberge, "Unravelling."

10. Various biographical materials, including typed memo, Box 9, Folder "1956–

57," BCHP; Heezen to Dennis Flanagan, 24 August 1960, Box 9, Folder "1960," BCHP; see also Heezen et al., "Submarine Topography."

11. Box 12, Folder "10/50–3/51," BCHP. In Box 69, Folder 9, "Heezen," SIO-Menard, there is an undated, unsigned page with a photo from 1947, describing the 1947 work: "The road to glory for Doc began with a hunk of Volcanic Rock dredged up from the bottom of the sea.... Bruce was disappointed not to be on the first cruise to the Mid-Atlantic Ridge in 1947 with this group."

12. Various letters, but see esp. Sue to Bruce, 14 June 1952, Box 13, Folder "52–53 R/S," BCHP.

13. Various documents, Box 13, Folder "12 53–55 S," and "12 53–55 Misc.," BCHP.

14. Arthur C. Zale to B. C. Heezen, 17 December 1951 and L. L. Walton to Bruce Heezen 2 June 1952, Box 13, Folder, "1952–53 HIJ," BCHP.

15. Memo, Worzel to Heezen, 5 February 1952, Box 13, Folder, "52–53 U-Z," BCHP.

16. This letter appears not to have been sent, Box 13, Folder "12/53–55 Misc.," BCHP.

17. Ibid.

18. B. C. Heezen to Dennis Flanagan, 24 August 1960, Box 9, Folder "1960," BCHP. This was recently confirmed to me by Walter Pitman, personal communication, New York City, 9 May 2016, who recalled that Heezen was "hardly ever around."

19. Johnson, *Submarine Canyons*, preface.

20. F. P. Shepard "Submarine Valleys"; Shepard et al., "Origins of Georges Bank"; Daly, "Origin of Submarine Canyons"; Daly, "Glaciation and Submarine Valleys."

21. Martin Rudwick has recounted the important role of greywackes in *The Great Devonian Controversy*, when scientists tried to sort out both the empirical question of what portion of the geological column certain Paleozoic rocks belonged to and the methodological question of whether the question should be answered on lithological or paleontological grounds. See Rudwick, *Great Devonian Controversy*.

22. Heezen and Ewing, "Turbidity Currents," 871.

23. Heezen and Northrop, "Eocene Sediment."

24. Ericson et al., "Deep-Sea Sands."

25. Heezen et al., "Submarine Topography."

26. Ericson et al., "Deep-Sea Sands"; Heezen and Ewing, "Turbidity Currents"; Heezen et al., "Submarine Topography"; Heezen and Northrop, "Eocene Sediment."

27. Heezen and Ewing, "Turbidity Currents," 849.

28. Kuenen to Heezen, 2 December 1953, Box 6, Folder "Kuenen," BCHP.

29. "Marie Tharp," Columbia 250, Columbians ahead of Their Time, http://c250 .columbia.edu/c250_celebrates/remarkable_columbians/marie_tharp.html; Tharp and Frankel, "Mappers of the Deep"; Tharp, "Mapping the Ocean Floor."

30. Oreskes, "Objectivity or Heroism"; Tharp to Gertler, 20 June 1960, Box 9, Folder "1960," BCHP. Oreskes, "Laissez-tomber."

31.   Tharp to Gertler, 20 June 1960, Box 9, Folder "1960," BCHP.

32.   Luskin et al., "Precision Measurement."

33.   On the role of the German findings in Ewing, Heezen, and Tharp's work, and whether they adequately credited them, see Theberge, "Discovering," "Unravelling," and "Seeking."

34.   Ewing, "Proposed Survey of the Middle Atlantic Ridge," in Box 38, Folder "Atlantis 150 Logbook, 1," BCHP.

35.   Ibid. This comment reveals just how ignorant Ewing was of geology. According to Daly's theory, the canyons were carved by submarine mudslides developing at the toes of deltas. Since no rivers ran on the mid-ocean ridges, it was not possible for canyons to develop by this means.

36.   Theberge, "Discovering the True Nature." Albert Theberge suggests that credit for the recognition that the ridge was in fact a rift belongs at least in part to the German oceanographer on the 1938 *Meteor* cruise, Gunter Dietrich, who first pointed out the "striking depressions" at the center of the Mid-Atlantic Ridge. See also Dierrsen and Theberge, "Bathymetry."

37.   Tharp and Frankel, "Mappers of the Deep"; Tharp, "Mapping the Ocean Floor"; Ewing and Heezen, "Reviews and Abstracts," 343; Heezen, "Geologie Sous-Marine." On the disappearing of women's contributions, see Rossiter, *Women Scientists*; Oreskes, "Objectivity or Heroism." Also relevant is the classic paper by Robert Merton on the so-called Matthew effect (Merton, "Matthew Effect"). In a notable irony, this paper was based largely on work done by Harriet Zuckerman. In the introduction Merton wrote: "The conception is based on analysis of the composite of experience reported in Harriet Zuckerman's interviews with Nobel laureates in the United States" (Merton seems not to have noticed the irony).

38.   Hess to Heezen, 8 February 1954, Box 13, Folder "12/53 1955 H," BCHP.

39.   Heezen to Hess, Box 13, Folder "12/53 1955 H," BCHP.

40.   Hess to Heezen, Box 13, Folder "12/53 1955 H," BCHP.

41.   Heezen et al., *Floors of the Oceans*, 3.

42.   Tharp and Frankel, "Mappers of the Deep," 53. See also Lawrence, "Mountains under the Sea."

43.   Heezen to John Northrop, 29 January 1959, Box 9, Folder "1959," BCHP.

44.   Heezen to Edwin Hamilton, 2 December 1960, Box 9, Folder "1960," BCHP.

45.   For the defense of qualitative field geology, see Pettijohn, *Memoirs*.

46.   Smith and Sandwell, "Conventional Bathymetry," point out that the maps were misleading in two ways. First, they created the illusion that the entire ocean floor had been mapped (which it had not). Second, the portrayal of the bottom shape was misleading: Heezen and Tharp, "Abstracts," 11, extrapolated from their detailed knowledge of the Atlantic, but the texture of the East Pacific Rise and Southeast Indian Ridge are quite a bit smoother, so their reliance on an analogy with the Atlantic produced a misleading result.

47.   Heezen to Iselin, 24 June 1959, Box 1, Folder, "Iselin," BCHP, including memo on classification of soundings. It might seem arrogant for a civilian scientist

to make assertions about naval needs, but like Hess, Revelle, Iselin, and most leading oceanographers of this time, Heezen was serving on several Navy committees convened to solicit technical input from scientists on operational matters, and like them, was frequently briefed on Navy operations. For a civilian, he knew quite a bit about those operations.

48. Heezen to Iselin, 24 June 1959, Box 1, Folder, "Iselin," BCHP, including memo on classification of soundings.

49. Collins, *Changing Order*; Shapin, *Social History of Truth*.

50. Heezen to Dill, c. 1953 (n.d.), Box 13, Folder "12/53-55 D," BCHP.

51. Memo, Heezen to Hess, 1954, Box 13, Folder "12/53-55, H" BCHP.

52. Heezen to Hess, 17 February 1954, Box 13, Folder "12/53-55, H" BCHP; Heezen to Kuenen, 28 October 1957, Box 9, Folder "1956–57," BCHP. In Heezen, Ewing and Ericson, "Submarine Topography," 1951, Heezen cites only Kuenen, *Marine Geology*, but none of his earlier works and misrepresents Kuenen's contribution: "Kuenen (1950), who was largely responsible for bringing turbidity currents to the attention of the geological profession, shows in his book that he did not suspect their vast significance in oceanic sedimentation." This was ridiculous; not surprisingly, Kuenen took offense at the characterization, for which Heezen subsequently apologized, weakly demurring that if he didn't cite the earlier work, it was only because he assumed everyone knew it. Yeah, right.

53. Agnes Creagh to Dr. Aldrich, summary of reviews of "Heezen paper" (GSA Special Paper 65), Box 16, unlabeled folders, BCHP.

54. Robert McAfee Jr. to Heezen, 25 May 1964, and Mildred Rippey to McAfee, 25 May 1964, Box 233, Folder "Heezen," 2 of 2, MEP.

55. W. A. Finck to Heezen, 27 May 1966, Box 233, Folder "Heezen," 2 of 2, MEP.

56. Memo, Ewing to Heezen and Tharpe [sic], 9 August 1954, Box 13, Folder "12/53–1955 E," BCHP.

57. Ibid.

58. Ibid.

59. On Revelle, this is based on my personal communication with Deborah Day; on Hess, this is based on the status of his papers at Princeton University Archives, which, when I first examined them in the 1980s, were completely unsorted, uncataloged, and contained classified documents mixed in with routine materials.

60. Memo, Heezen to Howard Davis, 11 May 1964, Box 10, Folder "1964, 1 of 3," BCHP.

61. Memo, Mildred Rippey to Alma Kesner, 11 May 1964, Box 10, Folder "1964, 1 of 3," BCHP.

62. There are many places where this comes up in the disciplinary proceedings correspondence, but see Ewing to Polycarp Kusch, 17 March 1970, Box 233, Folder "Heezen 2 of 2," MEP; Ewing to Heezen, 12 May 1970, Box 233, Folder "Heezen 2 of 2," MEP; Heezen to Ewing, 13 May 1970, Box 233, Folder "Heezen 2 of 2," MEP. See also Heezen telegram to Ewing, 24 June 1966, Box 234,

Folder "Heezen," MEP; memo, Harriet Bassett (later Ewing) to Heezen, 6 April 1962, Box 234, Folder "Heezen," MEP. See also Tolstoy and Ewing, "North Atlantic Hydrography." Ewing's policy was that any data collected by Lamont staff belonged to Lamont as an institution and not to the scientist who had collected it. This was justified in part by the argument that, in most cases, it was the Navy who had paid for it.

63.   Heezen to Ewing, 24 April 1963, Box 10, Folder, "1963," BCHP. Ultimately Heezen did work with Menzies; see Menzies et al., "Abyssal Fauna."

64.   Ewing was furious about this, and there is a great deal of correspondence that refers to it indirectly, but nothing that lays out exactly what the source of the dispute was, so I have had to read between the lines. Walter Pitman believes that Ewing was competitive with Heezen because Ewing considered the structure of the Atlantic seafloor to be his domain and because Heezen was the better scientist of the two. Walter Pitman, personal communication with the author, New York City, 9 May 2016.

65.   Draft memo, "A Proposed Solution to the Heezen Affair," 23 January 1970, Box 233 Folder "Heezen, 2 of 2," MEP.

66.   Undated draft letter from Heezen to Columbia President Grayson Kirk, c. 1966, sent to Menard by Marie Tharp, 8 November 1984, Box 96, Folder 6, "Heezen," SIO-Menard. Someone wrote in pencil on it, perhaps in 1966, but given events, it may more likely have been 1967.

67.   Draft memo, "A Proposed Solution to the Heezen Affair," 23 January 1970, Box 233 Folder "Heezen, 2 of 2," MEP. The issue apparently came to a head at a meeting in Moscow in 1966, where Heezen discussed the possible relation between magnetic reversals and the evolution of life. Ewing and others at Lamont were furious. Given that Heezen had done no work on the reversal question, they evidently viewed this as completely outside Heezen's domain. Whether the specific idea had already been proposed by others at Lamont, I do not know, but it would seem to be likely, given Ewing's reaction.

68.   Draft memo, "A Proposed Solution to the Heezen Affair," 23 January 1970, Box 233, Folder "Heezen, 2 of 2," MEP.

69.   Undated draft letter from Heezen to Columbia President Grayson Kirk, c. 1966, sent to Menard by Marie Tharp, 8 November 1984, Box 96, Folder 6, "Heezen," SIO-Menard.

70.   Ibid.

71.   Ibid.

72.   Stommel, notes for the receipt of the Ewing Medal, 1977, MC-06, Folder "National Medal of Science 1989," WHOI-Stommel. See also Carl Wunsch interview with Elizabeth Stommel, MC-06, Folder "1990," WHOI-Stommel, where the word *dictator* is used several times in reference to Ewing; see discussion in chapter 2.

73.   Ibid.

74.   Memo for the files, 19 March 1970, Box 233, Folder "Heezen 1 of 2," MEP.

75. This point was made in later years by Crane, *Sea Legs*.

76. Disclosure: Chuck Drake was a mentor and friend to me when I was an assistant professor in the Department of Earth Sciences at Dartmouth College. He passed away before I began the work on this book, so I was not able to discuss the material in this chapter with him. Drake was a thoughtful and judicious man with outstanding social skills and political instincts. He was a man to whom many would turn for counsel. In 1968 he was serving as chair of the Columbia Geology Department; in later years he would serve on the Council of Advisors on Science and Technology to President George H. W. Bush, and he would serve as president of the American Geological Society and the American Geophysical Union. For an oral history interview, see Interview of Charles Drake by Ronald Doel, 28 November 1995, Niels Bohr Library and Archives, American Institute of Physics, College Park, MD, https://www.aip.org/history-programs/niels-bohr-library/oral-histories/22583-1.

77. A similar issue arose at UCSD and Scripps, where academics on the "main campus"—up the hill from SIO—began to criticize classified research on the "lower campus." When I interviewed Bill Nierenberg about this as part of the 2000 ONR-Heinz oral history project, he became indignant, insisting that there was "no classified" research at Scripps and that his colleagues on upper campus were bloody minded. At the time, I did not know what I later learned about classified research at Scripps. It was true that most of the classified projects at Scripps were run through MPL, which was partly based at Point Loma, so in a strict sense, one could say that the classified research was done at Point Loma, not La Jolla. However, many Scripps faculty were engaged in classified projects, including John Isaacs and Ed Goldberg, who were not on the MPL staff. Isaacs in particular was known to have many locked files in his office, which colleagues assumed held classified materials. Some of this is discussed in the various ONR-Heinz interviews, as well as in various Scripps files. In this sense I believe that Nierenberg's position was disingenuous; having written *Merchants of Doubt*, I now believe that Nierenberg had an impressive capacity to persuade himself of the truth of his views, even when they were demonstrably false. See Scripps Biographical Files, SAC 5, Special Collections & Archives, UC San Diego, https://library.ucsd.edu/speccoll/findingaids/smc0087.html.

78. Claude ZoBell is a rare example of someone in this narrative who did chose to reject military projects and funding and looked instead to the petroleum industry (chapter 1).

79. Turner, "The Survey"; Hendrickson, "State Geological Surveys"; Lucier, "Geological Industries."

80. The word *enlightened* was repeatedly invoked by Lamont scientists to describe the US Navy in general and ONR in particular.

81. Memo: Classified Research-Geology/Lamont Drake to Ewing, 23 September 1968, Box 9, Folder "Classified Research," MEP. Drake was also ignoring or

dissenting from Vannevar Bush's argument that both military and industrial patrons were likely to neglect basic research in favor of projects with more obvious, short-term dividends. See Wolfe, *Competing with the Soviets*.

82.   See also Lowen, *Creating the Cold War*; Engerman, "Rethinking."

83.   Siever, "Doing Earth Science."

84.   Ibid., 161.

85.   A number of scientists who were at Woods Hole in the 1950s have told me they really did not know where the funds came from—Columbus Iselin took care of that. It was only later, with the increasing importance of the NSF, that grant writing became routine, even for younger scientists. Many scientists at Scripps in the 1990s and 2000s also told me that, when they were young scientists, they did not really know (or seek to know) where their advisers' funds were coming from, except insofar as from time to time they might be asked to give a "dog-and-pony show" to a visiting Navy officer.

86.   Drake, Memo: Classified Research, 27 September, 1968, Box 9, Folder "Classified Research," MEP.

87.   The budget was $9.3 million annually in 1968. See, e.g., Ewing to Cordier, 8 October, 1968, Box 9, Folder "Classified Research," MEP.

88.   ONR N0014-67-A-0108-004, for ship support, was worth $1.7 million. See Drake, Memo: Classified Research, 27 September 1968, Box 9, Folder "Classified Research," MEP. I communicated directly with Gordon Hamilton in August 2002; he confirmed many details in a letter to me on 23 August 2002, at which time he was retired in Rockville, Maryland.

89.   Gordon Hamilton to Lester R. Watson, re: ONR N6onr-27124, 2 February 1951, Box 72, Folder "Columbia Committee on Government Aided Research, 1951," MEP.

90.   Hamilton, personal communication with author, 23 August 2002.

91.   Hamilton to Ewing, 23 September 1968, Box 9, Folder "Classified Research," MEP. A duplicate copy can also be found in Box 223, Folder "Hamilton," MEP.

92.   Ibid.

93.   Ibid.

94.   J. Dorman, "Classified Research and the University," 2 October 1968, Box 9, Folder "Classified Research," MEP.

95.   Ibid.

96.   Ibid.

97.   Ibid.

98.   Ewing to Cordier, 8 October 1968, Box 9, Folder "Classified Research," MEP. Historian Charles Gillispie found a very similar argument—that building weapons of destruction was a moral good because it helped to support civilization against the forces of darkness—in the scientific weapons projects of revolutionary France. See his "Science and Secret Weapons."

99.   Worzel claimed in 1968 that "at least half "of the educational opportunities for graduate students came through government contracts," and one can safely assume that most of those were DOD. See Worzel to Warren Goodell,

15 October 1968, Box 9, Folder "Classified Research," MEP. This letter was written on the very day that the Columbia faculty voted to establish a policy on outside grants and contracts, following an earlier faculty recommendation that no new classified grants or contracts be accepted at Columbia. "Faculty to Study Columbia's Ties," *New York Times*, 16 October 1968, 17, press clipping in Box 9, Folder "Classified Research," MEP. According to the *New York Times*, there were "at least 10" classified contracts on campus. If so, then the nine at Lamont were a very substantial proportion!

100. Ultimately, Columbia did ban classified research on campus, and a group of Lamont faculty, including Ewing, his brother John, and others set up a separate institute, the Palisades Institute, to which they moved some of their classified contracts. However, by this time in the early to mid-1970s, Navy contracts in oceanography were somewhat in decline. On the creation of the Palisades Geophysical Institute, see Box 9, Folder "Classified Research," MEP. See also chapter 4.

101. This attitude continues to prevail in the scientific community today. When over one hundred members of the American Geophysical Union asked their society to reject funding from the oil and gas giant, ExxonMobil, as inconsistent with the society's bylaws that it will not accept funding from organizations that promote disinformation or misleading information about science, the AGU board and many members rejected that suggestion as "political" while refusing to accept the argument that accepting Exxon-Mobil funding was also political. See Margaret Leinen, "AGU Board Votes to Continue Relationship with ExxonMobil and to Accept Sponsorship Support," *From the Prow*, American Geophysical Union, 14 April 2016, https://fromtheprow.agu.org/agu-board-votes-continue-relationship-exxonmobil-accept-sponsorship-support/; Joe Romm, "American Geophysical Union Sells Its Scientific Integrity for $35,000 in ExxonMobil Money," *Think Progress*, 19 April 2016, http://thinkprogress.org/climate/2016/04/19/3770435/american-geophysical-union-exxon-mobil/. Although one board member recused himself from the vote on grounds of conflict of interest, he was not only permitted to participate in the discussions but also allowed to make a presentation on behalf of continued ExxonMobil support. In contrast, no one was invited to make the case for cessation of that support. According to *Inside Climate News*, AGU's executive secretary Christine McEntee, "No sponsor influences anything we do about the science." Yeah, right. See Phil McKenna, Zahra Hirji, and Lisa Song, "Exxons Donations and Ties to American Geophysical Union Are Larger and Deeper Than Previously Recognized," *Inside Climate News*, 26 May 2016, https://insideclimatenews.org/news/26052016/agu-american-geophysical-union-exxon-climate-change-denial-science-sponsorship. I think we can understand this as a form of status quo bias—sustaining the status quo requires no action and is therefore viewed as normative and apolitical; change is viewed as inspired by affirmative normative commitments and therefore political.

564 NOTES TO PAGES 220-223

102. Ewing appears to have been shocked when Columbia University did, in fact, ban classified research from campus. Unable or unwilling to compromise, he left Columbia and died soon thereafter of a cerebral hemorrhage at the age of sixty-eight. In his NAS memoir of Ewing, Teddy Bullard glossed over the reasons Ewing left Columbia but acknowledged that Ewing found the experience "bitter and deeply disturbing." See Bullard, "William Maurice Ewing."

103. Report of the Columbia University Committee on Relations with Outside Agencies, 31 May 1968, Box 9, MEP, photocopy of selected pages ii–iii and 7–21. On 10: "Accepting a grant from the CIA, moreover, might breed the suspicion that there are other university associations with it which are not revealed, and may create obstacles for faculty and students to carry on research and maintain intellectual contacts abroad." This, of course, was true, for Ewing and others at Lamont did do work that was not revealed, like Ewing's work for the State Department. The overall report concludes that "secrecy, particularly secrecy imposed from without, is contrary to the spirit of free inquiry and open exchange of information which the University exemplifies.... Government limitations on publication of the fruits of research interfere with the dissemination of knowledge and deprive the University of its autonomy in regard to one of its principal functions.... We therefore recommend that the University refrain from accepting new grants or contracts involving government classified research."

104. P. Kusch to Ewing, 4 April 1969, Box 233, Folder "Heezen 2 of 2," MEP.

105. Ewing, Worzel, Hamilton, and others set up a separate institute, the Palisades Institute, to which they moved some of their classified contracts.

106. Heezen was not the only one antagonized by Ewing. For some idea, consult the interview with Elizabeth Stommel by Carl Wunsch, MC-06, Folder "1990," WHOI-Stommel; see also chapter 2 of this book. People who admired him used terms like *focused, aggressive,* and *hard-driving.* Those who did not were less polite.

107. Laughton to Heezen, 28 November 1955, Box 1, Folder "Laughton," BCHP. See also Laughton to Heezen, 6 February 1956 and 11 April 1956, Box 1, Folder "Laughton," BCHP.

108. Laughton to Heezen, 11 April 1956, Box 1, Folder "Laughton," BCHP.

109. Heezen to Zoltan de Cserna, 3 August 1956, Box 9, Folder "1956–57," BCHP. see also Cserna to Heezen, 2 February 1956, Box 9, Folder "1956–57," BCHP, and Tolstoy to Heezen, 25 July 1956, Box 9, Folder "1956–57," BCHP.

110. There are more examples that I cannot cite without bogging down this already-long chapter, but they include US scientists who had clearances, as in January 1959 when Heezen wrote to sedimentologist Kenneth O. Emery (15 January 1959) that he had "only two contour charts" available to share "due to well-known Navy policy." Box 9, Folder "1959," BCHP.

111. Menard to Heezen, 4 April 1961, Box 69, Folder 6, "Heezen," SIO-Menard.

112. Ibid.

113. Heezen to Menard, 30 April 1968, Box 69, Folder 6, "Heezen," SIO-Menard

114. Menard to Heezen, 8 May 1968, Box 69, Folder 6, "Heezen," SIO-Menard.

115. William Dunkle to Heezen, 10 April 1957, Box 3, Folder "Dunkle," BCHP.

116. Heezen to Robert Dietz, 7 May 1956, Box 9, Folder "1956–57," BCHP.

117. Heezen to Warren Wooster, 18 May 1956, Box 9, Folder "1956–57," BCHP.

118. Doel et al., "Extending," 619, note that Heezen hoped his work would persuade his colleagues of the expanding Earth idea, but despite the attention the maps received, they were not interpreted this way. The authors stress the importance of Heezen's relationship with AT&T, but it is not clear how that relationship connected to his commitment to the expanding Earth theory, if it did.

119. Tharp and Frankel, "Mappers of the Deep," 53.

120. Ibid., 62.

121. King, "Origin and Significance."

122. Oreskes, *Continental Drift*.

123. Heezen to John Mann, 15 April 1957, Box 9, Folder "1956–57," BCHP.

124. Box 9, Folder "1956–57," BCHP. This exchange does not appear to have led to any discussion.

125. For a comprehensive history of the theories of continental drift and plate tectonics, see Frankel, *Drift Controversy*. For a detailed analysis of the American reaction to drift theory, including a discussion of Taylor, see Oreskes, *Continental Drift*. For the definitive biography of Alfred Wegener, see Greene, *Alfred Wegener*.

126. Heezen, "Submarine Geology and Continental Displacements," lecture, Box 9, Folder "1958," BCHP.

127. On Carey's work, see his *Continental Drift* and his more recent works: *Expanding Earth* and *Theories of the Earth*. For continued interest in the expanding Earth theory, see Hoshino, *Expanding Earth*; Scalera and Jacob, *Why Expanding Earth*. The expanding Earth theory was also seriously considered around 1960 by Canadian geophysicist Tuzo Wilson, in "Some Consequences of Expansion."

128. Carey, *Continental Drift*, 187.

129. Heezen to Holmes, 6 October 1959, Box 9, Folder "1959," BCHP.

130. Dicke, "Gravitation." See also Egyed, "Dynamic Conception." Keith Runcorn would later conclude that, while the argument might be correct, the amount of Earth expansion expected on Dirac's model was far too small to account for the geological separation of the continents. S. Keith Runcorn, in Carey, *Expanding Earth*, 327.

131. Heezen, "Deep-Sea," 394."

132. Heezen to A. C. Munyan, 15 June 1959, Box 9, Folder "1959," BCHP.

133. Press clipping, "Earth seem expanding like ripening orange," *New York Herald Tribune*, 8 September 1959, Box 69, Folder 11, SIO-Menard

134. "How Oceans Grew," *Time Magazine*, 14 September 1959, 46.

135. Heezen, "Ocean Floor."
136. Ibid.
137. Jacobs to Heezen, 17 October 1960, Box 9, Folder "1960," BCHP.
138. Heezen to Jacob, 17 May 1961, Box 9, Folder "1961," BCHP.
139. Ibid. In fact, Even the marine paleomagnetic data were not entirely helpful to his cause. The data collected by Lamont scientists over the Mid-Atlantic Ridge revealed major anomalies—later to be understood as the result of geomagnetic field reversal. Heezen explained them away as the effect of rocks of high magnetic intensity close to the surface at the ridge.
140. Heezen to Holmes, 6 October 1959, Box 9, Folder "1959," BCHP.
141. Ibid.; Ewing and Worzel, "Gravity Anomalies"; Talwani et al., "A Crustal Section"; Ewing and Heezen, "Reviews and Abstracts."
142. Heezen, "Concepts of the Expanding Earth and Continental Drift: A Symposium Held at Columbia University," lecture notes, 10 December 1959, Box 15, Folder "Misc. Meetings, 1959," BCHP.
143. Heezen to Holmes, 6 October 1959, Box 9, Folder "1959," BCHP.
144. Oreskes, *Continental Drift*. In 1960, Heezen received a letter from Arthur Meyerhoff, who would later be an opponent of plate tectonics, in which Meyerhoff argued that the rifts were compressional features. Heezen responded to this by calling Meyerhoff's view "astounding." Heezen to Meyerhoff, 6 June 1960, Box 9, Folder "1960," BCHP.
145. Worzel, "Configuration," 21. See also Worzel, *Pendulum Gravity*.
146. Ewing and Heezen, "Puerto Rico," 266; Ewing and Worzel, "Gravity Anomalies."
147. Ibid.
148. Oreskes, *Continental Drift*.
149. Especially Neil Opdyke, Walter Pitman, and Xavier Le Pichon. See Oreskes, *Plate Tectonics*.
150. Solomon, *Social Empiricism*.
151. Hess, "Ocean Basins"; R. S. Dietz, "Continent." Hess is credited with priority because his paper was widely circulated in preprint before Dietz's was submitted.
152. Work supported by ONR, Nonr 1858 (10) to Princeton University, Original Title "Nature of the Great Oceanic Ridges." This report acknowledges that it is continental drift revisited. In the published version of his now-famous "Essay in Geopoetry," Hess tried to argue that his ideas were substantively different from those of the 1920s and 1930s. However, in private he acknowledged that essentially this was the same idea Wegener had proposed before.
153. For summaries of this work, see Cox, *Plate Tectonics*; Glen, *Road to Jaramillo*; Frankel, "The Continental Drift Debate"; Frankel, "Jan Hospers"; Oreskes, *Plate Tectonics*, 21–22. The key primary references are Tarling, "Tentative Correlation"; McDougall and Tarling, "Polarity Zones"; Cox and Doell, "Pleistocene Geochronometry"; Cox et al., "Sierra Nevada II." On the importance

of rendering data visible and therefore potentially persuasive—a key element of the significance of Heezen and Tharp's physiographic maps—see Spanagel, "Utility."

154. The best summary of this is Vine, "Reversals of Fortune."

155. Hess, "Ocean Basins," 608.

156. Bullard et al., "Fit of the Continents."

157. Runcorn, "Paleomagnetic Comparisons."

158. Girdler, "New Oceanic Crust," 123.

159. Wilson, "Evidence from Ocean Islands."

160. Worzel to Heezen, 6 March 1964, Box 8, Folder "Worzel," BCHP.

161. Frankel, *Continental Drift*, 4:148–232.

162. Worzel to Ewing, 24 March 1964, Box 254, Folder "Worzel 1 of 2," MEP.

163. Ibid.

164. Heezen and Tharp, "Tectonic Fabric," 105.

165. Blackett, Bullard, and Runcorn to Heezen, 25 March 1964, Box 15, Folder "The Royal Society March 18, 20, 1964," BCHP.

166. Bullard to Munk, 6 April 1964, Box C18, Folder "Correspondence 1964," ECBP.

167. Indeed, Menard later discovered that Bruce Heezen was the principal citer of his own work. See Menard, *Science*, 116.

168. Heezen to Runcorn, 10 February 1964, Box 10, Folder "1964 2 of 3," BCHP.

169. Menard, application for federal employment, circa 1949, in Box 1, Folder 13, "Personnel Records 1947–48"—the application includes discussions of what he did in 1948–1949, so it appears to be written in 1949, SIO-Menard. Alternatively, it could be what he was planning to do—a letter from Dietz to Stetson dated 17 December 1948 asks for Menard to send in this form.

170. On Menard's thesis adviser, Kirtley Mather, see Bork, *Cracking Rocks*.

171. Application for federal employment, 1 September 1949, Box 1, Folder 14, SIO-Menard. See also Menard to E. L. Packard (rough draft), 9 January 1949, Box 1, Folder 13, SIO-Menard.

172. University of California Biography, 1956, Box 1, Folder 16, SIO-Menard.

173. Menard to George Thiel, 3 March 1952, Box 1, Folder 16, SIO-Menard.

174. He also had a strong interest in the history and sociology of science. His posthumously published *The Ocean of Truth: A Personal History of Plate Tectonics* (1986) documents his role in the plate tectonics revolution, particularly his personal relationships with the scientists responsible for developing the new paradigm.

175. A Guide to the Henry William Menard Papers, MC18, processed by Carol Lynn Flanigan, SIO Ref No. 92–27. 1992, https://library.ucsd.edu/speccoll /findingaids/Menard82-53.pdf.

176. See SIO "Henry William Menard Biography," http://scilib.ucsd.edu/sio/biogr /Menard_Biogr.pdf; see also Menard, *Ocean of Truth*, 315n2.

177. Menard to Bullard, 27 October 1966, Box C22, ECBP.

178. Menard, "Fractures," 41.

179. Ibid.
180. Bullard et al., "Heat Flow."
181. Menard, "Deformation," 1149, 1182.
182. Ibid. Similarly the Revilla Gigedo Islands and Volcanic Province of Central Mexico appeared to be an extension of the Clarion. Menard, *Deformation*.
183. Vacquier, "Many Jobs."
184. Menard and Vacquier, "Magnetic Survey," 5.
185. Menard and Fisher, "Fracture Zone," 252.
186. Menard, "East Pacific," 1745.
187. Ibid. This is particularly ironic, because in later years people would argue precisely the reverse: that land-based geologists were slow to accept plate tectonics because most of the convincing evidence came from the oceans.
188. Menard knew both men well—he and Dietz had worked together at NEL—and he corresponded with Hess in 1961. See Menard, *Ocean of Truth*, chap. 13. For Dietz's ideas, see Dietz, "Continent and Ocean Basin."
189. Menard, *Ocean of Truth*, 133.
190. Heezen and Menard, "Topography," 270.
191. Menard, *Ocean of Truth*. Although elsewhere Menard claims that up to 1968 it received thirty citations per year, suggesting that people did make use of its factual information, that many scientists were not paying attention to the forefront of the field, or that some people will cite anything. See Menard, *Science*, 114.
192. Menard, "Rise-Ridge System," 112.
193. Ibid., 121.
194. Menard and Chase, "Tectonic Effects," 34.
195. Ibid.
196. Menard, "Sea Floor," 362.
197. Press and Siever, *Earth*, 508, in the fourth edition.
198. Menard, *Ocean of Truth*, chap. 3.
199. Shor, *Scripps*, 92.
200. On leaving data analysis to women, see Oreskes, "Laissez-tomber."
201. On Revelle being a sailor, see Deborah Day, "Quotations from Roger Revelle," Compiled for the Scripps Institution of Oceanography Archives, http://scilib.ucsd.edu/sio/biogr/day_quotations-from-roger-revelle.pdf.
202. Menard, *Science*, 11.
203. Menard, *Ocean of Truth*, 205. Although Menard was well liked by students and colleagues, it was known that he could be possessive about his territory (an accusation he lodged, not incidentally, at Heezen). One oceanographer who joined Scripps in the late 1950s claims that Menard told him point-blank: "The Pacific is mine" (Tjeerd van Andel, personal communication with author, October 2002.). Of course, many scientists are competitive, but they cannot necessarily shield their data from others in the way that Menard, because of data classification, could.
204. Worzel to Heezen, 6 March 1964, Box 8, Folder "Worzel," BCHP.

205. In 1971, Ewing wrote to Walter Stern at Defense Contract Administration Services: "A continuing review has been undertaken to reduce the amount of classified material possessed at this facility. The destruction of records maintained at the observatory indicates that an active program of authorized disposal has been vigorously pursued." Box 107, Folder "LDGO Security and Safety Procedures," MEP.

206. Siever, "Doing Earth Science," 163.

207. See essays by Matthew Shindell and Erik M. Conway in Oreskes and Krige, *Cold War.*

208. See Pettijohn, *Unrepentant Field Geologist.*

209. Wood, *Dark Side of the Earth.*

210. From 1951 through 1994, the US Navy, under Project Magnet, continuously collected vector aeromagnetic survey data to support the US Defense Mapping Agency's world magnetic modeling and charting program. For the NOAA information on the program, see "Project Magnet Data," https://www.ngdc.noaa.gov/geomag/proj_mag.shtml; for information on marine geophysical data, in general, see "Marine Trackline Geophysical Data," https://www.ngdc.noaa.gov/mgg/geodas/trackline.html.

211. Mason, "Stripes on the Sea Floor," 33.

212. Bullard et al., "Heat Flow."

213. Mason, "Stripes on the Sea Floor," 33.

214. Bullard and Mason, "Magnetic Field," 194.

215. Fred Spiess, personal communication with author, 2002.

216. Mason, "Stripes on the Sea Floor," 36.

217. Morley, "Zebra Pattern," 68.

218. Ibid., 79.

219. Vacquier, "Horizontal Displacement," 1959; Vacquier et al., "Horizontal Displacement," 1961; see also Menard, "East Pacific Rise"; Vacquier, "Transcurrent Faulting."

220. Revelle, quoted in Shor, *Scripps,* 101.

221. It was known that some terrestrial rocks were magnetized in a direction opposite of today's prevailing field, but it was by no means accepted that reversely magnetized rocks indicated a reversely magnetized field at the time of the rocks' formation. Some geophysicists thought that some rocks might for some reason record a polarity opposite to the field prevailing when they formed, or that their polarity might be reversed during metamorphosis or deformation. The issue would be resolved by Cox et al., "Sierra Nevada II," in 1963, but in 1961 it was still in dispute.

222. Ron Mason, email communication to Dan McKenzie, 16 January 2001.

223. Vine, "Reversals of Fortune," 54.

224. Ibid., 55.

225. Vine and Matthews, "Magnetic Anomalies," 948.

226. Vine, "Reversals of Fortune," 58.

227. Ibid.

228. It was rejected by *Nature* and *Journal of Geophysical Research*. See Morley, "Zebra Pattern."

229. Meanwhile, at the Navy's Hydrographic Office there was an opposite problem: a man with the data who was denied the opportunity to explain it. In 2003, the American Geophysical Union gave a medal to a man named Robert H. Higgs for his work on seafloor magnetic stripes. Higgs worked for the Hydrographic Office collecting magnetic survey data at sea. In 1961, he analyzed unclassified data from the Pacific-Antarctic Ridge and recognized a linear pattern, noted its relation to the ridge axis, and suggested one might use the patterns to determine the age of the oceanic crust. This was written up in a report that apparently was widely distributed. As head of the Marine Section of the Geomagnetic Division of the Hydrographic Office from 1963 onward, Higgs promoted further magnetic surveys and encouraged their publication, but he was told to curtail operations by officials concerned that he was "deviating from the mission of the Office." In 1969, as plate tectonics broke open, Higgs presented a "sanitized" version of the data—that is to say, a degraded version in the form of "low resolution 'zebra charts.'" In accepting the AGU medal in 2003, Higgs noted that many scientists in the Navy "made significant contributions . . . but they often went anonymous and unrecognized," and so he accepted "on their behalf and on behalf of other scientists in government and private industry who have been restricted in publication of their work, but who have unselfishly found a way to share it with others and moved on." Korgen, "Higgs Receives." I first heard the Higgs story from Norman Sleep in an email; the historical record affirms it. For an example of the Navy advising scientists to "sanitize" charts by removing detail in specified regions to permit them to be released, see memorandum, Joseph E. King to Laboratory Director, 9 May 1963, AC 5, Box 11, Folder 367, "Henry William Menard," SBF.

230. Hess to Charles C. Bates, 8 December 1967, Box 25, Folder "US Navy Oceanographic Office 1967," HHP. For more information on continued discussions on classification of seafloor data, see memo, Research and Development Department, Code 70, 25 March 1966, CH/T #260, and attached consultants list, 8 July 1966, CH/T #270, Consultants List, Box 25, Folder "US NOO 1966," HHP.

231. Hess to H. P. Stockard, 6 January 1967, Box 25, Folder "US Navy Oceanographic Office 1967," HHP.

232. Transmittal and Transfer Record, "1 copy—Total Intensity Contour Charts of eastern and Western Sea [sic]," Box 25, Folder "US Navy Oceanographic Office 1967," HHP.

233. Leroy Dorman, Scripps Institution of Oceanography, personal communication, 24 September 2002.

234. O. D. Waters Jr., "Memorandum for Assistant Secretary of the Navy (Research and Development)," 14 November 1968, Box 31, Folder "Ocean Data Classification," HHP.

CHAPTER SIX

1.   Haymon et al., "Volcanic Eruption."
2.   Hessler and Kaharl, "Vent Community." While individual organisms had been recovered from depths on early oceanographic expeditions, the idea of rich communities was largely discounted.
3.   Hessler and Kaharl, "Vent Community"; see also Kaharl, *Water Baby*.
4.   German et al., "Mid-Ocean Ridges," 156.
5.   Cowan, "Hyperthermophilic Enzymes." For a review of hydrothermal sulfide deposits and metalliferous sediments, see Mills, "Marine Ecology"; Hannington et al., "Seafloor Mineralization."
6.   Historians of technology have of course challenged this—for example, in Kline, "Constructing Technology," and references cited therein. As Kline points out, whether or not the "linear model" accurately represents the historical facts, it has been widely accepted. In the conventional wisdom of scientists and their supporters, basic science yields application, not the other way around. Kline's discussion focuses on the early to mid-twentieth century in the United States; for critics who protest that no one believes this anymore, I offer as evidence a former president of the United States and the man who won the popular vote in 2000: William J. Clinton and Albert Gore Jr., "National Interest." For those who protest that this is merely rhetoric, fifty years after *Science: The Endless Frontier*, it remained the rhetoric driving US science policy (Sarewitz, *Frontiers of Illusion*) and arguably still does.
7.   In saying this, I do not mean to imply that *Alvin* was unique; on the contrary, the entire history of military-sponsored scientific research in physics and space supports an alternative reading of the relationship between science and technology, basic and applied research. See, e.g., Dennis, *Change of State*; Edwards, *Closed World*; Leslie, *Cold War*; Galison, *Image and Logic*; Hamblin, *Oceanographers in the Cold War*.
8.   On how Woods Hole director Columbus Iselin forged relations with the US Navy to support oceanographic research even before the outbreak of World War II, see Weir, *Ocean in Common*; Schlee, *Unfamiliar World*; Sears and Merriam, *Oceanography*.
9.   Weir, *Forged in War*, chap. 3; Weir, *Ocean in Common*, chap. 7.
10. Hill, *Afternoon Effect*. See also Weir, *Ocean in Common*.
11. Spiess, "Undersea Research," 46–50, in *Seeking Signals*.
12. For a concise review with references to original Navy sources, see Urick, "Sound Propagation." On the physics of underwater sound, see Kinsler and Frey, *Fundamentals of Acoustics*; Frosch, "Underwater Sound." Both Kinsler and Frey were professors at the US Navy Postgraduate School.
13. The thermocline beneath the surface layer is sometimes referred to as the "main thermocline," because other zones of strong temperature gradient may also occur. The main thermocline has no fixed depth; its location depends on the thermal conditions of the ocean in any given region, which

depend on climate and circulation patterns. See Iselin, *Application of Ocean-ography*, chap. 5; Spilhaus, "Bathythermograph."

14. Iselin, *Application of Oceanography*, 4–5. See Research Analysis Group, *Physics of Sound*; Eckart, "Refraction of Sound." Sound research was funded under Division 6, Subsurface Warfare, headed by John T. Tate.

15. Weir, *Ocean in Common*, 128.

16. Ibid., 130.

17. The velocity minimum occurs because sound velocity decreases with decreasing temperature but increases with increasing pressure. Near the surface where the temperature gradient is steep, velocity drops off sharply. However, at some point, the temperature effect tapers off, while pressure continues to increase with depth, and the net effect is that sound velocity begins to rise again. Where Ewing and colleagues were working, near Bermuda, the velocity minimum occurs at a depth of approximately 700 fathoms (4,200 feet, or about 1,500 meters). In the equatorial Pacific, the velocity minimum typically occurs between 500 and 1000 meters. The exact depth of the velocity minimum depends on the temperature profile; the overall variation in sound velocity is greater in the equatorial zones, where surface heating is great, and least in the polar regions, where surface heating is minimal.

18. Ewing and Worzel, "Sound Transmission," 3–4. For more details on the science of the sound channel and additional references, see Urick, *Sound Propagation*.

19. Ewing and Worzel, "Sound Transmission." Under Contract Nobs-2083, task no. 1 was to "study the basic phenomena involved in the propagation of underwater sound, including the transmission, reflection and scattering of sonic and supersonic sound." Task no. 1B was to "study the transmission of low-frequency sound at the greatest possible ranges, especially the transmission of explosive sound in sound channels."

20. The pilot would drop a charge set to go off at the appropriate depth (usually about 1300 m), and the sound waves could be picked up at listening stations hundreds of kilometers away. Given three stations, one could triangulate to determine the precise position of the pilot and rescue him. Conversely, a submerged submarine could accurately determine its position by triangulation using distant signals transmitted on the SOFAR channel; this application, known as RAFOS, became the basis for postwar submarine navigation systems. See Weir, *Forged in War*, 64. Note this is the same principle behind triangulating to find earthquake epicenters—a matter with which as a seismologist Ewing was very familiar.

21. Urick, *Sound Propagation*, 7–1.

22. For a convergence zone to occur, rays must bounce off the base of the sound channel, which means that the water depth must be greater than the depth to the base of the sound channel. Hence, bathymetry was also important—

see chapters 4 and 5. The best summary of the technical aspects is Eckart, *Underwater Sound*.

23.   Oreskes, "Laissez-tomber."

24.   WHOI, Biographical Sketch of Allyn C. Vine, in *Guide to the Collections*, Allyn Collins Vine, 1914–1994, Papers 1939–1980, http://dlaweb.whoi.edu/PHP /FAID/faids_files/MC-01_Vine.html.

25.   Weir, *Ocean in Common*, 148.

26.   Urick, *Sound Propagation*, emphasizes the work of NRL scientists in discovering and elucidating the afternoon effect, underscoring that the Navy was not helpless without academic scientists. It had considerable internal scientific resources of its own; the turn to academic scientists in the 1940s was pushed by the academics who ran the NDRC. On the history of the ONR, see Sapolsky, *Science and the Navy*. From UCDWR, the Navy also established the NEL at Point Loma, so UCDWR effectively became two labs: a military lab, NEL, and a civilian lab, MPL. Throughout their history MPL and NEL had close relations, and a number of scientists had either joint appointments (Robert Dietz) or moved from one lab to the other (Bill Menard).

27.   Geiger, "Seeking Signals," 118.

28.   Memo 00c-U-195, to CDR James B. Davidson, ONR Code 466, 23 October 1965, Box 4, Folder "Annual Reports, 1958–1964," SIO-MPL.

29.   Spiess, personal communications with author, 2002.

30.   Summary Report, Contract N6 Nonr 277, Task Order I, 2 May 1951, Box 20, Folder 80, WHOI-ODR; see also Contract Reports N6 Nonr 27701 (Basic task), 1950, Box 15, Folder 54, WHOI-ODR and 15 June 1951, Box 22, "Oceanography," WHOI-ODR. In 1951, institutional research funds at Woods Hole amounted to $140,000; government funds for that same year amounted to $1,500,000.

31.   15 June 1951, Box 22, WHOI-ODR.

32.   AC-09.5, Folder 15, WHOI-ODR. Vine to J. Donald Harris, Audiology Division Submarine Medical Center, New London, 18 May 1966, MC-01, Folder "Correspondence One of a Kind Jobs, 1964–1968," WHOI-Vine. While Vine was the author of many ingenious (or absurd?) suggestions to the Navy, he was by no means the only one; one Woods Hole memo on 28 February 1962, apropos of a visit to the NAS for a meeting of the deep-diving submarine group, describes an idea to use "dried chameleons as an additive to fuels—they would combine with water vapor in the exhaust gases and the exhaust would take on the color of the surroundings and thus be undetectable." I found no evidence that this idea was pursued.

33.   Memo, "IDA Meeting," 6 February 1961, MC-01, Folder 4, WHOI-Vine. I suppose Vine imagined himself among the "few others." He also noted that one would need to have studied in advance the "edible marine food supplies of the world." Vine felt that most people were pussy-footing around this issue and deemed it appropriate to plan intelligently for a nuclear exchange. In

May 1961, he wrote to Herman Kahn to congratulate him on his new book, *On Thermonuclear War*. Vine wrote, "It is seldom that one finds such intellectual honesty, particularly when dealing with such a tough and controversial subject. It was also a great relief to hear constructive thoughts about the people who [will] live instead of only lamenting about the ones that [will] die." In this sense, Vine can be seen as a "defense intellectual." Vine to Herman Kahn, 10 May 1961, MC-01 Folder 4, WHOI-Vine.

34.   Vine to Henry Kissinger Jr., 25 April 1969, MC-01, Folder 5, WHOI-Vine.

35.   Memorandum on the disposal of radioactive waste, 27 February 1956, AC-09.5, Folder "Personnel, Vine, Allyn, 1953–1959," WHOI-ODR.

36.   Vine to Hamilton Howze, Vice President of Bell Helicopters, MC-01, 31 May 1966, Folder "Correspondence: One of a Kind Jobs, 1964–1968," WHOI-Vine.

37.   Allyn C. Vine and W. E. Schevill, Memorandum on Submarine Rescue, 13 December 1945, MC-01, Folder "Sub Archives: Memos, Reports, Data, 1944–1961, 1 of 2," WHOI-Vine. Work done under Contract Nobs-2083, formerly OEMsr-31, Task No. 2, Problem 2A. Confidential, declassified 1967.

38.   Weir, *Ocean in Common*, 297.

39.   Shor, *Scripps*, 26–27; see also Hersey and Backus, "New Evidence"; Tucker, "Relation of Fishes"; Marshall, Bathypelagic Fishes." In 1957, Robert Gibbs and Bruce Collette of Woods Hole received a grant from NSF of $170,000 for a study of Gulf Stream fish, which would "incidentally" help determine the role of fish in the deep scattering layer. Box 22, Folder "Activities, NSF, 2 of 4, 1957–1958," WHOI-ODR. This is a good example both of how even "basic" research funded by NSF was influenced by military questions in a solicitous milieu that was imbued with awareness of and concern for military-scientific problems. For an early memo discussing the deep scattering layer and its potential military significance, see memorandum, 17 October 1947, From N428D/Gordon G. Lill to Dr. [Emanuel] Piore, et al., "Preliminary Draft for comment and Criticism 'Oceanography in the Navy Post War Program.'" Box 5, Folder "Operation Crossroads, 1947–47," 3, SIO-Revelle.

40.   On SOSUS, see Weir, *Forged in War*; Shor, *Seeking Signals*; Brown, *Means of War*. The first officially acknowledged detection of a Soviet submarine did not occur until 1962. See "First-Generation Installations and Initial Operational Experience," https://www.public.navy.mil/subfor/underseawarfaremagazine/Issues/Archives/issue_25/sosus2.htm; Declassification of SOSUS occurred in 1991. See Weir, *Forged in War*.

41.   The terminology can be confusing. SOFAR was the locational system first developed based on the sound channel. SOSUS was the submarine acoustic surveillance system that was based on monitoring in the SOFAR channel. Project CAESAR referred to the laying of the cables. SOSUS was designed under the guidance of a scientific group known as Project Hartwell, linked to MIT's Lincoln Laboratory (located on Hartwell Farm Road in Lexington, Massachusetts). The research contribution to SOSUS by Columbia University was Project MICHAEL; the contribution by AT&T Corporation

was Project JEZEBEL. For a discussion of the "cover story" for SOSUS, see "Unclassified Cover Story," https://www.iusscaa.org/coverstory.htm. For one scientist's perspective on this history, see Nierenberg, "Interview."

42. See "First-Generation Installations and Initial Operational Experience," https://www.public.navy.mil/subfor/underseawarfaremagazine/Issues /Archives/issue_25/sosus2.htm. On SOSUS generally, see Weir, *Forged in War*; Schwartz et al., *Atomic Audit*. For background on oceanographers' involvement in and contributions to SOSUS, see Eckart, "Refraction of Sound." On the physical principles behind SOSUS, see Urick, *Sound Propagation*. On Captain Paul Kelly, the manager of the SOSUS program, see "CAPT Joseph P. Kelly, USN (1914–1988)," http://www.public.navy.mil/subfor/cus/Pages /sosus_father.aspx, as well as the personal (and possibly slightly fictionalized) account by his daughter, Wilhelm, *Leviathan*.

43. Historian Cargill Hall suggests there may also have been concern about Soviet trawlers deliberately dragging the seafloor to attempt to break the SOSUS lines. Personal communication with author, 2001.

44. Office memorandum, 10 October 1960, Box 26, Folder 35, WHOI-ODR.

45. Vine to Buships, 4 October 1960, Box 26, Folder 35, WHOI-ODR.

46. For a brief but useful review, see Hornig, *Undersea Vehicles*, 81.

47. Memo, Robert Dietz to Bill Menard, "Program to Measure Deep Ocean Currents at NIO, 9 March 1955, Box 22, Folder "Activities, ONR, Washington, 2 of 2, 1954–1957," WHOI-ODR.

48. Von Arx, "Deep Sea Research," 181. Piccard and Dietz, *Seven Miles*, 134, emphasized that the direct military applications were nil, although the bathyscaphe might prove useful for deep salvage work.

49. The motion was put forward by Willard Bascom in von Arx, "Deep Sea Research," 174.

50. Piccard and Dietz, *Seven Miles*.

51. Ibid., 88.

52. Ibid., 181. See also Harold Froelich to Victoria Kaharl, 31 December 1988, Box 126, Folder 2, WHOI-ODR, which recounts the history of *Trieste* and the motivations behind *Alvin*. Froelich called *Trieste* "essentially an elevator."

53. MC-01, Folder 20, WHOI-Vine; Wenk, *Feasibility Studies*; see also Wenk et al., "Research Submarine," 1960. Wenk later worked at the White House Office of Science and Technology (see Folder "President's Science Advisory Committee," MEP) and at the University of Washington, where he helped to build its marine policy program (Keith Benson, personal communication, 2019).

54. MC-01, Folder 20, WHOI-Vine; Wenk et al., "Research Submarine."

55. Ballard, "Woods Hole's." Charles Momsen Sr. had served as assistant chief of naval operations for undersea warfare, and among other things developed the Momsen lung for diving. See Weir, *Oceans in Common*, 299. *Aluminaut* would be positively buoyant, diving by taking water into its air tanks and rising by jettisoning iron-pellet ballast.

56. Allyn C. Vine, "Proposed *Aluminaut* Program," WHOI Reference No. 60-19, 31 March 1960. This became ONR contract Nonr-3484(00) NR 261-140, 2–3.

57. Ballard, "Woods Hole's."

58. It was also suggested that submarine canyons and seamounts might provide places where submarines could hide, evading sonar detection by running close to the bottom or along the walls. See, e.g., Dolphin and Research Subs, US NEL, R. L. Waldie, "A Proposed Sonar Suit and Research Program for the Deep Diving Submarine (AGSS-555)," Report No. 081, p. 2, declassified 31 December 1969, Box 6, Folder 9, SIO-MPL. Seamounts were also used in tests of the Artemis system; an acoustic signal returned from the known position of a seamount could be used to determine propagation losses. See Berman, "Project Artemis."

59. Allyn Vine to Paul Fye, 21 June 1960, Box 28, Folder 11, WHOI-ODR.

60. Ibid.

61. Department of Navy, ONR (unsigned but presumably Coates) to Paul Fye, 15 June 1962, Box 28, Folder 14, WHOI-ODR.

62. A. C. Vine to Admiral Dwight Day and Reynolds Metal Company, 30 June 1960, Box 38, Folder 11, WHOI-ODR.

63. Fye to J. Louis Reynolds, 25 June 1962, Box 28, Folder 14, WHOI-ODR. This is a retrospective memo laying out the sequence of events leading up to June 1962. Woods Hole's involvement would be financed by ONR under contract Nonr-3484.

64. First quotation: Memo, Vine to Fye, 24 August 1960, Box 29, Folder 11, WHOI-ODR; second quotation: Letter from Director, WHOI, to Chief of Naval Research, 1 April 1960, Box 28, Folder 11, WHOI-ODR. The "preparation" costs mostly involved Woods Hole personnel salaries for time spent on the project. Once operations began, the main costs would be rental of *Aluminaut*, repairs, and maintenance.

65. Office memorandum, Vine to Fye, 2 December 1960, Box 28, Folder 11, WHOI-ODR.

66. Press release, Technical Information Office, ONR, 5 September 1961, Box 28, Folder 13, WHOI-ODR. In September, this was finally worked out and ONR issued a press release stating that a contract would soon be signed "between Woods Hole and Reynolds [that] will provide for lease and eventual transfer of title to the US Navy."

67. Vine to Paul Fye, 21 June 1960, Box 28, Folder 11, WHOI-ODR. See also Fye to Reynolds, 31 March 1960, and Day to Fye, 14 April 1960, Box 28, Folder 11, WHOI-ODR.

68. That, and depth capability, which in most cases probably did not exceed one thousand feet.

69. Vine to Admiral Dwight Day and Reynolds Metal Company, 30 June 1960, AC-09.5, Folder 11, WHOI-ODR. See also memorandum, "Visibility from *Aluminaut*," 16 February 1961, MC-01, Folder 21, WHOI-Vine.

70. Memo, William Schevill to Fye, 16 February 1961, Box 29, Folder 27, WHOI-

ODR. Towed cameras were the principal alternative conceptualized at the time for exploring the deep ocean, and scientists at the NRL and at Scripps were actively pursuing innovations in towed vehicles.

71. Press clipping, "Whale-Like Submarine to Explore Ocean's Depths," *New York Times*, 28 September 1961, Box 28, Folder 13, WHOI-ODR.

72. Reynolds Metals Company, press release, 16 August 1961, prepared for release September 1961, Box 28, Folder 13, WHOI-ODR. During World War II, most submarines operated at depths less than four hundred feet; by the end of the war, this had increased to around a thousand feet. In the 1960s, the Navy information on submarine operating depths was tightly guarded; in the early 2000s information on the *Thresher*'s failure depth was still classified. The declassified documents I examined at that time had all depth references obscured. Sontag and Drew, *Blind Man's Bluff*, 48, claim that *Thresher* failed at 1,300 feet but do not give a source.

73. Reynolds Metals Company, press release, 16 August 1961, prepared for release September 1961, AC-09.5, Folder 13, WHOI-ODR.

74. Reynolds Metals Company, press release, 16 December 1960, AC-09.5, Folder 11, WHOI-ODR. This claim was untrue, as *Alvin* was soon to be built of steel and was only slightly smaller. Unlike the *Trieste* sphere, *Aluminaut* was positively buoyant, but so were conventional submarines; both achieved negative buoyancy by taking on water.

75. Reynolds Metals Company, press release, 16 December 1960, AC-09.5, Folder 11, WHOI-ODR.

76. Ibid.; see also "*Aluminaut*: The Deep Diving Submarine," Reynolds Metals Company Brochure, MC-01, Folder 20, WHOI-Vine. The press releases accompanying the contract signing were followed by articles in the *Wall Street Journal* and the *New York Times*. In the latter, the intersection between the scientific and military questions was brought out explicitly. Fye cited the examination of the "deep scattering layers" of the sea as an example of what the *Aluminaut* could do: "These layers are of both scientific and military interest for they consist of oceanic life and they scatter the sound signals used to detect submarines. The layers form about 600 ft down in daylight, migrating to the surface at sundown. They appear to consist of plankton, the tiny drifting plants and animals of the sea, plus millions of fish feeding upon them. Another scientific problem high on the list of *Aluminaut*, Dr. Fye said, is the question of deep currents and their erosion of the bottom." Press clipping, "Whale-Like Submarine to Explore Ocean Depths," *New York Times*, 28 September 1961, Box 28, Folder 13, WHOI-ODR.

77. "*Aluminaut*: The Deep Diving Submarine," Reynolds Metals Company Brochure, MC-01, Folder 20, WHOI-Vine.

78. This was a common motif, repeating in many public statements about the importance of oceanographic research, later reiterated by the Stratton Commission.

79. Press clipping, Louis Reynolds, quoted in "Concerns Plan to Build Submarine

Capable of Cruising at Record Depth of Three Miles," *Wall Street Journal* [n.d., c. 1961], AC-09.5, Box 13, WHOI-ODR. Exploration of the deep sea proved far more difficult than suggested by its proximity: men would walk on the moon before they walked on the deep-ocean floor, and mining and farming the seafloor have yet to occur.

80.   Press release, Technical Information Office, ONR, 5 September 1961, AC-09.5, Folder 13, WHOI-ODR.

81.   William O. Rainnie Jr. to Robert Frosch, 20 July 1962, AC-09.5, Folder 19, WHOI-ODR. See also WHOI Technical Report 60-19, Proposed *Aluminaut* Program," Allyn C. Vine, 31 March 1960.

82.   A report on Security of Overseas Transport, 21 September 1950, vols. 1 and 2, Project Hartwell, Massachusetts Institute of Technology Archives. The original report was secret.

83.   See Nierenberg, "Interview."

84.   Alan Berman, "Introductory Speech for Artemis General Meeting," 27 September 1963, Box 2, Folder "Hudson Laboratories of Columbia University, Bora Bora Conference January 13–17, 1964, V. C. Anderson, 2 of 2," SIO-MPL. The project was funded primarily by ONR, under contract Nonr-266 (66).

85.   Beyer, *Sounds*, note 113 in chapter 9, calls Artemis an early version of SOSUS, but that is not correct.

86.   Very little has been published on Artemis; this summary is based on Anderson, "MPL and Artemis"; Berman, "Project Artemis"; and conversations and email correspondence with Robert Frosch, Alan Berman, and Fred Spiess. There are various materials dealing with specific aspects of the project, particularly signal processing, in SIO-MPL. For readers still wondering, Artemis is the Greek goddess of the hunt.

87.   Alan Berman, "Introductory Speech for *Artemis* General Meeting," 27 September 1963, Box 2, Folder "Hudson Laboratories of Columbia University, Bora Bora Conference January 13–17, 1964, V. C. Anderson, 2 of 2."

88.   Various documents speak of the project as "urgent" and stress that the Navy was anxious for the project to get started as soon as possible. See, e.g., Box 2, Folder "Nonr 2216(07) Task [Navy Project Artemis] 1958–1960, Office Memorandum, January 27, 1959," SIO-MPL. On the bravery of the scientists and the "battle for decibels," see Berman, "Project Artemis."

89.   Alan Berman, "Introductory Speech for *Artemis* General Meeting," 27 September 1963, Box 2, Folder "Hudson Laboratories of Columbia University, Bora Bora Conference January 13–17, 1964, V. C. Anderson, 2 of 2," SIO-MPL, 6–7, abstracts of papers.

90.   John C. Beckerle, 1969, "Final Report of Contract Nonr-2866, WHOI Reference No. 69-51, p. i. The project ran from March 1959 to June 1969. Most of the environmental measurements appears to have been made from surface ships.

91.   Box 2, Folder "Nonr 2216(07) Task: Navy Project Artemis: 1958–1960, Pro-

posal for participation in US Navy Project Artemis, October 30, 1958, 3 of 3,"
SIO-MPL. MPL participation as approved by C. B. Momsen Jr., head of the
Undersea Branch of ONR, 2 December 1958. Bermuda was the site of most
Navy acoustic experiments and its underwater terrain satisfied project requirements: a large area with a uniform steep slope on which a sonar array
could be placed appropriate depths within the sound channel. Early plans
called for the modules to be fitted by a remotely operated underwater vehicle, but later surveys showed that the terrain was in fact much rougher
than initially thought and unsuitable for the vehicle that had been designed
for the purpose. In the end, the Artemis cables were laid conventionally from
a ship.

92. Monthly Progress Report of the *Aluminaut* Project, Contract Nonr-3484, for
     January 1962, Box 28, Folder 15, WHOI-ODR; quotes from Progress Report,
     Electric Boat Division, General Dynamics Corporation, 11 September 1962,
     Box 28, Folder 15, WHOI-ODR.
93. Memo, Mavor to Fye, *Aluminaut,* Trip Report of 11 September 1962, Box 28,
     Folder 15, WHOI-ODR. This memo summarizes the events of the previous
     year.
94. Fye to Rear Admiral L. D. Coates, 11 April 1962, Box 28, Folder 14, WHOI-ODR.
     See also attached documents.
95. Memorandum to the files, 7 April 1962, Box 28, Folder 14, WHOI-ODR.
96. Department of the Navy, ONR (unsigned but replying to Fye's letter to
     Coates) to Fye, 15 June 1962, Box 28, Folder 14, WHOI-ODR.
97. Ibid.
98. Ibid.
99. Memorandum to the files, 7 April 1962, Box 28, Folder 14, WHOI-ODR.
100. Ibid.
101. Allyn C. Vine, "Proposed *Aluminaut* Program," WHOI Reference No. 60-19,
     31 March 1960. This became ONR contract Nonr-3484(00) NR 261-140, p. 6.
102. Fye to Raymond Stevens, 15 January 1962, AC-09.5, Folder 15, WHOI-ODR.
     Fye hoped that the opposition would fade when he brought in a proper head
     for the project, suggesting that the concern had to do with diffuse manage-
     ment. Hence the appointment a few months later of Earl Hays. However,
     given the larger context of Fye's leadership at Woods Hole, the opposition
     may well have included concerns on the part of Woods Hole scientists that
     the institution was moving too fast and too firmly in the direction of mili-
     tarily motivated science, what Fye called "need-to-know" science (chapter 2).
103. Fye to Raymond Stevens, 15 January 1962, AC-09.5, Folder 15, WHOI-ODR.
104. Jim Mavor to Paul Fye, 1 December 1961, AC-09.5, Folder 29, WHOI-ODR.
105. Memorandum to the files, Subject, meeting with Louis Reynolds, 6 April
     1962, AC-09.5, Folder 14, WHOI -ODR.
106. Term Contracts, Trip Report Memo, Earl Hays and William O. Rainnie Jr.,
     13 June 1962, AC-09.5, Folder 16, WHOI-ODR. There was considerable con-

cern about the safety of aluminum as a hull material for deep submergence; Jacques Cousteau had apparently tried aluminum and "gave up." Memo to the files from Fye, late January 1961, AC-09.5, Folder 27, WHOI-ODR.

107. Department of the Navy, ONR (unsigned but replying to Fye's letter to Coates) to Fye, 15 June 1962, AC-09.5, Folder 14, WHOI-ODR. See also memorandum to the files, Subject, meeting with Louis Reynolds, 6 April 1962, AC-09.5, Folder 14, WHOI-ODR. The specifications and list of invited bidders can also be found in Box 14, Folder 15, "Submersibles: Research Projects." The new contract was Nonr-3484.

108. Memo from Earl Hays to ONR, Code 466, "Informal Progress Report Letter, May, June, July 1962," 20 August 1962, AC-09.5, Folder 16, WHOI-ODR.

109. Ibid.

110. Box: 14, Folder 15, "Submersibles: Research Projects, May 18, 1962," 16, SIO-OCR. Specifications for the Design and Construction of a Research Submarine for Operation to a Depth of 6,000 ft. They settled on HY-100 steel.

111. This comment shows they already knew about deep currents, at least in some portions of the ocean (chap. 8).

112. Typescript, Summary of Bidders' Conference, Emphasis added, Box 14, Folder 15, "Submersibles: Research Projects, May 18, 1962," SIO-OCR.

113. Ibid., 7.

114. Memo from Earl Hays to ONR, Code 466, "Informal Progress Report Letter, May, June, July 1962," 20 August 1962, AC-09, Folder 16, WHOI-ODR.

115. Gary Hoglund to Director, SIO, 13 October 1961 (describing *Seapup*), Box 14, Folder 14, "Submersibles: Research Projects 1959–1966," SIO-OCR. See also Ballard, "Woods Hole's."

116. Buships memo, Paul Fye to Chief of Bureau of Ships, "Deep Submergence Research Program," 12 June 1962, AC-09.5, Folder 16, WHOI-ODR.

117. Office memorandum to Paul Fye from J. W. Mavor Jr. 18 September 1962, AC-09.5, Folder 15, WHOI-ODR.

118. Memo from Earl Hays to ONR, Code 466, "Informal Progress Report Letter, May, June, July 1962," 20 August 1962, AC-09.5, Folder 16, WHOI-ODR. There were tensions over Vine's freewheeling manner of operating. When Fye wrote a memo on 14 January 1962, outlining who was who on the project, Jim Mavor had complained that an earlier letter did not do the trick, because it did not "limit Al Vine's external activities at all.... Are you aware that Columbus Iselin once found it necessary to widely distribute a letter stating that Al Vine did not represent this institution?" Memo to Fye, 21 December 1961, AC-09.5 Folder 27, WHOI-ODR.

119. Fye to Michael Waller, 25 July 1962, AC-09.5, Folder 19, WHOI-ODR. Crewed dives were associated in many people's minds with public spectacle, such as the famous bathysphere dives of Beebe and Barton in the 1920s.

120. William O. Rainnie Jr. to Robert Frosch, 20 July 1962, AC-09.5, Folder 19, WHOI-ODR.

121. Ibid.

122. Kissinger, Frosch, and Spiess to Paul Fye, n.d., Box 14, Folder 15, "Submersibles: Research Projects 1959–1966," SIO-OCR.

123. Memo, William Rainnie Jr. to WHOI Files, 15 May 1962, AC-09.5, Folder 15, WHOI-ODR.

124. Kissinger, Frosch, and Spiess to Paul Fye, n.d., Box 14, Folder 15, "Submersibles: Research Projects 1959–1966," SIO-OCR.

125. The civilians were primarily members of the design and production divisions of the Portsmouth Naval Yard where it was built, plus representatives of the Sperry and Raytheon Corporations.

126. "Hope Abandoned," *New York Times*, 5 June 1963.

127. The fast attacks were not intended to carry or launch warheads but to find Soviet subs that did. However, much of the initial publicity in the press is ambiguous about this, suggesting that the public may have been confused about it as well. Since both fast-attack and boomers were nuclear powered, they were both referred to as nuclear submarines, but only the latter normally carried warheads.

128. US Navy Court of Inquiry Summary, 1963, item 49. That *Thresher* was "beautiful" was a view shared by many. See, e.g., Spiess and Maxwell, "Thresher." In their account of scientists' efforts in the search, they also note that despite the loss of *Thresher*, the nuclear submarines of the 1960s were far safer than those of World War II.

129. Copyright 1964, Barricade Music, Inc., reprinted with permission of Hal Leonard LLC.

130. When pressed under examination in the US Navy Court of Inquiry to hazard a guess as to what words preceded the words *test depth* in that penultimate message, Watson testified that both he and his commanding officer felt that they heard the word *exceeding*.

131. Testimony of Lt. James D. Watson, US Navy Court of Inquiry, 13 April 1963; declassified 16 February 1999, courtesy of David van Keuren, US NRL, Washington, DC.

132. "Loss at Sea of the USS *Thresher*, Summary of Events." US Navy Court of Inquiry, issued to the US NRL, Washington, DC, at the request of historian David van Keuren, 16 August 1999, Reference No. 5830, 354.1: 9200923. My summary is based on materials in the published congressional record and materials related to the US Navy Court of Inquiry record as released to van Keuren. These include a copy of the basic investigative report and the summary of events from the US Navy Court of Inquiry, approximately one hundred pages of documents. The remaining materials from the US Navy Court of Inquiry, approximately twelve boxes containing forty-two thousand pages of material, remained classified "under current Department of Defense directives" at the time that I did the research for this chapter (in the early 2000s). Memo, 16 August 1999, Judge Advocate General to David van Keuren.

133. By the end of its session on 5 June, the court had heard testimony from 179 witnesses, producing 1,700 pages of testimony. In June, Congress convened

hearings as well. A huge array of issues was raised, including quality control at the Portsmouth shipyard, deviations from design specification, failure of specific joints and piping, safety of the nuclear reactor compartments, and the effect of automation and mechanization on worker pride. The *Thresher* was part of a major expansion of the US submarine fleet, the first of a planned set of twenty-five nuclear-powered fast-attack submarines. The cost, scale, and complexity of building had accordingly increased; during the previous four years, the Navy shipbuilding budget had nearly doubled from $2.5 billion to $4.5 billion. Yet personnel in the Bureau of Ships had decreased by about 20 percent, even as the complexity of the tasks involved in designing, building, and testing nuclear submarines had dramatically increased, a backdrop implying problems in design, supervision, and quality control.

134. Because *Thresher* was a nuclear submarine, it also fell under the jurisdiction of the US Congressional Joint Committee on Atomic Energy, which heard testimony that quality control was difficult in Navy operations because of the constant rotation of personnel. Career advancement required experience in as wide a variety of areas as possible, so as soon as a man became experienced, he was rotated, sometimes leaving inexperienced personnel to inspect or supervise at critical times. Workers at the Portsmouth Naval Yard testified that specifications were often viewed as goals rather than requirements. They presented a worrisome example: of the 3,000 silver-brazed joints on the vessel, only 145 had been ultrasonically tested. Of those 145, 14 percent were found to be below standard, but the shipyard commandant decided not to pursue further tests, lest completion deadlines be missed. While these types of joints had been used for many years, they were being asked to contend with depths where their capabilities were unverified. In 1993, the Navy finally declassified the primary materials related to the cause of the loss of the *Thresher*, and the 1968 loss of the US Submarine *Scorpion*, in response to a Freedom of Information Act request from the *Chicago Tribune* and ABC News. The *Thresher* was found to have sunk due to "a faulty silver-braze piping joint," while the *Scorpion* was sunk by a hot running torpedo, which turned on the ship and destroyed it. William Broad, "Navy Says 2 Subs Pose No Hazard," *New York Times*, 7 November 1993; "Navy Says Whistle-Blower Lied about Dumping," *Chicago Tribune*, 6 June 1993. An excellent summary of relevant declassified documents and news reports related to *Scorpion* is given in Sontag et al., *Blind Man's Bluff*. In 1993, the Navy also released results of radiological surveys of the *Thresher* and *Scorpion* sites by *Alvin* in 1983 and 1986, as well as a survey of a sunken Russian sub site. The reports claimed "no significant impact" on the radioactivity of the environment surrounding either of the two US subs; this despite the fact that, against earlier denials, it was admitted that *Scorpion* carried two Mark 45 Astor torpedoes armed with nuclear warheads. On 27 October 1993, Jane Caruso, Woods Hole security officer, penned a memo reminding staff that, even though the Navy had declassified these materials, nondisclosure agreements were still in

effect. Thus, even with the end of the Cold War and declassification of military documents, Woods Hole scientists still operated under conditions that restricted their ability to discuss their work.

135. A. J. Hollings, 1963, "Narrative" [of the *Thresher* search], 12–20 April 1963 and 20 April–18 May 1963, typescript, NRLT.
136. Spiess and Maxwell, "Thresher," 1964; Memo from WHOI to ONR, Code 416, 24 May 1963, MC-01, Folder 39, WHOI-Vine. Fye later billed the Navy for costs lost due to the diversion of Atlantis II to the search: $236,649.
137. NRL Notice 3130, 8 May 1963, NRLT. This idea is also mentioned in Hollings, A. J., 1963, "Narrative" [of the *Thresher* search], 12–20 April 1963 and 20 April–18 May 1963, p. 7, typescript, NRLT. Hollings (8) also mentions the suggestion that the *Guardfish*, sunk off Long Island in torpedo trials, might be used as a sample target, but I have been unable to confirm if this suggestion was pursued. See also Spiess and Maxwell, "Thresher."
138. A. J. Hollings, 1963, "Narrative" [of the *Thresher* search], 12–20 April 1963 and 20 April–18 May 1963, typescript, NRLT. See especially the entry for 20 April 1963 at 9.
139. Dwight W. Batteau to Vine, 16 May 1963, MC-01, Folder 39, WHOI-Vine.
140. Minutes of the April 27 Meeting of the SUBLANT Technical Advisory Group for Thresher Search Operations at Woods Hole, 1 May 1963, ONR: 416: AEM: dmt, NRLT.
141. Once the hull was located, a further radiation survey concluded that "any release of fission products added less than 3% to the normal radioactivity of the water." Although the hull had been crushed, the reactor core had not been breached. Riel et al., *Radiation Survey*. The reference to "normal levels" reflects the fact that the ambient radionuclide levels in the world oceans were elevated compared to pre–World War II levels by contamination from Pacific nuclear tests. *Alvin* participated in several later surveys, which were declassified in the early 1990s.
142. Spiess and Maxwell, "Thresher," 1964.
143. Wakelin, "Lesson and Challenge"; Andrews, "Searching." The magnetic anomalies are discussed in the minutes of the 18 June 1963, TAG meeting, ONR: 416: AEM: mlg, 19 June 1963, NRLT.
144. "Largest Section of Lost Submarine Located," *Washington Post*, 4 March 1964.
145. On the history of *Mizar*, see Brundage, *Search Era*.
146. "NRL Research Reservists Participate in *Thresher* Search," typescript, NRLT.
147. Andrews, "Section II," 775.
148. Weir, *Ocean in Common*, 328.
149. Andrews, "Section I," 553.
150. Buchanan, "Love Affair."
151. Andrews, "Section I," 554.
152. Andrews, "Section II," 773.
153. Andrews, "Section I," 549–52. See also Hornig, *Undersea Vehicles*, 12.
154. "Navy names personnel to hold key posts in Deep Submergence Systems

Review Group," Department of Defense Office of Public Affairs News Release, 658–63, 9 May 1963, Box 14, Folder 15, "Submersibles: Research Projects 1959–1966," SIO-OCR. The task force was established in May 1963. One can only speculate from what other large bodies the Navy expected to rescue men on the seafloor.

155. Andrews, "Section I" and "Section II." The first chief scientist of DSSP was John Craven; on his work, see Sontag, Drew, and Drew, *Blind Man's Bluff*.

156. Fye to Admiral Stephan, 28 May 1963, AC-09.5, Folder 46, WHOI-ODR. See also Box 14, Folder 11, "Navy DSRV: Commemorating the Launching of the US Navy's First Deep Submergence Rescue Vehicle," SIO-OCR.

157. Stephan to Fye, 9 May 1963, AC-09.5, Folder 46, WHOI-ODR.

158. I do not know what happened to the penalty clause; presumably this was renegotiated.

159. On *Aluminaut*'s final costs, see Press clipping, "Research Subs: Sinking Business," *Los Angeles Times*, 13 December 1971, 12, Box 14, Folder 13, "Submersibles: Miscellaneous," SIO-OCR. On *Aluminaut*'s trial dives, see, "*Aluminaut* Operations Brief," 15 November 1965, Box 14, Folder 16, "Submersibles: Research Projects, 1964–1965," SIO-OCR. In November 1965, *Aluminaut* successfully dived to 6,250 feet in the Bahamas, and Reynolds announced that it was "ready to perform time charter contracts for the scientific and industrial communities to depths of 6,000 ft." (Apparently the 15,000-foot goal was not achieved.) Reynolds made a virtue of necessity and boasted that the vessel had been entirely paid for by Reynolds Metals, "a clear example of how our system of free enterprise works in the best interest of the public," September 1964, rpt., Box 14, Folder 16, "Submersibles: Research Projects, 1964–1965, *UST*," SIO-OCR. That same year, company representatives visited Scripps in an attempt to interest Fred Spiess in the vessel, but nothing came of it.

160. The upper six thousand feet includes only about 15 percent of the ocean by area.

161. This summary is taken primarily from WHOI, a guide to the records of the deep-submergence vehicle *Alvin*, 1949–1998 (bulk 1964–1988), "Operational Scientific Service Records, Historical Information," http://archives.mblwhoilibrary.org/repositories/2/resources/127. In later discussions, I have seen it claimed that *Alvin* was named after Alvin and the Chipmunks, or some combination of the Chipmunks and Al Vine, but I have seen no evidence for this interpretation in primary documents.

162. Memo, 15 June 1965, AC-09.5, Folder 25, WHOI-ODR. A copy of this can also be found in Memo, "DSRV Survey of Artemis Array," and attached "Planning Guide for Subject Survey," 15 June 1965, AC-18, Folder 28, WHOI-OSSR.

163. Alan Berman, "Introductory Speech for Artemis General Meeting," 27 September 1963, Box 2, Folder "Hudson Laboratories of Columbia University,

Bora Bora Conference January 13–17, 1964, V. C. Anderson, 2 of 2," SIO-MPL. See also memo, "DSRV Survey of Artemis Array," and attached "Planning Guide for Subject Survey," 15 June 1965, AC-18, Folder 28, WHOI-OSSR.

164. Alan Berman, "Introductory Speech for Artemis General Meeting," 27 September 1963, Box 2, Folder "Hudson Laboratories of Columbia University, Bora Bora Conference January 13–17, 1964, V. C. Anderson, 2 of 2," 3, SIO-MPL.

165. Memo, "DSRV Survey of Artemis Array"; and attached "Planning Guide for Subject Survey," 15 June 1965, AC-18, Folder 28, WHOI-OSSR.

166. Ibid.

167. Retrieval was not a trivial task. Each module was fifty-seven feet long, and maintained in a vertical position by an aluminum sphere, at the top of a forty-eight-foot mast and held down on the ocean floor by a cast steel base. For *Alvin* to retrieve a module required severing it from its base. A memo of 1 July 1965, discusses various options for doing this; I have not been able to determine how, if at all, a module was actually retrieved. ONR, Code 467, memo, "Underwater Module Recovery," 1 July 1965, AC-18, Folder 28, WHOI-OSSR.

168. Memorandum, 6 December 1963, J. C. Mumson to Artemis Systems Research Committee, "Conclusions and Recommendations regarding the Environmental Measurements Program, Box 3, Folder "Artemis System Research Committee, Minutes, December 18, 1963," SIO-MPL.

169. Memo, "DSRV Survey of Artemis Array"; and attached "Planning Guide for Subject Survey," 15 June 1965, AC-18, Folder 28, WHOI-OSSR. Discussion of temperature and current measurements in Section II, *Alvin* Tasks, ONR.

170. William Rainnie Jr., typescript, "*Alvin* Operations and the H-bomb Search," AC-18, Folder 17, WHOI-OSSR. Consistent with its designation—deep-submergence vehicle (DSV)—*Alvin* is called a research vessel, although at this point it had done no scientific research.

171. Schwartz, *Atomic Audit*.

172. Place et al., "Summary Report."

173. The debris and contaminated soil were packed into 4,810 fifty-five-gallon barrels and buried in an AEC facility in Aiken, South Carolina. See Lewis, *H-Bombs*, 135; Place et al., "Summary Report," 68.

174. WHOI, *Alvin* Materials (unprocessed), Press clipping, Russell Baker, "How to Find a Lost H-bomb," *Boston Herald*, 27 January 1966.

175. These articles ultimately became a book: Lewis, *H-Bombs*. For the cartoon see 91.

176. Lewis, *H-Bombs*, illustration opposite 135.

177. Baldwin, "Nuclear Mishap." This article cited a list of nuclear accidents—fifteen in all known to the public—a list that could either be viewed as "reassuring" or chilling.

178. Place et al., "Summary Report."

179. Ibid., DOD quote at 192, "reassuring" quote at 24. The military also later acknowledged that it was in fact possible "low-order detonation" had occurred; because the bomb was underwater, it might not be obvious. As far as I can tell, this was never publicly acknowledged at the time. It appeared in 1975 in the Air Force report. Ibid., 196.

180. Hayes, "Aircraft Salvage," 1.

181. Using the resources of the Sixth Fleet, the Navy mobilized what it would call Task Force 65. Its director would be Rear Admiral William S. Guest, deputy commander of naval strike and support forces, Southern Europe; the technical work would be coordinated by Rear Admiral Leroy V. Swanson, director of fleet operations.

182. Place et al., "Summary Report," 75.

183. William Rainnie Jr., Folder "Oceanus Article: H-bomb Search Manuscript"; William Rainnie Jr., "*Alvin* Operations in H-Bomb Search," typescript, n.d., AC-18, Folder 17, WHOI-OSSR.

184. Place et al., "Summary Report," 92–93; Swanson, "Aircraft Salvage," 5. On OBSS and Submarine, see also Lewis, *H-Bombs*, 135; Press clipping, *Time*, 19 January 1968, Box 14, Folder 13, "Submersibles, Misc.," SIO-OCR.

185. Place et al., "Summary Report," 100–106.

186. Earl Hays to Paul Fye, 8 March 1966, and William Rainnie Jr. to ONR Code 466, "Quarterly Informal Letter Status Report of Contract NONR 3484(00) Deep Submergence Research Vehicle Project, January 1, 1966 through April 9, 1966," 10 June 1966, AC-18, Folder 7, WHOI-OSSR.

187. Earl Hays to Paul Fye, 8 March 1966, AC-18, Folder 7, WHOI-OSSR. Fye wrote back, thanking the men for their hard work, and reminding them to write to their wives, who were "getting anxious." Fye to John Bruce, 8 March 1966, AC-18, Folder 7, WHOI-OSSR.

188. On *Mizar*'s role in the search and recovery, see Buchanan, "Search for the Scorpion," 20.

189. Place et al., "Summary Report," 203.

190. Szulc, "Cable Snaps."

191. Place et al., "Summary Report," 148.

192. Swanson, "Aircraft Salvage," 4.

193. Place et al., "Summary Report," 114.

194. William Rainnie Jr. to ONR Code 466, "Quarterly Informal Letter Status Report of Contract NONR 3484(00) Deep Submergence Research Vehicle Project, January 1, 1966 through April 9, 1966," 10 June 1966, AC-18, Folder 7, WHOI-OSSR.

195. Andrews, "Section II"; Hayes, "Aircraft Salvage"; Swanson, "Aircraft Salvage"; Place et al., "Summary Report."

196. Place et al., "Summary Report," 196. Photos were permitted the following day, after identifying marks were obscured. This was a break with prior policy that permitted no photographs of nuclear weapons of any kind, a departure deemed necessary to placate international concern. After this, the

regular policy permitting neither discussion nor pictures was restored. Ibid., 200.

197. Ibid., 141, gives the total cost as $10,230,744, or $126,305 per day, in 1966 dollars. This does not include other costs, such as administrative costs in Washington, or the claims later paid to Spanish citizens, many of which were still unsettled years later. According to the same report (181), over $7 million in damages were claimed, of which approximately $700,000 was paid.

198. Swanson, "Aircraft Salvage." See also Buchanan, "Scorpion Search."

199. William Rainnie Jr. to ONR Code 466, "Quarterly Informal Letter Status Report of Contract NONR 3484(00) Deep Submergence Research Vehicle Project, January 1, 1966 through April 9, 1966," 10 June 1966, AC-18, Folder 7, WHOI-OSSR.

200. Larry Megow, "Little *Alvin* and the Bomb," typescript, 2 May 1966, WHOI-OSSR. Megow was the vice president of Hahn and Clay's Steel Fabrication Division.

201. Swanson, "Aircraft Salvage," ix.

202. Ibid., 35.

203. Ibid., 4, 7.

204. Memorandum, Frank Omohundro to Paul Fye, 20 July 1966, AC-09.5 Folder 25, WHOI-ODR.

205. WHOI Technical Report, Arnold G. Sharp and Lawrence A. Shumaker, "DSRV *Alvin*: A Review of Accomplishments," 76–114, January 1977.

206. Fye to Captain Allan R. Davison, 24 June 1966, AC-09.5, Folder 25, WHOI-ODR.

207. "Meeting on Scientific Problems and Instrumentation for *Aluminaut*," 12 February 1962, AC-18, Folder 20, WHOI-OSSR.

208. "Desired Work on *Aluminaut*" questionnaire, MC-01, Folder 20, WHOI-Vine. Two replies were kept in this file, one from Richard Backus, who wished to work on fish and other animals in the upper thousand meters and on the bottom, and Hartley Hoskins, who was interested in examining outcrops in the Hydrographer and Oceanographer submarine canyons, at depths down to five hundred fathoms (three thousand feet). MC-01, Folder 21, WHOI-Vine. A reply by Carl Bowen describes his interest in "seeing and sampling the bottom to aid in structural and tectonic interpretations."

209. There is no sustained discussion of the Artemis program either, but that can be explained by classification. Moreover, enough is said about Artemis to reconstruct its specific goals, which is more than can be said about any intended scientific work.

210. Typescript, Staff meeting, 12 December 1961 [Fye speech to staff], MC-24, Folder "Palace Revolt, 1961," WHOI-von Arx.

211. An article in Nature in 1983 noted that the squid had still never been observed in their natural habitat; see Brix, "Giant Squids." Fifty years after Aldrich proposed his study—the giant squid remained mysterious. See Ellis, *Giant Squid*; see also Grann, "Squid Hunter."

212. Fye to Frederick Aldrich, 22 July 1965, Box 43, Folder 24, WHOI-ODR.
213. Allyn C. Vine, "Proposed *Aluminaut* Program," WHOI Report Reference No. 60-19, 31 March 1960.
214. Ibid. This became ONR contract Nonr 3583 (00) NR 261-140.

CHAPTER SEVEN

1. On Spiess, see "Obituary Notice Pioneer in Ocean Technology: Fred N. Spiess," https://scripps.ucsd.edu/news/2591.
2. For details of the Deep Tow System, see Spiess and Tyce, "Cruise Report." The earliest reference to the idea for Deep Tow that I have found is in Fred Spiess to Edward Wenk Jr., 31 October 1958, MC-249, Folder 20, WHOI-Vine. Spiess wrote that a group at Scripps—Vacquier, Menard, Allen, Shor, and Spiess— were interested in the "problem of studying the character of the deep ocean floor with considerable more detail than is possible with present instrumentation. Although our interest stems primarily from a desire to understand the ocean basins better, we have been further encouraged by the variety of military and economic uses to which this information can be put."
3. Research and Technology Resume, 17 June 1966, "Near Bottom Acoustic Properties," Box 4, Folder "Nonr 2216 (05) Class Amend. DD 1498s, 1965–June 1966," SIO-MPL. For readers who are innocent of such military matters (as I was when I began this project), fire control refers to the targeting of weapons, as in ready, aim, fire (not, for example, putting out fires on ships). The *Subroc* system came on line in 1965; see Spiess, "Motivating." It was guided by the Subroc Technical Advisory Group, or STAG. Spiess was the representative from Scripps; Al Vine from WHOI.
4. This ONR grant supplanted earlier support in the form of a general contract from the Bureau of Ships.
5. Nobsr 72512. See MPL Annual Summary Report, 1 November 1957–1958, PL-U-18/58, Box 4, Folder "Annual Reports 1958–1964," SIO-MPL. From 1958 to 1967, ONR's basic grant supplied approximately half the MPL annual budget; see Spiess, "Rough Draft," 7 December 1967, Box 4, Folder "Annual Reports 1965–68," SIO-MPL. In the early period, accountings seem to have been quite loose. A memo dated 21 September 1962, notes that from then on, "all procurement actions must be identified with a specific project," Box 7, Folder "MPL Memoranda, F. W. Outler to MPL Project Leaders, 21 September 1962," SIO-MPL.
6. Research and Technology Resume, 16 June 1965, "Marine Physics," Box 4, Folder "Nonr 2216 (05) Class Amend. DD 1498s, 1965–June 1966," SIO-MPL. Other documents in this folder discuss other Deep Tow activities, including work on behalf of the Bureau of Naval Weapons and work on behalf of ONR in helping to site SEALAB II.
7. Contract Nonr 2216(05), MPL Annual Summary Report, 1 November 1957– 1958, PL-U-18/58, Box 4, Folder "Annual Reports 1958–1964," SIO-MPL.

8.   Research and Technology Resume, 17 June 1966, "Sea Floor Properties," Box 4, Folder "Nonr 2216 (05) Class Amend. DD 1498s, 1965–June 1966," SIO-MPL.

9.   F. N. Spiess, typescript "Rough Draft," 7 December 1967, Box 4, Folder "Annual Reports 1965–68," SIO-MPL. See also Spiess to Admiral W. N. Dietzen Jr. (OP-23) 22 March 1973, Box 6, Folder "Project Deep Tow," SIO-MPL.

10.   Enclosure 1, to MPL file, 11 November 1971, Box 6, Folder "Project Deep Tow," SIO-MPL. This was part of a set of materials sent both to DSSP and Naval Operations accompanying a request for ongoing support for Deep Tow; Spiess noted that Enclosure 3 was intended as much for Naval Operations as for DSSP. Spiess to Admiral W. N. Dietzen Jr. (OP-23), 22 March 1973, Box 6, Folder "Project Deep Tow," SIO-MPL. Spiess was not claiming that the original intent of Deep Tow was to search for things on the seafloor—Deep Tow was *designed* to map the seafloor in greater detail and precision than could be obtained by echo sounding from surface ship. Rather, he was stressing that the same technology that could map the seafloor could also find something sitting on it.

11.   Atwater and Mudie, "Gorda Rise"; see also Spiess, "Origins"; Spiess, *Dr. Fred Noel Spiess*.

12.   Fye to Von Herzen and attached memo from Von Herzen, "*Alvin* and the Mid-Atlantic Ridge," 15 January 1968, AC-09.5, Folder 19, "Admin., DSRV *Alvin*, 1968, 1 of 3," WHOI-ODR.

13.   Phillips was best known for work on paleomagnetism in relation to plate tectonics. See Phillips and Forsyth, "Plate Tectonics."

14.   Memo, J. D. Phillips to E. Hays, 11 March 1968, AC-09.5, Folder 19, "Admin., DSRV *Alvin*, 1968, 1 of 3," WHOI-ODR.

15.   Typescript, Staff meeting [Fye comments to staff], 12 December 1961, MC-24, Folder 1, "Palace Revolt," WHOI-von Arx. See also chapter 3.

16.   F. N. Spiess, typescript "Rough Draft," 7 December 1967, Box 4, Folder "Annual Reports 1965–68," SIO-MPL. Other documents around this time refer to pending cuts to the ONR budget for the MPL. While the driving force behind the cuts was likely the Vietnam War, there is no explicit reference to this in these files.

17.   *Turtle* was the name of the first working submarine, built by David Bushnell and used during the American Revolutionary War. On Hahn and Clay, see Kaharl, *Water Baby*, 35–38.

18.   Swanson, "Aircraft Salvage," 10.

19.   Ibid., 13.

20.   Place et al., *Palomares Summary Report*, 106–7, summarized the Air Force's conclusions after Palomares: that remotely operated vehicles lacked adequate navigation and communication capabilities, adequate leverage for hoisting objects, and adequate escape mechanisms in the event of entanglement or entrapment. They also suffered from crew fatigue, and the long period (about eight hours) required to recharge the batteries after each dive. The report summarized: "Tethered vehicles provide maximum endur-

ance, safety and area coverage.... Manned submersibles provide maximum maneuverability and flexibility, but are endurance limited by life support systems, crew fatigue and power limitations" (138). This summary was released in 1975, but Frank Andrews had come to the same conclusion after the *Thresher* search: "It is the author's personal opinion ... that the manned vehicle is not as efficient in the deep ocean search as is the towed vehicle." See Andrews, "Section II," 776. Andrews believed that crewed vehicles had a role to play where visual inspection was required, but that rescue and salvage would be better done with remotely crewed equipment.

21.　Swanson, "Aircraft Salvage," 10.

22.　General Dynamics, press releases, 3 May 1966 and 5 January 1966, Robert West, Director of Customer Relations, General Dynamics, Electric Boat Division, to William Nierenberg, Box 14, Folder 10, "Submersibles," SIO-OCR. West notes that General Dynamics has made Asherah available to other academic institutions, including the Smithsonian and University of Rhode Island, and he suggested that Scripps might also make use of it. So far as I am aware, they never did.

23.　Box 14, Folders 9 and 10, "Submersibles," SIO-OCR. The depth capacities of these boats are often quoted variously; where possible, I have cited the manufacturer's claims; where this was unavailable, I have gone with the most frequently cited values. On NEL's use of *Deep Star*, see Box 14, Folder 14 "Submersibles: NEL," SIO-OCR.

24.　"Navy DSRV: Commemorating the Launching of the US Navy's First Deep Submergence Rescue Vehicle," n.d., c. January 1970, Box 14, Folder 11, SIO-OCR, and Box 14, Folder 12, "Press Clippings," SIO-OCR.

25.　"Navy DSRV: Commemorating the Launching of the US Navy's First Deep Submergence Rescue Vehicle," n.d., c. January 1970, Box 14, Folder 11, SIO-OCR. *Deep Quest* was billed as a prototype for *DSRV-1*. In this case, DSRV stood for deep-submergence rescue vehicle, not deep-submergence *research* vehicle. *DSRV-1* had a depth capacity of five thousand feet. In addition, Lockheed was building DSSV—deep-submergence search vehicle (sometimes referred to as *DSRV-2*) with depth capacity of twenty thousand feet. See also Press clipping, "Oceanology: Work beneath the Waves," *Time*, 19 January 1968; Press clipping, "Submersible Fleet Needs More Use," *UST*, April 1969 Box 14, Folder 13, "Submersibles, Misc.", SIO-OCR. *Time* reported that the Navy had commissioned "the world's first nuclear-powered research submarine, the NR-1, to be built by General Dynamics and able to "sustain a crew of seven underwater for at least thirty days." So far as I know, this was never completed. Corporate brochures Box 14, Folder 13, "Submersibles, Misc.", SIO-OCR. Submersibles were also being built in other countries, such as the Kawasaki Dockyard's *Shinkai*, in Japan, with a depth capacity of six hundred meters.

26.　Miscellaneous, clipping, "Submersible Fleet Needs More Use," *UST*, 36–40, April 1969, Box 14, Folder 13, "Submersibles," SIO-OCR. By 1969, this article was lamenting the overbuilding of submersible capacity, which it blamed on decreased government demand due to the costs of the Vietnam War. In 1971, a *Los Angeles Times* article discussed the financial losses among private contractors who had built submersibles. Certainly, scientists working in the federal government had encouraged the private sector to believe that there would be a large demand for such vessels, akin to the demand for rocketry associated with the space program. See, e.g., Press Release, Executive Office of the President, Office of Science and Technology, 12 January 1965, Box 14, Folder 16, "Submersibles: Research Projects, 1964–1965," SIO-OCR.

27.   Press clipping, "The Deep Sea ... Design's Newest Frontier," *Product Engineering*, 84–94, Box 14, Folder 13, "Submersibles: Miscellaneous," SIO-OCR. Why inner space never captured the public imagination as outer space did is an interesting question; for general background on deep-ocean exploration, see Rozwadowski, *Fathoming the Ocean*.

28.   In May 2002, *Dolphin* flooded and caught fire off the coast of San Diego. The crew was rescued; the boat was damaged but later repaired and remained in service until 2007. "Navy Sub Catches Fire," *CBS News*, 22 May 2002, https://www.cbsnews.com/news/navy-sub-catches-fire/; Jay Withgott, "Navy Research Sub Burns in Pacific," *Science*, 22 May 2002, http://www.sciencemag.org/news/2002/05/navy-research-sub-burns-pacific.

29.   Box 14, Folder 10, "Submersibles—General Dynamics and Convair," SIO-OCR. I have not seen this stated explicitly, but it seems to be implicit in activities taking place around this time and in statements made by Navy officials. For example, a *San Diego Union* article about *Asherah*, on 17 September 1966, reported that "a number of high ranking Naval Officers have made dives in the boat since its arrival in San Diego (for an Interior Department project). It was also used for a secret military project recently on the East Coast." With the development of additional submersibles, it was possible to consider subcontracting a specific submersible for a specific task, making it less desirable for the Navy to commit resources inflexibly to *Alvin*. In the fall of 1969, naval oceanographic office engineer Frank Busby, who had been extensively involved in the submersible program, including the use of *Aluminaut* for a bottom survey of the Navy Atlantic Fleet Weapons Range near the Virgin Islands, was quoted in the press opposing the idea of the Navy supporting an all-purpose submersible: "The concept of an all-purpose submersible is a delusion that only serves to frustrate the user and retard the development and application of manned submersibles." *Copley News Service*, "Oceanography Special: Oceanography Boom Hasn't Panned Out," 23 September 1969, Box 14, Folder 13, "Submersibles: Miscellaneous," SIO-OCR. Moreover, ONR was not the only branch of the Navy involved in deep submergence. Under Project GLAUCUS, NEL and the Bureau of Ships had developed a design for a deep-diving submarine to operate at four thousand feet, for "acoustic, weapon, and oceanographic research."

30.   Fye, memo to files on telephone conversation with James W. Smith, ONR, 6 March 1968, AC-09.5, Folder 21, WHOI-ODR. In conversation, many scien-

tists have told me that the decline in Navy funding for basic research was the result of the 1970 Mansfield Amendment, which required that any military funds dedicated to research were directly related to military projects. But the ONR had begun to back away from *Alvin* before this. In fact, already in the late 1960s, the Navy was concerned about overreliance on academics for operationally crucial matters. A 1969 memo from the deputy assistant oceanographer of the Navy to the "distribution list" noted: "The Chief of Naval Operations has stated that the Navy cannot depend in any large part on non-Navy oceanographic contributions to operational readiness because, it is believed, other ocean research or development programs will not be fully responsive to Navy requirements." Memo, 16 April 1969, Box 31, no folder, HHP.

31. Toye, "Deep Submergence."
32. Memo, Spiess to Nierenberg, "NSF Block Funding Ship Time for Deep Tow Work," MPL file 02-U-46, 15 March 1968, Box 7, Folder, "Deep Tow," SIO-MPL.
33. Sapolsky, *Science and the Navy*, 88–89.
34. There is a huge literature. One useful website is Jerry M. Lewis and Thomas R. Hensley, "The May 4 Shootings at Kent State University: The Search for Historical Accuracy," https://www.kent.edu/may-4-historical-accuracy.
35. Spiess, "Acoustics Community," 10. Other scientists have made the same comment to me, that the close relations with the Navy began to break down in the late 1960s and had greatly diminished by the mid-1970s. It appears that the legacy of World War II lasted a generation, but no more, which would tend to reinforce the importance of personal ties.
36. AUTEC was one of the Navy's most secret facilities; many documents from this period refer to it only obliquely, often referring to "work in Bermuda" without specifying AUTEC. Indeed, AUTEC was (and remains) so secret that conspiracy theorists call it the Navy's "Area 51," and have tied it to UFO activity and Bermuda Triangle disappearances. See the website Navy Secrets, at https://www.thelivingmoon.com/45jack_files/03files/AUTEC_Navys _Area_51.html; Matthew Stuart, "The Navy has its own Area 51 and its right in the middle of the Bahamas," *Business Insider*, 25 July 2019, https://www .businessinsider.com/navy-area-51-autec-in-the-bahamas-2016-10. I would suggest that the fact that large, important facilities were hidden has contributed to conspiracist ideation in contemporary American culture (see chapter 9).
37. Rainnie to Fye, n.d., c. 1968, AC-09.5, Folder 21, WHOI-ODR.
38. Ibid.
39. Memorandum, Chief of Naval Research to Chief of Naval Operations, 7 June 1968, Box 56, Folder 20, WHOI-ODR. In response, Brackett Hersey suggested developing a new acoustics initiative, see memorandum, Hersey to Fye, 2 July 1968, AC-09.5, Folder 20, WHOI-ODR.
40. Perhaps because of the other events overtaking US society in the spring and

summer of 1968, particularly the assassination of Robert F. Kennedy in early June, the *Scorpion* loss received far less attention than that of the *Thresher*. The *Scorpion* was finally found at the end of October by USS *Mizar*, in ten thousand feet of water, four hundred miles southwest of the Azores, through a combination of magnetometry and underwater photography. See Brundage, *Brief History*; Buchanan, "Search for Scorpio"; Sontag, Drew, and Drew, *Blind Man's Bluff*, chap. 6.

41. Memo to the files, 26 June 1968, AC-09.5, Folder 24, WHOI-ODR. The contemporary documents do not reveal what "the job" was, but since the *Scorpion* was not yet found, it seems evident that the job was to assist in the search. Subsequently declassified documents reveal that *Alvin* also participated in radiation surveys of the *Scorpion* site once the boat was located. There are many websites about *Scorpion* and even a Facebook page, "USS Scorpion (SSN-589)," https://www.facebook.com/groups/83467923449 /10152925916558450/.

42. Memo to the files, 3 July 1968, AC-09.5, Folder 24, WHOI-ODR. Upper case in the original.

43. Memorandum, Daubin to File, 5 July 1968, AC-09.5, Folder 24, WHOI-ODR.

44. Ibid.

45. Research and Technology Resume, 16 June 1965, "Marine Physics," Box 4, Folder "Nonr 2216 (05) Class Amend. DD 1498s, 1965–June 1966," SIO-MPL. Other documents in this folder discuss other Deep Tow activities, including work on behalf of the Bureau of Naval Weapons and on behalf of the ONR in helping to site SEALAB II.

46. Stephan to Fye, 9 May 1963, AC-09.5, Folder 46, WHOI-ODR.

47. William Rainnie Jr. to Director of Undersea Programs, ONR, 22 May 1962, "Research Submarine, Bid Specifications for," Box 14, Folder 14, "Submarines: Research Projects," SIO-OCR. One fascinating detail is that Admiral Swanson, when discussing Alvin's role at Palomares, did not use the designation DSRV, but referred to it as DSV *Alvin*. See, e.g., Swanson, "Aircraft Salvage," 6. Moreover, when the DSSP commissioned a rescue vehicle from Lockheed Missile and Space Corporation, they referred to these as DSRVs—deep-submergence rescue vehicles. For a retrospective on the rescue program, see Crawley, "Rescue Subs."

48. Sapolsky, *Science and the Navy*, 7.

49. Press release, Technical Information Office, Office of Naval Research, 5 September 1961, AC-09.5, Folder 13, WHOI-ODR. This is an interestingly parsed phrase, because it allows for the reading that, were the Navy interest to change so that it no longer embraced oceanography, then matters would be different.

50. James Wakelin Jr., typescript, Address at the Commissioning of *Alvin* at the Woods Hole Oceanographic Institution, Woods Hole Massachusetts, 5 June 1964, WHOI-OSSR.

51. On the ambiguity of the meaning of the word *research* and all its modifiers (e.g., *basic, fundamental*) in the specific case of the ONR, see Sapolsky, *Science and the Navy*, 6; see also discussion in the conclusion.

52. The Bureau of Ships could call *Dolphin* a research submarine because it was not intended to engage in combat operations, but this was not the same as it being a vessel for basic scientific research.

53. Alan Berman, "Introductory Speech for Artemis General Meeting," 27 September 1963, p. 4, Box 2, Folder "Hudson Laboratories of Columbia University, Bora Bora Conference January 13–17, 1964, V. C. Anderson, 2 of 2," SIO-MPL. For a summary of Woods Hole activities under this project, see John C. Beckerle, 1969, "Final Report of Contract Nonr-2866, WHOI Reference No. 69-51. The project ran at WHOI from March 1959 to June 1969.

54. Ballard, "Woods Hole's."

55. Robert Frosch, Personal communication, November 2001. Frosch was a visiting scholar at the Harvard Kennedy School of Government at that time.

56. R. L. Waldie, "A Proposed Sonar Suit and Research Program for the Deep Diving Submarine (AGSS-555)," p. 2, US NEL, Report No. 081, declassified 31 December 1969, Box 6, Folder 9, "Dolphin and Research Subs," SIO-MPL.

57. Arnold G. Sharp and Barrie B. Walden, "New Titanium Personnel Sphere for *Alvin*" (supported by ONR Contact N00014-71-C-01017), 8, in Earl E. Hays, ed., *Summary of Investigations Conducted in 1973*, WHOI Reference No. 74-18.

58. William Rainnie Jr. and Lawrence A. Shumaker, "The Deep Submergence Engineering and Operations Section," 3–7, in Earl E. Hays, ed., *Summary of Investigations Conducted in 1973*, WHOI Reference No. 74-18, p. 3.

59. Arnold G. Sharp and Lawrence A. Shumaker, "DSRV ALVIN, A Review of Accomplishments," Technical Report, January 1977, WHOI Reference No. 76-114, p. 3, cite the new capacity as twelve thousand feet; elsewhere it is ten thousand feet. Kaharl explains that it depends on which safety factor is used. Using 1.5, the engineers approved the retrofitted *Alvin* for work down to twelve thousand feet, but Fye wanted 1.8, so therefore quoted the new capability as ten thousand feet. See Kaharl, *Water Baby*, 146.

60. For Le Pichon's personal account of this work see his "Plate Tectonics."

61. Menard, *Ocean of Truth*.

62. Hess's hypothesis of seafloor spreading was critical in two ways: first, and widely recognized, it presented a mechanism for continental drift. But others had suggested this mechanism before. Equally, if not more important, was that Hess reopened the debate in the United States—where drift had been subject to hostile rejection in the 1920s—by being the first prominent US scientist to call attention to, and unequivocally embrace, the British land-based paleomagnetic work of Runcorn, Irving, Blackett, and others, which showed that the continents had moved independently of one another.

63. Oreskes, *Continental Drift*.

64. Shor, "Heat-Probe." On Bullard, see "Edward Crisp Bullard (1907–1980)," https://honors.agu.org/bowie-lectures/edward-crisp-bullard-1907-1980/.

65. Revelle and Maxwell, "Heat Flow"; and Bullard, Maxwell, and Revelle, "Heat Flow."

66. Bullard, "Comment."

67. Bullard, "Flow of Heat."

68. Von Herzen and Uyeda, "Heat Flow." See also Von Herzen, "Heat Flow."

69. Von Herzen, "Heat Flow."

70. On the Atlantic data, see Langseth et al., "Crustal Structure"; Le Pichon and Langseth, "Heat Flow"; Talwani et al., "Ridge Crest"; Lee and Uyeda, "Review." A slightly different pattern was observed over the Juan de Fuca Ridge in the northern Pacific. See Lister, "Thermal Balance." The Juan de Fuca values lacked the sporadic anomalously low values found elsewhere but were still lower overall than predicted by theoretical modeling. Lister therefore focused on the discrepancy between his measurements and McKenzie's prediction and concluded that an additional source of heat loss was required.

71. McKenzie, "Some Remarks."

72. Oreskes, *Plate Tectonics.*

73. McKenzie, "Some Remarks." McKenzie noted that the overall values were too low—lower than would be predicted on geophysical grounds for conductive cooling, and therefore that some other cause of heat loss was likely. See also Sclater, "Heat Flow."

74. Sclater, "Heat Flow," 140. In 1970, Sclater published an important paper with Jean Francheteau, "Terrestrial Heat Flow," analyzing heat flow through the continents and oceans. For retrospectives on objections to plate tectonics based on heat-flow data, see MacDonald, "Seafloor Spreading"; Sclater, "Heat Flow." At the time of his death, MacDonald considered that the heat-flow data had still not been adequately explained. The best contemporaneous discussion of the gap between the theoretical models and the measured values is given by Lister, "Thermal Balance."

75. The idea of cooling at the ridge crests by hydrothermal circulation was apparently first suggested by Pálmason, "On Heat Flow"; and by Talwani et al., "Ridge Crest." However, these workers seem to have been imagining flow through a porous layer, such as an individual basalt flow—a kind of submarine aquifer—in which seawater would circulate and soak up some of the missing heat. In 1972, Bodvarsson and Lowell noted that flow along submarine basalts would be unlikely to behave as a uniform porous medium and would likely be concentrated along irregular openings at the contacts of flows, open spaces between dikes and country rock, and/or vertical fractures, and presented a very simple quantitative model for fracture flow at a ridge crest.

76. Lister, "Thermal Balance," 528. Lister's work was funded by NSF and ONR.

77. Ibid.

78. Swallow and Crease, "Salty Water"; Degens and Ross, *Hot Brines*; Swallow, "Discovery Account," 3–9.

79. The first measurements of this are credited to scientists on the Swedish

Albatross expedition, Bruneau, Jerlov, and Koczy, "Expedition Report," but Rona et al., "Seafloor Spreading," 771, note that "the anomaly was not recognized at the time." They credit the recognition that the water column was stratified, with ponds of hot hypersaline fluid, to Charnock, "Energy Transfer," and to Swallow and Crease, "Salty Water." See also discussion in Degens and Ross, *Hot Brines*; Rona et al., "Seafloor Spreading."

80. John Mudie to Fred Spiess, memo, 23 July 1968, Box 7, Folder "Deep Tow," SIO-MPL.

81. Irving, "Review and Discussion"; Irving, Park, Haggerty, Aumento, and Loncarevic, "Mid-Atlantic Ridge."

82. Bostrom and Peterson, "Precipitates." They in turn credited the idea to Von Gümbel, "Stillen Ocean."

83. Scott et al., "Hydrothermal Field." See also Rona, "Comparison"; Rona, "New Evidence"; Rona and Scott, "Convenors"; Rona et al., "Water Temperatures"; Lowell and Rona, "Interpretation"; Rona, "Tag Hydrothermal Field"; Rona et al., "Seafloor Spreading." In hindsight, this can be viewed as the first hydrothermal field discovered, although it did not get as much attention as the Galápagos discoveries (then or now), no doubt for lack of spectacular biological discoveries.

84. Piper, "Metalliferous Sediments."

85. Corliss, *Ridge Basalts*.

86. Ibid. See "hydrothermal exhalations" on 23, "mobilized by dissolution" on xiv. See also Corliss, "Hydrothermal Solutions."

87. FAMOUS Report, "Planning Proposal for the Franco-American Mid-Ocean Undersea Study Program," n.d., but reports on a meeting held at Woods Hole on 30 November–2 December 1971, MC-01, unprocessed folder, WHOI-Vine.

88. Heirtzler and van Andel, "Project Famous." See also ARCYANA, "Transform Fault"; Ballard et al., "Submersible Observations"; Bellaiche et al., "Submersible Study"; Hekinian et al., "Rift Valley"; Renard et al., *Bathymétric détaillée*.

89. Heirtzler and van Andel, "Project Famous." The ships involved were the Navy vessels *Hayes*, *Mizar*, and *Lynch*, and Woods Hole's research vessel *Knorr*.

90. Brundage and Cherkis, *Cruise Results*. See also Laughton and Rusby, "Long Range Sonar." On the history of *Mizar* and LIBEC, see Brundage, *Brief History*; Brundage and Cherkis, *Cruise Results*. Between June 1964 and September 1974, *Mizar* participated in thirty-nine deep-sea search missions and took over 750,000 photographs. The photomosaic technique had previously been used to create a map of the debris field associated with the lost submarine *Scorpion*. See Buchanan, "Search for *Scorpion*."

91. Heirtzler, "Project Famous." This is the half-spreading rate; that is, each side moves laterally about this much each year, although one aspect of their discovery was the realization that the half-spreading rates were not symmetrical.

92. Heirtzler and van Andel, "Project Famous."

93. Heirtzler, "Project Famous."

94. Heirtzler and van Andel, "Project Famous."

95. Heirtzler, "Project Famous"; Ballard, "Project FAMOUS II."

96. Hoffert et al., "Hydrothermal Deposits."

97. Fye to Tjeerd van Andel, 11 September 1974, AC-09.5, Folder 4, WHOI-ODR.

98. For reference, the entire NOAA budget for underwater research in 1970–1971 was about $700,000, so NOAA was committing about 10 percent of its subsea research funds to *Alvin*. See Press clipping, "Research Subs: Sinking Business," *Los Angeles Times*, 13 December 1971, Box 14, Folder 13, "Submersibles: Miscellaneous," SIO-MPL.

99. G. J. Mischke to Art Maxwell, 22 July 1974, AC-09.5, Folder 4, WHOI-ODR.

100. Ibid.

101. "History of UNOLS," https://www.unols.org/what-unols/history-unols.

102. DSRV Alvin Support Agreement: Memorandum for the Distribution List, 11 November 1974, Department of the Navy, Oceanographer of the Navy, OCEANAV: ser 1032/N411, copy courtesy of the UNOLS office, Graduate School of Oceanography, University of Rhode Island, 2003. See also "UNOLS Charter," https://www.unols.org/what-unols/unols-charter.

103. Atwater, "Plate Tectonic Revolution," 246.

104. Atwater and Menard, "Magnetic Lineations."

105. F. N. Spiess, PI, "Proposal for Research to be conducted under the sponsorship of the ONR, Code 466, "Augmentation to the core contract of the MPL for analysis of natural background returns of the side-looking sonar," 11 November 1971, Box 6, Folder "Project Deep Tow," SIO-MPL.

106. Spiess to J. A. Cestone, DSSP Office, 20 June 1972, Box 6, Folder "Project Deep Tow," SIO-MPL.

107. Box 8, Folder 6, "Expeditions, South Tow, 1973," SIO-OCR.

108. On Sclater's earlier interpretation of the heat flow lows as a sampling artifact, see Sclater and Francheteau, "Terrestrial Heat Flow"; Sclater, Mudie, and Harrison, "Geophysical Studies"; discussion in Sclater et al., "Tag Hydrothermal."

109. Sclater and Klitgord, "Heat Flow," 6973.

110. Talwani et al., "Geophysical Study," 494–96. In an earlier paper, Langseth et al., "Crustal Structure," Lamont scientists had stressed that the lack of conspicuous difference between the heat flow over ridges and the rest of the ocean basins "precludes the possibility of continuous continental drift during the Cenozoic by the spreading-floor mechanism." The legacy of antidrift at Lamont may help explain why Lamont scientists were slower to interpret their heat-flow data in terms of hydrothermal venting than scientists elsewhere.

111. Sclater and Klitgord, "Detailed Heat Flow," and also work from the previous leg (Leg 6) of the South Tow expedition; see Klitgord and Mudie, "Geophysical Study." Although NSF provided the specific funding for the scientific data collection, MPL continued to receive ship support funds from ONR. On the shift here to metric—around this time, many American scientists began using metric units more consistently.

112. Glen, *Road to Jaramillo.*
113. South Tow Prospectus: being a sporadically published newsletter concerning the progress of an expedition aboard the R/V T. Washington," by John Mudie, vol. 1, no. 2, 18 January 1971, Box 8, Folder 6, "Expeditions, South Tow, 1973," SIO-OCR. Klitgord and Mudie, "Geophysical Study," discuss the magnetic as well as the heat-flow data.
114. Sclater et al., "Spreading Centre"; Williams et al., "Spreading Centre."
115. Williams et al., "Spreading Centre."
116. Detrick et al., "Spreading Centre"; Williams et al., "Spreading Centre."
117. Detrick et al., "Spreading Centre"; Williams et al., "Spreading Centre."
118. Scripps press release on conclusion of cruise, 21 February 1973, Box 8, Folder 6, "Expeditions, South Tow, 1973," SIO-OCR. This work became the basis of Williams's PhD dissertation and several papers: see Williams, "Heat Loss"; Williams et al., "Spreading Centre"; Sclater et al., "Spreading Centre"; Detrick et al., "Spreading Center"; as well as a number of other papers attempting to place these data in tectonics context. See Le Pichon and Langseth, "Heat Flow"; McKenzie and Sclater, "Heat Flow"; Sleep, "Sensitivity of Heat Flow"; Sclater and Francheteau, "Heat Flow"; Parker and Oldenburg, "Thermal Model"; Williams and Von Herzen, "Heat Loss."
119. Williams et al., "Spreading Centre," 599.
120. Ibid., 602.
121. Work on EPR at 21°N dates back at least to the 1967 Tow Mas cruise, an early Deep Tow study funded by ONR and DSSP.
122. MacDonald and Spiess, "Workshop Report," 1976, quote on 4.
123. Later this question would be answered using satellites; see Sandwell, "Martian View."
124. MacDonald and Spiess, "Workshop Report," 4.
125. This was done under the auspices of the International Decade of Ocean Exploration program. Until now the programmed decade scarcely features in the work we have been discussing, perhaps because it was mostly used to fund programs that the Navy did not fund.
126. MacDonald and Spiess, "Workshop Report," 9.
127. Ibid.
128. Lonsdale, *Samoan Passage*, vii.
129. Ibid.
130. See Ozima and Podosek, *Noble Gas Geochemistry*, for a comprehensive overview of helium isotopes and their applications.
131. Weiss, *Dissolved Gases.*
132. Various Documents, Box 8, Folder 1, "Pleiades Cruise Reports," SIO-OCR.
133. Ibid.
134. Weiss et al., "Spreading Center"; see also Craig et al., "Pacific Rise."
135. Leg II, 21 July 1976, Box 8, Folder 1, "Pleiades Cruise Reports," SIO-OCR; Crane ms., "The Best of Times, the Worst of Times"; and Peter Lonsdale, Personal communication with author, June 2002. See also Crane, *Sea Legs.* Crane went

on to a career as professor at Hunter College in New York and at NOAA, but she was very bitter at how she had been treated as one of the few women at Scripps when she was there, and, in particular, at having been excluded from the key paper first announcing the Galápagos Rift vent discovery. See Weiss et al., "Galapagos Rift."

136. Leg II, 21 July 1976, Box 8, Folder 1, "Pleiades Cruise Reports," SIO-OCR.
137. Ibid.
138. Ibid.
139. Ibid.
140. Corliss et al., "Sediment Mound"; Crane et al., "Structural Activity"; Weiss et al., "Spreading Center." These abstracts were followed by a series of papers over the next two years, including Lonsdale, "Clustering"; Lonsdale, "Galapagos Rift"; Weiss et al., "Galapagos Rift"; Corliss et al., "Galapagos Rift"; Crane, "Inner Rift."
141. Weiss et al., "Galapagos Rift"; Lupton, Weiss, and Craig, "Mantle Helium." Crane was particularly aggrieved that she was not included as an author on the first of these two papers. For some of the closely related papers published at this time, see also Lonsdale, "Deep-Tow Observations"; Lupton et al., "Hydrothermal Site"; MacDonald et al., "East Pacific Rise."
142. Weiss et al., "Galapagos Rift"; Lonsdale, "Clustering."
143. Lonsdale, "Clustering," 857–58. This paper was received at *Deep Sea Research* on 7 April 1977; in the paper, Lonsdale notes that taxonomic analysis of the organisms should await the result of the *Alvin* dives, presently taking place, which are "sampling the hydrothermal precipitates and benthic organisms that we photographed" (858).
144. MacDonald and Spiess, "Workshop Report," 11, says that "in 1977, an already funded submersible program will focus on hydrothermal problems on the Galapagos rise crest."
145. Ballard, "Deep Pacific."
146. The Pleiades Cruise Report records Crane as "in charge of transponder navigation." Her later publications dealt mainly with the structure of the ridge axis and interpretation of the underlying magma chamber.
147. Corliss and Ballard, "Cold Abyss."
148. Crane, *Sea Legs*.
149. German et al., "Hydrothermal Activity," 156.
150. Tjeerd van Andel, Personal Diary, "Atlantis II 1966; FAMOUS June–August 1974; and Galapagos Rift, February–March 1977," courtesy of Tjeerd van Andel, on 96–97.
151. Corliss and Ballard, "Cold Abyss"; Corliss et al., "Galapagos Rift"; Crane, "Galapagos Inner Rift"; Corliss et al., "Galapagos Rift."
152. Diary of Tjeerd van Andel, loaned to the author by Tjeerd van Andel prior to his death in 2010. His papers are at Stanford University, where he spent most of his later career, and where I knew him. "Guide to the Tjeerd van Andel Papers," https://oac.cdlib.org/findaid/ark:/13030/kt267nf0vb/.

153. For example, Sullivan, "Diving Scientists," although this particular piece ran before the biological discoveries.

154. Box 11, Folders 59–63, SIO-OCR.

155. The *New York Times*, to its credit, did mention Crane, Stakes, and Atwater but did not interview any of them; see "Diving Scientists," *New York Times*. The article was based on a presentation given by Bob Ballard.

156. Press clipping, David Perlman, "Astounding Undersea Discoveries," *San Francisco Chronicle*, 9 March 1977, Box 11, Folder 61, SIO-OCR.

157. Corliss et al., "Galapagos Rift."

158. Wolery and Sleep, "Hydrothermal Circulation."

159. MacDonald and Spiess, "Workshop Report."

160. Francheteau et al., "Deep-Sea Sulfide"; Hekinian et al., "Sulfide Deposits."

161. Press release, n.d., c. April 1978, Box 63, Folder 11, "Rise and Galapagos Rift File," SIO-OCR.

162. Ibid.

163. Haymon would get her PhD from Scripps in 1982 based on this work (Haymon, *Hydrothermal Deposition*) and be a coauthor on several papers. She would go on to become a professor at UC Santa Barbara.

164. Spiess et al., *Active Spreading Center*.

165. MacDonald and Spiess, "Workshop Report"; Spiess et al., "East Pacific Rise"; see also Spiess et al., *Active Spreading Center*; Various documents, Box 63, Folder 11, "Rise and Galapagos Rift File," SIO-OCR.

166. Ibid.

167. Spiess et al., *Active Spreading Center*. See also Spiess et al., "East Pacific Rise," 1980. The implications for ore formation were heavily emphasized in press reports. See, e.g., Sullivan, "Sea-Floor Geysers."

168. Peter Gwynn and Mary Hager, "Germs from the Deep," *Newsweek*, 14 May 1979, 129; Perlman, "Ocean Chemistry"; Sullivan, "Sea-Floor Geysers." Soon after this, a study of the flow rates by Converse et al., "Flow Rates," demonstrated that most of the mineral particles entrained in buoyant plumes are dispersed, and the rate at which the remaining particles settled and built up hydrothermal mounds was far too low to form large massive sulfide deposits, such as the famous Troodos deposits on Cyprus.

169. For example, Peter Gwynn and Mary Hager, "Germs from the Deep," *Newsweek*, 14 May 1979, 129, wrote that "scientists also believe that their new underwater finds may possess commercial potential" and quoted Bob Ballard describing the vents as containing a "bonanza of minerals." In reality, none of these scientists had experience in mineral exploration or development, as so far as the record reveals, none was in touch with any mining companies. Nor would any reasonable company at that time have been seriously interested in attempting to exploit the deposits. While the metal grades at some active hot springs compete with land-based ore deposits, the total resource is minuscule, and the technological and environmental difficulties associated with attempting to mine them overwhelming. Moreover, despite

the claims of the late 1970s, we know now that there really is no shortage of land-based mineral deposits; there are simply environmental, political, and social issues associated with our discovery and exploitation of them. But such difficulties would be no less at sea. On the alleged geothermal energy potential, see "Diving Scientists," *New York Times*.

170. Sullivan, "Sea-Floor Geysers."

171. This raises an interesting point: in the archives, I found almost no serious discussion of economic geology, underscoring the point that Navy funding led scientists to focus on matters of Navy concern. The one place where mineral deposits on the sea were seriously discussed was in the matter of mining manganese nodules, but that was a cover story for the recovery of a sunken Soviet submarine. Originally, I had hoped to have a chapter on this episode, but my Freedom of Information Act requests were denied, presumably because of Central Intelligence Agency involvement. In recent years, there has been some discussion. See, e.g., Nikola Budanovic, "Project Azorian: Howard Hughes' Secret Mission, Which Involved the CIA and a Missing Soviet Submarine Off the Coast of Hawaii," *Vintage News*, 25 April 2018, https://www .thevintagenews.com/2018/04/25/project-azorian/. It is not clear to me how accurate these popular accounts are. The story would still make for a good serious history, if one could access the classified documentary record.

172. "William Whewell," *Stanford Encyclopedia of Philosophy*, 23 December 2000, revised 22 September 2017, https://plato.stanford.edu/entries/whewell/.

173. Daubrée, *Études synthétiques*, 9, my translation. Daubrée lived from 1814 to 1896.

174. Lonsdale, "Submersible Exploration." See also "Marine Life Is Found," *La Jolla Light*, 21 April 1980, Box 11, Folder 59, SIO-OCR; Lonsdale et al., "Spreading Center."

175. Lonsdale, "Hot Vents"; see also Lonsdale et al., "Spreading Center."

176. Lonsdale, "Hot Vents," 1984. The Sea of Cortez is also known as the Gulf of California; see also Einsele et al., "Intrusion of Basaltic Sills"; Kennish et al., "Hydrothermal Activity"; Lutz and Kennish, "Review."

177. The bare ridge environment of the East Pacific Rise may be analogous to basalt-hosted massive sulfide deposits in ophiolite complexes, such as the famous Troodos deposits of Cyprus, mined since the Bronze Age. But volumetrically and commercially, such deposits are less significant than sediment-hosted massive sulfides.

178. Lipscomb, *Dive to the Edge*; Gilbert Grosvenor to John Steele, 8 January 1980, AC-09, Folder 20, WHOI-ODR.

179. Since 1980, hydrothermal vents have been found at spreading centers in all the world's major oceans, at temperatures as high as 400° Celsius. Typically, they occur in vent fields, a few tens to hundreds of meters across; exploration has suggested that fossil hot springs are very common. See German et al., "Regional Setting," and other papers in this volume. Recent work has shifted the focus of attention away from the mid-ocean ridges toward hydro-

thermal activity at a variety of plate boundaries, including diffuse venting; see, e.g., Stein and Stein, "Heat Flux."

180. Turner and Lutz, "Growth and Distributions." Accurate analyses have shown vent fluids to contain about 350 micromoles per liter H$_2$S at the Galápagos, 600 micromoles per liter at the East Pacific Rise.

181. For an excellent review of the toxicity problem, see Childress and Fisher, "Vent Animals."

182. Turner and Lutz, "Growth and Distributions." As Lutz and Kennish, "Review," note, terminology is not consistent. Jones, "Tube Worms," calls the bacteria *chemolithotrophic*, to emphasize that they exploit inorganic molecules; Somero, "Physiology and Biochemistry," prefers *chemolithoautotrophic*; Jannasch, "Nutritional Basis," likes *chemoautolithotrophic*. I use the term *chemoautotrophic* as clear and to the point: these bacteria are autotrophs, as opposed to heterotrophs, and they derive the energy to metabolize carbohydrates from chemical reactions rather than photosynthesis. Jannasch, "Nutritional Basis," notes that there are, in fact, bacteria that are chemoorganoautotrophic—that is, autotrophic but derive their energy from reactions involving organic molecules. Either way, it is interesting to note that while these communities do not require sunlight, they are not wholly independent of the Earth's surface, because the carbohydrate metabolizing reactions require oxygen, which is either dissolved in ocean water equilibrated with Earth's atmosphere or supplied by photosynthetic plankton.

183. Winogradsky, *Microbiologie*; see also Lechevalier and Solotorovsky, *Three Centuries*. Inspired by the work of Louis Pasteur and Robert Koch, Winogradsky devoted a lifetime to study of bacteria. As he later recalled, autotrophic bacteria were previously unknown and unstudied, and their characteristics incompatible with prevailing notions of physiology. Winogradsky coined the term *autotroph* to describe these organisms who generated their own food supply. In 1889, he demonstrated that some autotrophic bacteria were capable of fixing nitrogen in soils, including those involved in the nitrogen fixation by leguminous plants. It is sometimes suggested that scientists discovered a new form of life when they discovered these vents. Given Winogradsky's work, this claim is certainly overstated. However, it is true that the *Alvin* discoveries were unique in finding complex biotic communities, including higher-order species like clams and worms, in a food chain entirely dependent on chemosynthesis. Since the *Alvin* discoveries, chemosynthetic bacteria have been extensively studied in the hot springs of Yellowstone National Park. On the earlier assumption that they were ecologically unimportant, see Jannasch, "Nutritional Basis."

184. When first discovered, the giant tube worms appeared to be unlike any other known worm species—the closest known similar types were *Pogonophora*, who also lack guts and mouths but are much smaller—and in the mid-1980s, many biologists accepted the radical interpretation proposed by Smithsonian biologist Meredith Jones that the giant tube worms of the Galápagos

Rift were a new genus and species, *Riftia pachyptila*, of a new family, the
Riftiidae, of a new order, the Riftiida, of a new class Axonobranchia, of a new
phylum, the Vestimentifera. (Vestimentiferans were known since the mid-
twentieth century but previously considered an order, class, or subphylum).
Jones, "Introduction," and "Vestimentifera"; see also Wirsen et al., "Activities"; Wirsen et al., "Microbial Activity"; Shillito et al., "Composition." Others
disputed the phylum designation and considered the vestimentiferans as a
subclass of the pogonophorans; still others considered both vestimentiferans and pogonophorans classes of the Annelida. See discussions in Tunnicliffe, "Hydrothermal Vents," 377; Childress and Fisher, "Vent Animals," 339.
Erin McMullin, personal communication with the author, 2002. More recently, molecular data—plus the recognition that the giant tube worms are
segmented at their very back end, and earlier samples were all broken—has
supported this latter interpretation: that the vestimentiferans are part of the
Annelid worm phylum, family Siboglinidae, class Polychaeta. This molecular
evidence also argues against a very ancient origin and suggests instead that
Riftia (and other vent fauna) diverged from related shallow-water types only
twenty million to one hundred million years ago. Van Dover et al., "Evolution and Biogeography."

185. Wirsen and Jannasch, "Primary Production"; Somero, "Physiology and
Biochemistry"; Jannasch, "Nutritional Basis"; Jones, "Tube Worms"; Jones
"Vestimentifera"; and Rieley et al., "Lipid Characteristics," provide biochemical evidence for the diets of three major vent invertebrates: *Riftia pachyptila*
(tube worms), *Bathymodiolus thermophilus* (mussels), and *Halice hesmonecetes*
(amphipod crustaceans), which confirms earlier work that emphasized the
presence of these bacteria within the tissues of these organisms, and the
physiological features, which indicated reliance on endosymbionts.

186. A detailed summary of the modes of chemosynthetic bacterial life is given in
Wirsen et al., "Microbial Activity."

187. Excellent reviews are given in Tunnicliffe, "Hydrothermal Vents"; Childress
and Fisher, "Hydrothermal Vent"; Lutz and Kennish, "Review"; Wirsen et al.,
"Microbial Activity"; Hessler and Kaharl, "Overview"; Van Dover, "Hydrothermal Vents." The details of vent biology and primary production are complex
and appear to differ between Atlantic and Pacific sites, with the dominant
fauna at the Pacific sites relying on endosymbiosis and the dominant fauna
at the Atlantic grazing on free-living and epibiotic autotrophs. Wirsen, Tuttle, and Jannasch, "Activities"; Wirsen et al., "Microbial Activity"; Van Dover,
"Hydrothermal Vents." Van Dover emphasizes that many bacteria live epibi-
otically (on the surfaces of other organisms), and an unanswered question is
whether they live mutualistically, providing their hosts with an important
or exclusive food source, or in a fouling relationship, alternatives that have
very different evolutionary implications. It is now known that autotrophic
bacteria exploit a variety of reduced compounds, including methane and
hydrogen gas in vent fluids, sulfur in sulfide minerals, and ferrous iron and

manganese in iron and manganese oxide crusts. Methanotrophic bacteria have also been found in cold-water methane seeps throughout the world's oceans; Lutz and Kennish, "Review." Some vent fauna exploit more than one autotrophic source; *Bathymodiolid* mussels at the Atlantic Snake Pit site, for example, exploit two types of endosymbionts, one of which relies on the oxidation of methane (CH_4), the other H_2S. Van Dover, "Hydrothermal Vents"; see also discussion in Childress and Fisher, "Hydrothermal Vent."

188. Jones, "Tube Worms"; see also discussion in Hessler and Kaharl, "Overview."

189. German et al., "Mid-Ocean Ridges"; Tunnicliffe, "Hydrothermal Vents"; Lutz and Kennish, "Review."

190. Deming and Baross, "Black Smoker," Delaney et al., "Quantum Event."

191. On the 400°C springs, see Haymon et al., "Volcanic Eruption."

192. Baross and Deming, "Deep-Sea Smokers"; Delaney et al., "Quantum Event"; Wills and Bada, *Spark of Life*, chap. 8; Delaney et al., "Edifice Rex"; Chyba and Phillips, "Abode of Life." The proposal for life on Europa generated a fair bit of controversy; for a negative assessment, see Soare and Green, "Habitability of Europa." While the proposal for life on Europa is pushing the data fairly hard, it underscores that our understanding of seafloor and possible subseafloor biotic communities has generated important new ideas about where we might look for life, both on Earth and on other planets.

193. Cowan, "Hyperthermophilic Enzymes." See also Turner and Lutz, "Growth and Distributions"; Rhoads et al., "Growth of Bivalves"; Lutz et al., "Far-Infrared Reflection"; Childress and Fisher, "Vent Animals"; Lutz and Kennish, "Review"; Hessler and Kaharl, "Overview."

194. Corliss et al., "Hypothesis Concerning"; Baross and Deming, "Deep-Sea Smokers"; Converse et al., "Flow Rates"; Baross and Hoffman, "Hydrothermal Vents"; see review and discussion in Tunnicliffe, "Hydrothermal Vents."

195. Maher and Stevenson, "Impact Frustration."

196. Sousa et al., "Early Bioenergetic Evolution." In 1988, Stanley Miller, famous for his experiments demonstrating the formation of amino acids out of inorganic molecules, published an article in *Science*, coauthored with Scripps marine chemist Jeffrey Bada, refuting the suggestion on the grounds that the inherent instability of organic molecules at high temperatures would preclude organic synthesis. See Miller and Bada, "Hot Springs." In Miller and Bada's view, vent temperatures are far too high for sustained organic molecule synthesis. In the early 2000s, Bada and UCSD colleague Christopher Wills pursued this argument further in *Spark of Life* (101, 258) arguing that, rather than incubating incipient life, seafloor vents would have destroyed it: circulation of seawater through hydrothermal systems in Earth's early history would have destroyed proto-organic molecules in any primordial "soup," effectively sterilizing the oceans. In essence, the vents acted as an autoclave. They invert the question and ask, how did life manage to evolve despite the organic-molecule destroying effects of deep-sea hydrothermal circulation?

For true life to have begun, primitive organic molecules must have achieved the ability to replicate before they were destroyed by heating. It has been estimated that the entire ocean circulates through seafloor hydrothermal systems in about ten million years, so they use this to place an upper limit on the time required for life to begin. The argument has not been resolved; see Arunas L. Radzvilavicius, "We've Been Wrong about the Origins of Life for 90 Years," *The Conversation*, 15 August 2016, http://theconversation.com /weve-been-wrong-about-the-origins-of-life-for-90-years-63744; Rachel Brazil, "Hydrothermal Vents and the Origins of Life," *Chemistry World*, 16 April 2017, https://www.chemistryworld.com/feature/hydrothermal-vents -and-the-origins-of-life/3007088.article.

197. Sousa et al., "Early Bioenergetic Evolution."

198. US Department of Interior, Mineral Management Service, 1983. Draft Environmental Impact Statement: Proposed Polymetallic Sulfide Minerals Lease Offering: Gorda Ridge Area Offshore Oregon and Northern California, prepared by the Mineral Management Service Headquarters Office, Reston, Virginia. See also Van der Voort and Mielke, Paper Presentation, 1982, US Department of Interior, 1983, Symposium: A National Program for the Assessment and Development of Mineral Resources of the United States Exclusive Economic Zone. Symposium held November 15–17 in Reston, VA, USGS Circular 929; and US Department of Commerce, National Oceanic and Atmospheric Administration with the University of Maryland Sea Grant Program, 1983, Marine Polymetallic Sulfides: A National Overview and Future Needs. Proceedings of a Workshop Held 19–20 January, Maryland Sea Grant Publication UM-SG-TS-A3-04.

199. At the time many people believed that impending shortages would drive up commodity prices, making costly marine mining feasible, but such events have yet to develop.

200. I know, because I was one of them. For the science see Haymon et al., "Volcanic Eruption"; Haymon, Koski, and Sinclair, "Vent Worms."

201. Corliss et al., "Thermal Springs."

202. Wolery and Sleep, "Hydrothermal Circulation."

203. Tyler et al., "Deep Side," 210.

204. Ballard, "Woods Hole's" in the NSF report *Fifty Years of Ocean Discovery*. See also the article "Deep Submergence," by Sandra Toye, a former NSF program director, in that same volume. Of course, it is reasonable that an NSF program director might be most interested in *Alvin's* research (rather than military) work; what is not reasonable is to expunge its earlier history.

205. Malakoff, "Researchers." 2002.

206. Kaharl, *Water Baby*, 57.

207. Sapolsky, *Science*.

208. Spiess, personal communication with author, La Jolla, 26 March 2002; see also Spiess, "Origins and Perspectives"; Spiess, "East Pacific Rise."

209. 17 October 1960 Ewing to Lee Westrate, PSAC, discussing presentations made to ad hoc committee on 10 October, Box 157, Folder "PSAC Ad Hoc Committee on Basic Research (2 of 2)," MEP.
210. Forman, "Quantum Electronics."
211. On this point, see especially the work of historian Jacob Darwin Hamblin: *Oceanographers, Poison in the Well*, and *Arming Mother Nature*.
212. Oreskes and Krige, *Science and Technology*, conclusion.
213. Stommel to Fye, 13 February 1961, MC-06, Folder 2, "Corrp 1960–1966," 3, WHOI-Stommel.
214. Some went further: Maurice Ewing worked with the State Department to block the entry of the People's Republic of China into the International Union of Geodesy and Geophysics; unpublished notes from the Maurice Ewing papers.
215. This is not unique to military contexts. In other work, my colleagues and I have shown that being useful, as an expert, is a major motivation for scientists participating in scientific assessments for policy. Oppenheimer et al., *Discerning Experts*. Audra Wolfe also discusses the role of the NAS, in performing what might be considered governmental functions, in her book, *Freedom's Laboratory*.
216. Edwards, *Closed World*, 82.
217. "Current (5 to 10 years) Trends and Needs," memo, Deputy Assistant Oceanographer of the Navy to the Distribution List, 16 April 1969, Box 31, no folder, HHP.
218. Andrew Dickson, "A Book Review."
219. Mukerji, *Fragile Power*.

CHAPTER EIGHT

1. US Department of the Navy, *TENOC*, 1.
2. Ibid.
3. Ibid., 69.
4. Gary Weir, *Forged in War*, emphasizes the importance of personal ties in launching many of the Navy-military collaborations, particularly at Woods Hole.
5. US Department of the Navy, *TENOC*, 4.
6. Hollister, "Health of Basic Research in Oceanography," Testimony to the House Armed Services Committee, Subcommittee on Research and Development, 28 March 1985, MC-31, Folder "House Testimony," WHOI-Hollister.
7. Paul Fye to Bostwick Ketchum, 24 July 1974, AC-09.5, Folder 29, WHOI-ODR.
8. Hollister, "Health of Basic Research in Oceanography," Testimony to the House Armed Services Committee, Subcommittee on Research and Development, 28 March 1985, MC-31, Folder "House Testimony," WHOI-Hollister.
9. Hollister, Testimony before the Research and Development Sub-committee of the House Armed Services Committee, 28 March 1985, MC-31, Folder

"Reports/Statements," WHOI-Hollister. Navy funds still represented a very significant amount of Woods Hole funding: in 1984, $11.5 million out of a total Woods Hole research budget of $40.9 million, of which $3.1 million was in the form of an omnibus contract (not too different for the "basic task" grant of the post–World War II years; see chapters 2 and 5) to support "basic ocean science projects that supply environmental data and knowledge of the fundamental processes that govern the ocean and its surrounds. The results are broadly applicable to Navy problems." House Armed Services Committee, 28 March 1985, MC-31, Folder "Congressional Testimony," WHOI-Hollister.

10. For overview of the tensions over declining Navy funding in 1985, MC-31, Folder "Congressional Testimony," WHOI-Hollister, which includes his own testimony to the House Armed Services Committee and various materials related to it. The ONR denied that Navy commitment to oceanography had declined, but the numbers show that expenditures to academic oceanography—that is to say, direct grants to academic institutions like WHOI—were in decline.

11. Another branch of the family founded the town bearing their name. When asked why Charles Hollister never mentioned this, one colleague replied: Have you ever been to Hollister, California? (Larry Armi, personal communication with the author, May 2016.) Hollister is a town known primarily for a riot at a 1947 motorcycle rally, which became the basis of a Hollywood film. In contrast, the forty-thousand-acre Hollister ranch in Santa Barbara was, until its sale in 1964, one of largest privately owned coastal holdings in California. The family purchased the property in 1866 and ranched until 1964; in 1971 it was put up for development as a unique "unspoiled" natural portion of California coastline. In the early 2000s, a set of hundred-acre parcels were offered for sale. "About Hollister Ranch," http://www.hollister-ranch .com. The terms caused considerable controversy, as the owners, refusing public access to the coast, were accused of harassing the enterprising surfers and kayakers who, exercising their legal rights, arrive there by sea. See Dan Weikel, "Hollister Ranch Owners Are Fighting the State Again over Public's Right to Use Beach," *Los Angeles Times*, 8 August 2016. Since then, conservationists have worked to protect what is one of the last undeveloped stretches of Southern California coastline.

12. Heezen et al., "Shaping the Continental Rise." See also Heezen and Hollister, "Deep-Sea Current Evidence." Ruddiman would later become known for his advocacy of the "Early Anthropocene" concept, that the thoroughgoing human-impact on the environment begins with the advent of farming. Ruddiman, *Plows, Plagues, and Petroleum*. This view was rejected by the Anthropocene Working Group, in favor of a mid-twentieth-century onset of the Anthropocene, on the grounds that the early impacts were not globally synchronous and therefore do not offer a geological recognizable basis for a geological time unit. Zalasiewicz et al. (twenty-first of twenty-six authors), "Working Group on the Anthropocene."

13. Throughout his life, Hollister took credit for the discovery of contourites and rarely emphasized others whose work contributed to the discovery, but sedimentologist and historian of science Gerry Friedman has challenged Hollister's account. Friedman "The Concept of Deep-Sea Contourites — Discussion" responding to Hollister, "The Concept of Deep-Sea Contourites," argues that primary credit should go Heezen, who first recognized contour-following currents when he spent time in *Alvin*, drifting with the engines turned off. Hollister often suggested that geologists had no idea such currents might ever exist, but Friedman notes that Murphy and Schlanger, "Sedimentary Structures," did, describing them in the geological record, in a paper that strongly influenced Heezen. Columbia Geology Department chair Charles Drake (later a mentor and personal friend of mine) also worked and published extensively on bottom sediments and their relation to contour-following currents. See Burk and Drake, *Geology of Continental Margins*. In short, there is good evidence that both Heezen and Hollister had the tendency to exaggerate the novelty and originality of their work, but as a historian, I would say that is true of many famous scientists, and particularly those who are not quite as famous as they think they deserve to be.

14. Nowell and Hollister, "HEBBLE"; Nowell, Hollister, and Jumars, "HEBBLE."

15. Heezen and Hollister, *Face of the Deep*.

16. Department of Navy, Office of Naval Research, to Derek Spencer, 24 October 1978, MC-31, Folder "Proposal Index," WHOI-Hollister.

17. Larry Armi, personal communication with the author, May 2016.

18. Review of Proposal No. OCE7681491, MC-31, Folder "Proposal Index," WHOI-Hollister.

19. Note on terminology. The participants in this discussion mostly use the locution *subseabed*, but sometimes *sub-seabed*. I have used the former to be consistent with the participants, except in direct quotations.

20. Deese, "Seabed Emplacement." These wastes were along with a considerable amount of nonnuclear-waste disposal at sea, as well as nuclear waste disposal but other countries, chiefly the United Kingdom and Soviet Union.

21. Hagen, "Waste Disposal"; Other useful reviews include Holcomb, "Ocean Disposal"; Deese, *Nuclear Power*; Isaacs and Schmidt, "Ocean Energy"; Nuclear Energy Agency, *Seabed Disposal*. On the London Dumping Convention controls on radioactive waste-disposal at sea, see Sjöblom and Linsley, "Sea Disposal." For a contemporaneous discussion of the ban on all radioactive waste disposal at sea, see "Ocean Science News," 19 March 1984, MC-31, Folder "Allyn Vine Drawing," WHOI-Hollister.

22. AEC, *Annual Report*, 289; Calmet, "Ocean Disposal."

23. AEC, *Annual Report*; Hewlett and Anderson, *A History*; de la Bruhèze, *Political Construction*; de la Bruhèze, "Closing the Ranks"; Hamblin, "Light and Bread."

24. Hamblin, *Poison in the Well*. Stommel, "Anatomy of the Atlantic," suggested that radioactive waste disposal was the major motivation to study cur-

rents now that steamships had made the sailing issues obsolete. However, one should not make too large a point of this, because radioactive waste disposal was a motivation that could be discussed in public, whereas most of the Navy motivations were classified. See also memo by Robert Deitz to Bill Menard, "Program to Measure Deep Ocean Currents at NIO [National Institute of Oceanography]," 9 March 1955, AC-09, Folder 22, "Activities ONR, Washington (2/2) 1954–1957," WHOI-ODR.

25. Arons to Oreskes, 16 September 2000, personal communication with author.
26. William von Arx, "AEC Conference of the Disposal of Radioactive Wastes," June 1955, MC-24, Folder "Comments on AEC Conference on the Disposal of Radioactive Wastes," WHOI-von Arx.
27. Arons to Oreskes, 16 September 2000, personal communication with author.
28. Ibid.
29. Stommel, "The Abyssal Circulation."
30. Memo by Robert Deitz to Bill Menard, "Program to Measure Deep Ocean Currents at NIO [National Institute of Oceanography]," 9 March 1955, AC-09, Folder "Activities ONR, Washington (2/2) 1954–1957," WHOI-ODR. See also Kulp, "The Carbon 14 Method"; discussion in Plass, "Carbon-Dioxide Band."
31. Stommel, "Anatomy of the Atlantic." See also Joseph, "Report." At the time, considerable focus was placed on short-lived fission products (e.g., cesium 137, with a half-life of only thirty-three years), so it was plausible (on their calculations) that the wastes could decay to stability in the deeps before ocean circulation moved them back toward the surface. However, over time estimates continued to fall, further increasing the concept of an active rather than sluggish deepwater regime. In the 1980s, carbon 14 work through the Geochemical Sections program indicated replacement times of deep waters in the Pacific, Indian, and Atlantic Oceans between 250 and 500 years, with distinct values for the different oceans. Stuiver, Quay, and Ostlund, "Abyssal Water," suggested that while some parcels might have residence times along the lines earlier suggested by Kulp—up to 1,500 years—the overall average residence time for ocean deep waters was about five hundred years, a figure dominated by the large volume of the deep Pacific. More recently, Fine et al., "Using a CFC," have used CFC concentrations to estimate an age of only twenty years for the North Atlantic Deep Water at the equator. They conclude from this that the effects of global warming would be seen first in the deep waters of the North Atlantic. Stommel was well aware of these arguments and the significance of his work for them. In a 1957 paper he noted, "The average time of residence of a particle of water between its introduction into the deep water at its source and its eventual upwelling through the 200 m level is estimated at between 400 and 1200 years." See Stommel, "The Abyssal Circulation," 734. More recent work has suggested that Worthington's estimates were closer to the accurate figure than Kulp's, which I find interesting, because Kulp would later be involved in the National Acid Precipitation Assessment Program, where he would be

famous (or infamous) for his skepticism of the severity of the problem and optimistic views on how readily acid precipitation could be mitigated (Oppenheimer et al., *Discerning Experts*). See also Valentine Worthington, "North Atlantic Circulation and Water Mass Formation, Final Report," PO-75-PO-88, Woods Hole Oceanographic Institution Archives, Woods Hole, MA.

32.   Sandia Laboratories Contract, Document No. 13-2559, MC-31, Folder "Proposal Index," WHOI-Hollister. See also Hollister, Silva, and Driscoll, "Piston Corer."

33.   Hollister, Silva, and Driscoll, "Piston Corer."

34.   Draft Report, D. R. Anderson, C. D. Hollister, and D. M. Talbert, "A Logic and Decision Model for the Seabed Geologic Seclusion of Radioactive Waste," MC-87, Folder "Vaughan Bowen," Miscellaneous Manuscripts Collection, Woods Hole Oceanographic Institution, Woods Hole, MA (this is Bowen's copy, with his annotations).

35.   Ibid. Hollister would repeat these exact words in another later paper: Hollister, Bishop, and Deese, "Siting Considerations," 1979, MC-31, Folder "Siting Considerations Report," WHOI-Hollister.

36.   To be fair, the *Alvin* discoveries about benthic life around vents were brand new, and no one knew how extensive such life might be. Still, given that Hollister was a Woods Hole scientist, and Woods Hole was broadcasting this as a discovery of the century, it is noteworthy that Hollister should have brushed them aside. This issue would be raised later by George Woodwell.

37.   Charles Hollister, "The North Central Pacific Marine Desert: Geological and Geophysical Data Collection," a proposal to Sandia Laboratories/Atomic Energy Commission, 1973, MC-31, Folder "Proposal Index," WHOI-Hollister.

38.   Draft Report, D. R. Anderson, C. D. Hollister, and D. M. Talbert, A Logic and Decision Model for the Seabed Geologic Seclusion of radioactive waste, MC-87, Folder "Vaughan Bowen," Miscellaneous Manuscripts Collection, Woods Hole Oceanographic Institution, Woods Hole, MA.

39.   Hollister, "High Level Nuclear Wastes in the Sea Bed?"

40.   Vaughan Bowen to D. Richard Anderson, Sandia Laboratories, 6 February 1976, p. 2, MC-87, Folder "Vaughan Bowen," Miscellaneous Manuscripts Collection, Woods Hole Oceanographic Institution, Woods Hole, MA, 2. See also Bowen and Hollister, "Pre- and Post-dumping Investigations."

41.   See also Hollister, Anderson, and Heath, "Sub-seabed Disposal."

42.   Sandia Laboratories, "Sub-seabed Disposal of Radioactive Waste: An Overview of the Concept and Program," draft version in MC-31, Folder "Nuclear, Low-Level, 1980s," WHOI-Hollister.

43.   The contract, No. 13-2559, dated November 1978, was a continuation of two earlier contracts under the general title of *Deep Sea Geology Study Program* under the DOE's Seabed Disposal Program.

44.   Contract No. 13-2559, Sandia Laboratories Contract, Document No. 13-2559, 18 November 1978, MC-31, Folder "Proposal Index," WHOI-Hollister.

45.   Ibid.

46. Walsh, "Marine Pollution."

47. Hagen, "Waste Disposal"; Sjöbolm and Linsley, "Sea Disposal." For a contemporaneous discussion of the ban on all radioactive waste disposal at sea, see "Ocean Science News," 19 March 1984, in MC-31, Folder "Allyn Vine Drawing," WHOI-Hollister.

48. In 1993, annexes I and II to the London Convention would ban all dumping of radioactive waste at sea. See Convention on the Prevention of Marine Pollution by Dumping of Wastes and Other Matter, *Amendments to the Annexes to the Convention on the Prevention of Marine Pollution by Dumping of Wastes and Other Matter, 1972 Concerning Phasing out Sea Disposal of Industrial Waste*, Resolution LC.49 (16) (12 December 1993), http://www.imo.org/en /KnowledgeCentre/IndexofIMOResolutions/London-Convention-London -Protocol-(LDC-LC-LP)/Documents/LC.49(16).pdf.

49. "US DOE Office of Civilian Radioactive Waste Management," https://www .energy.gov/downloads/office-civilian-radioactive-waste-management. See also Appendix, B, History of the Civilian Radioactive Waste Management Program, in Civilian Radioactive Waste Management Program Plan, Revision 3, DOE/RW-0520, February 2000, US Department of Energy, Office of Civilian Radioactive Waste Management.

50. Ibid.

51. Hollister, "Responses to Follow-Up Questions Regarding Testimony of Dr Charles D Hollister Before the Senate Committee on Energy and Natural Resources," 17 July 1987, MC-31, Folder "November Meeting with Udall," WHOI-Hollister WHOI. See also John Kelly, Testimony of Dr. John E. Kelly on Behalf of the Seabed Association before the United States House of Representatives Committee on Interior and Insular Affairs, Subcommittee on Energy and Environment, 18 November 1987, MC-31, Folder "November Meeting with Udall," WHOI-Hollister.

52. D. R. Anderson to "Distribution," Sandia Laboratories Memo, 14 February 1986, MC-31, Folder "Sub-seabed 1985–1988," WHOI-Hollister. See also Ben C. Rusch to Hollister, 4 April 1986, MC-31, Folder "Sub-seabed 1985–1988," WHOI-Hollister.

53. Edward P. Boland to Hollister, 24 April 1986, MC-31, Folder "Sub-seabed 1985–1988," WHOI-Hollister.

54. See Appendix, B, History of the Civilian Radioactive Waste Management Program, in Civilian Radioactive Waste Management Program Plan, Revision 3, DOE/RW-0520, February 2000, US Department of Energy, Office of Civilian Radioactive Waste Management. See also the "Hess Report" of the US National Academy of Sciences: *The Disposal of Radioactive Waste on Land*, Report of the Committee on Waste Disposal of the Division of Earth Sciences, 1957, National Academy of Science–National Research Council Archives.

55. D. R. Anderson to "Distribution," Sandia Laboratories Memo, 14 February 1986, MC-31, Folder "Sub-seabed 1985–1988," WHOI-Hollister.

56. Rep. William J. Hughes, "H. Con. Res. 2350: A Concurrent Resolution regard-

ing the United States Position on the Subseabed Emplacement of High-Level
Radioactive Materials," introduced 18 November 1983, 98th Cong. (1983–
1984).

57.   *Congressional Record*, 18 November 1983, E5907, "Sub-seabed Emplacement
of High Level Radioactive Waste," Hon. William J. Hughes of New Jersey.

58.   Hughes and Studds to George Shultz, 17 February 1984, MC-31, Folder "Con-
gressional Liaison Memos," WHOI-Hollister. The letter was cosigned by
Gerry Studds, representative from Massachusetts and chairman of the House
Committee on Merchant Marine and Fisheries. The first openly gay member
of the House, Studds was censured in 1983 for a relationship with a teenaged
House page ten years earlier: Steven V. Roberts, "House Censures Crane and
Studds for Sexual Relations with Pages," *New York Times*, 21 July 1983, https://
www.nytimes.com/1983/07/21/us/house-censures-crane-and-studds-for
-sexual-relations-with-pages.html.

59.   Hughes and Studds to George Shultz, 17 February 1984, MC-31, Folder "Congressional Liaison Memos," WHOI-Hollister.

60.   W. Tapley Bennett to William Hughes, March 1984, MC-31, Folder "Congressional Liaison Memos," WHOI-Hollister. Hollister later described the panel
that met in 1985 as "poisoned" because it included environmental lawyers, as
opposed to only "experts"—that is, only scientists. See typed notes to "Russ"
from "Vik," 29 January 1986, MC-31, Folder "Science 86 Query Idea: Nuclear,"
WHOI-Hollister. (The issue was finally resolved in 1993 against any ocean
disposal of radioactive wastes.)

61.   Weart, *Global Warming*.

62.   Memo, 11 March 1987, John Kelly to Members of the Seabed Association, MC-
31, Folder "Sub-seabed 1985–1988," WHOI-Hollister.

63.   The Seabed Association also approached some utility representatives, on the
theory that it was in their interest to support a backup plan to land-based
disposal, but these approaches went nowhere. See MC-31, Folder "Letters,"
WHOI-Hollister.

64.   Now JK Research, based in Denver, and since that time a subcontractor to
the DOE for work on Yucca Mountain. After seabed disposal was itself dead
and buried, JK Associates took on work in support of Yucca Mountain and
in 2001 was accused of manipulating data to in support of the repository.
This led to a 2005 hearing before the Subcommittee on Federal Workforce
and Agency organization. Full report, "Yucca Mountain Project: Have Federal
Employees Falsified Documents," https://archive.org/stream/gov.gpo.fdsys
.CHRG-109hhrg23207/CHRG-109hhrg23207_djvu.txt.

65.   The classical definition of lobbying comes from Milbrath, *American Lobbyists*,
8, who defined it as "the stimulation and transmission of a communication,
by which someone other than a citizen acting on his own behalf, directed to
a governmental decision-maker, with the hope of influencing his decision."
In 1995, the Lobbying Disclosure Act (109 Stat. 691) read:

The term "lobbying contact" means any oral or written communication (including an electronic communication) to a covered executive branch official or a covered legislative branch official that is made on behalf of a client with regard to (i) the formulation, modification, or adoption of Federal legislation (including legislative proposals); (ii) the formulation, modification, or adoption of a Federal rule, regulation, Executive order, or any other program, policy, or position of the United States Government; (iii) the administration or execution of a Federal program or policy (including the negotiation, award, or administration of a Federal contract, grant, loan, permit, or license); or (iv) the nomination or confirmation of a person for a position subject to confirmation by the Senate.

Kelly's activities certainly fit these definitions.

The diagram was prepared for a DOE report on the Office of Subseabed Disposal. It was sent to Charles Hollister with the notation, "FYI Here is a briefing package I would use if I were called upon to brief senior management." "Office of Subseabed Disposal Research," ER-80, by Walter Warnick, Director of the DOE Office of Subseabed Disposal, MC-31, Folder "Miscellaneous Materials to Do with Seabed Disposal," WHOI-Hollister.

66.   "Office of Subseabed Disposal Research," ER-80, by Walter Warnick, Director of the DOE Office of Subseabed Disposal, MC-31, Folder "Miscellaneous Materials to Do with Seabed Disposal," WHOI-Hollister.

67.   Hearing on S. 801 Reauthorization of the National Science Foundation, 30 April 1985 in MC-31, Folder "Senator Edward M. Kennedy," WHOI-Hollister.

68.   Hollister to Kennedy, 26 November 1984, MC-31, Folder "Senator Edward M. Kennedy," WHOI-Hollister. See also Hollister to James Ling, 26 November 1984, Office of Science and Technology Policy; Hollister to Hugh Loweth, 26 November 1984, Office of Management and Budget, MC-31, Folder "Letters," WHOI-Hollister.

69.   Hollister, "Responses to Follow-Up Questions Regarding Testimony of Dr. Charles D Hollister before the Senate Committee on Energy and Natural Resources, 17 July 1987, MC-31, Folder "November Meeting with Udall," WHOI-Hollister.

70.   John Kelly to John Byrne, 2 September 1987, MC-31, Folder "November Meeting with Udall," WHOI-Hollister.

71.   Ibid.

72.   Ibid.

73.   Ibid. See also "Recommencing Research on Sub-seabed disposal of radioactive waste," which specifically identifies "FY-88 Appropriations" as the "vehicle" by which the goal would be obtained; see MC-31, unlabeled folder, WHOI-Hollister.

74.   John Kelly to Ross (Heath), 3 September 1987, MC-31, Folder "November

Meeting with Udall," WHOI-Hollister. See also materials in MC-31, Folder "Reports/Statements," WHOI-Hollister.

75.   "The Seabed Association," Bulletin, December 18 1987, MC-31, Folder "Reports/Statements," WHOI-Hollister.

76.   Testimony of Dr. John E. Kelly on Behalf of the Seabed Association before the United States House of Representatives Committee on Interior and Insular Affairs, Subcommittee on Energy and Environment, 18 November 1987, MC-31, Folder "November Meeting with Udall," WHOI-Hollister. In hindsight, we can say that if that was DOE's intent, it did not work.

77.   The focus on the US Exclusive Economic Zone was clearly designed to refute the position that the United States could not depend on using sites in international waters, but also clearly in conflict with the scientific argument for placing the wastes in the stable interiors of the oceanic plates.

78.   Testimony of Dr. John E. Kelly on Behalf of the Seabed Association before the United States House of Representatives Committee on Interior and Insular Affairs, Subcommittee on Energy and Environment, 18 November 1987, MC-31, Folder "November Meeting with Udall," WHOI-Hollister.

79.   Status report, Seabed Association, 21 October 1987, MC-31, Folder "November Meeting with Udall," WHOI-Hollister.

80.   Disclosure: In 1999 I testified to the Nuclear Waste Technical Review board on the topic of model validation in the context of nuclear waste disposal: US Nuclear Waste Technical Review Board, "Developing a Repository Safety Strategy with Special Attention to Model Validation," Washington, DC, 14 September 1999.

81.   Text of the Nuclear Waste Policy Act, as Amended, p. 61, MC-31, unlabeled folder, WHOI-Hollister.

82.   Derek Spencer to Edward M. Kennedy, 18 January 1988, MC-31, Folder "Kennedy," WHOI-Hollister.

83.   Memo, Dave Ross to Charles Hollister, 11 January 1988, 88 MC-31, Folder "Kerry," WHOI-Hollister and duplicate copy in MC-31, Folder "Sub-seabed 1995," WHOI-Hollister.

84.   Robert Roe to Elizabeth Smedley, 7 July 1988, MC-31, Folder "Congressional Liaison," WHOI-Hollister. On the establishment of the Office of Sub-seabed Disposal Research, as part of the Nuclear Waste Policy Act Amendments, see Congressional Record, 21 December 1987. The whole thing is a little weird— Congress passed the amendment establishing the office, but then allocated no funds for it.

85.   Email, Hollister to Kelly, 27 July 1988, on "Bureau briefing," MC-31, Folder "Utilities," WHOI-Hollister.

86.   Guy Nichols to Bennett Johnston, 26 May 1988, MC-31, Folder "Johnston/ Nichols Letter," WHOI-Hollister.

87.   Hollister to Edward Teller, 8 July 1988, and Edward Teller to Charles Hollister, 2 August 1988, MC-31, Folder "Edison Electric Institute," WHOI-Hollister.

88.   Hollister to McNeer, second draft letter, 26 July 1988, see also other materials

in MC-31, Folder "Utilities," WHOI-Hollister and MC-31, Folder "Edison Electric Institute," WHOI-Hollister.

89. It is a little hard to understand why Hollister had thought that the Utilities Group would not oppose him, given that it was well known that the electrical utilities were keen to see Yucca Mountain brought on line.

90. Draft letter, Hollister to McNeer, n.d., but second draft is 26 July 1988, MC-31, Folder "Utilities," WHOI-Hollister.

91. Ibid.

92. Sub-seabed disposal of high-level radioactive waste and spent fuel, UNWMG, n.d., MC-31, Folder "Utilities," WHOI-Hollister.

93. *Firing Line Newsletter*, 30 June 1991, in AC-09, Folder "Hollister," WHOI-ODR.

94. "Nuclear Waste: A Solution," William F. Buckley, "On the right," 7 June 1991 in AC-09, Folder "Hollister," WHOI-ODR

95. Buckley to Hollister, Single Page Fax Transmission, 5 June 1991, AC-09, Folder "Hollister," WHOI-ODR.

96. *Firing Line Newsletter*, 30 June 1991, AC-09, Folder "Hollister," WHOI-ODR.

97. Here I think Hollister was telling the truth. He had become so committed to the subseabed option that his advocacy was decoupled from his interest in research funding.

98. *Firing Line Newsletter*, 30 June 1991, AC-09, Folder "Hollister," WHOI-ODR.

99. Actually, at that time, not many people knew much about the hydrology of the unsaturated zone, which is one reason the Nuclear Waste Technical Review Board ultimately decided that the scientific basis for disposal at Yucca Mountain was "weak to moderate." See also US Department of Energy, *Yucca Mountain*. For background on the board, see "Geologic Disposal," https://www.nwtrb.gov/scope/geologic-disposal.

100. Hollister to Steve Frishman, 11 July 1991, MC-31, Folder "Radioactive Waste 1988–1992," WHOI-Hollister. The recipient was a state geologist, Steve Frishman, who was still active in 2015. See Steve Tretault, "Heller Pushes to Include State Geologist on Yucca Tour," *Pahrump Valley (NV) Times*, 8 April 2015, http://pvtimes.com/news/heller-pushes-include-state-geologist-yucca-tour.html.

101. Woods Hole Oceanographic Institution's position on Buckley editorial, 12 June 1991. In AC-09, Folder "Hollister-Ocean Dumping," WHOI-ODR.

102. Keith Schneider, "Scientists Suggest Dumping Sludge on Vast, Barren, Deep Sea Floor," *New York Times*, 2 December 1991, AC-09, Folder "Hollister-Ocean Dumping," WHOI-ODR; see also Walter Sullivan, "Schemes Are Debated for Dumping Sewage Deep beneath the Sea," *New York Times*, updated press clipping, MC-31, Folder "Short Articles," WHOI-Hollister.

103. Goldberg, "Waste Space," 2–9.

104. On the 1998 Ocean Dumping Ban Act, see "EPA History: Ocean Dumping Ban Act of 1988," https://archive.epa.gov/epa/aboutepa/epa-history-ocean-dumping-ban-act-1988.html.

105. Craig Dorman, "Not Advocates," *New York Times*, 23 December 1991, AC-09, Folder "Hollister-Ocean Dumping," WHOI-ODR.

106. The IRS website states: "An organization will be regarded as attempting to influence legislation if it contacts, or urges the public to contact, members or employees of a legislative body for the purpose of proposing, supporting, or opposing legislation, or if the organization advocates the adoption or rejection of legislation." See "Lobbying," https://www.irs.gov/charities-non -profits/lobbying. Nichols was an advocate of nuclear power who dismissed public concerns as "jitters." See Carlowicz, "Guy Nichols."

107. See "Lobbying," https://www.irs.gov/charities-non-profits/lobbying.

108. This would appear to be electioneering, which is explicitly prohibited to 501(c)3 organizations. See "The Restriction of Political Campaign Interven- tion by Section 501(c)(3) Tax-Exempt Organizations," https://www.irs.gov /charities-non-profits/charitable-organizations/the-restriction-of-political -campaign-intervention-by-section-501-c-3-tax-exempt-organizations.

109. "Don't rule out deep-sea disposal of sewage sludge," *USA Today*, date not fully legible but appears to be 6 or 8 December 1991, MC-31, Folder "Short Articles," WHOI-Hollister.

110. A similar issue has arisen with respect to geoengineering of the climate systems, also called "climate intervention." The US National Academy of Sciences has endorsed research into climate intervention options, but opponents have argued that studying it implies the possibility of doing it. The topic was sufficiently controversial, even among scientists, that the academy divided the subject into two parts, with one part (solar radiation management) viewed as more potentially problematic that the other (carbon dioxide removal and sequestration, also known as "negative emissions"). See National Research Council, *Climate Intervention: Reflecting Sunlight to Cool Earth* (Washington, DC: National Academies Press, 2015), http://www.nap .edu/catalog/18988/climate-intervention-reflecting-sunlight-to-cool-earth; National Research Council, *Climate Intervention: Carbon Dioxide Removal and Reliable Sequestration* (Washington, DC: National Academies Press, 2015), http://www.nap.edu/catalog/18805/climate-intervention-carbon-dioxide -removal-and-reliable-sequestration.

111. Woodwell to Dorman, 19 December 1991, AC-09, Folder "Hollister-Ocean Dumping," WHOI-ODR.

112. Vaughan Bowen to D. Richard Anderson, Sandia Laboratories, 6 February 1976, MC-87, Folder "Vaughan Bowen," Miscellaneous Manuscripts Collec- tion, Woods Hole Oceanographic Institution, Woods Hole, MA. Original emphasis.

113. Vaughan Bowen and John C. Burke, n.d., "The Sea Bottom in the OECD/NEA Solid Waste Dumping Site, submitted to Marine Pollution Bulletin," draft copy found in MC-87, Folder "Vaughan Bowen," Miscellaneous Manuscripts Collection, Woods Hole Oceanographic Institution, Woods Hole, MA.

114. Ibid.

115. Ibid. The irony, of course, was that Hollister was one of the people who had

116. Goldberg, "Waste Space," 2–9.
117. The idea is articulated, for example, in E. P. Odum's *Fundamentals of Ecology*.
118. William Nierenberg, "Current Problems in Oceanography, November 1966, Speech to the American College of Physicians, Box 165, Folder 16, SIO-Nierenberg.
119. Oreskes and Conway, *Merchants of Doubt*, chaps. 4 and 6.
120. Ibid.
121. Kamlet, "Waste Space: Rebuttal," 10.
122. Ibid., 14.
123. Beneath the issue of epistemic confidence was also a value issue: what rights does man have vis-à-vis marine organisms? Do our needs come first or does the preservation of other species merit equal consideration? There seems little doubt in this case that biologists were more likely to take the latter view; chemists and physicists the former.
124. Oppenheimer et al., *Discerning Experts*.
125. Craig Dorman, "Statement," 6 December 1991, and Statement, 19 December 1991, AC-09, Folder "Ocean Dumping," WHOI-ODR.
126. Full disclosure: That new chief of staff, in 1991, was Radford Byerly Jr. (1936–2016), who generously read and commented on an early draft of this chapter.
127. Notes from Trip, Charles D. Hollister and Derek W. Spencer, AC-09, Folder "Hollister," WHOI-ODR.
128. Hollister and Del Smith visit to Secretary Watkins Office, AC-09, Folder "Hollister," WHOI-ODR.
129. This would have been a nod to the "Waste Isolation Pilot Plant" being developed in New Mexico for disposal of high-level military wastes.
130. Confidential: Deep Ocean Isolation, n.d., AC-09, Folder "Hollister," WHOI-ODR.
131. Liberia is infamous for its corruption, and for ships flying under its flag to avoid regulation. Hollister apparently saw no irony in this. Julia Simon, "Liberia's Flags of Convenience Help It Stay Afloat," *All Things Considered*, 7 November 2014, https://www.npr.org/2014/11/07/362351967/liberias-flags-of-convenience-help-it-stay-afloat. Kongsli may be found today working as an adviser in the liquefied natural gas business: "Management Team," https://www.gls-ltd.com/team/.
132. Derek Spencer to Christian Kongsli, 13 February 1991, in AC-09, Folder "Ocean Dumping," WHOI-ODR.
133. Confidential: Deep Ocean Isolation, n.d., AC-09, Folder "Hollister," WHOI-ODR. The memo is filed under Hollister, but the writing style does not seem like Hollister's; it may have been written by Derek Spencer. See Derek Spencer to Christian Kongsli, 13 February 1991, in AC-09, Folder "Ocean Dumping," WHOI-ODR.

demonstrated the existence of deep currents, although he would have argued that they were not sweeping the abyssal plains.

134. Confidential: Deep Ocean Isolation, n.d., AC-09, Folder "Hollister," WHOI-ODR.

135. Memo, 11 April 1995, John Kelly to Hollister, MC-31, Folder "Sub-seabed 1995," WHOI-Hollister. Hollister and his colleagues preferred to use the term *subseabed* in response to legal opinions that seabed disposal would violate the Law of the Sea prohibition on ocean dumping. Hollister argued that his proposal was not dumping but "injection," because the wastes would be placed within the marine muds at the sea bottom rather than on top of them.

136. Hollister to Saalfeld, 6 September 1990, MC-31, Folder "Congressional Education Week," WHOI-Hollister.

137. Ibid.

138. Saalfeld to Hollister, 7 September 1990, MC-31, Folder "Congressional Education Week," WHOI-Hollister.

139. Title Code 31 U.S.C. § C, 1352, "Limitation on use of appropriated funds to influence certain federal contracting and financial transactions." For more on the Byrd Amendment, see "The Continued Dumping and Subsidy Offset Act ('Byrd Amendment')," https://www.everycrsreport.com/reports /RL33045.html; "OMB Circular A-21, https://www.whitehouse.gov/sites /whitehouse.gov/.../omb/circulars/A21/a21_2004.pdf; Defense appropriations bill language for 1990: "H.R. 3072- Department of Defense Appropriations Act, 1990," https://www.congress.gov/bill/101st-congress/house-bill /3072; and for 1991, "H.R. 2521-Department of Defense Appropriations Act, 1991-1992," https://www.congress.gov/bill/102nd-congress/house-bill/2521 /text.

140. Seminar on Ocean Science and National Security, Seminar Participants, n.d., MC-31, Folder "Congressional Education Week," WHOI-Hollister. See also "Draft: Ocean Science and National Security," MC-31, Folder "Congressional Seminar," WHOI-Hollister.

141. Seminar Report, "The Ocean, Climate Change and the Environment, and Participant List for the Same," MC-31, Folder "Congressional Education Week," WHOI-Hollister.

142. Fred Saalfeld to Hollister, 18 September 1991, MC-31, Folder "Congressional Seminar," WHOI-Hollister. How revealing that Saalfeld refers to the funds as "my funds," as opposed, for example, to taxpayer or even Navy funds!

143. Memo, Hollister to "The Directorate," 3 March 1981, MC-31, Folder "Congressional Liaison Memos," WHOI-Hollister.

144. Management and Disposition of Excess Weapons Plutonium, Prepublication Copy, MC-31, Folder "Reports/Statements," WHOI-Hollister. The committee did allow for the possibility that if subseabed disposal was reopened for radioactive waste, then it should be considered for excess plutonium as well, but they considered it unlikely that that would happen.

145. William J. Broad, "Scientists Fear Atomic Explosion of Buried Waste," *New York Times*, 5 March 1995; William J. Broad "Theory on Threat of Blast at Nu-

clear Waste Site Gains Support," *New York Times*, 23 March 1995; and David Applegate, "FORUM: Nuclear Explosions in a Geologic Repository? Peer Review Meets Politics and the Press," EOS, undated press clipping, all in MC-31, Folder "Short Articles," WHOI-Hollister.

146. Hollister to Frank Garman, 12 April 1995, MC-31, Folder "Letters," WHOI-Hollister. Hollister also presented a paper at the American Geophysical Union, in 1997, once again advocating seabed disposal and suggesting that the Russian ballistic missile submarine, *Yankee II*, sunk in 1986, could provide a site to study the dispersion of radionuclides in the ocean. It is interesting that Hollister said nothing about lost American nuclear materials, including the *Thresher* and *Scorpion* submarines, whose fate Woods Hole was continuing to track (chapter 2). See Hollister, "Nuclear Warheads and the Ocean Option," and "A Nuclear Waste Experiment at Sea?" MC-31, unlabeled folder, WHOI-Hollister.

147. Sub-seabed Disposal, "Background Blurbs," March 1995, MC-31, Folder "Sub-seabed 1995," WHOI-Hollister. See also Conference Call, 23 March 1995, MC-31, unlabeled folder, WHOI-Hollister.

148. "Recent activities that MAY relate to the Sub-seabed Disposal Concept," 15 March 1995, MC-31, unlabeled folder, WHOI-Hollister.

149. Various materials, MC-31, Folder "Sub-seabed 1995," WHOI.

150. Draft discussion paper, 25 September 1990, MC-31, Folder "Sub-seabed 1995," WHOI-Hollister. This was an earlier document that was recirculated in 1995.

151. Bennett Johnston to Guy Nichols, 21 July 1995, MC-31, Folder "Sub-seabed 1995," WHOI-Hollister.

152. On industry contacts, see various documents, MC-31, Folders "Edison Electric Institute" and "Utilities," WHOI-Hollister. In correspondence with industry representatives, Kelly insisted that he supported Yucca Mountain and advocated seabed disposal only as a backup, but Hollister had criticized Yucca Mountain loudly, directly, and frequently. See, e.g., Kelly to Charles McNeer, 30 June 1988, in MC-31, Folder "Edison Electric Institute," WHOI-Hollister. On outreach to the energy secretary, see fax from JK Research Associates to Henry W Kendall, 21 April 1995, MC-31, Folder "Sub-seabed 1995," WHOI-Hollister. Henry Kendall was a physics professor at MIT and a founder of the Union of Concerned Scientists, who opposed expansion of civilian nuclear power because of the risk of nuclear proliferation.

153. Memo, Bud Ris, Union of Concerned Scientists, to Hollister, 3 October 1995, MC-31, Folder "Sub-seabed 1995," WHOI-Hollister.

154. Hollister to Richard Stegemeir, 6 July 1995, MC-31, Folder "Sub-seabed 1995," WHOI-Hollister. Stegemeir was chairman and CEO of Unocal.

155. Memo, John Kelly to Hollister and Kendall, 30 September 1995, MC-31, Folder "Sub-seabed 1995," WHOI-Hollister.

156. See, e.g., various materials in MC-31, Folder "Edison Electric Institute," WHOI-Hollister. The position of industry was to oppose the diversion of

monies from the Nuclear Waste Fund from any activity that was not directly aimed at creating a functioning geological repository, or interim MRS system, as soon and as efficiently as possible.

157. Ris would later become the longtime head of the New England Aquarium. See Mark Shannon and Meredith Goldstein, "New England Aquarium CEO Bud Ris Stepping Down," *Boston Globe*, 2 April 2014, https://www .bostonglobe.com/lifestyle/names/2014/04/02/aquarium-ceo-bud-ris -steps-down-nigella-hillgarth-takes-over/etcWY7x236poCuHGKOCtdK /story.html.

158. Bud Ris to Charles Hollister, 3 October 1995, MC-31, Folder "Sub-seabed 1995," WHOI-Hollister.

159. Kelly to Hollister, email, 6 November 1996, MC-31, Folder "Sub-seabed 1996– 97," WHOI-Hollister.

160. Steven Nadis, "The Sub-seabed Disposal Solution," *Atlantic Monthly*, October 1996, MC-31, unlabeled folder, WHOI-Hollister.

161. "Edited Versions of Letters We Will Print in February," fax, Steve Nadis to Hollister, 22 November 1996, MC-31, Folder "Sub-seabed 1996–97," WHOI-Hollister.

162. Thorne-Miller, *Living Ocean*.

163. Ibid. See also "Edited Versions of Letters We Will Print in February," fax, Steve Nadis to Hollister, 22 November 1996, MC-31, Folder "Sub-seabed 1996–97," WHOI-Hollister.

164. Ibid.

165. Hollister and Nadis, "Radioactive Waste"; Vartanov and Hollister, "Nuclear Legacy."

166. Hollister and Nadis, "Radioactive Waste."

167. Merton, "Normative Structure."

168. Krimsky, *Science in the Private Interest*; Roosth, *Synthetic*.

169. See, e.g., Hollinger, "Money and Academic Freedom"; Healy, "Conflicting Interests," Press and Washburn, "Kept University"; Etzkowitz, "Conflicts of Interest."

170. Oreskes et al., "Why Disclosure Matters."

171. Oreskes, "Scientist as Sentinel"; Oppenheimer et al., *Discerning Experts*, chap. 6.

172. Oppenheimer et al., *Discerning Experts*, chap. 5.

173. Oreskes and Conway, *Merchants of Doubt*, esp. introduction and chapters 1, 2, and 5; Oreskes, *Why Trust Science?*

174. Weart, *Global Warming*.

175. Merton, *Sociology of Science*, 254; Shapin, *Social History*.

176. Oreskes, *Plate Tectonics*, chap. 1.

177. Mariette and Robinson, *Proceedings*, 3–4.

178. This was the meeting where Hollister scheduled the morning sessions to end in time for a half day skiing in the afternoon; Hollister apparently also took

the opportunity to show off his skiing prowess. Armi dryly recalled it as "an extremely effective format." Larry Armi, personal communication, May 2016.

179. Armi, "Dynamics."

180. While Armi made his measurements in the Atlantic, available evidence suggested that the bottom boundary layer was even larger in the Pacific, perhaps by a factor of four.

181. Garrett, "Isopycnal Surfaces," in Mariette and Robinson, *Proceedings*.

182. Bowen et al., "Fallout Radionuclides," 430.

183. Mullin, "Biology Report," in Mariette and Robinson, *Proceedings*.

184. Art Yayanos, email communication with the author, 28 September 2005.

185. Etzkowitz, "Conflicts of Interest."

186. Oppenheimer et al., *Discerning Experts*; Oreskes, *Why Trust Science?*

187. Makhijani and Saleska, *High-Level Dollars, Low-Level Sense*, Copy in MC-31, Folder "Radioactive Waste 1998–1992," WHOI-Hollister . The role of trust in public support for Yucca Mountain has also been explored by Flynn et al., "Trust."

188. Hollister was right that DOE and the US Congress created a situation in which there was no alternative to Yucca Mountain, and that when it failed the United States was left without a plan for long-term civilian radioactive waste storage (Hollister to Steve Frishman, 11 July 1991, MC-31, Folder "Radioactive Waste 1998–1992," WHOI-Hollister). But he was wrong in believing that the US would then turn to the sea. Since the 1990s, seabed storage has never been seriously reconsidered; in June 2018 it was reported that Congress was revisiting plans to open a waste facility at the site. See Michael Collins, "Congress Works to Revive Long-Delayed Plan to Store Nuclear Waste at Yucca Mountain," *USA Today*, 8 June 2018, https://www.usatoday .com/story/news/politics/2018/06/03/yucca-mountain-congress-works -revive-dormant-nuclear-waste-dump/664153002/. *Qui vivra verra.*

189. Hollister to Steve Frishman, 11 July 1991, MC-31, Folder "Radioactive Waste 1998–1992," WHOI-Hollister.

190. Sheila Jasanoff, *Fifth Branch*. Of course, this is not just true of the United States; see Pestre and Krige, *Science in the Twentieth Century*; Krige, *American Hegemony*.

191. The OMB guidelines around this time proposed that no scientist be allowed to consult on regulatory decisions if the scientist had received research funding from the agencies involved. That standard would have eliminated virtually all relevant experts in many areas; some skeptics suggested that was the aim. "Federal Agencies Subject to Data Quality Act," http:// corporate.findlaw.com/law-library/federal-agencies-subject-to-data-quality -act.html. The Data Quality Act, passed in 2001, did not instate such restraints, but similar proposals have been revisited under President Donald Trump. See Hannah Northey and Sean Reilly, "Proposal to Ban EPA Grantees from Agency Science Advisory Boards Stirs Controversy," *Science*, 18 October

2017, http://www.sciencemag.org/news/2017/10/proposal-ban-epa-grantees-agency-science-advisory-boards-stirs-controversy. Full disclosure: I served as a consultant to the US Environmental Protection Agency and to the US Nuclear Waste Technical Review Board, and as a member of National Academy of Science–National Research Council panel, which received funding from the US Department of Transportation. However, I personally have never received research funding from any of these agencies. As noted in the acknowledgments, I did receive seed money for this book project from the Scripps Institution of Oceanography, as well as research support from the University of California, San Diego, Faculty Senate grants. I leave it to my readers to judge whether I have pulled my punches or overcompensated.

192. For further discussion of this point, see Oppenheimer et al., *Discerning Experts*.

193. Here I am using the language of the Intergovernmental Panel on Climate Change, which, though imperfect, does get at an important distinction. See discussion in Oppenheimer et al., *Discerning Experts*.

CHAPTER NINE

1. E. Frieman, 2000, "Scripps: The End of the Cold War and New Beginnings," Speech dated 8 February 2000, typescript, Box 7, Folder 21, p. 1, SIO-Frieman.

2. Ibid., 5.

3. My own engagement with the issue of climate change began right here. In the early 2000s, when I first learned the context under which Ed Frieman had decided to shift the direction of Scripps from military matters to climate change, it struck me as troublingly opportunistic. However, as I dug into the scientific and historical materials, my view of the matter reversed: I now think that the Cold War focus on the Navy mission delayed Scripps from the sustained engagement with climate change that other institutions, such as National Center for Atmospheric Research and Geophysical Fluid Dynamics Laboratory, began in the 1970s, and that our understanding of key issues such as ocean heat update and acidification would have been better had Scripps and Woods Hole taken up climate change as a motivating context sooner than they did. I address this in chapter 10.

4. An interesting example of media attention is the 1958 Frank Capra film, "The Unchained Goddess," produced for the Bell Telephone Series and released 12 February 1958, https://www.youtube.com/watch?v=sqClSPWVnNE.

5. Speech to the Newcomen Society, 1957, "Matthew Fontaine Maury (1806–1873) Pathfinder of the Seas, Newcomen Society Meeting, 6 June 1957, Falmouth and Woods Hole, MA, AC-09.4, Box 24, Folder 19, "Personnel, Iselin COD, 1949–1959," WHOI-ODR.

6. Von Arx, "A Science in Bondage," February 1965, MC-24, Folder "A Science in Bondage," WHOI-von Arx, 3.

7. Friedman, *Appropriating the Weather*.

8. Pettersson, "Weathering the Storm."

9. Forman, "Into Quantum Electronics"; Forman, "Behind Quantum Electronics."

10. Von Arx, "A Science in Bondage," February 1965, MC-24, Folder "A Science in Bondage," WHOI-von Arx, 4. Paul Forman has made the same argument about twentieth-century physics, see Forman, "Into Quantum Electronics" and "Behind Quantum Electronics."

11. Von Arx, "A Science in Bondage," February 1965, MC-24, Folder "A Science in Bondage," WHOI-von Arx, 7. Original emphasis.

12. Fleming, *Historical Perspectives*; Weart, *Global Warming*; Howe, *Behind the Curve.*

13. Chamberlin, "Cause of Glacial Periods"; Arrhenius, "Influence of Carbonic Acid"; Arrhenius, *Lehrbuch der kosmischen Physik*; Arrhenius, "Temperatur der Erdoberfläche"; Arrhenius, "Ueber Die Wärmeabsorption"; Fleming, *Historical Perspectives.*

14. Sherlock, *Man as Geological Agent*, 302–3, 305.

15. Ibid., 302.

16. I think it may have been fairly widely read; when my husband became an assistant professor of earth science in 1990, he found a copy in the office of the late Noye Johnson, one of the pioneers of acid-rain studies. On Johnson, see Noye M. Johnson Dies; Acid Rain Researcher," *New York Times*, 30 December 1987, https://www.nytimes.com/1987/12/30/obituaries/noye-m-johnson -dies-acid-rain-researcher.html.

17. Hulburt, "Lower Atmosphere."

18. Callendar, "Amount of Carbon Dioxide"; see also Callendar, "Artificial Production"; Callendar, "Variations of the Amount"; and Callendar "Influence Climate." In hindsight his figures were pretty much correct, although scientists now put the preindustrial baseline at or about 280 parts per million. See "Noye M. Johnson Dies; Acid Rain Researcher," *New York Times*, 30 December 1987, https://www.nytimes.com/1987/12/30/obituaries/noye-m -johnson-dies-acid-rain-researcher.html.

19. Plass, "Infra-Red Cooling Rate"; Plass, "Carbon Dioxide Theory."

20. In 1956, he moved to the Ford Motor Company. It is not clear what connection there was, if any, between Ford and his work on carbon dioxide, but the possibility that Ford was attending to carbon dioxide and climate in the 1950s merits further investigation.

21. For details on the radiative transfer argument, see Weart, *Global Warming*, and the accompanying website at https://www.aip.org/history/climate /Radmath.htm. Weart makes a mistake, however, in saying that Plass's "specific numerical predictions for climate change made little impression on his colleagues." This might depend on how one interprets "specific numerical predictions," but it is certainly the case that Roger Revelle and Hans Suess were aware of, and cited, Plass's work, and it seems to have been very important in their thinking that anthropogenic climate change was likely to occur.

22.   Plass, "Infra-Red Cooling Rate."

23.   Hulburt, "Lower Atmosphere." See also Fleming, *Historical Perspectives*; Weart, *Global Warming*.

24.   Plass, "Carbon Dioxide Theory," 142.

25.   Ibid., 150.

26.   Plass wrote a popular version of this work, "Carbon Dioxide and Climate," for *Scientific American*. The piece focused primarily on the carbon dioxide theory of climate change in terms of ice ages but also has a well-developed discussion of anthropogenic climate change, stating that "during the past century a new geological force has begun to exert its effect upon the carbon dioxide equilibrium of the earth," and discussing the impacts of fossil fuels, land-use changes, and the prospect of ocean acidification. Like other scientists at this time, he noted the competing effects of dust and variations in solar radiation, but his final paragraph hints at the suggestion that in the long run carbon dioxide may prove dominant: "We shall be able to test the carbon dioxide theory against other theories of climatic change quite conclusively during the next half-century. Since we now can measure the sun's energy output independent of the distorting influence of the atmosphere, we shall see whether the earth's temperature trend correlates with measured fluctuations in solar radiation. If volcanic dust is the more important factor, then we may observe the earth's temperature following fluctuations in the number of large volcanic eruptions. But if carbon dioxide is the most important factor, long-term temperature records will rise continuously as long as man consumes the earth's reserves of fossil fuels" (47).

27.   Revelle and Suess, "Carbon Dioxide Exchange," 19–20.

28.   Ibid.

29.   Keeling, "Rewards and Penalties."

30.   This term was developed by Gordon MacDonald, as part of his work on the NAS Advisory Panel for Weather Modification. MacDonald introduced the phrase "inadvertent atmospheric modification" in the panel's 1966 report, the follow-up to an earlier report more narrowly concerned with the feasibility of deliberate "weather modification." See National Research Council, *Scientific Problems*, 1964; National Research Council, *Problems and Prospects*, 1966.

31.   Fleming, *Fixing the Sky*.

32.   In the early 1960s, several committees revisited the question, including the National Academy of Sciences Committee on Atmospheric Sciences (1961–1970), the President's Science Advisory Committee Panel on Atmospheric Sciences (1961–1964), and the US National Science Foundation Advisory Panel for Weather Modification (1964–1967).

33.   National Research Council, *Scientific Problems*, foreword.

34.   Munk, Oreskes, and Muller, *Gordon James Fraser MacDonald*.

35.   M. Pomeranz, personal communication with author, 2002. Comments at memorial for Gordon MacDonald, Scripps Institute of Oceanography, La

Jolla, CA. See also "Oral History Transcript." MacDonald's role is also noted in Rich, "Losing Earth."

36. National Research Council, *Weather and Climate Modification*, 1966, 10.

37. MacDonald, personal communication with author, 2001. Gordon and I discussed this and many other things while we were working on the book *Plate Tectonics*.

38. President's Science Advisory Committee, *Quality of Our Environment*, 9, 127. One should also note that Revelle and his colleagues made an early proposal for geoengineering: that one might counter act the effects of carbon dioxide by laying reflective particles on the sea surface.

39. On Smagorinsky and numerical weather prediction, see Harper, *Weather*. On the history of meteorology broadly, see Fleming, *Inventing Atmospheric Science*.

40. President's Science Advisory Committee, *Quality of Our Environment*, appendix Y4, 116.

41. Ibid., appendix Y4, 126–27.

42. MacDonald, "Climatic Consequences"; MacDonald, "Pollution, Weather and Climate"; MacDonald, *Long-term Impacts*; MacDonald, "Scientific Basis." MacDonald told me that he thought aerosols would lead to cooling, until he learned about Keeling's work. Then he thought it was a toss-up between the cooling effects of aerosols and the warming effects of carbon dioxide. But then, with the passage of the Clean Air Act (and other air-pollution control laws of the late 1960s and early 1970s), he concluded that we would clean up aerosols, and so the effect of carbon dioxide would come to dominate. He was right about that. On the history of the scientific input, including MacDonald's, into the Clean Air Act, see Oreskes et al., "Congressional Intent." The issue was also taken up at the Stockholm conference in 1972; see "United Nations Conference on the Human Environment (Stockholm Conference)," https://sustainabledevelopment.un.org/milestones/humanenvironment.

43. MacDonald, "Climatic Consequences."

44. White, "Oceans and Climate."

45. Schneider, "Climate Confusion"; Manabe and Wetherald, "Doubling the CO_2."

46. World Meteorological Organization, *Proceedings*, 1.

47. Ibid., 2.

48. MacDonald, *Long-Term Impacts*, 1–2, 25.

49. Verner E. Suomi in Charney et al., *Carbon Dioxide*, viii.

50. National Academy of Sciences, "An Evaluation of the Evidence for CO_2-Induced Climate Change," 1979, Film label: CO_2 and Climate Change: Ad Hoc, National Academy of Sciences Archives, Washington, DC.

51. Ibid.

52. Physicist William Nierenberg served on the JASON study group and was likely responsible for the addition of a dissenting voice suggesting that the changes would not be adverse. However, the bulk of reports, memos, and press releases from this time suggest a near consensus that the changes

would be adverse and difficult to manage. For an analysis of Nierenberg's influence, see Oreskes et al., "Chicken Little."

53.   Perry, "Energy and Climate."

54.   In White et al., *International Perspectives*; Miles, "International Negotiations."

55.   MacDonald, "Scientific Basis."

56.   Shabecoff, "Global Warming." For an interesting argument at that time over detection and attribution of specific droughts, see Hansen, "Thou Shalt Not."

57.   Wilford, "Bold Statement."

58.   Edwards, *Vast Machine*.

59.   Brysse et al., "Least Drama"; Hansen, "Scientific Reticence."

60.   Oreskes, *Rejection of Continental Drift*.

61.   Edward Frieman, personal communication with author, 16 March 2007.

62.   John Knauss to Richard Hallgren, 1971, MC-06, Folder "Correspondence, 1971," WHOI-Stommel.

63.   Revelle, "World Climate," 33.

64.   Solomon et al., *Fourth Assessment Report*, 5.

65.   Houghton et al., *Climate Change 1994*; Houghton et al., *Climate Change 2001*; Weart, *Global Warming*. Oreskes, "Scientific Consensus"; Edwards, *Vast Machine*.

66.   Frosch, "Underwater Sound"; Urick, *Sound Propagation*; Spiess, "Undersea Research."

67.   Munk, Testimony, US Commission on Ocean Policy. San Pedro, CA, 18 April 2002.

68.   On the history of the idea of temperature, see Chang, *Inventing Temperature*.

69.   Munk and Wunsch, "Acoustic Tomography."

70.   Ibid. See also Spiesberger et al., "Acoustic Mapping"; Spindel and Worcester, "Acoustic Tomography"; Munk et al., *Acoustic Tomography*.

71.   There is a long history of studying variation in the speed of sound over long distances in the ocean, see, e.g., Ewing and Worzel, *Propagation of Sound*; Ewing and Worzel, "Sound Transmission"; Ewing, Iselin, and Worzel, "Sound Transmission in Sea Water"; Johnson, "Synthesis of Point Data"; Del Grosso, "Speed of Sound"; Hamilton, "Time Variation." See also the discussion of comparable Russian work in Godin and Palmer, *Russian Underwater Acoustics*. There is nonetheless some ambiguity about how the idea to use acoustic tomography to detect climate change first developed. Like many scientific ideas, it seems to have been discussed by quite a few people before being taken up and advanced by a subset of scientists. In 1979, Munk and Wunsch, in "Acoustic Tomography," promoted the idea of using acoustic tomography for large-scale monitoring and later explicitly proposed monitoring ocean temperature related to anthropogenic climate change. See Munk and Forbes, "Global Ocean Warming"; Munk, "Global Ocean Warming." In 1983, John Spiesberger, T. G. Birdsall, and Kurt Metzger proposed developing an acoustic thermometer in a proposal to ONR. See Spiesberger et al., "Gulf Stream Meandering"; John Spiesberger, email communication, 15 February

2012. See also Spiesberger et al., "Ocean Acoustic Multipaths"; Spiesberger and Metzger, "Basin-Scale Tomography"; Spiesberger and Metzger, "Tomographic Section"; Spiesberger and Metzger, "New Algorithm"; Spiesberger and Metzger, "New Estimates"; Spiesberger et al., "Listening for Climatic Temperature."

72. Hamilton, "Time Variation"; Spiesberger et al., "Basin-Scale Ocean Monitoring with Acoustic Thermometers," typescript, MC-06, Folder 4, "Correspondence 1991, 1 of 2," WHOI-Stommel.

73. Hersey, "Introduction," 4.

74. Ibid.

75. House Armed Services Committee, 28 March 1985, MC-31, Folder "Congressional Testimony," WHOI-Hollister.

76. Spiesberger et al., "Acoustic Thermometer Proposal," MC-06, Folder 3, "Correspondence, 1983" WHOI-Stommel. Later referenced as ONR Research Contract N00014-82-C-0019, 1983. See Spiesberger and Metzger, "Basin-Scale Tomography."

77. Spiesberger et al., "Ocean Acoustic Multipaths." Robert C. Spindel was also at Woods Hole, and Kurt Metzger in the Department of Engineering Department at the University of Michigan.

78. Munk and Worcester, "Weather and Climate under the Sea."

79. The Advanced Research Projects Agency (ARPA) gained the letter *D* when it was renamed the Defense Advanced Research Projects Agency (DARPA) in 1972. The name briefly reverted to ARPA in 1993, only to have the *D* restored in 1996.

80. Behringer et al., "A Demonstration."

81. ATOC has had various names. Originally, it was called ocean acoustic thermometer, then acoustic tomography of ocean climate, and finally acoustic thermography of ocean climate in the draft environmental impact statement and all documents from about 1994 onward, presumably because this was more comprehensible to nonscientists. I have been unable to determine who was responsible for this change or precisely when it occurred. Spiesberger et al., "Acoustic Thermometer Proposal," MC-06, Folder 3, "Correspondence, 1983" WHOI-Stommel; Behringer et al., "A Demonstration," 121–25. In 1983, Spiesberger, by then at WHOI, along with T. G. Birdsall and Kurt Metzger submitted their own proposal to ONR for the development of an "acoustic thermometer." Later, they would develop a different technique from ATOC that they called global acoustic mapping of ocean temperature. See Spiesberger and Metzger, "Basin-Scale Ocean Monitoring"; Spiesberger et al., "Global Acoustic Mapping."

82. Spiesberger et al., "Acoustic Thermometer Proposal," MC-06, Folder 3, "Correspondence, 1983" WHOI-Stommel. Quotation from abstract. The work was funded by ONR and the Office of Naval Technology, so in this respect von Arx's belief that a shift in focus toward weather and climate would imply a shift in patronage was not borne out, although much later oceanographic

work in relation to global warming would be funded by other agencies, particularly NASA, NOAA, and DOE.

83. Spiesberger, et al. "Acoustic Thermometer Proposal," MC-06, Folder 3, "Correspondence, 1983" WHOI-Stommel. Many scientists involved in ATOC believe that, had they begun the measurements they envisaged in 1983 and continued them until now, some of the contentious debate over global warming might have been dampened. But that seems to me unlikely, given that most of the opposition to action in response to global warming has come not from scientists unsatisfied by the empirical evidence of global warming but from groups anxious to protect industrial interests from regulations that could affect their bottom lines. This point is discussed further later in this chapter. On the scientific consensus over global warming, see Oreskes, "Scientific Consensus." On the denial of global warming by interested parties, see Oreskes, "How Do We Know?"; Gelbspan, *Heat Is On*; Gelbspan, *Boiling Point*; Mooney, *Republican War on Science*; Oreskes and Conway, *Merchants of Doubt*.

84. Handwritten note, J. L. Spiesberger to H. Stommel on cover letter to Spiesberger et al., 3 May 1983, MC-06, Folder 3, "Correspondence, 1983" WHOI-Stommel.

85. Keeling, "Rewards and Penalties."

86. Edward Frieman to Richard Attiyeh, 26 January 1987, Box 15, Folder "Global Change, 16 January 1987–December 16, 1987," SIO-ODR (Frieman).

87. Munk and Forbes, "Global Ocean Warming"; Munk, "Global Ocean Warming."

88. Summary of "Acoustic Measurements," 4 October 1991, MC-06, Folder "Correspondence 1991, 1 of 2," WHOI-Stommel.

89. Munk et al., "Heard Island."

90. Potter, "Enviro-vandalism." See also Various documents, Box 12, Folder "Folders Applications for Permit, January 1991–August 1 1993," SIO-ATOC.

91. William W. Fox Jr. to Walter H. Munk, 7 December (not legible), 1990, Box 12, Folder "Applications for Permit, January 1991–August 1 1993," SIO-ATOC.

92. "Ann E. Bowles." Bowles got her PhD in 1994 from SIO, according to her biography on the website for the Hubbs-SeaWorld Research Institute, http://hswri.org/scientists/ann-e-bowles-phd/. Much of what followed was clearly not her fault: the ATOC lead scientists had hired a very young and inexperienced person to take on what turned out to be a very important task, which reveals something important about the attitude of the project leaders.

93. Munk et al., "Heard Island." The results were published in a special issue of the *Journal of the Acoustical Society of America* in 1994. Eighteen papers were published; one dealt with the biological results.

94. Munk et al., "Heard Island."

95. Sam Nunn, Strategic Environmental and Research Development Program," Senate Floor Speech, 28 June 1990, 101st Congress, Second Session. See Goodman, "Environment and National Security"; Potter, "Enviro-vandalism"; various documents, Box 12, Folder "Applications for Permit, January 1991–August 1 1993," SIO-ATOC.

96. On the use of the SOSUS system as the "bottom receiver," as well as the use of deliberately ambiguous language, see Munk et al., "Ocean Acoustic Tomography," appendix A.

97. Various documents in Box 12, Folder "Marine Mammal Correspondence Pre-September 1993," SIO-ATOC. This also included a study of humpback whales by Adam S. Frankel of the University of Hawaii at Manoa. See also various documents, Box 12, Folder "Applications for Permit, January 1991–August 1 1993," SIO-ATOC.

98. Craig Faanes to Marilyn Cox, 20 July 1994, Box 2, Folder, "California EIS 1994," SIO-ATOC.

99. Ann Bowles to Nancy Foster, National Marine Fisheries Service, 6 March 1992, Box 12, Folder "Marine Mammal Correspondence Pre-September 1993," SIO-ATOC.

100. Ibid.

101. Ann Bowles to David Hyde, 26 May 1993, Box 12, Folder "Marine Mammal Correspondence Pre-September 1993," SIO-ATOC.

102. Swartz and Hofman, *Mammal and Habitat Monitoring*, esp. 5. In Box 12, Folder "Correspondence September–October 1993," SIO-ATOC. The Marine Mammal Protection Act allows for accidental taking of small numbers of non-threatened species and populations during fishing and other commercial activities. In 1986, this was amended to permit the taking of depleted species and stocks as well, if it were shown to be inconsequential, which in turn required site-specific monitoring programs. For example, in 1991 the Fish and Wildlife Service had agreed to an unintentional take of small numbers of walruses and polar bears incidental to oil and gas exploration in the Chukchi Sea, near Alaska, so long as steps were taken to minimize losses, particularly in areas of traditional subsistence hunting, no walruses were taken during the spring migration period, and all of this was monitored and subject to peer review.

103. NOAA, "Marine Mammal Protection Act of 1972." See also discussion in Swartz and Hofman, *Marine Mammal and Habitat Monitoring*, in Box 12, Folder "Correspondence September–October 1993," SIO-ATOC.

104. Adam S. Frankel of the University of Hawaii at Manoa, for example, was given authorization for up to one thousand "takes," by harassment, of humpback whales, to study their response to low-frequency sound. The approval had come with strong caveats to "exercise extreme caution in approaching mother/calf pairs" and a requirement for a detailed report of all underwater approaches to whales, as well as the number and sex of whales involved. Michael F. Tillman to Adam S. Frankel, 1 February 1993, Box 12, Folder "Applications for Permit, January 1991–August 1, 1993," SIO-ATOC.

105. The historical record on Hyde's activities at SAIC is thin, but he evidently had a close involvement with Navy projects there; the lack of available records suggests they were likely classified projects. Hyde served on two NRC committees in 1988 addressing Navy-oceanography concerns. See Navy

Review Panel, *Oceanography and the Navy*; Naval Studies Board, *Implications of Advancing Technology*.

106. Ann Bowles to David Hyde, 26 May 1993, Box 12, Folder "Marine Mammal Correspondence Pre-September 1993," SIO-ATOC.

107. Ibid. If they had actually done this using federal funds, it would have been illegal. See chapter 8.

108. Ann Bowles to David Hyde, 26 May 1993, Box 12, Folder "Marine Mammal Correspondence Pre-September 1993," SIO-ATOC. The claim that industry routinely worked without permits could not have been entirely correct; elsewhere Christopher Clark noted that he frequently reviewed permits from industry. See Clark to Hyde, 27 July 1993, Box 12, Folder "Marine Mammal Correspondence Pre-September 1993," SIO-ATOC.

109. Ann Bowles to David Hyde, 26 May 1993, Box 12, Folder "Marine Mammal Correspondence Pre-September 1993," SIO-ATOC.

110. Ibid.

111. Daniel Costa to David Hyde, email, 13 August 1993, Box 12, Folder "Marine Mammal Correspondence Pre-September 1993," SIO-ATOC. The use of IUSS in Whales 93 is also discussed in Marine Acoustics Inc. Proposal MAI 208P, 28 September 1993, faxed to Cornell University Ornithology Lab, in Box 12, Folder "Correspondence September–October 1993," SIO-ATOC.

112. Christopher Clark, ATOC Task 7.0, "Acoustic Monitoring and Experimental Studies on the Effects of ATOC Transmissions on Marine Mammals," Box 13, Folder "Original Viewgraph Presentations, 7 June 1993, ARPA Review," SIO-ATOC.

113. Christopher Clark to David Hyde, 27 July 1993, Box 12, Folder "Marine Mammal Correspondence Pre-September 1993," SIO-ATOC.

114. Daniel Costa to David Hyde, email, 10 August 1993, 10:07 am, Box 12, Folder "Marine Mammal Correspondence Pre-September 1993," SIO-ATOC.

115. Daniel Costa to David Hyde, email 19 August 1993, 5:06 pm, in Box 12, Folder "Marine Mammal Correspondence Pre-September 1993," SIO-ATOC.

116. Daniel Costa to David Hyde, email, 13 August 1993, Box 12, Folder "Marine Mammal Correspondence Pre-September 1993," SIO-ATOC.

117. "Draft Study Plan, Acoustic Thermometry of Ocean Climate (ATOC)," Box 12, Folder "Marine Mammal Correspondence Pre-September 1993," SIO-ATOC. See also Christopher Clark to David Hyde, 6 July 1993, Box 12, Folder "Marine Mammal Correspondence Pre-September 1993," SIO-ATOC.

118. David Hyde to Nancy Foster, "Marine Mammal Permit, September 14, 1993, Current Version Submitted to NMFS," NMFS, 9 September 1993, Box 12, Folder "Marine Mammal Correspondence September- October 1993," SIO-ATOC.

119. See also Forbes California Coastal Commission, 29 November 1994, Box 2, Folder "California Coastal Commissions, November–December 1994," SIO-ATOC. This details the permit history. The permit signatories were Hyde, Bowles, Clark, William Kuperman—a professor of physical oceanography at Scripps and expert in noise and signal processing—and Sue Moore,

a master's-degree-level scientist employed by SAIC who had previously worked on small-take exemptions for the US Navy. The application was submitted with a cover letter signed by Hyde. See Bowles to Hyde, 26 May 1993, Box 12, Folder "Marine Mammal Correspondence Pre-September 1993," SIO-ATOC.

120. David Hyde to Nancy Foster, "Marine Mammal Permit, September 14, 1993, Current Version Submitted to NMFS," NMFS, 9 September 1993, Box 12, Folder "Marine Mammal Correspondence September–October 1993," SIO-ATOC. See also Christopher Clark, Proposal for ATOC-funded Marine Mammal Research Program, 3 February 1994, Box 12, Folder "Marine Mammal Research Program, February 1994–June 1994," SIO-ATOC. A copy of the cover letter accompanying the application can be found at https://groups.google .com/forum/#!topic/misc.activism.progressive/9xanKS9VFCA.

121. Munk and Baggeroer, "Heard Island Papers."

122. Christopher Clark to David Hyde, 27 July 1993, Box 12, Folder "Marine Mammal Correspondence Pre-September 1993," SIO-ATOC.

123. Years later, when Adam Frankel and Christopher Clark published an article in marine Mammal Science, based on work they had done through the MMRP, they noted that previous work had "indicated that whales responded to relatively low received levels of man-made sounds," and among the references they cited were ones that dated from 1983 and 1984. Frankel and Clark, "ATOC and Other Factors."

124. David Hyde to Nancy Foster, "Marine Mammal Permit, September 14, 1993, Current Version Submitted to NMFS," NMFS, 9 September 1993, Box 12, Folder "Marine Mammal Correspondence September–October 1993," SIO-ATOC. Italics added.

125. Oreskes et al., "Numerical Models."

126. Ibid.

127. Ibid. Considering the extensive study of marine mammals by the US Navy, this was a startling revelation.

128. Walter Munk and David Hyde to Nancy Foster, Chief, Permits Division, NMFS, 8 December 1993, Box 12, Folder "Correspondence November–December 1993," SIO-ATOC.

129. Marine Acoustics Inc. Proposal MAI 208P, 28 September 1993, fax to Cornell University Ornithology Lab, in Box 12, Folder "Correspondence September–October 1993," SIO-ATOC. See also Cornell, "IUSS Research."

130. Walter Munk and David Hyde to Nancy Foster, Chief, Permits Division, NMFS, 8 December 1993, Box 12, Folder Correspondence November–December 1993, SIO-ATOC.

131. David Hyde to Nancy Foster, "Marine Mammal Permit, September 14, 1993, Current Version Submitted to NMFS," NMFS, 9 September 1993, 10, Folder "Correspondence September–October 1993," SIO-ATOC.

132. ATOC NMFS permit, chart, copy in Box 2, Folder "CCC Correspondence November–December 1994," SIO-ATOC; Potter, "Enviro-vandalism."

133. Naomi Rose, telephone conversation with the author, 2001.
134. A search of this website on 16 January 2001, turned up 1,937 messages under the heading "ATOC."
135. Potter, "Enviro-vandalism."
136. Various documents, Box 12, Folder "MARMAM, Feb 1994–July 1995," SIO-ATOC.
137. Herman, "Hawaiian Humpback Whales."
138. Ibid., 65.
139. Edward Frieman, Draft of form Letter to The Honorable [blank], 18 March 1994, Box 3, Folder "Director's Office, Correspondence March 1994," SIO-ODR.
140. See, e.g., Potter, "Enviro-vandalism," 54.
141. Matthews, "Navy Whales."
142. Ibid.
143. Paddock, "Undersea Noise."
144. Ibid.
145. Potter, "Enviro-vandalism," 54.
146. Rozwadowski, *Fathoming the Ocean*; Burnett, *Sounding of the Whale*; Rozwadowski and van Keuren, *Machine*.
147. See, e.g., discussion by Lindy Weilgart of ATOC responses to MARMAM emails, 7 April 1994, Box 12, Folder "MARMAM, Feb 1994–July 1995," SIO-ATOC.
148. Perlman, "Undersea Plan." This and other press clippings in Box 2, Folder "News Clippings, March 1994–April 1994," SIO-ATOC.
149. The quotation is from the Executive Summary of National Research Council, *Low-Frequency Sound*, 1994. The NRC has reviewed the matter four times since 1994: in its publications *Progress since 1994*, 2000; *Ocean Noise*, 2003; *Biologically Significant Effects*, 2005; and "Approaches to Understanding the Cumulative Effects of Stressors on Marine Mammals," 2017. Also pertinent is Popper and Hawkins, *Effects of Noise*. While scientists continue to study and debate the issue, it is clear that various forms of ocean noise, including low-frequency active sonar, are likely to have some adverse effects, including a possible role in lethal stranding of deep-diving beaked whales. The 2005 NRC report tried to parse the matter in terms of effects that were "biologically significant," something they concluded was difficult to do, but the very framing was itself an acknowledgment that there were impacts, and the issue, now, was perceived to be differentiating between those that were biologically significant and those that were not. The 2017 report, chaired by WHOI's Peter Tyack, placed noise in the context of diverse stressors, including ocean acidification, overfishing, and pollution. Nonetheless, the report concluded that sound is a stressor, and that "additional research will be necessary to establish the probabilistic relationships between exposure to sound, contextual factors, and severity of response" (32). See also https://www.nrdc.org/issues/ocean-noise#priority-experts-resources.

150. National Research Council, *Low-Frequency Sound*, 1994.

151. Quoted in Ocean Science News, 22 March 1994, in Box 2, Folder "News Clippings, March 1994–April 1994," SIO-ATOC.

152. White, "Oceans and Climate," 1978; see also White et al., *International Perspectives*.

153. It is quite shocking that White's role is not more well known. I suspect this is because he worked mostly behind the scenes. For example, in the late 1960s he communicated extensively with members of the Nixon administration. See Oreskes et al., "Congressional Intent." Various White memos can also be found in the Nixon archives.

154. Statement by Robert M. White, 22 March 1994, Box 11, Folder "California Coastal Commission Public Hearing June 15, 1995, Agenda," SIO-ATOC. See also Ocean Science News, 22 March 1994, in Box 2, Folder "News Clippings, March 1994–April 1994," SIO-ATOC.

155. Statement by Robert M. White, 22 March 1994, Box 11, Folder "California Coastal Commission Public Hearing June 15, 1995, Agenda," SIO-ATOC.

156. Ocean Science News, 22 March 1994, in Box 2, Folder "News Clippings, March 1994–April 1994," SIO-ATOC.

157. Ibid.

158. Paddock, "Block Ocean Noise Tests," in Box 2, Folder "News Clippings, March 1994–April 1994," SIO-ATOC.

159. Various documents, Box 2, Folder "News Clippings, March 1994–April 1994," SIO-ATOC.

160. "Permits Are Delayed for Underwater Noise," *Pleasanton Valley (CA) Times*, 26 March 1994, Box 2, Folder "News Clippings, March 1994–April 1994," SIO-ATOC.

161. Patsy Mink et al. to Ron Brown, copy of faxed letter, 24 March 1994, AC-09136, Folder 2, "ATOC," WHOI-ODR.

162. Barbara Boxer to Ron Brown, copy of faxed letter, 23 March 1994, AC-09, Folder 2, "ATOC," WHOI-ODR. See also Box 2, Folder "News Clippings, March 1994–April 1994," SIO-ATOC. It is interesting that these members were all Democrats; I do not know if there is any significance to that, but perhaps because Scripps had such strong military ties, Republicans were less likely to question its activities than Democrats. Or perhaps because Brown was a Democrat, Democratic members felt comfortable approaching him.

163. Paddock, "Block Ocean Noise Tests," in Box 2, Folder "News Clippings, March 1994–April 1994," SIO-ATOC.

164. "Don't Deafen Whales," editorial in San Francisco *Examiner*, 24 March 1994, in Box 2, Folder "News Clippings, March 1994–April 1994," SIO-ATOC.

165. Various editorials, Box 2, Folder News clippings, March 1994–April 1994, Folder "News Clippings, March 1994–April 1994," SIO-ATOC.

166. Lankford, "Harpoons for Scripps Eggheads," *Santa Barbara (CA) News Press*, 30 March 1994, Box 2, Folder "News Clippings, March 1994–April 1994," SIO-ATOC.

167. "Backers Deflect," Box 2, Folder "News Clippings, March 1994–April 1994," SIO-ATOC.

168. "Lawmakers Protest Surge over Undersea Project," 25 March 1994, Box 2, Folder "News Clippings, March 1994–April 1994," SIO-ATOC.

169. "Backers Deflect," Box 2, Folder "News Clippings, March 1994–April 1994," SIO-ATOC.

170. "Sanctuary Panel Urges Delay," Box 2, Folder "News Clippings, March 1994–April 1994," SIO-ATOC.

171. "2 Environmental Camps," Box 2, Folder "News Clippings, March 1994–April 1994," SIO-ATOC.

172. Edward Frieman to [please fill in] (template) [sic], 28 March 1994, Box 3, Folder "Director's Office, Correspondence, E. A. Frieman, March 1994," SIO-ODR (Frieman).

173. Edward A. Frieman, "Dear SIO Associate," 29 March 1994, Box 3, Folder "Director's Office, Correspondence, E. A. Frieman, March 1994," SIO-ODR (Frieman).

174. Cindy Clark, memo and attachments sent to S. Lauzon at WHOI, SIO Communications Office, 2 April 1994, AC-09136, Folder 2, "ATOC," WHOI-ODR.

175. Ibid., unpaginated.

176. Ibid.

177. Ibid.

178. Tyack now has a TED talk, "The Intriguing Sound of Marine Mammals," April 2010, https://www.ted.com/talks/peter_tyack_the_intriguing_sound_of _marine_mammals.

179. Christopher Clark, comments on ATOC Program Description, forward to Shelley Lauzon by Peter Tyack, 2 May 1994, AC-09, Folder 2, "ATOC," WHOI-ODR. Lauzon forwarded to Bob Gagosian, with a sticky note saying, "Peter indicated I could share internally. Shelley." Emphasis in original.

180. Ibid. Emphasis in original.

181. Ibid.

182. Ibid. Emphasis in original.

183. Ibid.

184. It is telling that there is no evidence that any of the biologists at Scripps were involved in producing this fact sheet. The person whose name appeared at the bottom was Walter Munk. I found no archival evidence that he or others involved in ATOC had reached out to biologists at Scripps or at UCSD for advice.

185. Robert Gagosian to Shelley Lauzon, 8 May 1994, memo on subject "Your May 2 note on ATOC Questions," AC-09, Folder 2, "ATOC," WHOI-ODR.

186. The Advisory Board was headed by W. John Richardson, executive vice president of LGL Ltd., a private environmental consulting group based in Ontario, and a member of the US NRC committee on low-frequency sound and marine mammals. Other members included William Ellison, an expert in underwater acoustics at Marine Acoustics Inc.; Jeff Laake, a population

biologist at the National Marine Mammal Laboratory in Seattle who had developed techniques to estimate population densities; Jeannette Thomas, a biologist at Western Illinois University with expertise in cetacean and pinniped hearing; and Judy Zeh, a professor of statistics at University of Washington.

187. ATOC Marine Mammal Research Program Advisory Board, second meeting, 15 February 1994, Box 12, Folder "Marine Mammal Research Program February 1994–June 1994," SIO-ATOC.

188. Ibid.

189. Ibid.

190. Notes on Advisory Board meeting, 18 April 1994, Box 12, Folder "MMRP Research Protocols, April 1994–November 1995," SIO-ATOC.

191. Ibid.

192. Daniel Costa to David Hyde. Box 12, Folder "MMRP Research Protocols, April 1994–November 1995," SIO-ATOC.

193. "Acoustic Thermometry of Ocean Climate Program Update," press release, 13 May 1994, Box 6, Folder "Publications 1993–2002," SIO-ATOC. See also Scoping Process Summary, 15 June 1995, California Coastal Commission, Revised Staff Report and Recommendation, Box 2, Folder "California Coastal Commissions April–September 1994," SIO-ATOC; additional copy also in Box 2, Folder "California Coastal Commission June–November 1995," SIO-ATOC. Although initially Scripps scientists were reluctant to undertake the enormous effort involved in preparing an environmental impact statement, legal counsel advised that failure to do so would increase exposure to litigation and in the long run likely cost more time and money. See various documents in Box 12, Folder "Kauai DEIS Correspondence, May–June 1994," SIO-ATOC.

194. Various documents in Box 12, Folder "Kauai DEIS Correspondence, May–June 1994," SIO-ATOC. Naomi Rose, telephone conversation with author, 2001.

195. Robert J. Hofman to Clay H. Spikes, 15 June 1994, Box 12, Folder "Kauai DEIS Correspondence, May–June 1994," SIO-ATOC. It isn't clear which draft he read, but it may have been "Initial Study," 31 May 1994, Box 1, Folder "California EIR Initial Study May 31, 1994," SIO-ATOC.

196. John Turner to Ronald Schmitten, 4 May 1994, Box 2, Folder "California EIS (1994)," SIO-ATOC.

197. Ibid.

198. Ibid.

199. David R. Tomsovic to Marilyn Cox, 15 June 1994, Box 2, Folder "California EIS (1994)," SIO-ATOC.

200. Walter Munk to NOAH [*sic*], 21 April 1994, Box 2, Folder "California Coastal Commissions April–September 1994," SIO-ATOC; Peter Douglas to Ralph Alewine, 27 May 1994 Box 2, Folder "California Coastal Commissions April–September 1994," SIO-ATOC.

201. Peter Douglas to David Hyde and Terry Jackson, 14 July 1994, Box 2, Folder "California Coastal Commissions April–September 1994," SIO-ATOC. See also Peter Douglas to Marilyn Cox, 27 June 1994, in same. They had already been granted a coastal development permit for the shore end of the power and fiber-optic cables that would run from the source on Sur Ridge to the Naval Facility at Point Sur. Because the cable would run alongside an existing Navy pipeline, on land that was off limits to the public, the Coastal Commission had concluded that it was consistent with previously approved activities and approved it the previous January. But the debate had expanded considerably since January; now the question was the far larger one of potential impact to coastal resources—namely, marine life in the coastal zone. See Peter Douglas to P. Bosco, 18 January 1994, Box 2, Folder "California Coastal Commissions April–September 1994," SIO-ATOC.

202. Alan Waltner to Cindy Rogers, fax from law offices of Alan C. Waltner, to Cindy Rogers, 8 September 1994. Box 2, Folder "California Coastal Commissions April–September 1994," SIO-ATOC.

203. Milt Phegley to Russ Albertson, 10 June 1994, Box 2, Folder "California Coastal Commissions April–September 1994," SIO-ATOC.

204. Sometimes people called it the draft environmental impact statement; other times, the draft environmental impact report. Both are used in the printed version by Advanced Research Projects Agency, *Final Environmental Impact Statement*, 1995, www.dtic.mil/dtic/tr/fulltext/u2/a350539.pdf. The discussion presented here is based on the materials preserved at the time in the ATOC papers, SIO Archives.

205. Draft Environmental Impact Statement and various correspondences about it, Box 2, Folder "Records, California EIS, 1994," SIO-ATOC.

206. "Statement of Work, Cornell University," 28 July 1994, Box 12, Folder "MMRP Research Protocols, April 1994–November 1995," SIO-ATOC.

207. "Final Draft Environmental Impact Report," 16 August 1994, Box 2, Folder "California EIS, 1994," SIO-ATOC.

208. Paul Anderson, "An Update on the Acoustic Thermometry of Ocean Climate (ATOC) project," email, 30 January 1995, in Box 12, Folder "MARMAM February 1994–July 1995," SIO-ATOC.

209. Ibid.

210. Craig Faanes to Marilyn Cox, 14 July 1994, Box 2, Folder "California EIS, 1994," SIO-ATOC.

211. Paul Anderson, email, 16 June 1995, Box 12, Folder "MARMAM February 1994–July 1995," SIO-ATOC.

212. Ibid.

213. Walter Munk and David Hyde to John Richardson, 17 November 1993, Box 12, Folder "Correspondence November–December 1993," SIO-ATOC.

214. Cited in David R. Tomsovic to Marilyn Cox, 15 June 1994, Box 2, Folder "California EIS, 1994," SIO-ATOC. See also "40 CFR 1500-1508: CEQ Regulations Implementing Procedural Provisions of NEPA," https://www.energy

.gov/nepa/downloads/40-cfr-1500-1508-ceq-regulations-implementing
-procedural-provisions-nepa, specifically Sec. 1502.14, "Alternatives including the proposed action."

215. David R. Tomsovic to Marilyn Cox, 15 June 1994, Box 2, Folder "California EIS,
1994," SIO-ATOC.

216. Paul Anderson, email, 16 June 1995, Box 12, Folder "MARMAM February
1994–July 1995," SIO-ATOC. This point was also raised in several public
comments following the January 1995 California Coastal Commission public
hearings, where Walter Munk spoke, emphasizing the importance of ATOC
as a test of climate models rather than a proof of global warming.

217. Walter Munk to Carl L. Williams and Members, California Coastal Commission, 9 June 1995, Box 11, Folder ATOC, California Coastal Commission Public
Hearing June 15, 1995, Agenda," SIO-ATOC.

218. On the no-effects claim, see Forbes to Benoit, 13 January 1995, Box 2, Folder
"California Coastal Commissions (January–May 1995)," SIO-ATOC. For copies of public comments, see Box 11, various folders, SIO-ATOC.

219. James Hansen to NMFS, 5 May 1994, and Four Congressmen to Peter
Douglas, n.d., Box 11, Folder "ATOC, California Coastal Commission Public
Hearing June 15, 1995, Agenda," SIO-ATOC. This appears to be a form letter:
the signatures are on lines at the bottom with no typed attribution. It may
have been supplied by the SIO Communications Office, as the tone is similar to previous such "suggested letters." It is also possible that Hansen was
ambivalent, insofar as his own work had already demonstrated atmospheric
warming.

220. Box 11, Folder "ATOC, California Coastal Commission Public Hearing June 15,
1995, Agenda," SIO-ATOC.

221. Ibid.

222. Ibid.

223. Ibid. One need not belabor the irony of the emotional tone of Barry's remarks but merely note that NMFS was legally required to consider public
opinion, well informed or not.

224. Lindy Weilgart, Comment C27, Box 11, Folder "MMRP Draft EIS/EIR, Vol. 1,
January 1995, Comment received and responses C-1 to C-37," SIO-ATOC.

225. Ibid.

226. Ibid.

227. Center for Marine Conservation, 31 January 1995, Comments of the Center
for Marine Conservations on the Draft EIS/EIR, Box 11, Folder MMRP Draft
EIS/EIR, Vol. 1, January 1995, Comment received and responses C38 to C-61,"
SIO-ATOC.

228. Ibid.

229. Katherine Payne, Comment C-16, Box 11, Folder "MMRP Draft EIS/EIR, Vol. 1,
January 1995, Comment Received and Responses C-1 to C-37," SIO-ATOC.

230. Center for Marine Conservation, 31 January 1995, Comments of the Center
for Marine Conservations on the Draft EIS/EIR, Box 11, Folder "MMRP Draft

EIS/EIR, Vol. 1, January 1995, Comment Received and Responses C38 to C-61," SIO-ATOC.

231. Ibid.

232. Ibid.

233. Lawrence D. Six, Pacific Fishery Management Council, Comment C-55. Box 11, Folder "MMRP Draft EIS/EIR, Vol. 1, January 1995, Comment Received and Responses C38 to C-61," SIO-ATOC.

234. Munk et al., "Heard Island Feasibility Test."

235. Ibid.

236. Kauai Friends of the Environment, Comment C-56, Box 11, Folder "MMRP Draft EIS/EIR, Vol. 1, January 1995, Comment Received and Responses C38 to C-61," SIO-ATOC.

237. The paper was ultimately published as Spiesberger and Metzger, "Basin-Scale Tomography," 1991.

238. McCreary to Spiesberger, 5 November 1990, MC-06, Folder "Correspondence 1991 [sic] 1 of 2," WHOI-Stommel.

239. A bit more technical detail for the interested reader: ATOC scientists repeatedly stressed that the speed of sound in the ocean depended on temperature, but to be precise, it depended on density, which in turn depends on temperature, pressure, and salinity. Sound speed increases with increasing temperature, pressure, and salinity. To know the value of these variables with certitude would require measuring them at every point—a physical impossibility. Rather than attempt the impossible, oceanographers have measured the speed of sound in the laboratory at various temperatures, salinities, and pressures, then used those measurements to develop algorithms to relate the speed of sound to prevailing ocean conditions. Sound propagation is then modeled with ray theory, which predicts that an acoustic pulse emitted from a submerged source travels, like light, along multiple ray paths following Snell's law of refraction for the expected density field. Each potential path has a different travel time; acoustic tomography attempts to link actual arrival times to physically possible ray paths and so map the thermal structure of the ocean. For more details, see Munk and Wunsch, "Ocean Acoustic Tomography."

240. Appendix A, Travel time errors, unattributed typescript in MC-06, Folder "Correspondence 1990 1 of 1," WHOI-Stommel.

241. Review of "Acoustic Measures of Climate Variability, 19 October 1990, MC-06, Folder "Correspondence 1990 1 of 1," WHOI-Stommel. He judged the paper to be "less than a 'least publishable unit.'"

242. Review of Spiesberger and Metzger, "Basin-Scale Ocean Monitoring with Acoustic Thermometers," MC-06, Folder "Correspondence 1990 1 of 1," WHOI-Stommel.

243. Review of Spiesberger and Metzger, "Basin-Scale Ocean Monitoring with Acoustic Thermometers," MC-06, Folder "Correspondence 1991 [sic] 1 of 2," WHOI-Stommel. Inverse theory is a method in geophysics to determine the

(unknown) structure of an object—such as Earth—when you can measure some aspect of it, such as the propagation of seismic waves.

244. Review of Spiesberger, Metzger, and Furgerson, "Listening for Climatic Temperature Change in the Northeast Pacific, 1983–1989," 9 July 1991, in MC-06, Folder "Correspondence 1991, 1 of 2," WHOI-Stommel.

245. DEIS, p. ES-3, quoted by Linda S. Weilgart, Comment C-27, Box 11, Folder "MMRP Draft EIS/EIR, Vol. 1, January 1995, Comment Received and Responses C-1 to C-27," SIO-ATOC.

246. Mikolajewicz et al., "Acoustic Detection."

247. "Sea Creatures vs. Science," *Detroit Free Press*, 28 April 1994, Box 2, Folder "News Clippings, March 1994–April 1994," SIO-ATOC.

248. Munk et al., "Heard Island Feasibility Test."

249. Paul Anderson, email to MARMAM, 25 January 1995. Box 12, Folder "MARMAM B-B, February 1994–July 1995," SIO-ATOC.

250. No two scientific programs are ever exactly alike, so it was fair to say that ATOC would have provided distinctive and potentially very useful data. In contrast, other groups were also developing innovative means to detect global warming at that time. In 1996, scientists from the University Navstar Consortium, in Boulder, Colorado, working with colleagues at the University Corporation for Atmospheric Research, the University of Arizona, the Russian Institute of Atmospheric Physics, and Lockheed Martin published a paper on GPS sounding of the atmosphere from low Earth orbit in the *Bulletin of the American Meteorological Society*. The paper discussed a new technique using radio waves from global positioning system satellites to measure the temperature of the atmosphere. Radio waves are refracted as they travel through the atmosphere; that refraction depends on temperature and water vapor. If water vapor concentration is known independently or can be estimated from models, then the temperature of the atmosphere can be calculated. The 1996 paper demonstrated the technique's feasibility, and by the early 2000s there were scores of papers reporting the results of high-accuracy GPS sounding of the atmosphere to detect global warming. See Ware et al., "GPS Sounding." Some of the more highly cited include Duan et al., "GPS Meteorology"; Rocken et al., "Analysis and Validation"; Wickert et al., "Atmosphere Sounding"; Wickert et al., "Radio Occultation Experiment"; and Hajj et al., "CHAMP and SAC-C."

251. A point that, twenty-five years later, seems sadly prescient.

252. Rodney M. Fujita, Comment C-24, Box 11, Folder "MMRP Draft EIS/EIR, Vol. 1, January 1995, Comment Received and Responses C-1 to C-27," SIO-ATOC.

253. Ibid.

254. Ibid. For an analysis of these vast economic and political interests, see Oreskes and Conway, *Merchants of Doubt*. Fujita's analysis was crucial to my dawning realization of the motivations behind attacks on climate science.

255. Oreskes and Conway, *Merchants of Doubt*; Hoggan and Littlemore, *Climate Cover-Up*. As Rodney Fujita had rightly pointed out, scientifically compelling

was not at all the same as politically compelling. If the scientists involved were not asserting a naïve belief that facts alone could solve social problems, their actions and comments certainly seemed to imply something close to it. There was also an inverse assumption at play, that without good scientific data, governments would be unable to take action, as Ed Frieman argued in his "Dear SIO Associate" letter: "Until global warming is better understood, governments will not be able to take effective steps to counteract its negative impact," 29 March 1994, Box 3, Folder "Director's Office, Correspondence, E. A. Frieman, March 1994," SIO-ODR (Frieman). That is to say, scientists were assuming that compelling scientific evidence was both necessary and sufficient to lead to appropriate policy action. That was in the mid-1990s; events since have only underscored Fujita's point. Scientists are in near-unanimous agreement that the scientific evidence of anthropogenic climate change is compelling, definitive, even "unequivocal," yet political action has conspicuously failed to follow it its wake, particularly in North America, where the ATOC story took place. In Europe, however, under the precautionary principle, governments have acted on some environmental and health concerns without data that all (or even most) scientists agree are compelling. Scientific data alone are not sufficient to empower governments (or people) to act, and sometimes they are not even necessary.

256. Debby Molina, Comment C-52, Box 11, Folder "MMRP Draft EIS/EIR, Vol. 1, January 1995, Comment Received and Responses C38 to C-61," SIO-ATOC.

257. Ibid.

258. Ibid.

259. Derek J. Cole, Comment C-40, Box 11, Folder "MMRP Draft EIS/EIR, Vol. 1, January 1995, Comment Received and Responses C38 to C-61," SIO-ATOC.

260. Which indeed, scientists did, not only in the nineteenth century, but earlier too. In the seventeenth century, Stephen Hales famously stuck glass tubes in dogs' arteries while researching blood pressure; see Burgett, "Stephen Hales." Such experiments were routine in the eighteenth and nineteenth centuries, and continued in the twentieth century, when laboratory animals were routinely "sacrificed" in scientific experiments. For an entrée into the history of animal experimentation, see Guerrini, *Humans and Animals*.

261. Elaine Sohier, Comment C-35, Box 11, Folder "MMRP Draft EIS/EIR, Vol. 1, January 1995, Comment Received and Responses C-1 to C-37," SIO-ATOC; Peter Molitor, Comment C-57, Box 11, Folder "MMRP Draft EIS/EIR, Vol. 1, January 1995, Comment Received and Responses C38 to C-61," SIO-ATOC.

262. B. R. Harms, Comment C-19, Box 11, Folder "MMRP Draft EIS/EIR, Vol. 1, January 1995, Comment Received and Responses C-1 to C-37," SIO-ATOC.

263. Elaine Sohier, Comment C-35, Box 11, Folder "MMRP Draft EIS/EIR, Vol. 1, January 1995, Comment Received and Responses C-1 to C-37," SIO-ATOC.

264. On cetaceans in the twentieth century, see Burnett, *Sounding of the Whale*.

265. Debby Molina, Comment C-52, Box 11, Folder "MMRP Draft EIS/EIR, Vol. 1, January 1995, Comment Received and Responses C38 to C-61," SIO-ATOC.

266. Bob DeBolt, Comment C-37, Box 11, Folder "MMRP Draft EIS/EIR, Vol. 1, January 1995, Comment Received and Responses C-1 to C-37," SIO-ATOC.

267. Susanne Alterman, Comment C-14, Box 11, Folder "MMRP Draft EIS/EIR, Vol. 1, January 1995, Comment Received and Responses C-1 to C-37," SIO-ATOC.

268. The IPCC Second Assessment Report would come out later that same year, and declare that the balance of evidence suggested a "discernible human impact" on the climate system. See Intergovernmental Panel on Climate Change, *Climate Change 1995*, especially the first section; see also Oreskes and Conway, *Merchants of Doubt*, chap. 6. This argument had a personal impact on my own life. When I read these materials in the early 2000s, I was not fully conversant with the early history of climate science, so I dug in and learned much of the material presented in this chapter. That led me to publish my 2004 paper, "The Scientific Consensus on Climate Change," which in turn made me a target of the climate-change-denial network, and the rest is history.

269. Debby Molina, Comment C-52, Box 11, Folder "MMRP Draft EIS/EIR, Vol. 1, January 1995, Comment Received and Responses C38 to C-61," SIO-ATOC. Some environmentalists had already expressed this opinion. For example, Naomi Rose, a biologist for the Marine Mammal Commission who was interviewed by the *Orange County Register* in April 1994, said, "We already know that pumping man-made greenhouse gases into the atmosphere is probably bad news. Instead of studying things to death, we should act to reduce the amount of greenhouse gases that get released." This was also the view of the *San Francisco Examiner* in a March 14 editorial condemning the project: "The money could be far better spent on finding better and alternative energy sources—wind, solar, and even tidal—that would reduce humanity's dependence on fuels that could contribute to global warming." Box 2, Folder "News Clippings, March 1994–April 1994," SIO-ATOC. Other postings on MARMAM and letters submitted during public comment expressed the same point.

270. Derek J. Cole, Comment C-40, Box 11, Folder "MMRP Draft EIS/EIR, Vol. 1, January 1995, Comment Received and Responses C38 to C-61," SIO-ATOC.

271. Alan McGowan to MARMAM, email, 15 April 1994, Box 12, Folder "MARMAM, February 1994–July 1995," SIO-ATOC.

272. "Sound Science?" *Orange County Register*, Box 2, Folder "News Clippings, March 1994–April 1994," SIO-ATOC.

273. "Don't Deafen Whales," editorial, *San Francisco Examiner*, 24 March 1994, in Box 2, Folder "News Clippings, March 1994–April 1994," SIO-ATOC. For additional examples of this argument, see Alan McGowan to MARMAM, email, 15 April 1994, Box 12, Folder "MARMAM, February 1994–July 1995," SIO-ATOC.

274. "Outcry Could Give Scientists a Lesson in Public Relations," *Seattle Times*, 26 April 1994, Box 2, Folder "News Clippings, March 1994–April 1994," SIO-ATOC.

275. Flatté, personal communication with author, 2000.

276. Matt Hellman and Patti Kirby, Comment C-54, Box 11, Folder "MMRP Draft EIS/EIR, Vol. 1, January 1995, Comment Received and Responses C38 to C-61," SIO-ATOC.

277. Sarah Miquibas, handwritten letter in response to the draft environmental impact statement, for Kauai ATOC program, Box 11, Folder "DLNR Comments to EIS, March-May 1995," SIO-ATOC.

278. David Seielstad, 15 February 1995, written in response to the DEIS, for Kauai ATOC program, Box 11, Folder "DLNR Comments to EIS, March-May 1995," SIO-ATOC.

279. "Welcome to the Pacific Missile Range Facility Barking Sands," https://cnic.navy.mil/regions/cnrh/installations/pacific_missile_range_facility_barking_sands.html.

280. David Seielstad, 15 February 1995, written in response to the DEIS, for Kauai ATOC program, Box 11, Folder "DLNR Comments to EIS, March-May 1995," SIO-ATOC.

281. Natalie De Pasquale, 24 January 1995, written in response to the DEIS, for Kauai ATOC program, Comment C-41, Box 11, Folder "MMRP Draft EIS/EIR, Vol. 1, January 1995, Comment Received and Responses C38 to C-61," SIO-ATOC.

282. Dan Overmyer, 27 January 1995. Written in response to the DEIS, for Kauai ATOC program, Comment C-41, Box 11, Folder "MMRP Draft EIS/EIR, Vol. 1, January 1995, Comment Received and Responses C38 to C-61," SIO-ATOC.

283. Terry Jackson, 24 January 1995, written in response to the DEIS, for Kauai ATOC program, Comment C-41, Box 11, Folder "MMRP Draft EIS/EIR, Vol. 1, January 1995, Comment Received and Responses C38 to C-61," SIO-ATOC.

284. Matt Hellman and Patti Kirby, Comment C-54, Box 11, Folder "MMRP Draft EIS/EIR, Vol. 1, January 1995, Comment Received and Responses C38 to C-61," SIO-ATOC.

285. Michelle Waters, Comment C-36. Box 11, Folder "MMRP Draft EIS/EIR, Vol. 1, January 1995, Comment Received and Responses C-1 to C-37," SIO-ATOC.

286. Naomi Rose, personal communication with author, 2001.

287. One of the few supportive responses was a postcard from Ronald Peet and Sandra Castro of Seaside, California, 7 January 1995, Box 11, Folder "MMRP Draft EIS/EIR Vol. I Comments Received and Responses Jan 1995 C1-C37," SIO-ATOC.

288. California Coastal Commission, Revised Staff Report and Recommendation, 15 June 1995, Box 2, Folder "California Coastal Commissions April-September 1994 [sic]," SIO-ATOC. Quoting from Sylvia Earle's comments on the Draft EIS/R.

289. California Coastal Commission, Revised Staff Report and Recommendation, 15 June 1995, Box 2, Folder "California Coastal Commissions April-September 1994 [sic]," SIO-ATOC.

290. Peter Douglas to ATOC, file CC-110-94, Attachment 3, Hearing Modifications, Box 2, Folder "California Coastal Commissions June-November 1995,"

SIO-ATOC. Executive Summary, California Coastal Commission, Revised Staff Report and Recommendation, 15 June 1995, Box 2, Folder "California Coastal Commissions April–September 1994 [sic]," SIO-ATOC.

291. Peter Douglas to ATOC, file CC-110–94, Attachment 3, Hearing Modifications, Box 2, Folder "California Coastal Commissions June–November 1995," SIO-ATOC, 18.

292. Executive Summary, California Coastal Commission, Revised Staff Report and Recommendation, 15 June 1995, Box 2, Folder "California Coastal Commissions April–September 1994 [sic]," SIO-ATOC.

293. California Coastal Commission, Revised Staff Report and Recommendation, 15 June 1995, Box 2, Folder "California Coastal Commissions April–September 1994 [sic]," SIO-ATOC.

294. Executive Summary, California Coastal Commission, Revised Staff Report and Recommendation, 15 June 1995, Box 2, Folder "California Coastal Commissions April–September 1994 [sic]," SIO-ATOC; Andrew Forbes to Peter Douglas and Tami Grove, 8 May 1995, Box 2, Folder "California Coastal Commissions January–May 1995," SIO-ATOC.

295. Walter Munk to Carl L. Williams and Members, California Coastal Commission, 9 June 1995, Box 11, Folder "ATOC, California Coastal Commission Public Hearing June 15, 1995, Agenda," SIO-ATOC. Sylvia Earle evidently played a key role here, attending two meetings between the NRDC, Sierra Club Legal Defense, Save Our Shores, Friends of the Sea Otters, and others on the one hand and the ATOC project team on the other; the first on 13 May 1994 at the San Francisco airport Hilton and the second on July, 19, 1994, at the California Academy of Sciences. Scoping Process Summary, 15 June 1995, California Coastal Commission, Revised Staff Report and Recommendation, 15 June 1995, Box 2, Folder "California Coastal Commissions April–September 1994 [sic]," SIO-ATOC.

296. Au et al., "Acoustic Effects."

297. Klimley and Beavers, "Playback of Acoustic Thermometry," 2510.

298. National Research Council, *Marine Mammals and Low Frequency Sound*, 3.

299. Ibid., 73.

300. Although other data increasingly showed that the world was warming overall; see Houghton et al., *Climate Change 2001*.

301. Peter Worcester, written correspondence to the National Research Council, 1999, quoted in National Research Council, *Marine Mammals and Low-Frequency Sound*, 15n5, 21.

302. National Research Council, *Marine Mammals and Low Frequency Sound*, 5–6.

303. Ibid., 6–7.

304. Ibid., 82.

305. Anderson, 2000, email communication to listserv MARMAM@uvvm.uvic.ca and published on the MARAM website as of 2002, but no longer available.

306. Scripps Institution of Oceanography, "ATOC Source Recovery Operation Update, August 9, 2000." As of 6 July 2018, this link was still available at

https://scripps.ucsd.edu/news/2760. The winch operator's name was Ron Hardy.

307. Associated Press, "Halt to Sonic Blasts," 29 October 2002, Box 6, Folder "Publications Folder 1993–2002," SIO-ATOC; Malakoff, "Suit Ties Whale Deaths."

308. The Wikipedia page notes this connection, which is otherwise rarely mentioned. See https://en.wikipedia.org/wiki/Surveillance_Towed_Array _Sensor_System, last updated 30 September 2019.

309. "Mid and Low Frequency Sonar," https://www.justice.gov/enrd/mid-and-low -frequency-sonar.

310. For the 2003 report, see National Research Council, *Ocean Noise*, 2003; for the 2005 report, see National Research Council, *Marine Mammal Populations*, 2005, chaired by Peter Tyack, the Woods Hole scientist who sent the SIO Fact Sheet to Chris Clark for his comments. "Project Information," http:// www8.nationalacademies.org/cp/committeeview.aspx?key=49715.

311. National Research Council, *Ocean Noise*, 2003.

312. For a review of literature on impacts on anthropogenic sound on fish, see Popper and Hastings, "Anthropogenic Sources," and "Predicting and Miti-gating Hydroacoustic Impacts on Fish from Pile Installation: NCHRP Project 25–28," http://www.trb.org/main/blurbs/155418.aspx.

313. National Research Council, *Ocean Noise*, 2003, 5.

314. Ibid., 3.

315. Ibid.

316. Kenneth R. Weiss, "Navy Must Cut Sonar Exercises Off the Coast," *Los Angeles Times*, 4 January 2008, A1, A19. See also Steve Liewer, "Navy Ordered to Stop Using Sonar," *San Diego Union Tribune*, 7 August 2007.

317. "Groups Seek to Stop Navy from Blasting Marine Mammals with Sonar," http://www.nrdc.org/media/2012/121018. The list of plaintiffs included the NRDC, Humane Society of the United States, Cetacean Society Interna-tional, Ocean Futures Society, Jean-Michel Cousteau, and Michael Stocker. Defendants included Commerce Secretary Penny Pritzker, National Marine Fisheries Service, Eileen Sobeck (assistant admin for fisheries), Kathryn D. Sullivan (Admin NOAA). Ray Mabus (secretary of the Navy), and Jona-than Greenert (Admiral, Chief of Naval Ops). The Ninth Circuit overturned a lower-court decision in favor of the defendants, concluding: "The 2012 Final Rule did not establish means of 'effecting the least practicable adverse impact on' marine mammal species, stock and habitat, as was specifically required by the MMPA [Marine Mammal Protection Act]. The panel further held that the Fisheries Service impermissibly conflated the 'least practicable reverse impact' standard with the 'negligible impact' findings; and con-cluded that to authorize incidental take, the Fisheries Service must achieve the 'least practicable adverse impact' standard in addition to finding a negli-gible impact. The Panel held that the Fisheries Service did not give adequate protection to areas of the world's oceans flagged by its own experts as bio-

logically important, based on the present lack of data sufficient to meet the Fisheries Service's designation criteria." The decision continued:

> The Navy's plans for the use of LFA sonar, as approved by NMFS, have gone through several iterations, resulting in increased protection for marine mammals. We have every reason to believe that the Navy has been deliberate and thoughtful in its plans to follow NMFS guidelines and limit unnecessary harassment and harm to marine mammals. But the question is whether NMFS has satisfied the Congressional mandate that mitigation measures ensure the 'least practicable adverse impact' on marine mammals."
>
> In connection with peacetime activities such as use of LFA [low-frequency active] sonar for training, testing, and routine operations, Congress struck a balance to permit incidental take of marine mammals caused by deployment of LFA sonar or other techniques that might incidentally harm whales and other marine mammals, so long as the incidental take from the activity has a negligible impact on the species or stock involved, and so long as mitigation measures were fashioned to limit harm to the marine mammals to the 'least practicable adverse impact.' As the agency with delegate authority to implement the MMPA, NMFS is bound by these congressional mandates." In short, both Congress and the courts have at least implicitly acknowledged the LFA sonar may harm marine mammals, and that harm must be weighed against national security interests.

NRDC v. Pritzker, US Court of Appeals for the Ninth Circuit, No. 14-16375, DC No. 3:12-cv-05380-EDL, http://cdn.ca9.uscourts.gov/datastore/opinions /2016/07/15/14-16375.pdf.

318. "Mid and Low Frequency Sonar," https://www.justice.gov/enrd/mid-and-low -frequency-sonar.

319. "NOAA Fisheries Glossary," http://www.nmfs.noaa.gov/pr/glossary.htm#l, under *H* for *harassment*. As this book went to press, a new report on noise in the Arctic was released, which concluded, among other things, that several species of whales display avoidance behavior in response to underwater sound. Presumably if they avoid it, they don't like it. "Underwater Noise in the Arctic: A State of Knowledge Report," https://www.pame.is/index.php /document-library/pame-reports-new/pame-ministerial-deliverables/2019 -11th-arctic-council-ministerial-meeting-rovaniemi-finland/421-underwater -noise-report/file.

CONCLUSION

1. Wolfe, *Freedom's Laboratory*, 207. In fairness to Wolfe, this is not a major claim of her book, but the very fact that she makes this claim, en passant,

without argument or elaboration, shows how widespread this mythology is. Moreover, while she wants to argue that this period of allegedly unrestricted funding was an anomaly that lasted only twenty years, the work presented here shows that, at least in oceanography, it was *never* the case.

2. My discussion here overlaps with Frickel et al., *Undone Science*, but I wish to stress that the work that was left undone was of potential interest not just to civil society (their focus) but also to other expert communities and even to oceanographers themselves.

3. On the pure science ideal, see Rowland, "A Plea for Pure Science."

4. On the idea that NSF was established to be the nation's patron for pure science, see England, *Patron*.

5. A colleague of mine once criticized another colleague for getting research funding from the Department of Agriculture. When asked why that was a problem, she replied, "because it's government money." When further queried on where she thought this colleague should get his funding from, she replied unhesitatingly, "NSF." Clearly, this is related to the idea that NSF is the nation's patron for "pure science."

6. See the discussion in Oreskes, *Continental Drift*.

7. On glaciology, see Danielle Hallett Inkpen, "Frozen Icons."

8. In an interview in 2000, Bill Nierenberg insisted to me that there had been no classified research at Scripps. Nierenberg, William A. Interview by Naomi Oreskes and Ron Rainger, Scripps Institute of Oceanography, 10 February 2000. In a phone conversation a few years later, Alan Berman similarly insisted that *Alvin* had not done classified research and feigned surprise—as if he had entirely forgotten about it—when I first asked about Artemis. Quite possibly he really had forgotten—it was many years later—but perhaps the habits of denial were so great that they persisted long after they were unnecessary. I also learned from talking to these two men that often they were not aware of what projects had or had not been declassified. When I was able to demonstrate that I already knew about a particular project, they were quite willing to share what they remembered. I will never forget asking Bill Nierenberg about Project Michael. He replied, quite surprised, "You know about Project Michael?" Then he proceeded quite happily to talk about it.

9. On the belief in the freedom of science in relation to Cold War politics, see Wolfe, *Freedom's Laboratory*.

10. For a discussion of the demarcation of pure and applied science as a form of boundary work in ecology, see Lindseth, "Radioactive Fallout."

11. Glantz et al., *The Cigarette Papers*; Brandt, *Cigarette* ; Proctor, *Golden Holocaust*; Michaels, *Doubt*; Oreskes and Conway, *Merchants of Doubt*.

12. The editors of various publications under the *British Medical Journal* umbrella released a statement in 2013 on publication of research funded by the tobacco industry: "As editors of the *BMJ, Heart, Thorax,* and *BMJ Open* we have decided that the journals will no longer consider for publication any study that is partly or wholly funded by the tobacco industry. Our new policy is

consistent with those of other journals including *PLoS Medicine, PLoS One, PLoS Biology; Journal of Health Psychology*; journals published by the American Thoracic Society and the *BMJ*'s own *Tobacco Control*." Godlee et al., "Journal Policy," 8.

13. Bero, "Implications"; Oreskes et al., "Why Disclosure Matters." For a broader discussion of the adverse impacts of commercially motivated secrecy, see Davidoff et al., *Sponsorship*; Michaels and Vidmar, "Sequestered Science." On the adverse public health impacts of selective reporting of clinical trial data, see An-Wen Chan et al., "Selective Reporting."

14. Lundh et al., "Industry Sponsorship." See also Krimsky, *Science in the Private Interest*, and discussions in Michaels, "Sarbanes-Oxley," and Haack, "Scientific Secrecy."

15. Bero, "Tobacco Industry"; Bero, "Experimental Models"; Bero, "Bias Related to Funding."

16. Mirowski, *Science-Mart*.

17. Oreskes et al., "Why Disclosure Matters."

18. Michaels and Monforton, "Scientific Evidence"; Proctor, *Golden Holocaust*; Rosner and Markowitz, *Lead Wars*; Krosnick, "Replication"; Bollen et al., *Social, Behavioral, and Economic Sciences*; Dorothy Bishop, "The Why and How of Reproducible Science," talk presented as a part of the Oxford Reproducibility Lectures at the Autumn School in Cognitive Neuroscience, 28 September 2017; Munafò et al., "Manifesto."

19. Some commentators have suggested that the problem with directed research is that it removes the possibility of surprise. When one follows one's curiosity, one may discover unanticipated things (e.g., Mermin, "How Not to Create Tigers"); therefore the prospect of serendipitous discovery is often used as justification for basic research programs (Showstack, "Criteria"). The flaw in this argument is that there are surprises in mission-driven research as well; being mission driven does not mean being blinded. Moreover, if my argument in chapter 2 is correct, the Navy mission led Henry Stommel to pay attention to a phenomenon—the thermocline—that his earlier colleagues, undirected by military matters, had more or less ignored. Curiosity-driven scientists have certainly missed many things. A colleague of mine likes to say that if it's true that we never know where discoveries may come from, then we may as well work on something useful in the meantime.

20. William von Arx to Al Faller, 20 February 1958, MC-24, Folder "Correspondence 1958," WHOI-von Arx, and chapter 3.

21. The issue of scientific autonomy brings to mind the "social contract for science," which many commentators have said prevailed from the end of World War II until the late 1970s or early 1980s. Under this social contract, the US federal government funded scientists and permitted them a high degree of self-governance, in exchange for the promise of the social goods that basic science was presumed to yield. See Guston, *Between Scientists*, esp. chap. 2. There are many difficulties with this concept; to me the most acute is that

it ignores the lion's share of government support for science was channeled through the armed services; academic science misleadingly stands in for the whole scientific enterprise. Some colleagues have suggested that the emphasis on individuality and autonomy in the Cold War may have reflected the felt need to underscore a crucial distinction between the United States and the Soviet Union: Soviet communist totalitarianism was characterized by collectivism, American capitalistic democracy by individualism. One of the worst offenses of Soviet communism, from the US perspective, was the denial of individual freedom; therefore, it must be the case that in the United States we all have individual freedom. This argument resonates with the work I have done with Erik Conway examining the role of the ideology of personal freedom in American climate change denial (Oreskes and Conway, *Merchants of Doubt*; Oreskes and Conway, *Collapse of Western Civilization*), but I have not found evidence of it in the archives. Rather, it seems that the scientists I have studied take for granted that autonomy is a necessary element of science—it needs to be affirmed but not explained. This is consistent with the larger argument here that American scientists had been emphasizing individualism, autonomy, and competition well before the Cold War. In a sense, they are arguing that good scientists are good Americans: in the United States, we are all (meant to be) the author of our own destinies. But perhaps there is a deeper point: in Judeo-Christian traditions, to be a moral agent one must be capable of making independent decisions. To suggest that scientists had lost that capacity would be to suggest they had lost moral agency. All of which is to say the issue is complex; I have emphasized here the elements that seem clearest in this story.

22. Sarewitz, *Frontiers of Illusion*.
23. Gieryn, "Boundary-Work"; Gieryn, *Cultural Boundaries*.
24. Shapin, *Social History*.
25. Lucier, "The Professional," 713.
26. Hollinger, *Science, Jews, and Secular Culture*.
27. Bush, *Science*. Many commentators have argued that this document also provided the basis for the postwar social contract for science; see Guston, *Between Science and Politics*.
28. Bush, *Science*.
29. Dupree, *Science in the Federal Government*; Smith, *American Science Policy*; Fleming, *Meteorology*; Reingold, *American Style*; Slotten, *Patronage*; Oreskes, *Rejection of Continental Drift*.
30. Rowland, "A Plea"; Dennis, "Accounting for Research," 1987.
31. Merton, *Sociology*.
32. Ibid., 260.
33. For further background, see Merton, "Science and Technology"; Polanyi, "Autonomy"; Bush, *Science*; Shils, *Torment of Society*.
34. Schmalzer, "Self Reliant Science."
35. Kargon and Hodes, "Politics of Science," 1985.

36. In hindsight we can see how the word *Lysenko* has been used to cut off discussion of the political control of science in the West, but from the perspective of scientists at the time, concern over the fate of scientists in the Soviet Union was hardly unfounded. After all, how should a scientist have responded if in 1941 he encountered socialist J. G. Crowther's glowing description of science in the Soviet Union and blithe dismissal of its impact on individuals—as if mere growing pains—in *The Social Relations of Science*: "The numbers of new young scientists are still raw, and many troubles have arisen through immature or insufficiently trained men trying to accomplish too difficult tasks. The violent contests over political policy have been reflected in the life of the institutes, as in other departments of the state, many scientists have been imprisoned, and some have been shot.... These imperfections are due to the limitations of mankind" (556–57).

37. Aldrich had corresponded with Paul Fye about his proposal to study giant squid in 1965 (chapter 6). In 1968, he published an article based on stranded specimens. According to Google Scholar this paper has twelve citations. In 1991, he revisited the topic: again based on specimens found in and around Newfoundland. Aldrich's research did not obtain the level to which he had aspired; he was never able to study live squid in their habitat. He died in 1991. See Aldrich, "Distribution"; Aldrich "Some Aspects."

38. US Department of the Navy, *TENOC*, preface, 1.

39. It seems reasonable to suppose that this is still the case today and would be the case for the other armed services that support scientific research, so this helps to explain the larger phenomena of the close links between the sciences and national militaries, more broadly.

40. Michaels and Wagner, "Disclosure," argue that the right to publish free of sponsor control is essential to the integrity of science used in regulatory decisions. I concur. However, in regulatory environments the concern is that regulated parties may seek to suppress results contrary to their economic interests. Navy classification was different: it was based on real or perceived national-security concerns. Thus, without affirming the legitimacy of the Navy's classification schemes and decisions, I note that the Navy was not trying to squelch particular results or interpretations. Classification and suppression both lead to data sequestration, but they are not ethically equivalent. See also Michaels and Vidmar, "Sequestered Science."

41. Michaels, *Doubt*; Brandt, *Cigarette Century*; Oreskes and Conway, *Merchants of Doubt*; Proctor, *Golden Holocaust*. See also Haack, "Scientific Secrecy."

42. Shapin, *Never Pure*. See also Greenberg, *Politics of Pure Science*; Guston, *Between Politics and Science*.

43. On gentlemanly science, where scientists of independent means were their own patrons, see Morrell and Thackray, *Gentlemen of Science*; Rudwick, *Devonian Controversy*.

44. This was a phrase used by the late Allan Cox, geophysicist and dean of the School of Earth Sciences at Stanford in the 1980s. He urged us to work on

"the next most important question," and I remember thinking, "Well, how do we know what that is?"

45. Robert J. Hofman to Clay H. Spikes, 15 June 1994, Box 12, Folder "Kauai DEIS Correspondence, May–June 1994," SIO-ATOC. It isn't clear which draft he read, but it may have been "Initial Study," 31 May 1994, Box 12, "California EIR Initial Study May 31, 1994," SIO-ATOC.

46. Bowles et al., "Relative Abundance."

47. Edwards, *Closed World*.

48. Hamblin, "Pursuit of Science."

49. Paul Lucier, "The Professional and the Scientist," stresses the importance of public lectures in nineteenth-century American science.

50. Greene, *Alfred Wegener*.

51. "Oceanography: The Making of a Science, People, Institutions, and Discovery Collection, 2000 (SMC 0087)," https://library.ucsd.edu/speccoll/findingaids/smc0087.html. See also Oreskes and Conway, *Merchants of Doubt*.

52. Walter Munk made this point to me in a conversation in 2005.

53. See Rampton, *Trust Us*.

54. The most well documented is the *Glomar Explorer*, where the study and mining of manganese nodules was used as a cover story for an operation to salvage a sunken soviet submarine. Like all good cover stories, this one contained some truth: there are manganese nodules on the seafloor, and research science was funded to understand them. But the claim that the ship itself was built to mine them was a cover story. John Knauss once told me, "The best cover story is a true story"; this one was not. See Julia Barton, "Confirmed: The CIA's Most Famous Ship Headed for the Scrapyard," *The World* (PRI), 7 September 2015, http://www.pri.org/stories/2015-09-07/ship-built-cias-most-audacious-cold-war-mission-now-headed-scrapyard. See also Day, "Cover Stories," who notes that, among various projects where military and espionage were hidden under cover stories was the infamous U-2 aircraft, which was portrayed as a civilian weather reconnaissance project.

55. Nowotny et al., *Re-thinking*; Gibbons et al., *New Production of Knowledge*. See also discussion in Hessles and van Lente, "Re-thinking," for a review of the literature on mode 2 science. For an analysis of the effects in industry funding on scientific accountability, see Etzkowitz and Zhou, *Triple Helix*, 2017.

56. Certainly in the earth sciences, there has never been a time when earth scientists were not driven at least in part by the economic demand for resources, or the cultural demand for origins stories. See Oreskes and Doel, "Physics and Chemistry."

57. Norman, *The Design of Everyday Things*, 48–49.

58. American Academy of Arts and Sciences, *Perceptions of Science in America*, 2018.

59. Ibid.

60. La Porte and Metlay, "Technology Observed," 121–27. American views of technology have changed since the 1970s, and there is evidence that citi-

zens are in general favorably disposed toward many technologies, which spills over into science. However, citizens have concerns about some technologies—particularly the feeling that technology is changing too quickly—and this can be associated with negative views. So the overall point remains clear: to the extent that laypeople associated science with something other than the pursuit of knowledge, people's views of those other things will spill over into their views of science. See also American Academy of Arts and Sciences, *Perceptions of Science in America*, 2018.

61. "Acoustic Thermometry of Ocean Climate," 1994 Q&A press release, copy courtesy of Cindy Clark, SIO communications office. (She gave the copy to me in October 2005.) See also memo and attachments sent by Cindy Clark, SIO Communications Office to S. Lauzon at WHOI, 12 April 1994, AC-09, Folder 2, "ATOC," WHOI-ODR.

62. H. U. Sverdrup and Richard Fleming, "Memorandum on Post-war Studies of the Oceanography of the Surface Layers," 31 May 1945, Box 1, Folder 26, SIO-ODR (Sverdrup).

63. Box 33, Folder 87, "SIO Director July 31, 1947–Sept 24, 1947, various letters and memos," SIO-Hubbs. Lest my reader interpret this comment as endorsement of these men's views, they also liked Merriam for his pedigree: "He has good heritage, for he is a nephew of President Eliot and the son of a distinguished historical scholar of Harvard" (of dictionary fame). See SIO Faculty (Dayton Carritt et al.) to Robert G. Sproul, "Confidential Letter," 24 September 1947, SIO-Hubbs. See also Rainger, *Constructing a Landscape*.

64. Rainger, "Constructing a Landscape," 327–28. Revelle's definition of what was "basic" and what was not was of course convenient, if not self-serving, insofar as the US Navy had a far larger budget than the US Fish Commission (Revelle liked to call it the Fish Commission; it's official name was the Bureau of Fisheries). It is true, as we have seen, that the Navy through the ONR did take the long view, supporting many projects without expectation of immediate applicability. Still, in hindsight it strains credulity to claim that much of what was done at Scripps on behalf of the Navy was not equally applied as whatever might have been done on behalf of fish. Keith Benson (personal communication, June 2019) makes the point that marine biology was being pursued in other places, which is true, but does not negate the point that Revelle was consciously molding institutional priorities in light of military funding opportunities.

65. Revelle, 1948, quoted in Rainger, "Constructing a Landscape." As Rainger notes, Revelle later promoted the creation of the Institute of Marine Resources in La Jolla to address fisheries questions, but as a separate institution, not part of SIO. As if to prove the point, IMR ended up focusing on physical issues as well—beach erosion, ocean engineering, and underwater minerals—rather than fish.

66. AC-09, Box 20, Folder "Exec. Scientific Advisory Committee 1954–1959," WHOI-ODR.

67.   Proctor and Schiebinger, *Agnotology*. See also Frickel, Gibbon, and Howard, "Undone Science"; Lakoff, "Why It Matters."

68.   Historians will recognize this as the problem of Whig history, also known as history written by the victors (Butterfield, *Whig Interpretation*). Questions of why scientists failed to do X or to recognize Y have often been thought of as "whiggish" insofar they seem to imply an expectation of progress. The problem is closely related to "presentism," under which past actors are judged (anachronistically) by present standards. The agnotology framework helps reframe such questions in the "progress-neutral" framework of the production of both knowledge and ignorance, as well as without assuming that our present understandings are correct. See Proctor and Schiebinger, *Agnotology*; Oreskes, "Presentist."

69.   Oreskes, "Presentist."

70.   Revelle and Suess, "Carbon Dioxide Exchange," 19–20.

71.   National Science and Technology Medals Foundation, "Charles Keeling, National Medal of Science: Physical Science" (2001), http://www.nationalmedals.org/laureates/charles-d-keeling.

72.   That was not true. The Navy was very interested in weather but not yet very engaged in anthropogenic climate change. That would come later.

73.   Edwards, *Vast Machine*. For Charney report, see National Research Council, *Carbon Dioxide and Climate*. See also Harper, *Weather by the Numbers*. Kirk Bryan at Geophysical Fluid Dynamic Lab is a case in point: he was an oceanographer but working mostly with meteorologists. He was the first author on one of the most important early papers on the transient climate response to increasing carbon dioxide. Bryan et al., "Transient Climate."

74.   Edward Frieman to Richard Attiyeh, 26 January 1987, Box 15, Folder "Global Change, 16 January 1987–16 December 1987," SIO-ODR (Frieman).

75.   See, e.g., "Roger Revelle Centennial Symposium," held at Scripps Institution of Oceanography, 6 March 2009, proceedings accessed at https://scripps.ucsd.edu/revelle; Cindy Clark and Shannon Casey, "Around the Pier: Sustaining a Legacy," Scripps News, 1 June 2007, https://scripps.ucsd.edu/news/around-pier-sustaining-legacy.

76.   The IPCC WG1 Second Assessment Report (IPCC, 1996) (hereafter SAR) concluded that "the balance of evidence suggests that there is a discernible human influence on global climate." See also Santer, "Time-Dependent"; Santer et al., "Human Influences."

77.   This may also help explain why "climate scientists" were late to recognize the threat of ocean acidification in comparison to atmospheric temperature increase. Ocean chemistry was not something that most meteorologists thought much about, and this lack of early recognition also has important consequences. For example, some highly reputable scientists now advocate solar radiation management as a means to dampen climate change. The scientific evidence certainly suggests that it may be possible to counter the atmospheric warming effects of carbon dioxide and lower the global mean temperature by

putting aerosols into the atmosphere. However, that would do nothing to protect the ocean from acidification, and if it led to delay in controlling carbon dioxide emissions, its overall effect would likely be a net negative.

78.   "Ocean and Climate Change Institute: Understanding the Role of the Ocean in Global Climate," Woods Hole Oceanographic Institution, http://www.whoi.edu/main/occi.

79.   For some relevant insights related to this, see Genevieve Wanucha, "Where Time Stands Still, Ideas Travel Generations," *Oceans at MIT*, 9 September 2014, http://oceans.mit.edu/news/featured-stories/gfd-2014.

80.   The AEC funding a significant amount of work related to radionuclide uptake in fish. See NRC Effects of Atomic Radiation on Oceanography and Fisheries, 1957.

81.   Mitman, *Reel Nature.*

82.   Memo 00c-U-195 to CDR James B. Davidson, ONR Code 466, 23 October, 1965, Box 4, Folder "Annual Reports, 1958–1964," SIO-MPL. This is not to say that biological work was eliminated: Rainger also points out that even though Revelle had little interest in marine biology, he had to deal with the sardine program, which by 1948 was quite large, and a program that he was in charge of from 1948 to 1950, and which was part of the large California Cooperative Fisheries Investigations initiative. Revelle actively participated in the initiative, even as he was hoping to marginalize it. Ronald Rainger, email communication with author, 28 October 2005.

83.   Rozwadowski, *The Sea Knows.* Bizarrely, the right wing has rediscovered Pettersson and used him to try to discredit contemporary climate science. See Anthony Watts, "Who Knew? Rachel Carson—Climate Change Expert," *Watts Up with That?*, 15 July 2008, https://wattsupwiththat.com/2008/07/25/who-knew-rachel-carson-climate-change-expert/.

84.   Bigelow, *Oceanography*; Mills, *Biological Oceanography*; Rozwadowski, *The Sea Knows.*

85.   Alverson and Dunlop, *Status of World Marine Fish Stocks*; Arons et al., "Discussion," from the School of Marine Affairs 25th Anniversary Public Proceedings, Seattle: University of Washington, May 7–8, 1998.

86.   Pauly et al., "Fishing Down."

87.   Myers and Worm, "Rapid Worldwide Depletion." Their work garnered considerable media attention, culminating in a lead editorial in the *New York Times* on 27 May 2003, "Oceans in Peril," A26.

88.   Aron et al., "Discussion," 1998.

89.   Kay, *Book of Life*; Keller, *Century*; Landecker, *Culturing Life*; Roosth, *Synthetic.*

90.   Finley, *All the Fish*; Finley, *All the Boats.* See also Aron et al., "Discussion," 1998.

91.   McEvoy, *The Fisherman's Problem.*

92.   Finley, *All the Fish.*

93.   Roberts, "Deep Impact."

94.   There is a large literature on the Newfoundland cod collapse, for entry see Hubbard, *Science on the Scales*; Finlayson, *Fishing for Truth.* For other discus-

sions of fisheries collapse, see also McCay and Finlayson, "The Political Ecology"; Finley, "The Tragedy of Enclosure"; Finley and Oreskes, "Maximum Sustained Yield"; Finley, *All the Fish*, Finley, *All the Boats*.

95. Government of Newfoundland and Labrador-Canada, "Province Responds to Federal Government Closure of Cod Fisheries," *News Releases*, 23 April 2003, http://www.releases.gov.nl.ca/releases/2003/fishaq/0424n03.htm.

96. Vickers, *Farmers and Fishermen*, 1994.

97. European Commission, "Fishing Quotas: Commission Wants Better Protection for Fish Stocks to Avoid Closing Fisheries in 2004," 4 December 2003, http://europa.eu/rapid/press-release_IP-03-1656_en.htm.

98. Pew Oceans Commission, *America's Living Oceans: Charting a Course for Sea Change* (Philadelphia: Pew Oceans Commission, 2003), http://www.pewtrusts.org/en/research-and-analysis/reports/2003/06/02/americas-living-oceans-charting-a-course-for-sea-change.

99. "*Science* Study Predicts Collapse of All Seafood Fisheries by 2050," *Stanford News*, 2 November 2006, http://news.stanford.edu/news/2006/november8/ocean-110806.html.

100. Ban Ki-moon, "Secretary-General's Message on World Oceans Day," *United Nations Secretary-General*, 8 June 2009, https://www.un.org/sg/en/content/sg/statement/2009-06-08/secretary-generals-message-world-oceans-day; "Secretary-General's Message on the International Day for Biological Diversity," *United Nations Secretary-General*, 22 May 2012, https://www.un.org/sg/en/content/sg/statement/2012-05-22/secretary-generals-message-international-day-biological-diversity.

101. Lajus, "Foreign Science"; Lajus, "Linking People"; Doel et al., "Strategic Arctic Science."

102. Pew, "Landmark 'Census of Marine Life' Culminates Today," 5 October 2010, http://www.pewtrusts.org/en/research-and-analysis/analysis/2010/10/05/landmark-census-of-marine-life-culminates-today; Census of Marine Life, "First Census Show Life in Planet Ocean is Richer, More Connected, More Altered Than Expected," 23 September 2010, http://www.coml.org/press-releases-2010/.

103. "Look to the Sea, a Census Inspired Song," http://www.coml.org/census-arts/music/look-to-the-sea.html.

104. Jackson, *Shifting Baselines*; Knowlton and Jackson, "Shifting Baselines."

105. Jackson, *Shifting Baselines*; Knowlton and Jackson, "Shifting Baselines."

106. Census of Marine Life, "First Census Show Life in Planet Ocean is Richer, More Connected, More Altered Than Expected," 23 September 2010, http://www.coml.org/press-releases-2010/.

107. Fye to William Benson, 13 October 1959, AC-09.5, Folder "Activities, NSF, 3 of 4," WHOI-ODR.

108. Memo, Spiess to Nierenberg, "NSF block funding ship time for deep tow work, MPL File 02-U-46, Box 7, Folder "Deep Tow," SIO-MPL.

109. Roper and the Ocean Portal Team, "Giant Squid: *Architeuthis dux*," 2018. See also William J. Broad, "First Giant Squid Captured in Wild (on Film, That Is)," *New York Times*, 27 September 2005, http://www.nytimes.com/2005/09/27/science/first-giant-squid-captured-in-wild-on-film-that-is.html?_r=0; William J. Broad, "One Legend Found, Many Still to Go," *New York Times*, 2 October 2005, http://www.nytimes.com/2005/10/02/weekinreview/one-legend-found-many-still-to-go.html.

110. Broad, "One Legend Found."

111. I am not suggesting that creationists would have changed their views, but that scientists might have had better evidence with which to refute this particular claim, which, in my experience, many students find persuasive.

112. Vine to Hamilton Howze, Vice President of Bell Helicopters, 31 May 1966, MC-01, Folder "Correspondence: 'One of a Kind Jobs,' 1964–1968," WHOI-Vine. See chapter 6.

113. Charles Bates to John C. Maxwell, CC'ed to Harry Hess, 3 May 1948, Box 25, Folder "HO," HHP.

114. Vine, "Vehicles."

115. Memo, "Current Trends and Needs," Office of the Dept. Asst. Oceanographer of the Navy, Box 31, no folder, Hess Papers.

116. Spiess, "Motivating." See also *Dr. Fred Noel Spiess*, interview by Christopher Henke, La Jolla, CA, SIO Archives, University of California, San Diego, 2000, http://scilib.ucsd.edu/sio/oral/Spiess.pdf.

117. Bush, *Science*; Weinberg, *Reflections*; Kidd, "Basic Research."

118. The Navy supported this indirectly, with its continued funding of Alvin as a UNOLS platform, but the projects themselves were funded primarily by NSF.

119. Edward Frieman, "Oceanography: The Making of a Science," interview by Naomi Oreskes and Ron Rainger for the H. John Heinz III Center for Science, Economics, and the Environment, Scripps Institution of Oceanography (La Jolla: University of California, San Diego, 17 February 2000).

120. Haack, "Scientific Secrecy."

121. Edwards, *Closed World*; Mark Solovey, "Science and the State."

122. Kaiser, "Cold War Requisitions."

123. Edwin Hamilton to Bruce Heezen, 9 May 1967, Box 1, Folder "EL Hamilton (NEL)," BCHP. On Edwin Hamilton (not to be confused with Gordon Hamilton), see "Edwin Lee Hamilton Papers," http://www.oac.cdlib.org/findaid/ark:/13030/c8hh6pbc/admin/. The project was FASOR II.

124. This idea is also offered by seismologist Jack Oliver. Perhaps it was a truism among geophysicists, or perhaps Oliver got it from Cox or vice versa. Oliver, *Shocks and Rocks*.

125. Memo, Spiess to Nierenberg, "NSF Block Funding Ship Time for Deep Tow Work, MPL File 02-U-46," Box 7, Folder "Deep Tow," SIO-MPL.

126. Fye to William Benson, 13 October 1959, AC-09.5, Folder "Activities, NSF, 3 of 4," WHOI-ODR.

Bibliography

Advanced Research Projects Agency. *Final Environmental Impact Statement/Environmental Impact Report for the California Acoustic Thermometry of Ocean Climate Project and its associated Marine Mammal Research Project*. Silver Spring, MD: National Oceanic and Atmospheric Administration; La Jolla, CA: University of California, San Diego, April 1995.

Aldrich, Frederick A. "The Distribution of Giant Squids (Cephalopoda, Archxteuthidae) in the North Atlantic and Particularly about the Shores of Newfoundland." *Sarsia* 34 (1968): 393–98.

———. "Some Aspects of the Systematics and Biology of Squid of the Genus *Architeuthis* Based on a Study of Specimens from Newfoundland Waters." *Bulletin of Marine Science* 49 (1991): 457–81.

Alic, John A., Lewis M. Branscomb, Harvey Brooks, Ashton Carter, and Gerald L. Epstein. *Beyond Spin-Off Military and Commercial Technologies in a Changing World*. Cambridge, MA: Harvard University Press, 1992.

Allwardt, Alan O. "Evolution of the Tectogene Concept, 1930–1965." In *Oceanographic History: The Pacific and Beyond,* edited by Keith R. Benson and Philip F. Rehbock, 480–92. Seattle: University of Washington Press, 2002.

———. "The Roles of Arthur Holmes and Harry Hess in the Development of Modern Global Tectonics." PhD diss., University of California, Santa Cruz, 1990.

Alverson, D., and K. Dunlop. *Status of World Marine Fish Stocks: Technical Report*. St. Louis, MO: Washington University, School of Aquatic and Fishery Science, Fisheries Research Institute, 1998.

American Academy of Arts and Sciences. *Perceptions of Science in America: A Report from the Public Face of Science Initiative*. Cambridge, MA: American Academy of Arts and Sciences, 2018.

Anderson, Victor C. "'MPL and Artemis' in Seeking Signals in the Sea." In *SIO Reference No. 97-5*, 71–77. San Diego, CA: University of California, San Diego, Marine Physical Laboratory of the Scripps Institution of Oceanography, 1997.

Andrews, Frank A. "Searching for the Thresher." In *US Naval Institute Proceedings*, 69–77. Washington, DC: US Government Printing Office, May 1964.

———. "Search Operations in the Thresher Area—1964, Section I." *Naval Engineers Journal* (1965): 549–61.

———. "Search Operations in the Thresher Area—Section II." *Naval Engineers Journal* (1967): 769–79.

Appel, Toby. "Marine Biology/Biological Oceanography and the Federal Patron: The NSF Initiative in Biological Oceanography in the 1960s." In *The Pacific and Beyond: Proceedings of the Fifth International History of Oceanography Meeting*, edited by Keith R. Benson and Philip F. Rehbock, 38. Seattle: University of Washington Press, 2003.

ARCYANA. "Transform Fault and Rift Valley from Bathyscaph and Diving Saucer." *Science* 190 (1975): 108–16.

Armi, Laurence. "The Dynamics of the Bottom Boundary Layer of the Deep Ocean." *Elsevier Oceanography Series* 19 (1977): 153–64.

Arons, Arnold. "The Scientific Work of Henry Stommel." In *Evolution of Physical Oceanography*, edited by Bruce A. Warren and Carl Wunsch, xiv–xviii. Cambridge, MA: MIT Press, 1981.

Arons, Arnold, and Henry Stommel. "A Mixing-Length Theory of Tidal Flushing." *Eos* 32, no. 3 (1951): 419–21.

Arrhenius, Svante. *Lehrbuch der kosmischen Physik*. Leipzig: S. Hirzel, 1903.

———. "On the Influence of Carbonic Acid in the Air upon the Temperature of the Ground." *Philosophical Magazine and Journal of Science*, ser. 5, 41 (April 1896): 237–76.

———. "Ueber die Wärmeabsorption durch Kohlensäure und ihren Einfluss auf die Temperatur der Erdoberfläche." *Öfversigt af Kongl: Vetenskaps- Akademiens Föhandlingar* 58, no. 1 (1901): 25–58.

———. "Ueber die Wärmeabsorption durch Kohlensäure." *Annalen der Physik* 4 (1901): 690–705.

Atomic Energy Commission. *Annual Report to Congress for 1959*. Washington, DC: US Government Printing Office, 1960.

Atwater, Tanya M. "Implications of Plate Tectonics for the Cenozoic Tectonic Evolution of Western North America." *Geological Society of America Bulletin* 81 (1970): 3513–35.

———. "When the Plate Tectonic Revolution Met Western North America." In *Plate Tectonics: An Insider's History of the Modern Theory of the Earth*, edited by Naomi Oreskes, 243–63. Boulder, CO: Westview Press, 2001.

Atwater, Tanya M., and Henry W. Menard. "Magnetic Lineations in the Northeast Pacific." *Earth and Planetary Science Letters* 7 (1970): 445–50.

Atwater, Tanya M., and John D. Mudie. "Block Faulting on the Gorda Rise." *Science* 19 (1968): 729–31.

Au, Whitlow W. L., Paul E. Nachtigall, and Jeffrey L. Pawloski. "Acoustic Effects of the ATOC Signal (75 Hz, 195 dB) on Dolphins and Whales." *Journal of the Acoustical Society of America* 101, no. 5 (1997): 2973–77.

Baldwin, Hanson W. "Chances of a Nuclear Mishap Viewed as Infinitesimal." *New York Times*, March 27, 1966.

Ballard, Robert D. "The Exploits of Alvin and Angus: Exploring the Deep Pacific." *Oceanus* 37, no. 3 (1984): 7–14.

———. "The History of Woods Hole's Deep Submergence Program." In *50 Years of Ocean Discovery: National Science Foundation 1950–2000*, 67–84. Washington, DC: National Academies Press, 2000.

———. "Project FAMOUS II: Dive into the Great Rift." *National Geographic Magazine* 147, no. 5 (1975): 604–15.

Ballard, Robert D., W. B. Bryan, J. R. Heirtzler, G. Keller, J. G. Moore, and Tjeerd H. van Andel. "Manned Submersible Observations in the Famous Area." *Science* 190 (1975): 103–8.

Ballard, Robert D., and Will Hively. *The Eternal Darkness*. Princeton, NJ: Princeton University Press, 2000.

Baross, J. A. "Do the Geological and Geochemical Records of the Early Earth Support the Prediction from Global Phylogenetic Models of a Thermophilic Cenancestor?" In *Thermophiles: The Keys to Molecular Evolution and the Origin of Life*, edited by J. Wiegel and M. Adams, 3–18. London: Taylor & Francis, 1998.

Baross, John A., and Jody W. Deming. "Growth of 'Black Smoker' Bacteria at Temperatures of at Least 250°C." *Nature* 303, no. 5916 (1983): 423–26.

Baross, J. A., and S. E. Hoffman. "Submarine Hydrothermal Vents and Associated Gradient Environments as Sites for the Origin and Evolution of Life." *Origins of Life* 15 (1985): 327–45.

———. "Submarine Hydrothermal Vents and Associated Gradient Environments as Sites for the Origin and Evolution of Life." *Origins of Life* 15 (1985): 327–45.

Barton, Cathy. "Marie Tharp, Oceanographic Cartographer, and Her Contributions to the Revolution in the Earth Sciences." *Geological Society Special Publications* 192, no. 1 (2002): 215–28.

Barus, Carl. "Military Influence on the Electrical Engineering Curriculum since World War II." *IEEE Technology and Society Magazine* 6 (1987): 3–9.

Bass, Catherine. "ZoBell's Contribution to Petroleum Microbiology." In *Microbial Biosystems: New Frontiers, Proceedings of the 8th International Symposium on Microbial Ecology*. Halifax, NS: Atlantic Canada Society for Microbial Ecology, 1999.

Bates, Charles C., and John F. Fuller. 1986. *America's Weather Warriors, 1814–1985*. College Station: Texas A&M University Press.

Bates, Charles C., Thomas F. Gaskell, and Robert B. Rice. *Geophysics in the Affairs of Man: A Personalized History of Exploration Geophysics and Its Allied Sciences of Seismology and Oceanography*. Oxford, UK: Pergamon Press, 1982.

Behe, Michael. "Irreducible Complexity: Obstacle to Darwinian Evolution." In *Philosophy of Biology: An Anthology*, edited by Alex Rosenberg and Robert Arp. Malden, MA: Wiley-Blackwell, 2010.

Behringer, D., T. Birdsall, M. Brown, B. Cornuelle, R. Heinmiller, R. Knox, K. Metzger, W. Munk, J. Spiesberger, R. Spindel, D. Webb, P. Worcester, and C. Wunsch. "A Demonstration of Ocean Acoustic Tomography." *Nature* 299, no. 5879 (1982): 121–25.

Bekelman, Justin E., Yan Li, and Cary P. Gross. "Scope and Impact of Financial Conflicts of Interest in Biomedical Research: A Systematic Review." *JAMA* 289, no. 4 (2003): 454–65.

Bellaiche, G., J. L. Cheminee, J. Francheteau, R. Hekinian, X. Le Pichon, H. D. Needham, and Robert D. Ballard. "Inner Floor of the Rift Valley: First Submersible Study." *Nature* 250 (1974): 558–60.

Benson, Keith R. "Laboratories on the New England Shore: The 'Somewhat Different Direction' of American Marine Biology." *New England Quarterly* 61, no. 1 (1988): 55–78.

———. "Summer Camp, Seaside Station, and Marine Laboratory: Marine Biology and Its Institutional Identity." *Historical Studies in the Physical and Biological Sciences* 32, no. 1 (2001): 11–18.

———. "Why American Marine Stations?:The Teaching Argument." *Integrative and Comparative Biology* 28, no. 1 (1988): 7–14.

Benson, Keith R., and Philip F. Rehbock, eds. *Oceanographic History: The Pacific and Beyond.* Seattle: University of Washington Press, 2002.

Benson, William E., E. A. Frieman, Robert A. Frosch, Edward D. Goldberg, Gordon Hamilton, Feenan D. Jennings, John A. Knauss, et al. *Oceanography: The Making of a Science, People, Institutions and Discovery.* Arlington, VA: Office of Naval Research, 2000. VHS.

Berelson, Bernard. *Graduate Education in the United States.* New York: McGraw-Hill, 1960.

Berman, Alan. "Project Artemis—A Retrospective View. *Journal of the Acoustical Society of America* 110, no. 5 (2001): 2688.

Bernal, John Desmond. *The Social Function of Science.* Cambridge, MA: MIT Press, 1967.

Bernstein, Michael T. "The Cold War and Expert Knowledge: New Essays on the History of the National Security State." *Radical History Review* 63 (1995): 1–6.

Bero, Lisa. "Bias Related to Funding Source in Statin Trials." *British Medical Journal* 349 (2014): 9.

———. "Experimental Institutional Models for Corporate Funding of Academic Research: Unknown Effects on the Research Enterprise." *Journal of Clinical Epidemiology* 61, no. 7 (2008): 629–33.

———. "Implications of the Tobacco Industry Documents for Public Health and Policy." *Annual Review of Public Health* 24 (2003): 267–88.

———. "Tobacco Industry Manipulation of Research." In *Late Lessons from Early Warnings: Science, Precaution, Innovation,* 151–78. Copenhagen: European Environmental Agency, 2013.

Betz, Frederick, Jr., and H. H. Hess. "The Floor of the North Pacific Ocean." *Geographical Review* 32, no. 1 (1942): 99–116.

Beyer, Robert T. *Sounds of Our Times: Two Hundred Years of Acoustics.* New York: Springer Science & Business Media, 1999.

Biagioli, Mario. *Galileo, Courtier: The Practice of Science in the Culture of Absolutism.* Chicago: University of Chicago Press, 1993.

Bigelow, Henry B. *Oceanography: Its Scope, Problems, and Economic Importance*. Cambridge, MA: Riverside Press, 1931.

Bischoff, J. L., R. J. Rosenbauer, P. J. Aruscavage, P. S. Baedecker, and J. G. Crock. *Geochemistry and Economic Potential of Massive Sulfide Deposits from the Eastern Pacific Ocean*. Open File Report 83-324. Menlo Park, CA: US Geological Survey, 1983.

Blackett, P. M. S, Sir Edward Bullard, and S. K. Runcorn. *A Symposium on Continental Drift*. London: Royal Society, 1965.

Bolin, Bert, and Henry Stommel. "On the Abyssal Circulation of The World Ocean-IV: Origin and Rate of Circulation of Deep Ocean Water as Determined with the Aid of Tracers." *Deep-Sea Research* 8 (1961): 95–110.

Bollen, Kenneth, John T. Cacioppo, Robert M. Kaplan, Jon A. Krosnick, and James L. Olds. *Social, Behavioral, and Economic Sciences Perspectives on Robust and Reliable Science: Report of the Subcommittee on Replicability in Science Advisory Committee to the National Science Foundation Directorate for Social, Behavioral, and Economic Sciences*. Washington, DC: National Academy of Sciences, 2015.

Bork, Kennard Baker. *Cracking Rocks and Defending Democracy: Kirtley Fletcher Mather, Scientist, Teacher, Social Activist, 1888–1978*. San Francisco: Pacific Division, American Association for the Advancement of Science, 1994.

Bostrom, K., and M. N. A. Peterson. "Precipitates from Hydrothermal Exhalations on the East Pacific Rise." *Economic Geology* 61 (1966): 1258.

Bowen, V., and Charles Hollister. "Pre- and Post-dumping Investigations for Inauguration of New Low-Level Radioactive Waste Dump Sites." *Radioactive Waste Management* 1, no. 3 (1981): 235-239.

Bowen, Vaughan, Victor Noshkin, Hugh Livingston, and Herbert Volchok. "Fallout Radionuclides in the Pacific Ocean: Vertical and Horizontal Distributions, Largely from Geosecs Stations." *Earth and Planetary Science Letters* 49 (1980): 411–34.

Bowles, Ann E., Mari Smultea, Bernd Würsig, Douglas P. DeMaster, and Debra Palka. "Relative Abundance and Behavior of Marine Mammals Exposed to Transmissions from the Heard Island Feasibility Test." *Journal of the Acoustical Society of America* 96, no. 2469 (1994): 2469–84.

Boykoff, Maxwell T., and Jules M. Boykoff. "Balance as Bias: Global Warming and the US Prestige Press." *Global Environmental Change* 14 (2004): 125–36.

Brandt, Allan M. *Cigarette Century: The Rise, Fall, and Deadly Persistence of the Product That Defined America*. New York: Basic Books, 2009.

Bravo, Michael. "James Rennell Antiquarian of Ocean Currents." *Ocean Challenge* 4, nos. 1–2 (1993): 41–50.

Brennecke, Wilhelm. *Die ozeanographischen Arbeiten der Deutschen Antarktischen Expedition 1911–12*. Hamburg, DE: Druck von Hammerich & Lesser in Altona, 1921.

Briffa, Keith R., and Timothy J. Osborn. "Blowing Hot and Cold." *Science* 295, no. 5563 (2002): 2227–28.

Brix, Ole. "Giant Squids May Die When Exposed to Warm Water Currents." *Nature* 303 (1983): 422–23.

Broad, William J. "Navy Says 2 Subs Pose No Hazards." *New York Times*, November 7, 1993.

Broecker, W. S., R. D. Gerard, Maurice Ewing, and B. C. Heezen. "Geochemistry and Physics of Ocean Circulation." In *Oceanography*, edited by Mary Sears, 301–22. Washington, DC: American Association for the Advancement of Science, 1961.

Brown, Laurie, and Lillian Hoddeson, eds. *The Birth of Particle Physics*. Cambridge: Cambridge University Press, 1983.

Brown, Shannon A. *Providing the Means of War: Historical Perspectives on Defense Acquisition, 1945–2000*. Washington, DC: Department of Defense, 2005.

Browne, John. "Climate Change: The New Agenda." Speech. Stanford University, California, May 19, 1997.

Brundage, Walter L., Jr. *NRL's Deep Sea Floor Search Era—A Brief History of the NRL/MIZAR Search System and Its Major Achievements*. Naval Research Laboratory Memorandum Report No. 6208. Washington, DC: US Navy, 1988.

Brundage, Walter L., Jr., and N. Z. Cherkis. *Preliminary LIBEC/FAMOUS Cruise Results*. Naval Research Laboratory Report No. 7785. Washington, DC: US Naval Research Laboratory, 1975.

Bruneau, L., N. G. Jerlov, and F. F. Koczy. "Physical and Chemical Methods, Swedish Deep-Sea Expedition Report, V. III, Physics and Chemistry No. 4, Table I, Physical and Chemical Methods." *Physics and Chemistry. Fasc. 1–3. 1951–1956 Swedish Deep-Sea Expedition (1947–1948)*. Gothenburg: Elanders Boktryckeri Aktieboloag, 1951–1956.

Brysse, Keynyn, Naomi Oreskes, Jessica O'Reilly, and Michael Oppenheimer. "Climate Change Prediction: Erring on the Side of Least Drama?" *Global Environmental Change* 23, no. 1 (2013): 327–37.

Buchanan, C. L. "A Love Affair." In *NRL's Deep Sea Floor Search ERA: A Brief History of the NRL/MIZAR Search System and Its Major Achievements*, Naval Research Laboratory Memorandum Report No. 6208. Washington, DC: US Navy, 1973.

———. "Scorpion Search." In *NRL's Deep Sea Floor Search ERA: A Brief History of the NRL/MIZAR Search System and Its Major Achievements*, Naval Research Laboratory Memorandum Report No. 6208, 22. Washington, DC: US Navy, 1988.

———. "Search for the Scorpion: Organization and Ship Facilities." *Proceedings of the 6th Navy Symposium on Military Oceanography*, no. 1 (1969): 58–63.

———. "Strange Devices That Found the Sunken Sub Scorpion." *Popular Science*, April 1969, 66–71, 182.

Bullard, E. C. "Comment Following Letter by R. Revelle and A. E. Maxwell." *Nature* 170 (1952): 200.

———. "The Flow of Heat through the Floor of the Atlantic Ocean." *Proceedings of the Royal Society A, Mathematical and Physical Sciences* 222 (1954): 408–29.

———. "The Flow of Heat through the Floor of the Ocean." In *The Sea*, vol. 3, *The Earth Beneath the Sea: History*, edited by M. N. Hill, 218–32. New York: John Wiley, 1963.

———. "William Maurice Ewing, May 12, 1906–May 4, 1974." *Biographical Memoirs* 51. Washington, DC: National Academies Press, 1992.

Bullard, E. C., and A. Day. "The Flow of Heat through the Floor of the Atlantic Ocean." *Geophysical Journal of the Royal Astronomical Society* 4 (1961): 282.

Bullard, Edward, James E. Everett, and A. Gilbert Smith. "The Fit of the Continents around the Atlantic." *Philosophical Transactions of the Royal Society of London A: Mathematical, Physical and Engineering Sciences* 258, no. 1088 (1965): 41–51.

Bullard, E. C., and R. G. Mason. "The Magnetic Field over the Oceans." In *The Sea*, vol. 3, *The Earth beneath the Sea: History*, edited by M. N. Hill, 175–217. New York: John Wiley, 1963.

Bullard, E. C., A. E. Maxwell, and R. Revelle. "Heat Flow through the Deep Sea Floor." *Advances in Geophysics* 3 (1956): 153–81.

Bunge, Mario. "Technology as Applied Science." *Technology and Culture* 7 (1966): 329–47.

Bureau of Ships, Navy Department. *Prediction of Sound Ranges from Bathythermography Observations: Rules for Preparing Sonar Messages*. Washington, DC: National Defense Research Committee, 1944.

Burgett, G. E. "Stephen Hays (1677–1761)." *Annals of Medical History* 7, no. 2 (1925): 1–20.

Burk, Creighton, and Charles L. Drake. *The Geology of Continental Margins*. New York: Springer, 1974.

Burnett, D. Graham. *Masters of All They Surveyed: Exploration, Geography, and a British El Dorado*. Chicago: University of Chicago Press, 2000.

———. *The Sounding of the Whale: Science and Cetaceans in the Twentieth Century*. Chicago: University of Chicago Press, 2012.

Burstyn, Harold L. "Reviving American Oceanography: Frank Lillie, Wickliffe Rose, and the Founding of the Woods Hole Oceanographic Institute." In *Oceanography: The Past*, edited by Mary Sears and Daniel Merriam, 57–66. New York: Springer, 1980.

Bush, Vannevar. *Science: The Endless Frontier*. Washington, DC: National Science Foundation, 1945.

Butterfield, Herbert. *The Whig Interpretation of History*. New York: Norton Library, 1965.

Callendar, Guy Stewart. "The Artificial Production of Carbon Dioxide and Its Influence on Temperature." *Quarterly Journal of the Royal Meteorological Society* 64, no. 275 (1938): 223–40.

———. "Can Carbon Dioxide Influence Climate?" *Weather* 4, no. 10 (1949): 310–14.

———. "On the Amount of Carbon Dioxide in the Atmosphere." *Tellus* 10, no. 2 (1958): 243–48.

———. "Variations of the Amount of Carbon Dioxide in Different Air Currents." *Quarterly Journal of the Royal Meteorological Society* 66, no. 287 (1940): 395–400.

Calmet, Dominique P. "Ocean Disposal of Radioactive Waste: Status Report." *IAEA Bulletin* 4 (1989): 47–50.

Carey, S. Warren. *Continental Drift: A Symposium*. Hobart, Australia: Geology Department, University of Tasmania, 1958.

———. *The Expanding Earth*. Amsterdam: Elsevier, 1976.

———. *The Expanding Earth: A Symposium.* Sydney: University of Tasmania Press, 1981.

———. *Theories of the Earth and Universe.* Palo Alto, CA: Stanford University Press, 1988.

Carlowicz, Michael. "Guy Nichols: Transforming Institutions." *Oceanus* 44, no. 1 (2005): 35.

Cartwright, David Edgar. *Tides: A Scientific History.* Cambridge: Cambridge University Press, 1999.

Cartwright, Nancy. "Evidence-Based Policy: So, What's Evidence?" Lecture, Colloquium in Science Studies, University of California, San Diego, February 11, 2008.

Cartwright, Nancy, and Jeremy Hardie. *Evidence-Based Policy: A Practical Guide to Doing It Better.* New York: Oxford University Press, 2012.

Chamberlin, T. C. "An Attempt to Frame a Working Hypothesis of the Cause of Glacial Periods on an Atmospheric Basis." *Journal of Geology* 7, no. 6 (1899): 545–84.

Chan, An-Wen, Asbjørn Hròbjartsson, Mette Haahr, Peter Gotzsche, and Douglas Altman. "Empirical Evidence for Selective Reporting of Outcomes in Randomized Controlled Trials." *JAMA* 291, no. 20 (2004): 2457–65.

Chang, Hasok. *Inventing Temperature: Measurement and Scientific Progress.* Oxford: Oxford University Press, 2004.

Chang, Iris. *Thread of the Silkworm.* New York: Basic Books, 1995.

Chaplin, Joyce. "Knowing the Ocean: Benjamin Franklin and the Creation of Atlantic Knowledge." In *Science and Empire in the Atlantic World*, edited by James Delbourgo and Nicholas Dew, 73–96. New York: Routledge, 2007.

Chargaff, Erwin. *Heraclitean Fire: Sketches from a Life before Nature.* New York: Rockefeller University Press, 1978.

Charney, Jule G., Akio Arakawa, D. James Baker, Bert Bolin, Robert E. Dickinson, Richard M. Goody, Cecil E. Leith, Henry M. Stommel, and Carl I. Wunsch. *Carbon and Dioxide: A Scientific Assessment.* Washington, DC: National Academy of Sciences, 1979.

Charnock, Henry. "Energy Transfer by the Atmosphere and the Southern Ocean." *Proceedings of the Royal Society of London A: Mathematical, Physical and Engineering Sciences* 281, no. 1384 (1964): 6–14.

Chase, C. G., Henry W. Menard, R. L. Larson, G. F. Sharman III, and S. M. Smith. "History of Sea-Floor Spreading West of Baja California." *Geological Society of America Bulletin* 81 (1970): 491–98.

Childress, James J., and Charles R. Fisher. "The Biology of Hydrothermal Vent Animals: Physiology, Biochemistry, and Autotrophic Symbiosis." *Oceanography and Marine Biology: An Annual Review* 30 (1992): 337–441.

Chomsky, Noam, Ira Katznelson, R. C. Lewontin, David Montgomery, Laura Nader, Richard Ohmann, Ray Siever, Immanuel Wallerstein, and Howard Zinn, eds. *The Cold War & the University.* New York: New Press, 1997.

Chyba, Christopher F., and Cynthia B. Phillips. "Europa as an Abode of Life." *Origins of Life and Evolution of the Biosphere* 32 (2002): 47–68.

Clinton, William J., and Albert Gore Jr. *Science in the National Interest*. Washington, DC: Executive Office of the President, 1994.

Cloud, John. "Imagining the World in a Barrel." *Social Studies of Science* 31, no. 2 (2001): 231–52.

Cloud, John, and Keith C. Clarke. "Through a Shutter Darkly: The Tangled Relationships between Civilian, Military, and Intelligence Remote Sensing in the Early US Space Program." In *Secrecy and Knowledge Production*, edited by Judith Reppy, 35–56. Ithaca, NY: Cornell University, 1999.

Clowes, A. J., and G. E. R. Deacon. "The Deep Water Circulation of the Indian Ocean." *Nature* 136 (1935): 936–38.

Cohen, Jon. "Was Underwater 'Shot' Harmful to the Whales?" *Science* 252, no. 5008 (1991): 912.

Coles, James S. "Paul MacDonald Fye: An Appreciation." *Oceanus* 20, no. 3 (1977): 2–6.

Collins, Harry M. *Changing Order: Replication and Induction in Scientific Practice*. London: Sage Publications, 1985.

Committee on Oceanography, Committee on Undersea Warfare, Office of Naval Research. "Unedited Transcript of the Bathyscaph Conference." Washington DC: National Academy of Sciences, National Research Council, 1958.

Converse, D. R., H. D. Holland, and J. M. Edmond. "Flow Rates in the Axial Hot Springs of the East Pacific Rise (21°N): Implications for the Heat Budget and the Formation of Massive Sulfide Deposits." *Earth and Planetary Science Letters* 69, no. 1 (1984): 159–75.

Conway, Hannah C. "A Lively Stone: Industrial Microbes, Self-Healing Cement, and the Resilient Specter of Decay." Harvard University Seminar Paper, 2018.

Corliss, J. B. "Mid-Ocean Ridge Basalts, I: The Origins of Hydrothermal Solutions, II: Regional Diversity along the Mid-Atlantic Ridge." PhD diss., Scripps Institution of Oceanography, University of California, San Diego, 1970.

———. "The Origin of Metal-Bearing Submarine Hydrothermal Solutions." *Journal of Geophysical Research* 76 (1971): 8128–38.

Corliss, John B., and Robert D. Ballard. "Oases of Life in the Cold Abyss." *National Geographic* 152, no. 4 (1977): 441–53.

Corliss, J. B., J. A. Baross, and S. E. Hoffman. "An Hypothesis Concerning the Relationship between Submarine Hot Springs and the Origin of Life in Earth." *Oceanologica Acta* 4, suppl. (1981): 59–69.

Corliss, John B., Jack Dymond, Louis I. Gordon, John M. Edmond, Richard P. von Herzen, Robert D. Ballard, Kenneth Green, David Williams, A. E. Bainbridge, Kathy Crane, and Tjeerd H. van Andel. "Submarine Thermal Springs on the Galapagos Rift." *Science* 203 (1979): 1073–83.

Corliss, John B., Jack Dymond, Mitchell Lyle, and Kathy Crane. "The Chemistry of Hydrothermal Mounds near the Galapagos Rift." *Earth and Planetary Science Letters* 40 (1978): 10–24.

Corliss, John B., J. Dymond, M. Lyle, T. Doerge, K. Crane, P. Lonsdale, R. P. von

Herzen, and D. Williams. "Sediment Mound Ridges of Hydrothermal (?) Origin along the Galapagos Rift." *American Geophysical Union* 57, no. 12 (1976): 935–36.

Cowan, Don A. "Hyperthermophilic Enzymes: Biochemistry and Biotechnology." In *Hydrothermal Vents and Processes*, edited by L. M. Parson, C. L. Walker, and D. R. Dixon, 351–63. London: Geological Society 1995.

Cox, Allan. *Plate Tectonics and Geomagnetic Reversals*. New York: W. H. Freeman, 1973.

Cox, Allan, and Richard R. Doell. "Geomagnetic Polarity Epochs and Pleistocene Geochronometry." *Nature* 198 (1963): 1049–51.

Cox, Allan, Richard R. Doell, and G. Brent Dalrymple. "Geomagnetic Polarity Epochs: Sierra Nevada II." *Science* 142, no. 3590 (1963): 382–85.

Cox, C. S. "Ripples." In *The Sea: Ideas and Observations on Progress in the Study of the Seas*, edited by M. N. Hill, 720–30. New York: John Wiley & Sons, 1962.

Craig, H., W. B. Clarke, and M. A. Beg. "Excess 3HE in Deep Water on the East Pacific Rise." *Earth and Planetary Science Letters* 26 (1975): 125.

Crane, K. *Sea Legs: Tales of a Woman Oceanographer*. New York: Basic Books, 2004.

———. "Structure and Tectonics of the Galapagos Inner Rift, 86°10′W." *Journal of Geology* 86 (1978): 715–30.

Crane, K., P. Lonsdale, P. Weiss, and H. Craig. "Structural Activity at the Galapagos Spreading Center 86°W (Abs)." *American Geophysical Union* 57, no. 12 (1976): 935.

Crane, K., and W. Normark. "Hydrothermal Activity and Crestal Structure of the East Pacific Rise 21°N." *Journal of Geophysical Research* 82, no. 33 (1977): 5336–48.

Craven, John P. "Ocean Technology and Submarine Warfare." *Adelphi Papers* 8, no. 46 (1968): 38–46.

———. *The Silent War: The Cold War Battle beneath the Sea*. New York: Simon & Schuster, 2001.

Crawley, James W. "Anniversary of Rescue Subs Marked." *San Diego Union Tribune*, December 15, 1995.

Crosland, Maurice P. *Science under Control: The French Academy of Sciences, 1795–1914*. Cambridge: Cambridge University Press, 1992.

Crow, Michael, and Barry Bozeman. *Limited by Design: R&D Laboratories in the US National Innovation System*. New York: Columbia University Press, 1998.

Crowther, James Gerald. *The Social Relations of Science*. London: Macmillan and Co., 1941.

Cullen, Vicky. *Down to the Sea for Science: 75 Years of Ocean Research Education & Exploration at the Woods Hole Oceanographic Institution*. Woods Hole, MA: Woods Hole Oceanographic Institution, 2005.

Daly, R. A. "Glaciation and Submarine Valleys." *Nature* 149 (1942): 156–60.

———. "Origin of Submarine 'Canyons.'" *American Journal of Science* 31 (1936): 401–20.

Daubrée, Auguste. *Études synthétiques de géologie expérimentale*. Paris: Dunod, 1879.

Davidoff, F., C. D. DeAngelis, J. M. Drazen, J. Hoey, L. Hojgaard, R. Horton, S. Kotzin, et al. "Sponsorship, Authorship, and Accountability." *JAMA* 286, no. 10 (2001): 1232–34.

Day, Deborah. "Navy Support for Oceanography at SIO." Historical notes prepared for ONR/Heinz Center Symposium "Oceanography, Making of a Science," February 9, 2000.

Day, Dwayne A. "Cover Stories and Hidden Agendas: Early American Space and National Security Policies." In *Reconsidering Sputnik: Forty Years since the Soviet Satellite*, edited by Roger D. Lanius, John M. Logsdon, and Robert W. Smith, 161–91. London: Routledge, 2000.

Deacon, E. L., and E. K. Webb. "Small-Scale Interactions." In *The Sea: Ideas and Observations on Progress in the Study of the Seas*, edited by M. N. Hill, 43–87. New York: John Wiley & Sons, 1962.

Deacon, G. E. R. "The Hydrology of the Southern Ocean." *Discovery Reports* 15 (1937): 1–124.

———. "Note on the Dynamics of the Southern Ocean." *Discovery Reports* 15 (1937): 125–52.

Deacon, Margaret. *Scientists and the Sea, 1650–1900: A Study of Marine Science*. New York: Academic Press, 1971.

———. "Wind Power versus Density Differences: A 19th Century Controversy about Ocean Circulation." *Ocean Challenge* 4, nos. 1–2 (1993): 53–60.

Deese, Da. *Nuclear Power and Radioactive Waste: A Sub-seabed Disposal Option?* New York: Lexington Books, 1978.

———. "Seabed Emplacement and Political Reality." *Oceanus* (1997): 47–63.

Defant, Albert. *Physical Oceanography*. Vol. 1. New York: Pergamon Press, 1961.

Degens, E. T., and D. A. Ross, eds. Preface to *Hot Brines and Recent Heavy Metal Deposits in the Red Sea*, ix–x. New York: Springer, 1969.

De la Bruhèze, A. A. Albert. "Closing the Ranks: Definition and Closure of Radioactive Waste in the US Atomic Energy Commission, 1945–1960." In *Shaping Technology, Building Society: Studies in Sociotechnical Change*, edited by in W. E. Bijker and J. Law, 140–75. Cambridge, MA: MIT Press, 1992.

———. *Political Construction of Technology: Nuclear Waste Disposal in the United States 1945–1972*. Delft, The Netherlands: Eburon, 1992.

Delaney, J. W., D. S. Kelley, M. D. Lilley, D. A. Butterfield, J. A. Baross, R. W. Embley, and M. Summit. "The Quantum Event of Oceanic Crustal Accretion: Impacts of Diking at Mid-Ocean Ridges." *Science* 281 (1998): 222–30.

Delaney, J. W., Deborah S. Kelley, Edmond A. Mathez, Dana R. Yoerger, J. A. Baross, Matt O. Schrenk, Margaret K. Tivey, Jonathon Kaye, and Veronique Robigou. "'Edifice Rex' Sulfide Recovery Project: Analysis of Submarine Hydrothermal, Microbial Habitat." *Eos, Transactions, American Geophysical Union* 82, no. 6 (2001): 67, 72–73.

Del Grosso, V. A. "New Equation for the Speed of Sound in Natural Waters (with Comparisons to Other Equations)." *Journal of the Acoustical Society of America* 56, no. 4 (1974): 1084–91.

Demeritt, David. "The Construction of Global Warming and the Politics of Science." *Annals of the Association of American Geographers* 91, no. 2 (2001): 307–37.

Deming, Jody W., and John A. Baross. "Deep-Sea Smokers: Windows to a Subsurface Biosphere." *Geochimica et Cosmochimica Acta* 57 (1993): 3219–30.

Dennis, Michael A. "Accounting for Research: New Histories of Corporate Laboratories and the Social History of American Science." *Social Studies of Science* 17 (1987): 479–518.

———. "A Change of State: The Political Cultures of Technical Practice at the MIT Instrumentation Laboratory and the Johns Hopkins University." PhD diss., Johns Hopkins University, 1991.

———. "Historiography of Science: An American Perspective." In *Science in the Twentieth Century*, edited by John Krige and Dominique Pestre, 1–26. Amsterdam: Harwood Academic Publishers, 1997.

———. "Our First Line of Defense: Two University Laboratories in the Postwar American State." *Isis* 85 (1994): 427–55.

———. "Secrecy and Science Revisited: From Politics to Historical Practice and Back." In *The Historiography of Contemporary Science, Technology, and Medicine: Writing Recent Science*, edited by Ronald E. Doel and Thomas Söderqvist, 172–84. New York: Routledge, 2006.

Detrick, R. S., D. L. Williams, J. D. Mudie, and J. G. Sclater. "The Galapagos Spreading Centre: Bottom-Water Temperatures and the Significance of Geothermal Heating." *Geophysical Journal of the Royal Astronomical Society* 38 (1974): 627–36.

DeVorkin, David H. "The Military Origin of the Space Sciences in the American V-2 Era." In *National Military Establishments and the Advancement of Science and Technology*, edited by Paul Forman and José M. Sánchez-Ron, 233–60. Dordrecht, The Netherlands: Kluwer Academic Publishers, 1996.

———. *Science with a Vengeance: How the Military Created the US Space Sciences after World War II*. New York: Springer, 1992.

———. "Who Speaks for Astronomy? How Astronomers Responded to Government Funding after World War II." *Historical Studies in the Physical and Biological Sciences* 31, no. 1 (2000): 55–92.

Dexter, Ralph W. "History of American Marine Biology and Marine Biology Institutions Introduction: Origins of American Marine Biology." *Integrative and Comparative Biology* 28, no. 1 (1988): 3–6.

Dicke, Robert H. "Gravitation—An Enigma." *American Scientist* 47, no. 1 (1959): 25–40.

Dickson, Andrew G. "A Book Review of Chandra Mukerji, 1989: *A Fragile Power: Scientists and the State*." *Oceanography* (1990): 62–63.

Dierrsen, Heidi M., and Albert E. Theberge Jr. "Bathymetry: History of Seafloor Mapping." *Encyclopedia of Natural Resources* (2014). https://doi.org/10.1081/E -ENRW-120047531.

Dietz, Robert S. "Continent and Ocean Basin Evolution by Spreading of the Sea Floor." *Nature* 190, no. 4779 (1961): 854–57.

Dietz, Robert S., Henry W. Menard, and Edwin L. Hamilton. "Echograms of the Mid-Pacific Expedition." *Deep-Sea Research* 1 (1954): 258–72.

Dijkstra, Henk A. "Ocean Currents and Circulation and Climate Change." In *Global Environmental Change, Volume 1*, by B. Freedman, 85–95. New York: Springer, 2014.

Doel, Ronald E. "Constituting the Postwar Earth Sciences: The Military's Influence on the Environmental Sciences in the USA after 1945." *Social Studies of Science* 33, no. 5 (2003): 635–66.

———. "Defending the North American Continent." In *Exploring Greenland: Cold War Science and Technology on Ice*, edited by Ronald Doel, Kristine Harper, and Matthias Heymann, 25–46. New York: Palgrave Macmillan, 2016.

———. "Scientists as Policymakers, Advisors, and Intelligence Agents: Linking Contemporary Diplomatic History with the History of Contemporary Science." In *The Historiography of Contemporary Science and Technology*, edited by Thomas Söderqvist, 215–44. Paris: Harwood Academic Publishers, 1997.

Doel, Ronald, Robert Friedman, Julia Lajus, Sverker Sörlin, and Urban Wrakberg. "Strategic Arctic Science: National Interests in Building Natural Knowledge— Interwar Era through the Cold War." *Journal of Historical Geography* 44 (2014): 60–80.

Doel, Ronald E., Tanya J. Levin, and Mason K. Marker. "Extending Modern Cartography to the Ocean Depths: Military Patronage, Cold War Priorities, and the Heezen-Tharp Mapping Project, 1952–1959." *Journal of Historical Geography* 32, no. 3 (2006): 605–26.

Donnelly, J. D. "1967—Alvin's Year of Science." *Journal Naval Research Reviews* 21 (1968): 18–27.

Drew, Christopher. "Adrift 500 Feet under the Sea, a Minute Was an Eternity." *New York Times*, May 18, 2005.

———. "Submarine Crash Shows Navy Had Gaps in Mapping System." *New York Times*, January 15, 2005.

Duan, Jingping, Michael Bevis, Peng Fang, Yehuda Bock, Steven Chiswell, Steven Businger, Christian Rocken, et al. "GPS Meteorology: Direct Estimation of the Absolute Value of Precipitable Water." *Journal of Applied Meteorology* 35, no. 6 (1996): 830–38.

Dupree, Asa H. 1986. *Science in the Federal Government: A History of Policies and Activities*. Baltimore: Johns Hopkins University Press.

Eamon, William. *Science and the Secrets of Nature: Books of Secrets in Medieval and Early Modern Culture*. Princeton, NJ: Princeton University Press, 1994.

Easterling, D. R., G. A. Meehl, C. Parmesan, S. A. Changnon, T. R. Karl, and L. O. Mearns. "Climate Extremes: Observations, Modeling, and Impacts." *Science* 5487 (2000): 2068–74.

Eckart, Carl. "The Equations of Motion of Sea-Water." In *The Sea: Ideas and Observations on Progress in the Study of the Seas*, edited by M. N. Hill, 31–42. New York: John Wiley & Sons, 1962.

———. *Principles and Applications of Underwater Sound*. Washington, DC: Department of the Navy, 1968. Originally issued as *Summary Technical Report of Division 6, Volume 7, National Defense Research Committee*, 1946.

———, ed. "The Refraction of Sound." In *Principles and Applications of Underwater Sound*. Washington, DC: Department of the Navy, 1968. Originally issued as *Summary Technical Report of Division 6, Volume 7, National Defense Research Committee*, 1946.

Edgerton, D. E. H. "British Scientific Intellectuals and the Relations of Science, Technology and War." In *National Military Establishments and the Advancement of Science and Technology*, edited by Paul Forman and José M. Sánchez-Ron, 1–36. Dordrecht, The Netherlands: Kluwer Academic Publishers, 1996.

Edmond, John M. "The Carbonic Acid System in Sea Water." PhD diss., University of California, San Diego, 1970.

Edmond, John M., C. I. Measures, R. E. McDuff, L. H. Chan, R. Collier, and B. Grant. "Ridge Crest Hydrothermal Activity and the Balances of the Major and Minor Elements in the Ocean: The Galapagos Data." *Earth and Planetary Science Letters* 46 (1979): 1–18.

Edmond, John, and Karen von Damm. "Hot Springs on the Ocean Floor." *Scientific American* 248, no. 4 (1983): 78–93.

Edmond, John M., K. L. von Damm, R. E. McDuff, and C. I. Measures. "Chemistry of Hot Springs on the East Pacific Rise and Their Effluent Dispersal." *Nature* 297 (1982): 187–91.

Edwards, Paul. *The Closed World*. Cambridge, MA: MIT Press, 1997.

———. *A Vast Machine: Computer Models, Climate Data, and the Politics of Global Warming*. Cambridge, MA: MIT Press, 2010.

Egyed, László. "A New Dynamic Conception of the Internal Constitution of the Earth." *Geologische Rundschau* 46, no. 1 (1957): 101–21.

Einsele, Gerhardt, Joris M. Gieskes, Joseph Curray, David M. Moore, Eduardo Aguayo, Marie-Pierre Aubry, Daniel Fornari, et al. "Intrusion of Basaltic Sills into Highly Porous Sediments, and Resulting Hydrothermal Activity." *Nature* 283, no. 5746 (1980): 441–45.

Elder, J. W. "Physical Processes in Geothermal Areas." In *Terrestrial Heat Flow*, edited by W. H. K. Lee, 211–39. Washington, DC: Geophysical Union, National Academies of Science, National Research Council, 1965.

Ellis, Richard. *The Search for the Giant Squid*. New York: Lyons Press, 1998.

Elster, Jon. "Counterfactuals and the New Economic History." In *Logic and Society: Contradictions and Possible Worlds*, 175–232. Chichester, UK: John Wiley & Sons, 1978.

Emmery, Kenneth O., J. I. Tracey Jr., and H. S. Ladd. "Geology of Bikini and Nearby Atolls." *US Geological Survey Professional Paper* 260A (1954): 1–264.

Engerman, David. "Rethinking Cold War Universities." *Journal of Cold War Studies* 5, no. 3 (2003): 80–95.

England, J. Merton. "Dr. Bush Writes a Report: Science—the Endless Frontier." *Science* 191 (1976): 41–47.

———. *A Patron for Pure Science*. Vol. 1, *The National Science Foundation's Formative Years, 1945–1957*. Washington, DC: National Science Foundation, 1983.

Epstein, Steven. *Impure Science: AIDS, Activism, and the Politics of Knowledge*. Berkeley: University of California Press, 1996.

Ericson, D. B., Maurice Ewing, and B. C. Heezen. "Deep-Sea Sands and Submarine Canyons." *Bulletin of the Geological Society of America* 62 (1951): 961–66.

Escher, B. G. "On the Relation of the Volcanic Activity in the Netherlands East

Indies and the Best of Negative Gravity Anomalies Discovered by Vening Meinesz." *Proceedings of the Section of Sciences Koninklijke Nederlandse Akademie van Wetenschappen* 36 (1933): 677–85.

Esper, Jan, Edward R. Cook, and Fritz H. Schweingruber. "Low-Frequency Signals in Long Tree-Ring Chronologies for Reconstructing Past Temperature Variability." *Science* 295 (2003): 2250–53.

Etzkowitz, Henry. "Conflicts of Interest and Commitment in Academic Science in the United States." *Minerva* 34 (1996): 259–77.

Etzkowitz, Henry, and Chunyan Zhou. *Triple Helix: University-Industry-Government Innovation and Entrepreneurship.* Abingdon, UK: Taylor and Francis, 2017.

Everest, F. Alton, and E. M. Everest. *World War II Underwater, An Engineer Called to San Diego, CA for Wartime Research in Underwater Sound.* Santa Barbara, CA: Scripps Institution of Oceanography Archives, 1992.

Ewing, Maurice. "Gravity Measurements on the USS *Barracuda.*" *Transactions of the American Geophysical Union* 18 (1937): 66–69.

———. "Marine Gravimetric Methods and Surveys." *Proceedings of the American Philosophical Society* 79 (1938): 47–70.

Ewing, Maurice, A. P. Crary, and H. M. Rutherford. "Geophysical Investigations in the Emerged and Submerged Atlantic Coastal Plain, Part I: Methods and Results." *Bulletin of the Geological Society of America* 48, no. 9 (1937): 753–802.

———. "Geophysical Studies in the Atlantic Coastal Plain." *Lehigh University Publication* 11, no. 9 (1937): 753–812.

Ewing, Maurice, and B. C. Heezen. "Puerto Rico Trench Topographic and Geophysical Data." *Geological Society of America* 62 (1955): 255–67.

———. "Reviews and Abstracts." *Transactions of the American Geophysical Union* 37, no. 3 (1956): 343.

———. "Oceanographic Research Programs of the Lamont Geological Observatory." *Geographical Review* 46, no. 4 (1956): 508–35.

———. "Continuity of Mid-Oceanic Ridge and Rift Valley in the Southwestern Indian Ocean Confirmed." *Science* 131 (1960): 1677–79.

Ewing, Maurice, B. C. Heezen, and D. B. Ericson. "Significance of the Worzel Deep Sea Ash." *Proceedings of the National Academy of Sciences of the United States of America* 45 (1959): 355–61.

Ewing, Maurice, Columbus Iselin, and J. Lamar Worzel. "Sound Transmission in Sea Water." *Woods Hole Oceanographic Institution Report,* February 1941.

Ewing, Maurice, and J. Lamar Worzel. "Gravity Anomalies and Structure of the West Indies Part I." *Geological Society of America Bulletin* 65, no. 2 (1954): 165–74.

———. *Long Range Sound Transmission.* Interim Report No. 1, Contract Nobs-2083, August 25, 1945. Declassified March 12, 1946.

———. "Long Range Sound Transmission." *Geological Society of America Memoir 27* (1948).

Ewing, Maurice, J. Lamar Worzel, D. B. Ericson, and B. C. Heezen. "Geophysical and Geological Investigations in the Gulf of Mexico, Part I." *Geophysics* 20, no. 1 (1955): 1–18.

Falkowski, P., R. J. Scholes, E. Boyle, J. Canadell, D Canfield, J. Elser, N. Gruber, K. Hibbard, P. Hogberg, S. Linder, F. T. Mackenzie, B. Moore III, T. Pederson, Y. Rosenthal, S. Seitzinger, V. Smetacek, and W. Steffen. "The Global Carbon Cycle: A Test of Our Knowledge of Earth as a System." *Science* 5490 (2000): 291–96.

Field, Richard M., ed. "Geophysical Exploration of the Ocean Bottom." *Proceedings of the American Philosophical Society* 79, no. 1 (1938).

Field, Richard M. "The Importance of Geophysics to Submarine Geology." *Proceedings of the American Philosophical Society* 79, no. 1 (1938): 1–9.

Field, Richard M., T. T. Brown, E. B. Collins, H. H. Hess, and F. A. Vening Meinesz. *The Navy-Princeton Gravity Expedition to the West Indies in 1932*. Washington, DC: US Government Printing Office, 1933.

Fine, Rana, Monika Rhein, and Chantal Andrie. "Using a CFC Effective Age to Estimate Propagation and Storage of Climate Anomalies in the Deep Western North Atlantic Ocean." *Geophysical Research Letters* 29, no. 24 (2002): 2227.

Finlayson, Alan Christopher. *Fishing for Truth: A Sociological Analysis of Northern Cod Stark Assessments from 1977 to 1990*. St. Johns, NL: Institute of Social and Economic Research, Memorial University of Newfoundland, 1994.

Finley, Alan Christopher, and Naomi Oreskes. "Maximum Sustained Yield: A Policy Disguised as Science." *ICES Journal of Marine Science* 70, no. 2 (2013): 245–50.

Finley, Carmel. *All the Boats in the Ocean: How Government Subsidies Led to Overfishing*. Chicago: University of Chicago Press, 2017.

———. *All the Fish in the Sea: Maximum Sustainable Yield and the Failure of Fisheries Management*. Chicago: University of Chicago Press, 2011.

———. "The Tragedy of Enclosure: Fish, Fisheries Science, and US Foreign Policy, 1920–1960." PhD diss., University of California, San Diego, 2007.

Fisher, R. L., and H. H. Hess. "Trenches." In *The Sea: Ideas and Observations on Progress in the Study of the Seas*, edited by M. N. Hill, 411–36. New York: John Wiley & Sons, 1963.

Fleming, James R. "Climate Dynamics, Science Dynamics, and Technological Dynamics, 1804–2004." *From Beaufort to Bjerknes and Beyond—Critical Perspectives on Observing, Analyzing, and Predicting Weather and Climate*, edited by S. Emeis and C. Lüdecke, special issue, *Algorismus* 52 (2005): 9–16.

———. *First Woman: Joanne Simpson and the Tropical Atmosphere*. Oxford: Oxford University Press, 2020.

———. *Fixing the Sky: The Checkered History of Weather and Climate Control*. New York: Columbia University Press, 2012.

———. *Historical Perspectives on Climate Change*. Oxford: Oxford University Press, 2005.

———. *Inventing Atmospheric Science: Bjerknes, Rossby, Wexler, and the Foundations of Modern Meteorology*. Cambridge, MA: MIT Press, 2016.

———. *Meteorology in America, 1800–1870*. Baltimore: Johns Hopkins University Press, 2000.

Flynn, James, William Burns, C. K. Mertz, and Paul Slovic. "Trust as Determinant of

Opposition to a High-Level Radioactive Waste Repository: Analysis of a Structural Model." *Risk Analysis* 12, no. 3 (1992).

Fond, Eugene C. L. "Oceanography Researches at the US Navy Electronics Laboratory." *Transactions of the American Geophysical Union* 30 (1949): 594–96.

Forman, Paul. "Behind Quantum Electronics: National Security as Basis for Physical Research in the United States, 1940–1960." *Historical Studies in Physical and Biological Sciences* 18 (1987): 149–229.

———. "Into Quantum Electronics: The Maser as 'Gadget' of Cold-War America." In *National Military Establishments and the Advancement of Science and Technology*, edited by Paul Forman and José M. Sánchez-Ron, 261–326. Dordrecht, The Netherlands: Kluwer Academic Publishers, 1996.

Fox, Denis L. *Again the Scene*. Unpublished memoir. Scripps Institution of Oceanography Library, 1971.

Francheteau, J., H. D. Needham, P. Choukroune, T. Juteau, M. Séguret, R. D. Ballard, P. J. Fox, et al. "Massive Deep-Sea Sulphide Ore Deposits Discovered on the East Pacific Rise." *Nature* 277, no. 5697 (1979): 523–28.

Frankel, Adam, and Christopher Clark. "ATOC and Other Factors Affecting the Distribution and Abundance of Humpback Whales (*Megaptera novaeangliae*) off the Shore of Kauai." *Marine Mammal Science* 18, no. 3 (2002): 644–62.

Frankel, Henry. *The Continental Drift Controversy*. 4 vols. Cambridge: Cambridge University Press, 2012.

———. "The Continental Drift Debate." In *Scientific Controversies: Case Solutions in the Resolution and Closure of Disputes in Science and Technology*, by H. T. Engelhardt Jr. and A. L. Caplan. Cambridge: Cambridge University Press, 1987.

———. "Jan Hospers and the Rise of Paleomagnetism." In *History of Geophysics: Volume 4*, edited by C. Steward Gillmor, 156–60. Washington DC: American Geophysical Union, 1990.

Frickel, Scott, Sahra Gibbon, and Jeff Howard. "Undone Science: Charting Social Movement and Civil Society Challenges to Research Agenda Setting." *Science, Technology, and Human Values* 35, no. 4 (2010): 444–73.

Friedman, Gerald M. "The Concept of Deep-Sea Contourites—Discussion." *Sedimentary Geology* 88 (1994): 301–3.

Friedman, Robert M. *Appropriating the Weather: Vilhelm Bjerknes and the Construction of a Modern Meteorology*. Ithaca, NY: Cornell University Press, 1989.

———. "The Expeditions of Harald Ulrik Sverdrup: Contexts for Shaping an Ocean Science." William R. Ritter Memorial Lecture, Scripps Institution of Oceanography. Presented October 29, 1992; published by University of California, 1994.

Frosch, Robert A. "Underwater Sound: Deep-Ocean Propagation." *Science* 146 (1964): 889–904.

"The Future of American Science." *Science* 1, no. 1 (1883): 1–3.

Fyfe, W. S., and P. Lonsdale. "Ocean Floor Hydrothermal Activity." In *The Sea*, edited by Cesare Emiliani, 589–638. New York: John Wiley & Sons, 1981.

Galison, Peter. *Image and Logic*. Chicago: University of Chicago Press, 1997.

———. "Secrecy in Three Acts." *Social Research: An International Quarterly* 77, no. 3 (2010): 970–74.

Galison, Peter, and Bruce Hevly, eds. *Big Science: The Growth of Large-Scale Research.* Stanford, CA: Stanford University Press, 1992.

Geiger, Robert K. "*Seeking Signals in the Sea*: Address at the 30th Anniversary Observation of the MPL." In *SIO Reference No. 97-5.* San Diego, CA: University of California, San Diego, Marine Physical Laboratory of the Scripps Institution of Oceanography, November 11, 1976.

Gelbspan, R. *The Heat Is On: The Climate Crisis, the Cover-Up, the Prescription.* Cambridge, MA: Perseus Press, 1998.

Gemmell, Edgar M. "Harry Hess's War." *Princeton Alumni Weekly* 87, no. 17 (1987): 76.

German, C. R., E. T. Baker, and G. Klinkhammer. "Regional Setting of Hydrothermal Activity." In *Hydrothermal Vents and Processes*, edited by L. M. Parson, C. L. Walker, and D. R. Dixon, 3–15. London: Geological Society, 1995.

German, C. R., L. M. Parson, and R. A. Mills. "Mid-Ocean Ridges and Hydrothermal Activity." In *Oceanography: An Illustrated Guide*, edited by C. P. Summerhayes and S. A. Thorpe, 152–64. New York: John Wiley & Sons, 1996.

Gibbons, Michael, Camille Limoges, Helga Nowotny, Simon Schwartzman, Peter Scott, and Martin Trow. *The New Production of Knowledge: The Dynamics of Science and Research in Contemporary Societies.* Thousand Oaks, CA: Sage Publications, 1994.

Gieryn, Thomas F. "Boundary-Work and the Demarcation of Science from Non-Science: Strains and Interests in Professional Ideologies of Scientists." *American Sociological Review* 48, no. 6 (1983): 781–95.

———. *Cultural Boundaries of Science: Credibility on the Line.* Chicago: University of Chicago Press, 1999.

Gillispie, Charles Coulston. "Science and Secret Weapons Development in Revolutionary France." *Historical Studies in the Physical Sciences* 23, no. 1 (1992): 135–52.

Girdler, R. W. "The Formation of New Oceanic Crust." *Philosophical Transactions of the Royal Society of London A: Mathematical, Physical and Engineering Sciences* 258, no. 1088 (1965): 123–36.

Glantz, Stanton A., John Slade, Lisa A. Bero, Peter Hanauer, and Deborah E. Barnes. *The Cigarette Papers.* Berkeley: University of California Press, 1998.

Glass, B., D. B. Ericson, B. C. Heezen, N. D. Opdyke, and J. A. Glass. "Geomagnetic Reversals and Pleistocene Chronology." *Nature* 216 (1967): 437–42.

Glen, William. *The Road to Jaramillo: Critical Years of the Revolution in Earth Science.* Stanford, CA: Stanford University Press, 1982.

Godin, Oleg, and David R. Palmer. *History of Russian Underwater Acoustics.* Singapore: World Scientific, 2008.

Godlee, Fiona, Ruth Malone, Adam Timmis, Catherine Otto, Andy Busy, Ian Pavord, and Trish Groves "Journal Policy on Research Funded by the Tobacco Industry." *British Medical Journal* 347: f5193 (2013).

Goldberg, Edward D. "Accumulation of Radioiodine in the Thyroid Gland of Elasmobranches." *Endocrinology* 48 (1951): 485–88.

———. "Oceans as Waste Space: The Argument." *Oceanus* 24, no. 1 (1981): 2–9.

Goldberg, Edward D., and Douglas L. Inman. "Neutron Induced Quartz as a Tracer of Sand Movements." *Bulletin of the Geological Society of America* 66 (1955): 611–13.

Goodman, Sherri Wasserman. "The Environment and National Security." Speech at the National Defense University, Washington, DC, August 1996.

Gould, John. "James Rennell's View of Atlantic Circulation." *Ocean Challenge* 4, nos. 1–2 (1993): 26–32.

Graham, Loren. *Lysenko's Ghost: Epigenetics and Russia.* Cambridge, MA: Harvard University Press, 2016.

———. *What Have We Learned about Science and Technology from the Russian Experience.* Stanford, CA: Stanford University Press, 1998.

Grann, David. "The Squid Hunter." *New Yorker*, March 24, 2004, 57–71.

Grassle, J. F. "The Ecology of Deep-Sea Hydrothermal Vent Communities." *Advances in Marine Biology* 23 (1986): 301–62.

———. "Hydrothermal Vent Animals: Distribution and Biology." *Science* 229 (1985): 713–17.

Grassle, J. F., C. J. Berg, J. J. Childress, R. R. Hessler, H. J. Jannasch, D. M. Karl, R. A. Lutz, et al. "Galapagos '79: Initial Findings of a Biology Quest." *Oceanus* 22 (1979): 2–10.

Greatbatch, R. J., and J. Lu. "Reconciling the Stommel Box Model with the Stommel-Arons Model: A Possible Role for Southern Hemisphere Wind Forcing." *Journal of Physical Oceanography* 33, no. 8 (2003): 1618–32.

Greenberg, Daniel. *The Politics of Pure Science.* New York: World Publishing, 1967.

———. *Science, Money and Politics: Political Triumph and Ethical Erosion.* Chicago: University Chicago Press, 2001.

Greene, Mott T. *Alfred Wegener: Science, Exploration, and the Theory of Continental Drift.* Baltimore: Johns Hopkins University Press, 2015.

———. *Geology in the Nineteenth Century: Changing Views of a Changing World.* Ithaca, NY: Cornell University Press, 1985.

Griggs, D. T. "Creep of Rocks." *Journal of Geology* 47 (1939): 225–51.

———. "A Theory of Mountain Building." *American Journal of Science* 237, no. 9 (1939): 611–50.

Groves, Leslie R. *Now It Can Be Told: The Story of the Manhattan Project.* New York: Harper & Brothers, 1962.

Gudemann, Frances. "Dr. Joanne Malkus: Cloud Explorer." *Science World*, April 20, 1960, 28.

Guerrini, Anita. *Experimenting with Humans and Animals: From Galen to Animal Rights.* Baltimore: Johns Hopkins University Press, 2003.

Gummett, Philip, and Judith Reppy. *The Relations between Defence and Civil Technologies.* Dordrecht, The Netherlands: Kluwer Academic Publishers, 1988.

Gusterson, Hugh. *Nuclear Rites: A Weapons Laboratory at the End of the Cold War.* Berkeley: University of California Press, 1996.

———. "Secrecy and Science Revisited: From Politics to Historical Practice and

Back." In *Secrecy and Knowledge Production*, edited by J. Reppy, 57–76. Ithaca, NY: Cornell University, Peace Studies Program, 1999.

Guston, David H. *Between Politics and Science: Assuring the Integrity and Productivity of Research*. Cambridge: Cambridge University Press, 2000.

Haack, Susan. "Scientific Secrecy and 'Spin': The Sad, Sleazy Saga of the Trials of Remune." *Law and Contemporary Problems* 69, no. 3 (2006): 47–67.

Hacker, Barton. "Military Patronage and the Geophysical Sciences in the United States: An Introduction." *Historical Studies in the Physical and Biological Sciences* 30 (2000): 309–13.

Hacking, Ian. "Weapons Research and the Form of Scientific Knowledge." In *Nuclear Weapons, Deterrence, and Disarmament*, edited by David Copp, 237–60. Calgary, AB: University of Calgary Press, 1986.

Hagen, Amelia Ann. "History of Low-Level Radioactive Waste Disposal." *Environmental Science and Technology* (1983): 47–64.

Hajj, George A., C. O. Ao, B. A. Iijima, D. Kuang, E. R. Kursinski, A. J. Mannucci, T. K. Meehan, L. J. Romans, M. de La Torre Juarez, and T. P. Yunck. "CHAMP and SAC-C Atmospheric Occultation Results and Intercomparisons." *Journal of Geophysical Research: Atmospheres* 109, no. D6 (2004).

Hall, A. Rupert. "Engineering and the Scientific Revolution." *Technology and Culture* 2, no. 4 (1961): 333–41.

Hall, Cargill R. *Lunar Impact: A History of Project Ranger*. Washington, DC: National Aeronautics and Space Administration, 1977.

Hamblin, Jacob. *Arming Mother Nature: The Birth of Catastrophic Environmentalism*. New York: Oxford University Press, 2013.

———. "Let There Be Light … and Bread: The United Nations, the Developing World, and Atomic Energy's Green Revolution." *History and Technology* 25, no. 1 (2009): 25–48.

———. "The Navy's 'Sophisticated' Pursuit of Science." *Isis* 93 (2002): 1–27.

———. *Oceanographers and the Cold War: Disciples of Marine Science*. Seattle: University of Washington Press, 2005.

———. *Poison in the Well: Radioactive Waste in the Oceans at the Dawn of the Nuclear Age*. New Brunswick, NJ: Rutgers University Press, 2008.

Hamilton, G. R. "Time Variation of Sound Speeds over Long Paths in the Ocean." In *International Workshop on Low-Frequency Propagation and Wave Noise, Woods Hole, Massachusetts*. Washington, DC: Department of the Navy, 1977.

Hannington, Mark D., Ian R. Jonasson, Peter M. Herzig, and Sven Petersen. "Physical and Chemical Processes of Seafloor Mineralization at Mid-Ocean Ridges." *American Geophysical Union Geophysical Monograph* 91 (1995): 115–57.

Hansen, James E. "Scientific Reticence and Sea Level Rise." *Environmental Research Letters* 2, no. 2 (2007): 024002.

———. "Thou Shalt Not Mess with the Global Environment; Hot Summers in 1990s." *New York Times*, January 11, 1989. http://www.nytimes.com/1989/01/11/opinion/l-thou-shalt-not-mess-with-the-global-environment-hot-summers-in-1990-s-524589.html.

Hansen, James, Makiko Sato, Paul Hearty, Reto Ruedy, Maxwell Kelley, Valerie Masson-Delmotte, Gary Russel, et al. "Ice Melt, Sea Level Rise and Superstorms: Evidence from Paleoclimate Data, Climate Modelings, and Modern Observations that 2°C Global Warming Could Be Dangerous." *Atmospheric Chemistry and Physics* 16 (2016): 3761–3812.

Harper, Kristine C. *Weather by the Numbers: The Genesis of Modern Meteorology.* Cambridge, MA: MIT Press, 2012.

Hart, Stanley R., and Hubert Staudigel. "The Control of Alkalis and Uranium in Sea Water by Ocean Crust Alteration." *Earth and Planetary Science Letters* 58 (1982): 202–12.

Hayes, James F., and B. C. Heezen. "Quaternary Stratigraphy of Antarctic Ocean Bottom Sediments Based on Radiolaria." *Special Papers of Geological Society of America*, no. 76 (1964): 76–77.

Hayes, John D. "Aircraft Salvage Operation Mediterranean: Lessons and Implications for the Navy. Executive Summary of the Final Report Prepared for the Secretary of the Navy and the Chief of Naval Operation." Part 1. Washington, DC: Department of the Navy, 1967.

Haymon, R. M. "Hydrothermal Deposition on the East Pacific Rise at 21°N." PhD diss., University of California, San Diego, 1982.

Haymon, R. M., D. J. Fornari, K. L. von Damm, M. D. Lilley, M. R. Perfit, J. M. Edmond, W. C. Shanks III, R. A. Lutz, J. M. Grebmeier, S. Carbotte, D. Wright, E. McLaughlin, M. Smith, N. Beedle, and E. Olson. "Volcanic Eruption of the Mid-Ocean Ridge along the East Pacific Rise Crust at 9°45–52'N: Direct Submersible Observations of Seafloor Phenomena Associated with an Eruption Event in April, 1991." *Earth and Planetary Science Letters* 119 (1993): 85–101.

Haymon, R. M., R. A. Koski, and C. Sinclair. "Fossils of Hydrothermal Vent Worms from Cretaceous Sulfide Ores of the Samail Ophiolite, Oman." *Science* 223 (1984): 1407–9.

Healy, David I. "Conflicting Interests in Toronto: Anatomy of a Controversy at the Interface of Academia and Industry." *Perspectives in Biology and Medicine* 45, no. 2 (2002): 250–63.

Heath, G. R., and J. Dymond. "Genesis and Diagenesis of Metalliferous Sediments from the East Pacific Rise, Bauer Deep and Central Basin, Northwest Nazca Plate." *Geological Society of America Bulletin* 88, no. 5 (1977): 723.

Heezen, B. C. "The Deep-Sea Floor." In *Continental Drift*, edited by S. K. Runcorn, 235–88. New York: Academic Press, 1962.

———. "Deep-Sea Physiographic Provinces and Crustal Structure." *Transactions of the American Geophysical Union* 38 (1957): 394.

———. "Dynamic Processes of Abyssal Sedimentation: Erosion, Transportation, and Redeposition on the Deep-Sea Floor." *Geophysical Journal of the Royal Astronomical Society* 2 (1959): 142–63.

———. "Geologie sous-marine et deplacements des continents." In *La topographie et la geologie des profondeurs oceaniques*, 295–304. Paris: Centre National de la Recherche Scientifique, 1959.

————. "The Gulf Stream: A Physical and Dynamical Description." *Geographical Review* 50, no. 1 (1960): 144–45.

————. "Notes." *Scientist* 48, no. 3 (1960): 286A–287A.

————. "Review of Physics of the Earth's Interior." *Science* 131 (1960): 1216.

————. "Reviews and Abstracts." *Transactions, American Geophysical Union* 38, no. 3 (1957): 394.

————. "The Rift in the Ocean Floor." *Scientific American* 203 (1960): 99–110.

————. "The World Rift System." In *East African Rift System; Upper Mantle Committee-UNESCO Seminar, Nairobi, April 1965*, edited by UMC/UNESCO Seminar on the East African Rift System, 116–18. Nairobi: University College, 1965.

Heezen, B. C., E. T. Bunce, Brackett Hersey, and Marie Tharp. "Chain and Romanche Fracture Zones." *Deep-Sea Research* 11 (1964): 11–33.

Heezen, B. C., Roberta Coughlin, and Walter C. Beckman. "Equatorial Atlantic Mid-Ocean Canyon." *Geological Society of America Bulletin* 71 (1960): 1886.

Heezen, B. C., and Maurice Ewing. "The Mid-Oceanic Ridge." In *The Sea: Ideas and Observations on Progress in the Study of the Seas*, edited by M. N. Hill, 388–410. New York: John Wiley & Sons, 1963.

————. "The Mid-Oceanic Ridge and Its Extension through the Artic Basin." In *Geology of the Arctic*, edited by Gilbert O. Raash, 622–42. Toronto: University of Toronto Press, 1960.

————. "Turbidity Currents and Submarine Slumps, and the 1929 Grand Banks Earthquake." *American Journal of Science* 250 (1952): 849–73.

Heezen, B. C., Maurice Ewing, and D. B. Ericson. "Submarine Topography in the North Atlantic." *Bulletin of the Geological Society of America* 62 (1951): 1407–9.

Heezen, B. C., Maurice Ewing, and Edward Titus Miller. "Trans-Atlantic Profile of Total Magnetic Intensity and Topography, Dakar to Barbados." *Deep-Sea Research* 1 (1953): 25–33.

Heezen, B. C., and Paul J. Fox. "Mid-Oceanic Ridge." In *The Encyclopedia of Oceanography*, edited by Rhodes W. Fairbridge, 506–17. New York: Reinhold Publishing, 1966.

Heezen, B. C., R. D. Gerard, and Marie Tharp. "The Vema Fracture Zone in the Equatorial Atlantic." *Journal of Geophysical Research* 69, no. 4 (1964): 733–39.

Heezen, B. C., and Charles Hollister. "Deep-Sea Current Evidence from Abyssal Sediments." *Marine Geology* 1 (1964): 141–74.

————. *The Face of the Deep*. New York: Oxford University Press, 1971.

Heezen, Bruce C., Charles D. Hollister, and William F. Ruddiman. "Shaping of the Continental Rise by Deep Geostrophic Contour Currents." *Science* 152 (1966): 502–8.

Heezen, B. C., and C. L. Johnson. "Bathymetry." In *The Encyclopedia of Oceanography*, edited by Rhodes W. Fairbridge, 104–10. New York: Reinhold Publishing, 1966.

Heezen, B. C., and Henry W. Menard. "Topography of the Deep-Sea Floor." In *The Sea: Ideas and Observations on Progress in the Study of the Seas*, edited by M. N. Hill, 233–80. New York: John Wiley & Sons, 1963.

Heezen, B. C., and John Northrop. "An Outcrop of Eocene Sediment on the Continental Slope." *Journal of Geology* 59 (1950): 396–99.

Heezen, B. C., and Marie Tharp. *Physiographic Diagram of the Indian Ocean: The Red
Sea, the South China Sea, the Sulu Sea, and the Celebes Sea*. New York: Geological
Society of America, 1964.

———. "A Symposium on Continental Drift." *Philosophical Transactions of the Royal
Society* 258 (1965): 90–106.

———. "Tectonic Fabric of the Atlantic and Indian Oceans and Continental Drift."
Philosophical Transactions of the Royal Society of London 258, no. 1088 (1965): 90–
106.

Heezen, Bruce C., Marie Tharp, and Maurice Ewing. *The Floors of the Oceans*. New
York: Geological Society of America, 1959.

Heirtzler, J. R. "Project Famous." *Reviews of Geophysics* 13, no. 3 (1975): 542–42.

———. "Understanding the Mid-Atlantic Ridge." *Proceedings of the National Academy
of Sciences* (1972): 129.

Heirtzler, J. R., and X. Le Pichon. "FAMOUS: A Plate Tectonics Study of the Genesis
of the Lithosphere." *Geology* 2 (1974): 273–74.

Heirtzler, J. R., and Tjeerd H. van Andel. "Project Famous: Its Origins, Programs and
Setting." *Geological Society of America Bulletin* 88 (1977): 481–87.

Heiskanen, Weikko. "The Earth's Gravity." *Scientific American* 193, no. 3 (1955):
164–73.

Hekinian, R., M. Chaigneau, and J. L. Cheminee. "Popping Rocks and Lava Tubes
from the Mid-Atlantic Rift Valley at 36°N." *Nature* 245 (1974): 371–73.

Hekinian, R., M. Fevrier, J. L. Bischoff, P. Picot, and W. C. Shanks. "Sulfide Deposits
from the East Pacific Rise Near 21°N." *Science* 207, no. 4438 (1980): 1433–44.

Hendrickson, Walter. "Nineteenth-Century State Geological Surveys: Early Government Support of Science." *Isis* 52, no. 3 (1961): 357–71.

Herdman, Sir William. *Founders of Oceanography and Their Work*. London: Edward
Arnold & Co., 1923.

Herman, Louis M. "Hawaiian Humpback Whales and ATOC: A Conflict of Interests." *Journal of Environment & Development* 3, no. 2 (1994): 63–76.

Hersey, J. B. "Introduction: Low-Frequency Propagation and Noise Workshop."
From the International Workshop on Low-Frequency Propagation and Noise,
Woods Hole, MA, October 14–19, 1974.

Hersey, Brackett, and George Backus. "New Evidence That Migrating Gas Bubbles,
Probably the Swim Bladders of Fish, Are Largely Responsible for Scattering Layers
on the Continental Rise South of New England." *Deep-Sea Research* 1 (1954): 190–91.

———. "Sound Scattering by Marine Organisms." In *The Sea: Ideas and Observations
on Progress in the Study of the Seas*, edited by M. N. Hill. New York: John Wiley &
Sons, 1962.

Hersey, J. B., and R. H. Backus. "Sound Scattering by Marine Organisms." In *The Sea:
Ideas and Observations on Progress in the Study of the Seas*, edited by M. N. Hill, 398–
439. New York: John Wiley & Sons, 1962.

Hess, H. H. "Chemical Composition and Optical Properties of Common Clino-Pyroxenes, Part 1." *American Mineralogist* 34 (1949): 621–66.

———. "Drowned Ancient Islands of the Pacific Basin." *American Journal of Science* 244 (1946): 772–91.

———. "An Essay Review: The Petrology of the Skaergaard Intrusion Kangerdlugssuaq, East Greenland." *American Journal of Science* 238 (1940): 372–78.

———. "Extreme Fractional Crystallization of a Basaltic Magma: The Stillwater Igneous Complex." *Eos, Transactions, American Geophysical Union* (1939): 430–32.

———. "Further Discussion on Submerged Canyons." *Science* 85 (1936): 583.

———. "Geodesy and Geomorphology (2) a New Bathymetric Chart of the Caribbean Area." *Advanced Report of the Commission on Continental and Oceanic Structure* (1939): 29–32.

———. "Geological Interpretation of Data Collected on Cruise of USS *Barracuda* in the West Indies—Preliminary Report." *Eos, Transactions, American Geophysical Union* (1937): 69–77.

———. "Gravity Anomalies and Island Arc Structure with Particular Reference to the West Indies-Preliminary Report." *Proceedings of the American Philosophical Society* 79, no. 1 (1938): 71–96.

———. "History of the Ocean Basins." In *Petrologic Studies: A Volume to Honor A. F. Buddington*, edited by A. E. J. Engel, Harold L. James, and B. F. Leonard. Denver, CO: Geologic Society of America, 1962.

———. "Major Structural Features of the Western North Pacific, an Interpretation of HO 5485, Bathymetric Chart, Korea to New Guinea." *Bulletin of the Geological Society of America* 59 (1948): 417–46.

———. "Major Structural Trends of the Western North Pacific." *Transactions of the New York Academy of Sciences* 9, no. 2 (1947): 245–46.

———. "The Moore County Meteorite: A Further Study with Comment on Its Primordial Environment." *American Mineralogist* 34 (1949): 494–507.

———. *The Navy-Princeton Gravity Expedition to the West Indies in 1932*. Washington, DC: US Government Printing Office, 1933.

———. "Optical Properties and Chemical Composition of Magnesian Orthopyroxenes." *American Mineralogist* 25 (1940): 271–85.

———. "Orthopyroxenes of the Bushveld Type." *American Mineralogist* 23, no. 7 (1938): 450–56.

———. "Orthopyroxenes of the Bushveld Type, Ion Substitutions and Changes in Unit Cell Dimensions." *American Journal of Science* (1952): 173–87.

———. "Part III Geodesy and Geomorphology (2) a New Bathymetric Chart of the Caribbean Area." In *Advanced Report of the Commission on Continental and Oceanic Structure*, 29–32. Princeton, NJ: Princeton University, 1940.

———. "Part IV Gravity at Sea (2) Recent Advances in Interpretation of Gravity-Anomalies and Island-Arc Structure." *Proceedings of the American Geophysical Union* 18 (1938): 71–96.

———. "Primary Banding in Norite and Gabbro." *Transactions of the American Geophysical Union* (1938): 265–68.

———. "A Primary Peridotite Magma." *American Journal of Science* 35 (1938): 322–44.

———. "A Primary Ultramafic Magma." *Eos, Transactions, American Geophysical Union* (1937): 248–49.

———. "The Problem of Serpentinization." *Economic Geology* 30, no. 3 (1935): 320–25.

———. "Pyroxenes of Common Mafic Magmas (Part 1 and 2)." *American Mineralogist* 26 (1941): 515–35.

———. "Section of Geology and Mineralogy." *Transactions of the New York Academy of Sciences* 9, no. 7 (1947): 245–46.

———. "Seismic Anisotropy of the Uppermost Mantle under Oceans." *Nature* 203, no. 4945 (1964): 629–31.

———. "Submerged Valleys on Continental Slopes and Changes of Sea Level." *Science* 83, no. 2153 (1936): 332–34.

———. "Vertical Mineral Variation in the Great Dyke of Southern Rhodesia." *Transactions of the Geological Society of South Africa* 53 (1950): 159–66.

Hess, H. H., and M. W. Buell Jr. "The Greatest Depth in the Oceans." *Transactions of the American Geophysical Union* 31, no. 3 (1950): 401–5.

Hess, H. H., and A. H. Phillips. "Orthopyroxenes of the Bushveld Type." *American Mineralogist* 23, no. 7 (1938): 450–56.

Hessen, B. "The Social and Economic Roots of Newton's 'Principia,'" *Science at the Cross Roads: Papers Presented to the International Congress of the History of Science and Technology, Held in London from June 29–July 3, 1931 by the Delegates of the USSR*, 2nd ed.,152–212. London: Frank Cass & Co., 1971.

Hessler, Robert R., and Victoria Kaharl. "The Deep-Sea Hydrothermal Vent Community: An Overview." In *Seafloor Hydrothermal Systems: Physical, Chemical, Biological, and Geological Interactions*, edited by S. E. Humphris, R. A. Zeirenberg, and S. J. Mullineaux, 72–84. Geophysical Monographs. Washington, DC: American Geophysical Union, 1995.

Hessles, Lauren K., and Harro van Lente. "Re-thinking New Knowledge Production: A Literature Review and a Research Agenda." *Research Policy* 37, no. 4 (2008): 740–60.

Hevly, Bruce. "The Tools of Science: Radio, Rockets, and the Science of Naval Warfare." In *National Military Establishments and the Advancement of Science and Technology*, edited by Paul Forman and José M. Sánchez-Ron, 215–32. Dordrecht, The Netherlands: Kluwer Academic Publishers, 1996.

Hewlett, Richard, and Oscar Anderson Jr. *The New World: A History of the United States Atomic Energy Commission*. Vol. 1. Berkeley: University of California Press, 1990.

Hill, J. Warren. *"Afternoon Effect" Studies, Part 1*. RANRL Technical Note 3/82, Australia Department of Defense, April 1983.

Hill, M. N., ed. *The Sea: Ideas and Observations on Progress in the Study of the Seas*. Vol. 1, *Physical Oceanography*. New York: John Wiley & Sons, 1962.

———. *The Sea: Ideas and Observations on Progress in the Study of the Seas*. Vol. 3, *The Earth beneath the Sea History*. New York: John Wiley & Sons, 1963.

Hoffert, M. A., A. Perseil, R. Hekinian, P. Choukroune, H. D. Needham, J. Franche-
    teau, and X. Le Pichon. "Hydrothermal Deposits Sampled by Diving Saucer in
    Transform Fault 'A' near 37°N on the Mid-Atlantic Ridge, Famous Area." *Oceano-
    logica Acta* 1 (1978): 72–86.
Hoggan, James, and Richard Littlemore. *Climate Cover-Up: The Crusade to Deny Global
    Warming*. Vancouver, BC: Greystone Books, 2009.
Holcomb, W. F. "A History of Ocean Disposal of Packaged Low-Level Radioactive
    Waste." *Nuclear Safety* 23, no. 2 (1982): 183–97.
Holland, M. M., and C. M. Bitz. "Polar Amplification of Climate Change in Coupled
    Models." *Climate Dynamics* 21 (2003): 221–32.
Hollinger, David A. "Free Enterprise and Free Inquiry: The Emergence of Laissez-
    Faire Communitarianism in the Ideology of Science in the United States." In
    Science, Jews, and Secular Culture, 97–120. Princeton, NJ: Princeton University
    Press, 1996.
———. "Money and Academic Freedom a Half-Century after McCarthyism." In
    Unfettered Expression: Freedom in American Intellectual Life, edited by Peggy J.
    Hollingsworth. Ann Arbor: University of Michigan Press, 2000.
———. *Science, Jews, and Secular Culture: Studies in Mid-Twentieth-Century American
    Intellectual History*. Princeton, NJ: Princeton University Press, 1996.
Hollister, Charles D. "The Concept of Deep-Sea Contourites." *Sedimentary Geology* 82,
    no. 1 (1993): 5–11.
———. "The Seabed Option." *Oceanus* 20, no. 1 (1977): 18–25.
Hollister, Charles D., D. R. Anderson, and G. R. Heath. "Sub-seabed Disposal of Nu-
    clear Wastes." *Science* (1981): 1321–26.
Hollister, Charles D., and B. C. Heezen. "Ocean Bottom Currents." In *The Encyclope-
    dia of Oceanography*, edited by Rhodes W. Fairbridge, 576–83. New York: Rein-
    hold Publishing, 1966.
Hollister, Charles D., and Steven Nadis. "Burial of Radioactive Waste under the Sea-
    bed." *Scientific American* 278, no. 1 (1998): 60–65.
Hollister, Charles D., Armand Silva, and Alan Driscoll. "A Giant Piston-Corer." *Ocean
    Engineering* 2, no. 4 (1973): 159.
Holmes, Arthur. "Radioactivity and Earth Movements." *Transactions of the Geological
    Society of Glasgow* 18 (1929): 559–606.
Hornig, Donald. *Undersea Vehicles for Oceanography*. ICO Pamphlet 18, Federal Coun-
    cil for Science and Technology US Committee on Oceanography. Washington,
    DC: US Government Printing Office, 1965.
Hoshino, Michihei. *The Expanding Earth: Evidence, Causes, and Effects*. Tokyo: Tokai
    University Press, 1998.
Houghton, John T, et al., eds. *Climate Change 1994: Radiative Forcing of Climate Change
    and an Evaluation of the IPCC 1992 IS92 Emission Scenarios*. Cambridge: Cambridge
    University Press, 1995.
Houghton, John T, Y. Ding, D. J. Griggs, M. Noguer, P. J. van der Linden, X. Dai,
    K. Maskell, and C. A. Johnson, eds. *Climate Change 2001: The Scientific Basis*. Cam-
    bridge: Cambridge University Press, 2001.

Hounshell, David. "The Cold War, Rand, and the Generation of Knowledge, 1946–1962." *Historical Studies in the Physical and Biological Sciences* 27 (1997): 237–67.

———. "Edison and the Pure Science Ideal in 19th-Century America." *Science* 207, no. 4431 (1980): 612–17.

———. "Epilogue: Rethinking the Cold War; Rethinking Science and Technology in the Cold War; Rethinking the Social Study of Science and Technology." *Social Studies of Science* 31, no. 2 (2001): 289–97.

Hounshell, David A., and John Kenly Smith Jr. *Science and Corporate Strategy: Du Pont R&D, 1902–1980*. Cambridge: Cambridge University Press, 1988.

Howe, Joshua P. *Behind the Curve: Science and the Politics of Global Warming*. Seattle: University of Washington Press, 2014.

Hubbard, Jennifer. *A Science on the Scales: The Rise of Canadian Atlantic Fisheries Biology, 1898–1939*. Toronto: University of Toronto Press, 2006.

Hubbert, M. Kin. "Strength of the Earth." *Bulletin of the American Association of Petroleum Geologists* 29, no. 11 (1945): 1630–53.

Hughes, Thomas Parke. *American Genesis: A Century of Invention and Technological Enthusiasm, 1870–1970*. New York: Viking, 1989.

Hulburt, E. O. "Propagation of Radiation in a Scattering and Absorbing Medium." *Journal of the Optical Society of America* 33, no. 1 (1943): 42–45.

———. "The Temperature of the Lower Atmosphere of the Earth." *Physical Review* 38 (1931): 1876–90.

Humphris, Susan E., P. M. Herzig, D. J. Miller, J. C. Alt, K. Becker, D. Brown, G. Brügmann, et al. "The Internal Structure of an Active Sea-Floor Massive Sulphide Deposit." *Nature* 377, no. 6551 (1995): 713–16.

Inkpen, Danielle Hallet. "Frozen Icons: The Science and Politics of Repeat Glacier Photographs, 1887–2010." PhD diss., Harvard University, 2018.

Inman, Douglas. *Oral History Project of the H. John Heinz III Center for Science, Economics and the Environment*. La Jolla, CA: Scripps Institution of Oceanography, 2000.

Intergovernmental Panel on Climate Change. *IPCC Second Assessment: Climate Change 1995*. New York: United Nations, 1995.

Irving, E. "The Mid-Atlantic Ridge at 45°N. XIV. Oxidation and Magnetic Properties of Basalt; Review and Discussion." *Canadian Journal of Earth Sciences* 7 (1970): 1528–38.

Irving, E., J. K. Park, S. E. Haggerty, F. Aumento, and B. Loncarevic. "Magnetism and Opaque Mineralogy of Basalts from the Mid-Atlantic Ridge." *Nature* 228 (1970): 974–76.

Isaacs, J. D., and W. R. Schmitt. "Ocean Energy: Forms and Prospects." *Science* (1980): 265–73.

Iselin, Columbus O'Donnel. "National Defense Research Committee: Summary Technical Report of Division 6." In *The Application of Oceanography to Subsurface Warfare*. Washington, DC: Office of Scientific Research and Development, 1946.

Iselin, Gary E. *Sound Transmission in Sea Water: A Preliminary Report*. NDRC, Entry 29, 752. Woods Hole Oceanographic Institution, February 1, 1941.

Jackson, Jeremy. *Shifting Baselines: The Past and the Future of Ocean Fisheries*. Washington, DC: Island Press, 2011.

James, Harold L. "Harry Hammond Hess." In *Biographical Memoirs*, edited by National Academy of Sciences, 109–28. New York: Columbia University Press, 1973.

Jammer, Max. *The Philosophy of Quantum Mechanics: The Interpretations of Quantum Mechanics*. New York: John Wiley & Sons, 1974.

Jannasch, Holger W. "The Nutritional Basis for Life at Deep-Sea Vents." *Oceanus* 37, no. 3 (1984): 73–77.

Jasanoff, Sheila. *The Fifth Branch: Science Advisers as Policymakers*. Cambridge, MA: Harvard University Press, 1990.

———. "Transparency in Public Science: Purposes, Reasons, and Limits. *Law and Contemporary Problems* 69, no. 3 (2006): 21–45.

Jeffreys, J. Gwyn. "The Deep-Sea Dredging Expedition in H.M.S. *Porcupine*." *Nature* 1 (1869): 166–68.

Jewell, Ralph. "Vilhelm Bjerknes's Duty to Produce Something Clear and Real in Meteorological Science." In *The Earth, the Heavens, and the Carnegie Institution of Washington, History of Geophysics* 5, edited by G. Good, 37–46. Washington, DC: American Geophysical Union, 1994.

Johnson, Douglas. *The Origin of Submarine Canyons: A Critical Review of Hypotheses*. New York: Columbia University Press, 1939.

Johnson, G. Leonard, and B. C. Heezen. "The Artic Mid-Oceanic Ridge." *Nature* 215 (1967): 724–25.

Johnson, R. H. "Synthesis of Point Data and Path Data in Estimating SOFAR Speed." *Journal of Geophysical Research* 74 (1969): 4559–70.

Johnson, Stephen. "Sub Sank in 1968 after Skimpy Last Overhaul/USS *Scorpion* Was Lost with All on Board." *Houston Chronicle*, May 5, 1995.

Jones, Meredith L. "The Giant Tube Worms." *Oceanus* 27, no. 3 (1984): 47–52.

———. "Introduction." *Biological Society of Washington Bulletin* 6 (1985): 1–2.

———. "On the Vestimentifera, New Phylum: Six New Species and Other Taxa, from Hydrothermal Vents and Elsewhere." *Biological Society of Washington Bulletin* 6 (1985): 117–58.

Joravsky, David. *The Lysenko Affair*. Chicago: University of Chicago Press, 1970.

Joseph, Arnold B. *Report of Meeting on Ocean Disposal of Reactor Wastes*. Woods Hole, MA: Johns Hopkins University, 1955.

Kaharl, Victoria. "Physical and Chemical Processes of Seafloor Mineralization at Mid-Ocean Ridges." *American Geophysical Union, Geophysical Monograph* 91 (1995): 115–57.

———. *Water Baby: The Story of Alvin*. New York: Oxford University Press, 1990.

Kaiser, David. "Cold War Requisitions, Scientific Manpower, and the Production of American Physicists after World War II." *Historical Studies in the Physical and Biological Sciences* 33 (2002): 131–59.

Kamlet, Kenneth. "The Oceans as Waste Space: The Rebuttal." *Oceanus* 24, no. 1 (1981): 10.

Kargon, Robert, and Elizabeth Hodes. "Karl Compton, Isaiah Bowman, and the Politics of Science in the Great Depression." *Isis* 76 (1985): 301–18.

Kay, Lily E. *Who Wrote the Book of Life? A History of the Genetic Code*. Stanford, CA: Stanford University Press, 2000.

Kay, Marshall. "The Origin of Continents." *Scientific American* 193, no. 3 (1955): 62–66.

Keeling, Charles D. "Rewards and Penalties of Monitoring the Earth." *Annual Review of Energy and the Environment* 23, no. 1 (1998): 25–82.

Keller, Evelyn Fox. *The Century of the Gene*. Cambridge, MA: Harvard University Press, 2009.

Kelley, D. S., J. A. Baross, S. E. Hoffman, and J. R. Delaney. "Volcanoes, Fluids, and Life at the Mid-Ocean Ridge Spreading Centers." *Annual Review of Earth and Planetary Sciences* 30 (2002): 385–491.

Kennish, M. J., R. A. Lutz, and B. R. T. Simoneit. "Hydrothermal Activity and Petroleum Generation in the Guaymas Basin." *Reviews in Aquatic Science* 6 (1992): 467–77.

Kerker, Milton. "Science and the Steam Engine." *Technology and Culture* 2, no. 4 (1961): 381–90.

Kevles, Daniel J. "Cold War and Hot Physics: Science, Security and the American State, 1945–1956." *Historical Studies in the Physical and Biological Sciences* 20 (1990): 239–64.

———. "The National Science Foundation and the Debate over Postwar Research Policy, 1942–1945." *Isis* 68 (1977): 5–26.

———. *The Physicists: The History of a Scientific Community in Modern America*. 2nd ed. Cambridge, MA: Harvard University Press, 1995.

Kidd, Charles V. "Basic Research—Description versus Definition." *Science* 129 (1959): 368–71.

———. "Basic Research—Description versus Definition." In *Science and Society*, edited by Norman Kaplan, 146–55. New York: Arno Press, 1975.

King, L. C. "The Origin and Significance of the Great Sub-Oceanic Ridges." In *Continental Drift: A Symposium*, edited by S. W. Carey, 62–102. Hobart: University of Tasmania, 1958.

Kinsler, Lawrence E., and Austin R. Frey. *Fundamentals of Acoustics*. 2nd ed. New York: John Wiley and Sons, 1962.

Klimley, A. Peter, and Sallie C. Beavers. "Playback of Acoustic Thermometry of Ocean Climate (ATOC)-Like Signal to Bony Fishes to Evaluate Phonotaxis." *Journal of the Acoustical Society of America* 104, no. 4 (1998): 2506–10.

Kline, Ronald. "Constructing 'Technology' as 'Applied Science': Public Rhetoric of Scientists and Engineers in the United States 1880–1945." *Isis* 86 (1995): 194–221.

Klitgord, K. D., and J. D. Mudie. "The Galapagos Spreading Centre: A Near-Bottom Geophysical Study." *Geophysical Journal of the Royal Astronomical Society* 38 (1974): 563–86.

Knoll, A. H. "The Archean." *Science* 239 (1988): 199–200.

Knorr-Cetina, Karin. *The Manufacture of Knowledge: An Essay on the Constructivist and Contextual Nature of Science*. Oxford, UK: Pergamon Press, 1981.

Knowlton, Nancy, and Jeremy Jackson. "Shifting Baselines, Local Impacts, and Global Change on Coral Reefs." *PLoS Biology* 6, no. 2 (2008): e54.

Kohler, Robert E. *Landscapes and Labscapes: Exploring the Lab-Field Border in Biology*. Chicago: University of Chicago Press, 2002.

———. *Partners in Science: Foundations and Natural Scientists*. Chicago: University of Chicago Press, 1991.

Korgen, Ben J. "Higgs Receives 2003 Edward A. Flinn III Award." *Eos, Transactions, American Geophysical Union* 85, no. 8 (2004): 81.

Koski, R. A., et al. "Metal Sulfide Deposits on the Juan De Fuca Ridge." *Oceanus* 25, no. 3 (1982): 42–48.

Krause, D. C., and Henry W. Menard. "Depth Distribution and Bathymetric Classification of Some Sea-Floor Profiles." In *Marine Geology*, 169–93. Amsterdam: Elsevier, 1965.

Krige, John. *American Hegemony and the Postwar Reconstruction of Science in Europe*. Cambridge, MA: MIT Press, 2006.

———. "La science et la securité civile de l'Occident." In *Les sciences pour la guerre 1940–1960*, ed. A. Dahan and D. Pestre, 373–401. Paris: Éditions de l'École des hautes études en sciences sociales, 2004.

Krige, John, and Dominique Pestre, eds. *Science in the Twentieth Century*. New York: Routledge, 2014.

Krimsky, Sheldon. *Science in the Private Interest: Has the Lure of Profits Corrupted Biomedical Research?* Lanham, MD: Rowman and Littlefield, 2004.

Krosnick, Jon. "Replication." Talk at meeting of the Society for Personality and Social Psychology, 2015.

Kuenen, Philip H. *Marine Geology*. New York: Wiley, 1950.

———. "The Negative Isostatic Anomalies in the East Indies (with Experiments)." *Overdrukuit Leidsche Geologische Mededeelingen* 8 (1936): 169–214.

Kuenen, Ph. H., and Henry W. Menard. "Turbidity Currents, Graded and Non-Graded Deposits." *Journal of Sedimentary Petrology* 22, no. 2 (1952): 83–96.

Kuhlbrodt, T. A. Griesel, M. Montoya, A. Levermann, M. Hofman, and S. Rahmstorf. "On the Driving Processes of the Atlantic Meridional Overturning Circulation." *Reviews of Geophysics* 45, no. 2 (2007).

Kuklick, Henrika, and Robert E. Kohler, eds. *Science in the Field*. Vol. 11 of *Osiris*. Chicago: University of Chicago Press, 1996.

Kulp, J. L. "The Carbon 14 Method of Age Determination." *Scientific Monthly* 75, no. 5 (1952): 259–67.

Kundsin, Ruth B., ed. *Women and Success: The Anatomy of Achievement*. New York: William Morrow & Co., 1974.

Kuznick, Peter. *Beyond the Laboratory: Scientists as Political Activists in 1930s America*. Chicago: University of Chicago Press, 1987.

Lajus, Julia. "'Foreign Science in Russian Context': Murman Scientific-Fishery Expedition and Russian Participation in Early ICES Activity." *ICES Marine Science Symposia* (2002).

———. "Linking People through Fish: Science and Barents Sea Fish Resources in the Context of Russian-Scandinavian Relations." *Journal of Historical Geography* 42 (2014): 44–59.

Lakoff, George. "Why It Matters How We Frame the Environment." *Environmental Communication: A Journal of Nature and Culture* 4, no. 1 (2010): 70–81.

Landecker, Hannah. *Culturing Life: How Cells Became Technology*. Cambridge, MA: Harvard University Press, 2007.

Langseth, M. G., X. Le Pichon, and M. Ewing. "Crustal Structure of the Mid-Ocean Ridges, 5, Heat Flow through the Atlantic Ocean Floor and Convection Currents." *Journal of Geophysical Research* 71 (1966): 5321.

Langseth, M. G., and R. P. von Herzen. "Heat Flow through the Floor of the World Oceans." In *The Sea, Part I*, edited by A. E. Maxwell, 299. New York: Wiley-Interscience, 1971.

La Porte, Todd R., and Daniel Metlay. "Technology Observed: Attitudes of a Wary Public." *Science* 188, no. 4184 (1975): 121–27.

Laughton, A. S., and J. S. M. Rusby. "Long Range Sonar and Photographic Studies at the Median Valley in the Famous Area of the Mid-Atlantic Ridge near 37°N." *Deep-Sea Research* 22 (1975): 279–98.

Latour, Bruno. *Science in Action: How to Follow Scientists and Engineers through Society*. Cambridge, MA: Harvard University Press, 1987.

Latour, Bruno, and Steve Woolgar. *Laboratory Life: The Construction of Scientific Facts*. Beverly Hills, CA: Sage Publications, 1979.

Lawrence, David. "Mountains under the Sea." *Mercator's World* 4 (November 1999): 36.

———. *Upheaval from the Abyss: Ocean Floor Mapping and the Earth Science Revolution*. New Brunswick, NJ: Rutgers University Press, 2002.

Lawyer, L. C., Charles C. Bates, and Robert B. Rice, *Geophysics in the Affairs of Mankind: A Personalized History of Exploration Geophysics*. Tulsa, OK: Society of Exploration Geophysicists, 2000.

Lazcano, A., and S. L. Miller. "How Long Did It Take for Life to Begin and Evolve to Cyanobacteria?" *Journal of Molecular Evolution* 39 (1994): 546–54.

Lechevalier, Hubert, and Morris Solotorovsky. *Three Centuries of Microbiology*, 260–74. New York: McGraw-Hill, 1965.

Lederman, Leon M. "The Advancement of Science." *Science* 256 (1992): 1123.

Lee, W. H. K., and S. Uyeda. "Review of Heat Flow." In *Terrestrial Heat Flow*, Geophysical Monograph No. 8. Washington, DC: American Geophysical Union, 1965.

Legett, J. *The Carbon War: Global Warming and the End of the Oil Era*. New York: Routledge, 2001.

Le Grand, Homer. *Drifting Continents and Shifting Theories: The Modern Revolution in Geology and Scientific Change*. Cambridge: Cambridge University Press, 1988.

Lek, L. "Die Ergebnisse der Strom- und Serienmessungen auf den Ankerstationen." In *The Snellius Expedition in the Eastern Part of the Netherlands East-Indies, 1929–*

1930, under the Leadership of P. M. van Riel, edited by P. M. van Riel, 1938. Leiden: E. J. Brill, 1933.

———. "Tidal Phenomena in the Zuidersee Area During the Time of the Construction of the Closing Dam." PhD diss., Wilhelm Friedrich University, Berlin, 1934.

Lenz, Walter. "The Aspirations of Alfred Merz, Georg Wüst, and Albert Defant: From Berlin to Pacific Oceanography." In *Oceanographic History: The Pacific and Beyond*, edited by Keith R. Benson and Philip F. Rehbock, 118–23. Seattle: University of Washington Press, 2002.

Le Pichon, Xavier. "My Conversion to Plate Tectonics." In *Plate Tectonics: An Insider's History of the Modern Theory of the Earth*, edited by Naomi Oreskes, with Homer Le Grand, 201–26. Boulder, CO: Westview Press, 2001.

Le Pichon, X., and M. G. Langseth. "Heat Flow from Mid-Ocean Ridges and Sea-Floor Spreading." *Tectonophysics* 8 (1969): 319.

Leslie, Stuart W. *Cold War and American Science*. New York: Columbia University Press, 1993.

———. "Science and Politics in Cold War America." In *The Politics of Western Science 1640-1990*, edited by Margaret Jacob, 199–233. Atlantic Highlands, NJ: Humanities Press, 1992.

Lesney, Mark S. "Newton's Hair." *Today's Chemist at Work* (2003): 31–32.

Levin, Sheldon M. "Norwegians Led the Way in Training Wartime Weather Officers." *Eos* 78 (1997): 609–12.

Lewis, Flora. *One of Our H-Bombs Is Missing*. New York: McGraw-Hill, 1967.

Lewis, Sinclair. *Arrowsmith*. New York: Modern Library, 1925.

Lichte, H. "Über den Einfluss horizontaler Temperaturschichtung des Seewassers auf die Reichweite von Unterwasserschallsignalen." *Physikalische Zeitschrift* 17 (1919): 385–89.

Lindseth, Brian. "From Radioactive Fallout to Environmental Critique: Ecology and the Politics of Cold War Science." PhD diss., University of California, San Diego, 2013.

Lipscomb, Richard. *Dive to the Edge of Creation*. London: BBC TV.

Lipton, Peter. *Inference to the Best Explanation*. 2nd ed. New York: Routledge, 1991.

Lister, C. R. B. "On the Penetration of the Water into Hot Rock." *Geophysical Journal of the Royal Astronomical Society*. 39 (1974): 465–509.

———. "On the Thermal Balance of a Mid-Ocean Ridge." *Geophysical Journal of the Royal Astronomical Society* 26 (1972): 515–35.

Lizhi, Fang. *Bringing Down the Great Wall*. Edited by James Williams. New York: W. W. Norton & Co., 1990.

Long, Pamela and Alex Roland. "Military Secrecy in Antiquity and Early Medieval Europe: A Critical Reassessment." *History and Technology* 11 (1994): 259–90.

Longino, Helen E. *Science as Social Knowledge: Values and Objectivity in Scientific Inquiry*. Princeton, NJ: Princeton University Press, 1990.

Lonsdale, Peter. "Abyssal Geomorphology of a Depositional Environment at Exit of the Samoan Passage." PhD diss., University of California, San Diego, Scripps Institution of Oceanography, 1974.

————. "Abyssal Pahoehoe with Lava Coils at the Galapagos Rift." *Geology* 5 (1977): 147–52.

————. "Clustering of Suspension-Feeding Macrobenthos Near Abyssal Hydrothermal Vents at Oceanic Spreading Centers." *Deep-Sea Research* 24 (1977): 857–63.

————. "Deep-Tow Observations at the Mounds Abyssal Hydrothermal Field, Galapagos Rift." *Earth and Planetary Science Letters* 36, no. 1 (1977): 92–110.

————. "Hot Vents and Hydrocarbon Seeps in the Sea of Cortez." *Oceanus* 27, no. 3 (1954): 21–24.

————. "Structural Geomorphology of a Fast-Spreading Rise Crest: The East Pacific near 3°25'S." *Marine Geophysical Researches* 3 (1977): 251–93.

————. "Submersible Exploration of Guaymas Basin: A Preliminary Report of the Gulf of California, 1977 Operations of DSV-4 'Seacliff.'" In *SIO Reference 78-1* (1978).

Lonsdale, P. F., J. L. Bischoff, V. M. Burns, M. Kastner, and R. E. Sweeney. "A High-Temperature Hydrothermal Deposit on the Seabed at the Gulf of California Spreading Center." *Earth and Planetary Science Letters* 49 (1980): 8–20.

Lonsdale, Peter, and W. S. Fyfe. "Ocean Floor Hydrothermal Activity." *The Sea* 7 (1981): 589–638.

Lovell, Bernard. *P. M. S. Blackett: A Biographical Memoir.* London: Royal Society, 1976.

Lowell, R. P., and P. A. Rona. "On the Interpretation of Near-Bottom Water Temperature Anomalies." *Earth and Planetary Science Letters* 32 (1976): 18.

Lowen, Rebecca. *Creating the Cold War University: The Transformation of Stanford.* Berkeley: University of California Press, 1997.

Lucier, Paul. "Geological Industries." In *Cambridge History of Science,* ed. Peter J. Bowler and John V. Pickstone, 114–18. Cambridge: Cambridge University Press, 2009.

————. "The Professional and the Scientist in Nineteenth-Century America." *Isis* 100, no. 4 (2009): 699–732.

Lüdecke, Cornelia, and Colin Summerhayes. *The Third Reich in Antarctica: The German Antarctic Expedition, 1938–39.* Norwich, UK: Erskine Press, 2012.

Lundh, Andreas, J. Lexchin, B. Mintzes, J. B. Schroll, and L. Bero. "Industry Sponsorship and Research Outcome." *Cochrane Database of Systematic Reviews* 2 (2017).

Lupton, J. E., J. R. Delaney, H. P. Johnson, and M. K. Tivey. "Entrainment and Vertical Transport of Deep Ocean Water by Buoyant Hydrothermal Plumes." *Nature* 316 (1985): 621–23.

Lupton, J. E., G. P. Klinkhammer, W. R. Normark, R. Haymon, K. C. MacDonald, R. F. Weiss, and H. Craig. "Helium-3 and Manganese at the 21°N East Pacific Rise Hydrothermal Site." *Earth and Planetary Science Letters* 50, no. 1 (1980): 115–27.

Lupton, J. E., R. F. Weiss, and H. Craig. "Mantle Helium in Hydrothermal Plumes in the Galapagos Rift." *Nature* 267, no. 5612 (1977): 603–4.

Luskin, Bernard, B. C. Heezen, Maurice Ewing, and Mark Landisman. "Precision Measurement of Ocean Depth." *Deep-Sea Research* 1 (1954): 131–40.

Lutz, H. D., G. Schneider, and G. Kliche. "Far-Infrared Reflection Spectra, TO- and LO-Phonon Frequencies, Coupled and Decoupled Plasmon-Phonon Modes,

Dielectric Constants, and Effective Dynamical Charges of Manganese, Iron, and Platinum Group Pyrite Type Compounds." *Journal of Physics and Chemistry of Solids* 46, no. 4 (1985): 437–43.

Lutz, R. A. "Dispersal of Organisms at Deep-Sea Hydrothermal Vents: A Review." *Oceanologica Acta* 8 (1988): 23–29.

Lutz, R. A., D. Jablonski, and R. D. Turner. "Larval Development and Dispersal at Deep-Sea Hydrothermal Vents." *Science* 226 (1984): 1451–54.

Lutz, Richard A., and Michael J. Kennish. "Ecology of Deep-Sea Hydrothermal Vent Communities: A Review." *Reviews of Geophysics* 31, no. 3 (1993): 211–42.

MacDonald, Gordon. "Climatic Consequences of Increased Carbon Dioxide in the Atmosphere." In *Power Generation and Environmental Change*, edited by David Berkowitz and Arthur Squires, 246–62. Cambridge, MA: MIT Press, 1971.

———, ed. *The Long-Term Impacts of Increasing Atmospheric Carbon Dioxide Levels*. Cambridge, MA: Ballinger, 1982.

———. "Pollution, Weather, and Climate." In *Environment: Resources, Pollution, and Society*, edited by Walter F. Murdock, 326–35. Stamford, CT: Sinauer Associates, 1971.

———. "Scientific Basis for the Greenhouse Effect." *Journal of Policy Analysis and Management* 7, no. 3 (1988): 425–44.

MacDonald, K. C. "Seafloor Spreading: Mid-Ocean Ridge Tectonics." In *Encyclopedia of Ocean Sciences*, edited by J. Steele, S. Thorpe, and K. Turekian, 1798–1813. San Diego, CA: Academic Press, 2001.

MacDonald, Ken C., Keir Becker, F. N. Spiess, and R. D. Ballard. "Hydrothermal Heat Flux of the 'Black Smoker' Vents on the East Pacific Rise." *Earth and Planetary Science Letters* 48, no. 1 (1980): 1–7.

MacDonald, K. C., and J. D. Mudie. "Microearthquakes on the Galapagos Spreading Centre and the Seismicity of Fast-Spreading Ridges." *Geophysical Journal of the Royal Astronomical Society* 36, no. 1 (1974): 245–57.

MacDonald, K. C., and F. N. Spiess. "East Pacific Rise Submersible Program Workshop Report." In *SIO Reference 76-18*, 1–19. San Diego: University of California, San Diego, Marine Physical Laboratory of the Scripps Institution of Oceanography, 1976.

Mackenzie, Donald A. *Inventing Accuracy: An Historical Sociology of Nuclear Missile Guidance*. Cambridge, MA: MIT Press, 1990.

Mahan, Alfred Thayer. *The Influence of Sea Power upon History, 1660–1783*. New York: Little, Brown and Co., 1890.

Maher, K. A., and D. J. Stevenson. "Impact Frustration of the Origin of Life." *Nature* 331 (1988): 612–14.

Maienschein, Jane. *Transforming Traditions in American Biology, 1880–1915*. Baltimore: Johns Hopkins University Press, 1991.

Makhijani, Arjun, and Scott Saleska. *High Level Dollars, Low Level Sense: A Critique of Present Policy for the Management of Long-Lived Radioactive Wastes and Discussion of an Alternative Approach*. New York: Apex Press, 1992.

Malakoff, David. "Researchers Plunge into the Bait over New Sub." *Science* 297 (2002): 326–27.

Malkus, Joanne S. "Large-Scale Interactions." In *The Sea: Ideas and Observations on Progress in the Study of the Seas,* edited by M. N. Hill, 88–294. New York: John Wiley & Sons, 1962.

Malkus, Joanne S., and R. T. Williams. "On the Interactions between Severe Storms and Large Cumulus Clouds." In *Severe Local Storms,* edited by F. H. Ludlam, 59–64. Boston: American Meteorological Society, 1963.

Manabe, Syukuro, and Richard Wetherald. "The Effects of Doubling the CO_2 Concentration on the Climate of a General Circulation Model." *Journal of the Atmospheric Sciences* 32, no. 1 (1975): 3–15.

Marietta, M. G., and A. R. Robinson, eds. *Proceedings of a Workshop on Physical Oceanography Related to Seabed Disposal of High-level Nuclear Waste, Big Sky, Montana, January 14–16, 1980* (TM SAND-80-1776). Albuquerque, NM: Sandia National Labs, 1981.

Marshall, Norman Bertram. "Bathypelagic Fishes as Sounds Scatterers in the Ocean." *Journal of Marine Research* 10 (1951): 1–17.

Mason, Basil John. *Clouds, Rain and Rainmaking.* Cambridge: Cambridge University Press, 1962.

———, ed. *Climate Change 1994: Radiative Forcing of Climate Change and an Evaluation of the IPCC 1992 IS 92 Emission Scenarios.* Cambridge: Cambridge University Press, 1995.

Mason, R. G. "Stripes on the Sea Floor." In *Plate Tectonics,* edited by Naomi Oreskes, with Homer LeGrand, 31–45. Boulder, CO: Westview Press, 2001.

Matthews, Neal. "Sound Effects: Navy Whales Making Sure Noises Won't Hurt Their 'Civilian' Brothers." *San Diego Union-Tribune,* March 16, 1994.

Maury, Matthew Fontaine. *The Physical Geography of the Sea.* Edited by John Leighly. 8th ed. Cambridge, MA: Belknap Press of Harvard University Press, 1963.

Mayor, Adrienne. *Fossil Legends of the First Americans.* Princeton, NJ: Princeton University Press, 2005.

———. "Suppression of Indigenous Fossil Knowledge from Claverack, New York, 1705 to Agate Springs, Nebraska, 2005." In *Agnotology: The Cultural Production of Ignorance,* edited by R. Proctor and L. Schiebinger, 163–82. Stanford, CA: Stanford University Press, 2006.

McCay, Bonnie, and Alan Christopher Finlayson. "The Political Ecology of Crisis and Institutional Change: The Case of Northern Cod." Paper presented at the *Annual Meetings of the American Anthropological Association,* November 15–19, 1995, Washington, DC.

McConnell, Anita. *No Sea Too Deep: The History of Oceanographic Instruments.* Bristol, UK: Adam Hilger, 1982.

McDougall, Ian, and D. H. Tarling. "Dating of Polarity Zones in the Hawaiian Islands." *Nature* 200 (1963): 54–56.

McEvoy, Arthur. *The Fisherman's Problem: Ecology and Law in the California Fisheries, 1850–1980.* Cambridge: Cambridge University Press, 1986.

McKenzie, D. P. "Some Remarks on Heat Flow and Gravity Anomalies." *Journal of Geophysical Research* 72 (1967): 6261.

McKenzie, D. P., and M. J. Bickle. "The Volume and Composition of Melt Generated by Extension of the Lithosphere." *Journal of Petrology* 29, no. 3 (1988): 625–79.

McKenzie, D. P., and J. G. Sclater. "Heat Flow in the Eastern Pacific and Sea-Floor Spreading." *Bulletin Volcanologique* 33, no. 1 (1969): 101.

McNutt, Marcia, and Henry W. Menard. "Lithospheric Flexure and Uplifted Atolls." *Journal of Geophysical Research* 83, no. B3 (1978): 1206–12.

———. "Reply." *Journal of Geophysical Research* 84, no. B10 (1979): 5695–97.

———. "Reply." *Journal of Geophysical Research* 84, no. B13 (1979): 7698.

McVay, Scott. "In Appreciation of Harry Hammond Hess." *Princeton Alumni Weekly* 70, no. 6 (1969): 10–17.

Menard, Henry W. "Correlation between Length and Offset on Very Large Wrench Faults." *Journal of Geophysical Research* 67, no. 10 (1962): 4096–98.

———. "The Deep-Ocean Floor." *Scientific American* 221 (1969): 126–42.

———. "Deformation of the Northeastern Pacific Basin and the West Coast of North America." *Bulletin of the Geological Society of America* 66 (1955): 1149–98.

———. "The East Pacific Rise." *Science* 132, no. 3441 (1960): 1737–46.

———. "Elevation and Subsidence of Oceanic Crust." *Earth and Planetary Science Letters* 6 (1969): 275–84.

———. "Extension of Northeastern Pacific Fracture Zones." *Science* 155, no. 3758 (1967): 72–74.

———. "Fractures in the Pacific Floor." *Scientific American* 193 (1955): 36–41.

———. "Fracture Zones and Offsets of the East Pacific Rise." *Journal of Geophysical Research* 71, no. 2 (1966): 682–85.

———. "History of the Ocean Basins." In *The Nature of the Solid Earth*, edited by Eugene C. Robertson, James F. Hayes, and Leon Knopoff, 440–60. New York: McGraw-Hill, 1972.

———. "Minor Lineations in the Pacific Basin." *Bulletin of the Geological Society of America* 70 (1959): 1491–96.

———. *The Ocean of Truth: A Personal History of Global Tectonics.* Princeton, NJ: Princeton University Press, 1986.

———. "Permanency of Continents and Ocean Basin Lineations." *Nature* 210, no. 5037 (1966): 725.

———. *Science: Growth and Change.* Cambridge, MA: Harvard University Press, 1971.

———. "Sea Floor Relief and Mantle Convection." In *Physics and Chemistry of the Earth*, 315–64. London: Pergamon Press, 1966.

———. "Some Remaining Problems in Sea Floor Spreading." In *The History of the Earth's Crust*, edited by Robert A. Phinney, 109–18. Princeton, NJ: Princeton University Press, 1968.

———. "Topography and Heat Flow of the Fiji Plateau." *Nature* 216 (1967): 991–93.

———. "Transitional Types of Crust under Small Ocean Basins." *Journal of Geophysical Research* 72, no. 12 (1967): 3061–73.

———. "The World-Wide Oceanic Rise-Ridge System." *Philosophical Transactions of the Royal Society* 258 (1965): 109–22.

Menard, Henry W., and Tanya M. Atwater. "Changes in Direction of Sea Floor Spreading." *Nature* 219 (1968): 463–67.

———. "Origin of Fracture Zone Topography." *Nature* 222 (1969): 1037–40.

Menard, Henry W., and T. E. Chase. "Tectonic Effects of Upper Mantle Motion." *International Union of Geological Sciences, Upper Mantle Symposium, New Delhi* (1964): 29–36.

Menard, Henry W., T. E. Chase, and S. M. Smith. "Galapagos Rise in the Southeastern Pacific." *Deep-Sea Research* 11 (1964): 233–42.

Menard, Henry W., and Robert S. Dietz. "Mendocino Submarine Escarpment." *Journal of Geology* 60 (1952): 266–78.

Menard, Henry W., and Robert L. Fisher. "Clipperton Fracture Zone in the Northeastern Equatorial Pacific." *Journal of Geology* 66 (1958): 239–53.

Menard, Henry W., Jean Francheteau, and John Sclater. "Pattern of Relative Motion from Fracture Zone and Spreading Rate Data in the North-Eastern Pacific." *Nature* 226 (1970): 746–48.

Menard, Henry W., and Bruce C. Heezen. "Topography of the Deep Sea Floor." In *The Sea: Ideas and Observations on Progress in the Study of the Seas,* edited by M. N. Hill, 233–80. New York: John Wiley & Sons, 1963.

Menard, Henry W., and H. S. Ladd. "Oceanic Islands, Seamounts, Guyots and Atolls." In *The Sea: Ideas and Observations on Progress in the Study of the Seas,* edited by M. N. Hill, 365–87. New York: John Wiley & Sons, 1962.

Menard, Henry W., and Jacqueline Mammerickx. "Abyssal Hills, Magnetic Anomalies and the East Pacific Rise." *Earth and Planetary Science Letters* 2, no. 5 (1967): 465–72.

Menard, Henry W., and Stuart M. Smith. "Hypsometry of Ocean Basin Provinces." *Journal of Geophysical Research* 71, no. 18 (1966): 4305–25.

Menard, Henry W., and Victor Vacquier. "Magnetic Survey of Part of the Deep Sea Floor off the Coast of California." *Research Reviews, Office of Naval Research* (1958): 1–5.

Menzies, Robert J., John Imbrie, and Bruce C. Heezen. "Further Considerations Regarding the Antiquity of the Abyssal Fauna with Evidence for a Changing Abyssal Environment." *Deep-Sea Research* 8 (1961): 79–94.

Mermin, N. "How Not to Create Tigers." *Physics Today* 52, no. 8 (1999): 11–13.

Merton, Robert K. "The Matthew Effect in Science." *Science* 159, no. 3810 (1968): 56–63.

———. "Norms of Science Essay or Also Known as the Normative Structure of Science." In *The Sociology of Science: The Collected Work of Robert K. Merton,* 267–78. Chicago: University of Chicago Press, 1942.

———. "A Note on Science and Democracy." *Journal of Legal and Political Sociology* 1 (1946): 116.

———. "Science and Technology in a Democratic Order." *Journal of Legal and Political Sociology,* no. 1 (1942): 115–26.

Merz, Alfred, and Georg Wüst. "Die atlantische vertikalzirkulation." *Zeitschrift der Gesellschaft für Erdkunde: Berlin, Jahrgang* (1922): 1–35.

Michaels, David. "Foreword: Sarbanes-Oxley for Science." *Law and Contemporary Problems* 69, no. 3 (2006): 1–19.

———. *Doubt Is Their Product: How Industry's Assault on Science Threatens Your Health.* Oxford: Oxford University Press, 2008.

Michaels, David, and Celeste Monforton. "Scientific Evidence in the Regulatory System: Manufacturing Uncertainty and the Demise of the Formal Regulatory System." *Journal of Law and Policy* 13, no. 1 (2005): 17–41.

Michaels, David, and Neil Vidmar, eds. "Sequestered Science: The Consequences of Undisclosed Knowledge." *Law and Contemporary Problems* 69, no. 3 (2006).

Michaels, David, and Wendy Wagner. "Disclosure in Regulatory Science." *Science* 302, no. 5653 (2003): 2073.

Mikolajewicz, Uwe, Ernst Maier-Reimer, and Tim P. Barnett. "Acoustic Detection of Greenhouse-Induced Climate Changes in the Presence of Slow Fluctuations of the Thermohaline Circulation." *Journal of Physical Oceanography* 23, no. 6 (1993): 1099–1109.

Milbrath, Lester. *The Washington Lobbyists.* Chicago: Rand McNally, 1963.

Miles, Edward L. "Science, Politics, and International Ocean Management: The Uses of Scientific Knowledge in International Negotiations." *Policy Papers in International Affairs,* no. 3, 54–61. Berkeley: Institute of International Affairs, University of California, 1987.

Milgram, Stanley. *Obedience to Authority: An Experimental View.* New York: Harper & Row, 1817.

Miller, Benjamin L. "Geophysical Investigations in the Emerged and Submerged Atlantic Coastal Plain Part II: Geological Significance of the Geophysical Data." *Bulletin of the Geological Society of America* 48, no. 9 (1937): 803–12.

Miller, Howard Smith. *Dollars for Research: Science and Its Patrons in Nineteenth-Century America.* Seattle: University of Washington Press, 1970.

Miller, Stanley L., and Jeffrey L. Bada. "Submarine Hot Springs and the Origins of Life." *Nature* 334 (1988): 609–10.

Mills, Eric L. *Biological Oceanography: An Early History, 1870–1960.* Ithaca, NY: Cornell University Press, 1989.

———. "Book Review of Antarctic Science." *Science* 260 (1993): 1175–76.

———. "Book Review of Die Challenger-Expedition by Gerhard Muller." *History and Philosophy of the Life Sciences* 8 (1986): 165–67.

———. "Book Review of Frozen in Time." *Archives of Natural History* 15 (1988): 362–63.

———. "Book Review of 100 Years Exploring Life, 1888–1988." *Isis* 81, no. 4 (1990): 738–39.

———. "Book Review of the Secular Ark." *History and Philosophy of the Life Sciences* 8 (1986): 316–18.

———. "Book Review of Sir Joseph Banks." *History and Philosophy of the Life Sciences* 11 (1989): 349–51.

———. "Book Review of US Coastal Survey vs. Naval Hydrographic Office." *Archives of Natural History* 17 (1990): 378–79.

———. "Creating a Global Ocean Conveyor: George Deacon and the Hydrology of the Southern Ocean." In *Extremes: Oceanography's Adventures at the Poles*, edited by Keith R. Benson and Helen M. Rozwadowski, 107–32. Sagamore Beach, MA: Science History Publications, 2007.

———. "*De Motu Marium*: Understanding the Oceans before the Second Scientific Revolution." Annual Stillman Drake Lecture of the Canadian Society for the History and Philosophy of Science, Toronto, 1999.

———. "From Discovery to Discovery: The Hydrology of the Southern Ocean, 1885–1937." *Archives of Natural History* 32, no. 2 (2005): 246–64.

———. "From Marine Ecology to Biological Oceanography." *Helgoländer Meeresunter-suchungen* 49 (1995): 29–44.

———. "The Historian of Science and Oceanography after Twenty Years." *Earth Sciences History* 12, no. 1 (1993): 5–18.

———. "The History of Oceanography: Introduction." *Earth Sciences History* 12, no. 1 (1993): 1–4.

———. "Meeting the Demands of Historiographic Fashion: The Problem of Marine Science Archives." In *Oceans from a Global Perspective: International Cooperation in Marine Science Information Transfer*, edited by Carolyn P. Winn, 171–83. Woods Hole, MA: Woods Hole Oceanographic Institution, 1990.

———. "The Oceanography of the Pacific: George F. McEwen, H. U. Sverdrup and the Origin of Physical Oceanography on the West Coast of North America." *Annals of Science* 48 (1991): 241–66.

———. "Oceanography, Physical: Disciplinary History." In *Sciences of the Earth: An Encyclopedia of Events, People, and Phenomena*, edited by G. A. Good, 630–36. New York: Garland Publishing, 1998.

———. "The Ocean Regarded as a Pasture: Kiel, Plymouth, and the Explanation of the Marine Plankton Cycle." *Deutsche Hydrographische Zeitschrift Erganzungsheft*, series B, no. 22 (1990): 20–29.

———. "'Physische Meereskunde': From Geography to Physical Oceanography in the Institut für Meereskunde, Berlin, 1900–1935." *Historisch-Meereskundliches Jahrbuch* 4 (1997): 45–70.

———. "The Scripps Institution: Origin of a Habitat for Ocean Science." Inaugural lecture, University of California, San Diego, 1993.

———. "Socializing Solenoids: The Acceptance of Dynamic Oceanography in Germany around the Time of the 'Meteor' Expedition." *Historisch-Meereskundliches Jahrbuch* 5 (1998): 11–26.

———. "Useful in Many Capacities: An Early Career in American Physical Oceanography." *Historical Studies in the Physical and Biological Sciences* 20, no. 2 (1990): 265–311.

———. "A View of Edward Forbes, Naturalist." *Archives of Natural History* 11, no. 3 (1984): 365–93.

———. "Why Haven't Limnologists and Oceanographers Read Each Other More?" In *Communication to Members*, edited by American Society of Limnology and Oceanography. Waco, TX: American Society of Limnology and Oceanography, 1991.

Mills, Eric L., and Jacqueline Carpine-Lancre. "The Oceanographic Museum of Monaco." In *Ocean Frontiers: Explorations by Oceanographers on Five Continents*, edited by Elizabeth Mann-Borgese, 121–36. New York: Harry N. Abrams, 1992.

Mills, Rachel A. "Hydrothermal Deposits and Metalliferous Sediments from Tag, 26°N Mid-Atlantic Ridge." In *Hydrothermal Vents and Processes*, edited by L. M. Parson, C. L. Walker, and D. R. Dixon, 351–63: London: Geological Society 1995.

Mirowski, Philip. *Science-Mart: Privatizing American Science*. Cambridge, MA: Harvard University Press, 2011.

Mirowski, Philip, and Dieter Plehwe, eds. *The Road from Mount Pèlerin: The Making of the Neoliberal Thought Collective*. Cambridge, MA: Harvard University Press.

Mirowski, Philip, and Esther-Mirjam Sent, eds. *Science Bought and Sold: Essays in the Economics of Science*. Chicago: University of Chicago Press, 2001.

Mitman, Gregg. *Reel Nature: America's Romance with Wildlife on Film*. Seattle: University of Washington Press, 2009.

Molengraaff, G. A. F. "Modern Deep-Sea Research in the East Indian Archipelago." *Geographical Journal* (February 1921): 95–118.

———. "On Recent Crustal Movements in the Island of Timor and Their Bearing on the Geological History of the East-Indian Archipelago." *Koninklijke Akademie van Wetenschappen te Amsterdam* 2, no. September (1912): 224–35.

———. "Wegener's Continental Drift." In *Theory of Continental Drift: A Symposium*, edited by W. A. J. M. van Waterschoot van der Gracht, 90–92. London: Murray, 1928.

Mooney, Chris. *The Republican War on Science*. New York: Basic Books, 2005.

Morley, L. W. "The Zebra Pattern." In *Plate Tectonics*, edited by Naomi Oreskes, with Homer LeGrand, 31–45. Boulder, CO: Westview Press, 2001.

Morrell, Jack, and Arnold Thackray. *Gentlemen of Science: Early Years of the British Association for the Advancement of Science*. Oxford: Oxford University Press, 1982.

Mottl, M. J. "Submarine Hydrothermal Ore Deposits." *Oceanus* 23, no. 2 (1980): 18–27.

Moynihan, Daniel Patrick. *Secrecy: The American Experience*. New Haven, CT: Yale University Press, 1998.

Mukerji, Chandra. *A Fragile Power: Scientists and the State*. Princeton, NJ: Princeton University Press, 1989.

Mullarkey, William, and A. Donn Cobb. *Artemis Module Field Survey by Alvin*. Fort Trumbull, CT: US Navy Underwater Sound Laboratory, 1966.

Munafò, Marcus R., Brian A. Nosek, Dorothy V. M. Bishop, Katherine S. Button, Christopher D. Chambers, Nathalie Percie du Sert, Uri Simonsohn, Eric-Jan Wagenmakers, Jennifer J. Ware, and John P. A. Loannidis. "A Manifesto for Reproducible Science." *Nature Human Behavior* 1 (2017): art. 0021.

Munk, Walter. "The Circulation of the Oceans." *Scientific American* 193, no. 3 (1955): 96–104.

———. "Global Ocean Warming: Detection by Long Path Acoustic Travel Times." *Oceanography* 2 (1989): 40–41.

———. "Harald Ulrik Sverdrup: In Memoriam." *Journal of Deep Sea Research* 4 (1957): 289–90.

———. "Long Ocean Waves." In *The Sea: Ideas and Observations on Progress in the Study of the Seas*, edited by M. N. Hill, 647–63. New York: John Wiley & Sons, 1962.

———. "On the Wind-Driven Ocean Circulation." *Journal of Meteorology* 7, no. 2 (1950): 79–93.

———. "The Solitary Wave Theory and Its Application to Surf Problems." *Annals of the New York Academy of Sciences* 51, no. 3 (1949): 376–424.

———. Testimony, US Commission on Ocean Policy. San Pedro, CA, April 18, 2002.

Munk, Walter, and Arthur Baggeroer. "The Heard Island Papers: A Contribution to Global Acoustics." *Journal of the Acoustical Society of America* 96, no. 4 (1994): 2327–29.

Munk, Walter, and Deborah Day. "Harald U. Sverdrup and the War Years." *Oceanography: The Official Magazine of the Oceanography Society* 4 (2002): 7–29.

Munk, Walter, and A. M. G. Forbes. "Global Ocean Warming: An Acoustic Measure?" *Journal of Physical Oceanography* 13 (1983): 1765–77.

Munk, Walter, and Earl Gossard. "Gravity Waves in the Atmosphere," *Quarterly Journal of the Royal Meteorological Society* 81, no. 359 (1955): 484–87.

Munk, Walter, and Gordon MacDonald. *The Rotation of the Earth: A Geophysical Discussion*. Cambridge: Cambridge University Press, 1960.

Munk, Walter, Naomi Oreskes, and Richard Muller. *Gordon James Fraser MacDonald, 1930–2002: A Biographical Memoir*. Biographical Memoirs 84. Washington, DC: National Academies Press, 2004.

Munk, Walter H., and Rudolph W. Preisendorfer. "Carl Henry Eckart, 1902–1973." *Biographical Memoirs of the National Academy of Sciences* 47: 195–219.

Munk, Walter H., Robert C. Spindel, Arthur Baggeroer, and Theodore G. Birdsall. "The Heard Island Feasibility Test." *Journal of the Acoustical Society of America* 96, no. 4 (1994): 2330–42.

Munk, Walter, Peter Worcester, and Carl Wunsch. *Ocean Acoustic Tomography*. Cambridge: Cambridge University Press, 1995.

Munk, Walter, and Carl Wunsch. "Ocean Acoustic Tomography: A Scheme for Large Scale Monitoring." *Deep Sea Research Part A: Oceanographic Research Papers* 26, no. 2 (1979): 123–61.

Murton, B. J., Cindy Lee Van Dover, and Eve Southward. "Geological Setting and Ecology of the Broken Spur Hydrothermal Vent Field: 29°10'N on the Mid-Atlantic Ridge." In *Hydrothermal Vents and Processes*, edited by L. M. Parson, C. L. Walker, and D. R. Dixon, Geological Society Special Publications 87, 33–41. London: Geological Society, 1995.

Myers, Ransom, and Boris Worm. "Rapid Worldwide Depletion of Predatory Fish Communities." *Nature* 423, no. 6937 (2003): 280.

Nansen, F. *Fridtjof's Nansen's Farthest North: Being the Record of a Voyage of Exploration of the Ship "Fram" 1893–96 and of a Fifteen Months' Sleigh Journey by Dr. Nansen and Leiut. Johansen, with an appendix by Otto Sverdrup*. Westminster, UK: Archibald Constable, 1897.

National Defense Research Committee. *Physics of Sound in the Sea, National Defense Research Committee Summary Technical Reports, Originally Issued as Division 6, Vol-*

ume 8. Edited by J. R. Lyman Spitzer and Research Analysis Group, Committee on Undersea Warfare. Washington DC: National Research Council, 1969.

National Research Council. *International Perspectives on the Study of Climate and Society: Report of the International Workshop on Climate Issues*. Washington, DC: National Academies of Sciences, 1978.

National Research Council. *Low-Frequency Sound and Marine Mammals: Current Knowledge and Research Needs*. Washington, DC: National Academies Press, 1994.

———. *Marine Mammal Populations and Ocean Noise: Determining When Noise Causes Biologically Significant Effects*. Washington, DC: National Academies Press, 2005.

———. *Marine Mammals and Low-Frequency Sound: Progress since 1994*. Washington, DC: National Academies Press, 2000.

———. *Ocean Noise and Marine Mammals*. Washington, DC: National Academies Press, 2003.

National Research Council Committee on Undersea Warfare. *Proceedings of the Symposium on Aspects of Deep-Sea Research*. Edited by William S. Von Arx. Washington, DC: National Research Council, 1957.

National Research Council Panel on Weather and Climate Modification. *Scientific Problems of Weather Modification: A Report of the Panel on Weather and Climate Modification*. Publication No. 1236. Washington, DC: National Academies of Sciences, National Research Council, 1964.

———. *Weather and Climate Modification Problems and Prospects: Final Report of the Panel on Weather and Climate Modification, Vol. 1*. Publication No. 1350. Washington, DC: National Academies of Sciences, National Research Council, 1966.

National Science Foundation and Special Commission on Weather Modification. *Weather and Climate Modification*. Washington, DC: National Science Foundation, 1966.

Naval Studies Board, Commission on Physical Sciences, Mathematics and Resources, National Research Council. *Implications of Advancing Technology for Naval Operations in the Twenty-First Century*. Washington, DC: National Academies Press, 1988.

Navy Review Panel, Ocean Studies Board, Commission on Physical Sciences, Mathematics and Resources, National Research Council. *Oceanography and the Navy: Future Directions*. Washington, DC: National Academies Press, 1988.

Neushul, Peter, and Zuoyue Wang. "Between the Devil and the Deep Blue Sea: C. K. Tseng, Mariculture, and the Politics of Pure Science in Modern China." *Isis* 91 (March 2000): 59–88.

Nierenberg, William A. "Harald Ulrik Sverdrup." *National Academy of Sciences Biographical Memoirs* 69 (1996): 339–74.

Nils, Roll-Hansen. *The Lysenko Effect: The Politics of Science*. Amherst, NY: Humanity Books, 2005.

Nisbet, E. G. *The Young Earth*. Winchester, MA: Allen & Unwin, 1987.

Nye, Mary Jo. *Blackett: Physics, War, and Politics in the Twentieth Century*. Cambridge, MA: Harvard University Press, 2004.

Norman, Don. *The Design of Everyday Things*. New York: Basic Books, 2013.

Normark, William R. "Delineation of the Main Extrusion Zone of the East Pacific Rise at 21°N Latitude." *Geology* 4, no. 11 (1976): 681–85.

———. "Photographic Study of the Zone of Crustal Accretion of the East Pacific Rise Crest and the Mouth of the Gulf of California (Abs)." *Geological Society of America Bulletin* 7 (1975): 358.

Northrop, J., and B. C. Heezen. "An Outcrop of Eocene Sediment on the Continental Slope." *Geology* 59 (1951): 369–99.

Nowell, Arthur R. M., and Charles D. Hollister. "HEBBLE epilogue." *Marine Geology* 99 (1991): 445–60.

Nowell, Arthur R. M., Charles D. Hollister, and Peter A. Jumars. "High Energy Benthic Boundary Layer Experiment: HEBBLE." *Eos* 63, no. 31 (1982): 594–95.

Nowotny, Helga, Peter Scott, and Michael Gibbons. *Re-thinking Science: Knowledge and the Public in an Age of Uncertainty*. Cambridge, UK: Blackwell, 2001.

Nuclear Energy Agency. *Feasibility of Disposal of High-Level Radioactive Waste into the Seabed, Vol. 1*. Paris: Organization for Economic Co-operation and Development, 1984.

Odum, Eugene. *Fundamentals of Ecology*. Philadelphia: Saunders, 1953.

Offley, Ed. "Navy Says Sinking of the Scorpion Was an Accident; Revelations Suggest a Darker Scenario." *Seattle Post-Intelligencer*, May 21, 1998.

Oliver, Jack E. "Earthquake Seismology in the Plate Tectonics Revolution." In *Plate Tectonics*, edited by Naomi Oreskes, with Homer Le Grand, 155–68. Boulder, CO: Westview Press, 2001.

———. *The Incomplete Guide to the Art of Discovery*. New York: Columbia University Press, 1991.

———. *Shocks and Rocks: Seismology in the Plate Tectonics Revolution*. New York: Wiley, 2013.

Oppenheimer, Michael, Naomi Oreskes, Dale Jamieson, Keynyn Brysse, Jessica O' Reilly, Matthew Shindell, and Milena Wazeck, eds. *Discerning Experts: The Practices of Scientific Assessment for Environmental Policy*. Chicago: University of Chicago Press, 2019.

Oreskes, Naomi. "From Scaling to Simulation: Changing Meanings and Ambitions of Models in the Earth Sciences." In *Science without Laws: Model Systems, Cases, and Exemplary Narratives*, edited by Angela N. H. Creager, Elizabeth Lunbeck, and M. Norton Wise. Durham, NC: Duke University Press, 2007.

———. "Laissez-tomber: Military Patronage and Women's Work in the Mid-20th-Century Oceanography." *Historical Studies in Physical and Biological Sciences* 30, no. 2 (2000): 373–92.

———. "Objectivity or Heroism? On the Invisibility of Women in Science." *Osiris* 11 (1996): 87–113.

———. *The Rejection of Continental Drift: Theory and Method in American Earth Science*. New York: Oxford University Press, 1999.

———. "Science in the Origins of the Cold War." In *Science and Technology in the Global Cold War*, edited by Naomi Oreskes and John Krige, 11–30. Cambridge, MA: MIT Press, 2014.

———. "The Scientific Consensus on Climate Change." *Science* 306, no. 5702 (2004): 1686.

———. "Scientist as Sentinel." Plenary speech, American Association for the Advancement of Science annual meeting. Boston, February 17, 2017.

———. "Weighing the Earth from a Submarine: The *S-21* Expedition." In *The Earth, the Heavens, and the Carnegie Institution of Washington: Historical Perspectives after Ninety Years*, edited by Gregory Good, 53–68. History of Geophysics Series. Washington, DC: American Geophysical Union, 1994.

———. "Why I Am a Presentist." In "How and Why We Write History of Science," edited by Oren Harman and Alexandre Metraux, special issue, *Science in Context* 26 (2013): 595–609.

———. *Why Trust Science?* Princeton, NJ: Princeton University Press, 2019.

Oreskes, Naomi, Daniel Carlat, Michael E. Mann, Paul D. Thacker, and Frederick S. Vom Saal. "Viewpoint: Why Disclosure Matters." *Environmental Science and Technology* 49, no. 13 (2015): 7527–28.

Oreskes, Naomi, Colleen Lanier-Christensen, and Hannah Conway. "Congressional Intent in the 1970 Clean Air Act: What Congress Knew about Weather, Climate, and Carbon Dioxide." Manuscript in progress, 2019.

Oreskes, Naomi, and Erik M. Conway. *Merchants of Doubt: How a Handful of Scientists Obscured the Truth on Issues from Tobacco Smoke to Global Warming*. New York: Bloomsbury, 2010.

Oreskes, Naomi, Erik M. Conway, and Matthew Shindell. "From Chicken Little to Dr. Pangloss: William Nierenberg, Global Warming, and the Social Deconstruction of Scientific Knowledge." *Historical Studies in the Natural Sciences* 38, no. 1 (2008): 109–52.

Oreskes, Naomi, and Ronald E. Doel. "Physics and Chemistry of the Earth." In *The Cambridge History of Science*, vol. 5, *Modern Physical and Mathematical Sciences*, edited by Mary Jo Nye, 538–52. Cambridge: Cambridge University Press, 2002.

Oreskes, Naomi, and John Krige. *Science and Technology in the Global Cold War*. Cambridge, MA: MIT Press, 2014.

Oreskes, Naomi, with Homer Le Grand, eds. *Plate Tectonics: An Insider's History of the Modern Theory of the Earth*. Boulder, CO: Westview Press, 2001.

Oreskes, Naomi, and Ronald Rainger. "Science and Security before the Atomic Bomb: The Loyalty Case of Harald U. Sverdrup." *Studies in the History and Philosophy of Modern Physics* 31B (2000): 309–69.

Oreskes, Naomi, Kristin Shrader-Frechette, and Kenneth Belitz. "Verification, Validation, and Confirmation of Numerical Models in the Earth Sciences." *Science* 263, no. 5147 (1994): 641–46.

Ozima, M., and F. A. Podosek. *Noble Gas Geochemistry*. Cambridge: Cambridge University Press, 1983.

Paddock, Richard C. "Undersea Noise Test Could Risk Making Whales Deaf." *Los Angeles Times*, March 22, 1994.

Pálmason, Gudmar. "On Heat Flow in Iceland in Relation to the Mid-Atlantic

Ridge." In *Iceland and Mid-Ocean Ridges*, edited by S. Björnsson. Reykjavik: Societas Scientiarum Islandica, 1967.

Parker, R. L., and D. W. Oldenburg. "Thermal Model of Ocean Ridges." *Nature* 242, no. 122 (1973): 137–39.

Parson, L. M., C. L. Walker, and D. R. Dixon. *Hydrothermal Vents and Processes*. London: Geological Society, 1995.

Pauly, Daniel, Villy Christense, Johanne Dalsgaard, Rainer Froese, and Francisco Torres Jr. "Fishing Down Marine Food Webs." *Science* 279, no. 5352 (1998): 860–63.

Perlman, David. "New Theory of Ocean Chemistry." *San Francisco Chronicle*, November 10, 1979.

Perry, John S. "Energy and Climate: Today's Problem, Not Tomorrow's." *Climatic Change* 3, no. 3 (1981): 223–25.

Petit, J. R., J. Jouzel, D. Raynaud, N. I. Barkov, J. M. Barnola, I. Basile, M. Bender, J. Chappellaz, M. Davis, G. Delaygue, M. Delmotte, V. M. Kotyakov, M. Legrand, V. Y. Lipenkov, C. Lorius, L. Pepin, C. Ritz, E. Salzman, and M. Stievenard. "Climate and Atmospheric History of the Past 420,000 Years from the Vostok Ice Core, Antarctica." *Nature* 399 (1999): 429.

Peterson, R. G., L. Stramma, and G. Kortum. "Early Concepts and Charts of Ocean Circulation." *Progress in Oceanography* 37, no. 1 (1996): 1–115.

Petterssen, Sverre. 2001. *Weathering the Storm: Sverre Petterssen, the D-Day Forecast, and the Rise of Modern Meteorology*. Edited by J. R. Fleming. Historical Monograph Series. Boston: American Meteorological Society.

Pettersson, Hans. *The Ocean Floor*. New Haven, CT: Yale University Press, 1954.

Pettijohn, F. J. *Memoirs of an Unrepentant Field Geologist: A Candid Profile of Some Geologists and Their Science, 1921–1981*. Chicago: University of Chicago Press, 1987.

Pfeiffer, John E. "The Office of Naval Research." *Scientific American*, February 1949, 11–15.

Phillips, J. D., H. S. Fleming, R. Feden, W. E. King, and R. Perry. "Aeromagnetic Study of the Mid-Atlantic Ridge Near the Oceanographer Fracture Zone." *Geological Society of America Bulletin* 86 (1975): 1348–57.

Phillips, J. D., and D. Forsyth. "Plate Tectonics, Paleomagnetism, and the Opening of the Atlantic." *Bulletin of the Geological Society of America* 83, no. 6 (1972): 1579–1600.

Pianin, Eric. "Group Meets on Global Warming: Bush Officials Say Uncertainties Remain on Cause, Effects." *Washington Post*, December 4, 2002.

Piccard, Jacques, and Robert S. Dietz. *Seven Miles Down: The Story of the Bathyscaph "Trieste."* New York: G. P. Putnam's Sons, 1961.

Pickard, George, and William J. Emery. *Descriptive Physical Oceanography: An Introduction*. New York: Pergamon Press, 1990.

Piper, D. Z. "Origin of Metalliferous Sediments from the East Pacific Rise." *Earth and Planetary Science Letters* 19 (1973): 75–82.

Place, Commander W. M., Colonel F. C. Cobb, and Lt. Col. C. G. Defferding. *Palomares Summary Report*. Kirtland Air Force Base, NM: Field Command, Defense Nuclear

Agency, Technology and Analysis Directorate, Kirtland Air Force Base, January 15, 1975.

Plass, Gilbert N. "Carbon Dioxide and Climate." *Scientific American,* July 1959, 41.

———. "The Carbon Dioxide Theory of Climatic Change." *Tellus* 8, no. 2 (1956): 140–54.

———. "The Influence of the 15μ Carbon-Dioxide Band on the Atmospheric Infra-Red Cooling Rate." *Quarterly Journal of the Royal Meteorological Society* 82, no. 353 (1956): 310–24.

Polanyi, Michael. "The Autonomy of Science." *Scientific Monthly* 60, no. 2 (1945): 141–50.

———. "The Growth of Thought in Society." *Economica,* 8, no. 32 (1941): 428–56.

———. *Personal Knowledge: Towards a Post-critical Philosophy.* Chicago: University of Chicago Press, 1974.

———. "Rights and Duties of Science." Pamphlet, Society for Freedom in Science Occasional, 18, no. 2 (1945). Reprinted from Michael Polanyi *The Contempt of Freedom.* London: Watts, 1940.

———. *Science, Faith, and Society: A Searching Examination of the Meaning and Nature of Scientific Inquiry.* Chicago: University of Chicago Press, 1946.

Poldervaart, A., and H. H. Hess. "Pyroxenes in the Crystallization of Basaltic Magma." *Journal of Geology* 59, no. 5 (1951): 472–89.

Pollard, Raymond, and Gwyn Griffiths. "James Rennell, the Father of Oceanography." *Ocean Challenge* 4, nos. 1–2 (1993): 24–25.

Popper, A. N., and M. C. Hastings. "The Effects of Anthropogenic Sources of Sound on Fishes." *Journal of Fish Biology* 75 (2009): 455–89.

Porter, Theodore M. *Trust in Numbers: The Pursuit of Objectivity in Science and Public Life.* Princeton, NJ: Princeton University Press, 2005.

Potter, John R. "ATOC: Sound Policy or Enviro-Vandalism? Aspects of a Modern Media-Fueled Policy Issue." *Journal of Environment & Development* 3, no. 2 (1994): 47–62.

President's Science Advisory Committee and Environmental Pollution Panel. *Restoring the Quality of Our Environment.* Washington, DC: White House, 1965.

Press, Eyal, and Jennifer Washburn. "The Kept University." *Atlantic Monthly* 285, no. 3 (2000): 39–42, 44–52, 54.

Press, Frank, and Raymond Siever. *Earth.* 4th ed. New York: W. H. Freeman, 1985.

Proctor, Robert N., and Londa Schiebinger, eds. *Agnotology: The Making and Unmaking of Ignorance.* Palo Alto, CA: Stanford University Press, 2008.

Rahmstorf, Stefan. "The Current Climate." *Nature* 421 (2003): 699.

Rainger, Ronald. "Constructing a Landscape for Postwar Science Roger Revelle, the Scripps Institution and the University of California, San Diego." *Minerva* 39 (2001): 327–52.

———. *Palomares Summary Report.* Kirtland Air Force Base, NM: Field Command, Defense Nuclear Agency, Technology and Analysis Directorate, Kirtland Air Force Base, 1975.

———. "Patronage and Science: Roger Revelle, the Navy, and Biological Oceanography at the Scripps Institution." *Earth Sciences History* 19 (2000): 58–89.

———. "Science at the Crossroads: The Navy, Bikini Atoll and American Oceanography in the 1940s." *Historical Studies in the Physical and Biological Sciences* 30 (2000): 349–71.

———. "'A Wonderful Oceanographic Tool:' The Atomic Bomb, Radioactivity and the Development of American Oceanography." In *The Machine in Neptune's Garden: Historical Perspectives on Technology and the Marine Environment,* edited by Helen M. Rozwadowski and David K. van Keuren, 96–132. Sagamore Beach, MA: Science History Publications, 2004.

Rainger, Ronald, et al. *The American Development of Biology.* Philadelphia: University of Pennsylvania Press, 1988.

Raitt, Helen, and Beatrice Moulton. *Scripps Institution of Oceanography: First Fifty Years.* Los Angeles: W. Ritchie Press, 1967.

Rampton, Sheldon. *Trust Us, We're Experts! How Industry Manipulates Science and Gambles with Your Future.* New York: Putnam Press, 2001.

Reichenbach, Hans. *Experience and Prediction: An Analysis of the Foundations and Structures of Knowledge.* Chicago: University of Chicago Press, 1938.

Reilinger, R., et al. (2006). "GPS Constraints on Continental Deformation in the Africa-Arabia-Eurasia Continental Collision Zone and Implications for the Dynamics of Plate Interactions." *Journal of Geophysical Research* 111 (2006): B05411.

Reingold, Nathan. *Science, American Style.* New Brunswick, NJ: Rutgers University Press, 1999.

Renard, V., B. Schrumpf. *Bathymétrie détaillée d'une partie de vallée du rift et de faille transformante près de 36°50′N dans l'ocean Atlantique.* Paris: CNEXO, 1974.

Reppy, Judith, ed. *Secrecy and Knowledge Production.* Cornell University: Peace Studies Program, 1999.

Research Analysis Group. *Physics of Sound in the Sea.* Washington, DC: Department of the Navy, Headquarters Naval Material Command, 1969.

Revelle, Roger. "Alfred C. Redfield" In *Biographical Memoirs,* 315–29. Washington DC: National Academy of Sciences Press, 1995.

———. "Carbon Dioxide and World Climate." *Scientific American* 247, no. 2 (1982): 35–43.

———. "The Oceanographic and How It Grew." In *Oceanography: The Past,* edited by Mary Sears and Daniel Merriam, 10–24. New York: Springer, 1980.

Revelle, Roger, T. R. Folsom, E. D. Goldberg, and J. D. Isaacs. "Nuclear Science and Oceanography." Contribution from the Scripps Institution of Oceanography to the International Conference on the Peaceful Uses of Atomic Energy, June 30, 1955.

Revelle, R., and A. E. Maxwell. "Heat Flow through the Floor of the Eastern North Pacific Ocean." *Nature* 170 (1952): 199–200.

Revelle, Roger, and Walter Munk. "Harald Ulrik Sverdrup: An Appreciation." *Journal of Marine Research* 7 (1948): 127–38.

Revelle, Roger, and Hans E. Suess. "Carbon Dioxide Exchange between Atmosphere and Ocean and the Question of an Increase of Atmospheric CO_2 during the Past Decades." *Tellus* 9, no. 1 (1957): 18–27.

Rhoads, D. C., R. A. Lutz, R. M. Cerrato, and E. C. Revelas. "Growth of Bivalves at Deep-Sea Hydrothermal Vents along the Galápagos Rift." *Science* 214 (1981): 911–13.

———. "Growth and Predation Activity at Deep-Sea Hydrothermal Vents along the Galápagos Rift." *Journal of Marine Research* 40 (1982): 503–16.

Rich, Nathaniel. "Losing Earth: The Decade We Almost Stopped Climate Change." *New York Times Magazine*, August 1, 2018.

Richter, Catherine A., Linda Birnbaum, Francesca Farabollini, Retha Newbold, Beverly Rubin, Chris Talsness, John Vandenbergh, et al. "In Vivo Effects of Bisphenol A in Laboratory Rodent Studies." *Reproductive Toxicology* 24, no. 2 (2007): 199–224.

Riel, Gordon K., Donald G. Simons, and Commander P. V. Converse. "Dunc 'Thresher Radiation Survey' Noltr-64–21." White Oak, MD: US Naval Ordnance Laboratory, 1964.

Rieley, Gareth, Cindy Lee Van Dover, David B. Hedrick, David C. White, and Geoffrey Eglinton. "Lipid Characteristics of Hydrothermal Vent Organisms from 9°N, East Pacific Rise." In *Hydrothermal Vents and Processes*, edited by L. M. Parson, C. L. Walker, and D. R. Dixon, 351–63. London: Geological Society 1995.

Roberts, Callum. "Deep Impact: The Rising Toll of Fishing in the Deep Sea." *Trends in Ecology and Evolution* 17, no. 5 (2002): 242–45.

Robinson, Allan, and Henry Stommel. "The Oceanic Thermocline and the Associated Thermohaline Circulation." *Tellus* 11, no. 3 (1959): 295–308.

Rockefeller Foundation. "Oceanography." *A Digital History*, 2018. https://rockfound.rockarch.org/oceanography.

Rocken, C., R. Anthes, M. Exner, D. Hunt, S. Sokolovskiy, R. Ware, M. Gorbunov, et al. "Analysis and Validation of GPS/MET Data in the Neutral Atmosphere." *Journal of Geophysical Research: Atmospheres* 102, no. D25 (1997): 29849–66.

Rona, P. A. "Comparison of Continental Margins of Eastern North America at Cape Hatteras and Northwestern Africa at Cap Blanc." *American Association of Petroleum Geologists Bulletin* 54 (1970): 129–57.

———. "Hydrothermal Mineralization at Seafloor Spreading Centers." *Earth Science Reviews* 20 (1983): 1–104.

———. "New Evidence for Seabed Resources from Global Tectonics." *Ocean Management* 1 (1973): 145–59.

———. "Tag Hydrothermal Field: Mid-Atlantic Ridge Crest at Latitude 26°N." *London: Geological Society* 137 (1980): 385–402.

Rona, P. A., B. A. McGregor, P. R. Betzer, G. W. Bolger, and D. C. Krause. "Anomalous Water Temperatures over Mid-Atlantic Ridge Crest at 26°N Latitude." *Deep-Sea Research* 22 (1975): 611–18.

Rona, P. A., B. A. McGregor, P. R. Betzer, and D. C. Krause. "Anomalous Water Temperatures over Mid-Atlantic Ridge Crest at 26°N." *Eos, Transactions, American Geophysical Union* 55, no. 4 (1974): 293.

Rona, P. A., and R. B. Scott. "Convenors, Symposium: Axial Processes of the Mid-Atlantic Ridge." *Eos, Transactions, American Geophysical Union* 55, no. 4 (1974): 292–95.

Roosth, Sophia. *Synthetic: How Life Got Made*. Chicago: University of Chicago Press, 2017.

Rosenberg, Charles. "Martin Arrowsmith: The Scientist as Hero." In *No Other Gods: On Science and American Social Thought*. Baltimore: Johns Hopkins University Press, 1961.

Rosner, David, and Gerald Markowitz. *Lead Wars: The Politics of Science and the Fate of America's Children*. Berkeley: University of California Press, 2013.

Rossiter, Margaret. *Women Scientists in America: Before Affirmative Action, 1940–1972*. Baltimore: Johns Hopkins University Press, 1995.

———. *Women Scientists in America: Forging a New World since 1972*. Baltimore: Johns Hopkins University Press, 2012.

———. *Women Scientists in America: Struggles and Strategies to 1940*. Baltimore: Johns Hopkins University Press, 1982.

Rowland, Henry Augustus. *The Physical Papers*. 1883. Baltimore: Johns Hopkins Press, 1902.

Rozwadowski, Helen M. *Fathoming the Ocean: The Discovery and Exploration of the Deep Sea*. Cambridge, MA: Harvard University Press, 2005.

———. *The Sea Knows No Boundaries: A Century of Marine Science under Ices*. Seattle: University of Washington Press, 2002.

Rozwadowski, Helen, and David K. van Keuren, eds. *The Machine in Neptune's Garden: Historical Perspectives on Technology and the Marine Environment*. Sagamore Beach, MA: Science History Publications, 2004.

Ruddiman, W. F. *Plows, Plagues, and Petroleum: How Humans Took Control of Climate*. Princeton, NJ: Princeton University Press, 2005.

Rudwick, Martin J. S. *Bursting the Limits of Time: The Reconstruction of Geohistory in the Age of Revolution*. Chicago: University of Chicago Press, 2007.

———. *The Great Devonian Controversy: The Shaping of Scientific Knowledge among Gentlemanly Specialists*. Chicago: University of Chicago Press. 1985.

Runcorn, Stanley Keith. "Palaeomagnetic Comparisons between Europe and North America." *Philosophical Transactions of the Royal Society of London A: Mathematical, Physical and Engineering Sciences* 258, no. 1088 (1965): 1–11.

Sandwell, David T. "Plate Tectonics: A Martian View." In *Plate Tectonics: An Insider's History of the Modern Theory of the Earth*, edited by Naomi Oreskes, with Homer Le Grand, 331–45. Boulder, CO: Westview Press, 2001.

Santer, B. D. "Signal-to-Noise Analysis of Time-Dependent Greenhouse Warming Experiments." *Climate Dynamics* 9, no. 6 (1994): 267–85.

Santer, B. D., K. E. Taylor, T. M. L. Wigley, T. C. Johns, P. D. Jones, D. J. Karoly, J. F. B Mitchell, et al. "A Search for Human Influences on the Thermal Structure of the Atmosphere." *Nature* 382, no. 6586 (1996): 39.

Sapolsky, Harvey M. "Academic Science and the Military: The Years since the Second World War." In *The Sciences in the American Context: New Perspectives*, edited by Nathan Reingold, 379–99. Washington, DC: Smithsonian Institution Press, 1979.

———. *Science and the Navy: The History of the Office of Naval Research.* Princeton, NJ: Princeton University Press, 1990.

Sarewitz, Daniel. *Frontiers of Illusion: Science, Technology, and the Politics of Progress.* Philadelphia: Temple University Press, 1996.

Scalera, G., and Jacob, K.-H., eds. *Why Expanding Earth? A Book in Honour of O. C. Hilgenberg.* Rome: INGV, 2003.

Schlanger, S. O., and M. A. Murphy. "Sedimentary Structures in Ilhas and São Sebastião Formations (Cretaceous), Reconcavo Basin, Brazil." *AAPG Bulletin* 46, no. 4 (1962): 457–77.

Schlee, Susan. *On the Edge of an Unfamiliar World.* Boston: E. P. Dutton, 1973.

Schmalzer, Sigrid. *The People's Peking Man: Popular Science and Human Identity in Twentieth-Century China.* Chicago: University of Chicago Press, 2008.

———. *Red Revolution, Green Revolution: Scientific Farming in Socialist China.* Chicago: University of Chicago Press, 2016.

———. "Self-Reliant Science: The Impact of the Cold War on Science in Socialist China." In *Science and Technology in the Global Cold War*, edited by Naomi Oreskes and John Krige. Cambridge, MA: MIT Press, 2014.

Schmid, Sonja D. "Defining (Scientific) Direction: Soviet Nuclear Physics and Reactor Engineering during the Cold War." In *Science and Technology in the Global Cold War*, edited by Naomi Oreskes and John Krige. Cambridge, MA: MIT Press, 2014.

Schneider, Eric D., Paul J. Fox, Charles D. Hollister, H. David Needham, and Bruce C. Heezen. "Further Evidence of Contour Currents in the Western North Atlantic." *Earth and Planetary Science Letters* 2 (1967): 351–59.

Schneider, Stephen H. "On the Carbon Dioxide–Climate Confusion." *Journal of the Atmospheric Sciences* 32, no. 11 (1975): 2060–66.

Schneider, S. S. "A Constructive Deconstruction of Deconstructionists: A Response to Demeritt." *Annals of the Association of American Geographers* 91, no. 2 (2001): 338–44.

Schwartz, Stephen I., ed. *Atomic Audit.* Washington, DC: Brookings Institution Press, 1998.

Schweber, S. S. "The Empiricist Temper Regnant: Theoretical Physics in the United States 1920–1950." *Historical Studies in the Physical and Biological Sciences* 17, no. 1 (1986): 55–98.

———. *In the Shadow of the Bomb.* Princeton, NJ: Princeton University Press, 2000.

"Science: How Oceans Grew." *Time Magazine*, September 14, 1959.

"The Scientists: A Sympathetic Portrait." *Fortune* (1948): 106–12.

"Scientists Find Hot Springs on Pacific Ocean Floor Teeming With Life." *New York Times*, April 19, 1977.

Sclater, John. "Heat Flow under the Oceans." In *Plate Tectonics: An Insider's History of the Modern Theory of the Earth*, edited by Naomi Oreskes, 128–47. Boulder, CO: Westview Press, 2001.

Sclater, J. G., and J. Francheteau. "The Implications of Terrestrial Heat Flow Observations on Current Tectonics and Geochemical Models of the Crust and Upper Mantle of the Earth." *Geophysical Journal of the Royal Astronomical Society* 20 (1970): 509–42.

Sclater, J. G., and K. D. Klitgord. "A Detailed Heat Flow, Topographic and Magnetic Survey across the Galapagos Spreading Centre at 86°W." *Journal of Geophysical Research* 78 (1973): 6951.

Sclater, J. G., J. D. Mudie, and C. G. A. Harrison. "Detailed Geophysical Studies on the Hawaiian Arch Near 24°25'N, 157°40'W: A Closely Spaced Suite of Heat-Flow Stations." *Journal of Geophysical Research* 75 (1970): 333.

Sclater, J. G., R. P. Von Herzen, D. L. Williams, R. N. Anderson, and K. Klitgord. "The Galapagos Spreading Centre: Heat-Flow Low on the North Flank." *Geophysical Journal International* 38, no. 3 (1974): 609–25.

Scott, E. B., P. A. Rona, B. A. McGregor, and M. R. Scott. "The Tag Hydrothermal Field." *Proceedings of the National Academy of Sciences* 251 (1974): 301.

Scrutton, R. A., and M. Talwani, eds. *The Ocean Floor: Bruce Heezen Commemorative Volume*. Chichester, UK: Wiley, 1982.

Sears, Mary, and Daniel Merriam, eds. *Oceanography: The Past*. New York: Springer, 1980.

Seidel, Robert W. "From Glow to Flow: A History of Military Laser Research and Development." *Historical Studies in the Physical and Biological Sciences* 18 (1987): 111–47.

———. "A Home for Big Science: The Atomic Energy Commission's Laboratory System." *Historical Studies in the Physical and Biological Sciences* 16, no. 1 (1986): 135–75.

———. "The Postwar Political Economy of High Energy Physics." In *Pions to Quarks, Particle Physics in the 1950s*, edited by Laurie Brown, Max Dresden, Lillian Hoddeson, and May West, 497–507. Cambridge: Cambridge University Press, 1989.

Shabecoff, Philip. "Global Warming Has Begun, Expert Tells Senate." *New York Times*, June 24, 1988. http://www.nytimes.com/1988/06/24/us/global-warming-has-begun-expert-tells-senate.html.

Shapin, Steven. *A Social History of Truth: Civility and Science in Seventeenth-Century England*. Chicago: University of Chicago Press, 1994.

———. *Never Pure: Historical Studies of Science as if It Was Produced by People with Bodies, Situated in Time, Space, Culture, and Society, and Struggling for Credibility and Authority*. Baltimore, MD: Johns Hopkins University Press, 2010.

Shapin, Steven, and Simon Schaffer. *Leviathan and the Air-Pump: Hobbes, Boyle, and the Experimental Life*. Princeton, NJ: Princeton University Press, 1985.

Shepard, F. P. "The Underlying Causes of Submarine Canyons." *Proceedings of the National Academy of Sciences* 22 (1936): 496-502.

———. "Submarine Valleys." *Geographical Review* 23, no. 1 (1933): 77–89

Shepard, F. P., J. M. Trefethen, and G. V. Cohee. "Origins of Georges Bank." *Bulletin of the Geological Society of America* 45, no. 2 (1934): 281–302.

Sherlock, Robert Lionel. *Man as a Geological Agent: An Account of His Action on Inanimate Nature*. London: H. F. & G. Witherby, 1922.

Shillito, Bruce, Jean-Pierre Lechaire, Gèrard Goffinet, and Francoise Gaill. "Composition and Morphogenesis of the Tubes of Vestimentiferan Worms." In *Hydro-thermal Vents and Processes*, edited by L. M. Parson, C. L. Walker, and D. R. Dixon, 351–63: London: Geological Society, 1995.

Shils, Edward A. *The Torment of Secrecy*. Chicago: Elephant Paperbacks, 1956.

Shindell, Matthew. "From the End of the World to the Age of the Earth: The Cold War Development of Isotope Geochemistry at the University of Chicago and Caltech." In *Science and Technology in the Global Cold War*, edited by Naomi Oreskes and John Krige, 107–40. Cambridge, MA: MIT Press, 2014.

Shor, Elizabeth Noble. "E. C. Bullard's First Heat-Probe." *Eos, Transactions, American Geophysical Union* 65, no. 9 (1984): 73–74.

———. *Scripps Institution of Oceanography: Probing the Oceans 1936–1976*. San Diego: Tofua Press, 1978.

———, ed. *Seeking Signals in the Sea*. In *SIO Reference No. 97–5*. San Diego: University of California, San Diego, Marine Physical Laboratory of the Scripps Institution of Oceanography, 1997.

Shor, G. G., Jr., Henry W. Menard, and R. W. Raitt. "Structure of the Pacific Basins." In *The Sea*, edited by A. E. Maxwell, 3–27. New York: John Wiley & Sons, 1969.

Showstack, R. "Bush Administration Proposes Criteria for Basic Research." *Eos* 83, no. 12 (2002): 129.

Siever, Raymond. "Doing Earth Science Research during the Cold War." In *The Cold War & the University: Toward an Intellectual History of the Postwar Years*, edited by Andre Schiffrin, 147–70. New York: New Press, 1997.

Simpson, Joanne. "Meteorologist." In *Women and Success: The Anatomy of Achievement*, edited by Ruth B. Kundsin, 62–67. New York: William Morrow & Company, 1974.

Simpson, S. "Life's First Scalding Step." *Science News* 1999, 24–26.

Sjöblom, Kirsti-Liisa, and Gordon Linsley. "Sea Disposal of Radioactive Wastes: The London Convention 1982." *International Atomic Energy Agency Bulletin* 36, no. 2 (1994): 12.

Sleep, Norman H. "Sensitivity of Heat Flow and Gravity to the Mechanism of Sea-Floor Spreading." *Journal of Geophysical Research* 74, no. 2 (1969): 542–49.

Slotten, Hugh Richard. *Patronage, Practice, and the Culture of American Science: Alexander Dallas Bache and the US Coast Survey*. Cambridge: Cambridge University Press, 1994.

Smith, Bruce L. R. *American Science Policy since World War II*. Washington, DC: Brookings Institution, 1990.

Smith, W. H. F., and D. T. Sandwell. "Conventional Bathymetry, Bathymetry from Space, and Geodetic Altimetry." *Oceanography* 17, no. 1 (2004): 8–23.

Soare, Richard J., and David M. Green. "The Habitability of Europa: A Cautionary Note." *Eos, Transactions, American Geophysical Union* 83 (2002): 231.

Solomon, Miriam. *Social Empiricism*. Cambridge, MA: MIT Press, 2001.

———. "The Web of Valief." In *Out from the Shadows: Analytical Feminist Contributions to Traditional Philosophy*, edited by Sharon L. Crasnow and Anita M. Superson, 435–50. Oxford: Oxford University Press, 2012.

Solomon, Susan, ed. *Climate Change 2007: The Physical Science Basis: Working Group I Contribution to the Fourth Assessment Report of the IPCC*. Vol. 4. Cambridge: Cambridge University Press, 2007.

Solovey, Mark. "Science and the State during the Cold War: Blurring Boundaries and a Contested Legacy." *Social Studies of Science* 31 (2001): 165–70.

Solovey, Mark, and Hamilton Cravens. *Cold War Social Science: Knowledge Production, Liberal Democracy, and Human Nature*. New York: Palgrave McMillan, 2012.

Somero, George N. "Physiology and Biochemistry of the Hydrothermal Vent Animals." *Oceanus* 37, no. 3 (1984): 67–72.

Sontag, Sherry, Christopher Drew, and Annette Lawrence Drew. *Blind Man's Bluff: The Untold Story of American Submarine Espionage*. New York: Public Affairs, 1998.

Sousa, Filipa L., Thorsten Thiergart, Giddy Landan, Shijulal Nelson-Sathi, Inês A. C. Pereira, John F. Allen, Nick Lane, and William F. Martin. "Early Bioenergetic Evolution." *Philosophical Transactions of the Royal Society of London. Series B, Biological Sciences* 368, no. 1622 (2013): 1-30

Spanagel, David I. "Utility of Cartographic History to Historical Studies of the Earth Sciences." *History of the Earth Sciences Society International Commission on the History of Geological Sciences INHIGEO Meeting* 34, no. 2 (2015): 263–74.

Speer, Kevin G., and Karl R. Helfrich. "Hydrothermal Plumes: A Review of Flow and Fluxes." In *Hydrothermal Vents and Processes*, edited by L. M. Parson, C. L. Walker, and D. R. Dixon, 351–63: London: Geological Society 1995.

Spiesberger, J. L., T. G. Birdsall, K. Metzger, R. A. Knox, C. W. Spofford, and R. C. Spindel. "Measurements of Gulf Stream Meandering and Evidence of Seasonal Thermocline Development Using Long-Range Acoustic Transmissions." *Journal of Physical Oceanography* 13 (1983): 1836–46.

Spiesberger, J. L., et al. "Stability and Identification of Ocean Acoustic Multipaths." *Journal of the Acoustical Society of America* 67 (1980): 2011–17.

Spiesberger, John L., D. E. Frye, J. O'Brien, H. Hurlburt, J. W. McCaffrey, M. Johnson, and J. Kenny. "Global Acoustic Mapping of Ocean Temperatures (GAMOT)." In *Oceans '93: Engineering in Harmony with Ocean. Proceedings*, I253–I257. IEEE, 1993.

Spiesberger, John L., and Kurt Metzger. "Basin-Scale Ocean Monitoring with Acoustic Thermometers." *Oceanograpy* 5, no. 2 (1992), 92–98.

———. "A Basin-Scale (3000 km) Tomographic Section of Temperature and Sound Speed in the Northeast Pacific." *Transactions of the American Geophysical Union* 71 (January 7, 1990): 4869–89.

———. "Basin-Scale Tomography: A New Tool for Studying Weather and Climate." *Journal of Geophysical Research* 96, no. C3 (1991): 4869–89.

———. "A New Algorithm for Sound Speed in Seawater." *Journal of the Acoustical Society of America* 89 (1991): 2677–88.

———. "New Estimates of Sound-Speed in Water." *Journal of the Acoustical Society of America* 89 (1991): 1697–1700.

Spiesberger, John L., Kurt Metzger, and John A. Furgerson. "Listening for Climatic Temperature Change in the Northeast Pacific: 1983–1989." *Journal of the Acoustical Society of America* 92 (1992): 384–96.

Spiess, Fred Noel. "A Beginning in Undersea Research." In *Seeking Signals in the Sea*, edited by Elizabeth N. Shor, *SIO Reference No. 97-5*. San Diego: University of California, San Diego, Marine Physical Laboratory of the Scripps Institution of Oceanography, 1997.

———. *Dr. Fred Noel Spiess*. Interview by Christopher Henke. La Jolla, CA: SIO Archives, University of California, San Diego 2000.

———. "Motivating the Underwater Acoustics Community: 50's and 60's Version." *Acoustical Society of America* 110, no. 5 (2001): 2688–89.

———. "Some Origins and Perspectives in Deep-Ocean Instrument Management." In *Oceanography: The Past*, edited by Mary Sears and Daniel Merriam, 226–39. New York: Springer, 1980.

Spiess, F. N., P. Lonsdale, R. C. Tyce, and J. D. Mudie. "Shipboard Cruise Report on Leg 10 of Expedition South Tow." In *SIO Reference*. San Diego: University of California, San Diego, Marine Physical Laboratory of the Scripps Institution of Oceanography, 1973.

Spiess, Fred Noel, Ken C. Macdonald, T. Atwater, R. Ballard, A. Carranza, D. Córdoba, C. Cox, V. M. Díaz García, J. Francheteau, J. Guerrero, J. Hawkins, R. Haymon, R. Hessler, T. Juteau, M. Kastner, R. Larson, B. Luyendyk, J. D. Macdougall, S. Miller, W. Normark, J. Orcutt, and C. Rangin. "East Pacific Rise: Hotsprings and Geophysical Experiments." *Science* 207 (1980): 1421–23.

Spiess, F. N., K. C. MacDonald, and B. P. Luyendyk. *Mexican, American and French Near-Bottom Investigations of an Active Spreading Center Using Deep Tow, Angus, and Alvin, Expedition Rise Cruise Report at the 21°N East Pacific Rise, Marine Physical Laboratory Technical Memorandum, No. 313*. La Jolla, CA: Scripps Institute of Oceanography, March–May 1979.

Spiess, Fred Noel, and Albert Ernest Maxwell. "Search for the 'Thresher.'" *Science* 145 (1964): 349–55.

Spiess, F. N., and R. C. Tyce. "Deep Tow Instrumentation System." In *SIO Reference*. San Diego: University of California, San Diego, Marine Physical Laboratory of the Scripps Institution of Oceanography, 1973.

Spiess, Fritz. *The Meteor Expedition: Scientific Results of the German Atlantic Expedition, 1925–1927*. New Delhi: Amerind Publishing Company, 1985.

Spilhaus, Athelstan F. "A Bathythermograph." *Journal of Marine Research* 1 (1938): 95–100.

Spindel, Robert C., and Peter F. Worcester. "Ocean Acoustic Tomography." *Scientific American* 263, no. 4 (1990): 94–99.

Spooner, E. T. C., and W. S. Fyfe. "Sub-Seafloor Metamorphism, Heat and Mass Transfer." *Contributions to Minerology and Petrology* 42 (1973): 287.

Stainforth, D. A., T. Aina, C. Christensen, M. Collins, N. Faull, D. J. Frame, J. A. Kettleborough, S. Knight, A. Martin, J. M. Murphy, C. Piani, D. Sexton, L. A. Smith, R. A. Spicer, A. J. Thorpe, and M. R. Allen. "Uncertainty in Predictions of the Climate Response to Rising Levels of Greenhouse Gases." *Nature* 433 (2005): 403–6.

Staudenmaier, John M. "What Shot Hath Wrought and What Shot Hath Not: Reflections on Twenty-Five Years of the History of Technology." *Technology and Culture* 25, no. 4 (1984): 707–30.

Stein, C. A., and S. Stein. "Constraints on Hydrothermal Heat Flux through the Oceanic Lithosphere from Global Heat Flow." *Journal of Geophysical Research* 99 (1994): 3081–95.

Stocker, Thomas F. *The Ocean in the Climate System: Observing and Modeling Its Variability.* Bern: Climate and Environmental Physics Institute, University of Bern, 2000.

———. "Past and Future Reorganizations in the Climate System." *Quaternary Science Reviews* 19 (2000): 301–19.

Stommel, Henry. "The Abyssal Circulation." *Deep-Sea Research* 5 (1959): 80–82.

———. "The Abyssal Circulation of the Ocean." *Nature* 180, no. 4589 (1957): 733–34.

———. "The Anatomy of the Atlantic." *Scientific American* 192, no. 1 (1955): 30–35.

———. "Circulation in the North Atlantic Ocean." *Nature* 173 (1954): 886–88.

———. "An Example of Thermal Convection." *Transactions of the American Geophysical Union* 31, no. 4 (1950): 553–54.

———. "Future Prospects for Physical Oceanography." *Science* 168, no. 3939 (1970): 1531–37.

———. "The Gulf Stream: A Brief History of the Ideas Concerning Its Cause." *Scientific Monthly* 70, no. 4 (1950): 242–53.

———. "Horizontal Diffusion Due to Oceanic Turbulence." *Journal of Marine Research* 8, no. 3 (1949): 199–225.

———. "Is the South Pacific 3h Plume Dynamically Active?" *Earth and Planetary Science Letters* 61 (1982): 63–67.

———. "The Large-Scale Oceanic Circulation." In *Advances in Earth Science*, edited by P. Hurley, 175–84. Cambridge, MA: MIT Press, 1966.

———. "Note on the Deep Circulation of the Atlantic Ocean." *Journal of Meteorology* 7 (1950): 245–46.

———. "On the Abyssal Circulation of the World Ocean-III. An Advection-Lateral Mixing Model of the Distribution of a Tracer Property in an Ocean Basin." *Deep-Sea Research* 14 (1967): 441–57.

———. *Report on the Geomagnetic Electrokinetograph: The General Theory of the Electric Potential Field Induced in Deep Ocean Currents.* Woods Hole, MA: Woods Hole Oceanographic Institute, 1948.

———. *The Starbuck Essays of Henry Stommel.* Woods Hole, MA: Friends of Starbuck, 1992.

———. "A Survey of Ocean Current Theory." *Deep-Sea Research* 4 (1957): 149–84.

———. "The Theory of the Electric Field Induced in Deep Ocean Currents." *Journal of Marine Research* 7, no. 3 (1948): 386–92.

———. "Thermohaline Convection with Two Stable Regimes of Flow." *Tellus* 13, no. 2 (1961): 224–30.

———. "Trajectories of Small Bodies Sinking Slowly through Convection Cells." *Journal of Marine Research* 2, no. 1 (1949): 24–29.

———. "Varieties of Oceanographic Experience." *Science* 139, no. 3555 (1963): 572–76.

———. *A View of the Sea*. Princeton, NJ: Princeton University Press, 1987.

———. "The Westward Intensification of Wind-Driven Ocean Currents." *Transactions, American Geophysical Union* 29 (1948): 202–6.

Stommel, Henry, and Arnold Arons. "On the Abyssal Circulation of the World Ocean—I. Stationary Planetary Flow Patterns on a Sphere." *Deep-Sea Research* 6 (1960): 140–54.

———. "On the Abyssal Circulation of the World Ocean—II. An Idealized Model of the Circulation Pattern and Amplitude in Oceanic Basins." *Deep-Sea Research* 6 (1960): 217–33.

———. "On the Abyssal Circulation of the World Ocean—V. The Influence of Bottom Slope on the Broadening of Inertial Boundary Currents." *Deep-Sea Research* 19 (1972): 707–18.

Stommel, Henry, Arnold Arons, and A. J. Faller. "Some Examples of Stationary Planetary Flow Patterns in Bounded Basins." *Tellus* 10, no. 2 (1958): 179–87.

Stommel, Henry, Kim Saunders, William Simmons, and John Cooper. "Observations of the Diurnal Thermocline." *Deep-Sea Research* 16 (1969): 269–84.

Stommel, Henry, and George Veronis. "Steady Convective Motion in a Horizontal Layer of Fluid Heated Uniformly from Above and Cooled Non-Uniformly from Below." *Tellus* 9 (1957): 401–7.

Storch, Hans van, and Klaus Hasselman. *Seventy Years of Exploration in Oceanography: A Prolonged Weekend Discussion with Walter Munk*. New York: Springer, 2010.

Stuiver, Minze, Paul D. Quay, and H. G. Ostlund. "Abyssal Water Carbon-14 Distribution and the Age of the World Oceans." *Science* 219, no. 4586 (1983): 849–51.

Sullivan, Patricia. "Joanne Malkus Simpson, 86, Dies; Atmospheric Scientist." *Washington Post*, March 8, 2010.

Sullivan, Walter. "Sea-Floor Geysers May Be Key to Ore Deposits." *New York Times*, May 8, 1979.

———. "Superhot, Metal-Rich Water Sought by Diving Scientists." *New York Times*, February 17, 1977.

Summit, M., and J. A. Baross. "Thermophilic Sub-Seafloor Microorganisms from the 1996 North Gorda Ridge Eruption." *Deep-Sea Research Part II—Topical Studies in Oceanography* 45, no. 12 (1998): 2751–66.

Sverdrup, Harald U. *Among the Tundra People*. 1939. Translated by Molly Sverdrup. La Jolla: University of California, 1978.

———. "Dynamics of Tides in the North Siberian Shelf." *Geofysiske Publikasjoner* 4 (1921): 1–75.

———. "Informal Autobiography." Typescript prepared for the US National Academy of Sciences, February 6, 1948, Box 1, Folder 1: Biographical File, Scripps Institution of Oceanography Archives, Office of the Director (Sverdrup).

———. *The Norwegian North Polar Expedition with the "Maud" 1918–1925: Scientific Results*. Bergen: Geofysisk Institutt, 1933.

———. *Oceanography for Meteorologists*. New York: Prentice-Hall, 1942.

Sverdrup, H. U., M. W. Johnson, and R. H. Fleming. *The Oceans: Their Physics, Chemistry, and General Biology*. Englewood Cliffs, NJ: Prentice Hall, 1942.

Sverdrup, Harald U., and Walter Munk. *Wind, Sea, and Swell: The Theory of Relations for Forecasting*. Washington, DC: Hydrographic Office, 1947.

Swallow, J. C. "History of the Exploration of the Hot Brine Area of the Red Sea: Discovery Account." In *Hot Brines and Recent Heavy Metal Deposits in the Red Sea*, edited by E. T. Degens and D. A. Ross, 3–9. New York: Springer, 1969.

Swallow, J. C., and J. Crease. "Hot Salty Water at the Bottom of the Red Sea." *Nature* 205 (1965): 165–66.

Swallow, J. C., and L. V. Worthington. "Measurements of Deep Currents in the Western North Atlantic." *Nature* 179 (1957): 1183–84.

———. "An Observation of a Deep Countercurrent in the Western North Atlantic." *Deep-Sea Research* 8 (1961): 1–19.

Swallow, M. "Deep Currents in the Open Ocean." *Oceanus* 7, no. 3 (1961): 2–8.

Swanson, L. V. *Aircraft Salvage Operation Mediterranean: Lessons and Implications for the Navy. Executive Summary of the Final Report Prepared for the Secretary of the Navy and the Chief of Naval Operation*. Washington, DC: Department of the Navy, 1967.

Swartz, Steven L., and Robert J. Hofman. *Marine Mammal and Habitat Monitoring: Requirements; Principles; Needs; and Approaches*. Report No. PB-91-215046/XAB. Washington, DC: Marine Mammal Commission, 1991.

Szulc, Tad. "H-Bomb Searchers Fail Again as Sea Cable Snaps." *New York Times*, March 26, 1966.

Talwani, Manik, George H. Sutton, and J. Lamar Worzel. "A Crustal Section across the Puerto Rico Trench." *Journal of Geophysical Research* 64, no. 10 (1959): 1545–55.

Talwani, M., C. C. Windish, and M. G. Langseth. "Reykjanes Ridge Crest: A Detailed Geophysical Study." *Journal of Geophysical Research* 76 (1971): 473.

Tarling, D. H. "Tentative Correlation of Samoan and Hawaiian Islands Using 'Reversals' of Magnetization." *Nature* 196 (1962): 882–83.

Tharp, Marie. "Mapping the Ocean Floor—1947 to 1977." In *The Ocean Floor: Bruce Heezen Commemorative Volume*, 19–31. Chichester, UK: Wiley, 1982.

Tharp, Marie, and Henry Frankel. "Mappers of the Deep." *Natural History* 10 (1986): 49–62.

Theberge, Albert E. "Discovering the True Nature of the Mid-Atlantic Ridge: Part I." *Hydro International*, February 9, 2014.

———. "Seeking a Rift." *Hydro International*, December 15, 2014.

———. "Unravelling the Ridge and Rift." *Hydro International*, October 15, 2014.

Thomson, Sir George Paget. *J. J. Thomson and the Cavendish Laboratory in His Day*. Garden City, NY: Doubleday & Co., 1965.

Thomson, Sir J. J. *Recollections and Reflections*. New York: Macmillan, 1937.

Thorne-Miller, Boyce. *The Living Ocean: Understanding and Protecting Marine Biodiversity*. Washington, DC: Island Press, 1999.

Tolstoy, I., and M. Ewing. "North Atlantic Hydrography and the Mid-Atlantic Ridge." *Geological Society of America Bulletin* 60 (1949): 1527–40.

Toye, Sandra. "Deep Submergence: The Beginnings of Alvin as a Tool of Basic Research." In *50 Years of Ocean Discovery 1950–2000*, edited by National Science Foundation, 65–66. Washington, DC: National Academies Press, 2000.

Tucker, G. H. "Relation of Fishes and Other Organisms to the Scattering of Underwater Sound." *Journal of Marine Research* 10 (1951): 215–38.

Tunnicliffe, Verena. "The Biology of Hydrothermal Vents: Ecology and Evolution." *Oceanography and Marine Biology* 29 (1991): 319–407.

———. "Hydrothermal-Vent Communities of the Deep Sea." *American Science* 80 (1992): 336–49.

Tunnicliffe, V., et al. "Biological Colonization of a New Hydrothermal Vents Following an Eruption on Juan De Fuca Ridge." *Deep-Sea Research*, pt. 1, 44 (1997): 1627–43.

Turchetti, Simone, and Peder Roberts. *The Surveillance Imperative: Geosciences during the Cold War and Beyond*. New York: Springer, 2014.

Turner, J. S. "Buoyant Plumes and Thermals." *Annual Review of Fluid Mechanics* 1 (1969): 29.

Turner, R. D., and R. A. Lutz. "Growth and Distributions of Mollusks at Deep-Sea Vents and Seeps." *Oceanus* 27, no. 3 (1984): 55–62.

Turner, Stephen. "The Survey in Nineteenth-Century American Geology: The Evolution of a Form of Patronage." *Minerva* 25, no. 3 (1987): 282–330.

Tuve, Merle. "Is Science Too Big for the Scientist?" *Saturday Review* (1959): 48–52.

Tyler, P. A., et al. "A Walk on the Deep Side: Animals in the Deep Sea." In *Oceanography: An Illustrated Guide*, edited by C. P. Summerhayes and S. A. Thorpe, 195–211. New York: John Wiley & Sons, 1996.

Underwater Research Analysis Group Committee. *Principles of Underwater Sound: Originally Issued as Division 6, Volume 7 of NDRC Summary Technical Reports*. Washington, DC: National Research Council, 1955.

Urick, Robert J. *Sound Propagation in the Sea*. Washington, DC: Defense Advanced Research Projects Agency, US Government Printing Office, 1979.

US Department of Energy, Office of Civilian Radioactive Waste Management. *Yucca Mountain Science and Engineering Report: Technical Information Supporting Site Recommendation Consideration*. Washington, DC: US Department of Energy, Office of Civilian Radioactive Waste Management, 2002.

US Department of the Navy. *Department of the Navy Ten Year Program in Oceanography: Tenoc, 1961–1970*. Washington, DC: Department of the Navy, Office of the Chief of Naval Operations, 1961.

US Office of Scientific Research and Development, National Defense Research Committee. *Principles and Applications of Underwater Sound*. Washington, DC: Department of the Navy, Headquarters Naval Material Command, 1968.

Vacquier, Victor. "Many Jobs." *Annual Review of Earth and Planetary Sciences* 21 (1993): 1–17.

———. "Measurement of Horizontal Displacement along Faults in the Ocean Floor." *Nature* 183 (1959): 452–53.

———. "Transcurrent Faulting in the Ocean Floor." *Philosophical Transactions of the*

Royal Society of London A: Mathematical, Physical and Engineering Sciences 258, no. 1088 (1965): 77–81.

Vacquier, Victor, Arthur D. Raff, and Robert E. Warren. "Horizontal Displacements in the Floor of the Northeastern Pacific Ocean." *Geological Society of America Bulletin* 72, no. 8 (1961): 1251–58.

Vallis, Geoffrey K. *Atmospheric and Oceanic Fluid Dynamics: Fundamentals and Large-Scale Circulation.* Cambridge: Cambridge University Press, 2006.

———. "Large-Scale Circulation and Production of Stratification: Effects of Wind, Geometry, and Diffusion." *Journal of Physical Oceanography* 30 (2000): 933–54.

Vandenberg, Lauren, Theo Colborn, Tyrone B. Hayes, Jerrold J. Heindel, David R. Jacobs Jr., Duk-Hee Lee, John Peterson Myers, et al. "Regulatory Decisions on Endocrine Disrupting Chemicals Should be Based on the Principles of Endocrinology." *Reproductive Toxicology* 38 (2013): 1–15.

Van der Voort, J. L., and J. E. Mielke. Paper presented at the Marine Hydrothermal Metal Deposits. Congressional Research Service, Library of Congress, for Committee on Merchant Marine and Fisheries, House, 97th Congress, 2nd Session, 1982.

Van Dover, Cindy Lee. "Ecology of Mid-Atlantic Ridge Hydrothermal Vents." In *Hydrothermal Vents and Processes,* edited by L. M. Parson, C. L. Walker, and D. R. Dixon, 351–63. London: Geological Society, 1995.

Van Dover, Cindy Lee, C. R. German, Kevin G. Speer, L. M. Parson, and R. C. Vrijenhoek. "Evolution and Biogeography of Deep-Sea Vent and Seep Invertebrates." *Science* 295 (2002): 1253–56.

van Keuren, David K. "Cold War Science in Black and White: US Intelligence Gathering and Its Scientific Cover at the Naval Research Laboratory 1948–62." *Social Studies of Science* 31, no. 2 (2001): 207–29.

Vartanov, Raphael, and Charles D. Hollister. "Nuclear Legacy of the Cold War: Russian Policy and Ocean Disposal." *Marine Policy* 21, no. 1 (1997): 1–15.

Vening Meinesz, F. A. *Gravity Expeditions at Sea 1923–1930.* 3 vols. Vol. 1. Delft: Netherlands Geodetic Commission, 1932.

———. *Gravity Expeditions at Sea, 1934–1939.* 3 vols. Vol. 3. Delft: Netherlands Geodetic Commission, 1941.

Vening Meinesz, F. A., J. H. F. Umbgrove, and Ph. H. Kuenen. *Gravity Expeditions at Sea, 1923–1932.* 3 vols. Vol. 2. Delft: Netherlands Geodetic Commission, 1934.

Vening Meinesz, F. A., and F. E. Wright. "The Gravity Measuring Cruise of the US Submarine S-21." In *Publications of the United States Naval Observatory,* 2nd ser., vol. 13. Washington, DC: Government Printing Office, 1930.

Veronis, George. "A Theoretical Model of Henry Stommel." In *Evolution of Physical Oceanography,* edited by Bruce A. Warren and Carl Wunsch, xix–xxiii. Cambridge, MA: MIT Press, 1981.

Vickers, Daniel. *Farmers and Fishermen: Two Centuries of Work in Essex County, Massachusetts, 1630–1850.* Chapel Hill, NC: University of North Carolina Press, 1994.

Vine, Allyn C. "Vehicles as Instruments for Oceanographers." *NRC-NAS Committee* on *Undersea Warfare, Proceedings of the Symposium on Aspects of Deep Sea Research* (1956): 98–104.

Vine, Frederick J. "Reversals of Fortune." In *Plate Tectonics*, edited by Naomi Oreskes, with Homer Le Grand, 46–66. Boulder, CO: Westview Press, 2001.

Vine, Frederick John, and Drummond Hoyle Matthews. "Magnetic Anomalies over Oceanic Ridges." *Nature* 199, no. 4897 (1963): 947–49.

vom Saal, Frederick S., and Claude Hughes. "An Extensive New Literature Concerning Low-Dose Effect of Bisphenol A Shows the Need for a New Risk Assessment." *Environmental Health Perspectives* 113, no. 8 (2005): 926–33.

vom Saal, Frederick S., and W. V. Welshons. "Large Effects from Small Exposures. II. The Importance of Positive Controls in Low-Dose Research on Bisphenol A." *Environmental Research* 100 (2006): 50–76.

von Arx, William S. *An Introduction to Physical Oceanography*. 1962. Reading, MA: Addison-Wesley Publishing, 1974.

———. "On the Promise and Limitations of Ocean Model Experiments." In *Proceedings of the Symposium on Aspects of Deep Sea Research*, edited by William S. von Arx, 45–49. Washington, DC: National Academy of Sciences, 1957.

von Damm, Karen L. "Controls on the Chemistry and Temporal Variability of Seafloor Hydrothermal Systems." In *Seafloor Hydrothermal Systems: Physical, Chemical, Biological, and Geological Interactions*, edited by Susan E. Humphris, R. A. Zierenberg, S. J. Mullineaux, and Richard E. Thomson, 222–47. Washington, DC: American Geophysical Union, 1995.

———. "Systematics of and Postulated Controls on Submarine Hydrothermal Solution Chemistry." *Journal of Geophysical Research* 93 (1988): 4551–61.

von Damm, K. L., and J. L. Bischoff. "Chemistry of Hydrothermal Solutions from the Southern Juan De Fuca Ridge." *Journal of Geophysical Research* 92 (1987): 11334–46.

von Damm, K. L., J. M. Edmond, B. Grant, and C. I. Measures. "Chemistry of Submarine Hydrothermal Solutions at Guaymas Basin, Gulf of California." *Geochimica et Cosmochimica Acta* 49 (1985): 2221–37.

von Damm, K. L., J. M. Edmond, B. Grant, C. I. Measures, B. Walden, and R. F. Weiss. "Chemistry of Submarine Hydrothermal Solutions at 21°N, East Pacific Rise." *Geochimica et Cosmochimica Acta* 49 (1985): 2197–2220.

von Damm, Karen L., B. Grant, and J. M. Edmond. "Preliminary Report on the Chemistry of Hydrothermal Solutions at 21° North, East Pacific Rise." In *Hydrothermal Processes at Seafloor Spreading Centers*, edited by P. A. Rona, K. Bostrom, and K. L. Smith Jr., 369–89. New York: Plenum Press, 1983.

von der Borch, C. C., and R. W. Rex. "Amorphous Iron Oxide Precipitates in Sediments Cored During Leg 5, Deep Sea Drilling Project." In *Initial Reports of the Deep Sea Drilling Project*, edited by D. A. McManus et al., 541–44. Washington, DC: Government Printing Office, 1970.

von Gumbel, G. "Ueber die im Stillen Ocean auf dem meeresgrunde vorkommended Manganknollen." Munich: Sitz Berichte d. K. Bayerisched Akademie d. Matem-Physik Klasse, Wissenschaften München, 1878.

von Herzen, R. P. "Heat Flow Values from the South-Eastern Pacific." *Nature* 183 (1959): 882–83.

von Herzen, R. P., and M. G. Langseth. "Present Status of Oceanic Heat Flow Measurements." *Physics and Chemistry of the Earth* 6 (1966): 365.

von Herzen, R. P., and S. Uyeda. "Heat Flow through the Eastern Pacific Ocean Floor." *Journal of Geophysical Research* 68 (1963): 4219–50.

Wager, L. R., and G. M. Brown. *Layered Igneous Rocks*. Edinburgh: Oliver & Boyd, 1967.

Wager, L. R., and W. A. Deer. "The Petrology of the Skaergaard Intrusion." *Meddelelser om Grønland* 105, no. 4 (1940): 1–352.

Wakelin, James H., Jr. "Thresher: Lesson and Challenge." *National Geographic*, June 1964, 759–63.

Waldrop, Mitch. "Ocean's Hot Springs Stir Scientific Excitement." *Chemical and Engineering News*, March 10, 1980, 30–33.

Walsh, James. "US Policy on Marine Pollution: Changes Ahead." *Oceanus* 21, no. 1 (1981): 18.

Wang, Jessica. *American Science in the Age of Anxiety: Scientists, Anticommunism, and the Cold War*. Chapel Hill: University of North Carolina Press, 1999.

Ware, R., C. Rocken, F. Solheim, M. Exner, W. Schreiner, R. Anthes, D. Feng, et al. "GPS Sounding of the Atmosphere from Low Earth Orbit: Preliminary Results." *Bulletin of the American Meteorological Society* 77, no. 1 (1996): 19–40.

Warren, Bruce A. "Arnold B. Arons." *Eos* (2001): 328.

Warren, Bruce A., and Carl Wunsch. *Evolution of Physical Oceanography*. Cambridge, MA: MIT Press, 1981.

Weart, Spencer R. *The Discovery of Global Warming*. Cambridge, MA: Harvard University Press, 2003.

Weaver, Andrew, Jan Sedlacek, Michael Eby, Kaitlin Alexander, Elisabeth Crespin, Thierry Fichefet, Gwenaelle Phillippon-Berthier, et al. "Stability of the Atlantic Meridional Overturning Circulation: A Model Intercomparison." *Geophysical Research Letters*, October 24, 2012.

Weber, Max. "Objective Possibility and Adequate Causation in Historical Explanation." In *The Methodology of the Social Sciences*, edited by Edward A. Shils and Henry A. Finch, 164–88. Glencoe, IL: Free Press of Glencoe, 1949.

Weinberg, Alvin M. *Reflections on Big Science*. Cambridge, MA: MIT Press, 1967.

Weir, Gary E. "'Fashioning' Naval Oceanography: Columbus O' Donnell Iselin and American Preparation for War, 1940–1941." In *The Machine in Neptune's Garden: Historical Perspectives on Technology and the Marine Environment*, edited by Helen M. Rozwadowski and David K. van Keuren, 65–95. Sagamore Beach, MA: Science History Publications/USA, 2004.

———. *Forged in War: The Naval-Industrial Complex and American Submarine Construction*. Washington, DC: Naval Historical Center, 1993.

———. *An Ocean in Common: Naval Officers, Scientists, and the Ocean Environment*. College Station: Texas A&M University Press, 2001.

———. "Selling Bellevue: The Emergence of American Naval Oceanography." In

Oceanographic History: The Pacific and Beyond, edited by Keith R. Benson and Philip F. Rehbock, 320-331. Seattle: University of Washington Press, 2002.

———. "Surviving the Peace: The Advent of American Naval Oceanography, 1914–1924." *Naval War College Review*, Autumn 1997, 85–103.

Weisgall, Jonathan. *Operation Crossroads*. Annapolis, MD: Naval Institute Press, 1994.

Weiss, Ray F. "Dissolved Gases and Total Inorganic Carbon in Seawater: Distribution, Solubilities, and Shipboard Gas Chromatography." PhD diss., University of California, San Diego, Scripps Institution of Oceanography, 1970.

Weiss, Rudolf Fritz, P. Lonsdale, J. E. Lupton, A. E. Bainbridge, and H. Craig. "Hydrothermal Plumes in the Galapagos Rift." *Nature* 267 (1977): 600–603.

Weiss, Rudolf Fritz, J. E. Lupton, P. F. Lonsdale, A. E. Bainbridge, and H. Craig. "Hydrothermal Plumes on the Galapagos Spreading Center (Abs)." *Eos, Transactions, American Geophysical Union* 57 (1976): 935.

Welander, Pierre. "An Advective Model of Ocean Thermocline." *Tellus* 11, no. 3 (1959): 309–18.

Wenk, Edward, Jr. *Feasibility Studies of Pressure Hulls for Deep Diving Submarines*. Washington DC: National Academy of Sciences, National Research Council, 1958.

———. *The Politics of the Ocean*. Seattle: University of Washington Press, 1972.

Wenk, Edward, Jr., Robert C. Dehart, Philip Mandel, and Ralph Kissinger Jr. "An Oceanographic Research Submarine of Aluminum for Operation to 15,000 Ft." Paper No. 5 presented March 23, 1960, at Royal Institution of Naval Architects.

Westwick, Peter. "Secret Science: A Classified Community in the National Laboratories." *Minerva* 38, no. 4 (2000): 363–91.

White, Robert M. "Oceans and Climate: An Introduction." *Oceanus* 21 (1978): 2–3.

Wickert, Jens, Christoph Reigber, Georg Beyerle, Rolf König, Christian Marquardt, Torsten Schmidt, Ludwig Grunwaldt, et al. "Atmosphere Sounding by GPS Radio Occultation: First Results from CHAMP." *Geophysical Research Letters* 28, no. 17 (2001): 3263–66.

Wickert, Jens, Torsten Schmidt, Georg Beyerle, Rolf König, Christoph Reigber, and Norbert Jakowski. "The Radio Occultation Experiment Aboard CHAMP: Operational Data Analysis and Validation of Vertical Atmospheric Profiles." *Journal of the Meteorological Society of Japan* 82, no. 1B (2004): 381–95.

Wilford, John Noble. "His Bold Statement Transforms the Debate on Greenhouse Effect." *New York Times*, August 23, 1988. http://www.nytimes.com/1988/08/23/science/his-bold-statement-transforms-the-debate-on-greenhouse-effect.html.

Wilhelm, Mary Jo Kelly. *Listening for Leviathan*. Hagerstown, MD: Storyfest Press, 2003.

Williams, D. L. "Heat Loss and Hydrothermal Circulation Due to Sea-Floor Spreading." PhD diss., Woods Hole Oceanographic Institution, 1974.

Williams, D. L., and R. P. Von Herzen. "Heat Loss from the Earth; New Estimate." *Journal of Geology* 2 (1974): 327.

Williams, D. L., R. P. Von Herzen, J. G. Sclater, and R. G. Anderson. "The Galapagos

Spreading Centre: Lithospheric Cooling and Hydrothermal Circulation." *Geophysical Journal of the Royal Astronomical Society* 38 (1974): 587.

Wills, Christopher, and Jeffrey Bada. *The Spark of Life: Darwin and the Primeval Soup.* Cambridge, MA: Perseus Publishing, 2000.

Wilson, E. Bright. *An Introduction to Scientific Research.* New York: McGraw-Hill, 1952.

Wilson, J. Tuzo. "Evidence from Ocean Islands Suggesting Movement in the Earth." *Philosophical Transactions of the Royal Society of London A: Mathematical, Physical and Engineering Sciences* 258, no. 1088 (1965): 145–67.

———. "Some Consequences of Expansion of the Earth." *Nature* 185 (1960): 880–82.

Winogradsky, Sergei. In *Microbiologie du sol: Problèmes et méthodes*, edited by Masson et Cie, 7–9. Paris: Libraires de L'Academie de medécines, 1949.

Wirsen, Carl O., and Holger Wihdekilde Jannasch. "Deep-Sea Primary Production at the Galápagos Hydrothermal Vents." *Science* 207 (1980): 1345–47.

Wirsen, Carl O., H. W. Jannasch, and S. J. Mullineaux. "Chemosynthetic Microbial Activity at Mid-Atlantic Ridge Hydrothermal Vent Sites." *Journal of Geophysical Research* 98, no. B6 (1993): 9693–703.

Wirsen, Carl O., H. W. Jannasch, and J. H. Tuttle. "Activities of Sulfur-Oxidizing Bacteria at the 21°N East Pacific Rise Vent Site." *Marine Biology* 92 (1986): 449–56.

Wittje, Roland. *The Age of Electroacoustics: Transforming Science and Sound.* Cambridge, MA: MIT Press, 2016.

Woese, C. R., O. Kandler, and M. L. Wheelis. "Towards a Natural System of Organisms: Proposals for the Domains Archaea, Bacteria, and Eukarya." *Proceedings of the National Academies of Science* 87 (1990): 4576–79.

Wolery, T. J., and N. H. Sleep. "Hydrothermal Circulation and Geochemical Flux at Mid-Ocean Ridges." *Journal of Geology* 84 (1976): 249–75.

Wolfe, Audra J. *Competing with Soviets: Science, Technology, and the State in Cold War America.* Baltimore: Johns Hopkins University Press, 2013.

———. *Freedom's Laboratory: The Cold War Struggle for the Soul of Science.* Baltimore: Johns Hopkins University Press, 2018.

Wood, Robert Muir. *The Dark Side of the Earth.* Crows Nest, Australia: Allen and Unwin, 1985.

World Meteorological Organization. *Proceedings of the World Climate Conference.* WMO-No. 537. Geneva: World Meteorological Organization, February 1979.

Worthington, L. V. *On the North Atlantic Circulations.* Baltimore: Johns Hopkins University Press, 1976.

Worzel, J. Lamar. "The Configuration of the Earth." *Naval Research Reviews* (1949): 15–21.

———. *Pendulum Gravity Measurements at Sea 1936–1959.* New York: John Wiley & Sons, 1965.

Worzel, J. Lamar, and Maurice Ewing. *Propagation of Sound: Explosion Sounds in Shallow Water.* New York: Geological Society of America, 1948.

Wunsch, Carl. *Henry Stommel, 1920–1992*. National Academy of Science Biographical Memoir. Washington, DC: National Academies Press, 1997

———. "What Is the Thermohaline Circulation?" *Science* 298 (2002): 1179–81.

Wunsch, Carl, and Raffaele Ferrari. "Vertical Mixing, Energy, and the General Circulation of the Oceans." *Annual Review of Fluid Mechanics* 36 (2004): 281–314.

Wüst, George. "The Major Deep-Sea Expeditions and Research Vessels, 1873–1960, A Contribution to the History of Oceanography." *Progress in Oceanography* 2 (1964): 1–3.

Wright, Susan, and David A. Wallace. "Varieties of Secrets and Secret Varieties: The Case of Biotechnology." In *Secrecy and Knowledge Production*, edited by J. Reppy, 107–32. Ithaca, NY: Cornell University, 1999.

Yao, Shuping. "Chinese Intellectuals and Science: A History of the Chinese Academy of Sciences." *Science in Context* 3, no. 2 (1989): 447–73.

Zalasiewicz, J., et al. (21st of 26 authors). "The Working Group on the Anthropocene: Summary of Evidence and Interim Recommendations." *Anthropocene* 19 (2017): 55–60.

Zeller, Suzanne, and Christopher J. Reis. "Wild Men in and out of Science: Finding a Place in the Disciplinary Borderlands of Arctic Canada and Greenland." *Journal of Historical Geography* 43 (2014): 31–43.

Zilsel, Edgar. "The Sociological Roots of Science." *American Journal of Sociology* 47 (1941–1942): 544–62.

Index

Page numbers in italics refer to figures.

493; Ocean Bottom Scanning Sonar (OBSS), 292; salinity, 59, 83–85, 89, 259–63; sound channel, 59, 83–84, 409, 414, 424, 435, 441, 451; temperature variation, 257–61. *See also* Sound Fixing and Ranging (SOFAR); Sound Surveillance System (SOSUS)

sound refraction, 59, 260; shadow zones, 59, 83, 261, 479

Sound Fixing and Ranging (SOFAR), 84, 204, 230, 261, 343, 408–9, 411. *See also* sonar; Sound Surveillance System (SOSUS)

Sound Surveillance System (SOSUS), 84, 248, 263, 266, 273, 302, 342–43, 408, 411, 418, 422, 500; CAESAR, 263; IUSS, 422, 427. *See also* sonar; Sound Fixing and Ranging (SOFAR);

Soviet Union, 205, 364, 393, 477, 497, 608n20, 647n21

Space Science Board, 142

Spencer, Derek W., 368, 375, 379–81, 390

Spiesberger, John, 411–16, 444, 450, 452, 626n71

Spiess, Fred, 258, 262, 282, 288, 301–2, 304, 313, 323, 331–32, 340–41, 471, 473, 494, 498, 500, 502

Spilhaus, Athelstan, 112, 257, 528n7

Spitzer, Lyman, 52

Sproul, Robert, 19, 25, 35, 41–42, 44, 50

Stalin, Joseph, 2, 521n94

Stanford University, 20, 502, 511n8, 599n152, 649n44

Stefansson, Vilhjalmur, 40–41, 49

Stephan, Edward C., Admiral, 288, 307, 342

Stern, Melvin, 120, 131, 493, 569n205

Stevens, Ted, 385

Stevens Institute of Technology, 91

Stommel, Henry Melson: approach to science, 72–74, 100–101; on classified data, 195, 205, 277; education, 72–74, 91; experience at Harvard, 132; military funding of work, 11–12, 72, 91, 96, 98, 100–116, 118, 120–23, 125, 129, 131, 134–37, 214, 308–9, 341,

413, 464, 471–74, 486, 493, 498, 501; on ocean circulation, 11–12, 60–61, 70–96, 258, 471; personality, 70–71; relationship with Paul Fye, 71, 104–7, 111–16 (*see also* Fye, Paul; Woods Hole Palace Revolt); relationship with Harald Sverdrup, 73–74, 86–87 (*see also* Sverdrup, Harald Ulrik); *Science of the Seven Seas*, 73. *See also* ocean circulation; Stommel-Arons model

Stommel-Arons model, 11, 60–61, 75–77, 88, 90–96, 325, 350, 353. *See also* Arons, Arnold; ocean circulation; Stommel, Henry Melson

Strategic Environmental Research and Development Program (SERDP), 417–18, 485

Studds, Gerald, 361, 368, 433, 612n58

submarines, 4, 13, 29, 53–54, 83–84, 138, 171, 218, 248, 261, 263, 265, 272–74, 281, 288, 310, 320, 411, 490, 497; design, 58–59; nuclear, 58, 108; operation, 82; U-boats, 3, 59. *See also* *Aluminaut*

Suess, Hans, 401, 403, 408, 491–92, 527n199

Sverdrup, Einar, 48

Sverdrup, Gudrun, 32, 37, 39, 42

Sverdrup, Harald Ulrik: on circulation, 65, 69, 72–74, 86–87 (*see also* ocean circulation); citizenship, 35, 39–40, 50–51; family, 32, 36–37, 39, 42, 48–49; FBI investigation, 24, 30, 35–49; loyalty and effect on research, 49–57, 141; on military funding of scientific research, 116, 244, 346, 486–88, 491, 495; Nazism, suspected, 11, 30, 32; *Oceanography for Meteorologists*, 43; *The Oceans*, 22–23, 43, 45, 69, 86–87, 177, 263; personality, 31–32, 42–43, 221; testimonies against or in favor of, 36–37, 44–47. *See also* Scripps Institution of Oceanography

Sverdrup, Leif, 36, 39, 48–49

Swallow, John, 70, 72

Sykes, Lynn, 310